Y0-ABC-143

Electronic Circuits

Electronic Circuits

CLYDE N. HERRICK
San Jose City College

CHARLES E. MERRILL PUBLISHING COMPANY
Columbus, Ohio

A Bell & Howell Company

Library of Congress Catalog Card Number: 68-11092

Printed in the United States of America

1 2 3 4 5 6 7 8 9 10 11 12 13 14 15/76 75 74 73 72 71 70 69 68

Preface

Electronics is a new and expanding field of scientific endeavor that includes increasingly sophisticated solid-state devices. To keep abreast of the advances in this science, the format and content of textbooks at the introductory level must be periodically re-examined. *Electronic Circuits* is an up-to-date first course on electronics circuitry. This text treats electronics circuitry by subdividing the material according to three topics: components, networks and network theorems, and systems. *Strict* subdivision cannot be made at the introductory level, however, because components are meaningless in isolation from simple networks and basic electrical laws or theorems. In turn, simultaneous development is required from the academic point of view.

Electrical and electronic components comprise devices such as cells, resistors, switches, lamps, electromagnets, alternators, inductors, capacitors, transformers, vacuum tubes, gas tubes, transistors, thermistors, varistors, varactors, photocells, transducers, saturable reactors, and so on. Electrical and electronic instruments are not components, but specialized devices that operate as analog computers. An electric circuit is a conducting part (or associated conducting parts) through which an electric current flows. A magnetic circuit is analogous to an electric circuit. The distinction between circuits and networks is not sharply defined; similarly, the distinction between networks and systems is not sharply defined, but is a general distinction in complexity.

Signals are basically electric currents, but they are currents that have been processed in a manner to provide *information* transfer; a simple electric current provides only power transfer. Current processing is an extensive topic, and an introductory text is necessarily limited to analysis of the more basic processes. Development of fundamental principles proceeds step-by-step to general forms so that the student may perceive the unity that prevails throughout the sciences of electricity and electronics. Thus, the nonlinear circuits employed to convert electric currents into signals are shown to have a common denominator with linear circuits and to be subject to analysis on the basis of the same general forms.

This text stresses the conceptual approach and provides a firm foundation in the unifying concept of energy. Both intuitive and analytic approaches are provided to facilitate understanding by the beginning student. Discussion of "hardware" has been minimized and is introduced only when it contributes to necessary perspective. The student is encouraged to think in mathematical terms, and it is assumed that he has had courses in arithmetic, algebra, geometry, and trigonometry. The student will not be seriously hampered, however, if he is taking a concurrent course in trigonometry, and although a basic knowledge of analytic geometry is required, this knowledge can be obtained from this textbook by an otherwise qualified student.

Clyde N. Herrick

Contents

Electronic Circuits

CHAPTER 1

Power Sources

1–1. Introduction

Vacuum tubes, transistors, and various semiconductor devices used in electronic circuits require a source of electric power. Communications circuits utilize d-c power, with a few exceptions. On the other hand, industrial-electronics circuits generally utilize a-c power. Vacuum tubes in television receivers, for example, have heaters that are energized by a-c power, but the chief functions of the tubes are performed with respect to a d-c power source. Transistors in radio receivers or high-fidelity amplifiers are powered solely from d-c sources. Industrial-electronics devices such as automatic welders, motor controllers, photoelectric relays, and so on, are usually energized from a-c power sources.

Primary batteries, exemplified by the familiar dry cell, are an important source of d-c power for portable radio and television receivers, military communications equipment, and space-age devices [see Figs.

Courtesy of Burgess Battery Co.

Courtesy of P. R. Mallory and Co.

Figure 1–1. (a) Nickel-cadmium secondary cells; (b) Mercury primary cell.

1–1(a)-(c)]. Secondary batteries, or storage batteries, are also applied to these same requirements and are found in familiar household devices such as electric razors and toothbrushes. The basic distinction between a primary cell and a secondary cell is that the former cannot be recharged after its chemical energy has been depleted. In other words, the chemical reactions in a primary cell are not reversible.

Figure 1–1(c) depicts the construction of a conventional dry cell. Battery action of a cell is based upon the chemistry of dissimilar conductors immersed in an electrolyte. The electrolyte (solution) acts more on one of the conductors than on the other. For example, in a dry

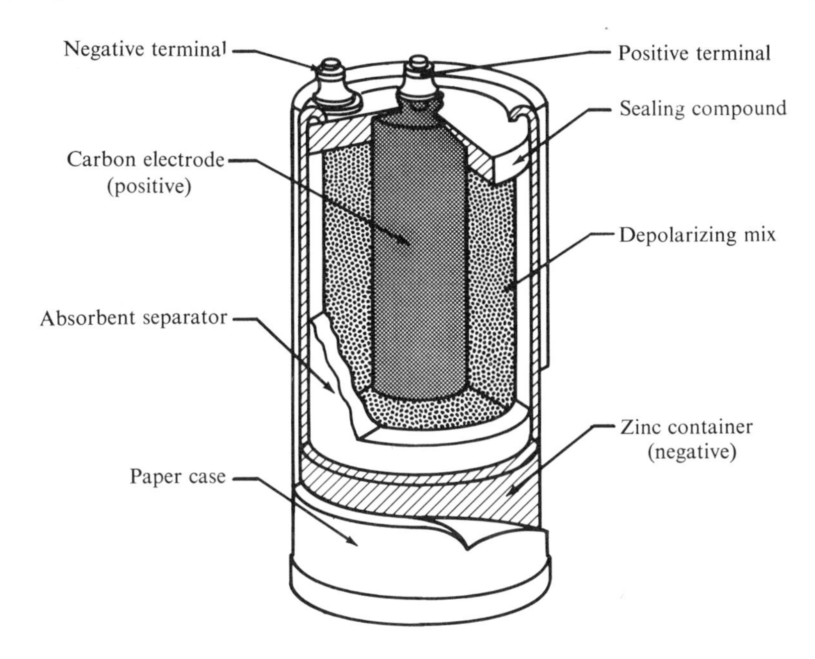

Figure 1-1. (c) Cutaway view of a conventional dry cell.

cell, the zinc is acted upon more than the carbon. The simplest primary cell consists of a strip of zinc and a strip of copper immersed in dilute sulphuric acid. However, this cell is impractical because the zinc is rapidly consumed by *local action* unless it has an extremely high degree of chemical purity.

Local action in a cell is defined as the loss of otherwise usable chemical energy by currents which flow within the cell regardless of its connections to an external circuit. Suppose, for example, that the zinc strip contains a few particles of copper. A copper particle makes electrical contact with the surrounding zinc, and if the particle is exposed on the surface of the zinc, both the zinc and copper can be acted upon by the acid. Evidently, the zinc surrounding the copper particle will be consumed as current flows from the zinc to the particle. That is, the presence of the copper particle establishes a small short-circuited cell. However, if the zinc strip has been amalgamated with mercury, local action is prevented. Let us analyze the effect of amalgamation.

When mercury is applied to the zinc surface, it dissolves a portion of the zinc to form a semiliquid alloy. Pure zinc is brought to the surface of the mercury, and particles of foreign matter are coated by the mercury. Furthermore, the mercury surface is extremely smooth, and a thin film of hydrogen gas clings to it. This gas film inhibits chemical action. However, when the external circuit of the cell is closed, comparatively large gas bubbles are formed that rise through the acid and escape. Thus, amalgamation does not prevent normal operation of the cell, but only inhibits local action. It has also been determined that loosened particles of foreign substances are expelled from the mercury due to its high density, whereupon the expelled particles fall to the bottom of the cell.

In the conventional dry cell, the zinc is not amalgamated; instead, zinc chloride is included to minimize local action. When current is drawn from a dry cell, hydrogen gas is liberated. Unless the hydrogen is removed, it collects as an insulating layer of bubbles on the carbon electrode, and the cell is said to be *polarized*. Hence, manganese dioxide is included to oxidize the hydrogen as rapidly as possible. Of course, this oxidation reaction cannot be entirely complete unless sufficient time is provided. Accordingly, the recovery from polarization in a dry cell occurs as depicted in Fig. 1–2. Note that after recovery is com-

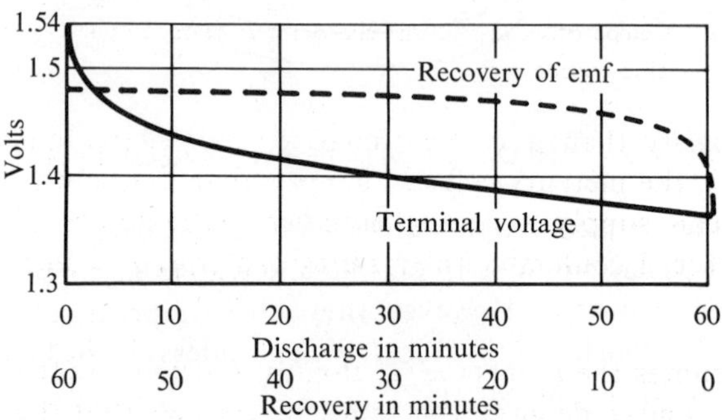

Figure 1–2. Recovery from polarization in a dry cell.

plete, the emf of the cell is slightly less than at the start of the discharge cycle.

Let us briefly note that electromotive force (emf) is the value of voltage generated by a cell or battery. When a cell is polarized, its internal resistance increases. Therefore, when current is supplied by the cell, an *IR* drop occurs across its internal resistance. In turn, the terminal voltage of the battery is decreased to a value equal to the emf of the cell minus its internal *IR* drop. We can measure the emf of a polarized cell with a voltmeter that draws negligible current. However, if an ordinary voltmeter is used that draws a significant amount of current, we will read a voltage value that is less that the emf of the cell. Of course, if a cell is not polarized, an ordinary voltmeter will indicate practically the emf value, because the internal resistance of the cell is very small. Details of internal-resistance measurement are reserved for subsequent discussion.

The familiar mercury cell, used to power portable radio receivers, is also a form of dry cell. It comprises zinc, zinc hydroxide, potassium hydroxide, mercuric oxide, and mercury. On discharge, a mercury cell maintains a more nearly constant terminal voltage than an ordinary dry cell, as depicted in Fig. 1–3. A mercury cell also has a greater

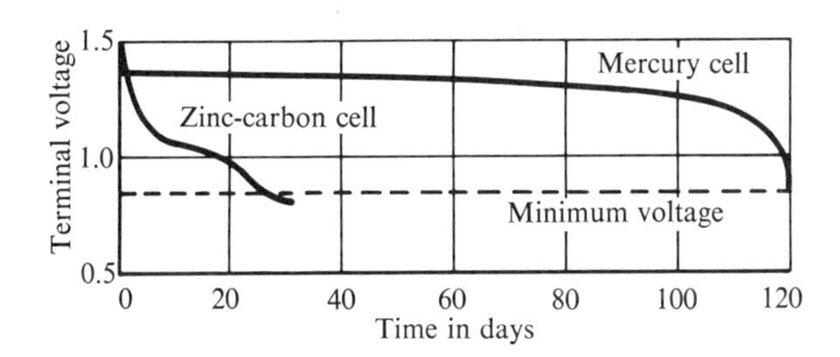

Figure 1–3. Comparative discharge characteristics of a mercury cell and a zinc-carbon cell.

current capacity than a conventional dry cell of the same weight. In other words, the mercury cell can supply a larger quantity of electricity, or it can supply a greater number of coulombs. When 1 amp flows for 1 sec, 1 coulomb of electricity has passed. Thus, we write

$$Q = It \tag{1–1}$$

where Q denotes the number of coulombs, I denotes a constant number of amperes, and t denotes the time in seconds that the current has flowed.

1–2. Internal Resistance and Battery Testing

Chemical action in a cell results in a practically constant value of generated voltage, called electromotive force (emf). However, the terminal voltage (Fig. 1–3) does not remain constant during discharge. In other words, a cell has internal resistance, denoted by R_{int} in the equivalent circuit shown in Fig. 1–4. When load resistor R is connected across the terminals of the cell, current flows and there is an IR drop across R_{int}. This IR drop subtracts from E, to give a difference equal to V. To measure the value of R_{int}, the circuit arrangement shown in Fig. 1–5 is convenient. When the switch is open, the voltmeter indicates the emf value of the battery for all practical purposes.

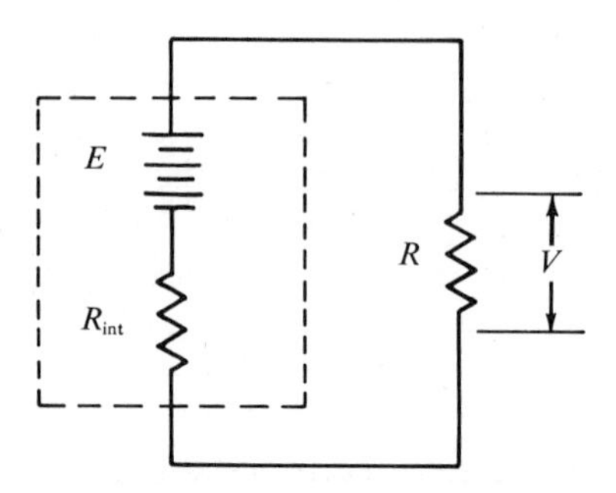

Figure 1-4. The equivalent circuit of the battery is enclosed by dotted lines.

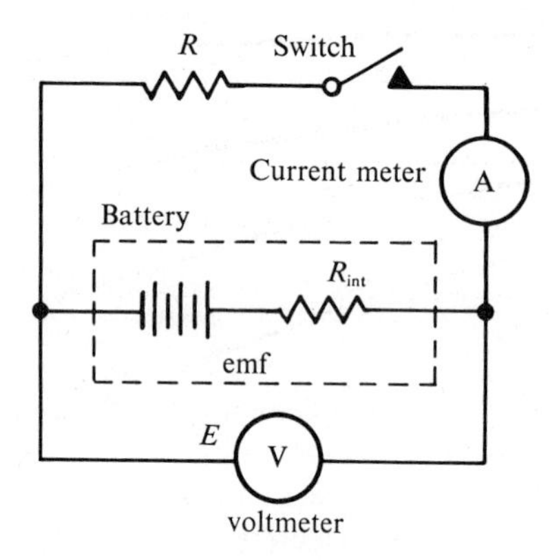

Figure 1-5. Test circuit for measurement of R_{int}.

When the switch is closed, the value of R_{int} is stated by the formula

$$R_{int} = \frac{E_{OC} - V_L}{I} \tag{1-2}$$

Since

$$I = \frac{V_L}{R_L}$$

then

$$R_{int} = \frac{(E_{OC} - V_L)R_L}{V_L} \tag{1-3}$$

where R_{int} is the value of the internal resistance for the selected value of current flow, E_{OC} is the voltage indication on open circuit, and V_L is the voltage indication on closed circuit.

Example 1. The open-circuit terminal voltage of a cell in a test circuit such as shown in Fig. 1-5 is 1.73 v and 1.2 v when a 12-Ω resistor is connected as the load. What is the internal resistance of the battery?

$$E_{OC} = 1.73 \text{ v}$$
$$V_L = 1.2 \text{ v}$$
$$R_{\text{int}} = \frac{(E_{OC} - V_L)R_L}{V_L} \tag{1-3}$$
$$= \frac{1.73 - 1.2}{1.2} \times 12$$
$$= \frac{0.53 \times 12}{1.2} = 5.3\ \Omega \qquad \text{(answer)}$$

Various types of storage batteries are used to power electronic devices. As previously noted, a storage battery is a secondary battery

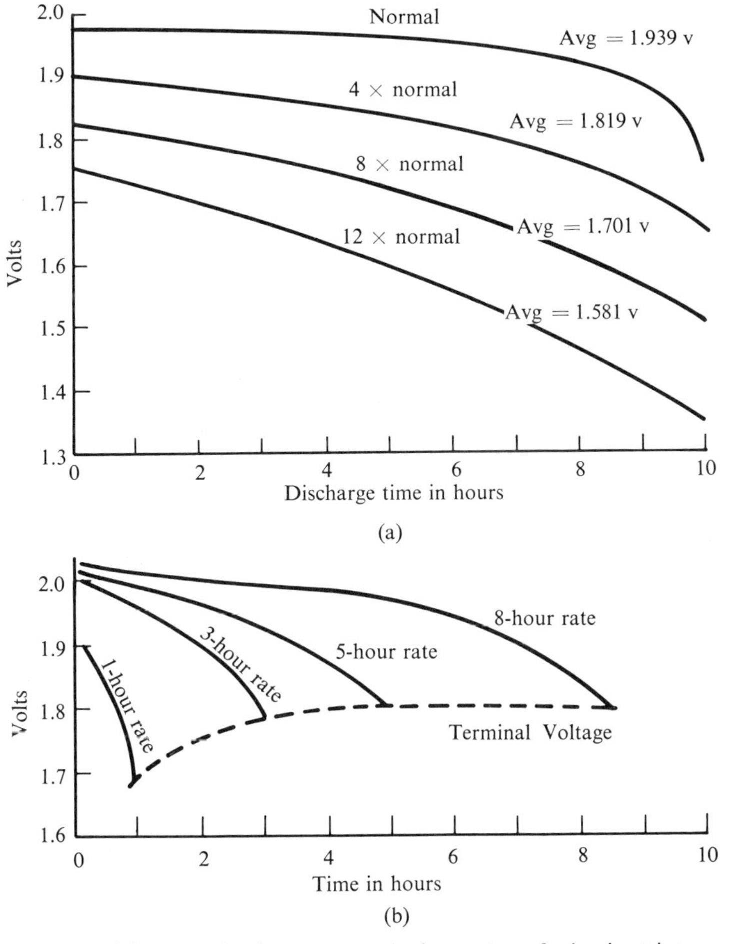

Figure 1-6. (a) Terminal voltage versus discharge time of a lead-acid storage cell; (b) Terminal voltage versus various rates of discharge of a lead-acid storage cell.

and can be recharged when its chemical energy is depleted. The most common type of storage battery is of lead-acid construction. Figure 1–6(a) depicts the decrease of terminal voltage for a storage cell versus discharge time. Observe that a greater current capacity is realized when the cell is discharged at a slow or "normal" rate. In other words, the efficiency of the cell is higher when the rate of discharge is slow. When a storage cell is recharged, its terminal voltage rises as shown in Fig. 1–7. The diagram shows that the terminal voltage reaches a constant value of 2.6 v when the cell is fully charged.

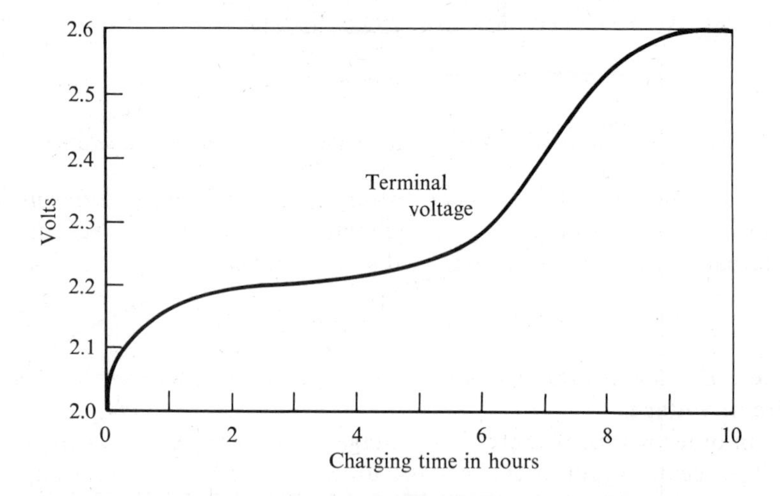

Figure 1–7. Terminal voltage versus charging time of a lead-acid storage cell.

Practically all storage batteries have a nominal rating based on an 8-hr discharge rate which is given in ampere-hours. A battery that will deliver 50 amp for 8 hr is given a rating of 400 amp-hr; however, this does not mean that a 400-amp-hr battery will be able to deliver 400 amp for 1 hr. Figure 1–6(b) illustrates the decrease of ampere-hour rating of a battery as the rate of discharge is increased.

The internal resistance of a storage battery is attributable to several factors. As the battery discharges, the electrolyte becomes more dilute. Note that the conductivity of a sulphuric acid solution varies nonlinearly with its concentration, as seen in Fig. 1–8. Internal resistance also depends on the condition of the lead compounds covering the surface of the lead plates. A fully charged storage battery has a low value of internal resistance compared with a primary battery. However, when a storage battery is discharged to the terminal value depicted in Fig. 1–6(a), its internal resistance is appreciable.

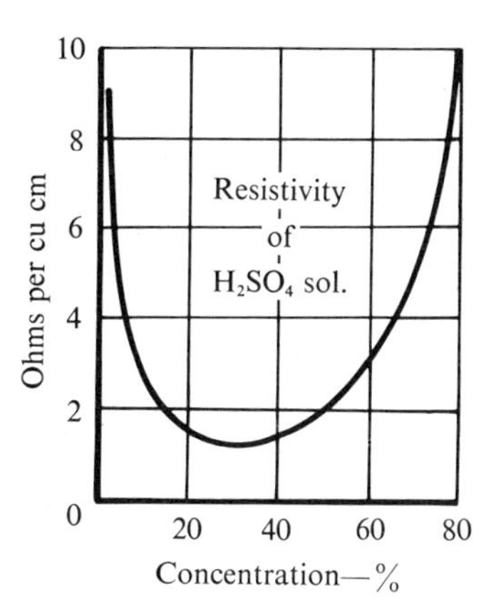

Figure 1-8. Resistivity versus concentration of a sulphuric-acid solution.

Any power supply, such as a battery, has a certain *regulation* under load. The regulation or lack of regulation is attributable chiefly to internal resistance and is expressed as a percentage

$$\% \text{ regulation} = \frac{E_{OC} - V_{CC}}{V_{CC}} \times 100 \qquad \textbf{(1-4)}$$

where E_{OC} denotes the open-circuit voltage, and V_{CC} denotes the closed-circuit voltage.

In other words, if there is no voltage drop under load, the regulation is 0 per cent. Again, if the voltage drops to one-half under load, the regulation is 100 per cent. If the voltage should drop to one-tenth under load, the regulation would be 900 per cent. Or, if the voltage approaches zero, under load, the regulation would approach infinity. The lower the regulation factor, the better will be the regulation of the power supply.

Example 2. The terminal voltage of a power supply is 300 v with no load and 270 v under full load. What is the per cent of regulation for the supply?

$$\text{Regulation} = \frac{300 - 270}{270} \times 100$$
$$= 11.1\% \qquad \text{(answer)}$$

Note that some engineers define power-supply regulation according to the following formula:

$$\text{Regulation} = \frac{E_{OC} - V}{E_{OC}} \times 100 \qquad \textbf{(1-5)}$$

Formula 1-5 states that if there is no voltage-drop under load, the regulation is 0 per cent. If the voltage drops to one-half under load, the regulation is 50 per cent. If the voltage should drop to one-tenth under load, the regulation would be 90 per cent; or, if the voltage should drop to zero under load, the regulation would be 100 per cent. Thus, if we use Formula 1-4, the regulation range is from zero to infinity. On the other hand, if we use Formula 1-5, the regulation range is from zero to unity. Therefore, whenever a regulation figure is stated, we must evaluate this figure according to the context of the discussion.

Example 3. Determine the regulation of the power supply in Example 2 using Formula 1-5.

$$\% \text{ Regulation} = \frac{300 - 270}{300} \times 100$$
$$= 10\% \qquad \text{(answer)}$$

Another important type of storage battery is the *nickel-cadmium* type, commonly used to power portable television receivers and other electronic devices. The nickel-cadmium storage battery is superior to the lead-acid storage battery in that it is much longer-lived and more rugged. For example, this type of battery is not damaged by complete discharge and subsequent storage for an extended period. Its internal resistance is comparatively low and in turn its regulation is good.

Chemical reaction in the nickel-cadmium storage battery is formulated:

$$2Ni(OH)_3 + Cd \rightleftarrows 2Ni(OH)_2 + Cd(OH)_2 + \text{electric charge} \qquad \textbf{(1-6)}$$

Discharge is from left to right in Formula 1-6, and charge is from right to left. Note that on discharge, two hydroxyl ions $(OH)^-$ combine with a cadmium atom with release of two electrons e^-. One of these electrons e^- forms a hydroxyl ion $(OH)^-$ to change one molecule of $Ni(OH)_3$ into a molecule of $Ni(OH)_2$. The other electron e^- produces a voltage across the battery terminals. One cell produces an emf of approximately 1.2 v. The electrolyte is potassium hydroxide, KOH. Concentration of the electrolyte does not change during charge or discharge.

1-3. Rectifiers and Power Supplies

Many d-c power supplies operate from an a-c source. One or more rectifiers are used to change the a-c source power into pulsating d-c power. In most cases, a low-pass filter follows the rectifier section to change the pulsating d-c power into substantially constant d-c power. *Half-wave rectifier action* is depicted in Fig. 1–9. The vacuum diode

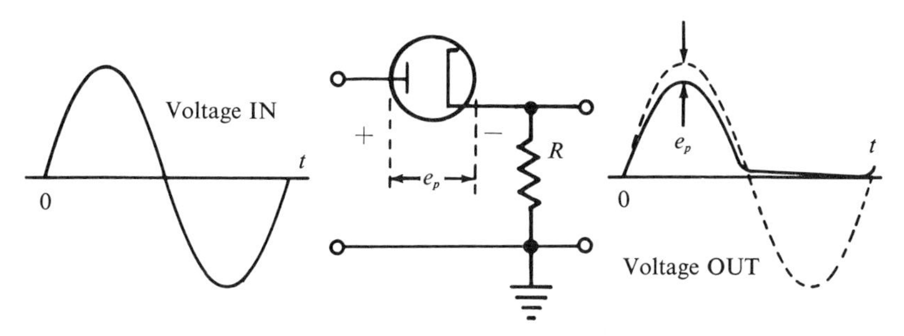

Figure 1–9. Half-wave rectifier action.

conducts on positive half cycles of the applied a-c voltage. Half-sine waves of current flow through the load resistor R. Since the diode has internal resistance, the voltage drop across the load resistor is less than the applied peak voltage. In other words, there is an IR drop across the internal resistance of the diode. This IR drop is denoted by e_p.

A vacuum diode exhibits a nonlinear relation between voltage and current, as seen in Fig. 1–10. In other words, the internal resistance of a diode is nonlinear. Child's law is a formula that expresses the relation between voltage and current in an ideal vacuum diode:

$$I = kE^{3/2} \tag{1–7}$$

We can compare the constant k in Child's law with the conductance parameter for a linear resistance. Thus, we may write

$$k = \frac{I}{E^{3/2}} \tag{1–8}$$

$$G = \frac{I}{E} \tag{1–9}$$

Conductance is a linear parameter, but k is a nonlinear parameter. Accordingly, we call k the *perveance* of the diode. The larger the value of k, the less is the voltage drop across the internal resistance of the diode. Observe that the internal d-c resistance of the diode has a value that depends upon what point we select on the curve depicted in Fig. 1–10.

Example 4. When 1 v is applied across the diode, a current of 1 amp flows and the internal resistance has a value of 1 Ω. On the other

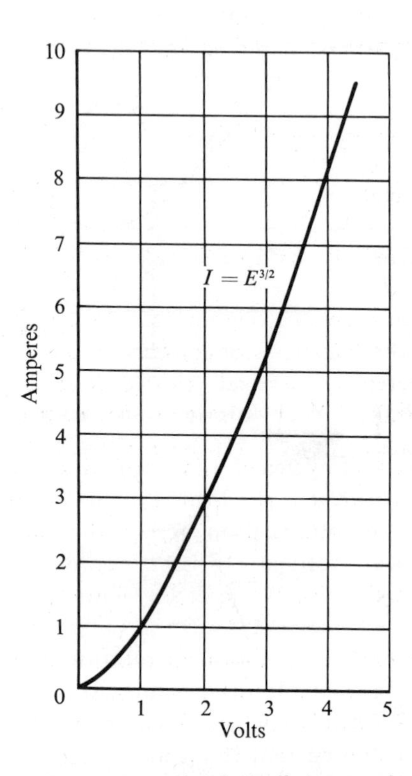

Figure 1–10. An example of Child's law in which $k = 1$.

hand, when 4 v are applied across the diode, a current of 8 amp flows and the internal resistance has a value of 0.125 Ω.

A sine wave has the values depicted in Fig. 1–11. These values are formulated as follows:

$$E_{\text{rms}} = \frac{1}{\sqrt{2}} E_p \cong 0.707 E_p \tag{1–10}$$

$$E_{pp} = 2E_p$$

$$E_{\text{avg}} = 0$$

The root-mean-square (rms) value of a sine wave is its equivalent d-c value insofar as power and energy are concerned. Its peak value is sometimes called its crest value; the positive-peak value is equal to the negative-peak value. The peak-to-peak value denotes the total excursion of the sine wave. Note that the average value of a sine wave over a complete cycle is zero, because the quantity of electricity in the negative half cycle is the same as in the positive half cycle.

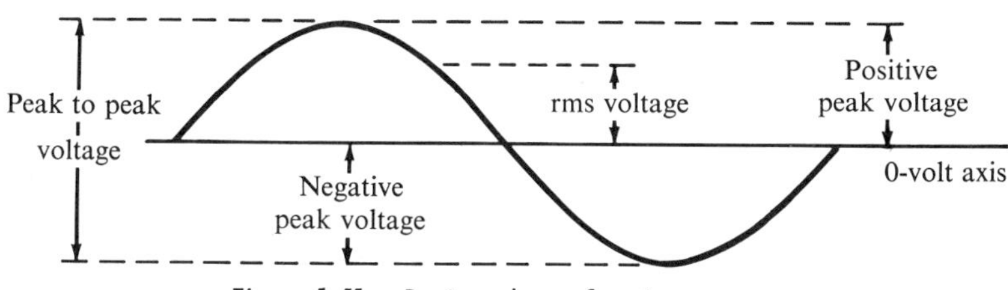

Figure 1-11. Basic values of a sine wave.

After a sine wave passes through a half-wave rectifier, the current that flows through the load resistor consists of half-sine waves. If a d-c voltmeter is connected across load resistor R in Fig. 1-9, the meter will read 0.318 of peak; an rms voltmeter will read 0.5 of peak.* The reason for this is seen in Figs. 1-12(a)-(c). Current flows for 180°, and

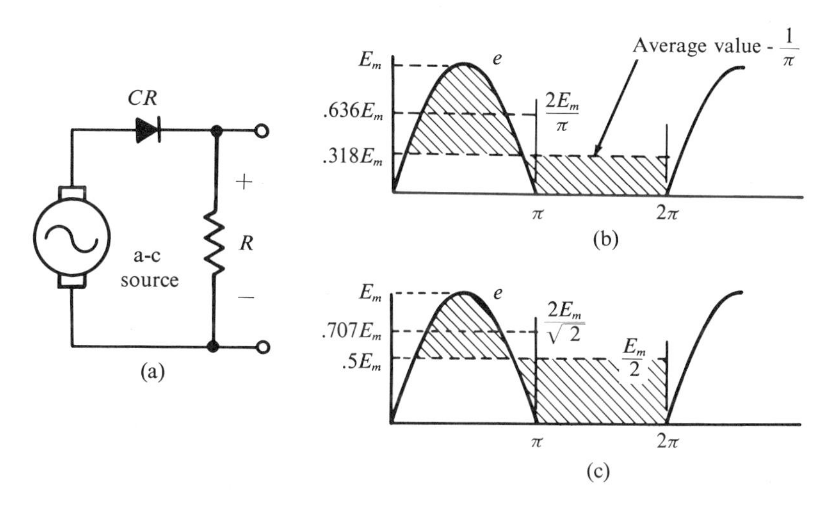

Figure 1-12. (a) A basic half-wave rectifier; (b) Instantaneous and average values; (c) Instantaneous and rms values.

then no current flows for the next 180°. The shaded area above the average level will exactly fill in the shaded area below the average level. It is proved in advanced mathematics that this average level is equal to $1/\pi$ times peak or 0.318 of peak. Thus, we formulate the average value of a half-rectified wave as shown at the top of page 14.

*The rms value of a half-rectified waveform over a half-cycle is not half of its full-cycle rms value.

$$E_{avg} = \frac{1}{\pi} E_p = 0.318\, E_p$$

$$I_{avg} = \frac{1}{\pi} I_p = 0.318\, I_p \qquad \textbf{(1–11)}$$

Example 5. In Fig. 1–12(b), the peak voltage of the half-rectified sine wave is 500 v. Therefore, the average value of the waveform is 159 v.

Note that the power in a half-rectified sine wave is equal to one-half of the power in a complete sine wave. Thus, the power in a waveform such as is shown in Fig. 1–12(c) is formulated

$$W = 0.5\, E_{rms} I_{rms} \qquad \textbf{(1–12)}$$

Example 6. Determine the load power of a half-wave rectifier with a peak-output voltage and current of 400 v and 2 amp, respectively.

$$\begin{aligned} E_{rms} &= \frac{1}{\sqrt{2}} E_p \\ &= 0.707 \times 400 \\ &= 282.8 \text{ v} \\ I_{rms} &= \frac{1}{\sqrt{2}} I_p \\ &= 0.707 \times 2 \\ &= 1.414 \text{ amp} \\ W &= 0.5\, E_{rms} I_{rms} \\ &= 0.5 \times 282.8 \times 1.414 \\ &= 200 \text{ w} \qquad \text{(answer)} \end{aligned}$$

Since an ordinary 60-Hz line provides only 115 v rms, we would use a step-up transformer as depicted in Fig. 1–13 to obtain a higher output voltage from a power supply. The extra winding is used to energize the heater of the rectifier diode.

If the secondary of the power transformer is center-tapped, we may employ a *full-wave rectifier*, as depicted in Fig. 1–14(a). It is evident that the average value of the full-rectified output must be twice that of a half-rectified output, or

$$E_{avg} = \frac{2}{\pi} E_p = 0.636\, E_p$$

$$I_{avg} = \frac{2}{\pi} E_p = 0.636\, E_p \qquad \textbf{(1–13)}$$

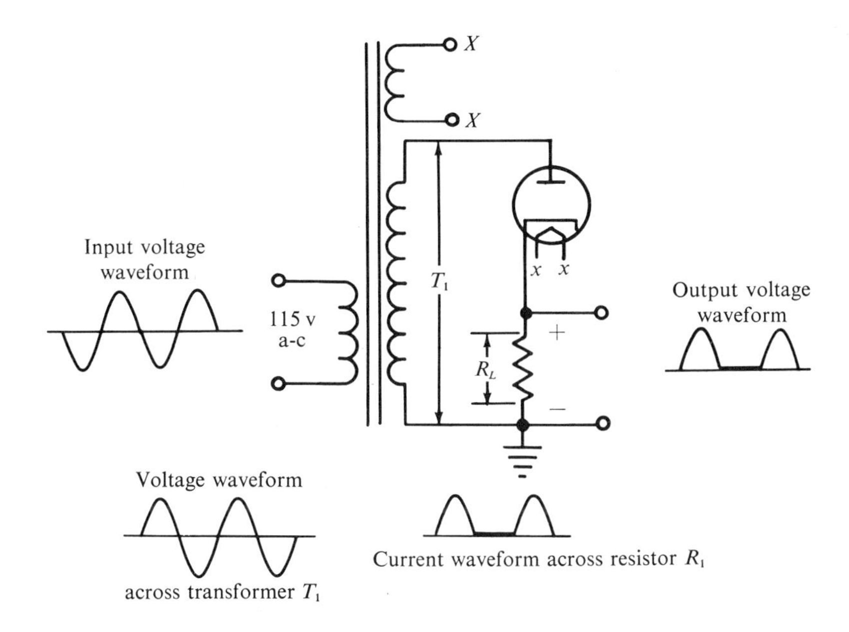

Figure 1–13. Half-wave rectifier with a step-up transformer.

In other words, if we connect a d-c voltmeter across R in Fig. 1–14(a), the voltmeter will read 0.636 of peak. We call the 0.318 value in Fig. 1–12(b) the d-c component of the rectified waveform. Similarly, we call the 0.636 value in Fig. 1–14(b) the d-c component of the rectified waveform. The half-sine waves are called the *ripple components* of the rectified waveform. Note that the power formula for the full-rectified sine wave depicted in Fig. 1–14(b) is the same as the power in a complete sine wave, or

$$W = E_{\text{rms}} I_{\text{rms}} \tag{1–14}$$

In the case of an a-c voltmeter with a half-wave instrument rectifier, a d-c voltage will be indicated as 2.22 times the d-c voltage value. In other words, we write

$$V_{\text{ind}} = \frac{0.707\, E_p}{0.318\, E_p} = 2.22 \text{ times} \tag{1–15}$$

In the case of an a-c voltmeter with a full-wave instrument rectifier, a d-c voltage will be indicated as 1.11 times the d-c voltage value, that is,

$$V_{\text{ind}} = \frac{0.707\, E_p}{0.636\, E_p} = 1.11 \text{ times} \tag{1–16}$$

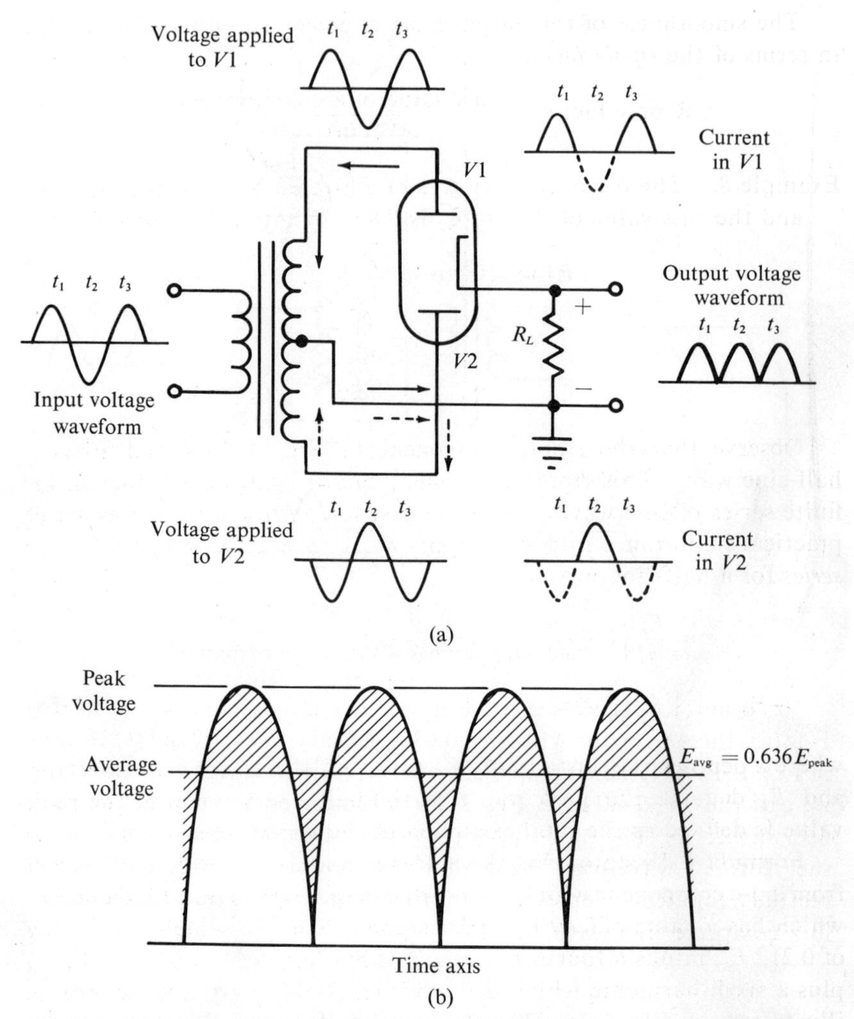

Figure 1-14. (a) A full-wave rectifier configuration; (b) The average value of a full-rectified sine wave is $2/\pi$ times peak or *0.636* of peak.

Example 7. Determine the reading of a d-c voltmeter connected across the output terminals of a full-wave power supply that delivers an rms voltage of 370 v.

$$V_{\text{d-c}} = \frac{0.636\ E_{\text{rms}}}{0.707}$$

$$= \frac{0.636 \times 370}{0.707}$$

$$= 333 \text{ v} \qquad \text{(answer)}$$

The smoothness of the output from a power supply is formulated in terms of the *ripple factor.*

$$\text{Ripple factor} = \frac{\text{rms value of a-c components}}{\text{average value}} \qquad \textbf{(1-17)}$$

Example 8. The d-c output voltage of a filtered power supply is 400 v and the rms value of the ripple is 0.8 v. What is the ripple factor?

$$\begin{aligned}\text{Ripple factor} &= \frac{E_{\text{rms}}}{E_{\text{d-c}}}\\ &= \frac{0.8}{400}\\ &= 0.002 \qquad \text{(answer)}\end{aligned}$$

Observe that the ripple component in Figs. 1-12(a) and (c) is a half-sine wave. This ripple component can be synthesized from an infinite series of sine waves. Only the first few terms in the series are of practical importance in power-supply work. Let us write the *Fourier series* for a half-sine waveform.

$$e = \frac{E_p}{\pi}\left(1 + \frac{\pi}{2}\sin\omega t + \frac{2}{3}\cos 2\omega t - \frac{2}{15}\cos 4\omega t + \frac{2}{35}\cos 6\omega t \cdots\right) \qquad \textbf{(1-18)}$$

where e denotes the instantaneous value of the half-sine waveform, and E_p denotes the peak value of the half-sine waveform; the peak value is defined as the total excursion of the waveform.

Formula 1-18 states that the half-sine waveform can be synthesized from a d-c component which has a value of 0.318 E_p, plus a fundamental which has a value of 0.49 E_p, plus a second harmonic which has a value of 0.212 E_p, minus a fourth harmonic which has a value of $-0.042\ E_p$, plus a sixth harmonic which has a value of 0.018 E_p, and so on, as illustrated in Fig. 1-15. The harmonic voltages decrease rapidly in amplitude, and we might consider only the d-c component, the fundamental, and the second harmonic in practical engineering work.

Next, let us consider the meaning of the "rms value of the a-c components" in Formula 1-17. When we have a mixture of sine waves of various frequencies, it can be shown that their rms value is equal to the square root of the sum of the squares of the individual rms sine-wave values.

$$E_{\text{rms}} = \sqrt{E_1^2 + E_2^2 + E_3^2} \cdots \qquad \textbf{(1-19)}$$

where $E_1 = 0.707$ of the peak value given by Formula 1-18 for $\cos\omega t$, and so on.

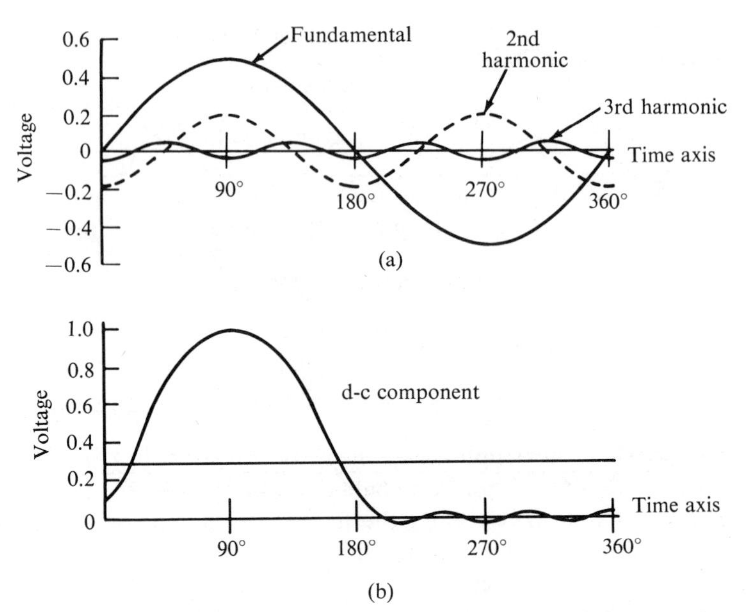

Figure 1–15. (a) Detail of d-c component, fundamental, and first two harmonics; (b) Approximate half-sine wave synthesized from its d-c component, fundamental, and first two harmonics.

If we calculate the rms value of the significant a-c components given in Formula 1–18 and divide this rms value by 0.318, we will find that the ripple factor of a half-rectified sine wave is equal to 121 per cent.

Note in passing that any waveform can be synthesized from an infinite series of sine waves. For example, some power supplies conduct current in pulses, which are narrower than sine waves. In turn, a pulse can be synthesized from sine waves in harmonic relation as depicted in Fig. 1–16. Again, some power supplies utilize a square waveform of voltage; a square wave can be synthesized from sine waves in harmonic relation as seen in Fig. 1–17(a) and (b) on page 20.

Now, let us consider the Fourier series for a full-rectified sine wave.

$$e = \frac{2E_p}{\pi}\left(1 + \frac{2}{3}\cos 2\omega t - \frac{2}{15}\cos 4\omega t + \frac{2}{35}\cos 6\omega t \cdots\right) \qquad \textbf{(1–20)}$$

where e denotes the instantaneous value, and E_p denotes the peak value of the full-rectified sine wave or its total excursion.

Formula 1–20 states that a full-rectified sine wave has a d-c component equal to 0.636 E_p, plus a second-harmonic component, minus a fourth-harmonic component, plus a sixth-harmonic component, and so on. If we apply Formula 1–19, we will find that a full-rectified sine

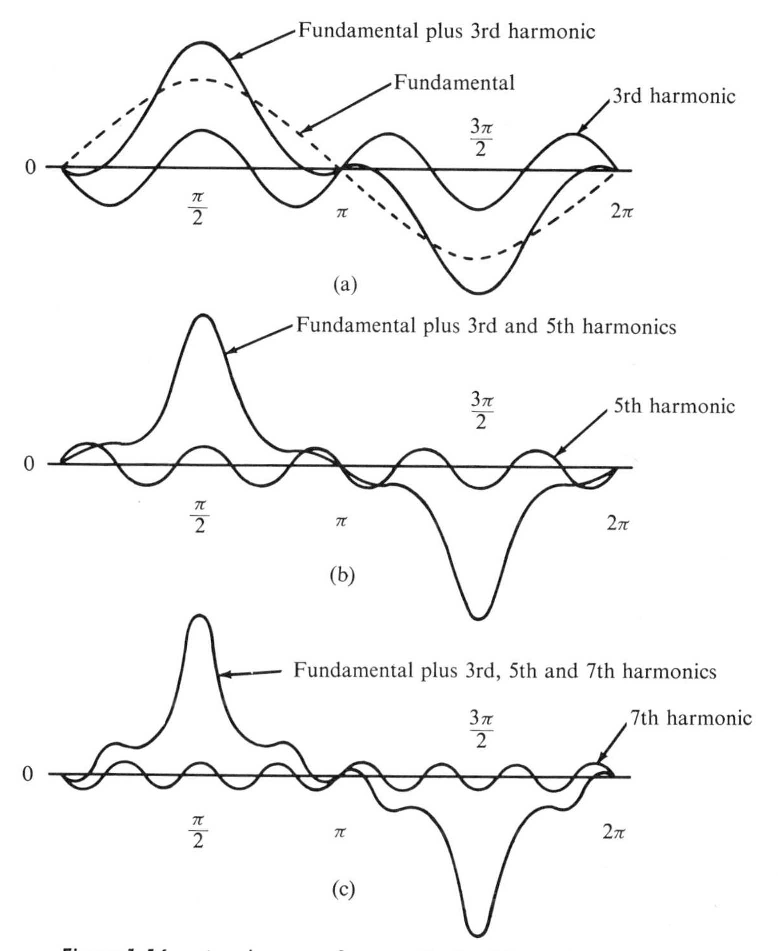

Figure 1-16. A pulse waveform synthesized from sine waves.

wave has a ripple factor equal to 0.48. The ripple factor is sometimes stated as a percentage.

$$\% \text{ ripple} = \text{ripple factor} \times 100 \qquad \textbf{(1-21)}$$

Example 9. The d-c output voltage of a power supply is 180 v, and the rms value of the ripple is 0.9 v. What is the per cent of ripple factor?

$$\begin{aligned} \% \text{ ripple} &= \frac{E_{\text{rms}}}{E_{\text{d-c}}} \times 100 \\ &= \frac{0.9}{180} \times 100 \\ &= 0.5\% \qquad \text{(answer)} \end{aligned}$$

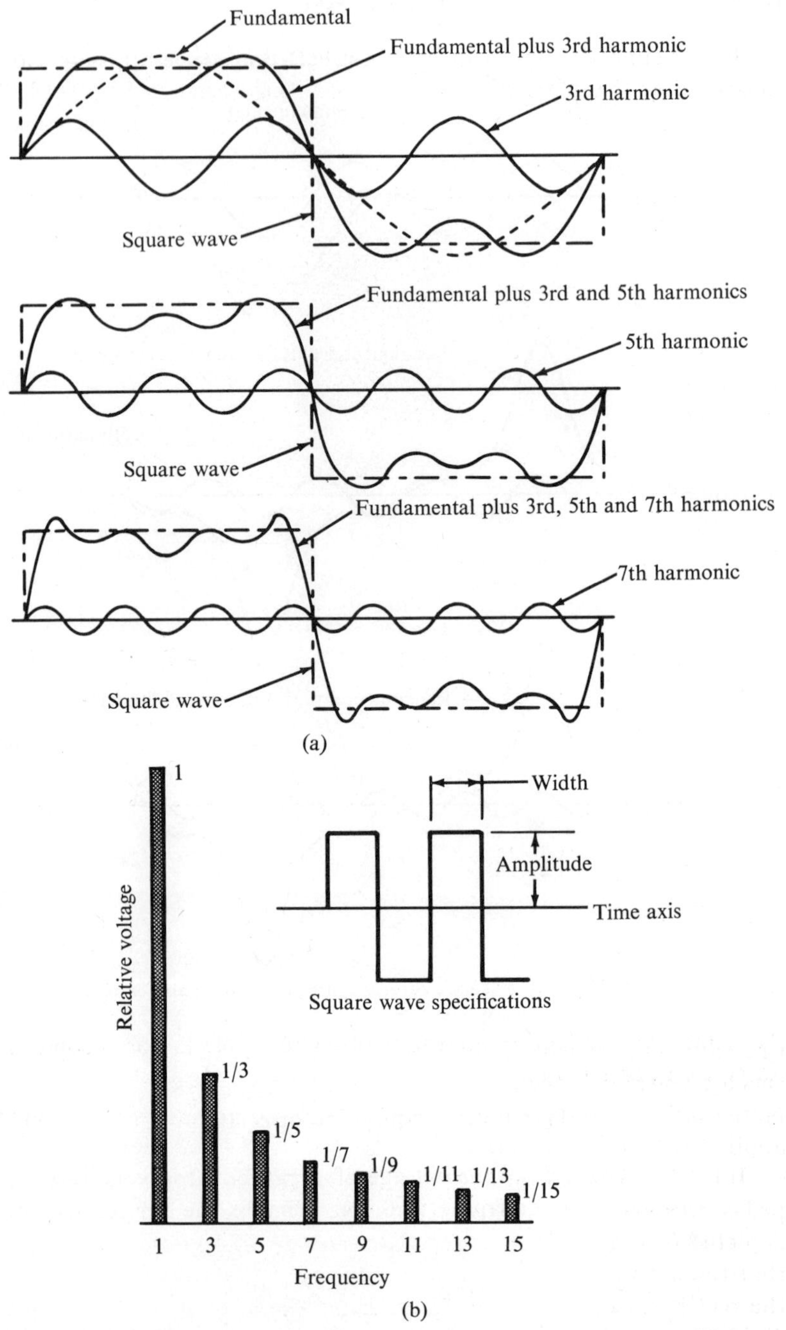

Figure 1-17. (a) A semi-square wave synthesized from sine waves; (b) Harmonic amplitudes and frequencies.

Power-supply ripple can be observed, and its amplitude can be measured with an oscilloscope such as illustrated in Fig. 1–18. The

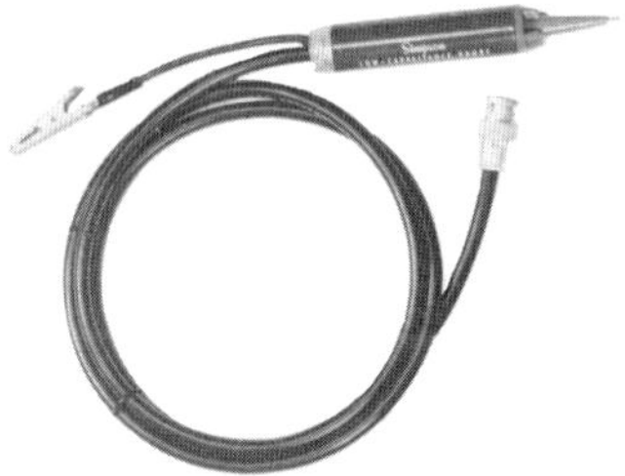

Courtesy of Simpson Electric Co.

Figure 1–18. A laboratory oscilloscope.

rms value of a nonsinusoidal waveform can be measured with a true rms a-c voltmeter, such as illustrated in Fig. 1–19 on the next page. Note that rectifier-type a-c voltmeters read correct values only when the applied voltage has a sine waveform.

If full-wave rectification is desired without the use of a center-tapped secondary [Fig. 1–14(a)], a bridge rectifier can be employed as depicted in Fig. 1–20 on page 23. Advantages of the bridge circuit are that the entire voltage from A to B in Fig. 1–20(a) is available to drive the rectifier section and the peak inverse voltage of the supply is divided across two rectifiers in series. On the other hand, only half the voltage from A to B in Fig. 1–14(a) is available to drive the rectifier

Courtesy of Ballantine Laboratories

Figure 1–19. A voltmeter that measures rms values of complex waveforms.

section. The disadvantage of the bridge circuit is that twice as many rectifiers must be employed for full-wave rectification.

1–4. Power-Supply Filters

A filter is used with a power supply to reduce the ripple amplitude of the pulsating d-c output. The simplest filter consists of a single shunt capacitor, as shown in Fig. 1–21(a). Observe that if the load resistance R_L were infinite, capacitor C would charge up to the peak voltage of the half-sine waveform, and the voltage across the capacitor would be pure d-c. On the other hand, if R_L were zero (a short-circuit), capacitor C would have no filtering action, and the current through the short-circuit would consist of half-sine waves.

In practical situations, R_L has a finite value, and the voltage output consists of pulsating d-c. The ripple waveform, or a-c component, has a peak-to-peak amplitude which is the difference of the crest voltage and trough voltage depicted in Fig. 1–21(b). Observe that the ripple waveform is not a half-sine wave when a half-sine voltage waveform is applied to C and R in parallel. The reason for this becomes apparent

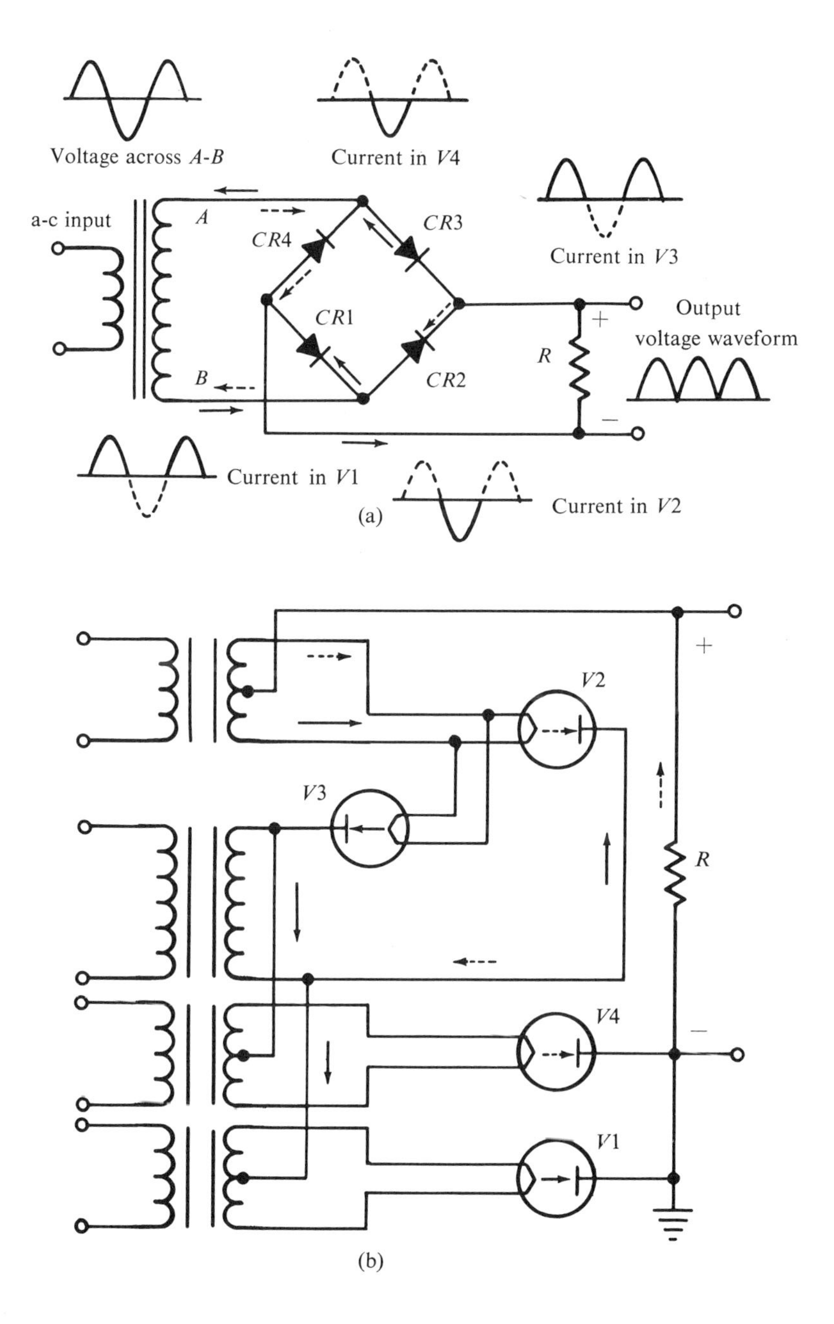

Figure 1-20. Full-wave bridge rectifier circuit. (a) Semiconductor circuit; (b) Vacuum tube circuit.

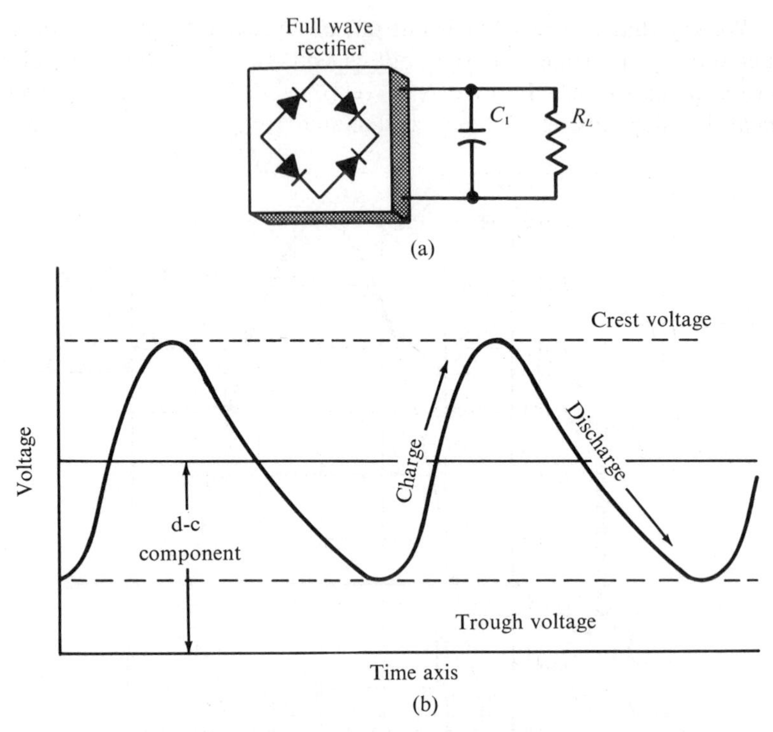

Figure 1–21. (a) Single capacitor filter; (b) Pulsating d-c voltage across load R_L.

when we consider the current flow from the rectifier to the filter. The shaded areas in Fig. 1–22 show that the current flows in pulses; in

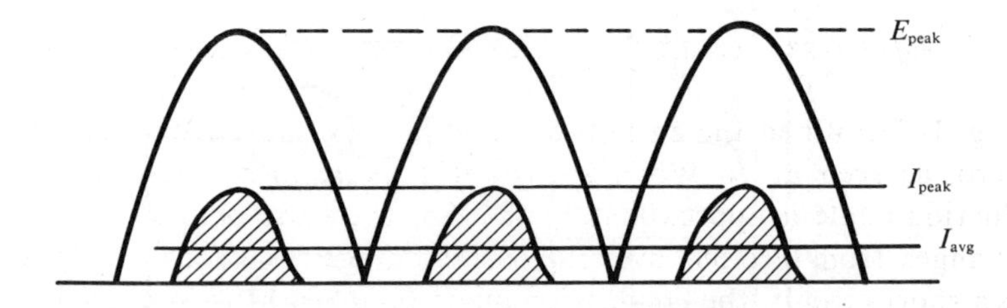

Figure 1–22. With moderate loads, pulses of current flow into the capacitor, as denoted by the shaded areas (I_{peak}).

other words, current cannot flow until the applied voltage exceeds the voltage across the capacitor. When the applied voltage falls below this level, the capacitor discharges through R_L and its voltage decays.

We say that the rectifier is cut off, or biased off, and cannot conduct during the time that the applied voltage is less than the voltage across the capacitor. In turn, the rectifier has a certain *conduction angle* that depends on the current demand from the filter, as shown in

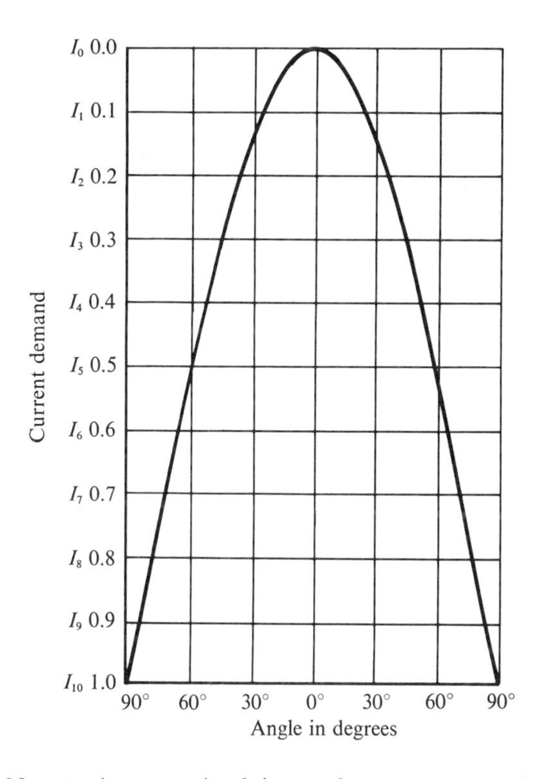

Figure 1–23. Conduction angle of the rectifier versus current demand.

Fig. 1–23. When the current demand is zero, the conduction angle is zero, as seen at I_0. When a current I_1 is taken by the load, the conduction angle is approximately 58°. So, the *width* of the current pulse changes from 0 to 58° over the current range from I_0 to I_1. If the load is a short-circuit, the conduction angle becomes 180°, as shown at I_{10}.

Another basic power-supply parameter is called the *peak inverse voltage* that appears across the rectifier. Let us consider the peak voltage that appears across rectifier *CR* in Fig. 1–24 at 270°; the rectifier does not conduct, or the rectifier is reverse biased to the peak inverse voltage. Since there is no current demand in this circuit, capacitor *C* charges up to the peak value of the applied voltage, 10 v,

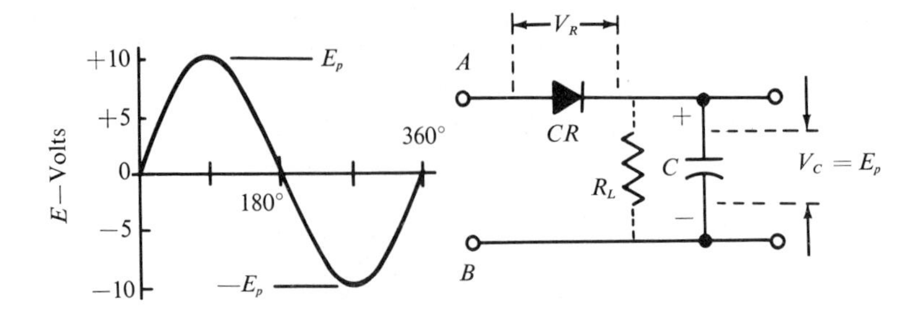

Figure 1–24. A peak inverse voltage $2E_p$ must be withstood by rectifier *CR*.

in the polarity indicated. At 270°, −10 v is applied to terminal *B*. Observe that this applied voltage is in series with voltage V_C. Therefore, rectifier *CR* must withstand 20 v at 270°. Thus, the maximum peak inverse voltage that must be withstood by rectifier *CR* is formulated

$$V_{PI} = 2\,E_p \qquad \textbf{(1–22)}$$

We perceive that if a load resistor were connected across *C* in Fig. 1–24, V_C would be less than E_p, because the capacitor would discharge through the resistor between pulses of current flow. Vacuum diodes and semiconductor diodes are rated by their manufacturers for maximum peak inverse voltage. The maximum rating must not be exceeded in a power-supply circuit or rectifier damage can be anticipated. Diodes are also rated for maximum peak anode current, and this rating must not be exceeded.

Analysis of circuits that entail nonsinusoidal waveforms is frequently made in terms of the *form factor* of a waveform. The form factor is the ratio of the rms voltage value to the average voltage value. For a half-rectified sine wave, the rms value is equal to 0.5 of its peak value, as the student may verify from Formulas 1–20 and 1–21. In other words, the rms value of the waveform is equal to the d-c component (0.318 peak), plus the *rms* value of the a-c components (0.182 peak); or, the rms value of the half-rectified sine wave is 0.5 of its peak value.

Thus, the form factor of a rectified half-sine wave is formulated

$$\text{Form factor} = \frac{\text{rms value (full cycle)}}{\text{average value (full cycle}} = \frac{0.5}{0.318} = 1.57 \qquad \textbf{(1–23)}$$

Example 10. The d-c output of a filter is 240 v and the rms value of the ripple is 0.96 v. What is the form factor of the filter?

$$\text{Form factor} = \frac{\text{rms value}}{\text{average value}}$$
$$= \frac{0.96}{240}$$
$$= 0.004 \qquad \text{(answer)}$$

Algebraic manipulation shows, in turn, that the ripple factor may be formulated

$$\text{Ripple factor} = \sqrt{\left(\frac{\text{rms}}{\text{d-c}}\right)^2 - 1} = \sqrt{F^2 - 1} \qquad \textbf{(1-24)}$$

where F denotes the form factor.

Example 11. The ripple factor for a half-wave rectifier is

$$\text{Ripple factor} = \sqrt{(1.57)^2 - 1} = 1.21 \qquad \text{(answer)}$$

Similarly, it can be shown that the form factor for a full-rectified sine wave is 1.11. In turn, the ripple factor for a full-rectified sine wave is 0.48 (from Formula 1-24), as previously obtained in the first analysis.

The form factor is also useful in analysis of rectifier-type voltmeters which do not respond to effective values. The voltmeters actually respond to the average output of a rectifier, and the a-c voltage scale is calibrated in terms of the rms value of one type of signal, usually a sine wave. Hence, any reading on the a-c scale is 1.11 times the average value to which the instrument responds, if the voltmeter employs a full-wave instrument rectifier. Again, a reading on the a-c scale is 2.22 times the average value to which the instrument responds, if the voltmeter employs a half-wave instrument rectifier.

This same basic principle is applied to filtered output waveforms such as that in Fig. 1-21(a) to calculate the form factor and the ripple factor. In practical engineering work, the situation is often simplified by assuming that the ripple waveform in Fig. 1-21(a) is a sawtooth waveform. The Fourier series for a sawtooth waveform is comparatively simple and is written

$$e = \frac{2E_p}{\pi}\left(\sin \omega t - \frac{1}{2}\sin 2\omega t + \frac{1}{3}\sin 3\omega t - \frac{1}{4}\sin 4\omega t \cdots\right) \qquad \textbf{(1-25)}$$

where e denotes the instantaneous value of the sawtooth wave, and E_p denotes one-half the peak-to-peak value of the sawtooth wave.

Another important parameter that characterizes a power supply is its *efficiency*. The efficiency of a power supply is simply the ratio of output power to input power. Thus, we may connect a wattmeter to the input of a power supply to measure its input power. An electrodynamometer-type wattmeter may also be connected to the output of the power supply to measure its output power. Or, if the ripple is very low in the output, the output power may be calculated as EI, where E is the d-c output voltage and I is the d-c output current.

Thus far, we have considered only the steady-state operation of a power supply. However, there is an important parameter called the *starting surge*. With reference to Fig. 1–24, it is apparent that when the a-c voltage is first switched into terminals A and B, capacitor C has zero charge. Suppose that the a-c voltage is switched at the 90° instant. Then, a pulse current flows, and this pulse current is not limited by any opposing voltage from the capacitor. The amplitude of current which flows is limited only by the internal resistance of rectifier CR and the internal resistance of the source transformer (which is negligible in many situations). Therefore, if C has a large capacitance, as is usually the case, the starting surge may exceed the peak-current rating of CR.

Accordingly, to avoid damage to the rectifier, a resistor is commonly connected in series between CR and C (as in Fig. 1–24). In a typical power supply, C has a value of 50 μf and is connected in series with a protective resistance of 120 Ω. Of course, the protective resistance reduces the efficiency of the power supply and causes poor regulation. The configuration in Fig. 1–24 is called a *capacitor-input filter*, because the rectifier feeds directly into the capacitor. If we connect a protective resistor between CR and C, the configuration is then called a *resistive-input filter*. We may also connect an inductor between CR and C, in which case the configuration is called a *choke-input filter*. Details of choke-input filters and elaborated filter configurations are explained in the next chapter.

1–5. Regulation versus Filter Capacitance

The regulation of a power supply can be improved by increasing the filter capacitance. For example, Fig. 1–25 depicts two simple filter configurations in which the parameters are the same except for the value of the filter capacitor. When the switch is closed in Fig. 1–25(a), the d-c output voltage falls by approximately 4.4 per cent. On the other hand, when the switch is closed in Fig. 1–25(b), the d-c output voltage falls by approximately 2.8 per cent. Thus, an *increase* of ten times in the value of the filter capacitor provides an increase of 1.6

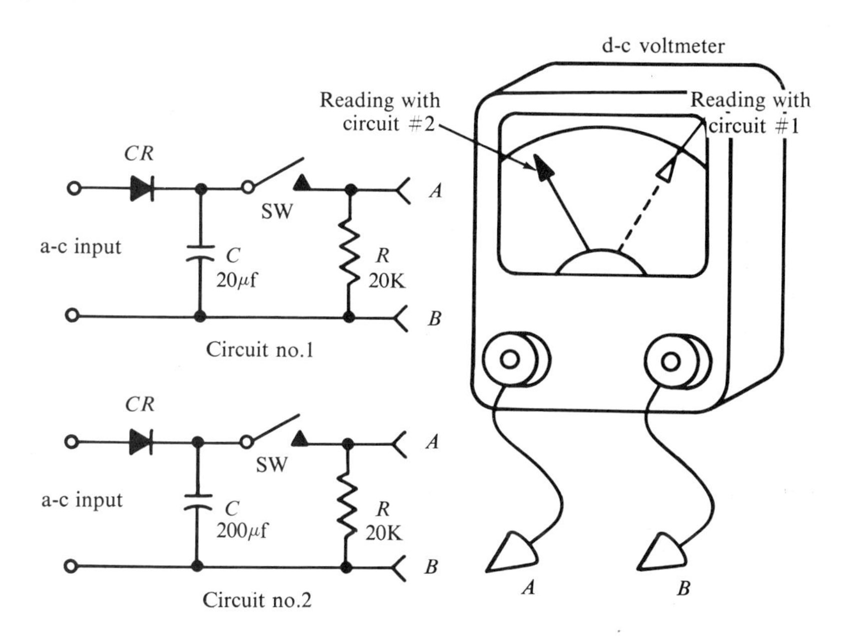

Figure 1-25. (a) Regulation is poor when the filter capacitor has a small value; (b) Regulation is much better when the filter capacitor has a large value.

per cent in d-c voltage under load and a 35.7 per cent decrease in the effective internal impedance of the supply.

Example 12. Determine the effective internal resistance of each power supply in Fig. 1-25 when the open-circuit voltage is 100 v. The open-circuit voltage of Fig. 1-25(a) reduces by 4.4 per cent when loaded by a 20-kΩ resistor.

$$E_{\text{load}} = 100 \times 0.956 = 95.6 \text{ v}$$

The voltage drop across the internal resistance is then 4.4 v and the internal resistance is

$$\frac{R_{\text{int}}}{R_{\text{load}}} = \frac{E_{\text{int}}}{E_{\text{load}}} \tag{1-26}$$

$$R_{\text{int}} = \frac{4.8}{95.2} \times 20 \text{ k}\Omega = 918 \ \Omega \qquad \text{(answer)}$$

The open-circuit voltage of Fig. 1-25(b) reduces by 2.8 per cent when loaded by a 20-kΩ resistor.

$$E_{\text{load}} = 100 \times 97.2 = 97.2 \text{ v}$$

The voltage drop across the internal resistance is 2.8 v and the internal resistance is

$$R_{\text{int}} = \frac{2.8}{97.2} \times 20 \text{ k}\Omega = 590 \ \Omega \qquad \text{(answer)}$$

Of course, the ripple amplitude is also reduced by the larger value of filter capacitance. Thus, if you connect an oscilloscope in place of the voltmeter depicted in Fig. 1-25, you will observe that the ripple amplitude is approximately 8.75 times greater when a 20-μf capacitor is used, than when a 200-μf capacitor is used. Thus, the ripple amplitude is greatly reduced while the regulation is improved to some extent. Note that in case the current demand from the filter is heavy, the improvement in regulation will be more evident; in turn the reduction in ripple amplitude will be less evident under heavy load.

1-6. Power-Supply Terminology

To review briefly, power supplies for vacuum-tube electronic equipment are chiefly classified as "A", "B", or "C" supplies. Thus, an "A" battery energizes the filament or heater of a tube, as seen in Fig. 1-26(a). A "B" battery energizes the plate of a tube. A "C" battery biases the grid of a tube. If a tube has a screen grid, the "B" battery energizes the screen grid. Heater-type tubes are usually powered by a-c, and we state, for example, that the heater or "A" supply provides 6.3 rms v. Plate-supply voltages are usually positive with respect to ground, and we state, for example, that the "B+" voltage is +250 d-c v. Grid-bias voltages are usually negative with respect to ground, and we state, for example, that the "C−" voltage is −4.5 d-c v. If the cathode of a tube is biased, the bias voltage is usually positive with respect to ground; we may state, for example, that the cathode bias voltage is +4.5 d-c v.

Power supplies for transistorized electronic equipment are commonly said to provide *source voltage* as bias voltage. Thus, a typical power supply for a transistor TV receiver is described as a −11.5 d-c v *source*, a −11.7 d-c v *source*, and a 12.3 d-c v *source*. TV receivers also supply from 7,000 v to 25,000 v to the second anode of the picture tube. Whether a receiver employs tubes or transistors, this circuit section is called the *high-voltage power supply*. Its output voltage is always positive with respect to ground. In the case of gas tubes, such as thyratrons used in industrial-electronics equipment, we speak of the *anode supply*, and the *filament* or *heater supply*.

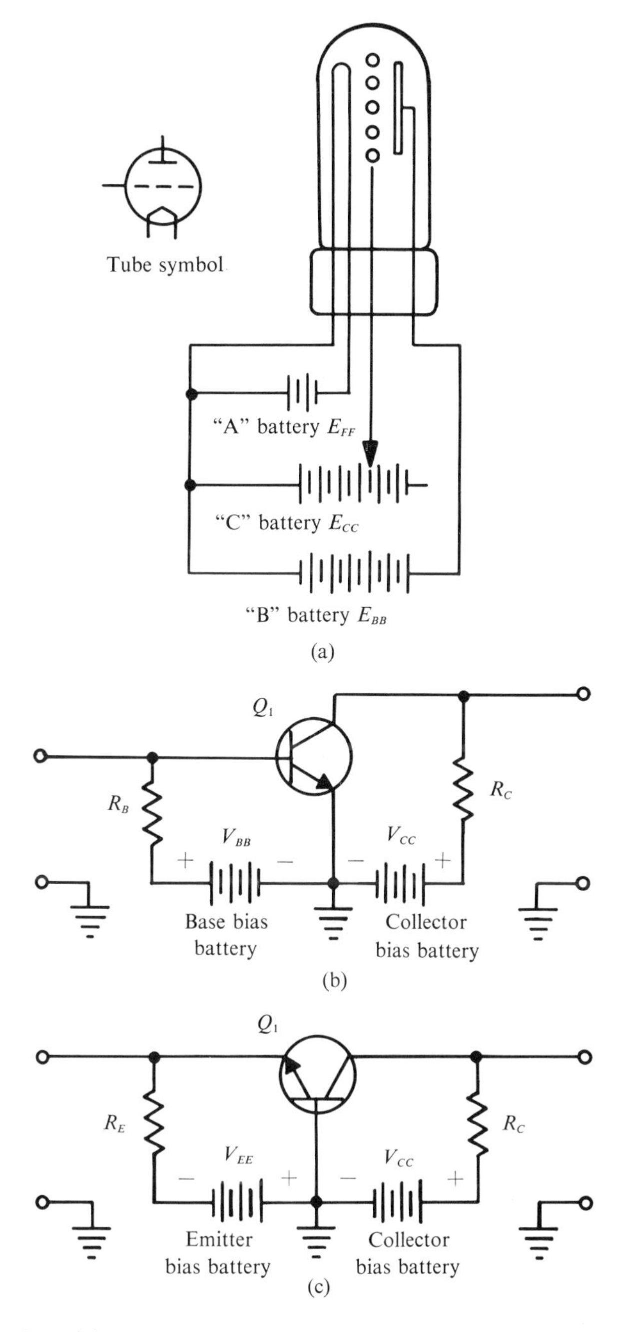

Figure 1-26. (a) "A," "B," and "C" battery terminology; (b) Collector and base bias battery terminology; (c) Collector and emitter bias battery terminology.

Power-supply terminology for transistor circuits is also somewhat different than that for vacuum-tube equipment. The collector supply in Fig. 1–26(b) is called the collector bias V_{CC}, and the base supply is called base bias V_{BB}. In Fig. 1–26(c), emitter supply is classified as the emitter bias V_{EE}. Power supplies or batteries for transistor circuits are usually much smaller than those used in similar vacuum-tube circuits.

1–7. Measurement of rms Values of Complex Waveforms

Since the current flow through a rectifier is nonsinusoidal, a-c circuit voltages and currents in power supplies cannot be measured accurately with a conventional volt-ohm-milliammeter. We can use an oscilloscope to measure peak-to-peak voltages of complex waveforms. A peak-to-peak indicating vacuum-tube voltmeter can also be utilized; however, to measure the rms value of a complex waveform, a true rms voltmeter must be used, such as illustrated in Fig. 1–19. The term "true rms" denotes that the meter movement or indicator responds to the rms value of a complex waveform—not to its average value nor to its peak-to-peak value.

There are two general types of true rms voltmeters. We recall that the electrodynamometer-type voltmeter indicates rms values of complex waveforms. This type of instrument requires comparatively large input power, and it is not suited to the measurement of small voltage values. The other type of true rms voltmeter is basically a hot-wire instrument; the hot-wire element is called a barretter and operates in a bridge circuit. Note that the input impedance of this voltmeter is 2 MΩ in parallel with 15 to 45 pf. Its voltage range is from 0.1 to 1199.9 v rms and operates at frequencies up to 20 kHz.

1–8. Basic Transformer Principles

Power transformers utilize iron cores that form a closed magnetic circuit. This design permits power transfer at high efficiency and minimizes the amount of copper wire that is required. A magnetic circuit is comparable in many respects to an electric circuit. We recall that Ohm's law is similar to the basic law for magnetic circuits.

$$I = \frac{E}{R} \tag{1–27}$$

$$\text{Flux} = \frac{\text{magnetomotive force}}{\text{reluctance}} \tag{1–28}$$

The unit of magnetic flux (ϕ) is 1 line. A magnetomotive force (mmf) of 1 amp-turn establishes 1 line of flux in a reluctance of 1 rel.* Note that a column of air 1 in. square and 3.19 in. long has a reluctance of 1 rel. Air has a reluctance approximately 2,000 times greater than that of iron. The reluctivity of a substance is equal to the reluctance of a unit cube of the substance—in this case, the reluctance of 1 cu in. On the other hand, the permeability of a substance is equal to the reciprocal of its reluctivity. A basic law of magnetic circuits states that

$$B = \mu H \tag{1-29}$$

wherein B denotes the flux density in lines per sq in., μ denotes the permeability of the core, and H denotes the magnetizing force in ampere-turns per in.

Note carefully that the magnetizing force is equal to the magnetomotive force divided by the length of the magnetic circuit.

$$H = \frac{\text{ampere-turns}}{L} \tag{1-30}$$

where L denotes the length around the magnetic circuit (core) in inches.

Iron is a nonlinear magnetic substance. Hence, core calculations are commonly based on BH charts, such as shown in Fig. 1-27. To find the required cross-section of core, we calculate the power to be supplied by the transformer; knowing the number of turns on the primary, we can calculate the mmf. Then, having chosen a certain core length, we calculate H. This value of H corresponds to a value of B [Fig. 1-27(a)] on the BH curve. If it should fall in the saturation region, we must increase the cross-section of the core sufficiently to place the operating point below the knee of the BH curve. The effect of core saturation on the *permeability* curves of iron samples is seen in Fig. 1-27(b).

Operation of a power transformer is based on the principle that electrical energy can be transferred efficiently by mutual induction from one winding to another. When the primary winding is energized from an a-c source, an alternating magnetic flux is established in the transformer core. This flux links the turns of both primary and secondary, thereby inducing voltages in them. Because the same flux cuts both windings, the same voltage is induced in each turn of both windings. Hence, the total induced voltage in each winding is proportional to the number of turns in that winding. That is,

---

*Note that there is no universally accepted unit of reluctance; however, many authors use the unit "rel."

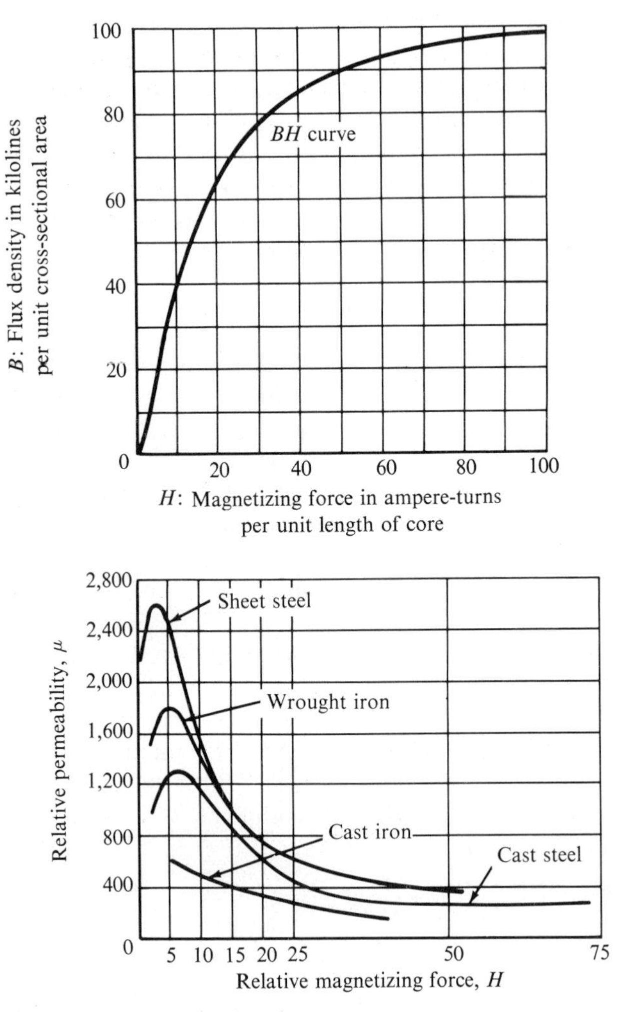

Figure 1-27. (a) A *BH* curve for a typical iron sample; (b) Permeability of various iron samples versus magnetizing force.

$$\frac{E_1}{E_2} = \frac{N_1}{N_2} \qquad \textbf{(1-31)}$$

where E_1 and E_2 are the induced voltages in the primary and secondary windings, respectively, and N_1 and N_2 are the number of turns in the primary and secondary windings, respectively.

In conventional transformers, the induced primary voltage is almost equal to the applied primary voltage; hence, the applied primary voltage and the induced secondary voltage are approximately proportional to the respective number of turns in the two windings. In other words, we wind a sufficient number of turns on the primary to obtain a high reactance at the operating frequency so that the no-load current of the transformer is negligible.

A constant-voltage single-phase transformer is represented in Fig. 1–28(a). For simplicity, the primary winding is shown as being on one leg of the core and the secondary winding on the other leg. The formula for induced voltage in one winding is written

$$E = \frac{4.44\ BSfN}{10^8} \qquad \textbf{(1-32)}$$

where E denotes the induced rms voltage, B the maximum value of magnetic flux density in lines per square inch in the core, S the cross-sectional area of the core in square inches, f the frequency in Hz, and N the number of turns in the winding.

For example, if the maximum flux density is 90,000 lines per sq in., the cross-sectional area of the core is 4.18 sq in., the frequency is 60 Hz, and the number of turns in a winding is 1,100, then the voltage rating of this winding is formulated

$$E_1 = \frac{4.44 \times 90{,}000 \times 4.18 \times 60 \times 1{,}100}{10^8} = 1{,}100 \text{ v} \qquad \textbf{(1-33)}$$

Let us suppose that E_1 is the voltage rating of the primary winding and that the secondary has 1/10 as many turns. Then the voltage rating of the secondary winding is formulated

$$E_2 = \frac{1{,}100}{10} = 110 \text{ v} \qquad \textbf{(1-34)}$$

Waveforms for an ideal power transformer at no load are shown in Fig. 1–28(b). When E_1 is applied to the primary winding N_1, with the switch S open, the resulting current I_0 is small and lags E_1 by almost 90° because the circuit is highly inductive. The no-load current is called the *exciting*, or *magnetizing* current, because it supplies the magnetomotive force that produces the transformer core flux ϕ. The flux produced by I_0 cuts the primary winding N_1 and induces a counter voltage E_c 180° out of phase with E_1 in this winding. The voltage E_2 induced in the secondary winding is in phase with the induced (counter) voltage E_c in the primary winding; both voltages lag the exciting current and flux, whose variations produce them, by an angle of 90°.

The foregoing relations are depicted in vector form in Fig. 1–28(c). Note that the values are not drawn to scale. When a load is connected to the secondary by closing switch S (Fig. 1–28(a), the secondary

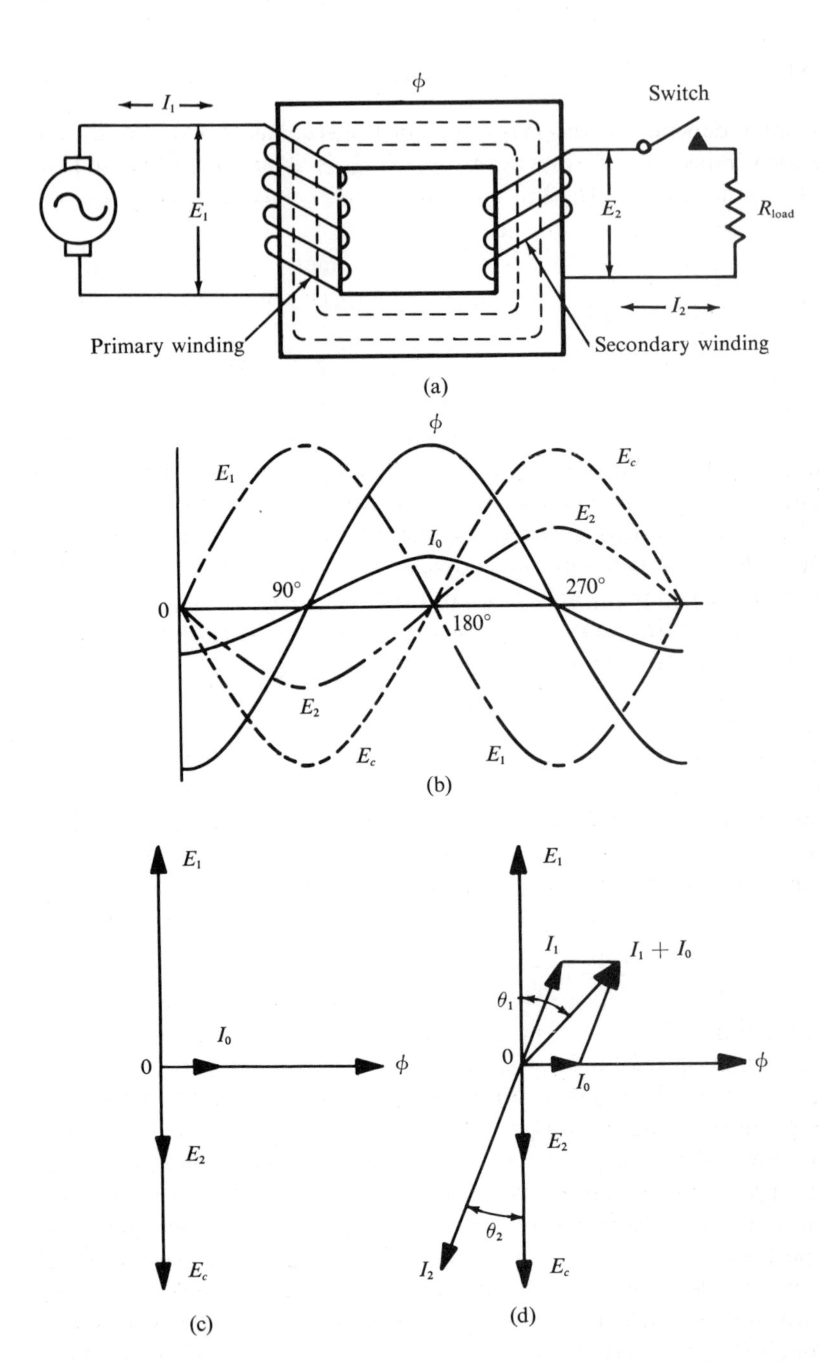

Figure 1–28. Physical arrangement and electrical parameter relations in a constant-voltage transformer. (a) Constant voltage transformer; (b) Transformer waveforms (no load); (c) 90° vector form; (d) Lagging power-factor load vectors.

current I_2 depends on the magnitude of the secondary voltage E_2, and the load impedance Z. For example, if E_2 is equal to 110 v, and the load impedance is 22 Ω, the secondary current will be

$$L_2 = \frac{E_2}{Z_2} = \frac{110}{22} = 5 \text{ amp} \tag{1-35}$$

If the secondary power factor is 86.6 per cent, the phase angle θ_2 between secondary current and voltage will be the angle whose cosine is 0.866, or 30°. The secondary load current flowing through the secondary turns comprises a load component of mmf, which is in such a direction (according to Lenz' law) to oppose the flux which is producing it. This opposition tends to reduce the transformer flux a slight amount. The reduction in flux is accompanied by a reduction in the counter voltage induced in the primary winding of the transformer. Because the internal impedance of the primary winding is low, and the primary current is limited principally by the counter emf in the winding, the transformer primary current increases when the counter emf in the primary is reduced.

The increase in primary current continues until the primary ampere turns are equal to the secondary ampere turns, neglecting losses. For example, in the transformer being considered, the magnetizing current I_0 is assumed to be negligible in comparison with the total primary current $I_1 + I_0$ under load conditions, because I_0 is small in relation to I_1 and lags it by an angle of 60°. Hence, the primary and secondary ampere turns are equal and opposite. In this example

$$N_1 I_1 = N_2 I_2 \tag{1-36}$$

$$I_1 = \frac{N_2 I_2}{N_1} = \frac{110 \times 5}{1100} = 0.5 \text{ amp} \tag{1-37}$$

Neglecting losses, the power demand of the primary is equal to the power supplied by the secondary to the load. If the load power is $P_2 = E_2 I_2 \cos \theta_2$, or $110 \times 5 \cos 30°$, or 476 w, the power supplied to the primary is approximately $P_1 = E_1 I_1 \cos \theta_1$, or 476 w. The load component of primary current I_1 increases with secondary loading and maintains the transformer flux at nearly its initial value. This action permits the transformer primary to take power from the source in proportion to the load demand and to maintain the terminal voltage approximately constant. The lagging power-factor load vectors are shown in Fig. 1-28(d). Note that the load power factor is transferred through the transformer to the primary and that θ_2 is approximately equal to θ_1. θ_1 is slightly larger than θ_2 because of the presence of the exciting current which flows in the primary winding but not in the secondary.

SUMMARY

1. Battery action of a cell is based upon the chemistry of dissimilar conductors immersed in an electrolyte.

2. Local action is caused by impurities in the zinc and may be controlled by coating the zinc with mercury.

3. Polarization results from production of hydrogen at the carbon electrode of a dry cell to form an insulating layer.

4. A mercury cell has a greater coulomb capacity than an ordinary dry cell.

5. The internal resistance of a cell, which does not remain constant during discharge, limits the maximum current by decreasing the terminal voltage when the cell is loaded.

6. The internal resistance of an energy source may be calculated from the formula

$$R_{\text{int}} = \frac{E_{\text{OC}} - E_L}{I}$$

7. Storage batteries are secondary cells and, as such, may be recharged when their chemical energy is depleted.

8. The efficiency of a storage cell is greater when discharged at a slower rate.

9. Any battery or power supply has a certain regulation under load, which is attributed chiefly to the internal resistance and may be calculated from the formula

$$\%\text{ regulation} = \frac{E_{\text{OC}} - E_{\text{CC}}}{E_{\text{CC}}} \times 100$$

10. A nickel-cadmium storage battery is superior to a lead-acid battery in that it may be stored discharged, and it has a longer life.

11. Child's law is a formula that expresses the nonlinear relationship between voltage and current in an ideal vacuum diode.

12. The important values of a sine wave are formulated

$$E_{\text{rms}} = \frac{1}{\sqrt{2}} E_p \cong 0.707\ E_p$$

$$E_{pp} = 2E_p$$

$$E_{\text{avg}} = 0$$

13. The relationships of the average current and voltage of a half-rectified sine wave are

$$E_{\text{avg}} = \frac{1}{\pi} E_p \cong 0.318\ E_p$$

$$I_{\text{avg}} = \frac{1}{\pi} I_p \cong 0.318\, I_p$$

14. The relationships of the average voltage and average current and peak voltage and peak current in a full-wave rectified sine wave are

$$E_{\text{avg}} = \frac{2}{\pi} E_p \cong 0.636\, E_p$$

$$I_{\text{avg}} = \frac{2}{\pi} I_p \cong 0.636\, I_p$$

15. The ripple factor of output voltage of a power supply is formulated

$$\text{Ripple factor} = \frac{\text{rms value of a-c components}}{\text{average value}}$$

16. Per cent ripple is equal to the ripple factor times 100.

17. The advantages of the bridge rectifier are (1) no center-tapped secondary is required, and (2) each rectifier must drop only half the inverse secondary voltage.

18. The peak-inverse voltage of a half-wave rectifier utilizing a capacitor input filter with no load is formulated

$$V_{PI} = 2E_p$$

19. The *form factor* of a nonsinusoidal waveform is the ratio of the rms voltage value to the average voltage value, and is formulated

$$\text{Form factor} = \frac{\text{rms value}}{\text{average value}}$$

20. The ripple factor may be formulated in terms of the form factor.

$$\text{Ripple factor} = \sqrt{F^2 - 1}$$

21. The form factor for a full-rectified sine wave is 1.11, and the ripple factor for a full-rectified sine wave is 0.48.

22. A disadvantage of a capacitive input filter is the possibility of a large surge current when the circuit is switched on.

23. Classifications of power supplies for vacuum-tube circuits originated with the names for batteries which were used for proper biasing of the tube elements. Likewise, the classifications of power supplies for transistor circuits are in accordance with their functions in biasing transistor circuits.

24. The rms value of a complex waveform must be measured with a special type of voltmeter called a *true rms voltmeter*.

Questions

1. Explain local action in a zinc-carbon cell. How is local action controlled in a conventional cell?
2. Explain polarization action in a dry cell. How can this action be minimized?
3. Why is the efficiency of a storage battery greater when the battery is discharged at a slower rate?
4. What factors contribute to the regulation of a battery or power supply?
5. What is the chemical reaction in a nickel-cadmium storage battery?
6. State the relationship of the peak voltage to the rms voltage of a sine wave.
7. Draw one cycle of a rectified-sine wave and indicate the value of the peak voltage, the rms voltage, and the average voltage.
8. What is the disadvantage of a capacitance-input filter, and what is the proper solution?
9. What are the limitations of batteries as power sources for electronics equipment? What are the advantages?
10. Why must a cell being charged have a vent hole?
11. What type of voltmeter may be used to measure complex waveforms?

PROBLEMS

1. The terminal voltage of a battery is 12.2 v on open circuit and 9.3 v with a 0.31-Ω load. What is the internal resistance of the battery?
2. The terminal voltage of a power source is 120 v on open circuit and 111 v at full load. What is the per cent regulation of the supply?
3. A 110-rms v a-c waveform is rectified by a half-wave rectifier. What is the d-c level of the output waveform?
4. The output of a power supply has an average value of 410 v and an rms value of 1.2 v. What is the ripple factor?
5. The secondary voltage from points *A* to *B* in Fig. 1-20 is measured at 420 v rms. What is the minimum peak-inverse breakdown voltage of each rectifier?
6. The d-c voltage output from the circuit in Fig. 1-25(b) is 300 v; the a-c input voltage is 420 v rms. What is the maximum peak-inverse voltage across the diode?

7. The output voltage of a power supply is 382 v on open circuit and 71 v when loaded by a 10-kΩ load. What is the effective internal resistance of the supply?

CHAPTER 2

Principles of Filtering

2–1. Introduction

Unless the output waveform from a rectifier is extensively filtered, the d-c output voltage will have an appreciable number of a-c components called *ripple*. Ripple voltage occurs because energy is supplied in pulses to the load by the rectifier. In most applications, the ripple amplitude must be reduced to a very low level. A simple filter consisting of a capacitor connected across the load reduces the ripple amplitude, because energy is stored in the capacitor during the time that the rectifier supplies a pulse output; the capacitor then discharges through the load during the time that the pulse output is absent.

Figure 2–1(a) depicts the output from a half-wave rectifier. Pulses of energy from the rectifier are applied to filter capacitor C in Fig. 2–1(b); in turn, the terminal voltage across C produces current flow through load resistor R. Capacitor C charges through a small series resistance (internal resistance of the rectifier diode and winding resis-

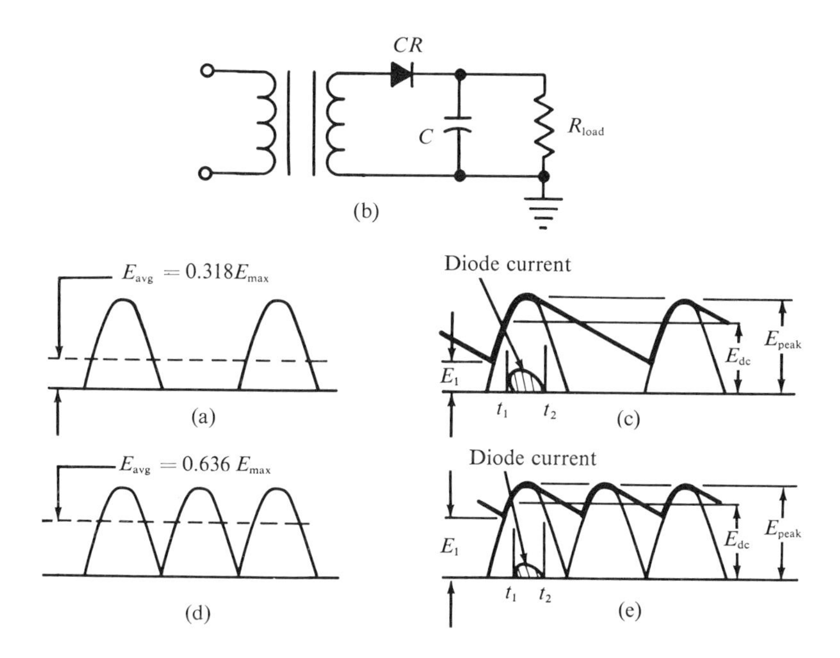

Figure 2-1. Simple capacitor filter with voltage and current waveforms. (a) Halfwave output without filtering; (b) Filter; (c) Filtered halfwave output; (d) Fullwave output without filtering; (e) Filtered fullwave output.

tance of the transformer secondary), and discharges through a much higher value of series resistance (resistance value of R). Figure 2-2 depicts the familiar universal RC time-constant chart which governs the basic waveshape of the ripple voltage.

We observe from Fig. 2-1(c) that there are two considerations that limit the rapidity with which capacitor C will charge. As previously noted, the internal resistance of the charging circuit is a limiting factor which may be evaluated from the charging curve in Fig. 2-2(a). Also, the rectifier current pulse has a slow rise, and the capacitor cannot charge faster than the pulse rises, even if the internal resistance of the charging circuit is negligible. When the conduction angle ends at t_2 in Fig. 2-1(c), the charge in the capacitor decays through R in accordance with the discharge curve in Fig. 2-2(b). In other words, this interval of the ripple waveform has an exponential shape. When a

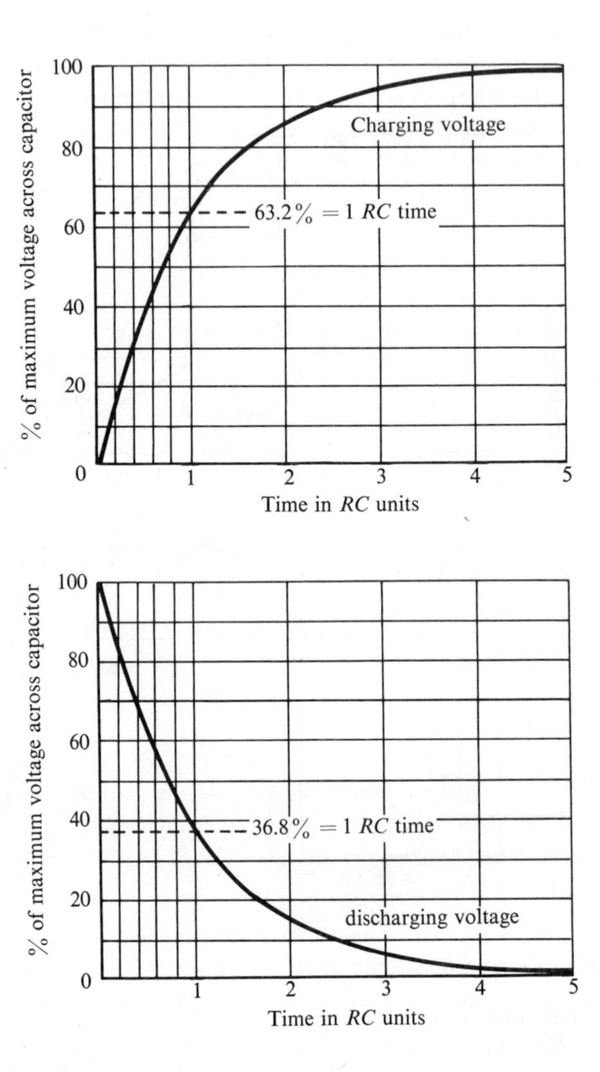

Figure 2–2. **Universal *RC* time-constant chart. (a) Charging curve; (b) Discharge curve.**

full-wave rectifier is connected to the filter, the voltage pulses are applied to the filter at twice the rate of the half-wave voltage pulses. The filter capacitor has less time to discharge between pulses, resulting in smaller conduction angle, less ripple amplitude, and double the ripple frequency.

If a full-wave rectifier were used, the ripple amplitude would be less for a given value of R and C. The charging pulses then occur twice as rapidly, and the filter capacitor does not discharge as extensively between pulses. Let us observe the d-c output voltage variation versus the load-resistance and capacitance values for a full-wave rectifier with a simple capacitor filter. This variation is shown in Fig. 2–3.

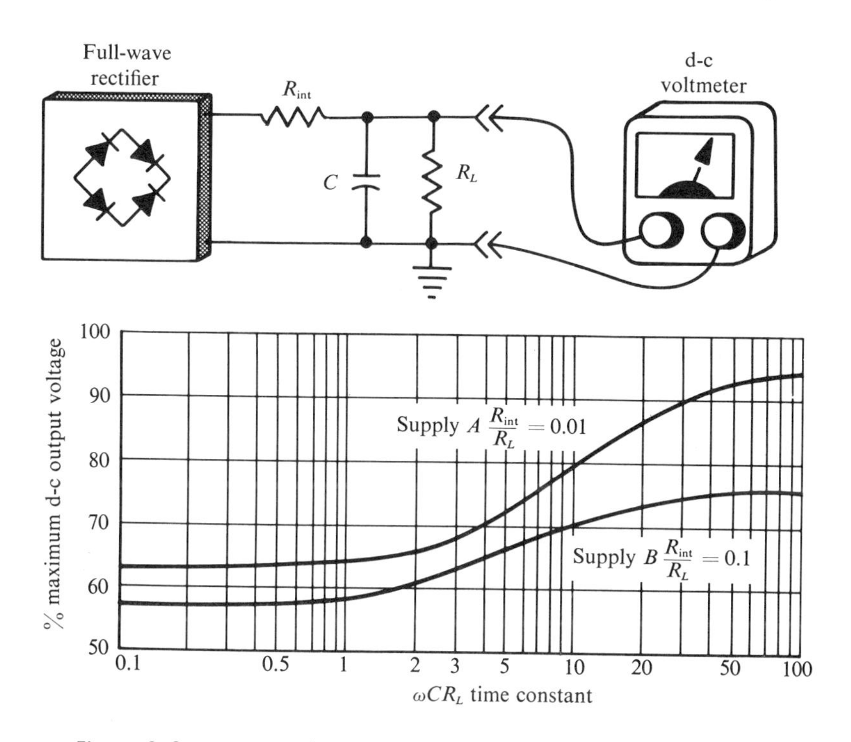

Figure 2–3. Output voltage versus load resistance and filter parameters.

Note that it was previously mentioned that the regulation depends considerably upon the internal resistance of the charging circuit.

Example 1. Determine from the graph in Fig. 2–3 the output voltage of each of the power supplies represented by the curves when the load resistance is 10 kΩ, the frequency is 60 Hz, the bypass capacitor is 10 μf, and the open-circuit voltage is 480 v.

Supply A: at $R_L = \omega C\, R_L = 37.7$, $E_{out} = 90\%\ E_{OC}$

$E_O = 0.8 \times 480 = 432$ v (answer)

Supply B: at $\omega C\, R_L = 37.7$, $E_O = 75\%\ E_{OC}$

$E_O = 0.7 \times 480 = 360$ v (answer)

It follows from previous discussion that the ripple amplitude will increase as the load current increases. Let us observe how the ripple percentage varies versus the load-resistance value for a full-wave rectifier with a simple capacitor filter. This variation is shown in Fig. 2–4. Note that if the capacitance value is doubled, the ripple percentage is substantially reduced.

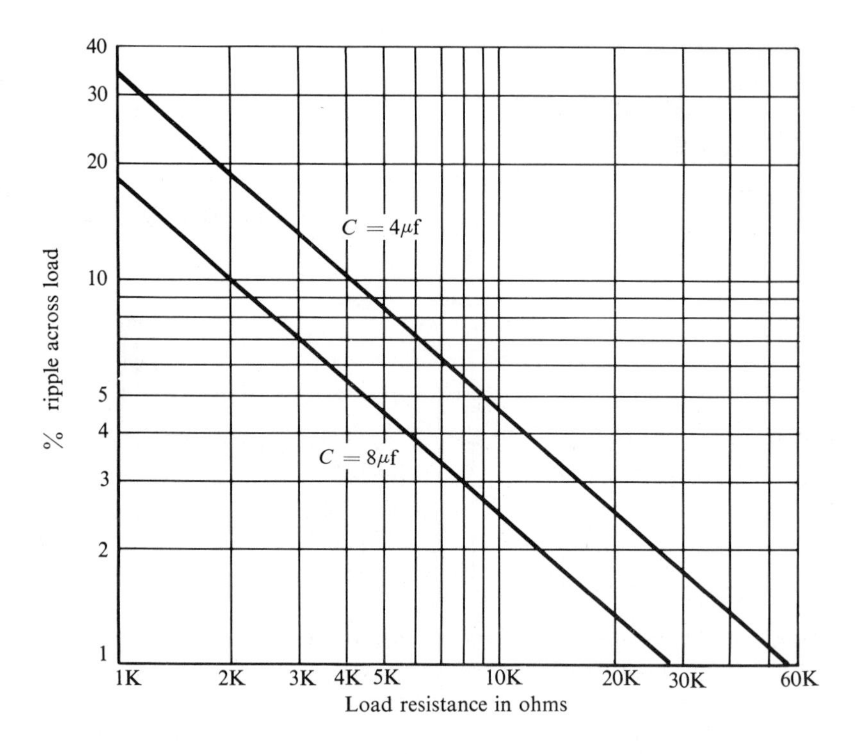

Figure 2–4. Per cent ripple versus load resistance and filter capacitor values.

Example 2. Determine the ripple for each of the filters represented in Fig. 2–4 with a 10-kΩ load.

$C = 8\ \mu\text{f},\ R_L = 10\ \text{k}\Omega,\ \text{ripple} = 3.9\%$ (answer)
$C = 4\ \mu\text{f},\ R_L = 10\ \text{k}\Omega,\ \text{ripple} = 4.6\%$ (answer)

A sensitive radio receiver may require that the ripple voltage be reduced to 0.01 per cent. On the other hand, a radio telephone transmitter operates satisfactorily if the ripple voltage is reduced to approximately 1 per cent. A radio telegraph transmitter requires that the ripple voltage be reduced to only 5 per cent.

As previously noted, the internal resistance of the charging circuit is the sum of the resistances in the circuit. The rectifier has a non-linear internal resistance. Figure 2–5 shows the forward and reverse

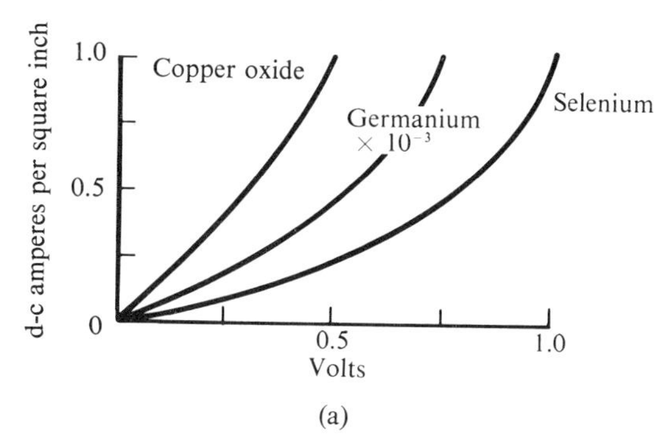

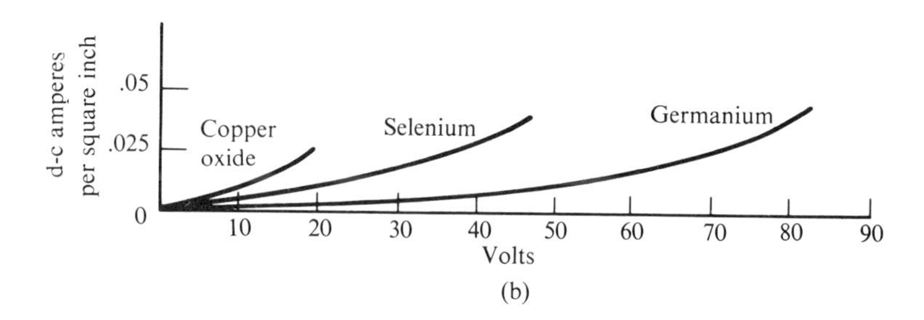

Figure 2–5. (a) Forward-current characteristics for three semiconductor rectifiers: note that germanium forward-current flow is approximately 1,000 times greater than either selenium or copper oxide; (b) Reverse-current characterisitics for copper-oxide, selenium, and germanium rectifiers.

characteristics for three types of semiconductor rectifiers. Note that germanium has much lower forward resistance than either selenium or copper oxide. If we plot the resistance of a semiconductor rectifier versus the applied voltage, we obtain a characteristic such as depicted in Fig. 2–6. It is also instructive to compare the internal resistance of a vacuum diode with that of a gas diode, such as a mercury-vapor rectifier tube. Figure 2–7 shows comparative voltage-current characteristics for a small vacuum diode and a small gas diode.

Note that a vacuum diode or a gas diode has an extremely high (practically infinite) reverse resistance. On the other hand, semiconductor power rectifiers may pass up to 1 per cent or more reverse

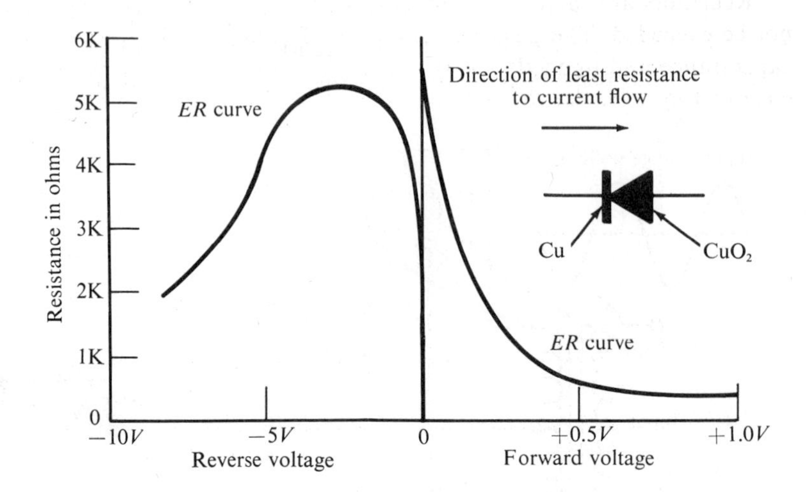

Figure 2-6. Resistance versus current value and direction for a copper oxide rectifier.

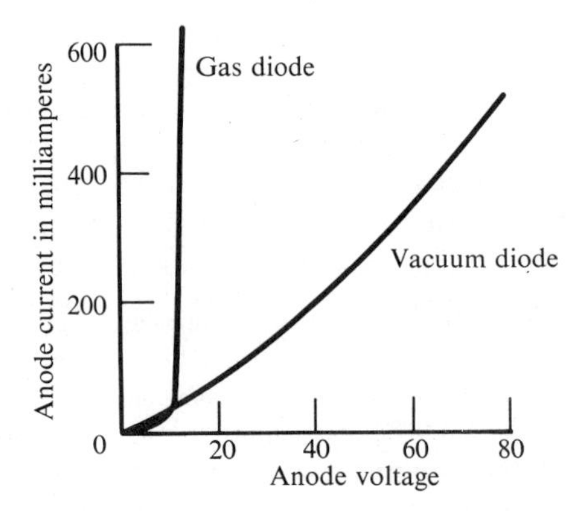

Figure 2-7. Comparative anode resistances of a vacuum diode and a gas diode.

current compared with their forward-current flow. The *front-to-back ratios* for typical power rectifiers are tabulated below:

Copper oxide at 6 v and 1/2 amp per sq in.	175 : 1
Copper oxide at 12 v and 1/2 amp per sq in.	100 : 1
Selenium at 15 v and 1/3 amp per sq in.	110 : 1
Selenium at 26 v and 1/3 amp per sq in.	66 : 1
Dot germanium at 65 v and 400 ma	133 : 1
Large-area germanium at 65 v and 75 amp	750 : 1

Rectifiers are rated for peak current flow, and this rating should not be exceeded. The peak-current value depends on the value of filter capacitance and upon the current demand from the filter. With reference to Fig. 2-8, the *RC* product is moderately large, and the conduc-

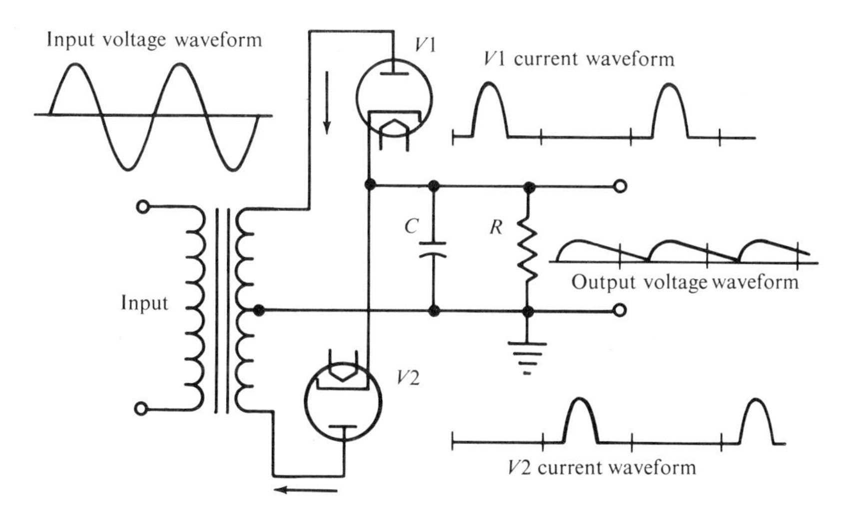

Figure 2-8. (a) Full-wave rectifier circuit; (b) Voltage and current waveforms.

tion angle is fairly small. Pulse currents flow through the rectifiers, and the pulse-current waveform has a peak-current value. If we wish to reduce the ripple amplitude, we may increase the value of *C*. However, the conduction angle then becomes smaller, and *the peak-current value of the pulse waveform increases.* Note that the value of *R* is not changed in this analysis.

Accordingly, an increase in the value of *C* in Fig. 2-8(a) may require that larger rectifier tubes be used, which are rated for adequate peak-current flow. If semiconductor rectifiers are used, the same rating considerations apply. Suppose that the value of *R* is reduced in Fig. 2-8 (a). Then the ripple amplitude increases and the conduction angle increases. Also, the peak-current value increases. In summary, the peak-current value increases as the filter capacitance is increased, and the peak-current value increases as the current demand from the filter is increased.

In practical work, it is often justified to assume that the current pulses depicted in Fig. 2-8(b) are rectangular pulses. In such case, the peak-current value is inversely proportional to the conduction angle:

$$I_p \simeq \frac{360^\circ}{\theta^0} \times I_L \tag{2-1}$$

where I_p is the peak current flow through a rectifier, theta (θ) is the conduction angle, and I_L is the load current.

Example 3. Determine the approximate peak current for one of the rectifiers in Fig. 2–8(a) when the load current is 100 ma and the conduction angle is 60°.

$$I_p \simeq \frac{360^\circ}{\theta^0} \times I_L$$
$$= \frac{360}{60} \times 100$$
$$= 600 \text{ ma} \qquad \text{(answer)}$$

Because exact calculation of peak-current values is not simple, engineers and technicians usually construct experimental circuits and measure the peak-current value. This measurement is made conveniently with a calibrated oscilloscope and a clip-on current probe. You will probably have an opportunity to measure peak-current values in the laboratory phase of your studies.

2–2. Resistive Input Filter

Instead of using excessively large values of filter capacitance to reduce the ripple amplitude to an acceptable level in the circuits of Figs. 2–1 and 2–8, we may use a resistive input filter circuit as shown in Fig. 2–9(a). Then, a larger value of filter capacitor may be used without exceeding the peak-current rating of the rectifier. In other words, the peak value of the charging current is reduced due to the resistance of R. This fact permits the value of C to be increased. Of course, the regulation becomes poorer when a resistive input filter is used; therefore, we use this type of filter only when R_L has a large value so that comparatively little current is demanded from the filter. The series R and shunt C components form an L section. Note that L sections may be connected in cascade.

To show the advantage of a resistive input filter, it is helpful to make the assumption that the ripple waveform consists of a sine wave with the fundamental ripple frequency, as that in Fig. 2–9(b). Filter action will actually be better than indicated by this simplified analysis, because harmonic frequencies "see" a lower reactance of C than the fundamental frequency. Now, in accordance with our simplified analysis, we have d-c voltage-divider action and a-c voltage-divider action [Fig. 2–10(a)]. It is evident that the capacitive reactance X_C

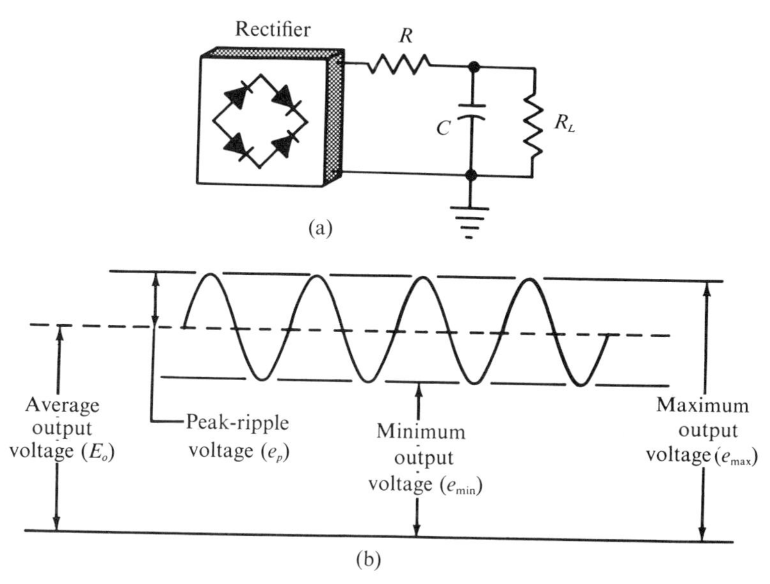

Figure 2–9. (a) Resistive input filter; (b) Simplified ripple voltage analysis (assumes that the entire ripple voltage occurs at the fundamental ripple frequency).

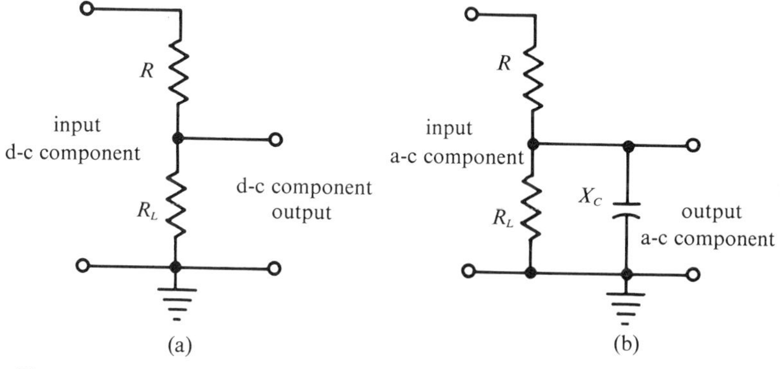

Figure 2–10. (a) The d-c component "sees" a resistive divider; (b) The a-c component "sees" an impedance divider.

has no divider action with respect to d-c, but that it provides divider action with respect to a-c.

In a practical situation, R_L might be ten times greater than R in the configuration of Fig. 2–10(a). Hence, 0.99 of the d-c input voltage appears across R_L. However, X_C in Fig. 2–10(b) might be only 0.001 of R_L; then X_C is 0.01 of R. Thus, approximately 0.0099 of the a-c

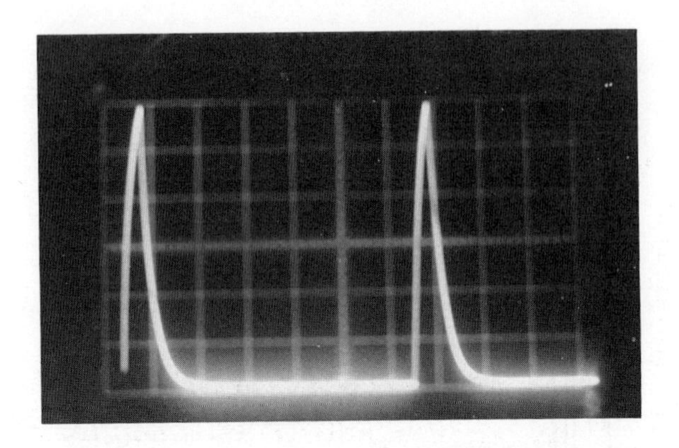

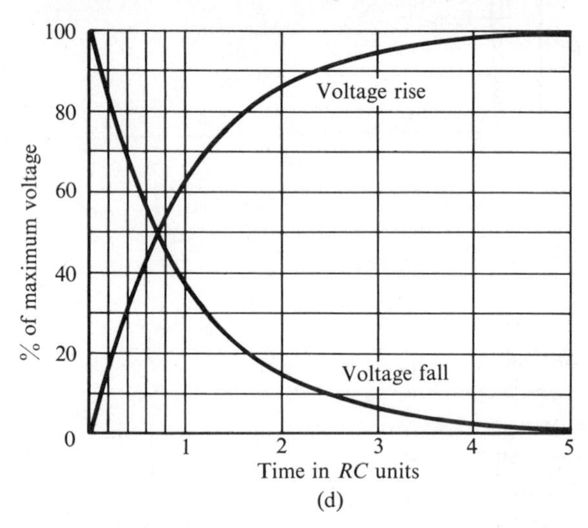

Figure 2–10. (c) Ripple waveform from a resistive-input filter with rectangular-pulse energization; (d) Rise of the ripple waveform is along curve *A*, fall is along curve *B*.

input voltage appears across R_L*. In summary, for this small-current situation, the use of a resistive input filter provides approximately 99 per cent reduction in ripple, with only 1 per cent reduction in d-c output voltage. This type of power supply is used, for example, in the picture-tube section of a television receiver, since a picture tube that operates at 20,000 v has a current demand of only a few hundred microamperes.

*We know that R and X_C may not be added directly; however, since $R \gg X_C$ the values are a close approximation.

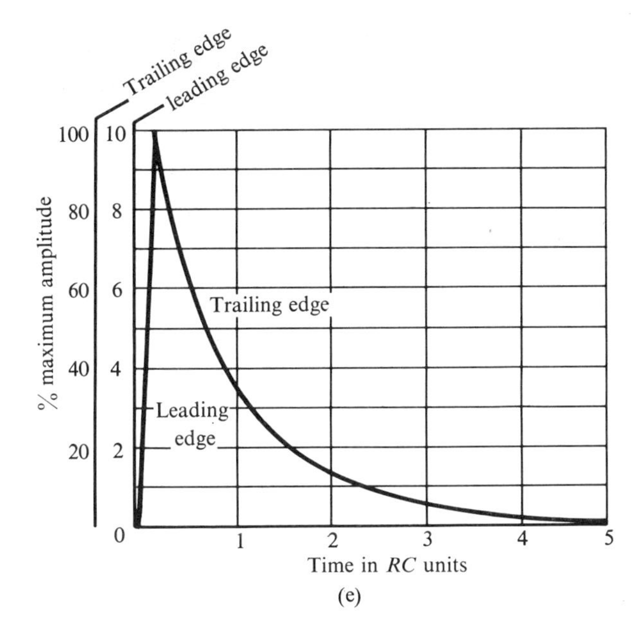

Figure 2–10. (e) Universal *RC* time-constant chart scaled for graphical solution of the ripple waveform.

Let us consider the ripple waveform in the output of a resistive-input filter, when the filter is energized by a rectangular pulse. This ripple waveform is illustrated in Fig. 2–10(c). Note that in this example the charge and discharge time constants were equal. Both the leading and trailing edges of the ripple waveform are exponential curves. However, the charge time is shorter than the discharge time because the charge time is basically limited by the duration of the charging pulse. On the other hand, the discharge time is determined solely by the pulse repetition rate. Thus, with reference to Fig. 2–10 (d), the leading edge of the ripple waveform corresponds to a short interval at the beginning of curve *A*; however, the trailing edge of the ripple waveform corresponds to a long interval on curve *B*.

Note that curve *B* in Fig. 2–10(d) must be scaled down to describe the waveform illustrated in Fig. 2–10(c). In other words, if the leading edge of the waveform attains 10 per cent of maximum amplitude, then the trailing edge of the waveform starts at 10 per cent of maximum amplitude [Fig. 2–10(e)]. Thus, 10 per cent of maximum amplitude along the leading edge of the ripple waveform corresponds to 100 per cent of maximum amplitude along the trailing edge of the ripple waveform.

2–3. Resistive π Filter

To reduce ripple amplitude, a resistive input filter [Fig. 2–11(a)] can be arranged as a resistive π filter [Fig. 2–11(b)]. Note that the

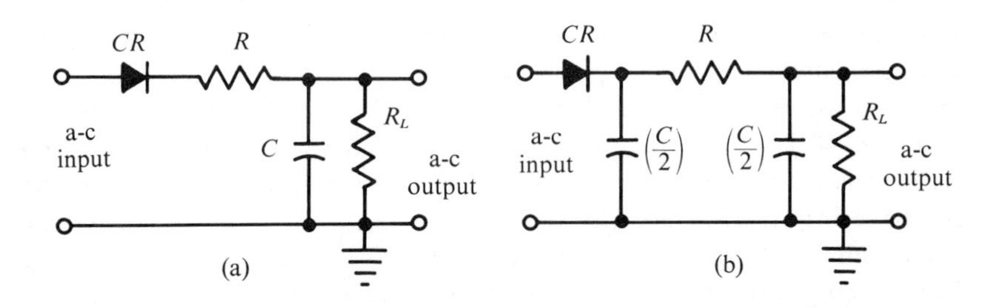

Figure 2–11. (a) Total filter capacitance connected across output end of filter; (b) Half of the filter capacitance connected at each end of the series resistor *R*.

total filter capacitance C has been split into $C/2$ and $C/2$ so that the circuit provides more a-c divider action. This increased a-c divider action is apparent in Fig. 2–12, which provides a functional presentation of the divider configurations. R_{int} denotes the internal resistance of the rectifier and is shown in Fig. 2–12 in order to include all para-

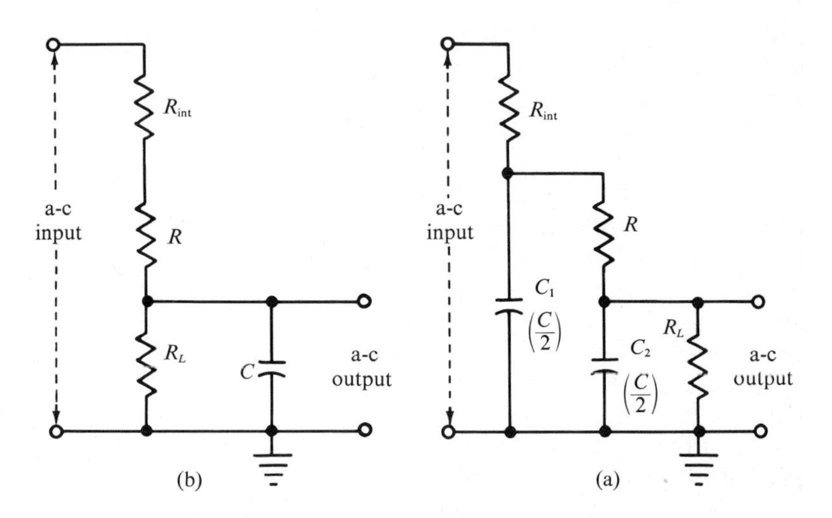

Figure 2–12. (a) Divider action of resistive input filter; (b) Divider action of resistive π filter.

meters which contribute to filter action. The d-c output is slightly greater in the π filter configuration. On the other hand, the a-c ripple voltage is considerably less in the π configuration.

An exact calculation of comparative a-c divider action for the circuits in Fig. 2–12 is somewhat involved, and it will be more instructive to compare typical ripple measurements for the two circuits. For example, suppose that R_{int} represents the internal resistance of a germanium diode (a very small resistance); $R = 44\ \Omega$, $R_L = 20{,}000\ \Omega$, and $C = 40\ \mu f$. Of course, $C/2$ then equals 20 μf. In this example, you will measure slightly more than double the ripple amplitude for the resistive input filter, compared with the resistive π filter. Thus, the π configuration provides a considerable advantage over the resistive input configuration when the same resistance and total capacitance values are used.

2–4. *LC* Filters

When a filter must supply substantial current with extensive reduction in ripple amplitude, an *LC* configuration is used. Inductance has a basic advantage over resistance as a filter component, because inductance may present negligible opposition to flow of d-c current and present great opposition to flow of a-c current. Thus, the regulation of an *LC* filter is much better than the regulation of an *RC* filter. Let us consider the configurations depicted in Fig. 2–13. The inductor and capacitor form an L-section filter. We will find that the filter action differs considerably when the L section is reversed. See Figs. 2–13(a) and (b) on the next page.

Observe that the configuration shown in Fig. 2–13(a) is a capacitor-input filter; stated otherwise, the 4-μf capacitor is connected directly across the rectifier output, and the capacitor charges to the peak value of applied voltage in the absence of a load. The regulation of this configuration is basically the same as that of the arrangement in Fig. 2–1. However, the ripple amplitude is considerably reduced by the reactance of the 5-h inductor. In summary, this *LC* arrangement is advantageous chiefly from the standpoint of ripple reduction. Its regulation is approximately the same as that of a simple capacitor filter if the inductor has a low-resistance winding. On the other hand, if the inductor has a high-resistance winding, the regulation becomes poorer than that of a simple capacitor filter.

Next, let us consider the action of the circuit shown in Fig. 2–13 (b). The rectifier output waveform has a d-c component and an a-c component, as seen in Fig. 2–14. We perceive that inductor L will

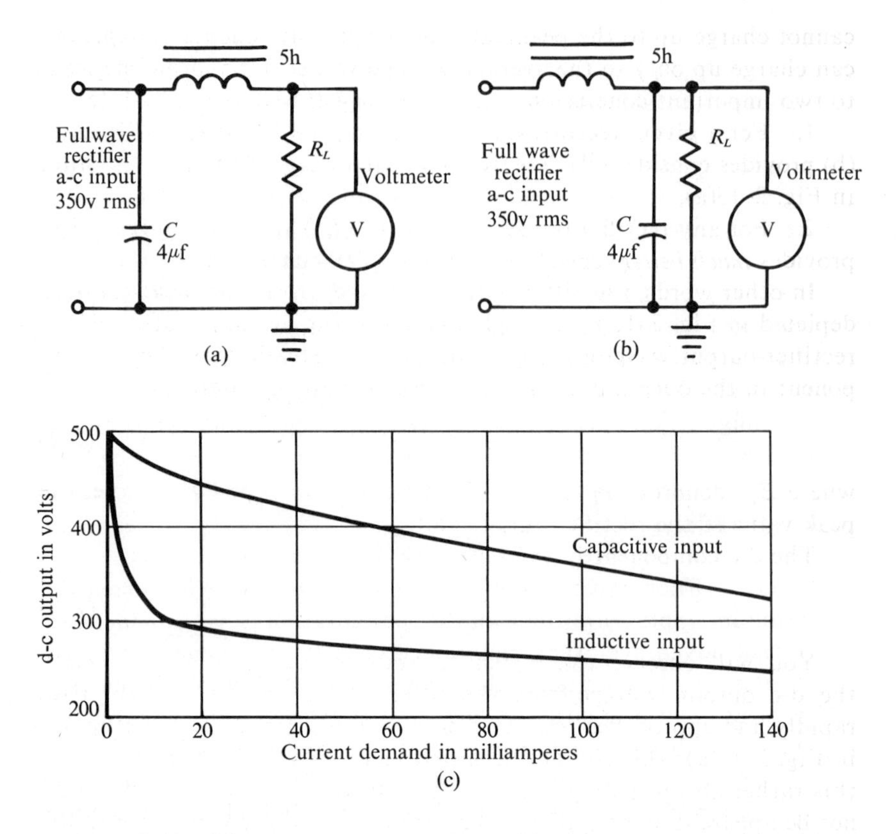

Figure 2-13. (a) Capacitive-input *LC* filter; (b) Inductive-input *LC* filter; (c) Comparative regulation.

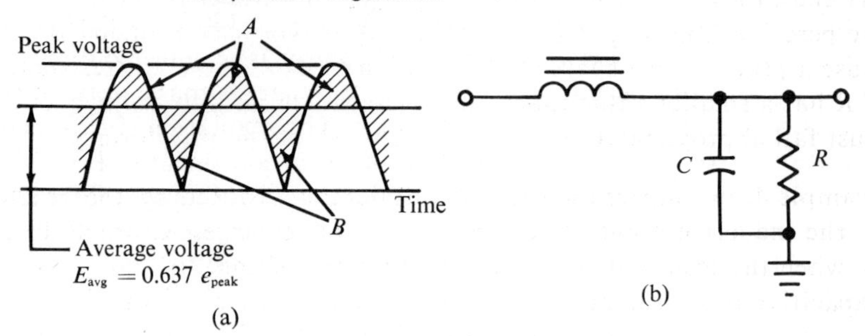

Figure 2-14. Inductor *L* opposes passage of excursions *A* and *B*.

oppose passage of the a-c excursions *A* and *B* into capacitor *C*, but will permit free passage of the d-c component. Therefore, *in principle*, *C* cannot charge up to the peak value of the rectifier output voltage, but can charge up only to the average voltage value. This principle leads to two important conclusions, which are illustrated in Fig. 2–13(c).

1. For a given rectifier-output voltage, the *LC* filter in Fig. 2–13 (b) provides considerably less filter-output voltage than the *LC* filter in Fig. 2–13(a).

2. For any rectifier-output voltage, the *LC* filter in Figure 2–13(b) provides *much better regulation* than the *LC* filter in Fig. 2–13(a).

In other words, the filter-output voltage from the configuration depicted in Fig. 2–13(b) is roughly equal to the d-c component of the rectifier-output waveform under varying load conditions. The d-c component in the output of a half-wave rectifier is formulated*

$$E_{\text{d-c}} = \frac{E_p}{\pi} \tag{2–2}$$

where $E_{\text{d-c}}$ denotes the value of the d-c component, and E_p denotes the peak value of the rectifier output voltage.

The d-c component in the output of a full-wave rectifier is

$$E_{\text{d-c}} = \frac{2E_p}{\pi} \tag{2–3}$$

You will observe that when the current demand is very small, the d-c output voltage from the filter shown in Fig. 2–13(b) rises rapidly and approaches the value of d-c voltage from the filter shown in Fig. 2–13(a); this rise is illustrated in Fig. 2–13(c). The reason for this rather abrupt voltage rise at very small current demand might not be apparent at first glance. This voltage rise is associated with the electronic switching action of the rectifiers. Because the rectifiers are switches, they permit no reverse d-c current flow through the inductor. We perceive from Fig. 2–14 that if *R* is disconnected, excursion *A* will cause a peak-voltage charge to build up in *C*. However, if the resistance of *R* has a smaller value than the reactance of *L*, the voltage of *C* must fall approximately to the average value of the waveform.

Example 4. Compare the regulation for the capacitive-input filter and the inductive-input filter represented by the curves in Fig. 2–13(c) when the load is increased from 60 ma to 100 ma.

Capacitive-input filter:

$$E_L \text{ at } 60 \text{ ma} = 400 \text{ v}$$

$$E_L \text{ at } 100 \text{ ma} = 360 \text{ v}$$

$$\text{Change} = \frac{40}{400} \times 100 = 10\%$$

*LC filters should not be used with half-wave rectifiers.

Inductive-input filter:

$$E_L \text{ at } 60 \text{ ma} = 275 \text{ v}$$
$$E_L \text{ at } 100 \text{ ma} = 260 \text{ v}$$
$$\text{Change} = \frac{15}{275} \times 100 = 5.37\%$$

Obviously, the inductive input filter is more stable with a changing load.

2–5. Basic Comparison of Linear and Nonlinear Systems

In general, analysis of nonlinear systems is much more difficult than analysis of linear systems. However, various simple nonlinear systems yield to elementary analysis. A rectifier is a nonlinear component and hence a filter is a nonlinear system. There are various situations in which a nonlinear system can be approximated by a linear analogy; in other situations, a linear analogy leads merely to an absurd answer. Engineers and technicians learn by experience to recognize when a linear analogy can be used to analyze a nonlinear system. For example, we used a linear analogy in Fig. 2–9 and regarded a ripple waveform as if it were produced by a sine-wave generator and a battery connected in series. This rough analogy serves satisfactorily to explain the comparative a-c and d-c divider action in a filter. On the other hand, this analogy fails completely to explain the voltage rise at very small current demand in an inductive-input filter.

2–6. Ripple Amplitude

In an inductive-input filter such as depicted in Fig. 2–13(b), the ratio of the amplitude of the rms ripple voltage to the rms a-c input voltage is formulated

$$\frac{\text{Output ripple voltage}}{\text{a-c input voltage}} = \frac{1}{(2\pi f)^2 LC} \simeq \frac{1.76 \times 10^{-6}}{LC} \text{ for 120 Hz} \qquad \textbf{(2–4)}$$

where f denotes the fundamental frequency of the rectifier output voltage in Hz, L denotes inductance in henrys, and C denotes capacitance in farads.

Formula 2–4 gives an *average value* for the ripple voltage under varying load conditions and remains approximately correct over considerable variation of load-current demand. To take a practical example, let us calculate the ratio of ripple voltage to the a-c input voltage for the filter depicted in Fig. 2–13(b). Since a full-wave rectifier is used, the fundamental frequency of the rectifier output voltage is 120 Hz. So we write

$$\frac{E_r}{E_{in}} = \frac{1}{(2\pi \times 120)^2 \times 5 \times 4 \times 10^{-6}} = 9\%, \text{ approximately} \qquad \textbf{(2–5)}$$

and

$$\text{Ripple voltage} = 0.09 \times E_{in} = 0.09 = 350 = 31.5 \text{ v (approximately)}$$

Now suppose that we add an inductive-input filter section, as depicted in Fig. 2–15. If both inductor values are the same and both

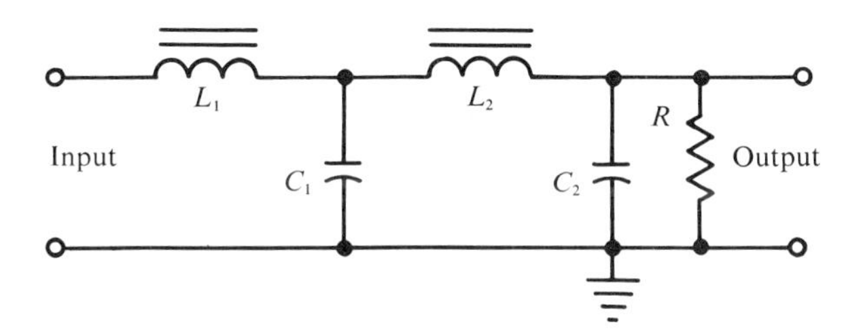

Figure 2–15. Two-section inductive-input filter.

capacitor values are the same, Formula 2–4 is modified to read

$$\frac{E_r}{E_{in}} = \frac{1}{(2\pi f)^4 L^2 C^2} \text{ or } \simeq \frac{3.1 \times 10^{-12}}{L^2 C^2} \qquad \textbf{(2–6)}$$

Example 5. Determine the ripple-output voltage of the circuit illustrated in Fig. 2–15 when $f = 120$ Hz, $C = 4\ \mu f$, $L = 5$ h, and $E_{in} = 350$ v.

$$\begin{aligned} E_r &= \frac{E_{in}}{(2\pi f)^4 L^2 C^2} \\ &= \frac{350}{(754)^4 (5)^2 (4 \times 10^{-6})^2} \\ &= \frac{350}{3.22 \times 10^{11} \times 2.5 \times 10^1 \times 1.6 \times 10^{-11}} \\ &= 2.72 \text{ v} \qquad \text{(answer)} \end{aligned}$$

Thus, cascaded inductive-input filter sections provide a large reduction in ripple amplitude. On the other hand, only a small change in regulation occurs. The foregoing discussion assumes that the inductance value is comparatively large so that we obtain normal inductive-input filter action. It is obvious that if the inductance value in Fig. 2–13(b) were very small, we would effectively have a capacitive-input filter. In any case, the inductance value should not be less than the *critical value*, which is formulated

$$L_{cr} = \frac{E_{d\text{-}c}}{I_{d\text{-}c}}, \text{ approximately} \qquad \textbf{(2–7)}$$

where L_{cr} is the critical value of inductance in henrys, $E_{d\text{-}c}$ is the d-c output voltage from the filter, and $I_{d\text{-}c}$ is the current demand in *milliamperes*.

Example 6. Determine the critical inductance value of the inductor in Fig. 2–13(b) when $E_{d\text{-}c} = 350$ v and $I_{d\text{-}c} = 175$ ma.

$$L_{cr} = \frac{E_{d\text{-}c}}{I_{d\text{-}c(ma)}}$$

$$= \frac{350}{175} = 2\text{h} \qquad \text{(answer)}$$

The significance of the critical inductance value is illustrated in Fig. 2–16. When L has a large value, the current through L fluctuates

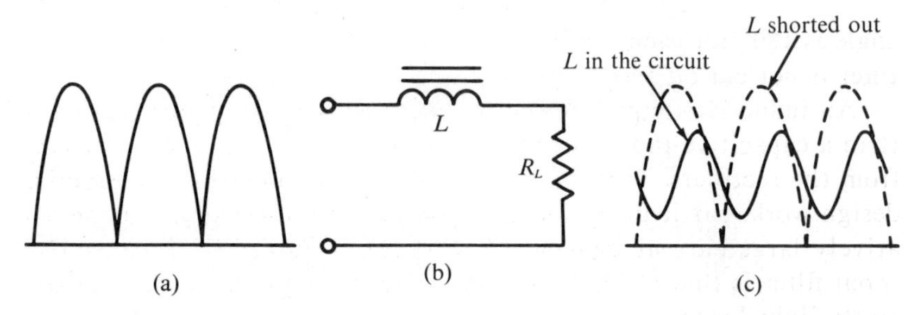

Figure 2–16. (a) Voltage waveform applied to *LR* circuit; (b) *LR* circuit; (c) Current flow with *L* in and out of the circuit.

more or less about the average-current value. On the other hand, when L is zero, the current flows in half-sine waves and passes through zero. Normal operation of an inductive-input filter requires that the rectifier output current never be cut off (even instantaneously) by the peak inverse voltage. In turn, the critical value of inductance is given by

Formula 2-7. It follows from Fig. 2-16 that the current flow through the rectifiers has a waveform that approximates a square wave when the inductance is large (see Fig. 2-17). Observe that the conduction

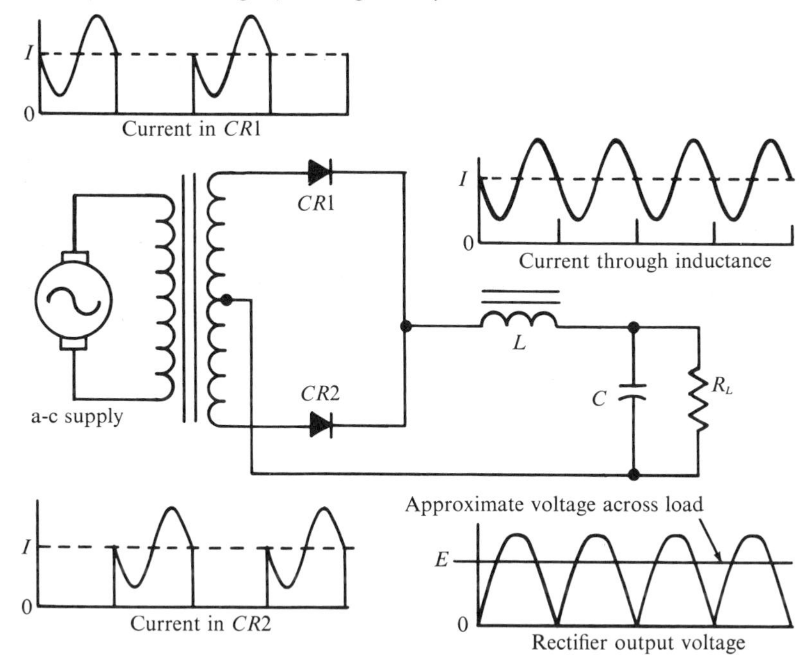

Figure 2-17. Voltage and current waveforms for an inductive-input filter.

angle is 180° for each rectifier and that output current from the rectifier is not cut off at any time.

An inductive-input filter not only provides much better regulation than a capacitive-input filter but also reduces the peak-current demand from the rectifiers. This is an important consideration in practical design work, particularly when a power supply must supply a comparatively large current demand. The only advantage of a capacitive-input filter is that of higher output voltage, particularly at comparatively light loads.

Let us consider the *LC* π configuration depicted in Fig. 2-18. As would be anticipated, this type of filter has characteristics which are intermediate to the capacitive-input and inductive-input arrangements. Regulation is comparatively poor, because the input filter capacitor tends to charge up to the peak value of the rectifier output voltage. Ripple amplitude is comparatively low because of the *LC* section following the input capacitor. Peak-current demand from the

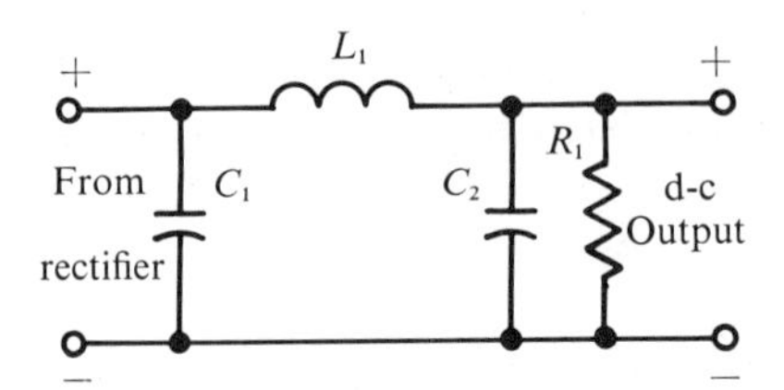

Figure 2–18. The LC π filter configuration.

rectifiers is practically the same as for a simple capacitor filter. Hence, the LC π configuration is used when current demand is moderate and comparatively high d-c output voltage is desired. The appearance of a small commercial power supply is seen in Fig. 2–19.

Courtesy of EICO, Electronic Instrument Co., Inc.

Figure 2–19. A small power supply.

2–7. Resonance in *LC* Filters

We observe that series resonance will occur at the ripple frequency in the configuration of Fig. 2–17 if certain values of L and C are used. Of course, resonance defeats the purpose of the filter because the amplitude of the ripple is magnified Q times across C. Therefore, a basic rule of filter design states that the value of L and the value of C should be chosen to be at least double the values which would produce resonance at the fundamental ripple frequency.

Example 7. Find the LC product for a filter such as the one shown in Fig. 2–13(b), so that the resonant frequency is less than 0.707 of the ripple frequency when the filter is used with a full-wave rectifier for a 60-Hz signal.

$$\text{Ripple} = 120 \text{ Hz}$$

$$120 = \frac{1}{2\pi\sqrt{LC}}$$

$$2LC = \frac{2}{\pi\sqrt{240}}$$

$$LC = 0.0367$$

$$LC \leq \frac{1}{753} = 0.0375 \qquad \text{(answer)}$$

2–8. Voltage-Multiplier Configurations

The voltage-multiplier type of power supply is restricted to small current demands since its regulation is comparatively poor. If a rectifier and a capacitor are connected in series with a source of a-c voltage, as in Fig. 2–20, the voltage will be doubled. When the input voltage has the polarity indicated in Fig. 2–20(a), the capacitor charges to the peak value of the line voltage. On the next half cycle, the condition depicted in Fig. 2–20(b) results. The full-peak voltage is retained across the capacitor, but because of polarity reversal of the source

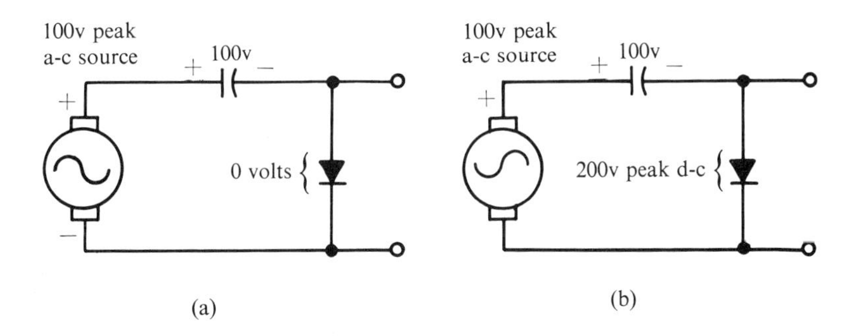

Figure 2–20. Basic voltage-doubling action. (a) Capacitor charging; (b) Capacitor voltage added to source voltage.

voltage, the rectifier no longer conducts; in turn, the voltage across the capacitor adds to the source voltage.

We know that the total voltage across the rectifier (peak inverse voltage) has a peak value twice that of the applied voltage. Thus, the output voltage varies between zero and twice the peak input value during each cycle. The output voltage can be maintained over the entire cycle if a second rectifier and capacitor are added, as shown in Fig. 2–21. The

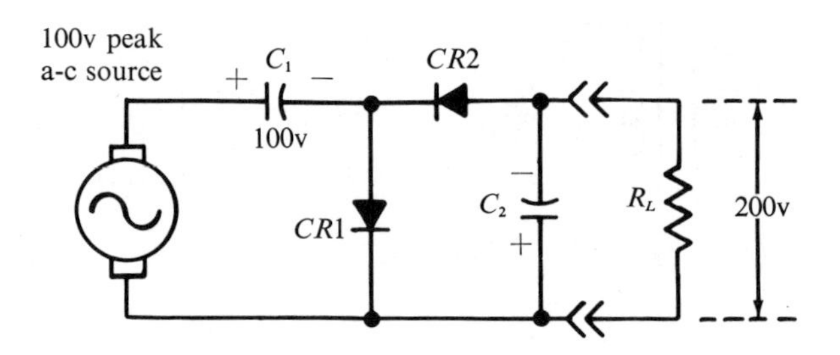

Figure 2–21. Half-wave voltage-doubler configuration.

second capacitor charges to double the peak input voltage when rectifier *CR*2 conducts, and this capacitor holds its charge during the time that *CR*2 is nonconducting.

Capacitor C_2 cannot, however, maintain the full output voltage over the complete cycle if there is a substantial current demand. This limitation results from the fact that when rectifier *CR*2 is nonconducting, no current is drawn from the input circuit. Thus, C_2 supplies the load current during discharge, and the output voltage falls in accordance with the discharge curve shown in Fig. 2–2. Because current is drawn from the source for only one-half cycle, this configuration is called a *half-wave doubler.* The ripple frequency is the same as the supply frequency.

The same components can be rearranged to form a *full-wave doubler* circuit (Fig. 2–22). In effect, we have connected a pair of half-wave rectifiers across the source so that the directions of their conducting paths are opposite. As a full-wave rectifier, this circuit draws current from the voltage source during both halves of the input cycle. For one-half cycle, C_1 charges to the peak source voltage through rectifier *CR*1, and for the next one-half cycle C_2 charges to the peak source voltage through rectifier *CR*2. The voltages impressed across C_1 and C_2 will therefore combine in series aiding to energize the load. The capacitors cannot, however, maintain the full output voltage from one conduction interval to the next if there is a substantial current demand. The ripple frequency, of course, is double the supply frequency.

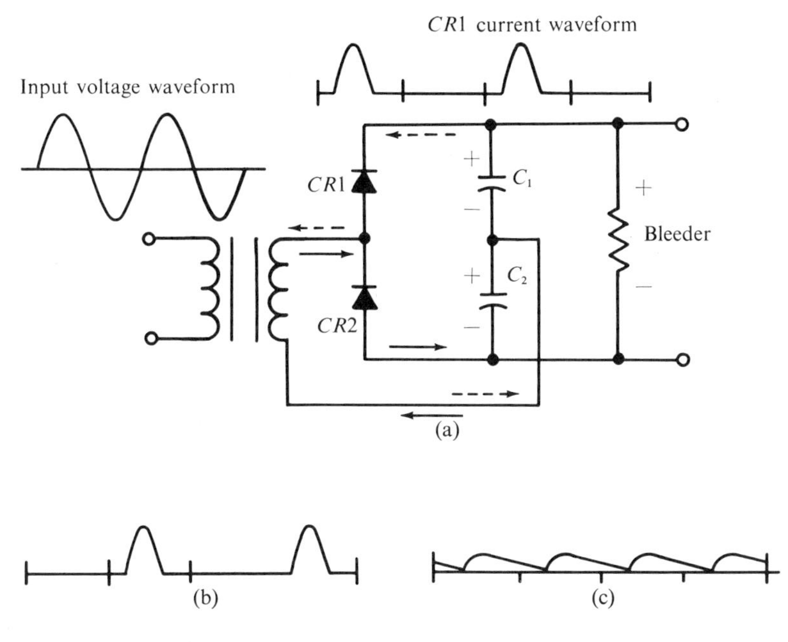

Figure 2–22. Voltage-doubler rectifier circuit; (b) CR2 current waveform; (c) Output voltage waveform.

This basic process of increasing a supply voltage can be done at higher levels of multiplication, such as tripling and quadrupling. In theory, there is no limit to the number of times that a supply voltage can be multiplied. However, practical considerations usually limit the multiplication to five times. With each additional section, the regulation becomes poorer. Larger capacitors must be used in turn to sustain a given current demand.

Let us briefly consider the operation of the seven-times multiplier depicted in Fig. 2–23. When the upper terminal is negative, electrons flow through all of the diodes charging C_1, C_3, C_5, and C_7 to the peak source voltage. When the upper terminal is positive, the charges stored in capacitors C_1, C_3 and C_5 act in series with the source voltage to charge C_2, C_4, and C_6 to twice the peak value of the source voltage. When the upper terminal again becomes negative C_1 charges to the peak value and C_2 and C_6 act in series with the source to charge C_3 and C_5 to twice the peak value. Capacitor C_6 and the source charge C_7, which has not discharged, to three times peak value. This analysis may be continued through the first seven cycles of operation, at which time the voltage across C_7 will have been built up to seven times the peak value of input voltage.

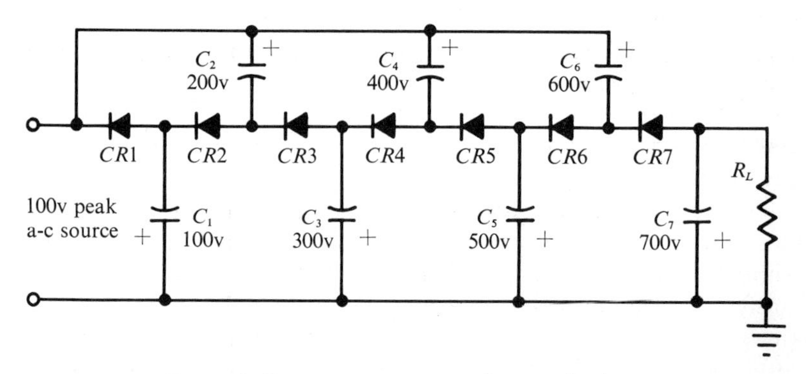

Figure 2-23. A times-seven voltage multiplier.

Contrary to what might be guessed at first glance, the inverse peak voltage across any one of the rectifiers does not increase with the number of sections that may be added. A voltage analysis through the circuit will show that the peak inverse voltage across each rectifier is double the peak value of the input voltage. Since the regulation of this circuit is very poor unless the capacitor values are extremely large, it is more economical to obtain a high output voltage from a step-up transformer and simple rectifier-filter configurations.

2-9. Physical Units

In analysis of power-supply circuitry, we determine relations among various physical units, such as voltage, current resistance, time, power, energy, inductance, capacitance, reactance, impedance, magnetomotive force, magnetizing force, magnetic flux, flux density, reluctivity, and permeability. Each of these physical units can be stated in terms of a *definitive dimensional formula.* A definitive dimensional formula defines a physical unit on the basis of force, distance, time, and charge. Let us develop the definitive dimensional formulas that are of interest in analysis of power-supply circuitry.

Voltage, or emf, is a familiar unit that is not always correctly understood. We often speak loosely of voltage as an electrical pressure that causes current to flow. Although current flow is indeed associated with voltage in a closed circuit, this is not a complete accounting of the situation. To clarify this point, let us consider the charging of a capacitor. If a capacitor has a capacitance of 1 f and is connected to a voltage source of 1 v, the charging action transfers 1 coulomb from the positive plate to the negative plate of the capacitor. One coulomb of charge comprises 6.25×10^{18} electrons. We recognize that work has been done on the capacitor; the charged capacitor contains 1 joule of potential energy (1 w-sec of potential energy). If the voltage source is disconnected from the capacitor and the capacitor terminals are connected to a resistor, the charge in the capacitor will produce 1 j of heat in the resistor.

This is just another way of saying that a complete accounting of the situation entails a description of voltage as a measure of work accomplished per unit charge. After the 1-f capacitor was charged by transfer of 1 coulomb of charge, a potential difference of 1 v then existed across the capacitor terminals. In other words, the volt is the unit of potential difference established between two points when one unit of work has been done in moving one unit of electrical charge from one point to another. In summary, voltage is fundamentally a measure of work done per unit charge that is moved.

Electric current comprises charge flow past a reference point. Thus, if a d-c voltage causes charges to flow past a point, and 1 coulomb of charge has passed at the end of 1 sec, the rate of charge flow (current value) is 1 amp. In terms of electronic charges, 1 coulomb consists of 6.25×10^{18} electrons. We may now express the physical units of voltage and current as definitive dimensional formulas.

$$E = \frac{FL}{Q} \tag{2-8}$$

$$I = \frac{Q}{T} \tag{2-9}$$

where E denotes voltage, F force, L distance, Q charge in coulombs, I current in amperes, and T time in seconds.

That is, Formula 2–8 states that voltage is equal to the work performed per unit charge. Formula 2–9 states that current is equal to the charge that flows per unit time. Resistance is defined as a voltage/current ratio, in accordance with Ohm's law. Accordingly, we may express the physical unit of resistance in a definitive dimensional formula.

$$\mathrm{R} = \frac{E}{I} = \frac{FLT}{Q^2} \tag{2-10}$$

Capacitance is related to charge and voltage by Coulomb's law for capacitance.

$$Q = CE \tag{2-11}$$

Hence, we may express the physical unit of capacitance as in Eq. 2–12.

$$C = \frac{Q}{E} = \frac{Q^2}{LF} \tag{2-12}$$

We recall that the time constant of an RC circuit is expressed in seconds. Let us make a dimensional analysis to establish this relation between the physical units.

$$T = RC = \frac{FLTQ^2}{Q^2LF} = T \tag{2-13}$$

Formula 2–13 states that the product of resistance and capacitance is equal to time, as derived by substituting dimensional units for R and C. Next, we recall that energy is numerically equal to work, and that power is the rate of performing work. Therefore, we write

$$P = EI = \frac{FLQ}{QT} = \frac{FL}{T} \tag{2-14}$$

$$W = EIT = FL \tag{2-15}$$

Let us next analyze the physical unit of inductance. A henry is defined as the inductance that will develop 1 v across its terminals when the current flow changes at the rate of 1 amp per second. That is,

$$E = \frac{LI}{T} \tag{2-16}$$

It follows that the definitive dimensional formula for inductance is written

$$L = \frac{ET}{I} = \frac{FLT^2}{Q^2}{}^{*} \tag{2-17}$$

It has been stated that the time constant of an LR series circuit is equal to L/R seconds. Let us verify this statement by means of a dimensional analysis.

$$\frac{L}{R} = \frac{FLT^2Q^2}{Q^2FLT}{}^{*} = T \tag{2-18}$$

Faraday's law states that an induced voltage is proportional to the rate of change of magnetic flux that links the turns of a coil. Thus, we may write

$$E = \frac{\phi}{T} \tag{2-19}$$

---

*L denotes inductance in the lefthand member; L denotes length in the righthand member.

or,

$$\phi = TE = \frac{FLT}{Q} \tag{2-20}$$

Magnetomotive force is defined as the number of ampere-turns that are producing magnetic flux. Since a pure number has no physical dimension, magnetomotive force is evidently expressed by the same dimensional formula as current.

$$\text{mmf} = \frac{Q}{T} \tag{2-21}$$

Magnetizing force is equal to magnetomotive force per unit length of core, from which it follows that

$$H = \frac{Q}{LT} \tag{2-22}$$

Since flux density is equal to flux per unit area, we write

$$B = \frac{FT}{LQ} \tag{2-23}$$

And since $B = \mu H$, it follows that permeability is dimensionally expressed

$$\mu = \frac{FT^2}{Q^2} \tag{2-24}$$

Ohm's law may be written: $I = E/R$, $I = E/X$, and $I = E/Z$. Accordingly, we perceive that reactance and impedance have the same physical dimensions as resistance.

$$X = \frac{FLT}{Q^2} \tag{2-25}$$

$$Z = \frac{FLT}{Q^2} \tag{2-26}$$

SUMMARY

1. For given sizes of filter components and loads, a full-wave rectifier will produce less ripple and more output voltage than a half-wave rectifier.
2. The regulation of a power supply depends considerably upon the internal resistance of the charging circuit.
3. The per cent ripple of a power supply generally increases as the load (current requirement) increases.
4. The internal resistance of a germanium diode is much lower than that of either selenium or copper-oxide diodes.

5. Vacuum-tube and gas diodes pass almost no reverse current, whereas a semiconductor rectifier may have a reverse current of 1 per cent of the forward current. However, the advantages of semiconductor rectifiers, such as longer life, small size, rugged construction, and lack of heater power, make them the logical choice as rectifiers in most applications.

6. The internal resistance of a gas diode is much less than that of a vacuum diode at high current levels.

7. Rectifiers are rated for peak inverse voltage values, average current values, and peak current values. The peak current value of a rectifier depends upon the peak applied voltage, the internal resistance of the filter and supply components, and the current demand from the supply.

8. The ripple of a supply and the peak rectifier current both increase as the load current increases.

9. In a practical "L" RC filter, a small value of resistance is placed in series with the load and the load bypass capacitor to reduce the ripple and the surge current.

10. A resistive input filter may be arranged in a π circuit to further decrease the ripple amplitude.

11. A capacitive input LC filter gives a higher output voltage but a less stable output voltage with a varying load.

12. The ripple voltage for an inductive input-filter circuit, such as shown in Fig. 2–13(b), may be formulated

$$\text{Ripple voltage} = \frac{\text{a-c input voltage}}{(2\pi f)^2 LC}$$

13. The ripple voltage for a double LC inductive input filter, such as is depicted in Fig. 2–15, is formulated

$$E_{\text{ripple}} = \frac{E_{\text{in}}\ \text{a-c}}{(2\pi f)^4 L_1 L_2 C_1 C_2}$$

14. The minimum value of inductance for an inductive input LC filter is

$$L_{\text{cr}} = \frac{E_{\text{d-c}}}{I_{\text{d-c}}}, \text{ approximately}$$

15. An inductive-input filter provides much better regulation than a capacitive-input filter and also reduces the peak current demand from the rectifiers.

16. To prevent series resonance from occuring at the ripple frequency of a power supply, the LC components are chosen to resonate at least 0.707 below ripple frequency.

17. A voltage-multiplier type of power supply is a simple, inexpensive method of developing a high value of d-c voltage; its application is limited to small current demands as its regulation is poor.

Questions

1. Compare a full-wave rectifier circuit to a half-wave rectifier circuit.
2. Explain each of the following terms as applied to power supplies and filter circuits: open-circuit voltage, regulation, internal resistance, filter circuit.
3. Which of the three diode materials—copper-oxide, germanium, or selenium—is the most desirable for use in a medium-current power supply? Why?
4. Give the advantages and disadvantages of a semiconductor rectifier in comparison to a vacuum-tube rectifier.
5. From the graphs in Fig. 2-4, draw a conclusion as to the relationship of ripple component versus filter capacitor value and load resistance value.
6. Explain the filtering action of a resistive-input filter, such as the one shown in Fig. 2-9(a).
7. Discuss the advantages gained with arranging a resistive-input filter as a π network.
8. Prove that the inductive-input *LC* filter gives a more stable output voltage than the capacitive-input *LC* filter under a changing load. See Fig. 2-13.
9. Define the terms: *LC* filter, *RC* filter, choke-input filter, capacitive-input filter, "*L*" filter, and π filter.
10. Explain the term *critical inductance* as applied to an inductive-input *LC* filter.
11. Compare the properties of the inductive-and capacitive-input filters.
12. When can an *LC* π filter circuit be used as a power supply filter?
13. How is resonance prevented in an *LC* filter circuit?
14. Determine the minimum *LC* ratio so that the resonant frequency of an *LC* filter, shown in Fig. 2-13(b), is less than 0.707 the ripple frequency when an 800-Hz signal is rectified by a full-wave rectifier.
15. Draw a voltage-doubler circuit schematic and explain the circuit operation through 2 cycles of input voltage.
16. What is the effect on a half-wave rectifier, such as the one shown in Fig. 2-1, if the diode shorts?
17. What is the effect on the half-wave rectifier shown in Fig. 2-1 if the capacitor shorts?
18. What is the effect on the output of a full-wave rectifier circuit if one diode opens? shorts?

19. What is the effect on the output of a circuit, such as the one shown in Fig. 2-11(b), if capacitor C_1 opens?

20. What is the effect on a circuit, such as the one in Fig. 2-13(b), if the inductor shorts?

21. What is the effect on the output of a full-wave bridge rectifier if one diode shorts? opens?

22. Why is the diode current in form of sharp pulses when a capacitor filter is used?

23. What effect does the d-c resistance of an inductor have on the output of a filter?

24. What is the effect on the output of a full-wave voltage doubler if one diode shorts? opens?

PROBLEMS

1. From the curves shown in Fig. 2-3, determine the output voltage of supply *A* and supply *B* when $R_L = 1\ \text{k}\Omega$.

2. From the curve shown in Fig. 2-4, determine the per cent of ripple for each supply for each of the load resistances: 2 kΩ, 6 kΩ, and 20 kΩ.

3. Determine the internal resistance of the gas diode and the vacuum diode in Fig. 2-7 when the current through each is 400 ma.

4. What is the approximate peak current of the rectifier in the circuit shown in Fig. 2-8(a) when the load current is 300 ma and the conduction angle is 50°?

5. The filter circuit in Fig. 2-13(b) has the following values: $L = 6$ h, $C = 6\ \mu$f, $f = 120$ Hz, and a-c input voltage of 680 rms v. What is the ripple voltage?

6. What minimum value of capacitance will allow a ripple voltage of 3.5 v or less in the circuit illustrated in Fig. 2-13(b)?

7. What is the ripple voltage of the filter shown in Fig. 2-15 when $E_{in} = 500$ rms v, $L_1 = 4$ h, $L_2 = 10$ h, $C_1 = 4\ \mu$f, $C_2 = 10\ \mu$f, and the ripple frequency $= 120$ Hz?

8. Determine the critical inductance value for the circuit illustrated in Fig. 2-13(b) when $I_{d\text{-}c} = 250$ ma and $E_{d\text{-}c} = 400$ v.

9. What is the critical inductance value for the circuit shown in Fig. 2-13(b) when $I_{d\text{-}c} = 400$ ma and $E_{d\text{-}c} = 300$ v?

10. Determine the ripple voltage of the filter shown in Fig. 2-15 when $E_{in} = 750$ rms v, $L_1 = L_2 = 12$ h, $C_1 = C_2 = 8\ \mu$f, and the line frequency is 400 Hz.

11. What is the conduction angle of the rectifier circuit in Fig. 2-8(a) when the peak current is 200 ma and the load current is 120 ma?

12. From Fig. 2-24, what is the output ripple at 320 ma load current when $R_L = 1\text{k}\Omega$ and E_{a-c} input $= 377$ v.

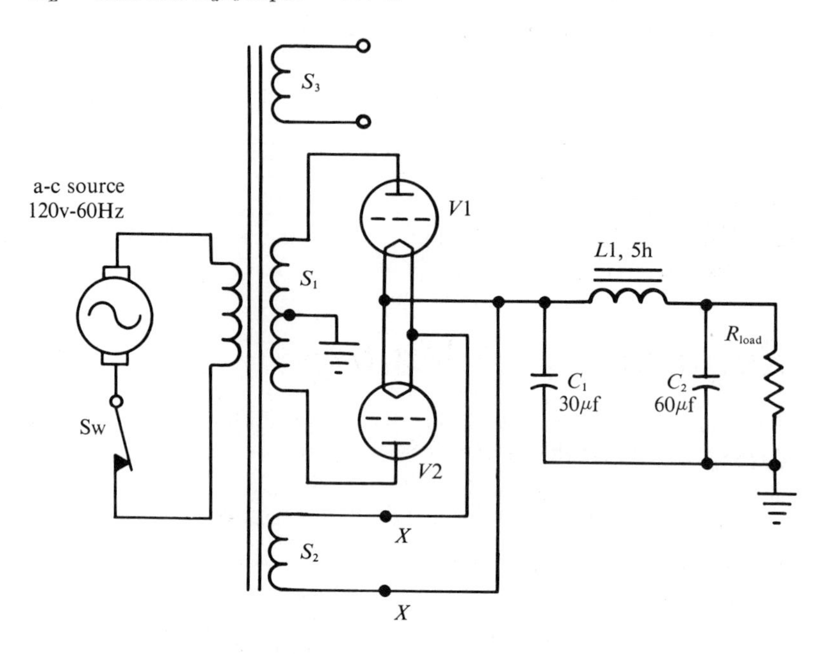

Figure 2-24. Full-wave rectifier and π-filter circuit.

13. The normal output of the circuit in Fig. 2-24 is 250 v at 100 ma; if the output drops to 159 v, what is most likely the trouble?

14. With the load removed, the output of the circuit in Fig. 2-24 is 314 v d-c. What is the output if C_1 opens?

15. If the output ripple of the circuit in Fig. 2-24 is 15 v at 350 ma load current, what is the most probable trouble?

16. What is the most probable voltage between points *X-X* in Fig. 2-24?

CHAPTER 3

Regulation of Voltage and Current

3–1. Introduction

Many modern electronic units and instruments are critical with respect to variation in supply voltage, and even a slight deviation from normal supply voltage will cause unsatisfactory operation in some cases. Other electronic devices or units are critical with respect to variation in supply current. Hence, voltage-regulating or current-regulating devices are widely utilized, either as part of a power supply or as an associated unit. A view of small bench-type regulated power supply is seen in Fig. 3–1.

Both gas tubes and semiconductors are used in voltage-regulating and current-regulating circuits. Vacuum tubes are commonly employed to extend the capabilities of gas tubes in regulator circuits. In a glow-discharge tube, such as a neon bulb, the voltage drop across the tube changes but slightly as the current flow through the bulb varies over a fairly wide range. This property is attributed to the proportionality

Courtesy of Heath Company

Figure 3–1. A bench-type regulated power supply.

of ionization in the gas with the value of current flow through the bulb. When a large current flows, the gas is very highly ionized, and the internal resistance of the bulb is low. On the other hand, when a small current flows, the gas is less extensively ionized, and the internal resistance of the bulb is high. Over the normal operating range of the bulb, the product of current flow and internal resistance is essentially constant.

A simple glow-tube regulator circuit is depicted in Fig. 3–2(a). Current flow through the series resistor R branches into the glow tube and into the load resistance. If the supply voltage decreases, the voltage across the glow tube tends to decrease. In turn, the gas in the glow tube deionizes slightly, and less current flows through the glow tube. Current flow through R is decreased by the amount of current decrease through the glow tube. Because the current flow through R is less, the voltage drop across R is less. If R has a suitable value with respect to the glow-tube resistance and to the load resistance, the

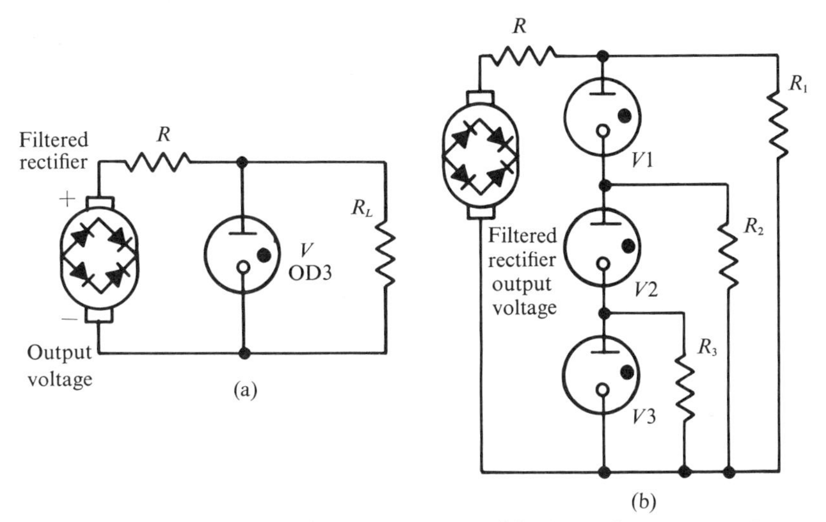

Figure 3–2. (a) Simple glow-tube regulator; (b) Glow tubes connected in series for higher output voltages.

voltage across the load remains reasonably constant as the supply voltage varies over a substantial range.

Of course, the value of R in Fig. 3–2(a) must never be chosen so large that the glow tube fails to ionize. Glow tubes are designed to operate at various useful values of voltage. These values are commonly denoted by the tube-type number. Note that the value of R must not be chosen so small that excessive current flows through the glow tube. Excessive current flow will change the characteristics of the tube, or if greatly excessive, will cause the tube to explode.

Example 1. The load current in Fig. 3–2(a) is 12 ma at 150 v. What value of R is necessary to allow 15 ma through the $OD3$ when the rectifier output is 239 v? What is the change of current through the $OD3$ when the load is removed?

$$I_R = I_{VR} + I_{\text{load}}$$
$$= 15 \text{ ma} + 12 \text{ ma} = 27 \text{ ma}$$
$$E_R = E_T - E_{VR}$$
$$= 239 - 150 = 89 \text{ v}$$
$$R = \frac{E}{I}$$
$$= \frac{89\ V}{27 \text{ ma}} = 3.3 \text{ k}\Omega \qquad \text{(answer)}$$
$$I_{VR} = \frac{89 \text{ v}}{3.3 \text{ K}} = 27 \text{ ma} \qquad \text{(answer)}$$

3–2. Analysis of Nonlinear Systems

A glow tube has a nonlinear voltage-current characteristic, as seen in Fig. 3–3. The most practical approach to analysis of the series

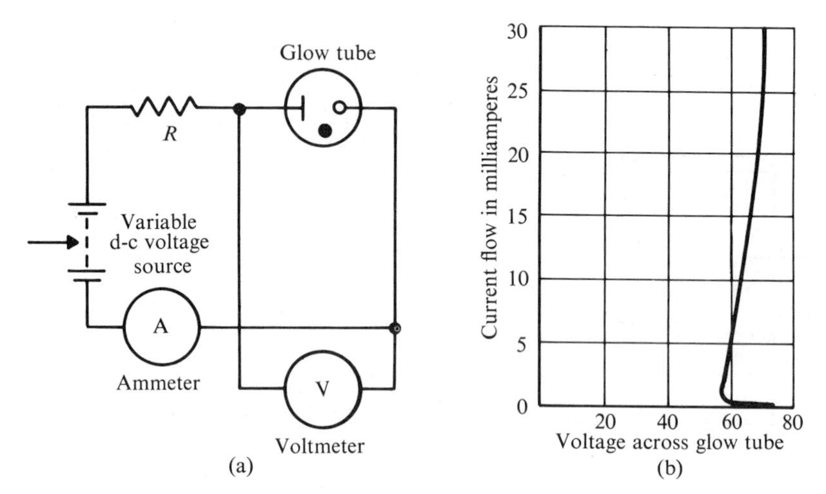

Figure 3–3. (a) Glow-tube test circuit; (b) Voltage-current characteristic.

circuit is a graphical analysis, as illustrated in Fig. 3–4. A constant-voltage source of 80 v is used. The load line for the 3,200-Ω series

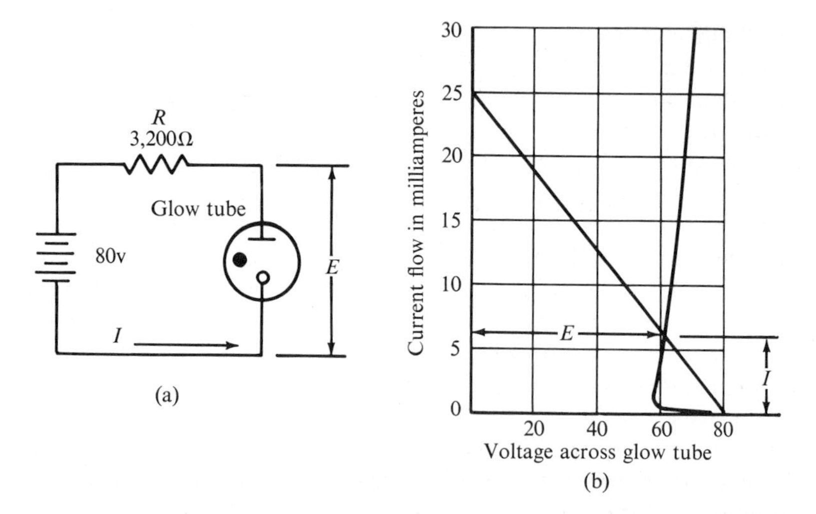

Figure 3–4. (a) Series circuit with glow tube; (b) Graphical solution of circuit.

resistor is drawn from the 80-v point to the 25-ma point on the coordinate system. In turn, the operating point is given at the intersection of the load line with the tube characteristic. At this point, the current flow is approximately 6 ma, and the voltage drop across the tube is approximately 60.1 v.

A graphical analysis for a parallel circuit is illustrated in Fig. 3–5.

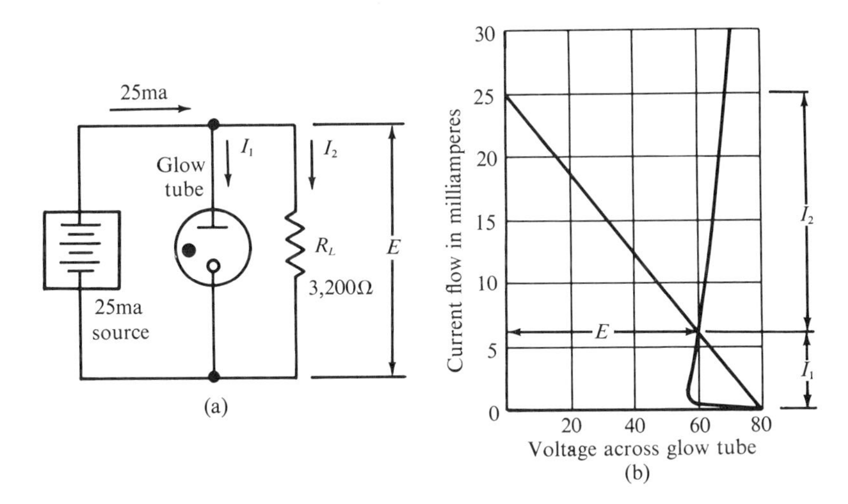

Figure 3–5. (a) Parallel circuit with glow tube; (b) Graphical solution of circuit.

A constant-current source of 25 ma is used. We could not employ a constant-voltage source, because the glow tube would explode. The load line is drawn as before. Note that if the glow tube is open-circuited, the voltage drop across the resistor is equal to 80 v; but if the glow tube is short-circuited, the voltage drop across the resistor is zero. The operating point is given at the intersection of the load line with the tube characteristic. At this point the current flow through the tube is approximately 6 ma, the current flow through the resistor is approximately 19 ma, and the voltage drop is approximately 60.1 v.

Next, if we wish to analyze the series-parallel configuration depicted in Fig. 3–2(a), we can work with neither a constant-voltage source nor a constant-current source. The voltage across all components is not the same; again, the current through all components is not the same. In turn, a graphical analysis is not straightforward, and it is advisable to approach the problem from the standpoint of linearity assumption. For example, Fig. 3–6 shows that the average dynamic resistance of the glow tube illustrated is approximately 400 Ω, and the

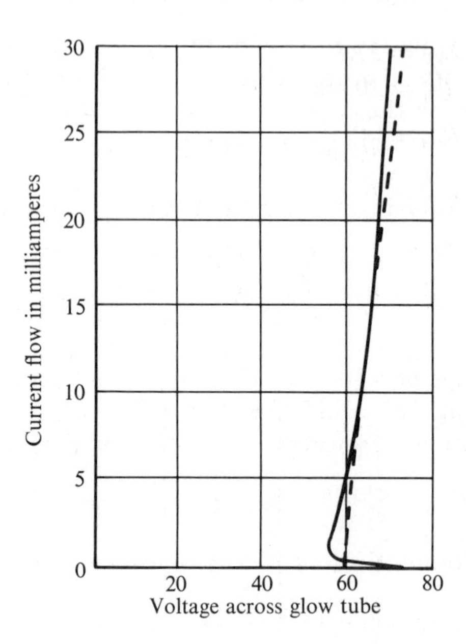

Figure 3–6. Average dynamic resistance of the glow tube is approximately 400 Ω.

static voltage is approximately 57.5 v at the point where static current is 10 ma. Accordingly, we can replace the glow tube in Fig. 3–2(a) with a 6-kΩ resistor and solve the series-parallel circuit for an approximate answer.

Example 2. When the load is removed, voltage across R and the current through R must remain the same for the output voltage to be 150 v. Therefore, the $OD3$ carries the total current of 27 ma. The dynamic resistance of the glow tube represented in Fig. 3–6 is

$$r = \frac{\Delta E}{\Delta I} = \frac{E_{\text{max}} - E_{\text{min}}}{I_{\text{max}} - I_{\text{min}}}$$

$$r = \frac{59\ V - 65\ V}{0 - 15\ \text{ma}}$$

$$r = \frac{16}{15\ \text{ma}} = 400\ \Omega \qquad \text{(answer)}$$

Example 3. The output of the rectifier in Fig. 3–2(a) is 100 v, and the output across a 3.6-kΩ load is 57.5 v. What is the value of R? From the graph in Fig. 3–6:

$$E_{VR} = 57.5 \text{ v}$$
$$I_{VR} = 10 \text{ ma}$$
$$R_{VR} = \frac{57.5}{10 \text{ ma}} = 5.75 \text{ k}\Omega$$
$$I_{\text{load}} = \frac{E_L}{R_L} = \frac{57.5}{3.6 \text{ k}} = 15 \text{ ma}$$
$$I_R = I_L + I_{VR} = 15 \text{ ma} + 10 \text{ ma} = 25 \text{ ma}$$
$$R = \frac{E_R}{I_T} = \frac{E_T - E_{\text{load}}}{25 \text{ ma}} = \frac{42.5}{25} = 1.7 \text{ k}\Omega \qquad \text{(answer)}$$

Note: A change of load or rectified input voltage (within limits) will cause a change of resistance of the *VR* tube and therefore a change of *VR* current. However, the total current will remain the same.

3–3. Series Connection of Glow Tubes

When a regulated voltage value is required that is in excess of the rating of a single glow tube, two or more glow tubes may be connected in series, as shown in Fig. 3–2(b). This arrangement permits several regulated voltages with comparatively small current drains to be obtained from a single regulator circuit. The individual tubes may have various voltage ratings. For example, a 75-v tube might be connected in series with a 150-v tube and a 105-v tube. In such a case, we would obtain regulated voltages with values of 105, 255, and 330 v in respect to the common point *B*.

Observe that it would *not* be permissible to replace one or more of the tubes in Fig. 3–2(b) with fixed resistors. The voltage drop across a resistor, of course, is unregulated. Accordingly, it would defeat the purpose of a voltage regulator to use a voltage divider between the glow tube and its load. Applications that require a regulated voltage supply that cannot be provided by glow tubes are implemented by other types of voltage-regulator circuits, discussed subsequently.

3–4. Semiconductor Voltage Regulation

Certain types of semiconductor diodes exhibit a practically constant voltage drop under suitable operating conditions. Comparative diode characteristics are shown in Fig. 3–7. If a small reverse bias voltage is applied across the *PN* junction, the height of the potential barrier is increased. Only a small reverse current flows, because the

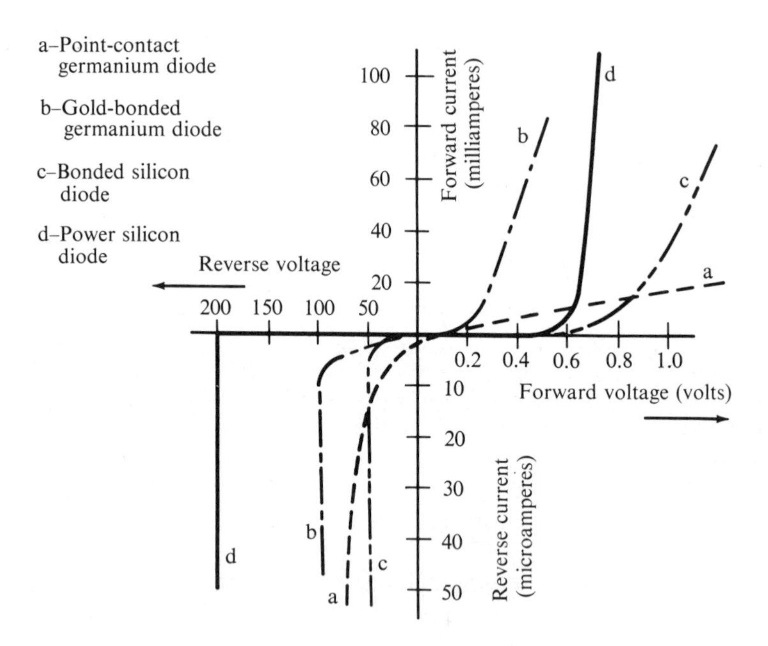

Figure 3–7. Semiconductor diode characteristics.

charge carriers are swept away from both sides of the junction. A space-charge depletion layer is established through the junction.

When a larger reverse bias voltage is applied across the diode, the velocity of the minority carriers in the depletion layer is increased. Some of these charge carriers collide with covalent-bond electrons and release these electrons as additional charge carriers. At a certain critical value of reverse bias voltage, this release action becomes self-sustaining and "runs away." This "snowballing" action is called *avalanche breakdown.* It is accompanied by a rapidly rising reverse-current flow that, unless limited by a series resistor, may destroy the diode. The value of reverse bias voltage at which the avalanche effect appears is called the reverse breakdown voltage, and it is denoted by BV_R.

A Zener-diode voltage-regulator circuit is depicted in Fig. 3–8. The symbol B denotes that the diode (CR) is being operated in the reverse breakdown-voltage region. When operated in this mode, the voltage across the diode and load is held practically constant because the voltage drop across the diode remains essentially constant over wide ranges of diode-current flow. Semiconductor diodes operated in

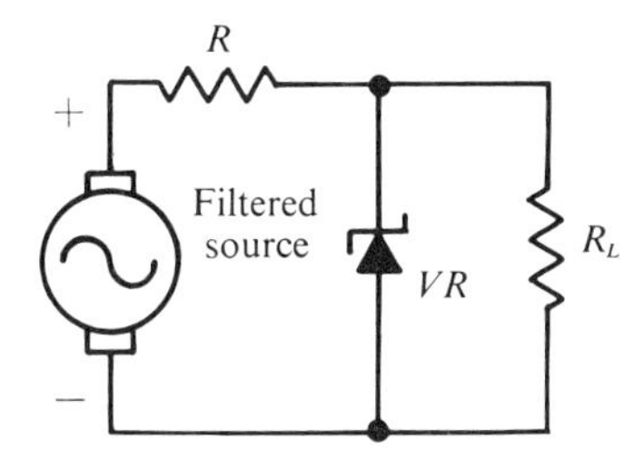

Figure 3-8. Zener diode voltage-regulator circuit.

this manner are in a class called breakdown diodes, as they operate either by Zener action (below approximately 8.6 v) or avalanche action (above approximately 8.6 v). The name *Zener* is universally used as this action was first described as a "field emission" effect by Clarence Zener. A family of Zener diode curves is shown in Fig. 3-9. Observe

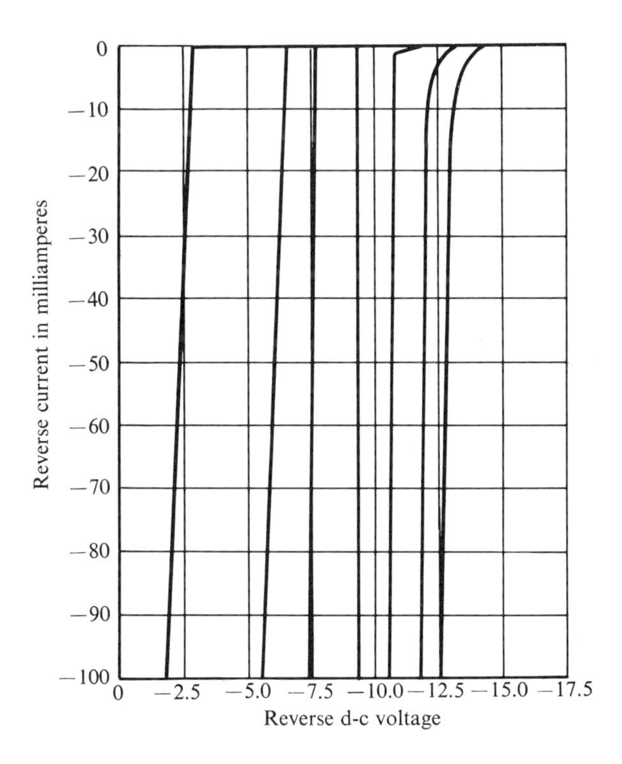

Figure 3-9. Reverse current curves of typical Zener diodes.

that the knee is rounded for low-voltage diodes and sharp for higher-voltage diodes. This would suggest that perhaps there are two different actions involved in breakdown diodes. These two actions are called Zener breakdown for low-voltage breakdown diodes and avalanche for higher-voltage breakdown diodes. Breakdown diodes present a 0° temperature coefficient at about 6 v, a negative temperature coefficient below, and a positive temperature coefficient above this voltage. This 0 temperature coefficient and the minimum Zener resistance coincide with a breakdown voltage of about 6 v.

If the supply voltage decreases in Fig. 3-8, the reverse voltage across the Zener diode *VR* will tend to decrease. In turn, the speed of the charge carriers within the crystal lattice will decrease, and the number of collisions with covalent-bond electrons will decrease; thus, the reverse current flow through the diode will decrease. Current flow through *R* is decreased by the amount of current decrease through *VR*. Because the current through *R* is decreased, the voltage drop across *R* is proportionally less. If *R* has a suitable resistance value with respect to the load and to *VR*, the voltage drop across the load will remain essentially constant. Of course, the value of *R* must not be so large that the avalanche action does not occur, nor so small that either the maximum current rating or the maximum power dissipation of the diode is exceeded. The value of *R* is limited by the formulas

$$R_{\text{minimum}} = \frac{E_{\text{in max}} - V_L}{I_{Z\ \text{max}}} \tag{3-1}$$

$$R_{\text{maximum}} = \frac{E_{\text{in min}} - V_L}{I_{Z\ \text{min}} + I_{L\ \text{max}}} \tag{3-2}$$

where R is the series resistance, $E_{\text{in max}}$ is the maximum value of input voltage, V_L is the maximum value load voltage, $I_{Z\ \text{max}}$ is the maximum Zener current ($P_{Z\ \text{max}}/V_Z$), $I_{Z\ \text{min}}$ is the minimum allowable Zener current, and $E_{\text{in max}}$ and $E_{\text{in min}}$ are the limits of the source voltage and $I_{L\ \text{max}}$ is the maximum load current.

Example 4. Determine the range of resistance for *R* in the regulator circuit shown in Fig. 3-8 to establish 25 v across a 10-kΩ load using a 25-v, 1/4-w Zener which must have at least 200 μa of current for stable operation. The maximum value of input voltage is to be 60 v, and the minimum value of input voltage is to be 40 v.

$$I_{Z\ \text{max}} = \frac{P_{Z\ \text{max}}}{V_Z} = \frac{0.25}{25} = 10\ \text{ma}$$

$$I_L = \frac{V_L}{R_L} = \frac{25}{10 \times 10^3} = 25\ \text{ma}$$

$$R_{\text{min}} = \frac{60 - 25}{10 \text{ ma}} = 3.5 \text{ k}\Omega$$

$$R_{\text{max}} = \frac{40 - 25}{0.2 \text{ ma} + 2.5 \text{ ma}} = 5.56 \text{ k}\Omega$$

$$3.6 \text{ k} < R < 5.56 \text{ k}\Omega \qquad \text{(answer)}$$

A regulator, such as illustrated in Example 8 of this chapter, would probably be used with a battery pack where $E_{\text{int max}}$ is the fully charged voltage of the pack, and $E_{\text{int min}}$ is the discharged voltage of the pack.

Zener diodes are designed to operate at various breakdown voltages and various power ratings. When a regulated voltage is required that exceeds the rated voltage value of a single Zener diode, two or more diodes may be connected in series as previously explained for glow-tube regulators. Note that commerical tolerances on Zener breakdown-voltage values are fairly wide. Accordingly, you may have to make a number of measurements on a group of diodes to select a diode that has a precise breakdown-voltage value. A practical circuit to measure the breakdown voltage of a Zener diode is shown in Fig. 3–10.

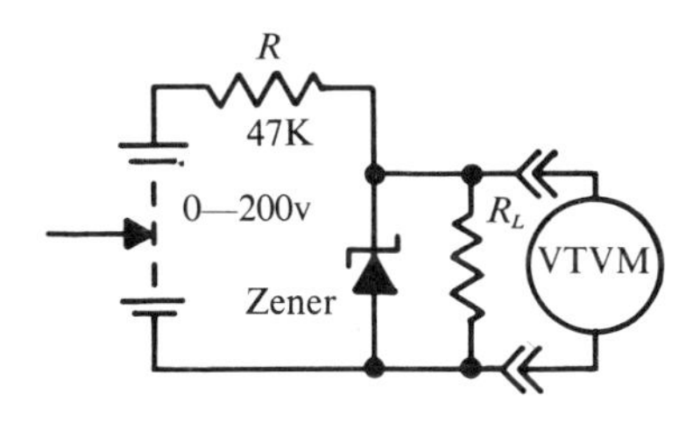

Figure 3–10. Zener diode test circuit.

3–5. Metallic Current Regulator

A *ballast tube* is basically a current regulator that maintains a fairly constant current flow over an appreciable range of input voltage variation. A ballast tube is also called a barretter. It consists of a length of iron wire enclosed in a hydrogen atmosphere. A ballast tube is utilized as a series component, as shown in Fig. 3–11(a). The resistance of the iron wire varies as the current flow changes. If the input voltage increases, more current starts to flow through the ballast tube. In turn, the resistance of the iron wire increases, and more volt-

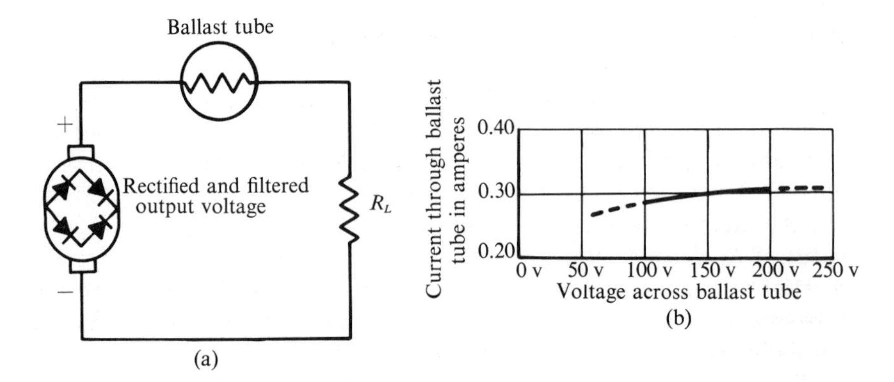

Figure 3–11. (a) A ballast-tube regulator circuit; (b) Voltage-current characteristic of a typical ballast tube.

age is dropped across the ballast tube; the current flow increases but slightly. Hydrogen is used to provide a high rate of heat removal. Iron has a high temperature coefficient of resistance, as seen in the following tabulation.

Temperature Coefficients of Metals

Metal	*Temperature coefficient per °C*
Aluminum	0.0043
Copper	0.0040
German Silver	0.0004
Iron	0.0062
Manganin	0.00002
Platinum	0.00366

A typical ballast tube passes 1.24 amp at 105 v and 1.36 amp at 125 v. Thus, a change of 19 per cent in applied voltage produces a change of 9.7 per cent in current flow. Ballast tubes are commonly designed to operate in the ranges from 5 to 8, 3 to 10, 15 to 21, 40 to 60, and 105 to 125 v. Since there is a thermal lag in a ballast tube, it cannot respond instantly to an input voltage change. It is evident from Fig. 3–11 that a ballast tube also operates as a voltage regulator in a series circuit. In other words, the voltage across the load is maintained as constant as the current which flows through it. Note that the current value will be changed, and the load voltage will also be changed if you vary the value of the load resistance. A ballast tube operates correctly only if the load resistance does not vary.

3–6. Electron-Tube Voltage Regulator

For our present viewpoint, an electron tube may be regarded as a variable resistance. When the tube is conducting a direct current, its resistance value is equal to the ratio of plate-to-cathode voltage and current flow. This is the d-c plate-resistance value and is denoted by R_p. An electron tube is basically a nonlinear device, and in turn, the value of R_p depends upon the amount of current flow; the current flow, of course, depends upon the value of grid bias. Since an electron tube operates as a variable resistance, it finds useful application in voltage-regulator circuits.

With reference to Fig. 3–12(a), the effective plate resistance of $V1$

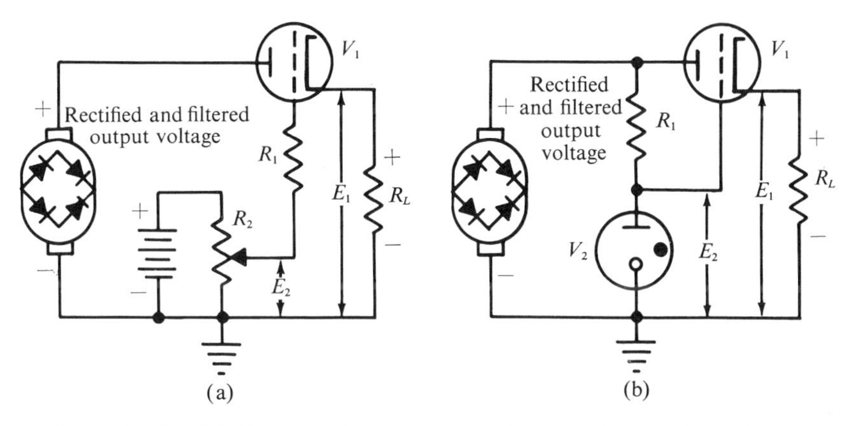

Figure 3–12. (a) Electron-tube voltage regulator with fixed bias from a battery; (b) A glow tube will supply the fixed bias.

is established initially by the grid-bias voltage. Let us assume that the voltage drop across the load is the desired value. Then, the cathode is positive with respect to ground by this load drop, E_1. The grid can be made positive with respect to ground by a voltage E_2 that is less than E_1. Potentiometer R_2 is adjusted to make the bias voltage (grid-to cathode voltage), which is $E_1 - E_2$, sufficient to pass a current through $V1$ equal to the desired load current. With this bias voltage, the resistance of $V1$ is established at the proper value to reduce the rectifier output voltage to the desired load voltage.

If the rectifier output voltage increases, the voltage at the cathode of $V1$ tends to increase. As E_1 increases, the negative bias on the grid increases, and the effective plate resistance of the tube becomes greater. Consequently, the voltage drop across $V1$ increases with the rise of input voltage. If the circuit is suitably designed, the increased voltage drop across $V1$ is approximately equal to the increase in voltage at the input to the regulator. Hence, the voltage across the load remains essentially constant.

Resistor R_1 is used to limit grid-current flow. This limitation is necessary in this particular configuration because the battery is not disconnected when the power is turned off. However, the battery can be eliminated from the circuit by use of a glow tube, $V2$, as shown in Fig. 3–12(b). The glow tube provides a fixed bias voltage for the grid of $V1$ whenever the circuit is operating. Thus, the circuit action is the same as if a battery were used to obtain the grid-bias voltage. The output voltage from the simple regulator circuits in Fig. 3–12 cannot remain absolutely constant. As the rectifier output voltage increases, the voltages on the cathode of $V1$ must rise slightly if the regulator is to function.

Note that the voltage regulators shown in Fig. 3–12 compensate not only for changes in the output voltage from the rectifier, but also for changes in the load. For example in Fig. 3–12(b), if the load resistance decreases, the load current will increase. The load voltage will tend to fall because of the increased voltage drop across $V1$. The decrease in load voltage is accompanied by a decrease in grid-bias voltage on $V1$. The bias voltage on $V1$ is equal to $E_1 - E_2$. Thus, the effective resistance of $V1$ is reduced at the same time that the load current is increased. The IR drop across $V1$ increases only a slight amount because R decreases about as much as I increases. Therefore, the tendency for the load voltage to drop when the load is increased is compensated by the decrease in effective resistance of the series triode.

3–7. Improved Voltage Regulation

A very stable voltage regulator can be obtained by taking advantage of the high amplification provided by a pentode tube. A voltage regulator employing a pentode is shown in Fig. 3–13. The output voltage is more nearly constant than that of the triode circuit in Fig. 3–12. The output voltage from the pentode regulator circuit is developed across bleeder resistors R_3, R_4, and R_5 in parallel with the load resistance. These resistors make up the resistance of one part of the total voltage divider. The other resistance, through which all of the load current must flow, is the effective plate-to-cathode resistance of $V1$. The other components of the circuit are used to control the resistance of $V1$ and thereby maintain constant voltage across the load.

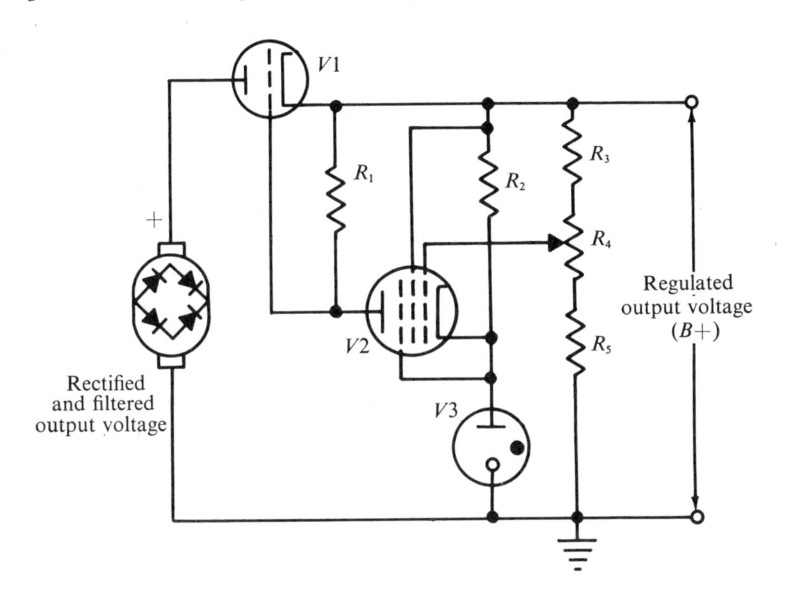

Figure 3-13. A voltage regulator circuit with a pentode d-c amplifier.

Note that the potential of the cathode (*V*2) is held at a constant positive value with respect to ground by glow tube *V*3. In other words, current flowing from ground through *V*3 causes an *IR* drop across *V*3 that maintains the cathode of *V*2 positive with respect to ground. The grid potential of *V*2 is a voltage that is selected by setting potentiometer R_4. This potentiometer is adjusted to make the grid-to-ground voltage less positive than the cathode-to-ground voltage by an amount equal to the bias that causes *V*2 to pass a certain value of plate current. Otherwise stated, the *IR* drop between the arm of R_4 and ground is less than the *IR* drop across *V*3 by an amount that is equal to the bias on *V*2. The plate current of *V*2 flows through R_1 and produces a voltage drop. Note that the magnitude of the voltage across R_1 is the bias voltage for *V*1. Therefore, adjustment of potentiometer R_4 establishes the normal resistance of *V*1. This adjustment is used to set the value of load voltage that the regulator is to maintain.

If the load voltage tends to rise, whether from a decrease in load current or from an increase of input voltage, the voltage between the arm on R_4 and ground will increase. The difference in this voltage and the fixed voltage across *V*3 decreases. Observe that these two voltages are opposed, and the voltage between the arm on R_4 and ground is less than the fixed voltage across R_3. Hence, the grid bias on *V*2 decreases, and the plate current of *V*2 increases through R_1. The increase in voltage across R_1 increases the effective resistance of $V1$. If the load voltage tends to rise because of an increase in input voltage, this increase is accompanied by an increase in voltage across $V1$ and the rise in load voltage is compensated. Again, if the rise in load voltage is caused by a decrease in load current, this rise is compensated because the IR drop across $V1$ remains constant; the decrease in I is accompanied by an equal increase in R.

A pentode is used in the $V2$ position because it provides high amplification. In turn, the output voltage is held more nearly constant; small variations in load voltage are sufficiently amplified to provide close regulation. The anode of glow tube $V3$ is connected to the cathode of $V2$ and to the positive terminal of the regulated output voltage ($B+$) through resistor $R2$. It is necessary to connect the glow tube to $B+$ in this manner in order to produce ionization of the gas in $V3$ when the power is first turned on.

All of the load current must flow through V1; therefore, this tube must be rated for sufficiently large current flow. In some regulators, a single tube does not have sufficient current capability to conduct the required value of current. In such case, several identical tubes may be

Courtesy of Simpson Electric Co.

Figure 3–14. A pulse generator that contains a built-in regulated power supply.

connected in parallel. The regulator configuration depicted in Fig. 3–13 is widely used because of its comparatively good performance. As would be anticipated, an electronic voltage regulator is also an efficient *filter*. In other words, ripple in the input voltage is largely removed by regulator action, just as if it were due to line-voltage fluctuation. However, to obtain maximum smoothness of the regulated output, it is customary to energize the regulator circuit from a filtered power supply. Many electronic instruments have built-in regulated power supplies. For example, Fig. 3–14 illustrates a pulse generator that is powered by a regulator section.

3–8. Transistor Voltage Regulator

Transistors are often used in place of tubes to obtain voltage regulation. Compare the circuit in Fig. 3–15(a) with that in Fig. 3–13.

Transistor *T*1 operates as a controlled variable resistor in series with the load. Variations in the output voltage level are applied to the base of *T*1 as negative feedback to increase or decrease the resistance of *T*1. For example, suppose the voltage at the base of *T*2 decreases, causing an increase of its collector voltage and hence an increase of the base voltage of *T*1. The increase of *T*1 base voltage *reduces* the *resistance* of *T*1 and increases the output voltage across the load. An *increase* in the output voltage would result in a similar regenerative action to increase the resistance of *T*1, thereby decreasing the load voltage.

The function of the Zener diode is to establish a d-c voltage level and to compensate for drift in the resistance of *T*1 due to variation in temperature. We will discuss temperature stability for transistors in Chapter 9.

Transistor regulators must be protected from load malfunctions which cause drastic increases of load current; for example, a short circuit. The simplest form of current protector places a resistor in series with the power supply which would limit the value of maximum current. However, the power dissipation of such a resistor at maximum current would be great, and this would reduce the efficiency of the supply.

Figure 3–16 depicts a current limiter circuit employed in a regulator circuit. In normal current generation, transistor *T*2 appears as a very low value resistance in series with R_E ($R_E \simeq 0.5\Omega$). The sum of the voltage drops across the emitter-base junction of *T*2 (V_{EB}) and across R_E are equal to the drop across diode *CR*1. Diode *CR*1 conducts establishing a constant voltage of about 0.6 v for a silicon diode. The

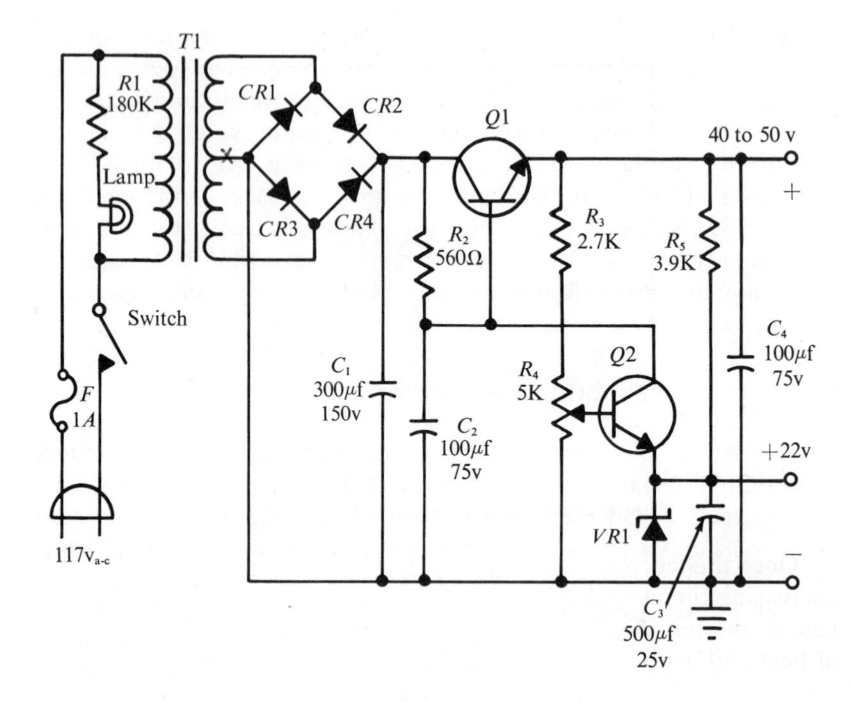

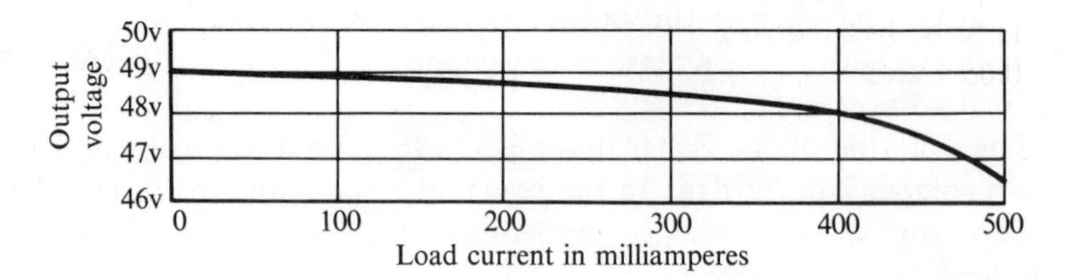

Courtesy of General Electric Co.

Figure 3-15. **(a) Transistor voltage-regulator configuration; (b) Regulation characteristic.**

difference between the voltage V_{CR1} and V_{EB} and the value of R_E determine the maximum current.

$$I_E = \frac{V_E}{R_E} \tag{3-3}$$

$$I_{E\ \max} = \frac{V_{D1} - V_{EB}}{R_E} \tag{3-4}$$

then,

$$R_E = \frac{V_{D1} - V_{EB}}{I_{E\ \max}} \tag{3-5}$$

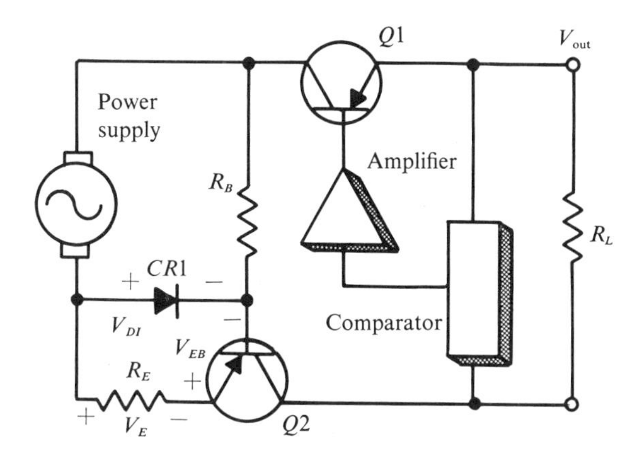

Figure 3-16. Power supply with current limiter.

Once this current value is reached, transistor $Q2$ goes into cutoff, decreasing the output current, and the resistance of $Q2$ increases. Of course, the transistors must be capable of dissipating great quantities of heat and heat sinks must be used.

Example 5. Suppose the current in the circuit depicted in Fig. 3-16 is to be limited to 2 amp. What is the value of R_E when $V_{D1} = 0.65$ v and $V_{EB} = 0.2$ v?

$$R_E = \frac{V_{D1} - V_{EB}}{I_{max}}$$

$$R_E \simeq \frac{0.65 - 0.2}{2}$$

$$R_E \simeq 0.225\ \Omega \qquad \text{(answer)}$$

The regulation characteristics for the transistor regulator are shown in Fig. 3-15(b). Uniform regulation is obtained for load currents from 0 to 350 ma, and 2 per cent regulation appears at 400 ma. The peak-to-peak output ripple from this regulator circuit is less than 0.3 v at 400 ma load current, and 0.01 v at no load. All voltage regulators have a very low output impedance; the output impedance of the transistor regulator is less than 1 Ω at 200 ma load current and is less than 2 Ω over the entire operating range.

Example 6. The output voltage of a voltage regulator, as in Fig. 3-15, changes from 180 v to 180.2 v with a decrease in load current of 100 ma to 80 ma. What is the internal impedance of the regulator?

$$R_{\text{int}} = \frac{\Delta E}{\Delta I} = \frac{180.2 - 180}{(100 - 80)\text{ ma}}$$
$$= \frac{0.2}{.02} = 10\,\Omega \qquad \text{(answer)}$$

3-9. Voltage-Stabilizing Transformer

When an a-c voltage is to be stabilized, a voltage-stabilizing transformer is commonly utilized, as shown in Fig. 3-17. The design

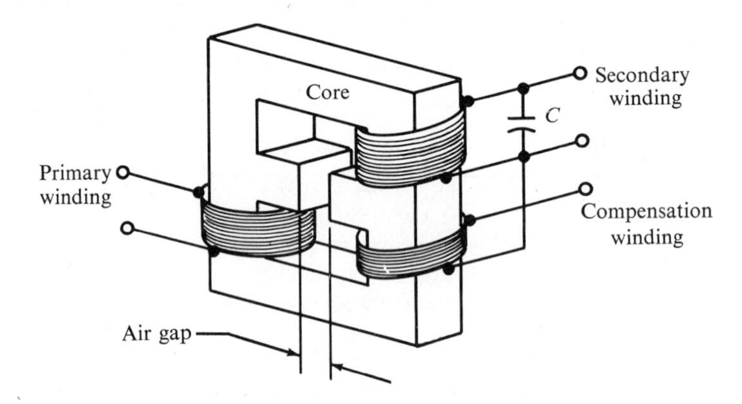

Figure 3-17. Arrangement of typical voltage-stabilizing transformer.

is directed to maintaining an approximately constant output voltage from the secondary, although the voltage applied to the primary varies over a considerable range. We know that iron provides a nonlinear magnetic circuit, as shown by the *BH* curve in Fig. 3-18. The operating point of a voltage-stabilizing transformer is placed on the curved portion of the core's *BH* characteristic. In turn, the inductance of the secondary winding changes when the load current changes.

A capacitor C of suitable value is connected across the secondary winding. When the secondary current demand has a certain value, the inductance of the compensating winding forms a series-resonant circuit with the capacitance; there is a Q magnification of secondary voltage. This circuit action is basically responsible for stabilization of the secondary voltage as the current demand increases. Thus, if the load across the secondary is changed, the operating point shifts farther away from, or closer to resonance, and the voltage across the load remains essentially constant.

Let us consider the transformer action when the input voltage to the primary changes in value. A magnetic shunt is provided by the

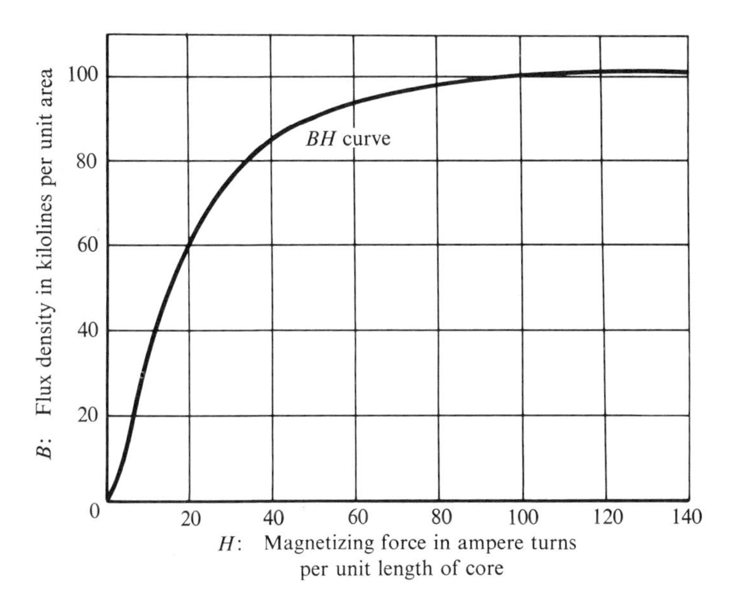

Figure 3-18. A *BH* curve for iron.

center leg of the transformer, and the amount of magnetic flux diverted by the shunt is determined by the design spacing of the air gap. Note that the capacitor connected across the secondary draws a leading current. Then, the secondary produces an out-of-phase flux that opposes the primary flux in its passage through the shunt. Consequently, the secondary portion of the core operates at greater magnetic saturation than the primary portion of the core.

It is evident that if the input voltage to the primary is increased, the additional primary current flow produces flux that passes chiefly through the shunt, because the secondary portion of the core is more highly saturated. This circuit action minimizes the voltage change across the secondary load. Further stabilization is provided by the compensating winding shown in Fig. 3-17. The induced voltage in the compensating winding opposes the secondary voltage. Therefore, if the input voltage to the primary increases, the tendency of the secondary voltage to increase is further opposed by increased output voltage from the compensating winding. The combined result of these circuit actions provides effective voltage stabilization. A typical transformer provides a secondary output voltage that varies $\pm$ 1 per cent when the input voltage to the primary varies $\pm$ 20 per cent.

The time-constant of a voltage-stabilizing transformer is such that abrupt changes in primary voltage cannot be compensated instantaneously; however, after several cycles of operation, the secondary voltage will return to almost its original value. Note that if better stabilization is desired than can be provided by a single transformer, two or more transformers may be connected in cascade. Of course, a voltage-stabilizing transformer is ineffective if operated at very light loads or at excessively heavy loads; the operating point must fall on a suitable portion of the *BH* core characteristic. Hence, a voltage-stabilizing transformer is rated for a certain range of power output, and should not be operated out of this range.

3–10. Pure and Applied Mathematics

Previous discussion has introduced us to the concepts of physical units and definitive dimensional formulas. We are now in a good position to consider the distinction between pure numbers and concrete numbers. For example, when we state that "$2 + 2 = 4$", we are not referring to any objects or physical concepts; we are stating a pure number relation. Note that the meaning of the first number "2" is the same as the meaning of the second number "2" in the expression "$2 + 2 = 4$". Since the meanings of the first and second numbers are exactly the same in every respect, each of the "2's" is said to be *indistinguishable*. Note in passing that scientists apply the principle of indistinguishability to electrons, as well as to pure numbers.

If we state that "2 v + 2 v = 4 v," we are now referring to a physical concept: the volt. We call "2 v" a *concrete number*. Note that the number "2" tells us *how many* and the physical unit "volt" tells us *what*. Other examples of concrete numbers are 3 amp, 6 Ω, 5 w, 0.001 μf, and 1.5 h. It is evident that any concrete number has two parts; the first part has the form of a pure number, and the second part has the form of a physical unit. We will find that the number "2" in the concrete number "2 v" is not actually a pure number—it merely has the appearance of a pure number. The number "2" is not a pure number in this form, because this number is a *measurement*, or it is a voltmeter *scale indication*. In other words, the number "2" in this example is a *scalar*.

Let us clearly observe the distinction between a pure number and a scalar (a measured value). A pure number is absolutely exact; for example, we never state that "2 plus 2 is 4, approximately"; this would be an absurd statement, because logic dictates that "2 plus 2 is 4, exactly." On the other hand, we never state that "the voltmeter reading is 2 v, exactly"; this would be an absurd statement, because no one can read a pointer indication exactly. The observational error might be very small, but it is inevitably present. Therefore, the true statement of the situation is "the voltmeter reading is 2 v, approximately," or, "the voltmeter reading is 2 v, with an unstated observational error present."

Moreover, no voltmeter is 100 per cent accurate. In turn, no voltage value can be measured exactly, even if we could eliminate observational error. In precise scientific work, we state the accuracy of voltmeter response in two ways: (1) instrument accuracy may be stated in terms of percentage of full-scale indication; (2) instrument accuracy may be stated in terms of percentage of the scale reading. For example, an ordinary 20,000 Ω per volt meter is rated for a typical full-scale accuracy of $\pm$ 2 per cent. This means that if we are operating on the 100-v scale, any reading on this scale has a rated accuracy of ± 2 v. Again, a laboratory-type hand-calibrated voltmeter might be rated for an accuracy of ± 0.5 per cent of the scale reading. And so if we are operating on the 100-v scale and the pointer indicates 60 v, the rated accuracy of the reading is 60 v ± 0.3 v.

We recognize that "60 v ± 0.3 v" is not a pure number but a concrete number that comprises a scalar and a physical unit. We state that a concrete number is a mathematical model of a physical measurement; this means that a concrete number describes a physical measurement in mathematical form. It implies that the mathematical model entails an experimental error that may be minimized but that cannot be completely eliminated. A concrete number is the simplest type of mathematical model.

The next mathematical model is a basic formula, such as Ohm's law. For example, we write

$$2 \text{ amp} = \frac{6 \text{ v}}{3\,\Omega}$$

Note carefully that the above formula entails a relation among scalars and a relation among physical units. The relation among scalars is written

$$2 = \frac{6}{3}$$

The relation among physical units is written

$$\text{amp} = \frac{\text{volts}}{\text{ohms}} \tag{3-6}$$

Furthermore, Formula 3-6 entails a dimensional analysis.

$$\text{amperes} = \frac{Q}{T} = \frac{\text{volts}}{\text{ohms}} = \frac{FLQ^2}{FLTQ} = \frac{Q}{T} \tag{3-7}$$

In summary, pure mathematics states relations among pure numbers, and these relations are absolutely exact. On the other hand, applied mathematics states relations among concrete numbers, and these relations are never exact, although they might be stated to a high order of precision. Relations among pure numbers are called equations; an equation expresses pure-number relations only. On the other hand, relations among concrete numbers are called formulas; a formula expresses scalar relations and physical-unit relations. A formula is also called a mathematical model.

Physical-unit relations, such as expressed by the foregoing dimensional Formula 3–4, are concerned solely with relations of physical concepts. In turn, dimensional formulas contain no numbers other than those which denote physical relations. For example, "2 v" and "3 v" both refer to the physical concept of "volt," and the scalars "2" and "3" are beside the point in a dimensional analysis. Dimensional analysis of "2 v" states the conclusion: FL/Q; similarly, dimensional analysis of "3 v" states the same conclusion: FL/Q. Therefore, we strike out all numerical coefficients when we write a dimensional formula.

On the other hand, when physical units are multiplied, a distinctive physical concept results. For example, one of the power laws states:

$$P = I^2 R \tag{3–8}$$

We note that this is an abbreviated writing of

$$P = I \cdot I \cdot R \tag{3–9}$$

Observe that we are forbidden to strike out an I in Formula 3–8 when we make a dimensional analysis. If we should make this error, we would write $P = IR$, which erroneously states that power is equal to voltage. Therefore, we are forbidden to strike out exponents when we make a dimensional analysis. To make certain that we clearly understand this principle, let us consider the following example. We will make a dimensional analysis of the power formula.

$$4 \text{ w} = (2 \text{ amp})^2 (1\ \Omega) \tag{3–10}$$

We note that this is an abbreviated way of writing

$$4 \text{ w} = 2 \text{ amp} \cdot 2 \text{ amp} \cdot 1\ \Omega \tag{3–11}$$

All numerical coefficients are inconsequential and are struck out.

$$\text{watts} = \text{amperes} \cdot \text{amperes} \cdot \text{ohms} \tag{3–12}$$

If we abbreviate our relation among physical units, we write

$$\text{watts} = \text{amperes}^2 \cdot \text{ohms} \tag{3–13}$$

Our final step in this dimensional analysis reveals an identity:

$$\frac{FL}{T} = \frac{Q^2 FLT}{T^2 Q^2} = \frac{FL}{T} \tag{3-14}$$

SUMMARY

1. A simple regulator using a parallel gas-regulator tube can be used for relatively constant loads with low current demands.

2. Gas-regulator tubes may be connected in series to develop several regulated voltages with small current drains.

3. The Zener or breakdown diode is especially suited for simple regulator circuits with low-voltage and low-current requirements.

4. Zener action in a diode was named for Clarence Zener, who explained the breakdown action as a field emission.

5. Breakdown diodes (Zener) of less than 6 v breakdown have a negative temperature coefficient; those of greater than 6 volts have a positive temperature coefficient.

6. Zener diodes are used extensively for regulators in portable or space-age equipment to establish a regulated voltage from a battery pack over an extended period of time.

7. A ballast tube or barretter consists of a length of iron wire enclosed in a hydrogen atmosphere which utilizes its positive temperature coefficient of resistance property to regulate current.

8. A ballast tube series regulator operates correctly only with a constant load.

9. The electron tube may be considered as a voltage variable resistor and, as such, finds useful application in voltage-regulator circuits.

10. Control devices such as electron tubes or transistors are used in regulators for applications which require very stable operating voltage levels.

11. Employment of more than one control device in a voltage regulator, as seen in Fig. 3-13 and 3-15, results in a very stable output and a method to control the output voltage level within very close limits.

12. A basic characteristic of a voltage regulator is that its output impedance appears as a very low value.

13. A voltage-stabilizing transformer is normally used when an a-c voltage is to be stabilized.

Questions

1. How does a gas glow tube maintain a relatively constant voltage in a simple voltage regulator, such as the one shown in Fig. 3-2(a)?
2. What are the factors determining the series resistance value in the filter in Fig. 3-2(a)?
3. How is the dynamic resistance of a voltage regulator determined?
4. Explain the difference between a current regulator and a voltage regulator.
5. What are the limitations of a regulator such as the one in Fig. 3-2(b)?
6. What changes would occur in load voltages and currents if the *VR*5 tube in the circuit illustrated in Fig. 3-2(b) were to open? If the *VR*5 were to short?
7. Explain the regulation action of a simple regulator as in Fig. 3-2(a).
8. What factors limit the high and low values of series resistance in a simple regulator as the one in Fig. 3-2(a)?
9. What is the name of the action whereby reverse "breakdown" occurs in a semiconductor diode?
10. What factors limit the maximum value of series resistance in a simple Zener regulator as illustrated in Fig. 3-8?
11. What factors limit the minimum value of series resistance in a simple Zener regulator as illustrated in Fig. 3-8?
12. How is a ballast tube used to regulate current, and why is a constant load a requirement?
13. Explain the operation of the voltage regulator in Fig. 3-12(b).
14. Explain why a voltage regulator must have a very low output impedance.
15. Explain the operation of a voltage-stabilizing transformer.
16. What is the effect on the output of the circuit in Fig. 3-13 if *V*3 shorts?
17. What is the function of the 5-kΩ potentiometer in Fig. 3-15(a)?
18. What effect would a shorted capacitor (C_3) have on the circuit in Fig. 3-15(a)?
19. What action would occur in the regulator circuit in Fig. 3-15(a) if the Zener diode opened?
20. What is the basic type of filter and rectifier shown in the circuit of Fig. 3-15(a)?

PROBLEMS

1. Determine the voltage across each load in Fig. 3-2(b) when the three voltage regulator tubes are type *VR*105 which regulates at 105 v.

2. What is the supply-voltage in Problem 1 when load 1 = load 2 = load 3 = 100 kΩ, R = 2.2 kΩ, and the current through $VR3$ is 10 ma?

3. Select the value of series resistance for the voltage regulator in Fig. 3-2(a) that will give a load voltage of 90 v at 10 ma and a regulator tube current of 12 ma when the regulator is connected to a 250-v d-c source.

4. Using the circuit in Fig. 3-2(a) and the graph in Fig. 3-6, determine the series resistance (R) necessary to establish a load voltage of 66 v across a 6.6-kΩ load when the applied voltage is 150 v.

5. For the circuit in Fig. 3-2(b) (using the graph in Fig. 3-6), determine the current of each load and VR tube, and the rectified voltage to establish voltages of 64 v, 130 v, and 198 v when load 1 is 50 kΩ, load 2 is 21.7 kΩ, load 3 is 14.2 kΩ, and R is 3.3 kΩ.

6. Determine the limits of the series resistance for the Zener regulator depicted in Fig. 3-10 when

$$V_L = 10 \text{ v}$$
$$R_L = 4.7 \text{ k}\Omega$$
$$E_{\text{in max}} = 30 \text{ v}$$
$$E_{\text{in min}} = 20 \text{ v}$$
$$P_{Z \text{ max}} = 1/4 \text{ w}$$
$$I_{Z \text{ min}} = 200\ \mu\text{a}$$

7. Determine the internal resistance of a voltage regulator when a change of load current from 200 ma to 100 ma produces a change of load voltage from 120 v to 119.6 v.

8. In reference to Fig. 3-16, calculate the ohmic value of resistance and the wattage rating of the resistor R_E when $V_{EB} = 0.15$ v, $V_{D1} = 0.7$ v and $I_{\text{max}} = 5$ amp.

CHAPTER 4

Special-Purpose Power Supplies

4–1. Introduction

Previous discussion has considered the more generalized forms of power supplies. We are now in a good position to analyze some of the special-purpose power supplies that are used extensively in electronics technology. For example, Fig. 4–1 illustrates a small power supply for transistorized equipment; the circuit diagram for the unit is shown in Fig. 4–2. Output voltage is adjustable from 0 to 24, and current demands up to 100 ma can be supplied. Half-wave rectification is used, with a selenium diode followed by an $RC\pi$ filter. A stiff bleeder (213 Ω) is used to provide adequate regulation. Recall that percentage regulation is commonly formulated

$$\% \text{ regulation} = \frac{E_{OC} - E_{CC}}{E_{CC}} \times 100 \qquad \textbf{(4–1)}$$

where E_{OC} denotes the open-circuit voltage from the power supply, and E_{CC} denotes the closed-circuit voltage when a load is connected.

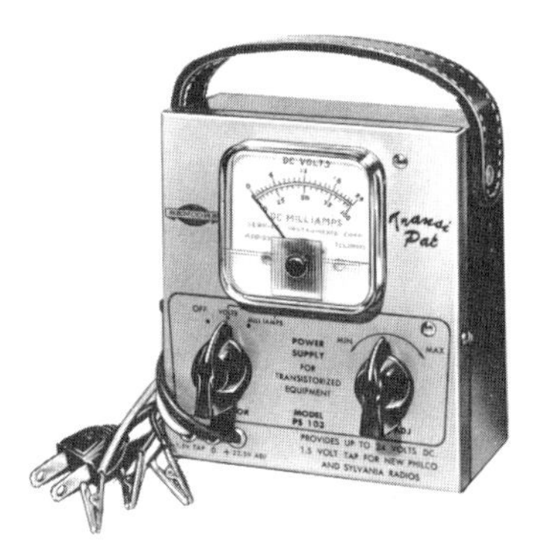

Courtesy of Sencore

Figure 4-1. A small auxiliary power supply for transistorized equipment.

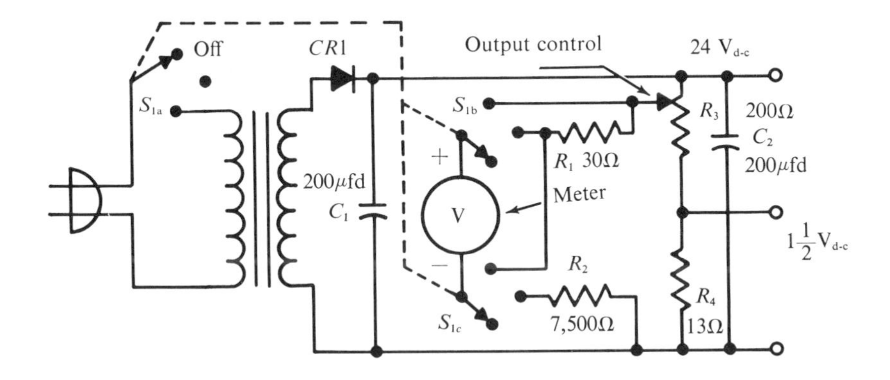

Figure 4-2. Configuration of the power supply illustrated in Figure 4-1.

Regulation of the power supply depicted in Fig. 4-2 is approximately 20 per cent under average load. The ripple waveform is illustrated in Fig. 4-3. At 24 v d-c output, the ripple amplitude is

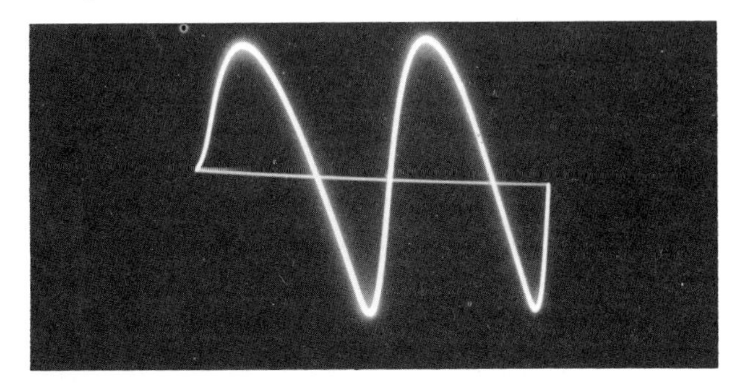

Figure 4-3. Ripple waveform from power supply depicted in Figure 4-2.

approximately 1.5 v peak-to-peak. However, when the output from the power supply is passed through the filter in a typical transistor radio receiver, the ripple amplitude is reduced to approximately 0.05 v peak-to-peak. This type of power supply is also used to recharge nickel cadmium batteries, such as used to power portable transistor television receivers.

Courtesy of EICO Electronic Instrument Co., Inc.

Figure 4-4. A power supply for use as a storage-battery eliminator or charger.

A heavier-duty low voltage power supply is illustrated in Fig. 4-4. Output voltage is adjustable from 0 to 12 v d-c. Maximum current drain at 6 v is 20 amp, or 12 amp at 12 v output. The circuit diagram for this power supply is depicted in Fig. 4-5. Full-wave rectification is used, followed by a simple capacitor filter. Since the power supply is designed to operate under comparatively heavy current demands, the filter capacitor has a very large value of 6,000 μf. The fundamental ripple frequency is 120 Hz. Filtering is not required, of course, when the power supply is used to charge a storage battery. On the other hand, when the power supply is used to eliminate a storage battery that powers electronic equipment, a filtered output is required.

Let us consider the ripple amplitude from the power supply depicted in Fig. 4-5. If there is no load on the power supply, the ripple amplitude is obviously 0. However, suppose that we draw 6 amp at 6 v from the power supply. Then, the load resistance is 1 Ω, and the time constant of the filter output circuit becomes 6 msec. Since the fundamental ripple frequency is 120 Hz, it follows that the filter

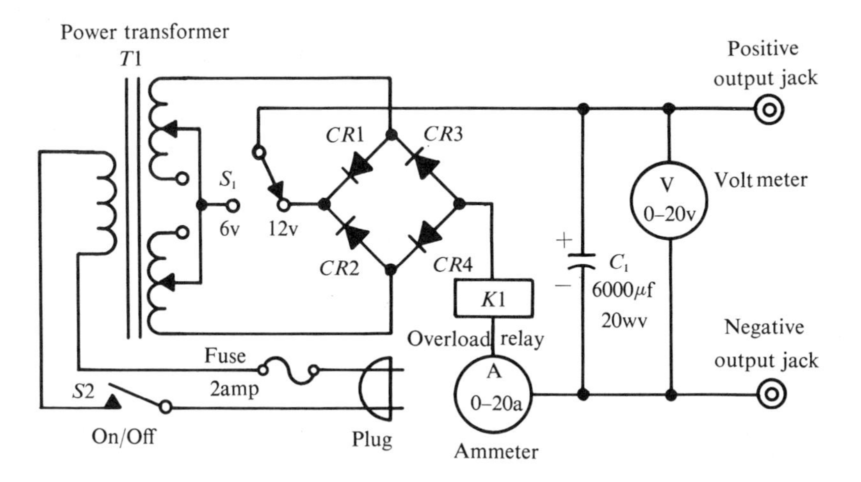

Figure 4-5. **Configuration of the power supply illustrated in Figure 4-4.**

discharges for 1.4 time constants between current pulses. For the reader's convenience, the universal *RC* time-constant chart is shown in Fig. 4-6. Inspection of curve *B* shows that the ripple amplitude

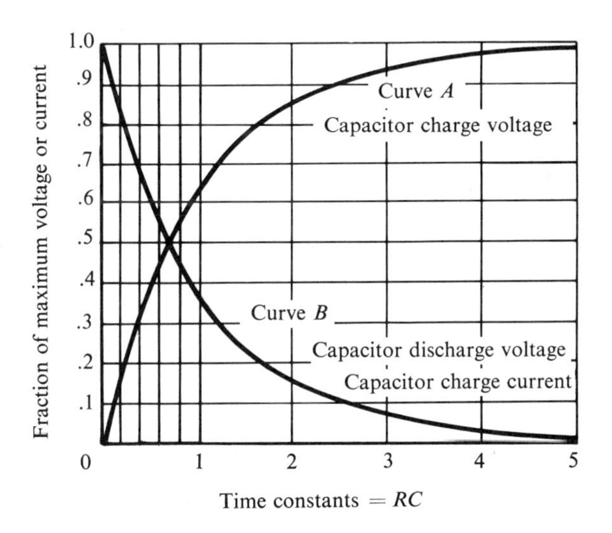

Figure 4-6. **Universal RC time-constant chart.**

in this example will be approximately 70 per cent of 6 v, or 4.2 v peak-to-peak.

4–2. Basic Ripple Waveforms

The ripple waveform from the circuit depicted in Fig. 4–5 is evidently an exponential waveform. In other words, when a pulse of charging current is applied to the filter capacitor, the terminal voltage of the capacitor rises rapidly to maximum amplitude. Immediately after application of the pulse current, the terminal voltage of the filter capacitor decays along curve *B* in Fig. 4–6. This is the simplest type of ripple waveform that we encounter in a power-supply output.

Let us consider next the ripple waveforms produced by resistive-input filters of various types. When a resistive-input filter is used, the pulses of charging current do not produce an immediate voltage rise across the terminals of the filter capacitor. Observe the basic configuration for an L-section resistive-input filter depicted in Fig. 4–7(a). A square-wave input is used to demonstrate the circuit action for ideal rectangular pulses of charging current. Evidently, capacitor *C* will charge with comparative slowness, due to the presence of input resistor *R*. Thus, the leading edge of the ripple waveform for the filter circuit depicted in Fig. 4–7(a) follows curve *A* in Fig. 4–7(d). Of course, this is the same curve as depicted at *A* in Fig. 4–6. Since the square-wave generator has negligible internal resistance, resistor R_1 in Fig. 4–7(a) operates as a load resistor across *C* after the charging current falls to zero. Therefore, the trailing edge of the ripple waveform is described by curve *A* "turned upside down" in Fig. 4–7(d). Of course, this is the same as curve *B* in Fig. 4–6.

Figure 4–7 is a universal time-constant chart for symmetrical L sections. In other words, ripple waveforms are shown for filter sections that have the same values of *R* and the same values of *C* throughout. With reference to Fig. 4–7(b), the leading edge of the output ripple waveform is depicted at *B* in Fig. 4–7(d). Note that the *RC* product is equal to the number of ohms for one resistor, multiplied by the number of farads for one capacitor. Curve *B* rises more slowly than curve *A*, because two capacitors must now be charged, instead of one. The waveshape of *B* is different from the waveshape of *A* because the second L section is not energized by an instantly-rising voltage, but by an exponentially rising voltage. Furthermore, the second *RC* section loads the first *RC* section, which also changes the output waveshape somewhat.

With reference to Fig. 4–7(c), the leading edge of the output ripple waveform is depicted at *C* in Fig. 4–7(d). Curve *C* rises more slowly than curve *B*, because three capacitors must now be charged, instead of two. The waveshape of *C* is different from the waveshape of *B* because

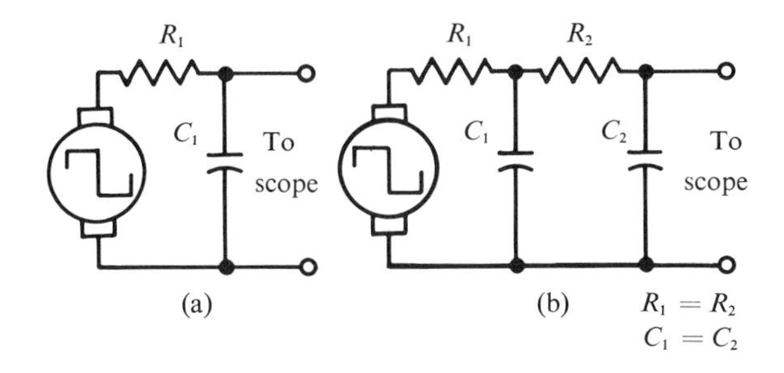

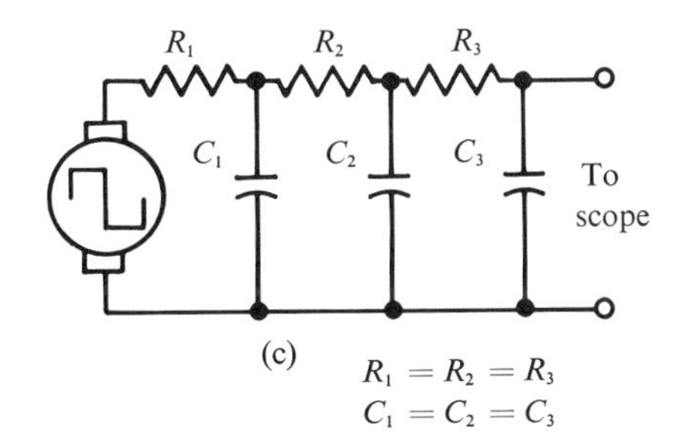

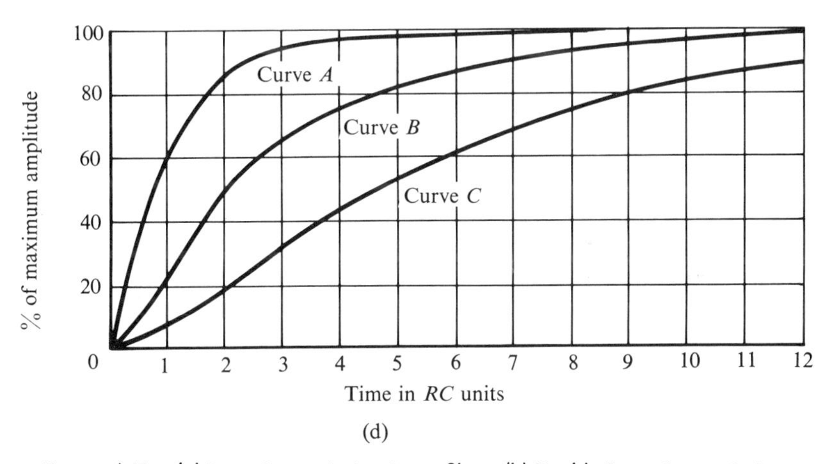

Figure 4-7. (a) L-section resistive-input filter; (b) Double L-section resistive-input filter; (c) Triple L-section resistive-input filter; (d) Leading edges of output ripple waveforms.

the first section of the filter is energized by an instantly-rising voltage, the second section is energized by a modified exponentially-rising voltage, and the third section is energized by a further modified exponentially-rising voltage. Moreover, the second *RC* section loads the first section, and the third section loads the second section. This loading action changes the output waveshape appreciably.

When the square-wave voltage falls to zero, the input resistor of each filter depicted in Fig. 4–7 then acts as a load resistor which discharges the filter capacitors. Thus, the trailing edges of the ripple waveforms in these demonstration circuits are represented by turning

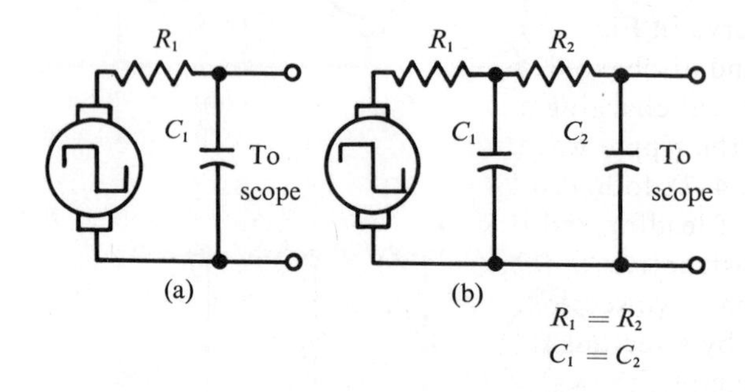

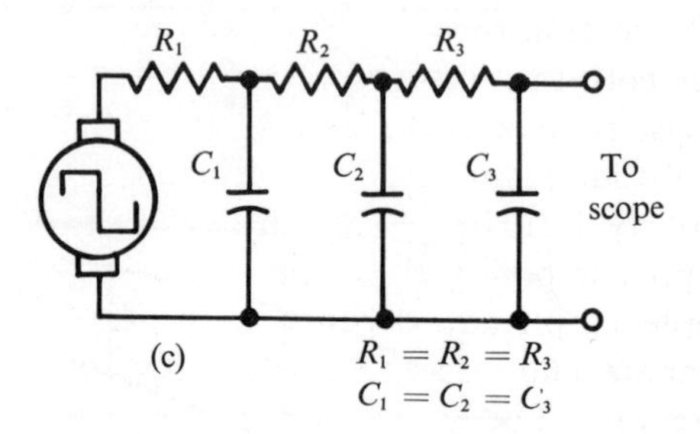

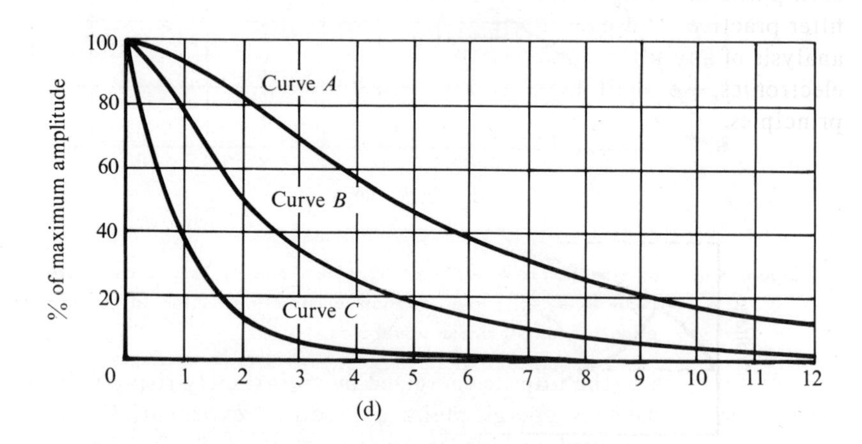

Figure 4–8. (a) L-section resistive-input filter; (b) Double L-section resistive-input filter; (c) Triple L-section resistive-input filter; (d) Trailing edges of output ripple waveforms.

the curves in Fig. 4-7(d) "upside down." This inversion is depicted in Fig. 4–8. To clearly recognize why the simple inversion principle applies, observe in Fig. 4–8(b) that an $RC\pi$ section is alternately charged and discharged through the input resistance R. Otherwise stated, this is a symmetrical charging and discharging situation. Similarly, observe in Fig.4–8(c) that cascaded $RC\pi$ sections are alternately charged and discharged through the input resistance R. Again, this is a symmetrical charging and discharging situation.

Thus, the ripple waveform is developed by a portion of a leading edge (Fig. 4–7), followed by a portion of a trailing edge (Fig. 4–8). The sequence of leading and trailing edges is depicted in Fig. 4–9(a).

It is instructive to observe the comparative ripple waveforms for one-section, two-section, and three-section symmetrical *RC* filters energized by a rectangular pulse source, as illustrated in Fig. 4–9(b). In this example, the values of R and C are identical throughout, and the effective load on the filter is also equal to R. Note that each additional *RC* section reduces the amplitude of the ripple waveform and increases both the rise time and the decay time of the waveform. Recall that rise time is measured between the 10 per cent and the 90 per cent amplitude points along the leading edge, as depicted in Fig. 4–9(c). Similarly, fall time is measured between the 10 per cent and 90 per cent amplitude points along the trailing edge. We will recognize that the ripple amplitude depends upon the *RC* product, upon the conduction angle, and also upon the elapsed time between successive charging intervals. The foregoing explanation of ripple waveforms has been purposely restricted to the most basic situations encountered in filter practice. However, these basic principles underlie the waveform analysis of any filter configuration. As we proceed in our study of electronics, we shall have occasion to build on these fundamental principles.

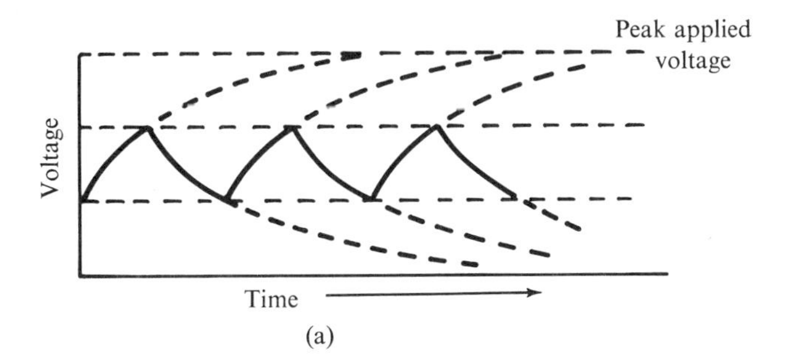

Figure 4–9. (a) Development of a ripple waveform.

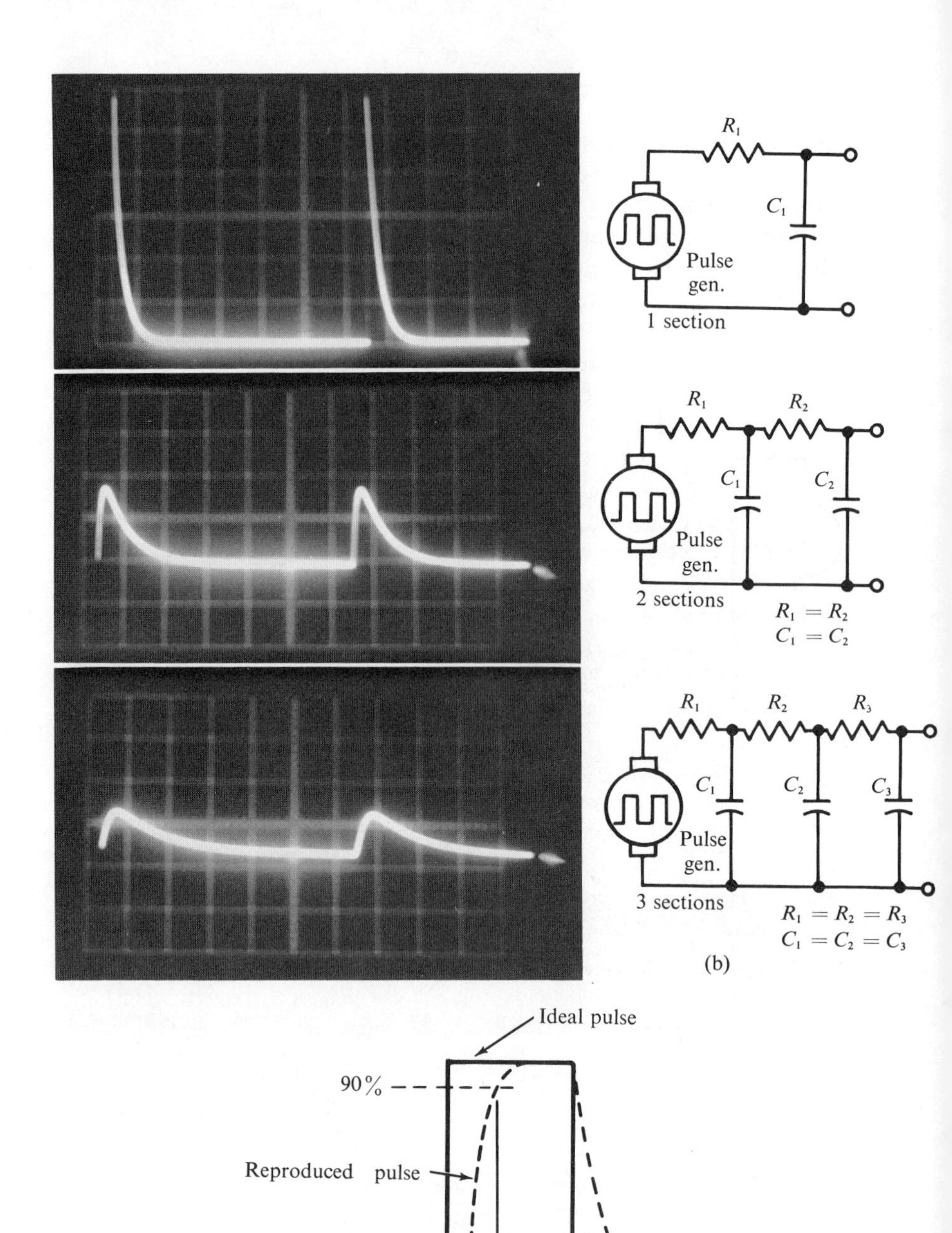

Figure 4-9. (b) Comparative ripple waveforms for one-, two-, and three-section RC filters; (c) Determination of pulse rise time.

4-3. Grid-Bias Voltage Supplies

In modern electronic equipment, grid-bias voltage is often derived from the plate power-supply voltage. Three common methods are seen in Fig. 4-10. The configurations shown in Figs. 4-10(a) and (b) are

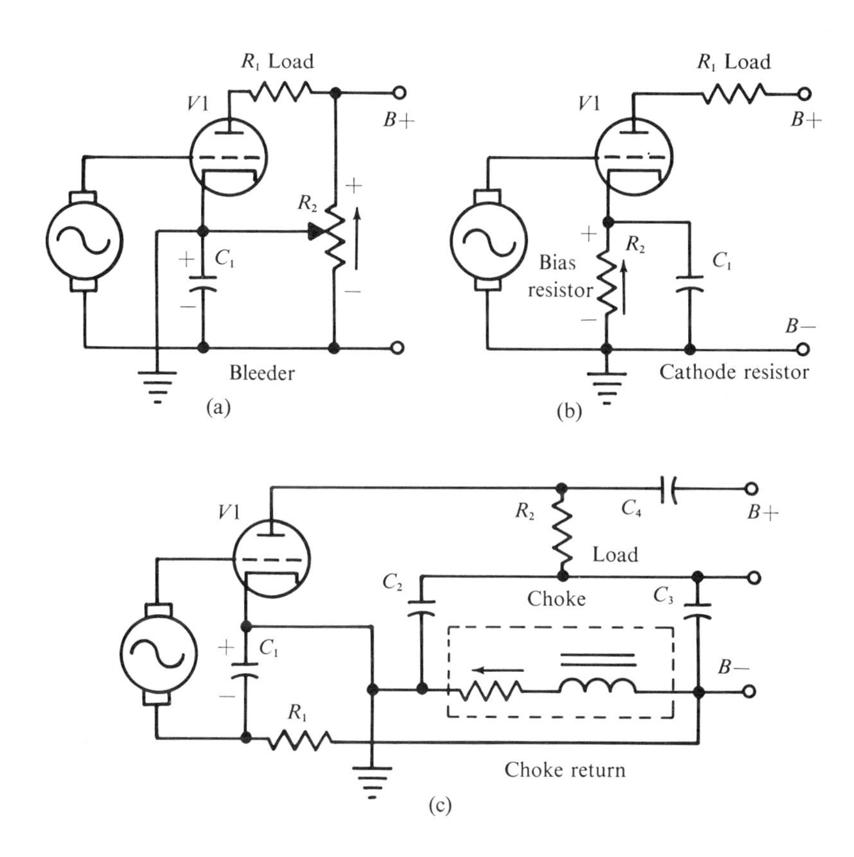

Figure 4-10. Methods of obtaining grid-bias voltage from the plate power-supply voltage.

commonly used in low-power amplifiers. Here, the prime consideration is that the bypass capacitor be chosen sufficiently large that a steady d-c voltage and practically zero a-c voltage is maintained across the resistor at the lowest operating frequency of the amplifier. That is, the reactance of the capacitor should be very small, less than one-tenth of the value of the resistance of the bias resistor at the lowest frequency of amplifier operation. Note that these bias arrangements reduce the effective plate-to-cathode voltage by the absolute value of the bias voltage.

The purpose of bypassing the a-c component around the bias resistor is to avoid *degeneration*, or *negative feedback*. As explained in a subsequent chapter, negative feedback reduces the gain of an amplifier. In this discussion, we assume that maximum gain is desired. A brief consideration of current flow through the bleeder and through the tube in Fig. 4–10(a) will show how the bias voltage is established. Bleeder current flows from $B-$ to $B+$, and the no-signal (bias) voltage between grid and cathode is determined by the position of the arm. We perceive that the grid is made negative with respect to the cathode. The d-c component of plate current also flows through the lower portion of the bleeder.

Capacitor C in Fig. 4–10(a) offers much less opposition to the signal component of the plate current than does the lower portion of the bleeder. Therefore, the signal component flows from $B-$ through C to the cathode, and back to $B+$. Because the signal component does not flow through the bleeder, the grid-bias voltage is d-c only. In Fig. 4–10(b), grid-bias voltage is developed by flow of the d-c component of plate current through the bias resistor. In turn, the a-c component is bypassed around the bias resistor by the shunt capacitor. Thus, the grid-bias voltage is d-c only.

The method shown in Fig. 4–10(c) is often used in power amplifiers. This arrangement makes use of the resistance in the power-supply filter choke to provide the voltage drop used as grid bias. In this circuit, the grid is connected to $B-$, and the signal component of the plate current flows from $B-$ through the low reactance of C to the cathode, and back to $B+$ via the tube and load resistor. Batteries are sometimes used for grid-bias supply when absolute stability of the bias

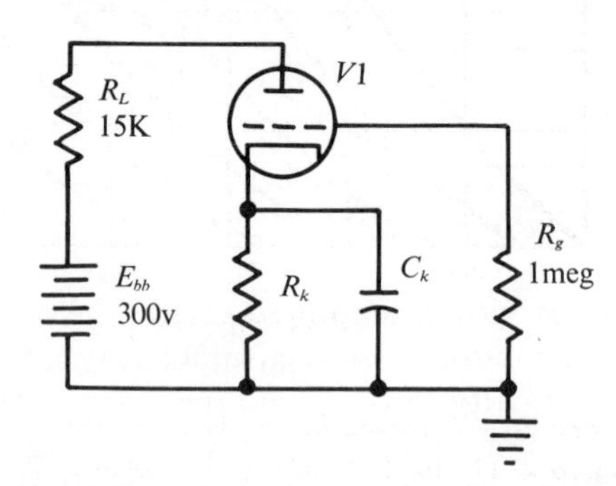

Figure 4–11. Cathode-bias circuit for analysis.

voltage is necessary. Under these conditions, the battery supplies no grid current and its effective life is the same as its "Shelf life." Similarly, Zener diodes may be used for absolute stability of the bias voltage.

Let us briefly review the method for calculation of a cathode-bias resistor value. With reference to Fig. 4–11, we will find the value of R_k that will provide 8 v bias. Using the family of $E_b - I_b$ curves depicted in Fig. 4–12(a), we locate the intersection of the $E_b - I_b$ curve

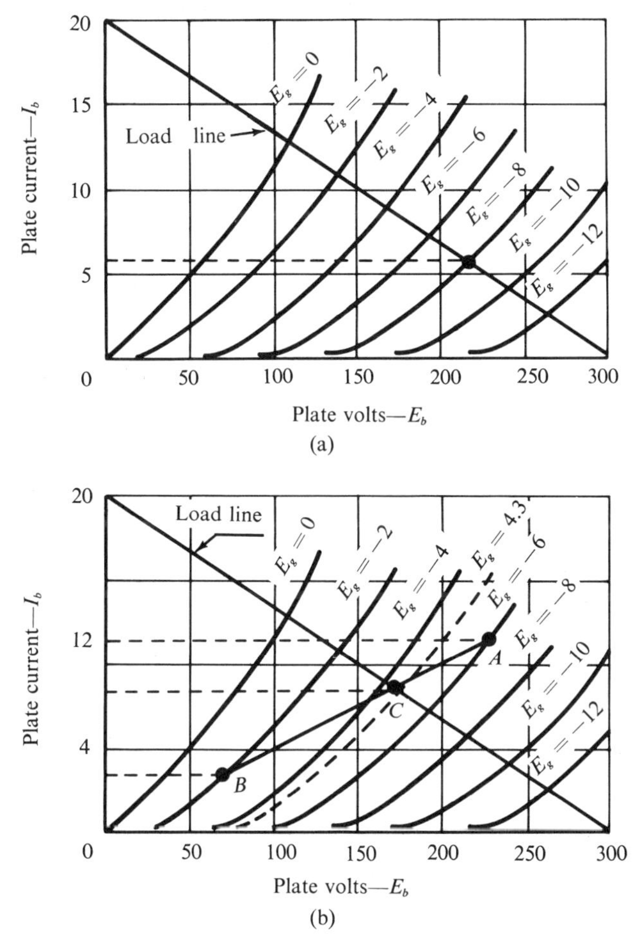

Figure 4–12. Triode $E_b - I_b$ curves for analysis of the circuit depicted in Figure 4–11. (a) Determining the value of R_k for the circuit in Figure 4–11 to establish 8 v bias; (b) Determining the cathode bias of Figure 4–11 when R_k is 500 Ω.

marked 8v and the load line.* This gives us point P, which is projected on the vertical axis at point S to determine I_b, which is 5.5 ma. Then, we determine the value of R_k from Ohm's law:

$$R_k = \frac{E_k}{I_b} = \frac{8}{5.5 \times 10^{-8}} = 1{,}450\ \Omega \qquad \textbf{(4–2)}$$

Next, with reference to Figs. 4–11 and 4–12(b), let us determine the cathode bias if $R_k = 500\ \Omega$. We may start by assuming that $I_b = 4$ ma. Then, the drop across R_k will be 2 v in accordance with Ohm's law. We locate the point on the $E_g = -2$ curve which is opposite 4 ma on the vertical axis. This gives us point B. Next, we may assume that $I_b = 12$ ma, and locate a corresponding point in a similar manner (point A). Then, we connect these two points by a straight line. This straight line intersects the load line at point C. The point C lies between the E_{gk} bias curves marked -4 and -6. The approximate value is determined by the proportional distance from the two curves. In this example, the value is about -4.3 v. Thus, the cathode is 4.3 v more positive than the grid. Note that the d-c load lines in Fig. 4–12(a) have not taken R_k into account; however, the error is small if R_k is small in relation to R_L.

High-powered transmitters require a fixed-bias source to protect the tubes and other circuit components as a fail-safe feature. Failure of a self-bias supply would cause a large increase of current flow through the associated tube and components, with resulting damage. Thus, we find self-bias supplies "backed up" by fixed bias supplies. A typical fixed-bias circuit combined with a conventional power supply is shown in Fig. 4–13. Full-wave rectification is followed by an $LC\pi$ filter to supply positive high d-c voltages for the plates and screens of the tubes. The fixed-bias circuit is enclosed in dotted lines, and comprises a half-wave rectifier followed by a resistive-input $RC\pi$ filter. Note that the same power transformer is used to energize both rectifier-filter sections.

When the lower end of the secondary winding (Fig. 4–13) is negative, electrons flow from the cathode to the anode of $CR3$, down through R_3 to ground, and thence back to the center tap on the secondary winding. Thus, the output voltage is negative with respect to ground, in contrast with the $B+$ supply circuit. An adjustable tap on R_3 permits the bias voltage to be set to the desired value. R_3 is called the bias resistor, and usually has a comparatively low resistance value to provide a "stiff" source. Thus, voltage variation is minimized due

---

*Vacuum tube voltages have been conventionally labeled (E). We will follow this convention even though it does not follow previous logic.

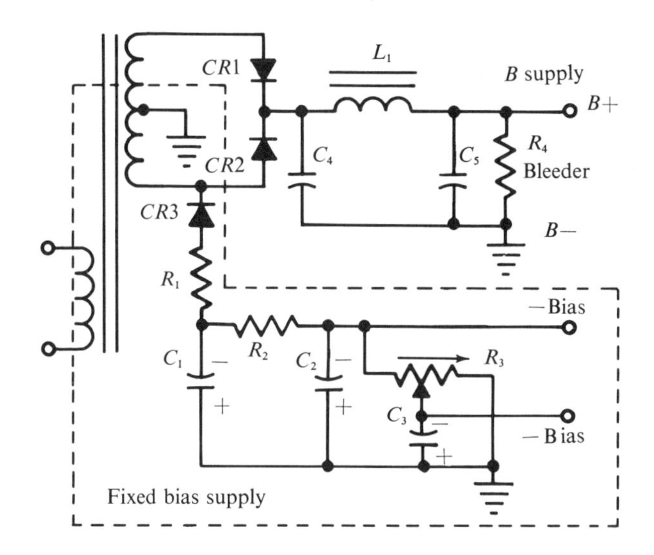

Figure 4-13. Hybrid $B+$ and fixed $C-$ supply.

to the conduction angle of grid-current flow in class C amplifiers, overdriven class A amplifiers, or class AB_2 amplifiers.

Capacitor C_3 is provided in the configuration of Fig. 4-13 to improve the regulation. If grid current flows in a tube connected to the fixed-bias supply, the current demand during the conduction angle partially discharges C_3. In turn, the "dip" in supply voltage is less than if C_3 were omitted. The foregoing configuration is an example of a hybrid $B+$ and $C-$ supply.

You will also find hybrid A and $C-$ supplies, as depicted in Fig. 4-14. This circuit operates in the following way. Assume that point 1 is instantaneously negative, and that point 2 is instantaneously positive. Electrons flow from point 1 through the diode, and charge the right-hand plate of C_1 negatively. During this conduction angle, resistor R_1 is short-circuited by the low plate resistance of the diode. On the following half cycle, when point 1 is positive and point 2 is negative, electrons cannot flow through the diode. Instead, C_1, which is charged to nearly the peak voltage of the applied a-c waveform, now partially discharges through R_1, thus developing the bias voltage. The time constant of $R_1 C_1$ is long, so that the voltage across C_1 remains reasonably constant during the nonconducting period. R_2 and C_2 provide filter action to further smooth the output voltage.

Another type of grid-bias supply is depicted in Fig. 4-15. It is a grid-leak resistance bias circuit which utilizes signal-developed d-c

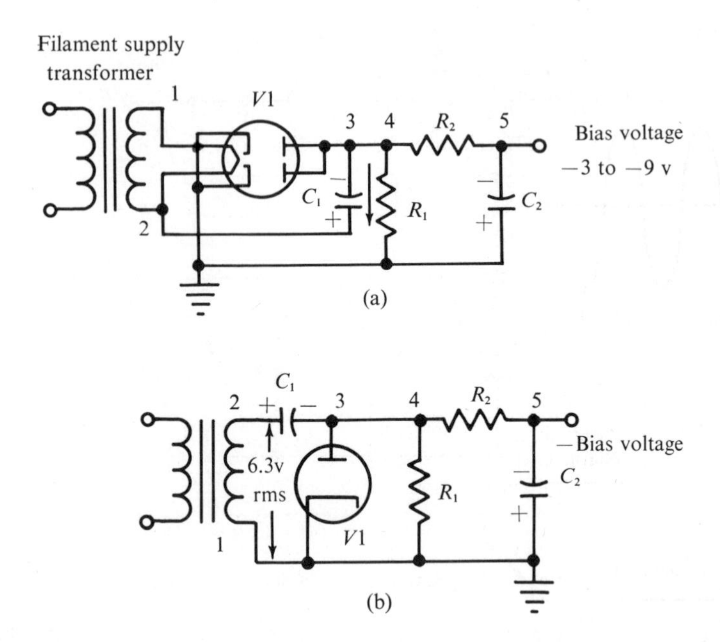

Figure 4-14. (a) Hybrid C— supply; (b) Simplified diagram of the shunt-fed rectifier bias supply.

voltage. Thus, it is a form of self-bias configuration. As shown in Fig. 4-15(b), the grid normally rests at cathode potential, via grid-leak resistance R_g. When an a-c signal voltage is applied, grid current flows on the positive peaks of the signal waveform. Otherwise stated, the grid-cathode circuit operates as a diode peak rectifier. In turn, the grid captures or collects electrons during the conduction angle, and these electrons "pile up" on the right-hand plate of the grid-coupling capacitor. Between consecutive conduction intervals, the capacitor charge slowly decays through R_g. Thus, the grid capacitor and R_g form a simple filter circuit. We perceive that the value of bias voltage depends both on the amplitude of the drive waveform and also on the time constant of the grid circuit.

The chief disadvantage of this grid-bias method is that the grid voltage falls to zero in the event that signal drive fails. If the tube is operated at comparatively high power level, failure of grid bias can result in tube damage. For example, the horizontal scanning tube in a television receiver may be damaged because of signal drive failure. Hence, some circuit designers "back up" grid-leak resistance bias with

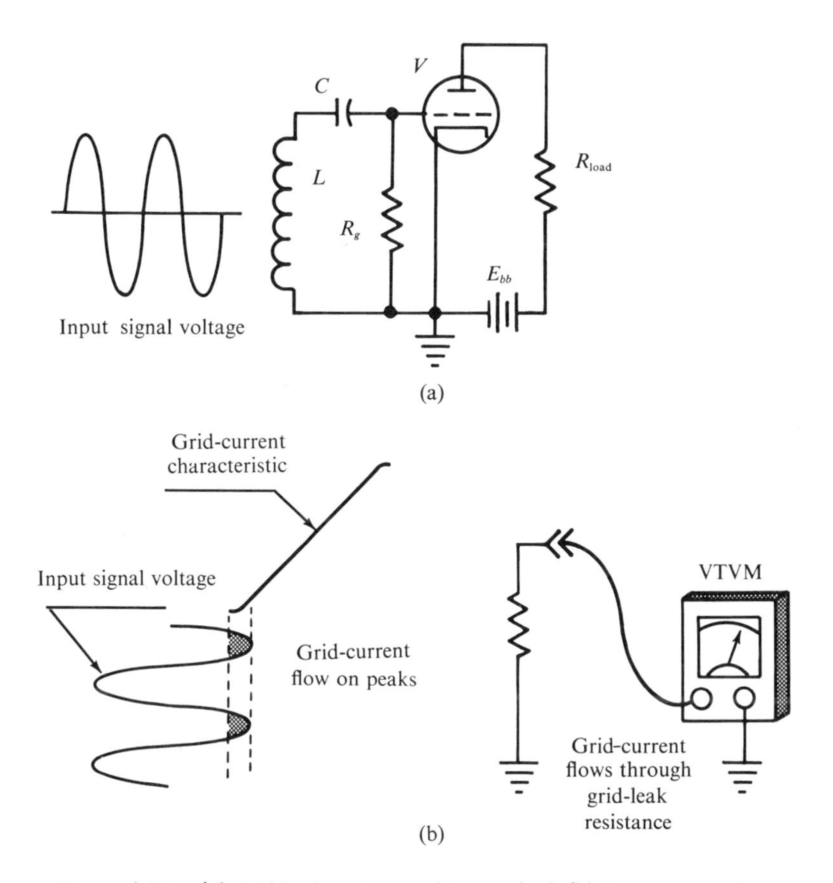

Figure 4-15. (a) Grid-leak resistance bias method; (b) Circuit operation.

cathode bias as depicted in Fig. 4-11. If the cathode resistor has a sufficiently large value, the bias system becomes virtually fail-safe. However, this fail-safe consideration is sometimes omitted in view of efficiency considerations, that is, there is an I^2R power loss in the cathode resistor, which must be supplied by the tube. In turn, a larger tube is required than if the cathode-bias resistor were omitted.

4-4. Electromechanical Power Supplies

Aircraft electronic equipment is commonly powered by a rotating machine called a *dynamotor*. The dynamotor performs the functions of a motor and a generator. It converts the 24-v d-c supply of the aircraft into a much higher voltage for utilization by the plates and screens of electron tubes. A dynamotor commonly employs two windings on a single armature. The windings occupy the same set of armature slots and terminate in two or more separate commutators. The armature (see Fig. 4–16) rotates in a single field frame that has a conventional field winding to provide excitation for both the motor and the generator.

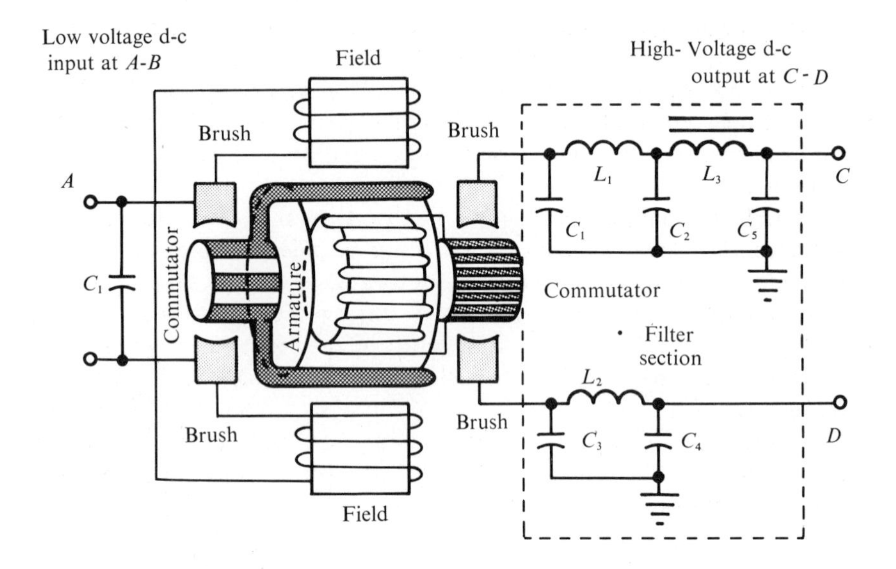

Figure 4–16. Functional diagram of a dynamotor power supply.

Figure 4–16 shows how the motor armature winding is connected to the 24-v power source and thereby develops driving torque. This portion of the system operates as a conventional motor. The generator winding has a comparatively large number of turns, and develops an adequately high value of d-c voltage for application to the plates and screens of the associated equipment. You will recall that any d-c generator has commutator ripple. Therefore, an *LC* filter section is provided for smoothing the output voltage.

Comparatively high current is drawn through the motor circuit (heavy lines) in Fig. 4–16. In other words, the power input must be at least as great as the power output, and the system is necessarily less than 100 per cent efficient. The field winding is also energized from the 24-v source. This type of generator is a shunt-wound configuration. The high-voltage winding (thin lines) delivers its induced voltage to the right-hand commutator. Since a shunt-wound motor has a fairly constant speed with changes in load, the output voltage from the generator section remains reasonably constant despite changes in current demand by the associated electronic equipment.

The comparatively high current value required by the motor in Fig. 4–16 necessitates a correspondingly large size of commutator, brushes, and armature wire, compared with the same components in the generator section. It is interesting to note that the motor commutator is larger in diameter but has fewer segments than the generator commutator. A large number of segments increases the ripple frequency and reduces the commutator ripple across the filter. Since the current demand from the generator is comparatively small, the generator armature wire has a proportionally smaller size than the motor armature wire.

The filter configuration shown in Fig. 4–16 suppresses sharp pulse voltages that are produced by sparking between the brushes and the rotating commutator segments. Inductors *L*1 and *L*2 (chokes) present a high impedance to high-frequency current flow or to sharp pulses. Capacitors *C*1, *C*2, *C*3, and *C*4 present a low impedance to ground for sharp pulses or high-frequency currents. The comparatively low ripple frequency is filtered by the iron-core inductor (choke) *L*3, and capacitor *C*5. Note that a capacitor is also connected across the brushes of the motor commutator; this capacitor reduces sparking between the brushes and commutator.

4–5. Vibrator Power Supplies

Another type of power supply, the vibrator converter, is also used to obtain a high a-c or d-c voltage from a comparatively low d-c source voltage. The vibrator converter has certain advantages over the dynamotor arrangement: it is lighter in weight, is less expensive, and has higher efficiency. However, a vibrator can be used only when a limited current demand is imposed at high voltage. Vibrator contacts have a comparatively short life, and sparking contacts generate *r-f* interference (hash) which is sometimes difficult to suppress. We find vibrator power supplies used extensively in "power packs" for lightweight mobile equipment.

Note that a vibrator alone is not a power supply, just as a dynamotor alone is not a power supply. They are merely essential components of complete power supplies. Both are the means whereby low-voltage direct current is converted to high-voltage pulsating direct current. Let us briefly review the meaning of pulsating d-c voltage. Three key waveforms are illustrated in Fig. 4–17; note that a pulsating d-c waveform does not cross the zero axis.

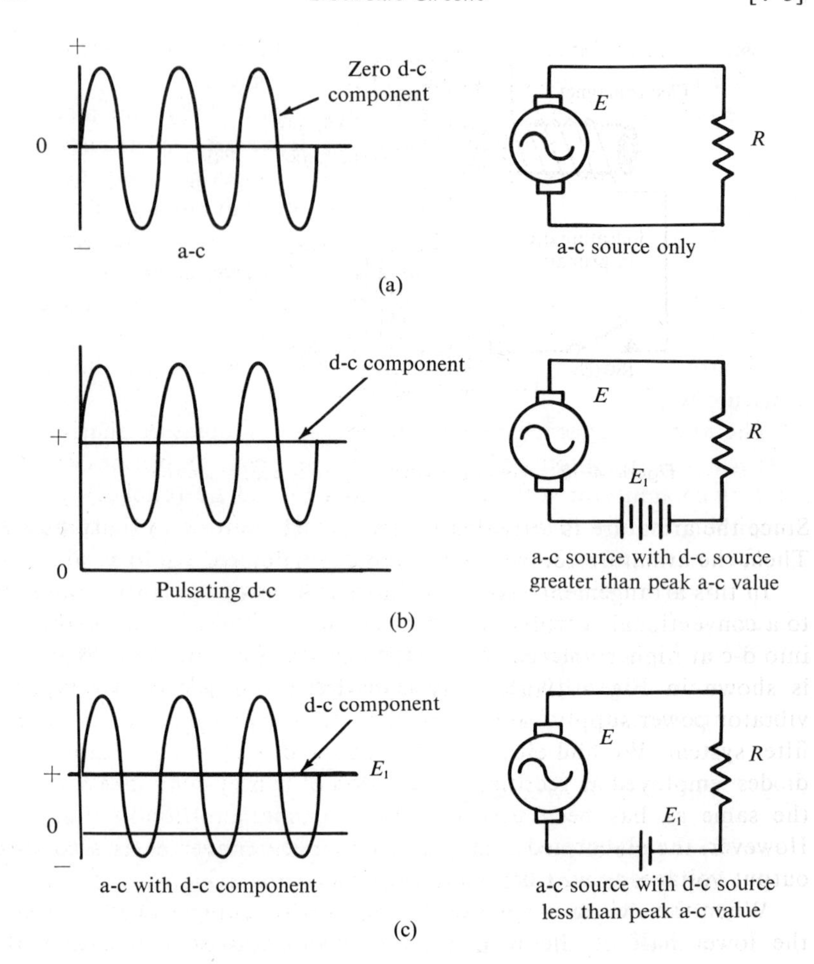

Figure 4-17. **Distinctions among (a) a-c; (b) pulsating d-c; and (c) a-c with d-c component.**

The configuration for a simple vibrator power supply is depicted in Fig. 4–18. Note that the vibrator operates as an interrupter, the action of which can be compared with that of an ordinary buzzer or doorbell. Pulsating d-c thus energizes the primary winding of the transformer, which in turn induces an a-c voltage in the secondary winding. A step-up transformer is utilized to obtain the desired value of secondary output voltage. When the switch is closed in Fig. 4–18, current flows from the battery through the electromagnet, then through contact *B*, armature *A*, primary winding *P*, and back to the battery.

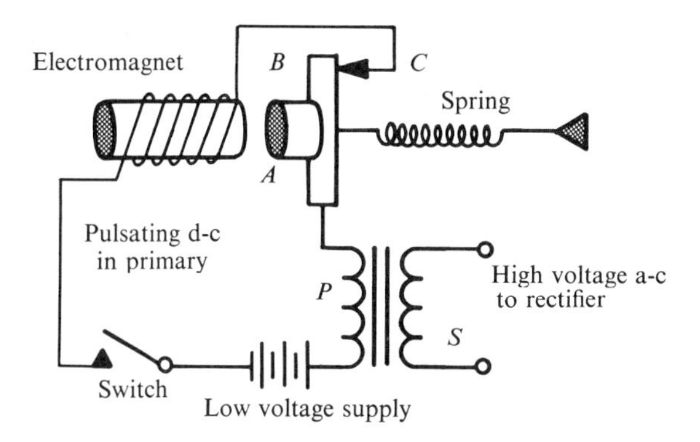

Figure 4-18. Plan of a vibrator power-supply system.

Since the armature is attracted to the core, the circuit is broken at *B*. Then, the armature returns and the circuit is closed again at *B*.

In this arrangement, output voltage from the secondary is applied to a conventional rectifier and filter system which changes the a-c into d-c at higher voltage. A variation of the foregoing configuration is shown in Fig. 4-19(a). This is called a *nonsynchronous* type of vibrator power supply, and it requires a conventional rectifier and filter system. We find either high-vacuum diodes or cold-cathode gas diodes employed as rectifiers. Operation of this system is essentially the same as has been described for the configuration in Fig. 4-18. However, the elaborated primary circuit provides a greater secondary output voltage, as will be explained.

When the switch is closed in Fig. 4-19(a), current flows through the lower half of the primary, the electromagnet, and back to the battery, which establishes a magnetic field in the transformer core. The armature is attracted down, contacts terminal 1, and temporarily short-circuits the electromagnet. Thereupon, the armature returns and contacts terminal 2. Current now flows through the upper half of the primary and back to the battery. At this time, the magnetic field previously established is collapsing, while the new field adds to the collapsing field. In turn, the induced voltage in the secondary winding is comparatively great.

Next, let us observe the *synchronous* vibrator power supply depicted in Fig. 4-19(b). This arrangement does not require a rectifier device. It is called a synchronous vibrator because two additional contacts are provided, which are connected to the ends of the secondary winding. Opening and closing of the pairs of contacts take place simulta-

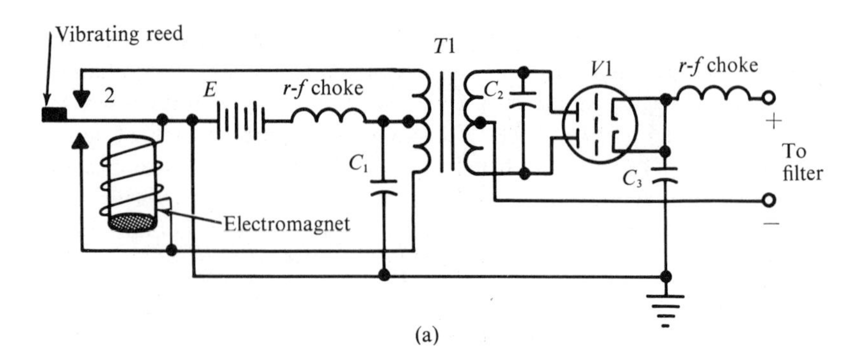

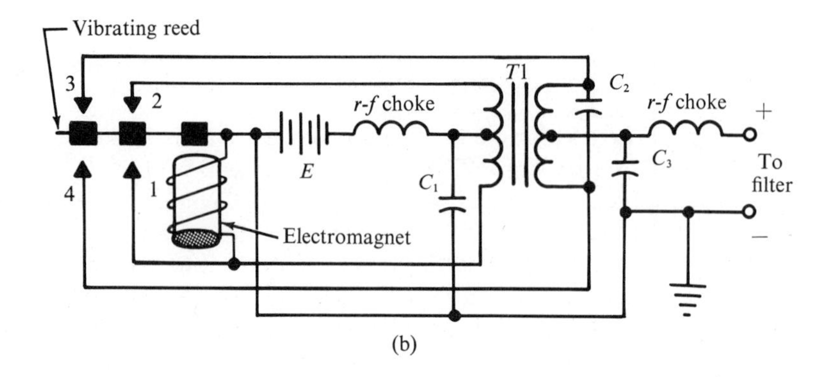

Figure 4–19. (a) Nonsynchronous vibrator power supply; (b) Synchronous vibrator power supply.

neously, and the secondary output voltage is rectified. In other words, contacts 3 and 4 perform the function of rectification. Thus, pulsating d-c is applied to the filter.

4–6. Inverters and Converters

In various aircraft installations, 115 v a-c is often derived from the 24-v d-c source by use of a device called an *inverter*. An inverter is a rotating machine consisting of a motor that operates from the 24-v d-c source, which drives an alternator that generates 115 rms at a frequency of about 800 Hz. This 115-v power source is often utilized by rectifier-type power supplies to energize high-power radar radio transmitters, units, and equipment.

An *electromechanical* system that changes a-c into d-c is called a *converter;* on the other hand, an electromechanical system that changes d-c into a-c is called an *inverter*. Inverters commonly operate at 400 Hz or 800 Hz. This comparatively high frequency permits use of small inductive components, which minimizes the weight of the system. *Semiconductor* systems are also used in inverter and converter configurations. A simple semiconductor d-c to d-c *converter* is depicted in

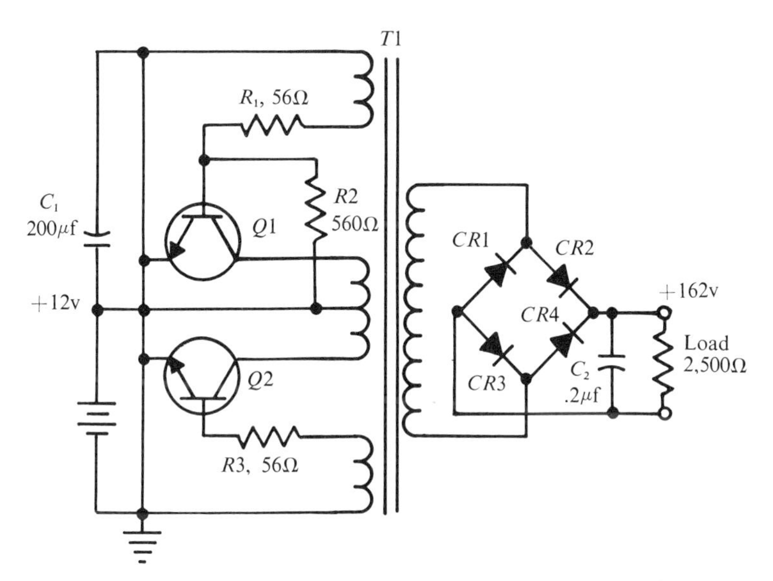

Courtesy of General Electric Co.

Figure 4-20. A d-c to d-c semiconductor converter.

Fig. 4-20. The transistors oscillate at about 8.6 kHz; oscillators are explained in detail subsequently. The efficiency of this system is about 80 per cent, and it can supply 10 w at 162 v from a 12-v source. Semiconductor *inverters* change d-c into square-wave a-c. Note that if we omit the bridge rectifier in Fig. 4-20, the system operates as a semiconductor inverter.

4-7. A-C/D-C Power Supplies

Radio receivers are often energized from a-c/d-c power supplies, as seen in Fig. 4-21. As indicated by its name, this type of power supply operates from either a d-c source or an a-c source (usually 117

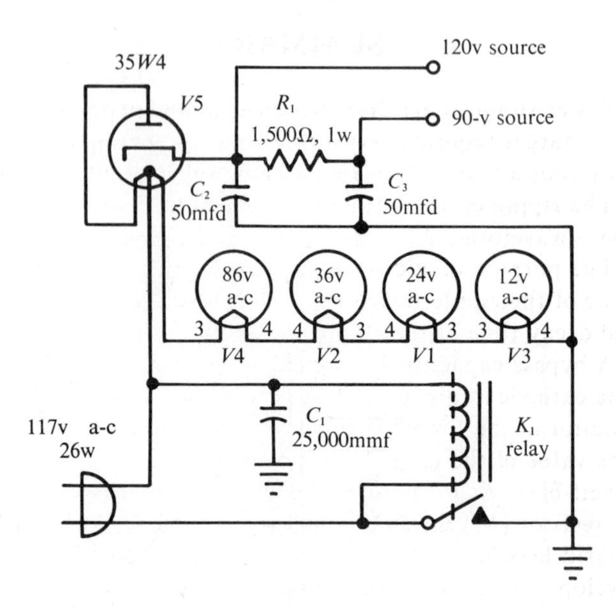

Figure 4–21. An a-c/d-c power supply.

rms v a-c, or 117 v d-c). It is a hybrid type of power supply which combines the functions of both *A* and *B*+ supplies. When energized from an a-c line, the half-wave rectifier *V*5 functions as in conventional power supplies discussed previously. However, when energized from a d-c line, the rectifier tube merely serves as a continuously conducting component. Observe that the power plug must then be inserted so that the upper prong is connected to the positive terminal of the d-c outlet.

Recall that the heaters of *V*1 through *V*5 are connected as a *series string*. Thus, voltage division occurs along the string, and the individual heater characteristics are chosen so that each obtains correct terminal voltage. Note that the left-hand section of the heater in *V*5 conducts 50 ma more current than the right-hand section. Accordingly, we find by reference to a tube manual that the heater sections of a 35W4 tube are designed to operate with corresponding unequal current values in the individual sections. It is evident that an a-c/d-c power supply cannot be operated from a transformer. The power supply must be energized directly from a line, because a transformer will block the flow of direct current. One of the disadvantages of this type of power supply is that there is a possibility for live plugs to be connected, and the chassis to be very negative in respect to ground (hot). Therefore, care must be taken to prevent a shock to oneself or damage to the test equipment by employing an isolation transformer which isolates the live "ground" from the equipment.

SUMMARY

1. Power supplies are designed for specific purposes. For example, a battery charger requires very little filtering; whereas, a supply for a frequency counter requires a very stable, well-regulated power supply.
2. The ripple on the output of an *RC* filter is best represented by a sawtooth waveform.
3. The purpose of adding L-section *RC* filters is to reduce the amplitude of the ripple waveform. This also increases both the rise time and decay time of the ripple waveform.
4. A bypass capacitor is shunted across a cathode-bias resistor to place the cathode at a-c ground. A general rule is that the reactance of the capacitor at the lowest frequency to be amplified should be one-tenth the value of the cathode-bias resistor.
5. Self-bias of a vacuum-tube circuit can be established by a cathode resistor (R_k) through which plate current (I_b) flows.
6. Grid-leak bias results from the charge on the coupling capacitor that develops when the grid is driven positive.
7. The chief disadvantage of the grid-leak bias method is that the tube may be destroyed in the event that signal drive fails.
8. A dynamotor is a motor-generator combination. The motor operates from a low d-c voltage and drives a generator which produces a higher d-c voltage.
9. The efficiency of a dynamotor is determined by the energy converted to heat in the system.
10. All generators produce alternating current and voltage; however, a commutator may be used to convert a generator's output to direct current and voltage.
11. A vibrator may be used to convert d-c voltage to a-c voltage by making and breaking d-c current in the primary of a transformer.
12. We may say that a vibrator operates as an interrupter to produce d-c pulses.
13. A nonsynchronous type of vibrator power supply requires a rectifier and filter to produce d-c voltage and current.
14. A synchronous type of vibrator power supply does not require a rectifier, because two additional contacts on the vibrator are provided to rectify the voltage from the transformer's secondary.
15. An inverter is a device used to convert d-c voltage and current to a-c voltage and current.
16. Portable equipments, which are designed to operate on either a-c or d-c voltages, utilize a-c/d-c power supplies.
17. An isolation transformer should be used when testing equipment employing an a-c/d-c power supply.

Questions

1. List five factors that determine the type of power supply required for a specific application.
2. Would the power supply illustrated in Fig. 4-2 be used as a laboratory supply voltage for testing oscillator circuits?
3. Identify the type of rectifier circuit and the type of filter circuit shown in Fig. 4-2.
4. Identify the type of rectifier and the type of filter circuit illustrated in Fig. 4-5.
5. Why isn't the waveform in Fig. 4-3(a) a sine wave?
6. What is the effect on the output ripple waveform with more L sections added in the *RC* filter?
7. What is the function of the cathode-bypass capacitor, and what determines its capacitance value?
8. Explain the terminologies $B+$ and $C-$ as applied to power supplies.
9. Explain "grid-leak" bias.
10. What is the disadvantage of grid-leak bias, and how may this disadvantage be overcome?
11. What is the advantage of grid-leak bias?
12. Explain the terminology: dynamotor, converter, and inverter.
13. What is the function of a commutator in a generator?
14. How does a vibrator operate to produce a-c voltage from a transformer?
15. Explain what is meant by the statement that a vibrator operates as an interrupter.
16. Compare the synchronous and nonsynchronous types of vibrators.
17. Explain the operation of the a-c/d-c power supply shown in Fig. 4-21.
18. What is the function of an isolation transformer?

PROBLEMS

1. The circuit in Fig. 4-11 has a load resistance of 20 kΩ; determine the value of R_K to establish a bias of -5 v, and the values of I_b and E_b under these conditions when a 6SN7 (Fig. A-1 in Appendix) is employed.
2. The circuit in Fig. 4-11 has a load resistance of 2 kΩ. Determine the values of R_k to establish -10 volts bias, and the values of I_b and E_b under these conditions when using the 6V6 tube represented by the characteristics in Fig. A-2.

3. Determine the value of bias for the circuit represented in Problem 1 when R_k is 800 Ω.

4. Determine the value of bias for the circuit represented in Problem 2 when R_k is 200 Ω.

5. The load (R_L) of the circuit shown in Fig. 4-11 is 4 kΩ, the supply voltage is 450 v, and the tube used is a 6V6 (see Fig. A-2 in Appendix). Determine the value of R_k to establish a static plate voltage of 250 v.

6. The values of the components for the circuit in Fig. 4-11 are $E_{bb} = 280$ v, $R_L = 10$ kΩ, and $R_k = 400$ Ω; the tube is the 6SN7 represented in Fig. A-3. What are the values of E_C, I_b, and V_b?

CHAPTER 5

Basic Measurements

5–1. Introduction

We are familiar with the measurement of d-c and a-c voltage, current, and power values. At this time, it is advisable to review briefly certain types of measurements, such as decibel, power-factor, phase-shift, and percentage distortion values. A decibel meter is illustrated in Fig. 5–1(a). We recall that a decibel value is equal to ten times the common logarithm of a power ratio. Thus, a bel value is the logarithm of a power ratio, and a decibel (db) is equal to 0.1 bel. And so we write

$$\text{bels} = \log \frac{P_2}{P_1} \tag{5–1}$$

$$\text{decibels} = 10 \log \frac{P_2}{P_1} \tag{5–2}$$

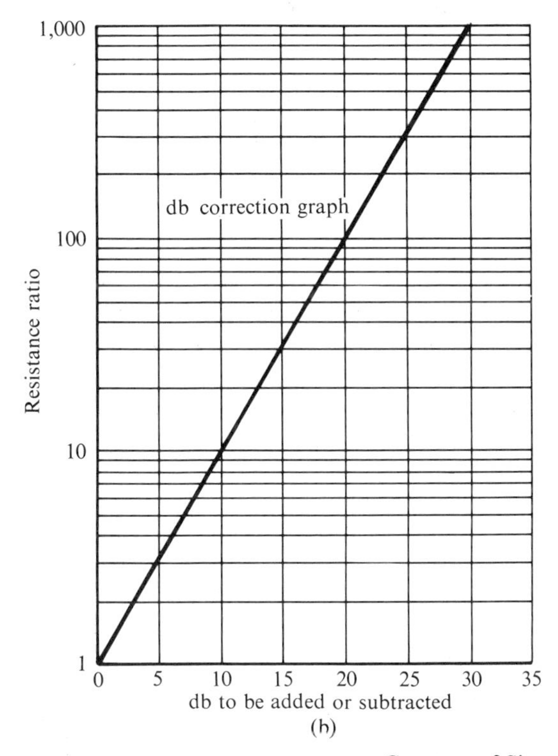

Courtesy of Simpson Electric Co.

Figure 5-1. (a) A typical decibel meter; (b) Load-correction chart.

where P_1 and P_2 denote power values with P_2 conveniently being the *largest power*, and a power loss being indicated by a minus sign. This may be demonstrated mathematically.

Example 1. Measurements on an audio amplifier with equal input and output resistances indicate an input of 10 μw and an output of 6 w. What is the power gain in db?

$$\text{db} = 10 \log \frac{P_{\text{out}}}{P_{\text{in}}}$$

$$= 10 \log \frac{6}{10 \times 10^{-6}}$$

$$= 10 \log 6 \times 10^{+5}$$

$$\text{Power gain} = 57.8 \text{ db} \qquad \text{(answer)}$$

Example 2. The power into a 300-Ω transmission line is 300 μw, and the power at the other end across the 300-Ω load is 1.5 nw. What is the power loss in db?

$$\text{db} = 10 \log \frac{P_2}{P_1}$$

$$= 10 \log \frac{3 \times 10^{-6}}{1.5 \times 10^{-9}}$$

$$\text{Power loss} = -23 \text{ db} \qquad \text{(answer)}$$

It is informative to measure audio power levels in db, because the db unit is proportional to ear response. Voltage, current, or power levels are not proportional to ear response. Note that a gain of 1 db is just perceptible to the ear; similarly, a loss of 1 db is just preceptible to the ear. In turn, 1 db is called a just-noticeable-difference, or a JND unit. A change of 2 db in an audio signal level is thus a slight audible increment. A change of 3 db is a half-power change. Again, an increase of 6 db is commonly judged to be a sound level that is twice as loud. A reduction of 6 db is often judged as a sound level that is half as loud.

Decibel values are additive and subtractive. For example, if a loss of 20 db is followed by a gain of 30 db in a system, the *system gain* has a value of 10 db. Again, if a loss of 30 db is followed by a gain of 20 db in a system, the *system loss* has a value of 10 db. We say that the system output is 10 db down with respect to the system input. If we have a number of stages in an amplifier, for example, and we know the db gain of each stage, the total gain is equal to the sum of the individual stage gains, expressed in db.

Example 3. Calculations of db gain in an *i-f* amplifier indicate a 2 db loss, a 31 db gain, a 3.5 db loss, a 25 db gain, and a 2.8 db loss. What is the overall power gain in db?

$$\text{Total db gain} = \text{db1} + \text{db2} + \text{db3} \cdots$$
$$= -2 + 31 - 3.5 + 25 - 2.8$$
$$\text{Power gain} = 47.7 \text{ db} \quad \text{(answer)}$$

Conventional db meters respond to a-c voltage values. Accordingly, the meter is an a-c voltmeter with a scale calibrated in db values. Let us see how this fact affects applications of the instrument. Since db values are being determined on the basis of voltage values, we make use of the power law and write

$$\text{db } 10 \log \frac{E_2^2}{E_1^2} \tag{5-3}$$

Formula 5–3 is commonly written in the form

$$\text{db } 20 \log \frac{E_2}{E_1} \tag{5-4}$$

Example 4. A signal voltage of 2 mv is measured across the 1-kΩ input of an amplifier, and a signal voltage of 3 v is measured across the 1-kΩ output of the amplifier. What is the power gain in db?

$$\text{db} = 20 \log \frac{E_2}{E_1}$$
$$= 20 \log 1.5 \times 10^3$$
$$= 20 \times 3.176$$
$$\text{Power gain} = 63.52 \text{ db} \quad \text{(answer)}$$

Note carefully that since $P_2 = E_2^2/R$, and $P_1 = E_1^2/R$, Formula 5–4 contains an implied assumption. It is implied that both E_2 and E_1 are measured across the *same* value of resistance. If we measure E_2 and E_1 across different values of resistance, the db values that we read on the meter scale will be incorrect as they stand. Unless such values are corrected with respect to the resistance values across which the readings were taken, we will arrive at incorrect conclusions. This is perhaps the most common mistake made by beginners in the measurement of db values.

It follows from the power law that db values can also be measured in terms of current values.

$$\text{db} = 10 \log \frac{I_2^2}{I_1^2} \tag{5-5}$$

or,

$$\text{db} = 20 \log \frac{I_2}{I_1} \tag{5-6}$$

Example 5. An audio preamplifier with a 2-kΩ input and output resistance has an input signal current of 10 μa and an output signal current of 0.5 ma. What is the power gain in db?

$$\text{db} = 20 \log \frac{I_2}{I_1}$$
$$= 20 \log 5 \times 10^1$$
$$= 20 \times 1.699$$
$$\text{Power gain} = 33.98 \text{ db} \qquad \text{(answer)}$$

Again, we must note that Formula 5–6 contains an implied assumption. It is implied that both I_2 and I_1 denote values of current flow through the *same* value of resistance. This implication follows from the power law: $P_2 = I_2^2R$, and $P_1 = I_1^2R$. A short table of decibel values and corresponding power, voltage, and current ratios is shown in Table 5–1. Observe that a power ratio of 10/1, for example, corresponds to 10 db, and that a power ratio of 1/10 will also correspond to 10 db. In the former case, we state that there is a 10-db gain, and in the latter case, we state that there is a 10-db loss.

With reference to Fig. 5–1, we observe that the number of db indicated by a meter might be either positive or negative. The algebraic sign of the db value depends solely upon the reference power level which might have been chosen by the instrument designer.

Example 6. In Fig. 5–1, zero db has been assigned a power value of 6 mw in 500 Ω. Therefore, in this case, zero db is assigned a voltage value of 1.73 v rms across 500 Ω.

$$P = \frac{E^2}{R} = \frac{(1.73)^2}{500} = 6 \text{ mw} \qquad \text{(answer)}$$

Therefore, if there is less than 1.73 rms v across 500 Ω, the meter will read a negative value of db. Suppose that we have an amplifier which has an input resistance of 500 Ω and an output resistance of 500 Ω. If we measure −6 db at the input and 25 db at the output of the amplifier, the gain is equal to 31 db. Many volt-ohm-milliammeters have a single db scale which is calibrated for operation on the lowest a-c voltage range (see Fig. 5–2). Accordingly, when the meter is operated on a higher a-c voltage range, we must add a suitable number of db to the scale reading. The number of db to be added will be noted on the meter dial, or in the instrument instruction manual.

When you measure db values across resistances other than that for which the db scale of the instrument has been calibrated, there are two essential points to keep in mind. If the resistances are different from the reference value specified for the meter, but are equal in value,

TABLE 5-1
Decibel Table

Power ratio	*Voltage and current ratio*	*Decibels*	*Power ratio*	*Voltage and current ratio*	*Decibels*
1.0000	1.0000	0			
1.0233	1.0116	0.1	19.953	4.4668	13.0
1.0471	1.0233	0.2	25.119	5.0119	14.0
1.0715	1.0315	0.3	31.623	5.6234	15.0
1.0965	1.0471	0.4	39.811	6.3096	16.0
1.1220	1.0593	0.5	50.119	7.0795	17.0
1.1482	1.0715	0.6	63.096	7.9433	18.0
1.1749	1.0839	0.7	70.433	8.9125	19.0
1.2023	1.0965	0.8	100.00	10.0000	20.0
1.2303	1.1092	0.9	158.49	12.589	22.0
1.2589	1.1220	1.0	251.19	15.849	24.0
1.3183	1.1482	1.2	398.11	19.953	26.0
1.3804	1.1749	1.4	630.96	25.119	28.0
1.4454	1.2023	1.6	1000.0	31.623	30.0
1.5136	1.2303	1.8	1584.9	39.811	32.0
1.5849	1.2589	2.0	2511.9	50.119	34.0
1.6595	1.2882	2.2	3981.1	63.096	36.0
1.7328	1.3183	2.4	6309.6	79.433	38.0
1.8198	1.3490	2.6	10^4	100.000	40.0
1.9055	1.3804	2.8	$10^4 \times 1.585$	125.89	42.0
1.9953	1.4125	3.0	$10^4 \times 2.512$	158.49	44.0
2.2387	1.4962	3.5	$10^4 \times 3.981$	199.53	46.0
2.5119	1.5849	4.0	$10^4 \times 6.31$	251.19	48.0
2.8184	1.6788	4.5	10^5	316.23	50.0
3.1623	1.7783	5.0	$10^5 \times 1.585$	398.11	52.0
3.5480	1.8836	5.5	$10^5 \times 2.512$	501.19	54.0
3.9811	1.9953	6.0	$10^5 \times 3.981$	630.96	56.0
5.0119	2.2387	7.0	$10^5 \times 6.31$	794.33	58.0
6.3096	2.5119	8.0	10^6	1,000.00	60.0
7.9433	2.8184	9.0	10^7	3,162.3	70.0
10.0000	3.1623	10.0	10^8	10,000.0	80.0
12.589	3.5480	11.0	10^9	31,623.0	90.0
15.849	3.9811	12.0	10^{10}	100,000.0	100.0

you can observe the scale readings and find the db gain or loss by taking the difference between the two readings. Thus, the reference value of the scale for taking db readings might be 600 Ω. If the input and output resistances of an amplifier are both 75 Ω, you can calculate

(a)

Courtesy of
Triplett Electrical Instrument Co.

RESISTIVE LOAD AT 1000 CPS	DBM*
600	0
500	+0.8
300	+3.0
250	+3.8
150	+6.0
50	+10.8
15	+16.0
8	+18.8
3.2	+22.7

*DBM IS THE INCREMENT TO BE ADDED ALGEBRAICALLY TO THE DBM VALUE READ FROM THE GRAPH.

(c)

By Permission
Radio Corporation of America

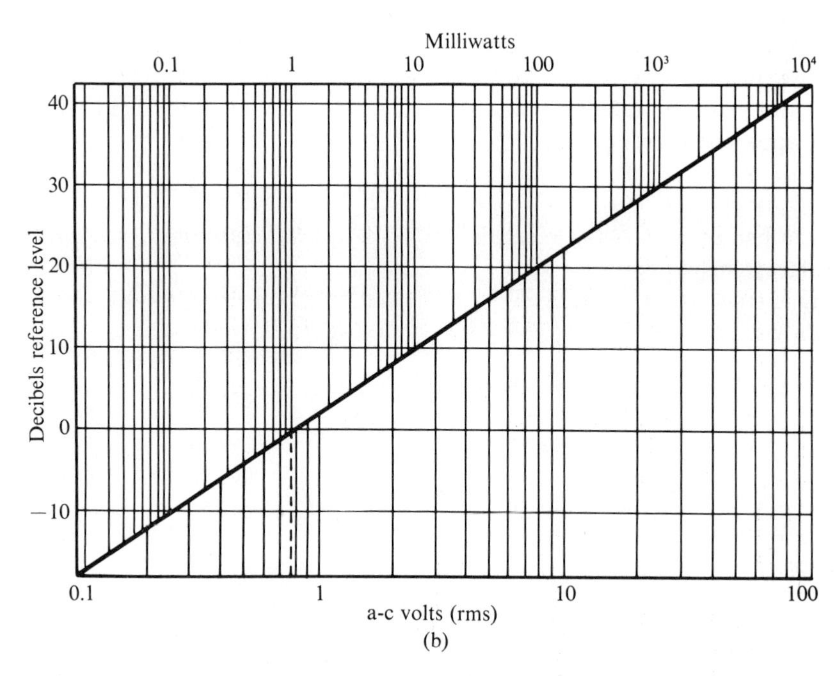

(b)

Figure 5-2. A VOM that has a db scale calibrated for 0 db = 1 mw in 600 Ω; (b) Graph for conversion of rms voltages to dbm values (1 mw in 600 Ω resistive load); (c) DB correction values for dbm measurements made across loads other than 600 Ω.

the db gain of the amplifier by taking the difference of the db readings at input and output terminals. Note that neither of these readings is correct of itself; nevertheless, the difference between the two readings is equal to the db gain of the amplifier. This fact follows from the previous formulations of db values.

Next, if the input and output resistances of the amplifier differ from the reference value of the meter, and these two resistance values are unequal, then the difference between db readings at input and output will not be the correct db gain figure for the amplifier. However, the correct db gain figure can be found from the chart in Fig. 5–1(b). When the output resistance is smaller than the input resistance, the corrected db value is added, and when the output resistance is larger than the input resistance, the corrected db value is subtracted.

Example 7. Suppose that a meter is calibrated for a reference level of 1 mw in 600 Ω. We will also suppose that the input resistance of the amplifier is 100,000 Ω, and that the output resistance is 500 Ω. If we measure 1 db at the amplifier-input terminals and 1 db at the amplifier-output terminals, the difference is 0 db, and this figure requires correction. The input/output resistance ratio is 200/1, and from Fig. 5–1(b), we observe that 23 db must be added to the apparent gain figure.

$$1 - 1 + 23 = 23 \text{ db gain} \qquad \text{(answer)}$$

When db gain must be calculated from voltage measurements across different resistance values, and a correction chart such as illustrated in Fig. 5–1(b) is not available, the formula for db may be derived as follows.

$$\begin{aligned} \text{db} &= 10 \log \frac{(E_0)^2/R_0}{E_{\text{in}}^2/R_i} \\ &= 10 \log \left(\frac{E_0}{E_{\text{in}}}\right)^2 \times \frac{R_i}{R_0} \end{aligned}$$

$$\text{db} = 20 \log \frac{E_0}{E_{\text{in}}} + 10 \log \frac{R_i}{R_0} \tag{5-7}$$

and

$$\begin{aligned} \text{db} &= 10 \log \frac{I_0^2 R_0}{I_{\text{in}}^2 R_i} \\ &= 10 \log \left(\frac{I_0}{I_{\text{in}}}\right)^2 \times \frac{R_0}{R_i} \\ &= 20 \log \frac{I_0}{I_{\text{in}}} + 10 \log \frac{R_0}{R_i} \end{aligned} \tag{5-8}$$

Example 8. Measurements with a voltmeter on an amplifier indicate 20 mv input across 1 kΩ and 35 v output across 20 kΩ. What is the db power gain?

$$\text{db} = 20 \log \frac{35}{20 \times 10^{-3}} + 10 \log \frac{1}{20}$$

$$\text{db} = 20 \times 3.243 + 10(-1.3)$$

$$\text{db} = 51.8 \text{ db} \qquad \text{(answer)}$$

A good sine waveform is required for accurate db measurements with a rectifier-type meter. If an audio oscillator does not provide a good waveform, or if an amplifier introduces waveform distortion, db measurements will be in error. Since an audio amplifier has a limited frequency-response range, it is customary to state its db gain figure at 400 Hz. Of course, we may also be interested in the amplifier gain figure at 60 Hz, and at 15 kHz. Unless the meter has full response at 15 kHz, the latter measurement would be in error. Therefore, do not attempt to make db measurements at frequencies outside the rated range of the instrument.

Example 9. The output of a power amplifier measures 10 w at 400 Hz and 4 w at 10 kHz. With 400 Hz as the reference, what is the power loss at 10 kHz?

$$\text{db} = 10 \log \frac{10}{4}$$

$$= 10 \log 2.5$$

$$\text{Power loss} = -3.98 \text{ db} \qquad \text{(answer)}$$

Note that if we read db values across a resistance with a value equal to the reference value for the meter, we can convert these db values into corresponding power values by calculation or by consulting a chart. These are called *absolute* db readings. On the other hand, suppose that the meter has a reference value of 500 Ω, and we make a pair of db readings across 1,200 Ω; these are called *relative* db readings. Neither reading corresponds directly to power values in 500 Ω, but it is nevertheless true as we have seen, that the difference between the relative readings is the actual number of db difference between the two power levels.

A dbm measurement is a decibel measurement of a sine-wave voltage made with a meter having a reference level of 1 mw in 600 Ω. Again, a VU (volume unit) measurement is a decibel measurement of a complex voltage waveform. It is made with a db meter [see Fig. 5-3(a)] that has a reference level of 1 mw in 600 Ω, and with a damping such that pointer overshoot is not greater than 1.5 per cent in

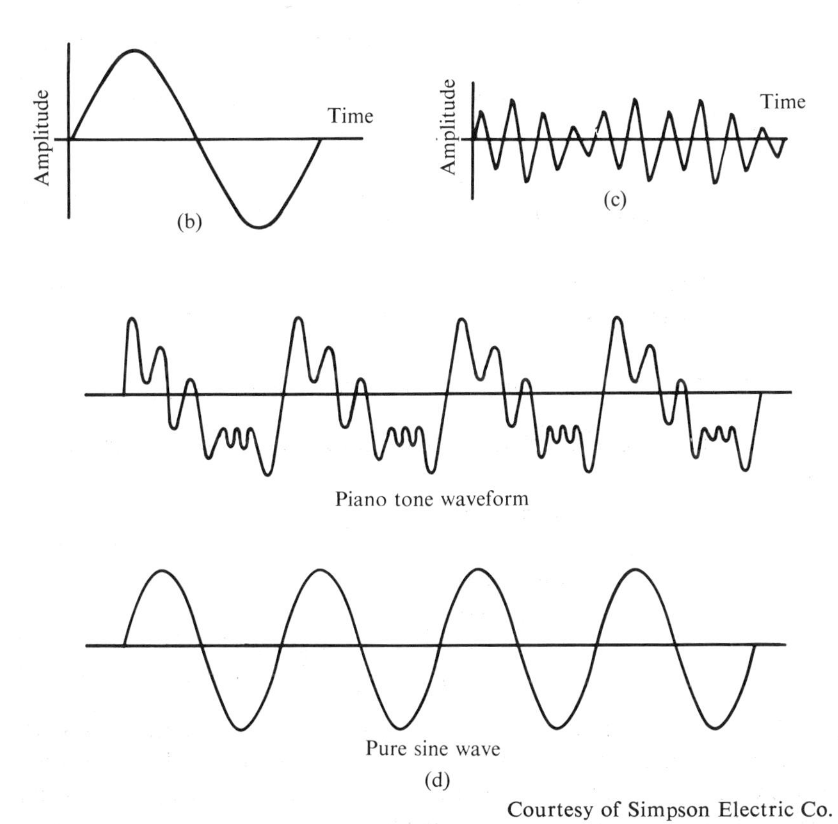

Courtesy of Simpson Electric Co.

Figure 5-3. (a) A typical VU meter; (b) Sine waveform; (c) Random voice waveform; (d) Comparison of a sine waveform with a piano tone waveform.

response to a suddenly applied voltage. That is, a dbm measurement is made on a steady sine-wave signal, where damping is inconsequential; on the other hand, a VU measurement is made on musical or vocal waveforms, and damping must be suitably controlled to obtain a valid indication of the power level. Figures 5-4(a), (b) and (c) show the difference between these types of waveforms. Figure 5-2(a) is a convenient chart that relates dbm values to power values and to voltage values across a 600-Ω load.

Example 10. A 400-Hz signal voltage of 20 mv is read across a 600-Ω earphone for a radio receiver. What is the dbm power level at the earphone?

From Fig. 5-2(b) we read 20 mv along the horizontal and +28 dbm along the vertical lines.

$$\text{Power level} = +28 \text{ dbm} \qquad \text{(answer)}$$

If the load has a lesser value, correction factors for dbm readings may be found in Fig. 5-2(c).

Example 11. A signal of 2 v is measured across a 3.2-Ω load. What is the dbm power level of the signal?

1. From Fig. 5-2(b), 2 v across 600 Ω gives +8 dbm.
2. From the conversion chart in Fig. 5-2(c), a resistive load of 3.2 Ω gives a dbm increment of +22.7 db to be added to the dbm value from Fig. 5-2(a).

$$\therefore \quad \text{Power level in dbm} = 30.7 \text{ dbm} \qquad \text{(answer)}$$

The following are other reference levels which have been established.

dbk: 1 k
dbv: 1 v
dbw: 1 w
dbvg: voltage gain
dbrap: db above a reference acoustical power of 10^{-16} w

Decibel measurements are often made across impedances. In such case, the db values correspond to *apparent-power levels*. The real power is equal to the power in the resistive component of the impedance, and the reactive power is equal to the power in the reactive component of the impedance. The power relations in an impedance are formulated

$$P_R^2 + P_X^2 = P_A^2 \tag{5-9}$$

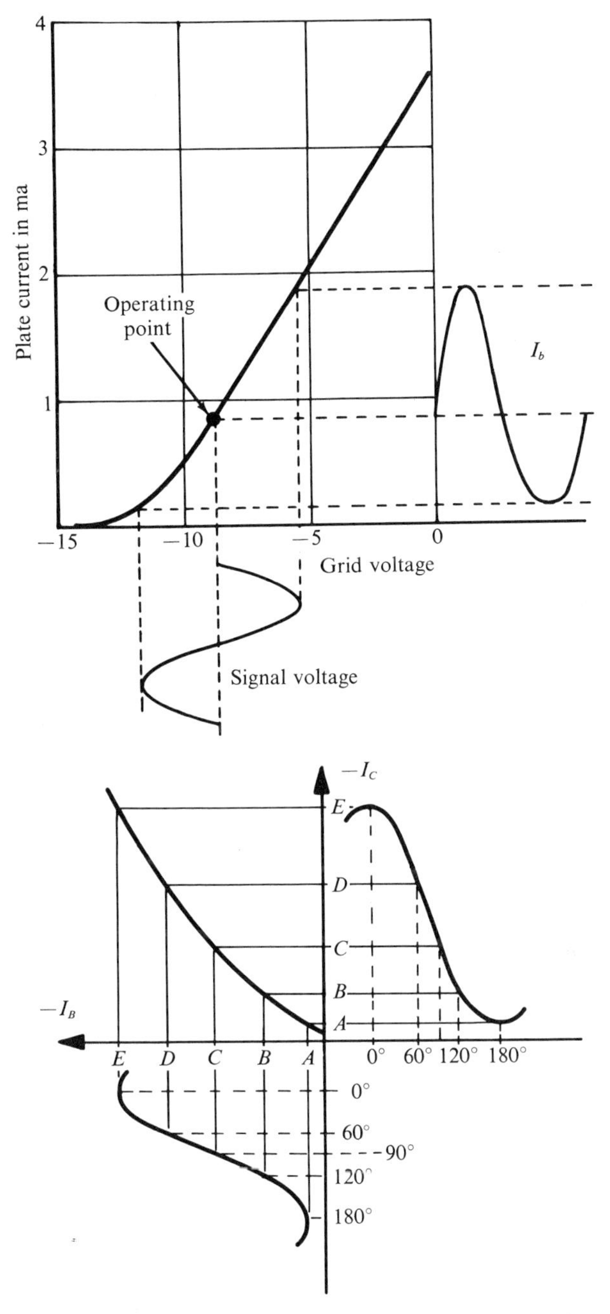

Figure 5-4. (a) Harmonic distortion produced by operation over the nonlinear portion of the vacuum-tube transfer characteristic; (b) Harmonic distortion produced by the nonlinear input of the transistor.

where P_R is the real power in watts, P_X is the reactive power in vars, and P_A is the apparent power in volt-amperes.

To take an extreme case, let us consider an amplifier which is loaded by a capacitor with a reactance of 600 Ω. In such a case, we might measure 25 db across the capacitor. It is evident that this is a reactive db value, and that it represents wattless power which merely surges into and out of the capacitor on alternate half cycles. When db values are measured across an impedance, it is necessary to determine the power-factor angle and calculate the real power component; it is this power value, for example, which is available to drive a loudspeaker.

5–2. Percentage Harmonic Distortion

Another basic measurement is a percentage harmonic distortion measurement. This measurement is usually made with a harmonic-distortion meter. Harmonic distortion is produced by operation over the nonlinear portion of a tube's transfer characteristic, as depicted in Fig. 5–4(a), or the nonlinear input characteristic of a transistor as depicted in Fig. 5–4(b). Harmonic distortion can also occur when an amplifier tube is driven into grid-current flow, as shown in Fig. 5–5.

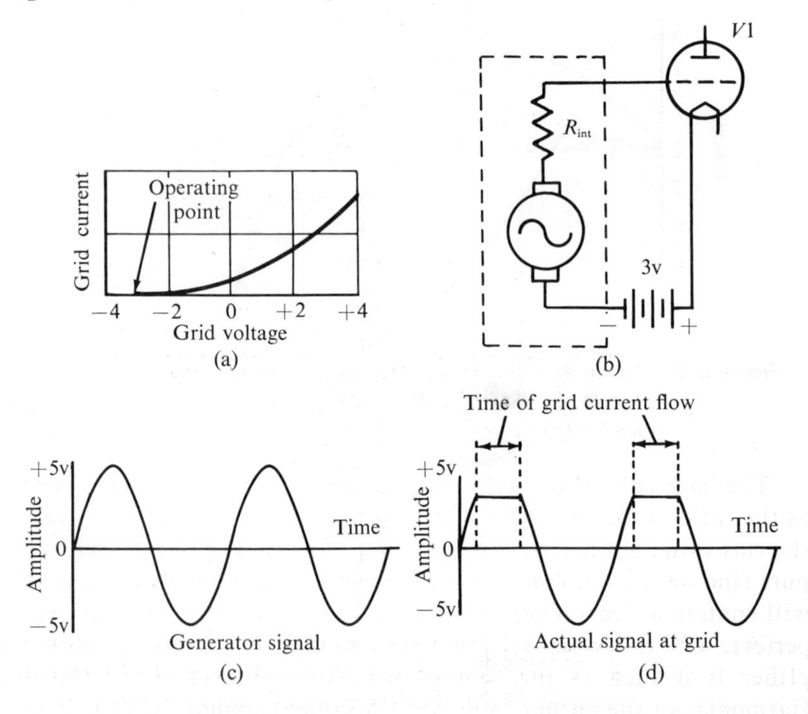

Figure 5–5. **Harmonic distortion can occur when an amplifier tube is driven into grid current flow.**

Harmonic distortion occurs because the modified sine waveform then consists of a series of harmonically related frequencies, as we can see in Fig. 5–6. For example, a half-sine wave is formulated

$$e = \frac{E_p}{\pi}\left(1 + \frac{\pi}{2}\cos\omega t + \frac{2}{3}\cos 2\omega t - \frac{2}{15}\cos 4\omega t + \frac{2}{35}\cos 6\omega t \cdots\right) \quad \textbf{(5–10)}$$

where e denotes instantaneous voltage of the half-sine wave, and E_p denotes its peak value.

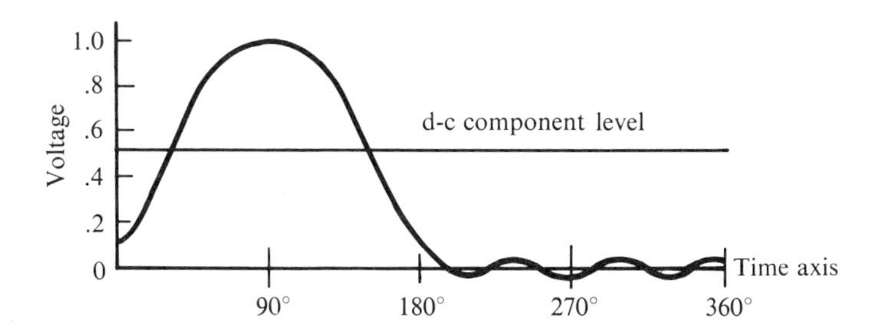

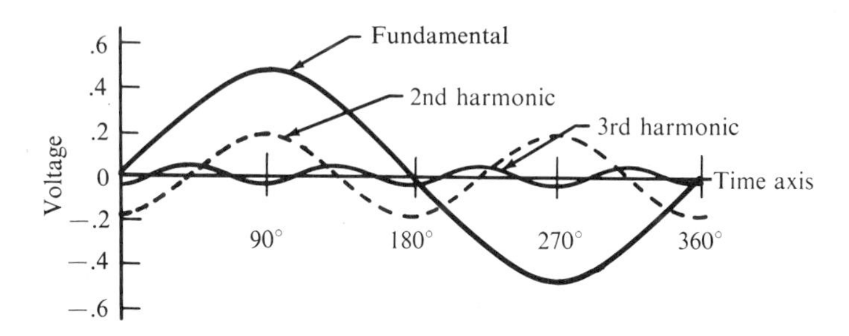

Figure 5–6. Synthesis of a half-sine wave from harmonically related sine waves. (a) Harmonically related frequencies; (b) How the harmonically related frequencies combine.

The harmonic distortion in the output from an amplifier is defined as the ratio of the rms value of all the significant output *harmonics* to the rms value of the sine wave applied to the amplifier. Although a pure sine wave is applied to an amplifier, the output from the amplifier will contain at least a slight harmonic value, because no amplifier is perfect. The rms value of the harmonics in the output from an amplifier is defined as the root-mean-square value of the harmonics. Harmonics in the output from an amplifier combine to form a complex waveform as depicted in Fig. 5–7(b). If we square the instantaneous values of the complex waveform in Fig. 5–7(a), we obtain the

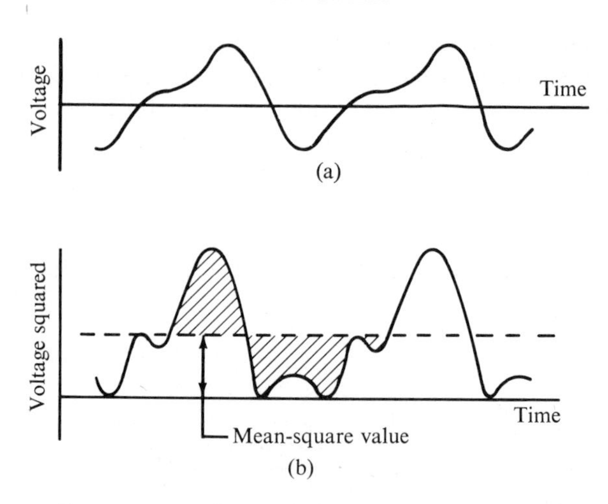

Figure 5–7. Illustration of the mean-square value. (a) Non-sine waveform; (b) The area above the line equals the area below the line.

waveform shown in Fig. 5–7(b). Equal shaded areas above and below the dotted line define a mean-square value for the waveform. The square root of this mean-square value is the rms value of the harmonics.

This calculation is automatically performed by a harmonic-distortion meter and its scale is calibrated in percentage harmonic-distortion values.

$$\%\,\text{HDM} = \frac{\text{rms harmonic voltage}}{\text{rms input voltage}} \times 100 \qquad \textbf{(5–11)}$$

You will probably have an opportunity to measure percentage harmonic-distortion values in the laboratory phase of your studies. High-fidelity amplifiers normally have a harmonic distortion of less than 1 per cent. Utility amplifiers, on the other hand, may have a harmonic distortion of 5 per cent, 10 per cent, or even more. Percentage harmonic-distortion measurements are customarily made at full rated power output from an amplifier, because the distortion value increases as the power output is increased.

Discussion of other basic measurements must be deferred until we have learned additional electrical principles and characteristics of circuit action.

5–3. Basic Measuring Instruments

We are now in a good position to consider the operating principles of representative basic measuring instruments. It is instructive to start with an analysis of d-c voltmeter action. Most d-c voltmeters employ a d'Arsonval movement, as depicted in Fig. 5–8(a). A permanent magnet (PM) is utilized (often in the form of a horseshoe), and the magnetic circuit also comprises a pair of pole pieces (P) and a soft-iron core (C). The small air gap provides a typical clearance of less than 0.1 in. In this gap, a moving coil (*A*) rotates on jeweled bearings. A strong magnetic field (typically 5,000 gauss) in sensitive instruments traverses the air gap. Construction of the moving coil is seen in Fig. 5–8(b). Approximately 1,500 turns of wire less than

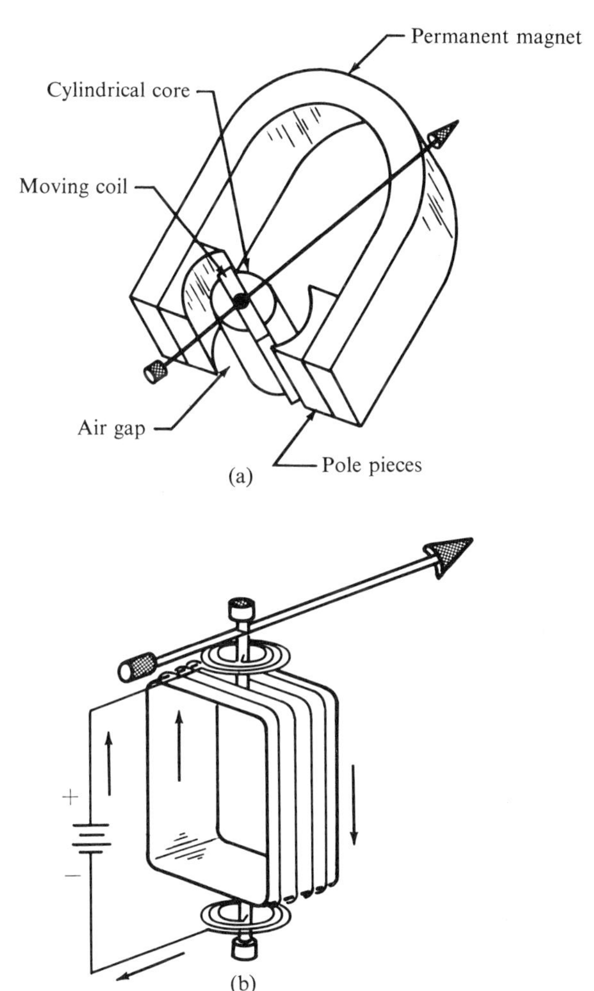

Figure 5–8. (a) Arrangement of a d'Arsonval movement; (b) Construction of the moving coil.

0.001 in. in diameter may be wound on the *bobbin*. Current to the coil is conducted through the spiral phosphor-bronze springs. Adjustable balance weights are provided for mechanical balancing of the assembly.

A d'Arsonval movement operates on the motor principle. When current flows through the moving coil, a torque is generated which is opposed by tension of the spiral springs. In a sensitive movement, a current of 10 μa will produce full-scale deflection. Most utility voltmeters used in school laboratories have a full-scale current value of 50 μa. The resistance of the moving coil is typically 2,000 Ω. Thus, the voltage drop across the moving coil at full-scale deflection is 100 mv in this example. Accordingly, the movement can be used directly as a d-c voltmeter with a full-scale indication of 100 mv. Any voltmeter has a sensitivity that is stated in ohms-per-volt. This ohm-per-volt value is equal to the reciprocal of the full-scale current.

$$\Omega/\text{v} = \frac{1}{I_{\text{fs}}} \qquad \textbf{(5–12)}$$

Example 12. The full-scale current of a meter is 50 μa; what is the ohms-per-volt rating?

$$\Omega/\text{v} = \frac{1}{I_{\text{fs}}}$$

$$= \frac{1}{50 \times 10^{-6}} = 20{,}000 \qquad \text{(answer)}$$

This figure can also be calculated as the quotient of the internal resistance of the movement and the full-scale voltage.

$$\Omega/\text{v} = \frac{\text{internal resistance}}{\text{full-scale voltage}} \qquad \textbf{(5–13)}$$

Example 13. The internal resistance of a meter is 2,000 Ω, and the full-scale voltage is 100 mv. What is the ohms-per-volt rating of the meter?

$$\Omega/\text{v} = \frac{R_{\text{int}}}{V_{\text{fs}}}$$

$$= \frac{2{,}000}{0.1} = 20 \text{ k}\Omega \text{ v} \quad \text{(answer)}$$

Most voltmeters have more than one range. To obtain another range with a greater full-scale value, a *multiplier resistor* is connected in series with the movement, as shown in Fig. 5–9. Let us assume that we wish to obtain a full-scale indication of 1 v, with the movement previously described. This is just another way of saying that

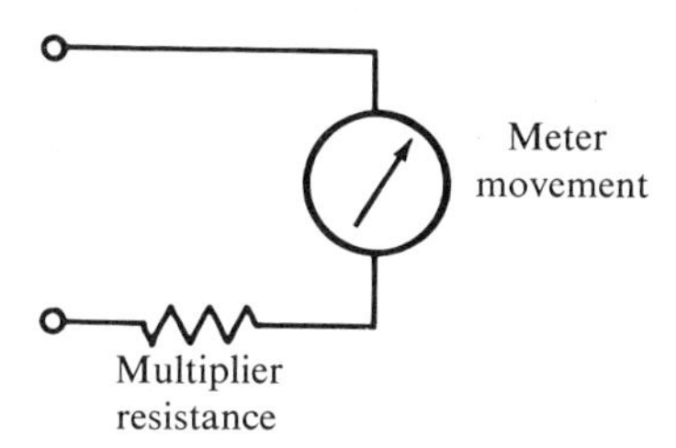

Figure 5-9. The full-scale voltage indication is determined by the value of the multiplier resistance.

50 μa of current must flow through the circuit in Fig. 5-9 when 1 v is applied. We write

$$\text{Full-scale current} = \frac{1}{R_M + R_{int}}$$
$$50 \times 10^{-6} = \frac{1}{R_M + 2{,}000} \qquad \textbf{(5-14)}$$

where R_M is the required resistance value for the multiplier, and R_{int} is the internal meter resistance.

It is evident that R_M must have a value of 18,000 Ω. Other full-scale ranges are calculated in the same manner. For example, suppose that we wish to use the foregoing movement in a d-c voltmeter to provide a choice of 2.5, 10, 50, 250, 1,000, and 5,000 v full-scale indication. The student may verify that the multiplier resistance values shown in Fig. 5-10 will provide these ranges. Note that the 5-kv

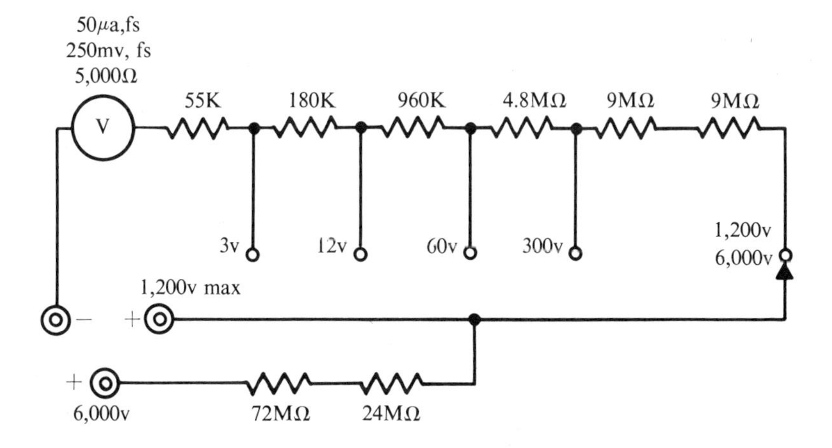

Courtesy of Triplett Electrical Instrument Co.

Figure 5-10. D-C voltmeter configuration with six ranges.

multiplier resistor (80 meg) is brought out to a separate terminal. This design feature is required because conventional rotary switches are rated to withstand less than 5,000 v. If it is desired to measure still higher voltage values, necessary insulation-resistance requirements are met by use of an external multiplier resistor. Figure 5–11 illustrates a 20,000-v d-c probe.

Courtesy of Precision Apparatus,
Division of Dynascan Corp.

Figure 5–11. A high-voltage d-c probe.

Current values can also be measured with the movement previously described. For example, the movement can be provided with a 50-μa full-scale indication. Of course, we are often required to measure much higher current values also. Hence, higher current ranges are provided by utilizing a meter *shunt*, as depicted in Fig. 5–12. The value of a meter shunt may be formulated

$$R_S = \frac{\text{full-scale meter voltage}}{\text{shunt current}} = \frac{V_{\text{fs}}}{I_S} \tag{5–15}$$

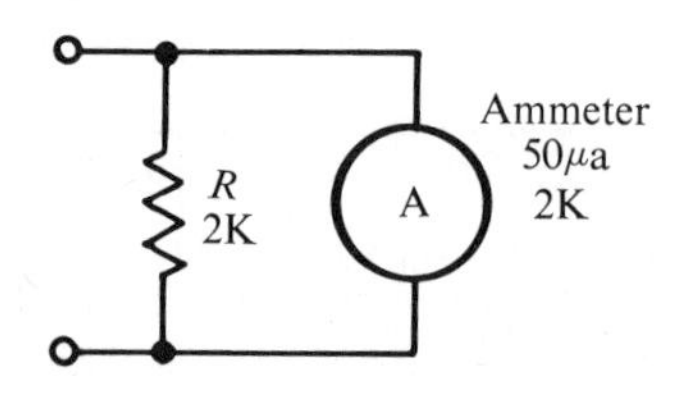

Figure 5–12. A shunt that provides a full-scale current indication of 100 μa.

Example 14. The full-scale current and voltage of a meter movement are 50 μa and 0.1 v respectively. What value of shunt resistance is required for the meter to read 1 ma full scale?

$$R_S = \frac{V_{fs}}{I_S}$$

$$= \frac{0.1}{0.0950 \text{ amp}} \simeq 14.2\ \Omega \qquad \text{(answer)}$$

If the shunt has the same resistance value as the movement, half of the applied current will flow through the shunt, and half will flow through the movement. Accordingly, the full-scale current indication in the configuration of Fig 5–12 is 100 μa.

Observe the shunt arrangement depicted in Fig. 5–13. Four d-c current ranges are provided. This is called a *ring-shunt* configuration.

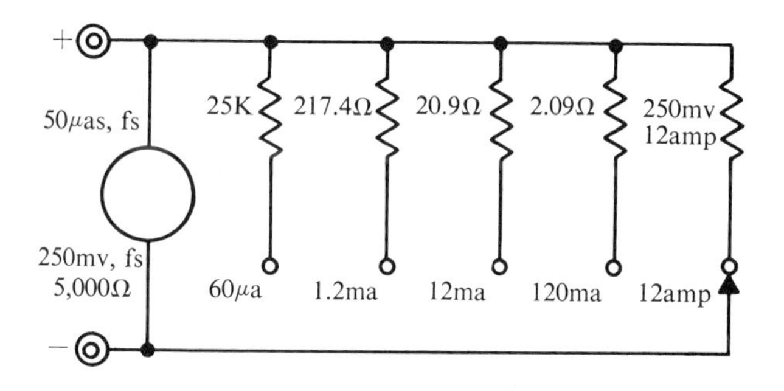

Courtesy of Triplett Electrical Instrument Co.

Figure 5–13. A ring-shunt configuration.

Ring shunts are used so that the meter circuit remains closed while switching from one range to the next; in this manner, the movement is not subject to damage when the range switch is changed with the terminals connected to a current source. Of course, the movement may be damaged nevertheless, if the range switch is set to too low a position. The student may verify the shunt-resistance values noted in Fig. 5–13. This is a series-parallel circuit arrangement, and all of the resistors must be taken into account on any position of the range switch.

Many utility meters provide for resistance measurement, as well as d-c voltage and current measurement. An *ohmmeter* employs the basic configuration depicted in Fig. 5–14. When a resistor under test is connected to the instrument terminals, current flows through the

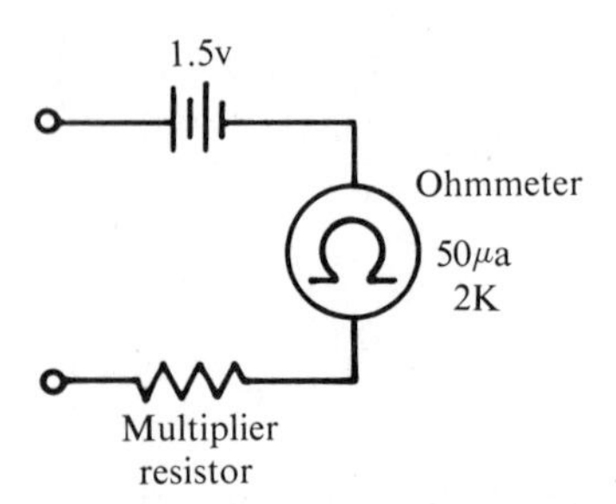

Figure 5-14. **Basic ohmmeter configuration.**

movement and the pointer is deflected. The value of the multiplier resistor is chosen to obtain full-scale deflection when the instrument terminals are short-circuited. This is a circuit in which the value of the multiplier resistor remains fixed, while various values of resistance may be connected to the instrument terminals for test. Therefore, the ohmmeter scale is nonlinear, as seen in Fig. 5-15.

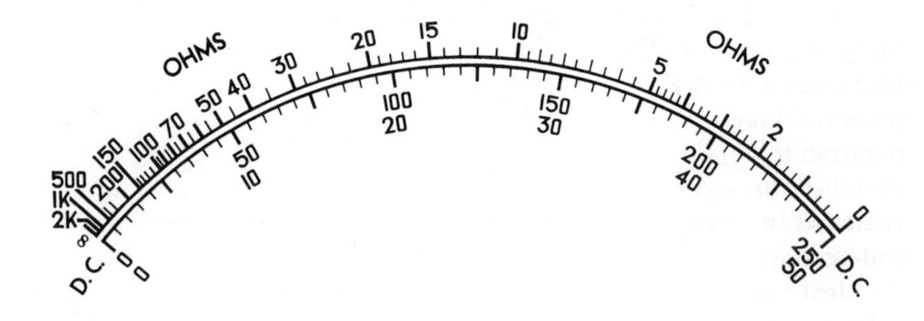

Courtesy of Simpson Electric Co.

Figure 5-15. **Comparison of ohms scale with d-c voltage and current scales.**

Since the internal resistance of the instrument depicted in Fig. 5-14 is 30,000 Ω, it is evident that half-scale deflection will occur when a resistance of 30,000 Ω is connected across the instrument terminals. Of course, a wide range of resistance values must be measured in practice. An ohmmeter scale is highly cramped at the left-hand end. Therefore, several resistance ranges are commonly provided. A typical configuration is shown in Fig. 5-16. This is a series-parallel arrangement; indication accuracy on the $R \times 1$ range requires that the test-lead resistance and the internal resistance of the 1.5-v battery be taken into account, and the scale is calibrated accord-

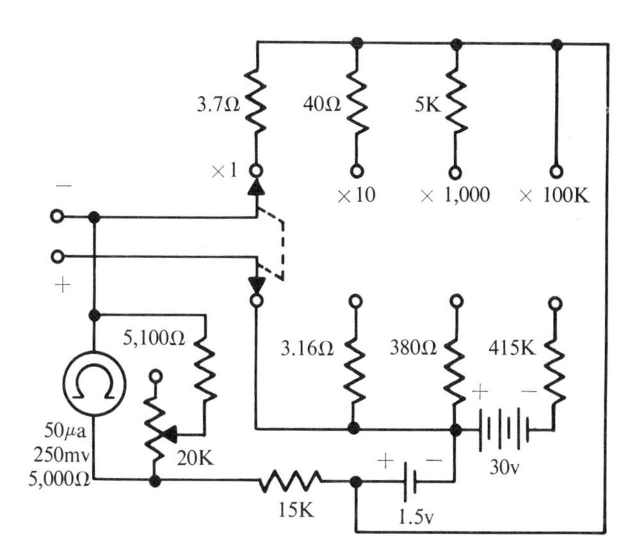

Courtesy of Triplett Electrical Instrument Co.

Figure 5-16. Four-range ohmmeter configuration.

ingly. In turn, a different value of resistor is switched into the test-lead circuit on each range to maintain circuit symmetry and accurate scale-tracking. On the $R \times 100$ K range, a higher battery voltage is required to obtain 50 μa full-scale current flow, and a 30-v battery is switched in series with the 1.5-v battery. A 415 K-Ω resistor is also switched in series with the test leads to maintain circuit symmetry and accurate scale-tracking.

Most volt-ohm-milliammeters also provide for measurement of a-c voltage values. The basic configuration is depicted in Fig. 5-17. Half-sine waves of current flow through the movement when an a-c voltage

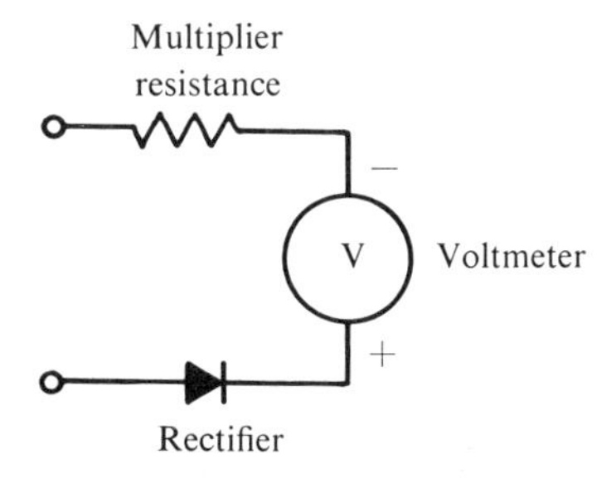

Figure 5-17. Basic a-c voltmeter configuration.

is applied to the instrument terminals. The movement responds to the average value of the rectified waveform (0.318 of peak for a sine wave). However, the scale is calibrated in terms of rms values for sine waves. Therefore, the scale indicates 0.707/0.318 times the average value of the rectified wave-form, or 2.22 times the d-c component value. This means that if you apply 1 d-c v to the configuration in Fig. 5–17 (in the conducting direction), the scale reading will be 2.22 v. This is a simple example of system analysis entailing nonlinear resistance, a-c calibration, and d-c response.

Since a semiconductor rectifier presents a nonlinear resistance to the meter circuit, the left-hand end of the a-c scale is somewhat cramped, as seen in Fig. 5–18. A complete a-c voltmeter configuration is shown in Fig. 5–19. Only the series rectifier passes current into the

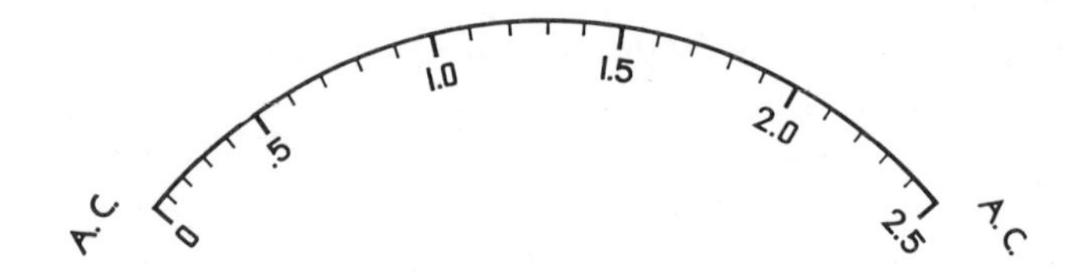

Figure 5–18. The a-c scale is somewhat cramped at the left-hand end.

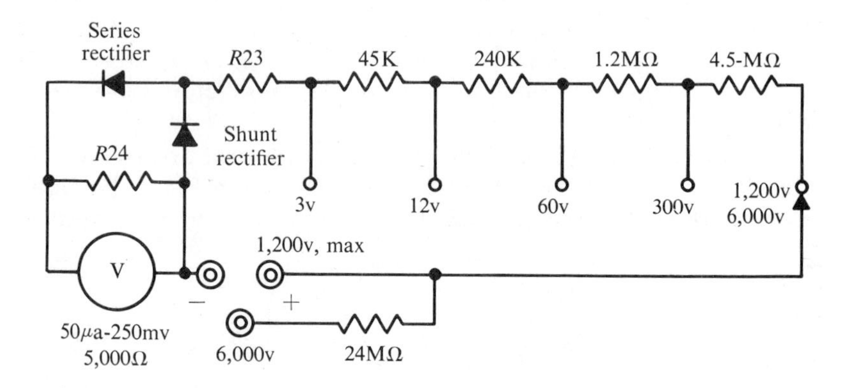

Courtesy of Triplett Electrical Instrument Co.

Figure 5–19. A complete a-c voltmeter configuration.

movement. The shunt rectifier serves merely to bypass the opposite half cycle around the movement. Accuracy is thereby improved, because the small reverse-current flow that would otherwise take place through the series rectifier is practically eliminated. R_{23} and R_{24} are calibrating resistors that are adjusted to take rectifier tolerances into account. You will note that the input resistance of this a-c voltmeter is less than that of the previously described d-c voltmeter configuration. The a-c voltmeter has a sensitivity of 1,000 Ω per volt. This reduced sensitivity on a-c voltage measurement is due to the fact that the movement responds to the average value of the rectified waveform and the presence of the "swamping" resistor R_{24}.

Note that the swamping resistor, R_{24} (which also serves as a calibrating resistor), makes the current demand of the circuit greater. This is a desirable condition, because a semiconductor characteristic becomes more nearly linear at high current values. In turn, a swamping resistor assists in partial linearization of the a-c scale. It also provides improved indication accuracy as aging changes the semiconductor characteristic.

Finally, let us consider the operation of a *watt-hour meter*. A watt-hour meter measures electrical *energy*. It integrates, or sums up, all the instantaneous rates of energy flow, so that the total energy utilized over a given interval can be measured. A watt-hour meter, in effect, multiplies the power in watts by the time in hours. Figure 5–20

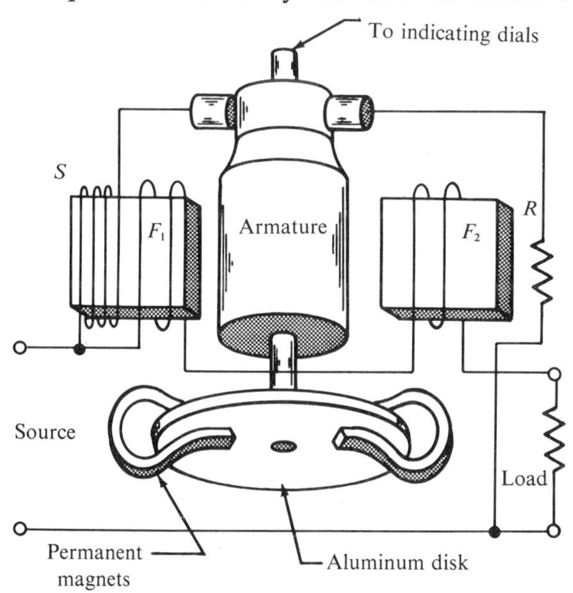

Figure 5–20. Simplified diagram for a Thompson watt-hour meter.

depicts a simplified diagram of a Thompson watt-hour meter. This is a d-c type of meter. We perceive that the instrument is designed around a series motor. The series combination of coil *S*, the armature, and resistor R is connected across the load. However, the field coils F_1 and F_2 are wound with a few turns of heavy wire, through which the load current flows. In turn, a magnetic field is established in which the armature rotates.

The operation of the d-c watt-hour meter may be outlined as follows.

1. The current that flows through the load also flows through F_1 and F_2, producing a field that is proportional to the load current.

2. The voltage that is applied to the load is also applied to the armature, coil S, and resistor R. In turn, the current that flows through the armature is proportional to this voltage.

3. Torque is generated that causes the armature to revolve. This torque is proportional to the product of the current through the load and the voltage across the load; in other words, the torque is proportional to the number of watts consumed by the load.

4. Some form of resistance (or opposing torque) must be provided that will vary directly with the speed of the armature in order that the motor speed will be directly proportional to the power consumed in the load. This opposing torque is obtained by mounting an aluminum disk at the lower end of the armature shaft, so that it will rotate in the two magnetic fields established by the permanent magnets. Eddy currents are induced in the disk as it rotates through the fields. These eddy currents produce a retarding force on the disk that is proportional to the speed of the disk. When the line power increases, the watt-hour meter speeds up until the disk-load torque increases to the same value as the increased driving torque. At this condition of equilibrium, the speed is constant and is proportional to the increased load power.

5. For any given load, the amount of energy indicated on the meter dials is proportional to the product of the speed of the disk and the time during which the power is being integrated.

The function of coil S in Fig. 5–20 is to produce just enough field flux to interact with the armature flux so that static friction is overcome. Thus, when a small load is present, the meter will immediately rotate at a speed proportional to this load. The full-load adjustment is made by moving the position of the drag magnets on the disk. Moving the magnets toward the center of the disk will lessen the drag and allow the meter to speed up for a given electrical load in the line.

As we have learned in our previous study of the series motor, this arrangement can be used also with alternating current, because both the armature flux and the field flux reverse at the same time, and the armature rotates continuously in the same direction. However, the induction type of watt-hour meter has certain advantages over the Thompson watt-hour meter and is commonly used to measure a-c energy.

The single-phase induction watt-hour meter includes a simple induction-drive motor, which consists of an aluminum disk, moving magnetic field, drag magnets, current and voltage coils, integrating dials, and associated gears. A simplified sketch of an induction watt-hour meter is seen in Fig. 5–21. The potential coil is connected across the load and comprises many turns of comparatively small wire. It is

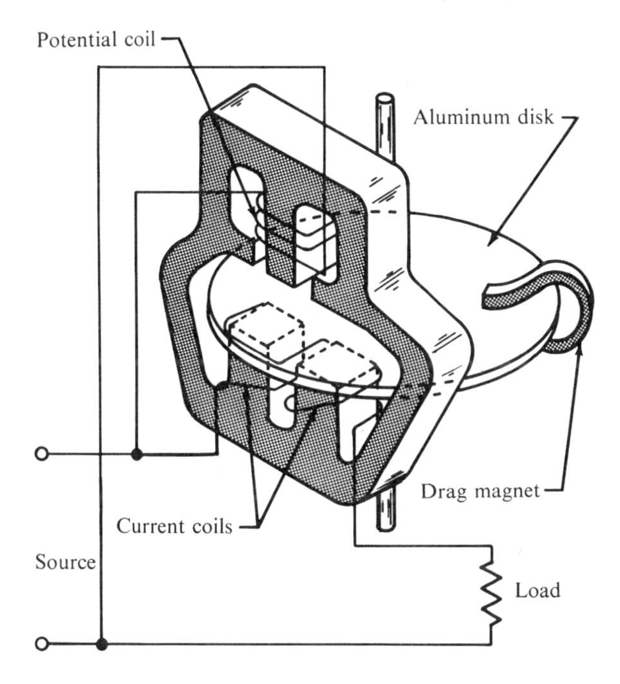

Figure 5–21. Simplified diagram for an induction watt-hour meter.

wound on one leg of the laminated magnetic circuit. Because of the large number of turns, this voltage coil has a high impedance and high inductance; therefore, the current through the coil lags the applied voltage by nearly 90°. The two current coils connected in series with the load comprise a few turns of heavy wire. They are wound on two legs of the laminated magnetic circuit. Because only a few turns are utilized, the current coils have small inductance and small impedance.

The arrangement of the potential coil, the aluminum disk, the current coils, and one of the drag magnets is depicted in Fig. 5–21.

Figure 5–22 shows the potential, current, and flux curves associated with the induction watt-hour meter. Also shown are the instantaneous directions of the flux through the disk at four instants during the cycle. To understand the operation of the induction watt-hour meter, it is necessary to consider how the force F is generated that causes the aluminum disk to rotate.

This force F, that causes the aluminum disk to move in a clockwise direction (looking down on the disk from above), is outlined as follows.

1. At instant 1, assuming a load of unity power factor, the current i_{pc}, and the flux ϕ_{pc} of the potential coil, are both at maximum. The line current, i_{line} (shown as i_{cc} in the figure), and the flux ϕ_{cc}, of the current coils are zero, but they are both *changing* at their maximum rate. As a result of the rapid flux change through the disk, eddy currents are generated in the disk in two distinct paths above the current coils. These two currents join under the potential coil and flow toward the edge of the disk (toward the observer). The right-hand rule for motor action applied to this area of current flow indicates a force acting on the disk that tends to turn the disk clockwise.

2. At instant 2, i_{line} and Φ_{cc}, the current and flux respectively of the current coils, are at a maximum. The current i_{pc} and the flux Φ_{pc} of the potential coil are zero, but they are both *changing* at their maximum rate. As a result of the rapid flux change through the disk, eddy currents are generated in the disk in the region below the potential coil. The current path is shown as a circle around the area covered by the potential coil. On the right-hand side, the current is moving toward the observer and directly through the flux established by the right-hand current coil. On the left-hand side, the current is moving away from the observer and directly through the flux established by the left-hand current coil. As may be seen by application of the right-hand motor rule, the force F, acting on both areas of the disk, tends to turn the disk clockwise.

3. Similar reasoning may be applied at instants 3 and 4. At each instant, the direction of the eddy currents and the flux through which the currents flow are such that the resultant forces tend to turn the disk clockwise.

From the foregoing discussion, we perceive that there are established in the disk induced currents that flow about each polar region in circles. The induced currents due to the potential coil are 90° out of phase with those due to the current coils when the load power factor is unity. Part of the current induced by the current coils passes under the pole of the potential coil and therefore through a magnetic field in phase with it. For power factors less than unity, the phase

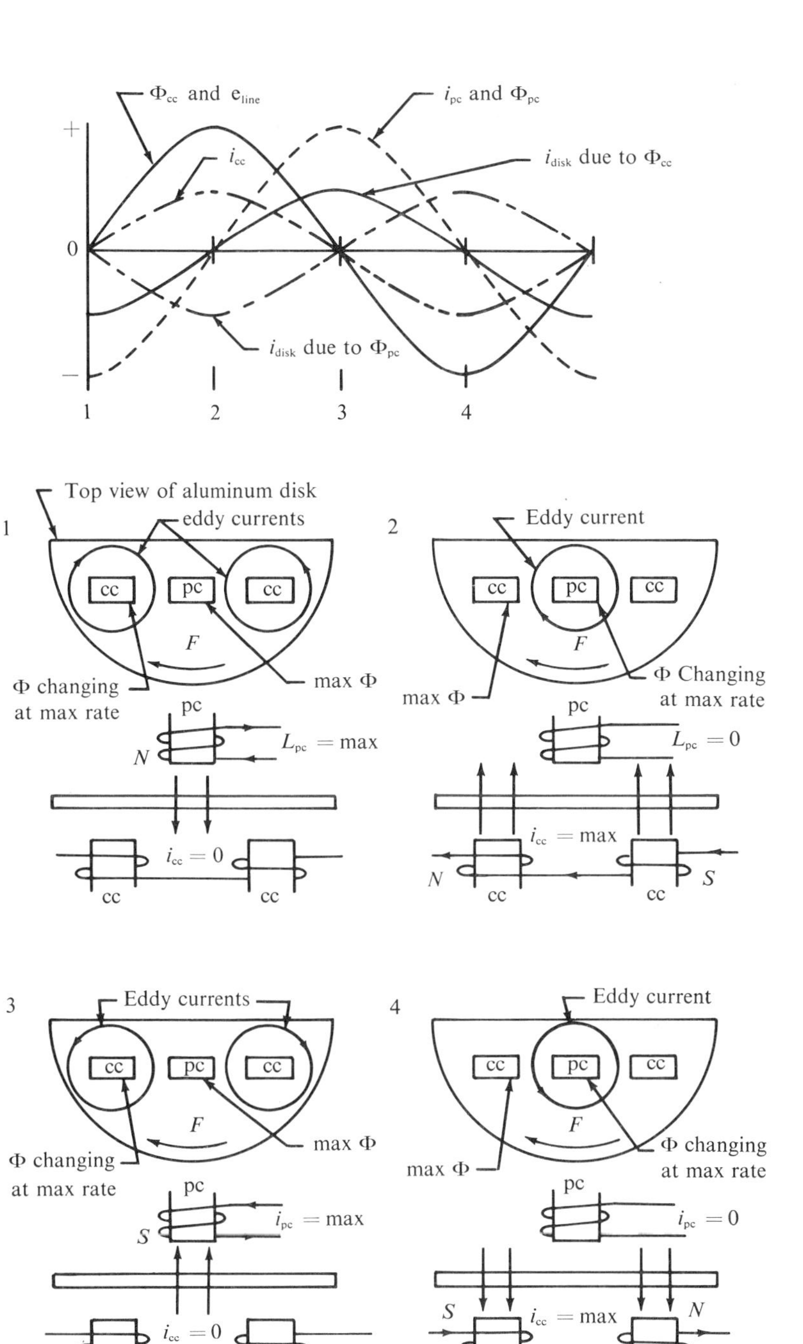

Figure 5-22. Operation of an induction watt-hour meter.

difference between these currents and the magnetic fields through which they pass is the same as the difference in phase between the load current and the load voltage. Hence, the driving torque on the disk is proportional to the true power supplied through the line to the load. A braking system similar to that of the Thompson watt-hour meter causes the speed of the disk to be proportional to the average torque, and therefore, the number of revolutions per second is proportional to the energy supplied to the load.

A small copper shading disk (not shown in the figure) is placed under a portion of the potential pole face in an adjustable mounting to develop a torque in the disk to counteract static friction. This disk has the effect of a shading pole and provides a small load adjustment for the meter. The two drag magnets provide the counter torque against which the aluminum disk acts when it turns. The drag is increased (speed of motor reduced) by moving the magnets toward the edge of the disk. Conversely, the drag is decreased and the speed increased by moving the magnets toward the center of the disk.

5–4. Introduction to the Oscilloscope

The oscilloscope is an electronic instrument which allows us to "see" electrical waveforms. In most common applications, the oscilloscope presents information on the screen of a cathode-ray tube (CRT) in the form of a graph. The horizontal (X) axis represents time, and the vertical (Y) axis represents amplitude. The signal to be observed on the oscilloscope is connected through a probe, to eliminate circuit loading, to the input of the vertical amplifier. The horizontal signal is normally an internally generated sawtooth waveform to develop the horizontal sweep. The resultant of the vertical and horizontal signals is impressed on the phospor coating of the CRT.

The success in our analysis of electronic circuits will depend, to a large extent, upon our ability to operate the controls of the oscilloscope and to evaluate the electrical waveforms which we observe on the CRT.

Figure 5–23(a) illustrates the action of the horizontal and vertical controls on the electron beam. Figure 5–23(b) depicts the action of the same controls with a horizontal sawtooth signal, but no vertical signal. Figure 5–24 depicts the front panel controls of a basic oscilloscope with illustrations of the action of each control.

The student should make a list of the controls and become familiar with their functions. One should, of course, refer to an instruction book when using an unfamiliar piece of test equipment.

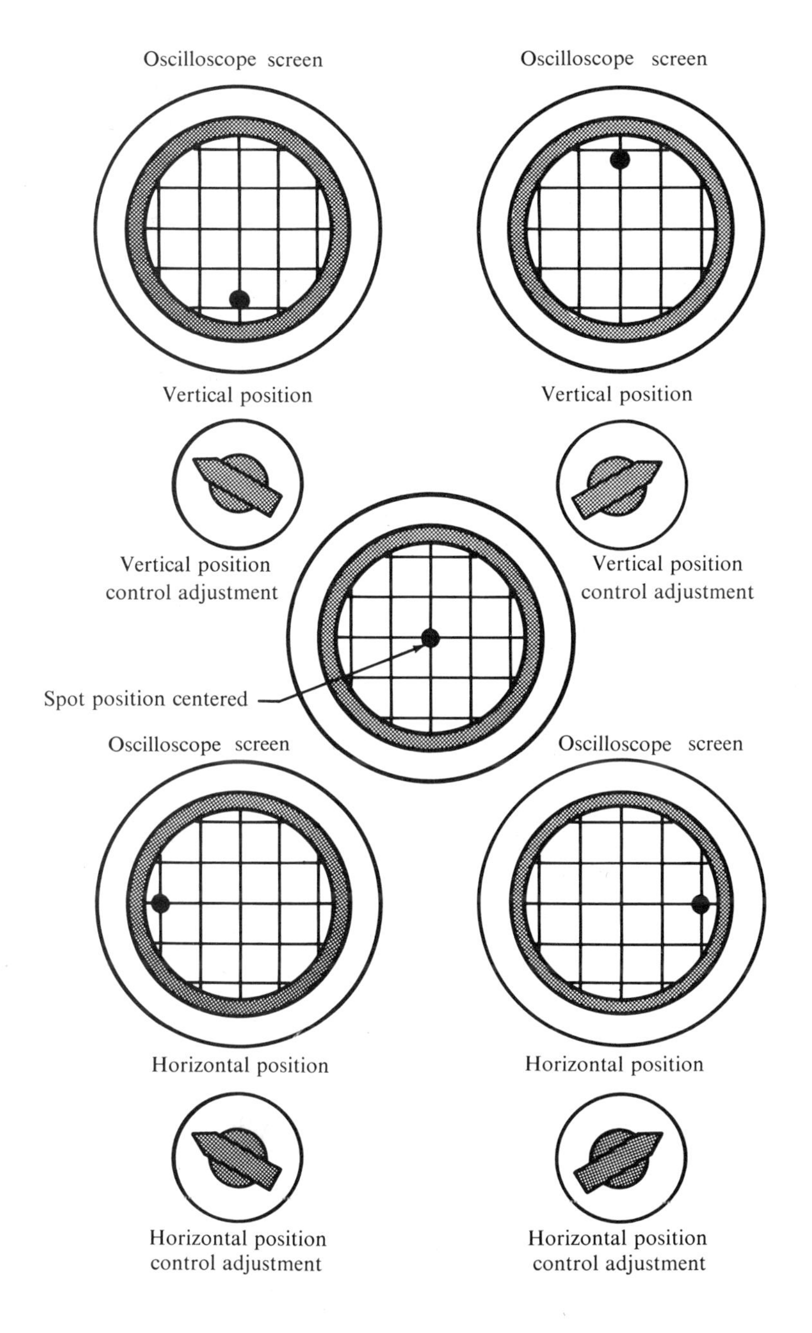

Figure 5-23. (a) Oscilloscope screen with no vertical or horizontal signal applied.

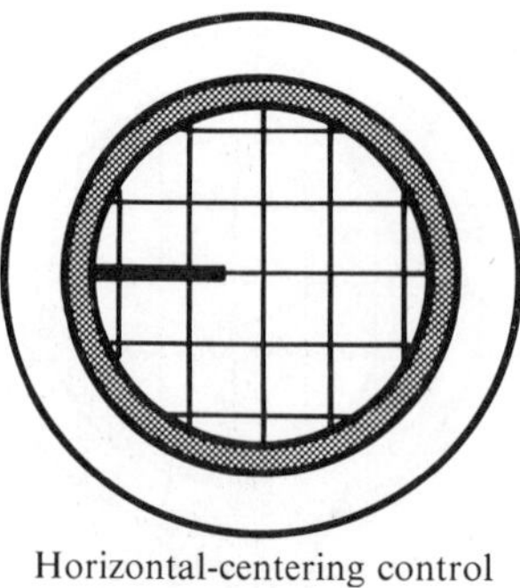

Horizontal-centering control
turned too far left

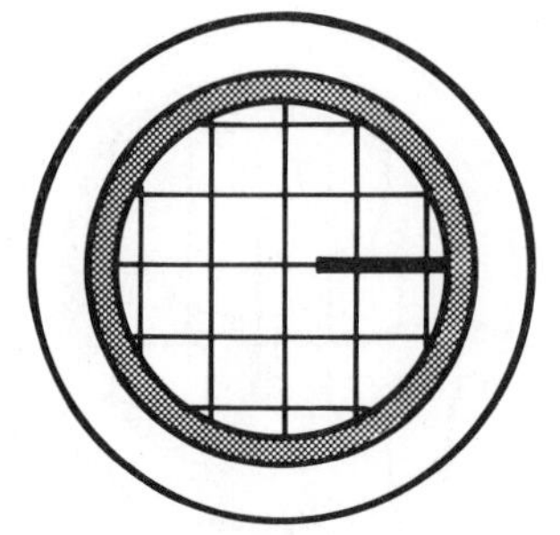

Horizontal-centering control
Turned too far right

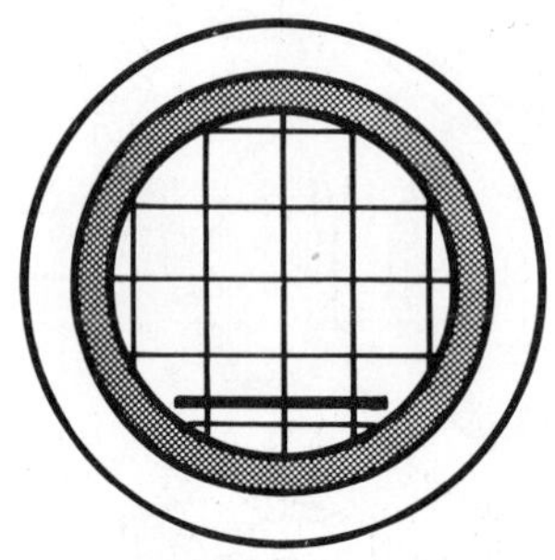

Vertical-centering control
turned too far left

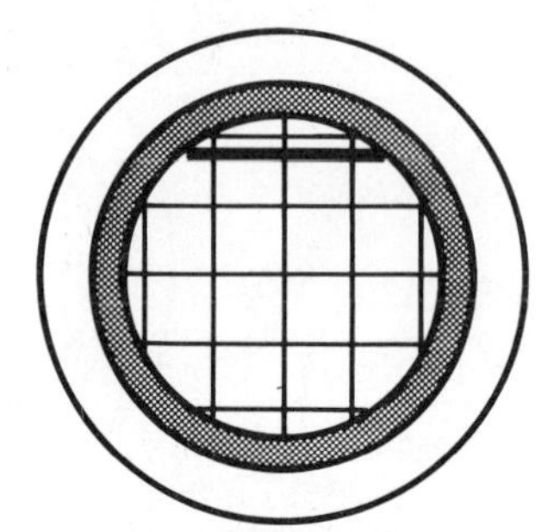

Vertical-centering control
turned too far right

Figure 5–23. (b) Oscilloscope screen with no vertical-input signal applied, time base operating, control of horizontal-trace position by the centering controls.

SUMMARY

1. A bel value is the logarithm of a power ratio, and a decibel (db) is 0.1 bel. Audio power levels are measured in db, because the decibel unit is proportional to ear response.

2. A change of 1 db is slightly noticeable; a change of 6 db is often judged as one-half or double the sound level.

3. A change of 3 db is a change of one-half or double the power level.

4. Decibel values are logarithms of power ratios and therefore are additive and subtractive.

5. Calculation of decibel gain or loss by measurement of input voltage or current in relation to output voltage or current is correct

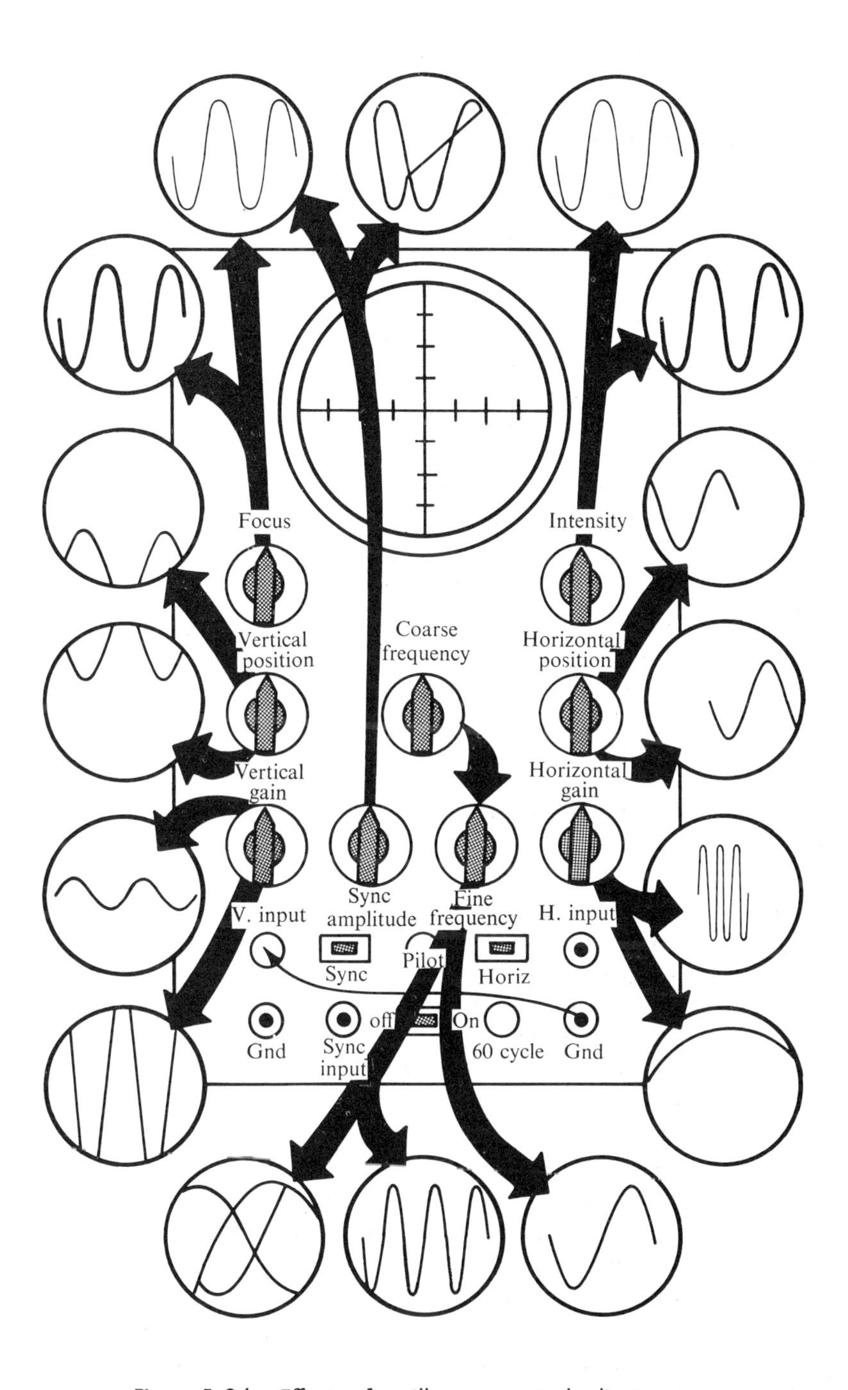

Figure 5-24. Effects of oscilloscope control adjustments.

only if the measurements are made across the same values of resistance. In other words, the input and output resistances are equal.

6. The db gain of an audio amplifier is customarily given at 400 Hz.

7. Absolute db values are those db values read across a resistance equal to the reference resistance value of the meter.

8. Relative db values are those db values read across resistances not equal to the reference resistance value of the meter. However, two relative db measurements made across the same value of resistance give the true db level between the two power levels.

9. A dbm measurement is a decibel measurement of a sine-wave voltage made with a db meter having a reference level of 1 mw across 600 Ω.

10. A volume unit (VU) measurement is a decibel measurement made on a complex waveform, such as music or voice on a db meter with a reference level of 1 mw in 600 Ω.

11. Reference levels other than dbm and VU have been established.

12. Apparent power levels correspond to db values measured across impedances.

13. Harmonic distortion is produced by operating an amplifier device over a region where the gain varies, as this causes nonlinearity of the output signal. Any distortion of a sine wave produces harmonics.

14. The total distortion of an amplifier is equal to the rms harmonic voltage divided by the rms input voltage and is measured by a distortion analyzer.

15. The ohms-per-volt rating of a voltmeter may be used to determine the resistance of the meter on any range and thereby determine the relative loading on the circuit.

16. A multiplier resistor is placed in series with a meter movement to increase the voltage range of the instrument. The value of the multiplier plus the value of the internal meter resistance limits the maximum meter current to full-scale current for a particular voltage range.

17. A shunt resistor is placed in parallel with a meter movement to increase the current range of the meter.

18. An ohmmeter uses an internal energy source to force current through an external resistance, thereby indicating the value of the external resistance by a magnitude of current.

19. An a-c signal is rectified by an a-c voltmeter and the average d-c value of the rectified a-c current produces a meter deflection which is indicated on a calibrated scale to read rms a-c values.

20. The low ohms-per-volt rating of an a-c voltmeter produces more loading on a circuit under test than the higher ohms-per-volt rating of a d-c meter.

Questions

1. What is the mathematical expression for the bel?
2. Why do we make calculations in decibel values rather than bel values? What is the psychological interpretation of the physical sound level of an increase of 1 decibel?
3. What approximate power level in decibels is necessary to double the average apparent level of a sound?
4. What decibel increase is necessary to double the power level of a signal?
5. Why may decibel levels be added and subtracted while power levels must be multiplied?
6. What are the constants implied when using the decibel power formulas with voltages or currents?
7. The db gain of an audio amplifier is given for what customary reference frequency?
8. List three things that must be taken into consideration when determining the db power gain of a system by measurements.
9. State the general equation for power gain or loss in db. How is a power loss indicated?
10. Explain the difference between relative db values and absolute db values.
11. What is the reference power level for a dbm measurement?
12. What type of voltage waveforms may be measured in dbm?
13. What is the VU unit, and upon what type of waveform is this unit measured? What are the references?
14. Give three reference levels other than dbm and VU units.
15. Explain apparent power db levels.
16. What are two possible reasons for harmonic distortion in a vacuum-tube amplifier?
17. What type of instrument is used to measure distortion in an amplifier?
18. When do we desire the meter movement to be lightly damped?
19. What are two possible reasons for harmonic distortion in a transistor amplifier?
20. Why are correction factors necessary when using a decibel meter?
21. Explain the operation of a d'Arsonval meter movement.
22. What significance does the ohms-per-volt rating have on the accuracy of the voltage measurement across a given value of resistance?
23. What is the function of a high-voltage probe for a voltmeter?

24. What is the function of a shunt resistor in an ammeter?
25. Explain the operation of an ohmmeter.
26. How is a d'Arsonval meter movement made to measure an a-c voltage?
27. What are the functions of a swamping resistor in an a-c voltmeter?
28. What causes the lower value of ohms-per-volt rating of a multimeter on the a-c function?

PROBLEMS

1. The output of a high fidelity amplifier is 2.8 w. What decibel increase is needed to double the power level? What power increase in db is needed to double the average apparent sound level?

2. A radar receiver delivers 1 w from an input of 100 nw. What is the power gain of the receiver in db?

3. The input and output resistances of an amplifier are 2.2 kΩ. What is the power gain in db of the amplifier when $E_{in} = 300\ \mu$v and $E_0 = 3$ v ?

4. A coupling circuit with an input of resistance of 2 kΩ, and an output resistance of 2 kΩ, has a −2.2 db gain. What is the output power when the input is 100 mw?

5. The power gain of an amplifier which delivers 6 w into 3.2 Ω is 62 db. What is the input voltage necessary to produce this output if the input resistance is 120 kΩ?

6. What input current is necessary to produce 6 w in Problem 5 if the input resistance is 20 kΩ?

7. What is the correction factor for a dbm measurement of 4 w across 300 Ω?

8. A phonograph pickup produces 20 mv across a 1-kΩ input. What power gain is necessary in an amplifier to deliver 30 w across an 8-Ω speaker?

9. A preamplifier produces 1 mw from a 200-nw input. What is the db gain if the input and output resistances are equal? What is the db gain if the input is 2.8 kΩ and the output is 3.3 kΩ?

10. The insertion loss of a filter circuit is 8 db. What is the power input if the output power is 2 w?

11. An amplifier is to be driven with a microphone rated at −68 db. What is the necessary power gain of the amplifier to deliver 10 w output?

12. The db gains of a system are: microphone −76 db, preamplifier +35 db, and power amplifier +48 db. What input power is necessary to deliver an output of 4 w?

13. An amplifier that has an input impedance of 1.1 kΩ and an output impedance of 40 Ω delivers 6-w output with an input of 50 μa. What is the db gain? What is the input power?

14. An amplifier with an input resistance of 60 kΩ, and an output of 100 Ω, has an input voltage of 3 v and an output voltage of 2.8 v. What is the db gain or loss?

15. With a reference of 1 mw, determine the output power for a gain of 68 db, 32 db, −15 db, and 22 db.

16. With a reference of 1 mw, determine the db gain for a power output of 3 w, 120 nw, 60 w, and 100 μw.

17. A field-effect transistor circuit has equal input and output resistances. What is the db gain when an input of 15 mv produces 18-v output?

18. A wide-band amplifier has an input resistance of 100 MΩ and an output resistance of 1 Ω. What is the db gain if the voltage gain is 1?

19. With the input signal level constant, the following outputs were measured from an audio amplifier: 20 Hz, 1 w, 100 Hz, 1.2 w, 400 Hz, 1.5 w, 2 kHz, 1.55 w, 1 kHz, 1.7 w, and 30 kHz, 1.3 w. What is the db gain or loss of each frequency in relation to 1 kHz?

20. Determine the ohms-per-volt rating of a meter movement in which the full-scale current is 40 μa.

21. What is the internal resistance of the meter in Problem 20 if the full-scale voltage is 0.1 v?

22. Two 2-MΩ resistors are connected in series across 20 v. What value of voltage will a 20,000-ohm-per-volt multimeter indicate on the 10-v scale when placed across one of the resistors?

23. A 100 μa, 1000-Ω meter movement is to be used as a voltmeter for the following voltage ranges: 1 v, 3 v, 10 v, 15 v, 50 v, 300 v, and 1,000 v. Calculate the multiplier resistance value for each range.

24. A multimeter with a 50 μa meter movement is designed to read 30,000 v on the 3-v scale with an external high-voltage probe. What is the value of the internal meter resistance and the probe resistance?

25. A 40-μa meter movement with an internal resistance of 2,000 Ω is to be used to measure the following full-scale currents: 1 ma, 5 ma, 10 ma, 25 ma, 100 ma, 250 ma, 500 ma, and 1 amp. What is the shunt resistance value for each scale?

CHAPTER 6

Amplifier Principles

6–1. Introduction

Engineers generally consider that the two most important circuit systems in electronic equipment are amplifiers and oscillators. In this chapter, we will consider basic principles of amplifier operation. A prototype amplifier is a device or a system of components which provides an output signal that is an enlarged reproduction of the essential characteristics of the input signal. In other words, a prototype amplifier is the most basic form of amplifier. As we proceed in our study of electronics, we will find that prototype amplifiers may be evolved into operational amplifiers, which perform mathematical operations on the input signal. Operational amplifiers find wide application in electronics technology and are second in importance only to prototype amplifiers.

The required power output from an audio amplifier ranges from a few milliwatts to 1 Mw or more in high-power electronic equipment.

Ordinary table-model radios have an audio output of about 1 w. A high-fidelity audio amplifier has a typical power output of 20 w. An amplifier is required to provide a comparatively large amount of audio output when the loudspeaker efficiency is low. In high-fidelity systems, a loud-speaker may be designed for optimum acoustical characteristics, at the cost of efficiency. Large exponential horns have maximum efficiency but do not provide a high-fidelity acoustical characteristic.

A prototype amplifier consists essentially of a vacuum tube with three or more elements, or a transistor, associated with suitable circuit components. Note in passing that devices other than vacuum tubes and transistors are utilized in some types of amplifiers to obtain an enlarged reproduction of the input signal. All amplifiers are *active* configurations; that is, all amplifiers contain a source of power, such as a battery or a rectifier-filter system. The fundamental principle of amplifier action is a "valving" of the source power in correspondence with the instantaneous amplitude of the input signal. It is for this reason that the English call a vacuum tube a valve. A high-fidelity transistor amplifier is illustrated in Figure 6–1(a).

Let us consider the single-stage vacuum-tube amplifier circuit depicted in Fig. 6–1(b). Input signal voltage is applied in series with the grid-cathode circuit and the grid-bias voltage supply. Changes in the grid-to-cathode voltage due to changes in the input signal voltage cause corresponding changes in plate-current flow. These plate-current changes flow through R_L, which is connected in series with the plate-cathode circuit and the plate-voltage supply. Resistor R_L, we recall, is termed the plate *load* resistor. The voltage drop across the plate-load resistor varies in accordance with the plate-current flow.

In most applications, the plate-load resistor has a fairly large resistance value; thus, its resistance value might be five times the internal (dynamic plate) resistance of the tube. Hence, the varying voltage drop across R_L is greater in magnitude than the varying input signal voltage. In turn, the varying input signal voltage is said to be *amplified.* This, of course, is voltage-amplification action. We may note in passing that other amplifier configurations can provide current-amplification action, or power-amplification action. However, we will restrict our present discussion to the basic principles of voltage amplification.

To formulate the voltage gain of a triode stage, we recall that the dynamic plate resistance of the tube (r_p), the plate-load resistance (R_L), and the voltage amplification factor of the tube (μ), are related as in Formula 6–1 at the top of page 168.

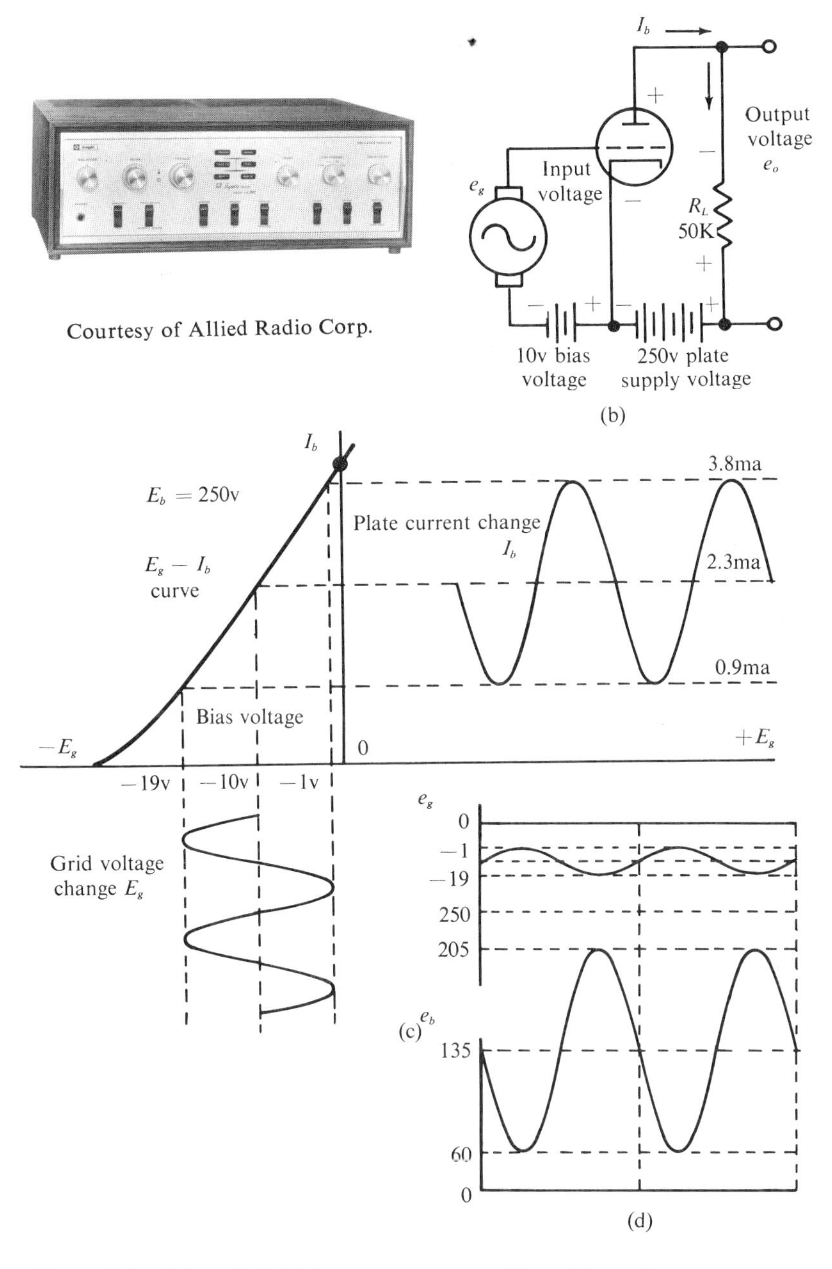

Figure 6–1. (a) A high-fidelity transistor amplifier; (b) Configuration of a simple vacuum-tube amplifier; (c) Visualization of amplifier action; (d) Comparative input and output signal voltages.

$$\text{Voltage gain} = A_v = \left(\frac{\mu R_L}{r_p + R_L}\right) \tag{6–1}$$

When e_g, the signal voltage applied to the grid, is multiplied by the gain figure A_v, the result expresses the number of times that e_g is amplified across R_L.

Example 1. Determine the voltage gain of the circuit in Fig. 6–1 when $\mu = 20$, $e_g = 1$, $R_L = 20\ \text{k}\Omega$, and $r_p = 10\ \text{k}\Omega$.

$$A_v = \frac{20 \times 20 \times 10^3}{10 \times 10^3 + 20 \times 10^3}$$
$$= 13.3 \qquad \text{(answer)}$$

Note that the voltage gain of the circuit (A_v) is always less than the voltage gain of the tube (μ). Figure 6–1(c) presents a visualization of amplifier action in terms of grid voltage versus plate current; Fig. 6–1(d) shows a comparison of grid-input and plate-output signal voltages.

R_L and r_p form a voltage divider, and the stage gain is increased when the value of R_L is increased, as depicted in Fig. 6–2. As R_L is increased, the voltage gain approaches the amplification factor of the tube. Of course, as the plate-load resistance is increased, the d-c plate current is correspondingly decreased for a fixed value of plate-supply voltage (250 v in Fig. 6–1). Since a triode is a nonlinear device, the amplification factor, dynamic plate resistance, and mutual conductance values vary with the d-c plate-current value as shown in Fig. 6–3. This simply means that the output waveform becomes more distorted as the signal voltage applied to the grid is increased.

We will observe that the grid of the amplifier tube is normally biased so that it remains negative with respect to the cathode, regardless of the input-voltage amplitude. This mode of operation is said to be in Class A when the angle of plate current flow is 360°. As long as the grid is negative with respect to the cathode, the grid does not attract electrons. When no electrons are attracted by the grid, no current flows in the grid circuit. Under this condition, the grid consumes virtually no power. But, if the grid should be driven positive with respect to the cathode, grid current then flows and the grid circuit consumes appreciable power.

In the analysis of amplifier principles, knowledge that an amplifier produces a large output voltage when a small input voltage is applied is necessary, but often is not sufficient. It is also important to know just how much and how faithfully the input voltage is increased by

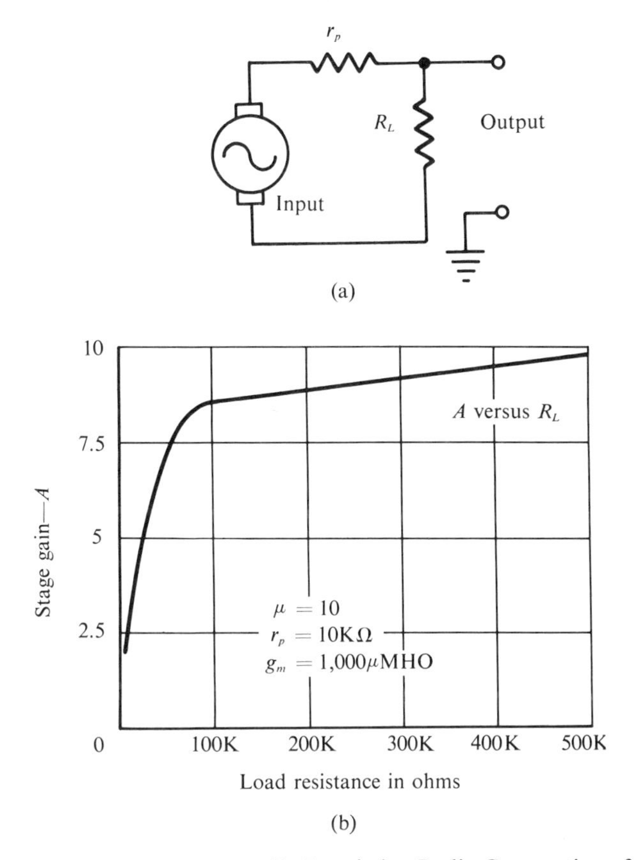

By Permission Radio Corporation of America

Figure 6-2. (a) Equivalent circuit of a vacuum-tube amplifier; (b) Stage gain versus load-resistance value.

amplification action. Therefore, let us briefly review the use of characteristic curves, such as shown in Fig. 6-4. Appropriate characteristic curves which can be used for calculating voltage amplification in vacuum-tube amplifiers are the grid-voltage/plate-current characteristics, and the plate-voltage/plate-current characteristics. Although we will start our discussion with respect to the grid-voltage/plate-current characteristics, we will learn subsequently that it is more practical to use the plate-voltage/plate-current characteristics, since it reduces the number of required steps in calculation.

Notice the grid-voltage/plate-current curve in Fig. 6-1(c) below the amplifier circuit in Fig. 6-1(b). This curve shows us the plate-

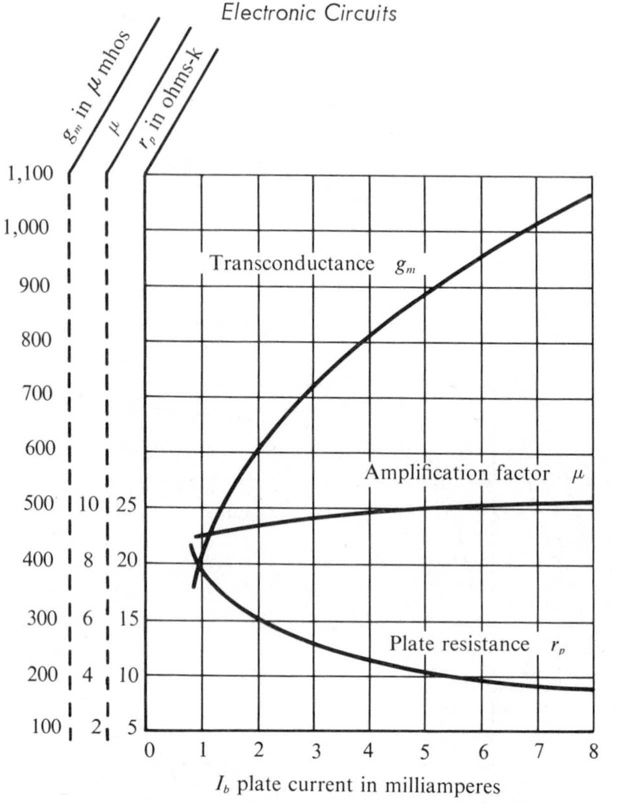

Figure 6–3. Variation of dynamic plate resistance, amplification factor, and mutual conductance versus d-c plate current for a typical triode.

current value for any grid-voltage value within the useful range of tube operation. Let us assume, for example, that the input voltage to the grid of the tube is a sine wave with a peak-to-peak amplitude of 18 v (about 6.3 v rms), and that the bias voltage applied to the tube is -10 v. In turn, the average grid voltage will be -10 v over a complete cycle of operation. We say that the operating point of the tube is -10 v.

To determine the plate-current flow for a chosen value of grid voltage in Fig. 6–1(c), we follow a vertical projection from this value of grid voltage up to the $E_g - I_b$ curve, and then move horizontally to the plate-current axis. This point of intersection denotes the value of plate current which corresponds to our chosen value of grid voltage. Observe from the curve that as the grid voltage varies from -1 to -19 v, the resulting plate-current flow varies from 3.8 ma to 0.9 ma. The average plate-current flow is 2.3 ma. At the no-signal condition (when no signal voltage is applied to the grid) the value of plate-

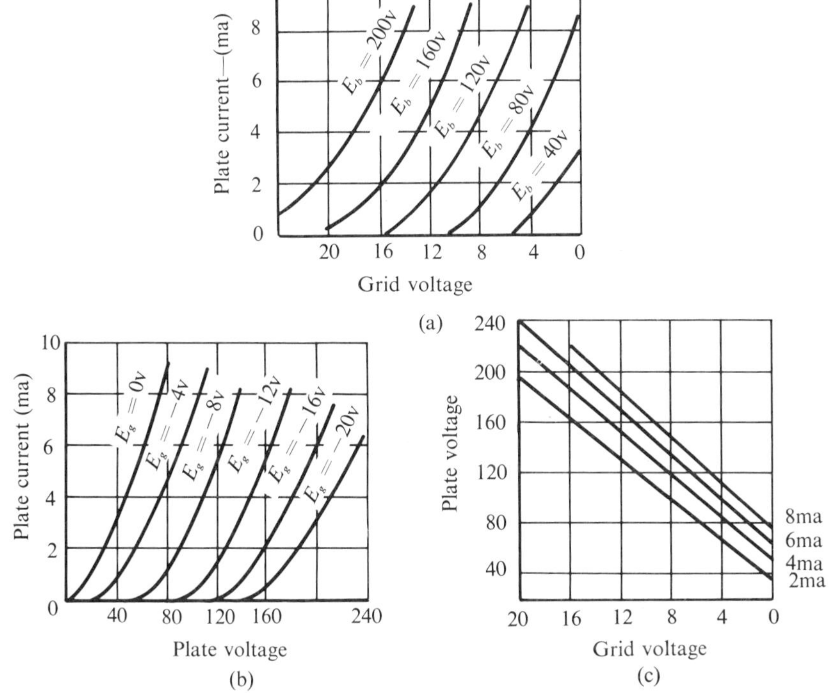

Figure 6–4. Characteristic curves are used to calculate the gain of an amplifier stage. (a) E_g — I_b curves of a triode vacuum tube; (b) E_b — I_C curves of a triode vacuum tube; (c) E_b — E_g curves of a triode vacuum tube.

current flow is likewise 2.3 ma. This is a basic characteristic of amplification in Class *A*.

The voltage "swing" across the plate-load resistor represents the useful output signal voltage from the amplifier. We can calculate the voltage output by multiplication of the plate-load resistance and the plate-current values that correspond to chosen grid-voltage values along the curve in Fig. 6–1(c). For example:

1. Grid potential — −1 v

$$E_o = I_b R_L = 3.8 \text{ ma} \times 50 \text{ k}\Omega = 190 \text{ v}$$

2. Grid potential = −10 v

$$E_o = 2.3 \text{ ma} \times 50 \text{ k}\Omega = 115 \text{ v}$$

3. Grid potential = −19 v

$$E_o = -0.9 \text{ ma} \times 50 \text{ k}\Omega = 45 \text{ v}$$

These calculations have not taken the plate-supply voltage into account. Note that any voltage drop across the plate-load resistor will subtract from the 250 v of the plate-supply voltage, and thus change the actual output voltage (E_o) of the amplifier stage. The amplitude of the IR_L drop is the same as that of the output voltage. Nevertheless, to find the exact value of E_o, add the 250 v to the voltage drop across the plate-load resistor with reference to the previous calculations as follows:

1. $-190 + 250 = 60$ v
2. $-115 + 250 = 135$ v
3. $-45 + 250 = 205$ v

Observe the curves in Fig. 6–1(d), which show the grid- and plate-voltage curves for the values chosen in the example. These curves depict correct amplitude and time relationships. Note that the input voltage (e_g) varies from -1 to -19 v; that is, the variation of the input signal is 18 v. The plate-current/grid-voltage curve represents the voltage variation at the top (plate end) of resistor R_L. Its range is from 60 to 205 v, or the total variation in output voltage is $205 - 60 = 145$ v. From the foregoing discussion, it is evident that the output voltage has considerably greater amplitude than the input voltage. The ratio between these two voltages is called the voltage *gain* of the amplifier stage.

$$A_v = \frac{e_o}{e_{\text{in}}} \qquad \textbf{(6–2)}$$

Example 2. Determine the voltage gain of the circuit in Fig. 6–1(b).

$$\begin{aligned} A_v &= \frac{e_o}{e_{\text{in}}} \\ &= \frac{145}{18} = 8.1 \qquad \text{(answer)} \end{aligned}$$

The power output of an amplifier may be formulated

$$\begin{aligned} P_o &= \frac{(I_{\text{max}} - I_{\text{min}})}{2\sqrt{2}} \cdot \frac{(V_{\text{max}} - V_{\text{min}})}{2\sqrt{2}} \\ &= \frac{I_{pp}\,V_{pp}}{8} \qquad \textbf{(6–3)} \end{aligned}$$

Example 3. Determine the power output of the amplifier represented in Fig. 6–1(b) when

$$V_{pp} = 145 \text{ v}$$

$$I_{pp} = \frac{V_{pp}}{R_L} = \frac{145}{50 \times 10^3} = 2.9 \text{ ma}$$

$$P_o = \frac{145 \times 2.9 \times 10^{-3}}{8} = 525 \text{ mw}$$

We conclude that this stage provides a fairly large voltage gain at a rather low power level. Since the circuit action entails appreciable voltage gain at a low power level, the configuration is called a voltage amplifier. Nevertheless, the power gain of the circuit is very large. Note that the grid circuit is practically an open circuit. The current flow in the circuit is extremely small at audio frequencies because the small series capacitance presented by the grid and cathode permits virtually zero current flow, and this extremely small current encounters practically zero resistance through the grid-circuit components. Therefore, the grid consumes virtually zero driving power, while the plate circuit provides a power level of approximately 54 mw.

Similarly, the current gain of the amplifier stage depicted in Fig. 6–1(c) is very large, in spite of the fact that the current level in the output (plate) circuit is comparatively small. Although virtually zero current flows in the grid circuit, the plate current swings from 3.8 ma to 0.9 ma. Thus, the current gain is very large. Power gain and current gain are formulated in the same manner as voltage gain.

$$\text{Current gain} = \frac{I_{\text{out}}}{I_{\text{in}}} \qquad \textbf{(6–4)}$$

where I_{in} denotes the peak-to-peak value of the current flow in the grid circuit, and I_{out} denotes the peak-to-peak value of the current flow in the plate circuit.

$$\text{Power gain} = \frac{V_{\text{out}} I_{\text{out}}}{V_{\text{in}} I_{\text{in}}} \qquad \textbf{(6–5)}$$

Note that the amplifier stage *reverses* the phase of the signal from the grid circuit to the plate circuit. Figure 6–5 illustrates the waveforms for an amplifier stage in which the grid signal is positive-going from −4 to 0 v; over this period, the plate current waveform is positive going from 12.75 ma to 18.25 ma, and the plate voltage is negative going from 198 v to 154 v. Thus, e_b is reversed in phase with respect to e_g.

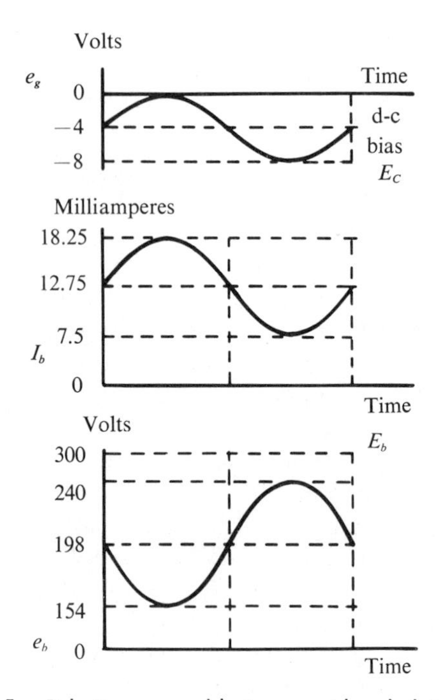

Figure 6-5. Polarity reversal between grid and plate voltages.

6-2. Transistor Characteristic Curves

Vacuum tubes, junction transistors, and field-effect transistors all function as electronic amplifying devices, even though they differ greatly in their basic mechanism of operation. The amplifying action of the vacuum tube is based upon the control of electrons through a vacuum by a control grid. The amplifying effect of the junction transistor is based upon the control of minority carriers through a semiconductor solid by a base current. Finally, the amplifying effect of the field-effect transistor is based upon the control of majority carriers through a semiconductor solid by a gate voltage. Figure 6-6 compares the basic vacuum-tube amplifier circuit to the basic junction transistor and field-effect transistor circuits.

In most applications, where the junction transistor is used in an amplifier, it is operated with the input circuit (base to emitter) forward biased, and the output circuit (collector to emitter) reverse biased as shown in Fig. 6-6.

Transistor characteristics are similar to those of vacuum tubes except they are more complicated. This complication is caused by the

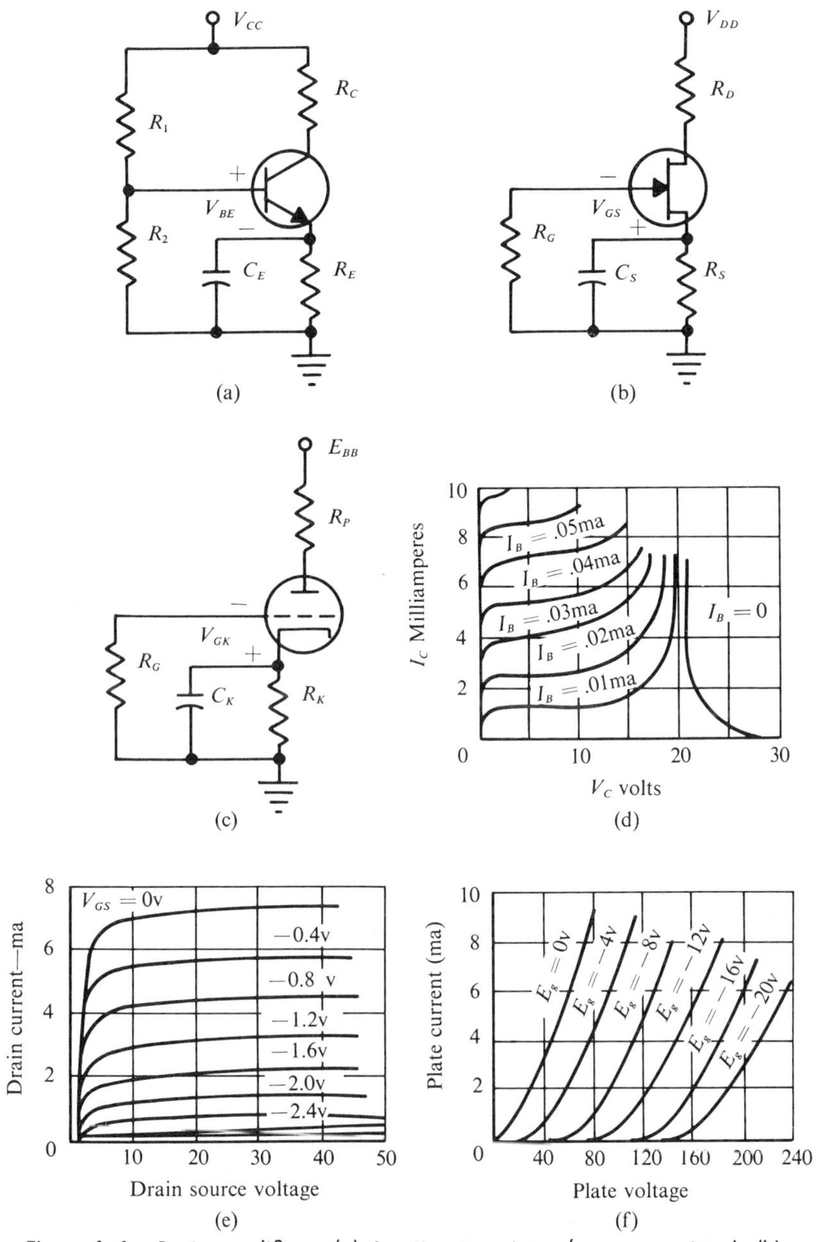

Figure 6–6. Basic amplifiers. (a) Junction transistor (common emitter); (b) Field-effect transistor (common source); (c) Vacuum tube (common cathode); (d) Junction-transistor characteristics; (e) Field-effect transistor characteristics; (f) Vacuum-tube characteristics.

input current flow which cannot be neglected due to its substantial value. As indicated in Fig. 6–7(a), a very small change in base voltage causes a large change in base current. Therefore, a value of base current is much easier to set accurately than a value of base voltage.

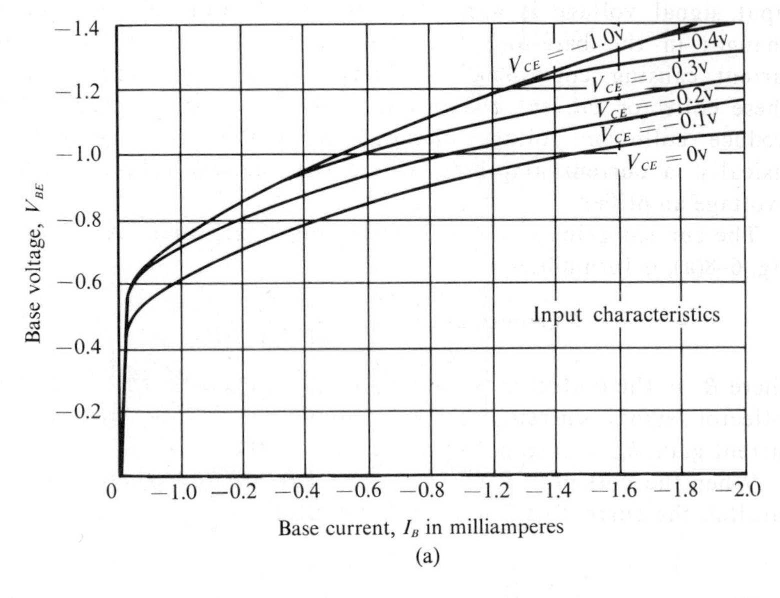

(a)

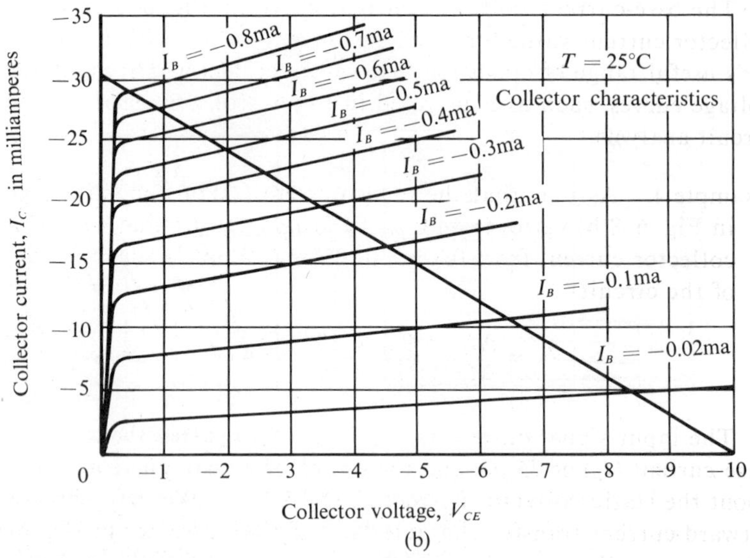

(b)

Figure 6–7. **(a) Input characteristics; (b) Collector characteristics for a junction transistor.**

For this reason, the base current (I_B) is normally taken as one of the parameters for analysis of the transistor [Fig. 6–7(b)] rather than base-to-emitter voltage (V_{BE}).

Consider the single stage transistor amplifier depicted in Fig. 6–8(a). Input signal voltage is applied across the base-to-emitter junction. Changes in the base-to-emitter voltage produce changes in base current causing corresponding changes in collector current flow. These collector current changes through the collector resistor (R_C) produce collector voltage changes. Note that the transistor is basically a current amplifier, where the vacuum tube is basically a voltage amplifier.

The current gain of a single transistor stage, such as shown in Fig. 6–8(a), is formulated

$$\text{Current gain} = \frac{i_C}{i_b} = \frac{h_{fe}}{(1 + h_{oe})R_L} \tag{6–6}$$

where R_L = the collector resistor R_C in parallel with the load, i_c = collector signal current, i_b = base signal current, h_{fe} = transistor current gain, h_{oe} = internal transistor admittance.

When the collector load (R_C and load) are less than 10 kΩ in parallel, the current gain may be formulated

$$A_i \cong h_{fe} \cong \beta \tag{6–7}$$

The base-current/collector-current curve in Fig. 6–8(c) shows the collector current value for any base current value within the transistor's useful range of operation. However, collector-current/collector-voltage curves, such as illustrated in Fig 6–8(b), are usually used for circuit analysis.

Example 4. Assume that the input current (I_B) on the curve depicted in Fig. 6–8(b) varies from 0 μa to 210 μa and produces a change of collector current from 0.3 ma to 9.8 ma. What is the current gain of the circuit?

$$A_i = \frac{\Delta I_C}{\Delta I_B} = \frac{9.5 \times 10^{-3}}{210 \times 10^{-6}} = 45.2 \qquad \text{(answer)}$$

The input signal current (i_b) in Fig. 6–8(b) varies about the static base current (I_B) of 75 μa, and the collector signal current (i_c) varies about the static collector current (I_C) of 5.2 ma. We may observe the forward-current transfer characteristic of the transistor in Fig. 6–8(c). The curves illustrate that the transistor as the vacuum tube is a nonlinear amplifier.

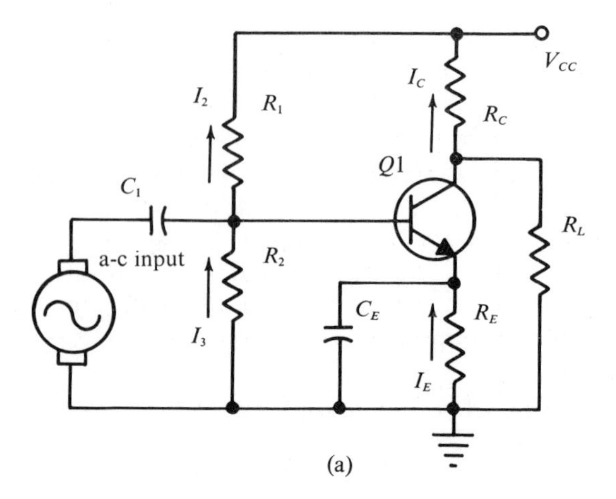

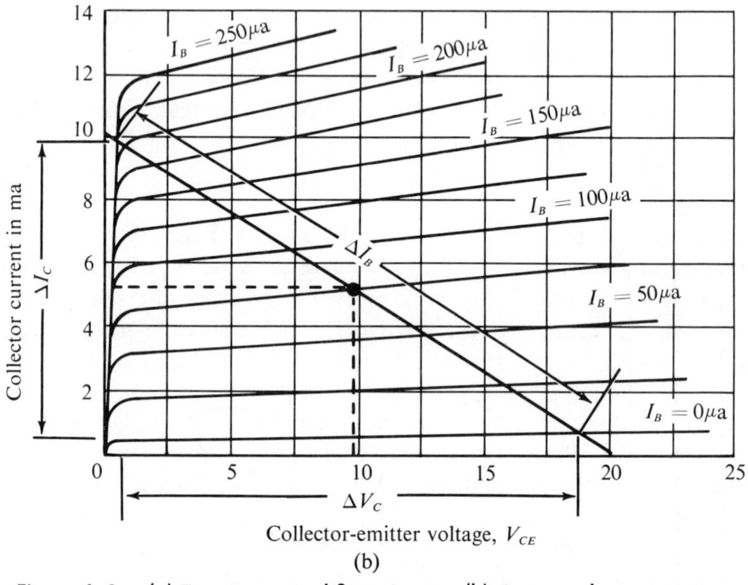

Figure 6–8. (a) Transistor amplifier circuit; (b) Output characteristics.

As may be expected from our discussion of vacuum-tube operation, the voltage and current gain of a transistor amplifier varies with the value of the load resistance. The current gain of a stage approaches the current gain of the transistor as the value of the load resistance decreases, as illustrated in Fig. 6–9 where the internal resistance of the transistor is approximately 30 kΩ. The voltage gain of a transistor amplifier with a load of less than 10 kΩ may be formulated as in Formula 6–8 at the top of page 180.

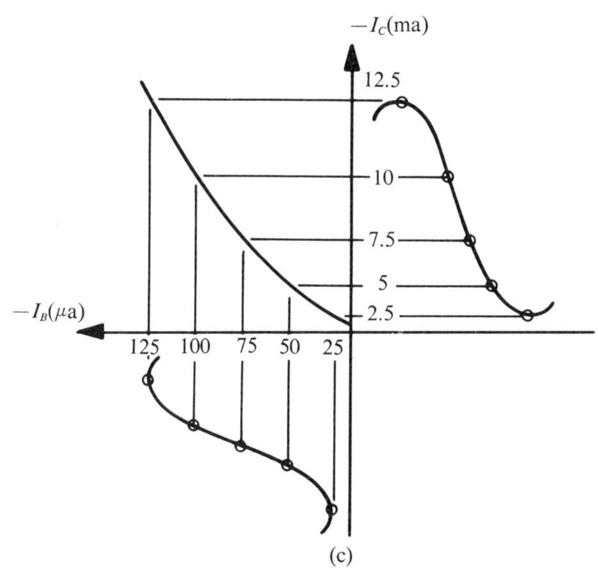

Figure 6-8. (c) $I_B - I_C$ characteristic curve.

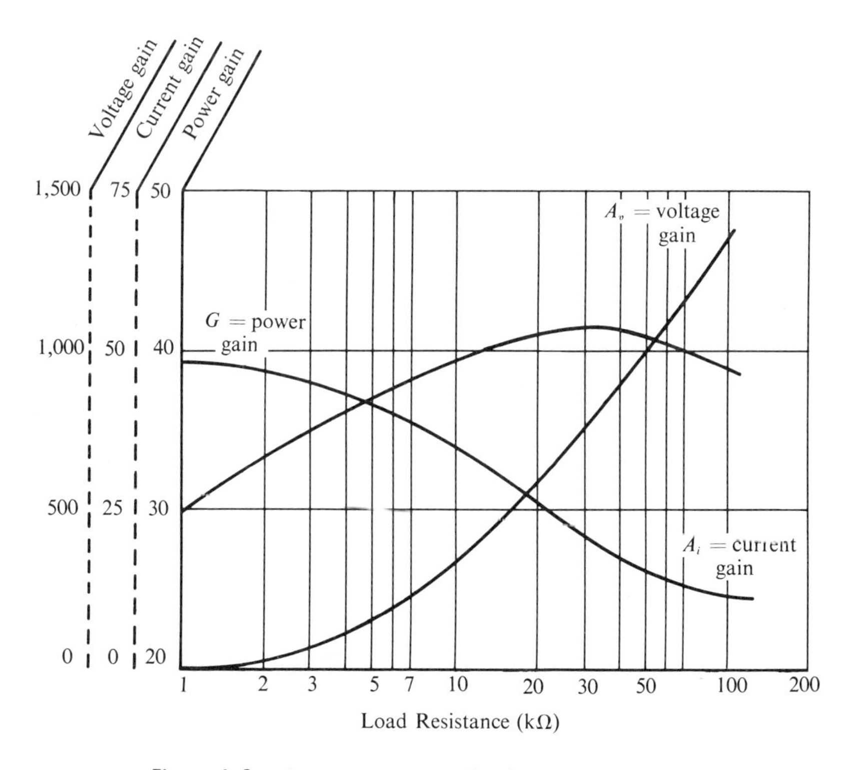

Figure 6-9. Junction transistor load-gain relationships.

$$A_v \cong A_i \frac{R_L}{h_{ie}} \tag{6-8}$$

where h_{ie} is the transistor input impedance.

Junction transistors inherently have much lower input and output impedances than vacuum tubes and are more sensitive to changes of temperature. However, they have advantages over vacuum tubes in their small size, longer life, greater efficiency, and ruggedness.

6-3. Field-Effect Transistors

Field-effect transistors (FET) have the advantages of both junction transistors and vacuum tubes in that they have high input and output impedance, small size, ruggedness, long life, and efficiency. In fact, the characteristics of the FET so closely match those of the vacuum tube that the design and circuit analysis formulas are similar for both devices. However, the mechanism of operation for the FET differs greatly from both that of the vacuum tube and the junction transistor. The FET is a majority carrier device which operates on the principle of the control of majority carriers through a semiconductor solid by the field established by a reverse-biased junction. On the other hand, the junction transistor operates on a principle of the control of minority carriers by a forward-biased junction, and the vacuum tube operates on a principle of the control of electrons by an electrostatic field in a vacuum.

The basic elements of the FET are the drain (D), the source (S), and the gate (G), corresponding to the plate, the cathode, and the control grid of the vacuum tube. Basically, the FET, illustrated in Fig. 6-10, consists of a semiconductor bar with a source attached to one end and the drain attached to the other end. The two junctions located at the center of the bar are called the gate, and the area between the gates is called the channel. FET's are classified as N-channel or P-channel depending upon the type of semiconductor material used in the construction of the bar.

Application of a voltage between the source and the drain causes carriers to flow through the resistance of the semiconductor bar. These carriers are controlled by application of a reversed bias between the gate and the source. This reverse bias causes a field which forms a depletion region around each gate as shown in Fig. 6-11. The depletion region has the effect of decreasing the cross-sectional area of the channel, thereby decreasing the source-to-drain conductance. The penetration of the depletion region into the channel follows a

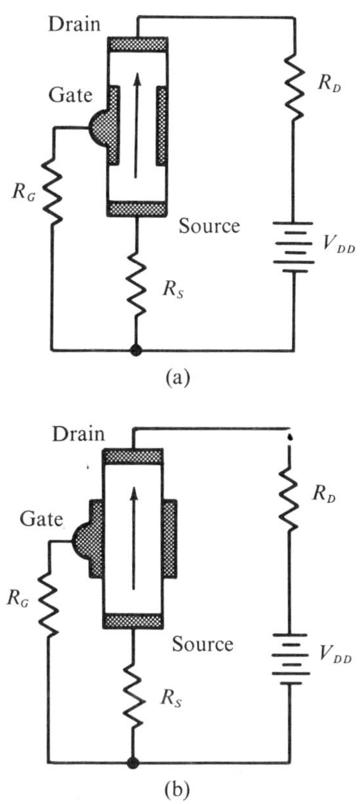

Figure 6–10. Schematic representation of a field-effect transistor. (a) Junction field-effect transistor (JFET); (b) Isolated-gate field-effect transistor (ISFET).

square-law characteristic. When the reverse bias on the gate causes the depletion layers to touch, as depicted in Fig. 6–11(c), the channel is said to be pinched off.

The characteristic curves for an FET are shown in Fig. 6–12. Area (a) is the region in which the device operates as a variable resistance, with the output current being controlled by the source to drain voltage. Area (b) is called the pinch-off point or saturation region, and it is the edge of the region where an increase in the source-to-drain voltage does not increase the drain current. Beyond the pinch-off region in area (c), the FET operates as a constant-current device, in which the drain current is controlled only by the gate bias voltage. The pinch-off voltage (V_p) can be deduced in two ways from the drain characteristics in Fig. 6–12: either as the voltage at which the zero gate-to-drain voltage (V_{GS}) curve saturates to a constant pinch-off cur-

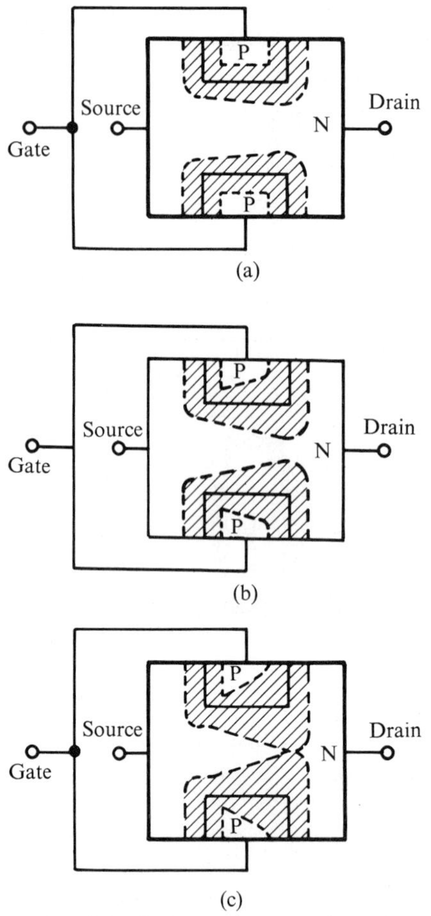

Figure 6–11. Schematic of the depletion region as the gate voltage increases. (a) Zero gate bias; (b) Medium gate bias. (c) Gate bias pinch-off.

rent called I_{DSS}, or the lowest V_{GS} value for which the drain current is shut off.

The expression for the voltage gain of the FET voltage amplifier depicted in Fig. 6–13 on page 184 is formulated

$$A_v = \frac{\mu R_L}{r_{ds} + R_L} \quad \text{or,} \quad \frac{g_m r_{ds} R_L}{r_{ds} + R_L} \tag{6–9}$$

where r_{ds} = dynamic drain to source resistance of the FET, μ = the voltage amplification factor of the FET, g_m = the transconductance of the FET. Then,

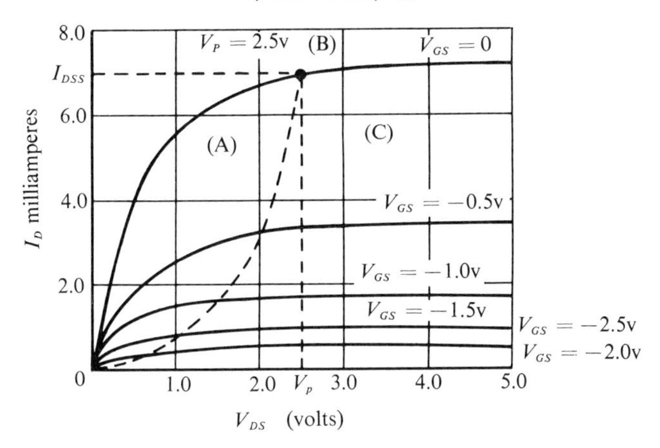

Figure 6–12. Drain characteristics of a field-effect transistor in the low-voltage region. (a) Ohmic region; (b) Pinch-off point; (c) Pentode region.

$$\mu = g_m r_{ds}$$

and

$$r_{ds} = \frac{v_{ds}}{i_{ds}}$$

If $r_{ds} >> R_L$, the expression for voltage gain at low frequencies becomes

$$A_v = g_m R_L \tag{6–10}$$

The forward-current transfer ratio (transconductance) is the ratio of output current versus input voltage with the output voltage held constant.

Example 5. Determine the voltage gain of the amplifier in Fig. 6–13 when $g_m = 1{,}000\ \mu$mho.

$$\begin{aligned} A_v &\simeq g_m R_L \\ &\simeq 1 \times 10^{-3} \times 15 \times 10^3 \\ &\simeq 15 \qquad \text{(answer)} \end{aligned}$$

The bias resistor (R_S) for the FET is determined as the bias resistor for the vacuum tube.

$$R_S = \frac{V_{GS}}{I_S} \tag{6–11}$$

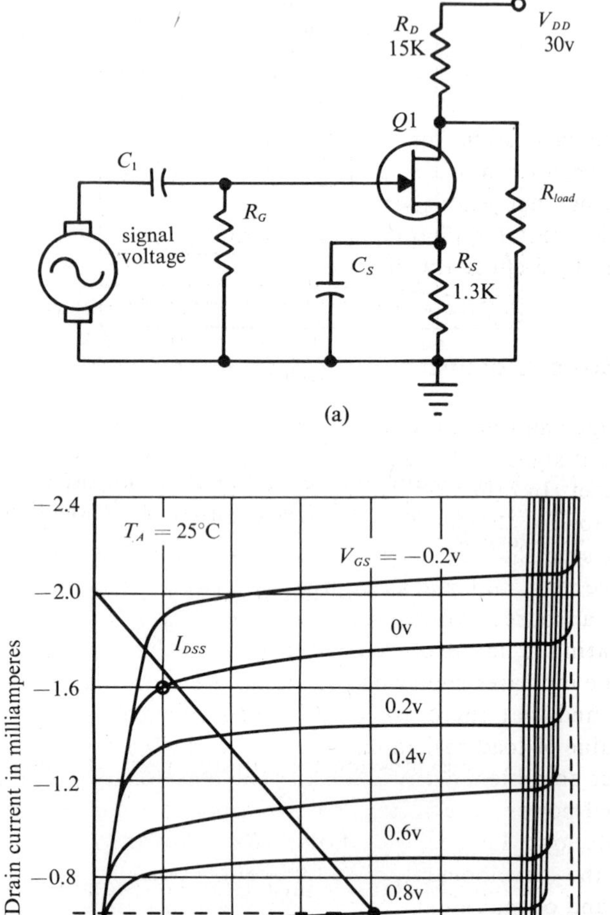

Figure 6–13. FET voltage amplifier. (a) Schematic; (b) Output characteristics with load line drawn.

Example 6. Determine the value of the source resistor in Example 5 to establish a bias of 0.8 v on the load line in Fig. 6–13(b).

$$R_S = \frac{0.8 \text{ v}}{0.6 \text{ ma}}$$
$$= 1.5 \text{ k}\Omega \qquad \text{(answer)}$$

As we have seen, junction transistors, vacuum tubes, and field-effect transistors may all be utilized as amplifying devices even though the internal operations are somewhat different. We will return to this subject in later chapters and introduce other types of vacuum tubes and field-effect transistors.

6-4. Basic Classification of Amplifiers

Voltage amplifiers are primarily intended to increase the amplitude of an input signal voltage and usually operate at a comparatively low power level. In other words, the design of a voltage-amplifier stage is directed toward the development of maximum output signal voltage. We have seen that to accomplish this objective, the value of the load resistance must be made as high as possible. It is also desirable to use as high a source voltage as is practical, provided the amplifying device rating is not exceeded. (Recall that the amplification factor of a triode increases somewhat as the plate-supply voltage is increased.) We will find that there are practical considerations which limit the upper value of load resistance that may be utilized.

Power amplifiers are primarily intended to provide maximum output signal power; the voltage gain of a power amplifier may be comparatively low. That is, the design of a power amplifier is directed toward the development of a maximum *product* of output signal current and output signal voltage. In a vacuum tube, this maximum product occurs when the plate-load resistance has a value equal to the dynamic plate resistance of the tube. We recall the maximum power-transfer theorem, illustrated in Fig. 6-14. Of course, the efficiency of a power amplifier is less than that of a voltage amplifier. When the plate-load resistance is selected for maximum power output, half of the d-c energy from the power supply is dissipated as heat within the power-amplifier tube.

Note carefully that a power-amplifier stage seldom works into a load resistor as such. Instead, the stage works into an *effective* resistance, as exemplified by a loudspeaker. The d-c resistance of the speaker is small as indicated by an ohmmeter measurement. On the other hand, when a-c signal voltage is applied to the speaker, power is demanded to move the air surrounding the diaphragm or cone. Thus, a speaker has an *effective* resistance to an a-c signal, and this effective

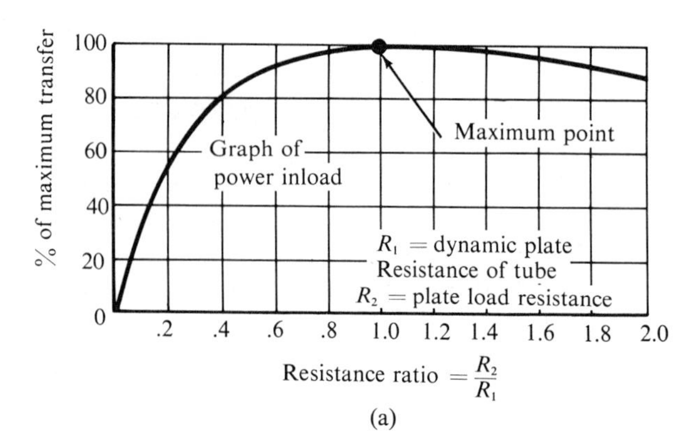

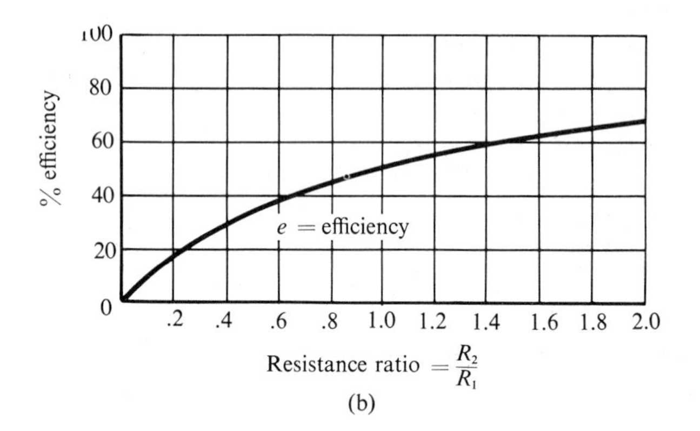

Figure 6-14. (a) Maximum power is developed when the plate-load resistance equals the internal resistance of the tube; (b) System efficiency is 50 per cent at maximum power transfer.

resistance value has no direct relation to the d-c resistance of the speaker.

Current amplifiers are primarily intended to provide large output signal currents. Thus, the voltage gain of a current-amplifier stage is often less than unity. Maximum power is seldom developed by a current-amplifier stage. When a vacuum tube is used as a current amplifier, a cathode-follower configuration is commonly utilized, as explained subsequently. When a transistor is used as a current amplifier, an analogous emitter-follower configuration is commonly utilized. It is instructive to compare the voltage gain, current gain, and power gain values that are typically obtained in the three basic transistor-amplifier stage configurations, as tabulated in Table 6-1.

TABLE 6–1
Comparative Voltage, Current, and Power Gains Figures for Transistor-Amplifier Stages

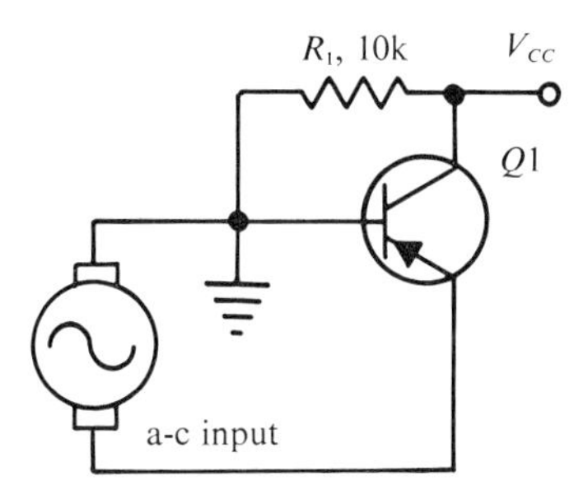

Voltage gain: 380 times
Current gain: 0.98
Power gain: 26 db
Input resistance: 35 Ω
Output resistance: 1 MΩ

Common-base stage
Generator internal resistance is equal to 1 kΩ

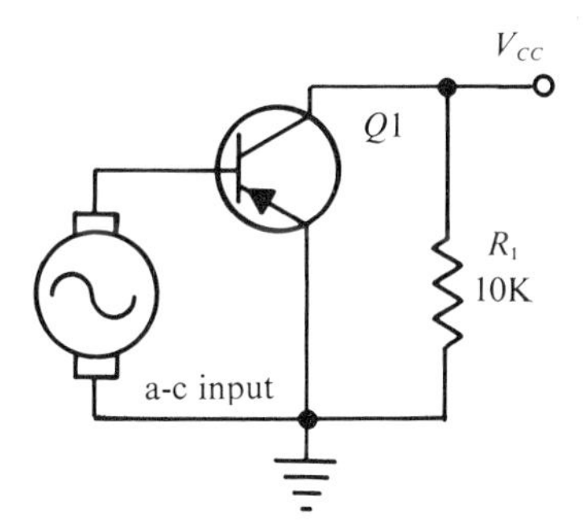

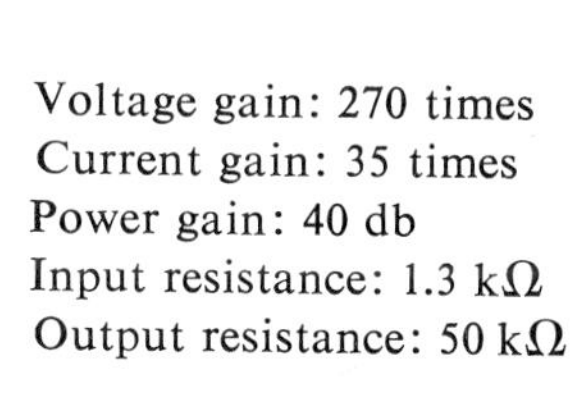
Voltage gain: 270 times
Current gain: 35 times
Power gain: 40 db
Input resistance: 1.3 kΩ
Output resistance: 50 kΩ

Common-emitter stage
Generator internal resistance is equal to 1 kΩ

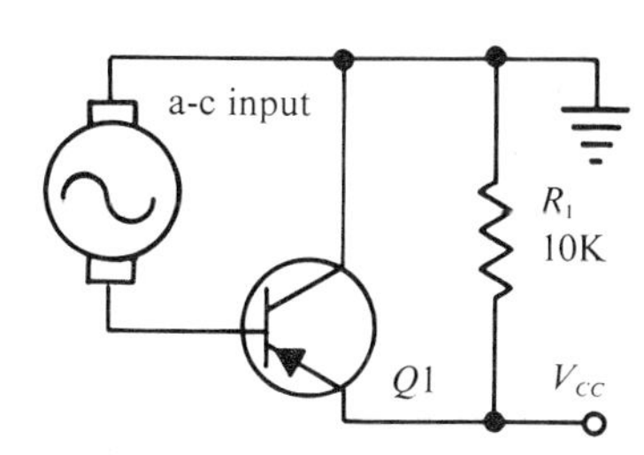

Voltage gain: 1
Current gain: 36 times
Power gain: 15 db
Input resistance: 350 kΩ
Output resistance: 500 Ω

Common-collector (emitter-follower) stage
Generator internal resistance is equal to 1 kΩ

We recall that a Class-*A* amplifier stage utilizes a bias value with respect to the input signal voltage, such that output current flows continuously throughout the cycle of the input signal and never reaches zero (except perhaps as an instantaneous point). Class-*A* amplifiers are characterized by low efficiency (ratio of signal-power output to d-c power input). For practical purposes, the efficiency of a Class-*A* amplifier stage ranges between 20 per cent and 25 per cent. Theoretically, it has a maximum efficiency with a resistive load of 25 per cent. However, in a vacuum tube, the power consumed by the heater in the tube reduces the overall efficiency, and distortion with any of the amplifying devices becomes objectionable if the Class-*A* stage is driven to maximum output.

Class-*B* amplifiers are biased approximately to output-current cutoff. In a vacuum-tube amplifier, plate current flows only during the positive half cycle of the input signal voltage. The d-c current consumption is less than that of Class-*A* stage. Power loss in a Class-*B* stage is low, because plate current does not flow when no input signal voltage is applied. In turn, little power is wasted during non-operating periods. Moreover, plate current flows only during the positive half of the input signal cycle, which means that the average current flow is only 32 per cent, approximately, of the peak current value in the tube. Figure 6–15 shows the relation between grid voltage and plate current in a tube that is operated in Class *B*. Note that

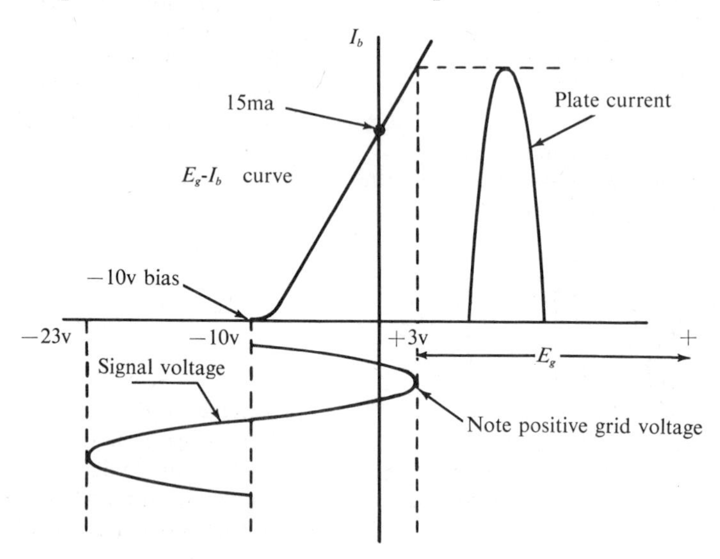

Figure 6–15. Waveforms in a Class-*B* amplifier stage.

the plate current flows only during the positive half cycle of the signal voltage. Grid current flows only during the time that the grid is positive. The maximum theoretical efficiency of a Class-*B* stage is 78 per cent.

Class-*B* amplifier stages are used chiefly as power amplifiers. Their power output is proportional to the square of the grid-input signal voltage. The best bias value for a Class-*B* stage is that which corresponds approximately to the cutoff bias value that would be obtained if the main part of the characteristic curves shown in Fig. 6–16 were

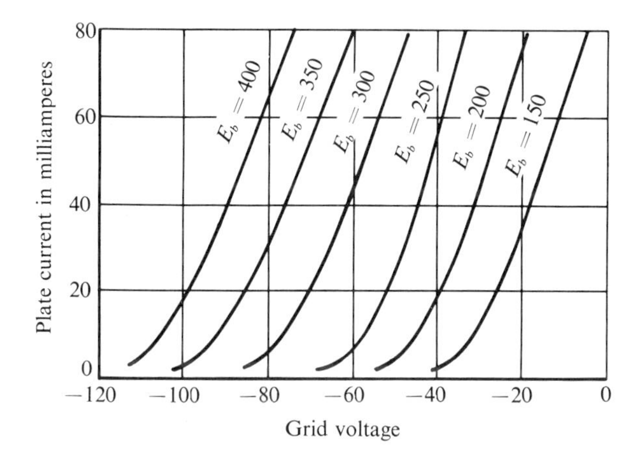

Figure 6–16. **Determination of correct bias voltage for a Class-*B* amplifier stage.**

projected as straight lines. Note the dotted line which extends from the straight-line part of the characteristic curve marked $E_b = 300$. The point at which this line intersects the grid-voltage axis (approximately −75 v) denotes the cutoff bias for 300-v plate potential. Curves of this type provide a convenient means of determining grid-bias values for various plate-supply voltages.

Distortion occurs in Class-*B* amplifier stages in the same general manner as in Class-*A* stages. Harmonic distortion in the output signal of a Class-*B* stage which has a normal value of load resistance depends upon the departure of the characteristic curves from a straight line, and also upon the operating point which is chosen for the tube. With reference to Fig. 6–15, the half-cycle output waveform is evidently quite different from a distortionless reproduction of the full-cycle input waveform. One-half of the original cycle is missing. In practice, this missing half cycle is recovered by one of two methods: an addi-

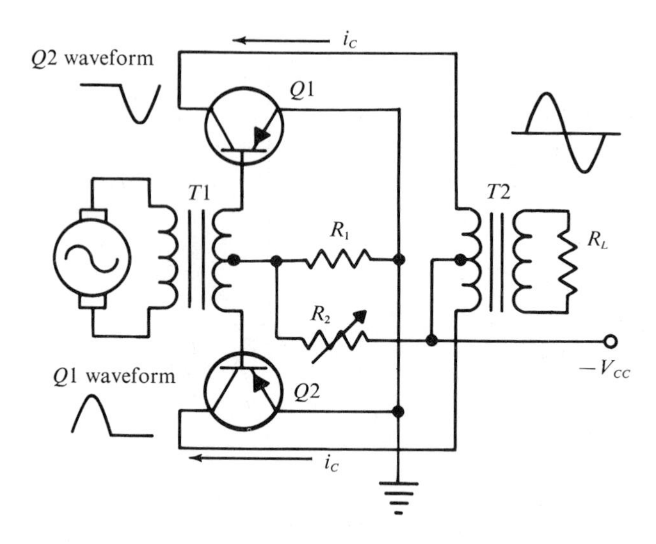

Figure 6–17. Class-*B* transistor amplifier; each transistor operates in push-pull for one-half cycle to supply the full cycle of output.

tional Class-*B* stage may be used to provide the missing half cycle (see Fig. 6–17), or the *flywheel action* of a resonant plate-load circuit may be employed. These elaborated configurations are analyzed subsequently.

A Class-*C* amplifier stage is one in which the grid-bias value is appreciably greater than the cutoff value. When no a-c input signal voltage is applied to the grid, the plate current is zero. With an a-c drive signal present, plate current flows for less than one-half cycle; the peak value of the grid-drive signal must exceed the grid-bias value before plate current can start to flow. Except for the comparatively high grid-bias voltage utilized, a Class-*C* amplifier stage operates in much the same way as a Class-*B* stage. Output waveform distortion is greater, because plate current flows for less than one-half cycle, as depicted in Fig. 6–18. A Class-*C* amplifier is also characterized by its comparatively low ratio of output power to input power, or power amplification. The grid of a Class-*C* stage is normally driven positive, and the tube is driven into plate-current saturation (see Fig. 6–19). In turn, harmonic distortion is high in the output signal.

In most applications, means are employed in a Class-*C* stage to minimize harmonics in the output signal; that is, circuit means are included in the plate-load circuit to restore the sine waveform. A common arrangement is seen in Fig. 6–20. The tuned-tank circuit, as we know, has maximum impedance at resonance, as shown in Fig.

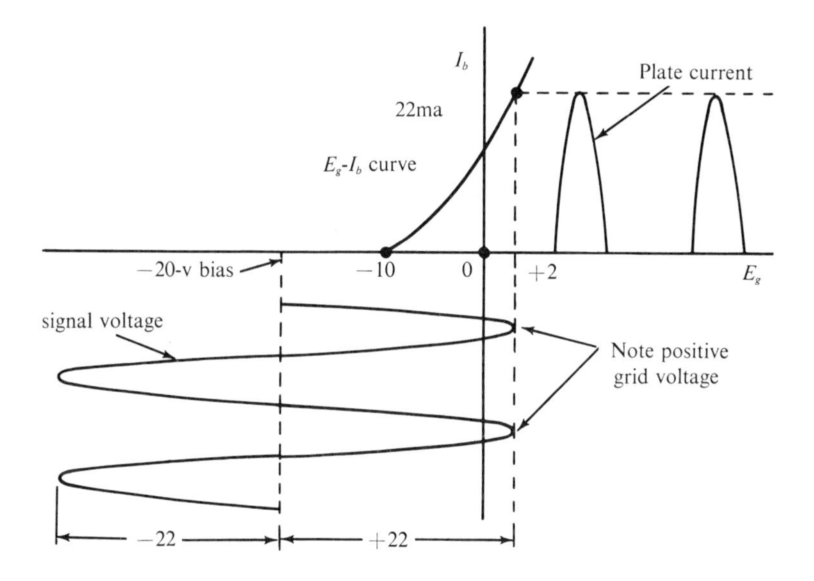

Figure 6–18. Waveforms in a Class-*C* amplifier stage.

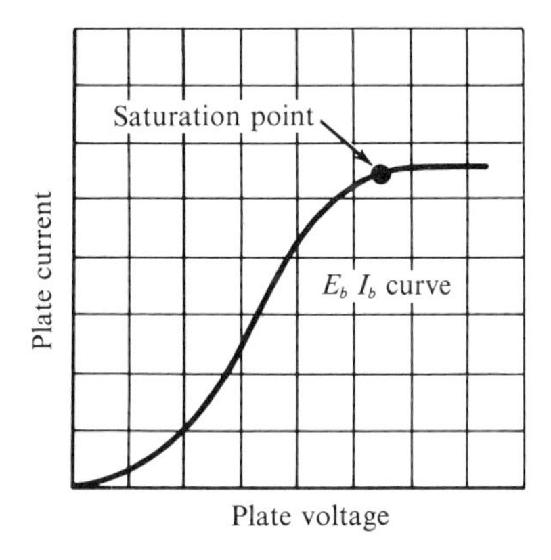

Figure 6–19. Saturation point on an F_b I_b curve.

6–21 on page 194. Accordingly, the resonant-tank circuit tends to bypass the harmonic components in the output signal. Furthermore, we recall that the power factor of an *LCR* parallel-resonant circuit is unity only at its resonant frequency. Although the tuned tank is pulsed by the Class-*C* amplifier, the flywheel effect of the resonant circuit develops a virtually pure sine waveform, provided its *Q* value is comparatively high ($Q \geq 10$).

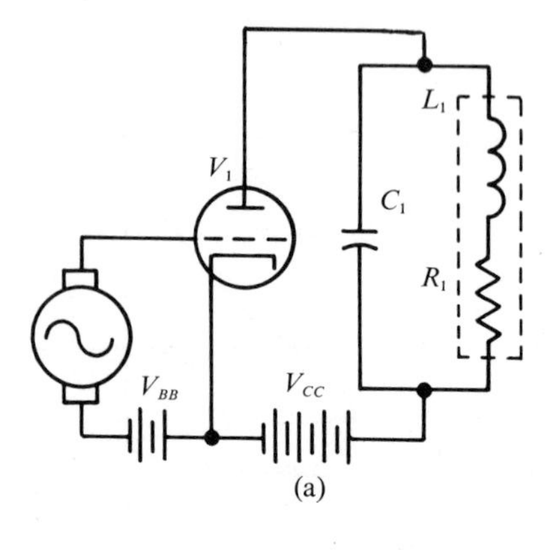

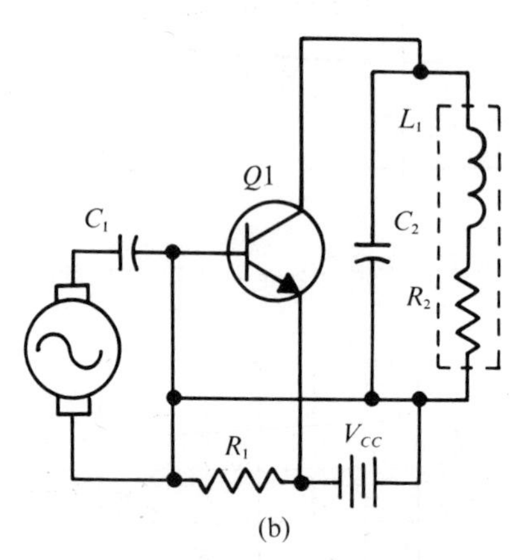

Figure 6–20. An *LCR* resonant circuit provides harmonic suppression and good output waveform. (a) Vacuum-tube circuit; (b) Transistor circuit.

It is evident that the high impedance and unity power factor of the tuned-output tank, when tuned to the input-signal frequency, provide much greater amplification of this fundamental-frequency signal than of harmonic-frequency signals (see Table 6–2). A Class-*C* amplifier has high efficiency because when plate current is permitted to flow, the instantaneous potential of the plate is low compared with the value of the plate-supply voltage. Thus, energy is supplied to the plate circuit only when most of the plate-supply voltage drops across the plate tank. In turn, most of the energy is delivered to the plate circuit instead of being consumed as heat by the tube. The efficiency of a Class-*C* amplifier usually ranges from 60 per cent to 80 per cent.

TABLE 6–2

Characteristics of Series and Parallel Resonant Circuits

Quantity	*Series circuit*	*Parallel circuit*
At resonance: Reactance ($X_L - X_C$)	Zero; because $X_L = X_C$	Zero; because nonenergy currents are equal
Resonant frequency	$\frac{1}{2\pi\sqrt{LC}}$	$\frac{1}{2\pi\sqrt{LC}}$
Impedance	Minimum; $Z = R$	Maximum; $Z = \frac{L}{CR}$, approx.
I_{line}	Maximum	Minimum value
I_L	I_{line}	$Q \times I_{\text{line}}$
I_C	I_{line}	$Q \times I_{\text{line}}$
E_L	$Q \times E_{\text{line}}$	E_{line}
E_C	$Q \times E_{\text{line}}$	E_{line}
Phase angle between E_{line} and I_{line}	0°	0°
Angle between E_L and E_C	180°	0°
Angle between I_L and I_C	0°	180°
Desired value of Q	10 or more	10 or more
Desired value of R	Low	Low
Highest selectivity	High Q, low R, high $\frac{L}{C}$	High Q, low R
When f is greater than f_o: Reactance	Inductive	Capacitive
Phase angle between I_{line} and E_{line}	Lagging current	Leading current
When f is less than f_o: Reactance	Capacitive	Inductive
Phase angle between I_{line} and E_{line}	Leading current	Lagging current

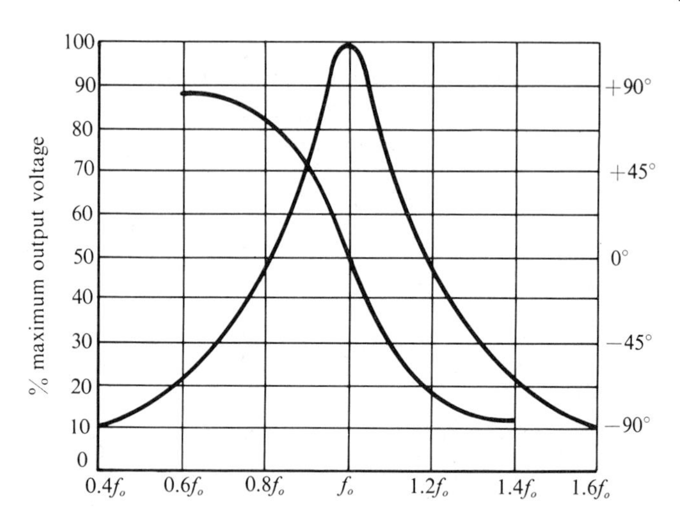

Figure 6–21. Electrical characteristics of an *LCR* parallel-resonant circuit.

6–5. Distortion Considerations

When the plate-current or collector-current waveshape in an amplifier stage differs from the input-voltage waveshape, the output waveform is accordingly distorted. We will find that there are three types of distortion to be considered. These are called *frequency distortion*, *harmonic* (or nonlinear, or amplitude) *distortion*, and *phase-shift distortion.* Any of these types of distortion may be present in the output signal from an amplifier stage, either individually or in combination. The basic distinctions between these distortions are depicted in Fig. 6–22.

Observe in Fig. 6–22(a) that a typical pair of input-drive signals are shown that might be applied to the tube or transistor. This pair of signals includes a sine wave that has a certain frequency, and another sine wave that has three times the frequency of the first wave. These two waveforms are combined in the input circuit and are then amplified by the device. If both frequencies are not amplified equally, it is clear that the waveform in the plate circuit will have a different waveshape. This will lead to frequency distortion in the output signal, as detailed next.

Again, observe in Fig. 6–22(b) that the higher-frequency signal in the distorted output waveform is only about 1/8 the amplitude of the

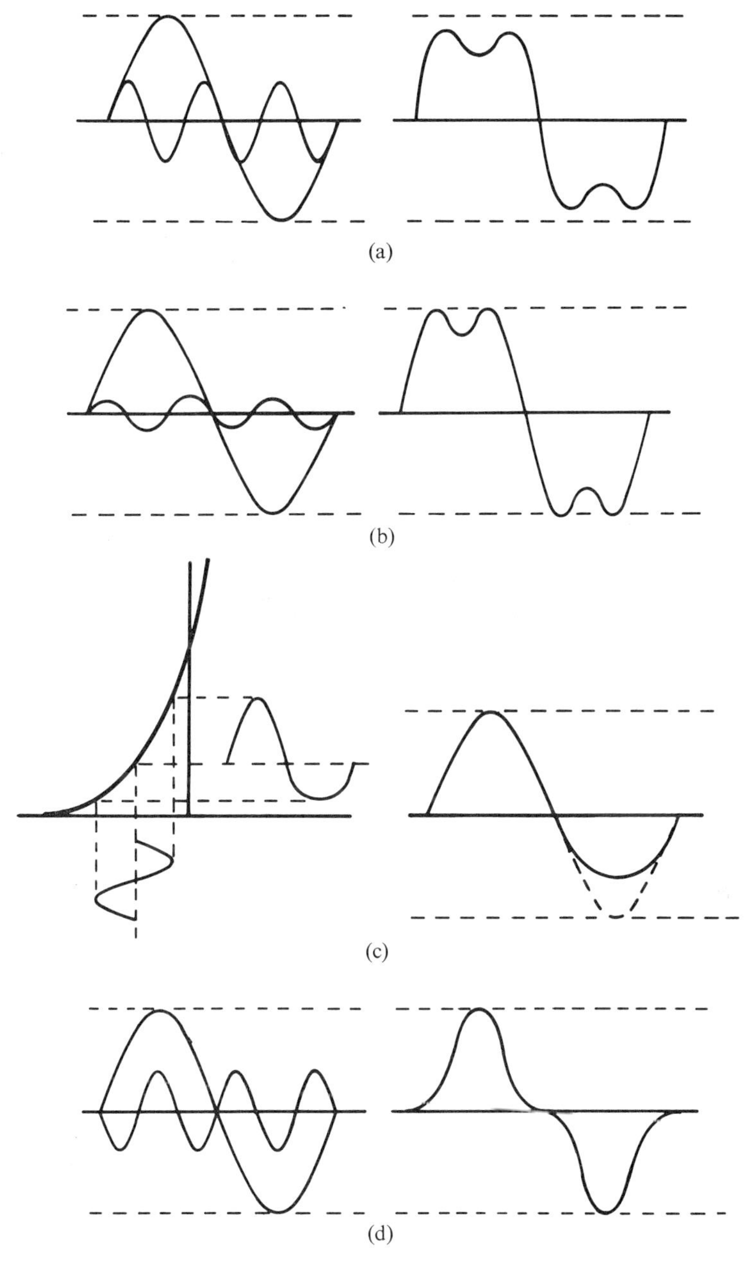

Figure 6-22. Three basic types of amplifier distortion that occur in amplifying devices. (a) Equal amplification; (b) Frequency distortion; (c) Harmonic distortion; (d) Phase shift distortion.

lower-frequency signal, although it started with an amplitude about 1/4 as large. However, the lower-frequency signal did not undergo this comparative attenuation. This is an example of frequency distortion. We will later learn how to determine the frequency-response curve of an amplifier stage (see Fig. 6–23). A frequency-response curve

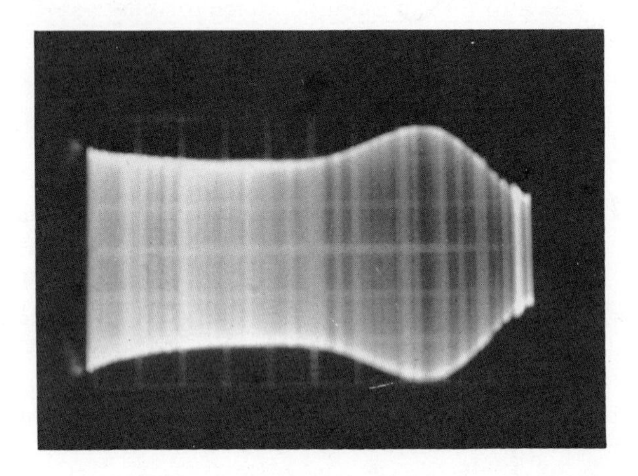

Figure 6–23. A frequency-response curve for an amplifier stage.

permits us to easily assess the frequency distortion of an amplifier stage. We commonly state that the response varies by plus or minus an associated number of db at a given frequency, with respect to the response at the mid-band frequency.

Harmonic distortion affects the amplitude of the output waveform. This type of distortion is often caused by a large plate-current change for a small grid-voltage change near zero grid voltages and with a small plate-current change for a large grid-voltage change near cutoff grid voltages in a vacuum tube. In a transistor, this type of distortion is produced by a large collector-current change for a small base-current change near maximum base current and with small collector-current changes for large base-current changes near zero base current. The effect of this nonlinear distortion is that one part of the output waveform does not have as great an amplitude as another part. In Fig. 6–22 (c), the negative portion of the output waveform is not as large as it should be. The dotted line shows how the negative portion would appear if amplitude distortion were not present. We will learn how to determine the amplitude characteristic of an amplifier stage, and thereby to evaluate the harmonic distortion of the stage.

Phase-shift distortion occurs when reactive components are present in the amplifier stage. Under these conditions, there is a phase shift of one or more of the signals comprising the input-drive waveform. In consequence, the output-signal waveform is not a true replica of the input-drive waveform. Figure 6–22(d) shows the high-frequency component of the signal shifted 180°, with resulting phase-shift distortion. We will learn how to test an amplifier stage for phase-shift distortion with a sine-wave oscillator and oscilloscope and with a square-wave generator and oscilloscope in later chapters.

6-6. D-C Amplifiers

Direct-current amplifier stages which are used to amplify very low frequency signals or d-c voltage levels may also be required to amplify frequencies in the MHz range. With reference to Fig. 6–1(a), it is evident that if the a-c generator is replaced by a short-circuit, and the value of the *C*-battery voltage is changed to −11 v, the plate-to-cathode voltage will increase. Conversely, if the value of the *C*-battery voltage is changed to −9 v, the plate-to-cathode voltage will decrease. It is left as an exercise for the student to calculate the approximate amount of increase and decrease in the plate-to-cathode voltage values. It follows from Table 6–1 and Fig. 6–8 that a transistor amplifier stage also operates as a d-c amplifier.

In practical application, the *C* battery might be replaced by a photocell, and the plate-load or collector resistor might be replaced by a relay (Fig. 6–24). Then, when a sufficient change occurs in the light

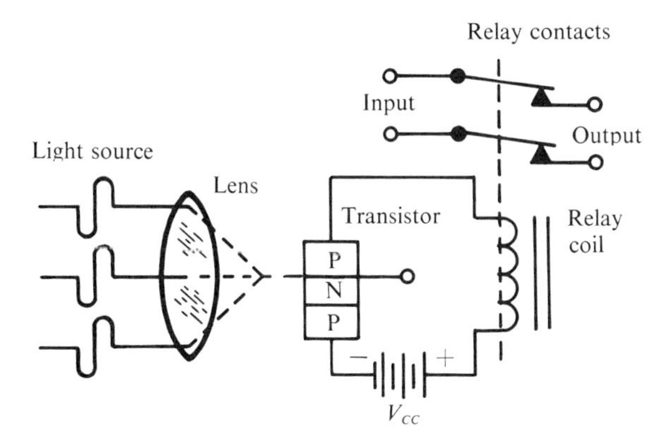

Figure 6–24. Phototransistor driving a relay coil as a d-c amplifier.

level at the photocell, the d-c output current of the amplifier stage causes the relay to pull in or drop out. In turn, associated heavy-current equipment can be turned on and off by the switch contacts in the relay. In summary, a slight amount of energy in a light beam can be used to control the operation of heavy electrical equipment. Transistor d-c amplifiers may be connected in many configurations, as we will subsequently discover.

Various types of electronic instruments, such as the vacuum-tube or transistor voltmeters illustrated in Fig. 6–25, also use d-c amplifiers.

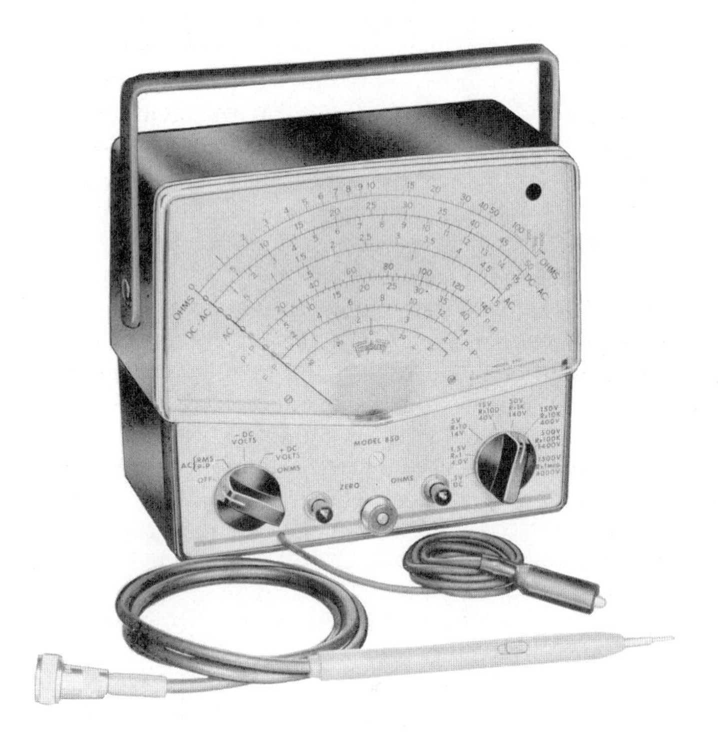

Courtesy of Triplett Electrical Instrument Co.

Figure 6–25. A typical service-type VTVM.

The chief advantage of a VTVM, or TVM, transistor voltmeter in comparison with a VOM is that the input of a d-c amplifier tube draws practically no current from the circuit under test. Accordingly, minimum loading is imposed on circuits that have a high internal resistance, and the voltage indication is more accurate.

SUMMARY

1. The two most important circuit systems in electronics equipment are considered to be oscillators and amplifiers.
2. A prototype amplifier consists of a vacuum tube, a junction transistor, or a field-effect transistor associated with suitable circuit components.
3. Power output from an audio amplifier ranges from a few milliwatts to 1 Mw, or more.
4. The equivalent circuit of a vacuum tube amplifier is formed by a voltage generator ($\mu\, e_g$) driving the internal resistance (r_p) and the load resistance (R_L) in series. The signal is divided between r_p and R_L.
5. Distortion of the output signal of a vacuum-tube voltage amplifier may be due to the nonlinear characteristics of the tube or the input signal overdriving the grid into grid-current flow.
6. The characteristic curves of either the plate-voltage/plate-current or the grid-voltage/plate-current values may be used to determine the voltage gain of a vacuum-tube amplifier.
7. The common-cathode vacuum-tube amplifier produces 180° phase shift between the input and output signal at low frequencies.
8. Amplification by a junction transistor is based upon the control of minority carriers through a semiconductor solid by a base current.
9. In a conventional amplifier, the input of a transistor is forward-biased and the output is reverse-biased.
10. Junction transistors are characterized by much lower input impedances than vacuum tubes or field-effect transistors.
11. In a transistor amplifier, the current gain approaches h_{fe} as the value of the load resistance decreases.
12. A field-effect transistor has the advantages of a transistor plus the advantages of a vacuum tube.
13. The field-effect transistor has input and output impedance characteristics very similar to those of vacuum tubes.
14. An FET operates as a variable resistance in its ohmic region with characteristics very similar to a triode vacuum tube.
15. An FET operates as a constant-current source in its pentode region with characteristics very similar to those of a pentode vacuum tube.
16. The transconductance of a device is the ratio of the output current to the input voltage.
17. A field-effect transistor is normally reverse-biased and may,

therefore, utilize the voltage developed across the source resistor as the bias.

18. Amplifiers may be classified as power amplifiers, voltage amplifiers, and current amplifiers.

19. The amplifying device in a Class-*B* amplifier conducts for less than 180° of the input signal; this is accomplished by biasing the device at, or near, its cutoff point.

20. In a Class-*C* amplifier, the device conducts for less than 90° of the input signal, usually into a tuned-tank circuit which produces an output by a flywheel effect.

21. Three basic types of distortion are phase shift, harmonic, and frequency distortion.

22. Phase-shift distortion occurs when reactive components are present in the amplifier stage.

23. Harmonic distortion affects the shape of the output waveform and is caused by overdriving the input of an amplifier or by driving the output toward cutoff or saturation.

24. Frequency distortion is the result of more amplification of the waveform at one frequency than at another frequency.

25. To amplify very low frequency signals, d-c amplifiers may utilize vacuum tubes, junction transistors, or field-effect transistors.

Questions

1. What are the three basic types of amplifiers?
2. Explain why the gain of a circuit is always less than the amplification of the device.
3. What factors may cause distortion in the output of a vacuum-tube voltage amplifier?
4. Why is the control grid of a vacuum tube normally biased negative?
5. What characteristic curves are most convenient to determine the voltage gain of an amplifier?
6. Why is base current, rather than base voltage, taken as a parameter for analysis of the junction-transistor amplifier?
7. Why is the input impedance of a junction transistor very low?
8. What are the advantages of transistors over vacuum tubes?
9. Identify the elements of an FET, and explain the operation of each element.
10. Explain the theory upon which the current flow through an FET is controlled.
11. Define transconductance of a vacuum tube.
12. Compare the three basic transistor amplifier configurations.
13. Compare the operating time and efficiency of Class-*A*, Class-*B*, and Class-*C* amplifiers.
14. Why is the efficiency of Class-*B* and Class-*C* amplifiers higher than that of Class-*A* amplifiers?
15. Why is the efficiency of a transistor amplifier higher than the efficiency of a corresponding vacuum-tube amplifier?
16. What is the relationship between the value of the plate- or collector-load resistor and the voltage gain? Why?
17. Explain the operation of the tuned-plate or collector-tank circuit in a Class-*C* amplifier.
18. What is the basic definition of distortion as referred to an amplifier?
19. Explain each of the three basic types of distortion.
20. What is the function of a d-c amplifier?

PROBLEMS

1. Determine the voltage gain of a vacuum-tube amplifier, such as illustrated in Fig. 6-1, when $\mu = 22$, $R_L = 20\text{ k}\Omega$, and $r_p = 5\text{ k}\Omega$.

2. Determine the voltage gain of the circuit in Problem 1 when the load resistance value is increased to 47 kΩ.

3. What practical factors limit the value of the load resistor?

4. Determine from the graph in Fig. 6-4(c) the values of r_p, g_m, and μ at the plate currents 3 ma, 5 ma, and 7 ma.

5. Determine the voltage gain for each of the plate-current values in Fig. 6-4(c) when a load of 22 kΩ is used with $E_g = -8$ v.

6. Calculate the power output of a stage when there is a voltage swing of 120 v peak-to-peak across 25 kΩ.

7. Determine the power output of a stage when the signal swing across a 68-kΩ load resistor is from 120 v to 280 v.

8. Determine the current gain of a transistor amplifier stage when $h_{fe} = 99$, $h_{oe} = 20$ kΩ, and $R_L = 2.5$ kΩ.

9. Determine the output current of the circuit in Fig. 6-8 when I_B varies from 50 to 100 μa.

10. What is the current gain of the circuit in Problem 9?

11. What is the output power of the circuit in Problem 9 if R_L has a value of 2 kΩ?

12. What is the voltage gain of the amplifier in Problem 9 if the transistor's input resistance is 1.2 kΩ?

13. What is the voltage gain of an amplifier with a 47-kΩ load resistor using an FET with a g_m of 2,000?

14. Determine the bias value (V_{GS}) of the amplifier in Fig. 6-13(a) when R_S is 1 kΩ and I_D is 0.4 ma.

CHAPTER 7

Fundamentals of RC-Coupled Amplifiers

7–1. Introduction

We recall that amplifiers designed to step up signals with frequencies from approximately 15 to 15,000 Hz are called audio-frequency amplifiers. To obtain a required amount of gain, it is often necessary to use several stages connected in cascade. When one stage of audio amplification is coupled to a following stage, it is common practice to utilize *RC* coupling, as depicted in Fig. 7–1. The equivalent circuits representing the amplifiers in Fig. 7–1 are presented in Fig. 7–2. Two other means of blocking d-c voltage from the input of a stage entail *LC* (impedance) coupling, or transformer coupling, as depicted in Figs. 7–3(a) and (b). At this point, we will consider the fundamentals of *RC*-coupled amplifiers.

An *RC*-coupling circuit transfers the varying signal voltage from one stage to the next, and supplies the necessary d-c voltages and currents for normal operation of the amplifying devices. Observe that

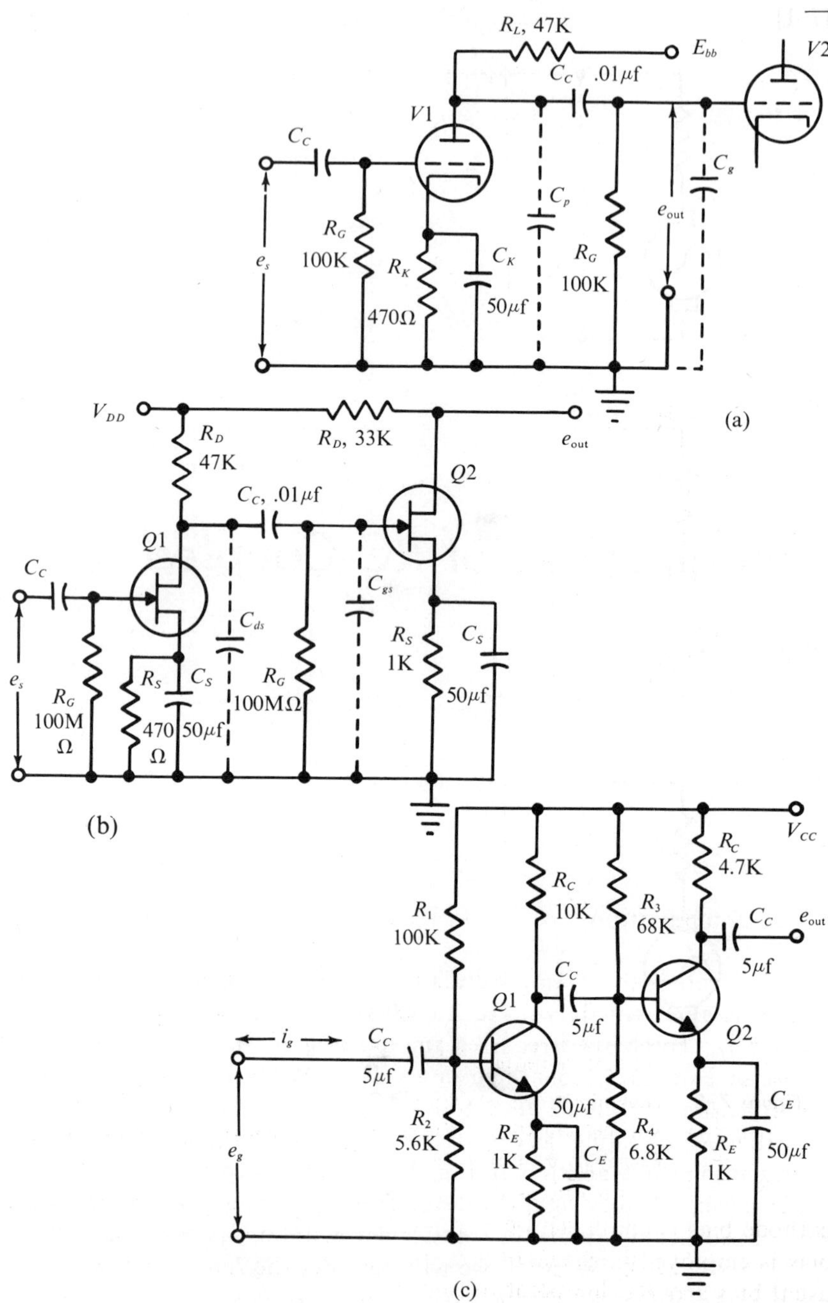

Figure 7-1. RC-coupled amplifiers. (a) Vacuum tube; (b) Field-effect transistor; (c) Junction transistor.

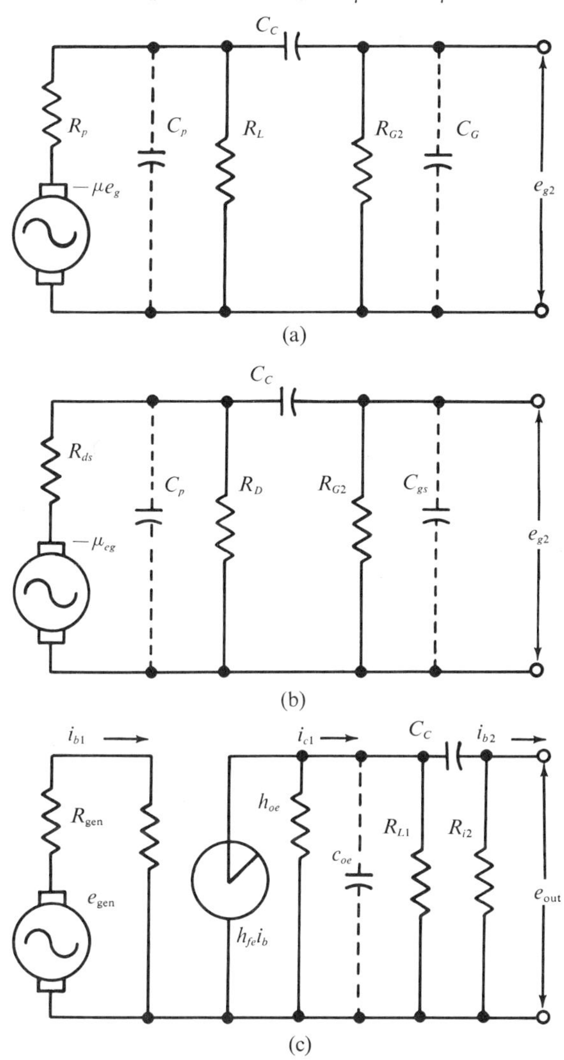

Figure 7–2. Equivalent circuits for RC-coupled amplifiers. (a) Employing a vacuum tube; (b) Employing a field-effect transistor; (c) Employing a junction transistor.

cathode bias is employed in the configuration of Fig. 7–1(a), and source bias is employed in the configuration of Fig. 7–1(b). This is the most usual bias arrangement for *RC*-coupled vacuum tubes or field-effect transistor amplifiers. Capacitor C_C is called a *coupling capacitor;* it provides transfer of a-c signal voltage from the source to the control

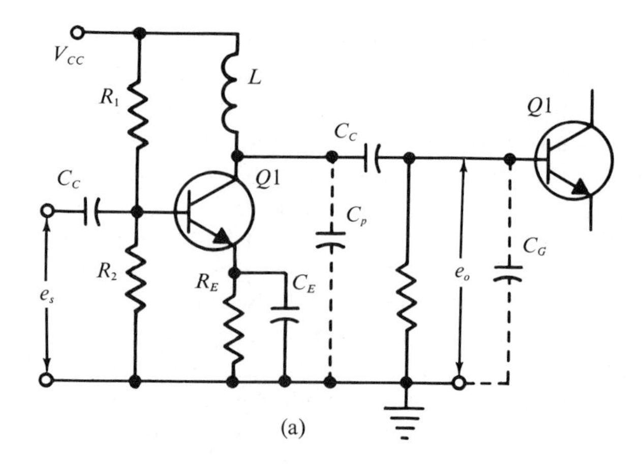

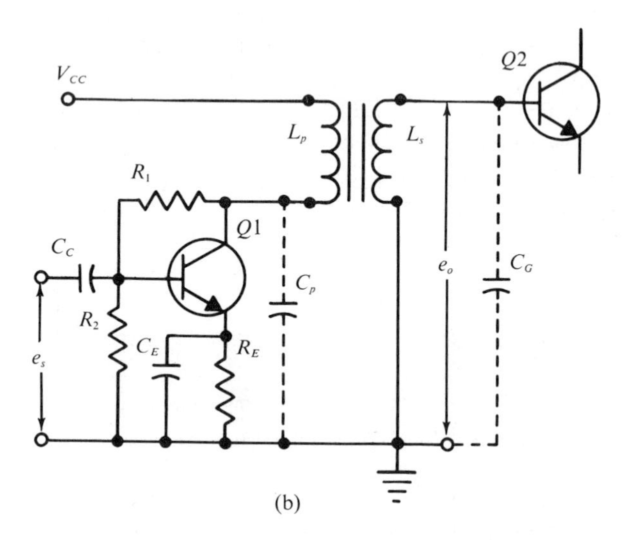

Figure 7-3. Two other basic amplifier coupling arrangements. (a) Impedance coupling; (b) Transformer coupling.

element of the first stage. A similar capacitor provides transfer of a-c signal voltage to the input of the second stage.

In the vacuum-tube amplifier depicted in Fig. 7-1(a), resistor R_L is called a *coupling resistor*. Its resistance value is chosen as high as possible in view of the required high-frequency response and the required d-c plate voltage value, as noted in the previous chapter. Resistor R_G is called a grid-return resistor, or a grid-leak resistor. Its purpose is to provide a d-c path between grid and ground, and thereby to maintain the grid at d-c ground potential in this configuration. Thus, R_G is a part of the grid-cathode bias circuit. The resistance value of R_G is chosen as high as possible in view of the necessity for stable tube operation. This consideration is explained in greater detail subsequently.

Observe that coupling capacitor (C_C) in Fig. 7-1(a) functions also as a blocking capacitor. That is, C_C blocks the application of d-c plate voltage from the first tube to the grid of the second tube. In turn, the grid-bias voltage is determined solely by the value of the cathode-bias resistor. There is more than meets the eye in a conventional audio-amplifier circuit diagram. For example, stray capacitance c_p is inevitably present (see Fig. 7-2). Though the value of c_p is small compared with the value of C_C, this stray capacitance has a dominant effect on amplification at high audio frequencies, as will be explained.

Amplitude variation produced in the plate current of a vacuum tube when a signal voltage is applied to the grid is exactly the same as would be produced by a generator developing a voltage of $-\mu e_g$ in a circuit consisting of the tube's plate resistance connected in series with the load impedance, as depicted in Fig. 7-2(a). Recall that the minus sign denotes the fact that the signal phase is inverted from grid to plate. The gain is formulated

$$A_V = \frac{-\mu e_g R_L r_p}{R_L + r_p} \tag{7-1}$$

The effect on the plate current of a signal voltage e_g applied to the grid is exactly as if the plate-cathode circuit were a generator developing a voltage and having an internal resistance equal to the dynamic *plate resistance* of the tube.

In turn, the *RC*-coupled amplifier depicted in Fig. 7-1(a) can be represented by the equivalent circuit shown in Fig. 7-2(a). This equivalent circuit makes it easier to visualize amplifier operation. Note that the cathode bypass capacitor and biasing resistor are omitted in the equivalent circuit. This omission is permissible because the cathode bypass capacitor ordinarily has a comparatively large value and is practically a short circuit for a-c signals.

Stray capacitance c_p and c_G in Fig. 7-2(a) have negligible effect on low-frequency response, but they have a dominant effect on high-frequency response, as we shall see. Both the stray capacitances and the coupling capacitance have negligible effect on mid-band response. At low and medium audio frequencies, c_p and c_G are practically open circuits; and at medium and high audio frequencies, C_C is practically a short-circuit. Again, at low frequencies, C_C has substantial capacitive reactance; at high frequencies, c_p and c_G have low reactance and tend to shunt the audio signal to ground.

An important fact to recall from the previous chapter is that the output from an audio stage is always less than μe_g because of r_p, the dynamic plate resistance of the tube. In turn, the ratio of output voltage to input voltage is formulated

$$\text{Voltage gain} = \frac{\text{output voltage}}{\text{input voltage}} = \frac{\mu \dfrac{R_L R_G}{R_L + R_G}}{r_p + \dfrac{R_L R_G}{R_L + R_G}} \tag{7-2}$$

The field-effect transistor amplifier shown in Fig. 7–1(b) employs source bias very similar to cathode bias in the vacuum-tube amplifier. Again, capacitor C_C is called the coupling capacitor and is approximately the same value as C_C for a similar vacuum-tube circuit. Resistor R_D is called the drain resistor. Its resistance value is selected as high as possible, compatible with high frequency response and bias voltage, for maximum voltage gain. Resistor R_G, the gate-return resistor, provides a d-c path from gate to ground; it thereby prevents the gate from becoming negative due to the small current through the back-biased source to gate junction. Capacitor c_{dg} is the internal capacitance of the unipolar device and as the corresponding vacuum-tube capacitance (c_p), is a limiting factor for amplification at high frequencies.

Amplitude variations on the gate of an *FET* produce an output voltage variation exactly as if a generator ($-\mu e_g$) were connected in series with the internal resistance (r_{ds}) and the load impedance. This equivalent circuit produces a voltage gain.

$$A_V = \frac{\mu \dfrac{R_L R_G}{R_L + R_G}}{r_{ds} + \dfrac{R_L R_G}{R_L + R_G}} \tag{7-3}$$

The above analysis using the voltage gain factor is similar to that of a vacuum tube. However, most writers prefer to analyze the FET amplifier as a transconductance device in terms of $g_m e_{gs}$ (or a vacuum-tube pentode); in which case the voltage gain is formulated

$$A_V = \frac{g_m R_L r_{ds}}{r_{ds} + R_L} \tag{7-4}$$

The *RC*-coupled amplifier illustrated in Fig. 7–1(b) can be represented by the equivalent circuit shown in Figure 7–2(b). Note that the equivalent circuit corresponds almost exactly to the equivalent circuit of a vacuum-tube *RC*-coupled amplifier, as seen in Fig. 7–2(a). A somewhat more exact equivalent circuit will be presented subsequently.

The three *RC*-coupled amplifiers depicted in Fig. 7–1 perform the same function: they amplify an audio frequency signal. However, the component values in the junction transistor amplifier are quite different from those in the vacuum tube and the FET amplifier due to the much lower input impedance of the junction transistor.

Resistors R_1 and R_2 form a bleeder circuit to establish a forward bias on the base of $Q1$, and resistors R_2 and R_3 perform this function for $Q2$. Resistors R_E in the emitter leads of $Q1$ and $Q2$ are stability resistors which limit the collector operating point drift with a change of temperature. We will return to the subject of temperature stability in later chapters.

The junction transistor amplifier in Fig. 7–1(c) may be represented by the equivalent circuit in Fig. 7–2(c). This equivalent circuit acts as if a current $(h_{fe}i_b)$ were placed across the internal transistor admittance (h_{oe}) in parallel with the load impedances (R_L of $Q1$ and r_i of $Q2$). The term h_{fe} is defined as the forward current gain factor. When the load impedance is much less than the internal impedance of the transistor, the current gain of the amplifier may be formulated

$$A_i \simeq h_{fe} \tag{7–5}$$

and the voltage gain is formulated

$$A_V \simeq \frac{h_{fe}R_L}{h_{ie}} \tag{7–6}$$

where h_{ie} is the input resistance to the transistor.

The power gain is the product of the current gain and voltage gain.

$$A_p \simeq A_i A_v \tag{7–7}$$

Example 1. Evaluate the current, voltage, and power gain of the output stage of Fig. 7–1(c) when a transistor with a forward-current gain (h_{fe}) of 100 and an input resistance (h_{ie}) of 2 kΩ is used in the circuit.

$$A_i \simeq 100 \qquad \text{(answer)}$$

$$A_v \simeq 100\frac{4.7\text{K}}{2\text{K}} \simeq 235 \qquad \text{(answer)}$$

$$A_p \simeq 100 \times 235 \simeq 23{,}500 \qquad \text{(answer)}$$

7–2. Low-Frequency Response

At low audio frequencies, the response of an *RC*-coupled stage is determined by the reactance of the coupling capacitor C_C. This response can be calculated from our knowledge of a-c circuit action.

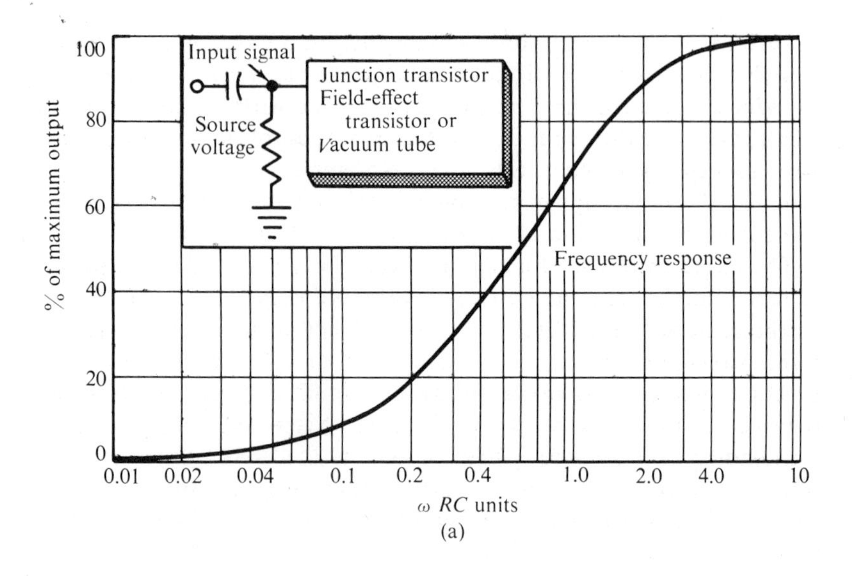

(a)

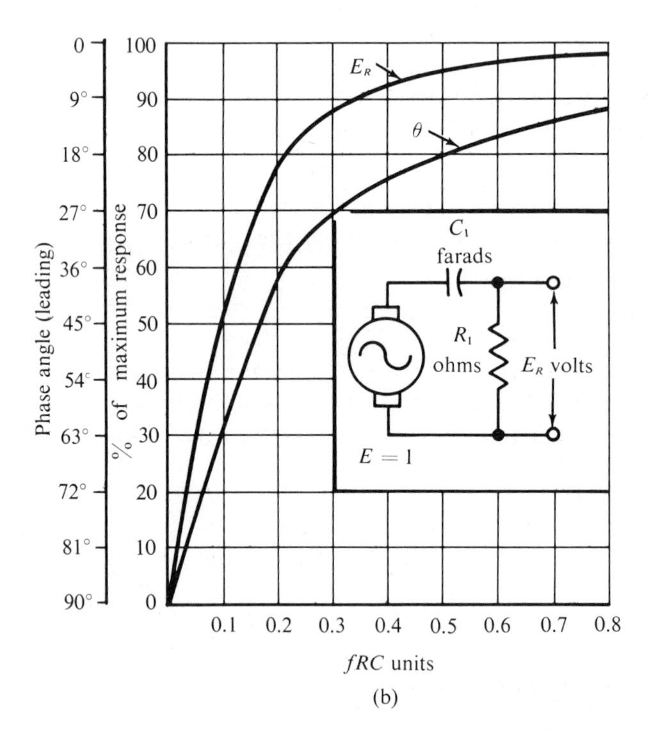

(b)

Figure 7-4. Universal *RC* frequency-response chart for one-stage amplifier at low frequencies.

However, it is instructive to visualize the low-frequency response with the aid of a universal RC frequency-response chart, as depicted in Fig. 7–4. We perceive that the input signal is practically zero when $\omega RC = 0.01$, and that the input signal has practically the same amplitude as the source signal when $\omega RC = 10$. In between these two frequency extremes, the input signal voltage is attenuated by the reactance of the capacitor to a greater or lesser extent.

The curve shown in Fig. 7–4 is very useful, because it illustrates the fact that the frequency-response curve of any RC-coupling circuit has the same shape. Regardless of the R and C values, a plot of ωRC versus percentage of maximum output gives us the same curve. Since universal charts are of great utility in amplifier analysis and circuit analysis, let us pause for a moment to derive the formula for the curve shown in Fig. 7–4(a). With respect to the insert diagram in Fig. 7–4(a), we perceive that the input signal voltage is applied across an impedance consisting of R and C connected in series. This impedance, of course, is formulated

$$Z = \sqrt{R^2 + X_C^2} \qquad \textbf{(7–8)}$$

In turn, the voltage drop across R represents the signal voltage applied to the grid of the tube in Fig. 7–4(a). Therefore, the ratio of input-driving voltage to the source-driving voltage is formulated

$$\% \, E_{\text{max}} = \frac{R}{\sqrt{R^2 + X_C^2}} \times 100 \qquad \textbf{(7–9)}$$

Formula 7–9 simply states that R and C form a voltage divider with respect to the input signal. A bit of algebraic manipulation permits us to write Formula 7–9 in the form

$$\% \, E_{\text{max}} = \frac{1}{\sqrt{1 + \dfrac{1}{R^2\omega^2C^2}}} \times 100 \qquad \textbf{(7–10)}$$

Observe that Formula 7–10 explicitly relates the per cent of maximum output voltage to the value of ωRC. In turn, we plot these related values in Fig. 7–4(a) and obtain a universal RC frequency response chart. A similar approach is used to derive a universal chart for any amplifier or circuit situation. Note that in the case of a junction transistor amplifier, the input is a low impedance and must be taken in parallel with the base resistor to arrive at a value of R in Formulas 7–9 and 7–10.

Example 2. Suppose the amplifier ($V1$) in Fig. 7–1(a) has a maximum output voltage of 30. What is the output voltage at 50 Hz?

The grid resistor is 100 kΩ, and the value of C is 0.01 μf. At 50 Hz,

$$\begin{aligned}\omega RC &= 2\pi f\, RC \\ &= 2\pi \times 5 \times 10^5 \times 10^{-8} \\ &\simeq 0.314\end{aligned}$$

From the chart in Fig. 7–4

$$E = 0.21\ E_{\max}$$

$$\therefore \quad E \text{ at } 50 \text{ Hz} = 0.21 \times 30 = 6.30 \qquad \text{(answer)}$$

a universal *RC* frequency-response chart can be plotted in terms of *fRC* units instead of ωRC units, as shown in Fig. 7–4(b). The phase angle of the circuit versus *fRC* units is also depicted in Fig. 7–4(b).

Next, suppose that we have two identical *RC*-coupled amplifier stages connected in cascade. This is called a symmetrical two-stage amplifier, because *R* and *C* values are the same in each stage. With reference to the insert in Fig. 7–5, it is evident that the source signal is attenuated by the coupling circuit of the first stage and is again at-

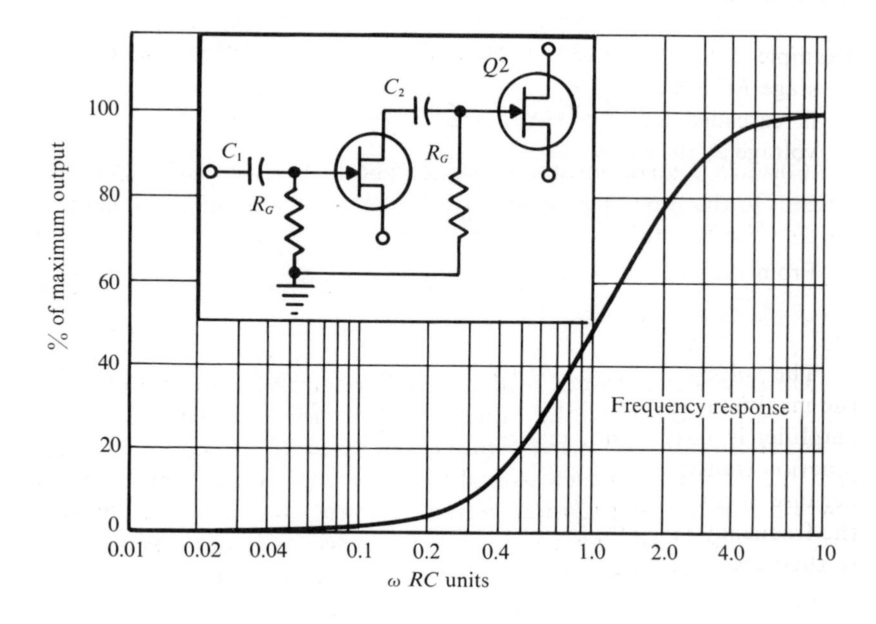

Figure 7–5. Universal *RC* frequency-response chart for symmetrical two-stage amplifier at low frequencies.

tenuated by the coupling circuit of the second stage. In effect, the output signal has a ratio to the source signal that is computed by applying Formula 7–10 twice; we multiply the source signal by the square of Formula 7–10. The result is the curve depicted in Fig. 7–5. This is a universal *RC* frequency-response chart for any symmetrical two-stage amplifier.

There are some essential facts to observe in Fig. 7–5. First, the low-frequency cutoff characteristic is more abrupt when the stages are connected in cascade. The *RC*-coupling circuits are frequency-selective filters from one basic viewpoint, and we know that when filter sections are connected in cascade the cutoff characteristic becomes more abrupt. Second, there is a difference in the response characteristic when *RC* sections are connected by wire leads and when they are "connected" by grid-plate action in a tube. Since the grid-input impedance of a Class-*A* tube amplifier, or the gate input of a Class-*A* FET amplifier, is extremely high, the *RC* sections in Fig. 7–5 are effectively isolated from each other. This is just another way of saying that the second *RC* section does not load the first *RC* section. Thus, the frequency response of a two-section *RC* integrator is somewhat different from the frequency response of a two-stage *RC*-coupled amplifier.

Example 3. As an example of a more rapid decrease in gain of a two-stage *RC* amplifier, let us assume that the amplifier in Example 2 is cascaded and determine the percentage of maximum output voltage at the same frequency (50 Hz).

$$\omega RC = 2\pi \times 10^5 \times 10^{-8}$$
$$= 0.314$$

From the chart in Fig. 7–5

$$E = 0.09\ E_{max} \qquad \text{(answer)}$$

The voltage at 50 Hz for the one-stage amplifier is 21 per cent of the maximum value; whereas, the voltage at 50 Hz for the two-stage amplifier is only 9 per cent of the maximum value. The two-stage junction transistor amplifier cannot be analyzed one stage at a time because of interaction between the collector and the base. However, the frequency response for two stages of amplification is very similar to the curve in Fig. 7–5.

7–3. Midband-Frequency Response

At medium audio frequencies, the reactance of the coupling capacitor(s) becomes sufficiently small so that we can replace the capacitor(s) with short circuits for purposes of signal analysis. The stray capacitances have such high reactances at medium audio frequencies that we can regard them as open circuits. In turn, amplifier voltage gain is calculated as if the source and amplifying devices were d-c coupled. Thus, the maximum available gain of the amplifier is realized at midband frequencies. Recall that the formula for a vacuum tube or FET stage gain under these circumstances is formulated

$$A_V = \frac{v_0}{v_g} = \frac{\mu r_p}{r_p + R_L} \quad \textbf{(7–11)}$$

where v_o is the signal voltage across the plate-load resistor, v_g is the value of the signal voltage, r_p is the dynamic plate resistance of the device, and R_L is the load resistance.

If two stages are connected in cascade, the voltage gains are multiplied. Thus, A_V or v_o/v_g denotes the voltage gain of one stage, and A_V^2 or $(v_o/v_g)^2$ denotes the gain of two identical stages connected in cascade. If we have three identical stages connected in cascade, the voltage gain of the amplifier becomes A_V^3 or $(v_o/v_g)^3$, where v_o and v_g refers to the same stage when each stage is isolated from the next.

Example 4. Three identical *RC*-coupled FET amplifiers, each with a voltage gain of 30, are connected in cascade. Assuming no losses in the coupling circuits, what is the overall voltage gain?

$$\begin{aligned} A_{V\text{total}} &= A_{V1} \times A_{V2} \times A_{V3} \\ &= (30)^3 \\ &= 27{,}000 \qquad \text{(answer)} \end{aligned}$$

7–4. High-Frequency Response

Amplification at high audio frequencies is limited by the shunting action of c_p and c_G depicted in Fig. 7–2. Of course, we can replace C_C by a short circuit, insofar as signal analysis is concerned. If R_G has a very high value (as is usually the case), it can simply be neglected. For an exact analysis, we can combine R_G in parallel with R_L. Of course, we can combine c_p and c_G in parallel and add their values to calculate the total shunt capacitance in the equivalent circuit. Then we have the simplified equivalent circuit shown in Fig. 7–6(a).

The alert student will perceive that the configuration depicted in Fig. 7–6(a) can be further simplified by applying Thevenin's theorem. That is, we will regard the shunt capacitance $c_p + c_G$ as the load component in the circuit. If we disconnect the capacitor, the equivalent source voltage becomes reduced by resistive voltage-divider action, the

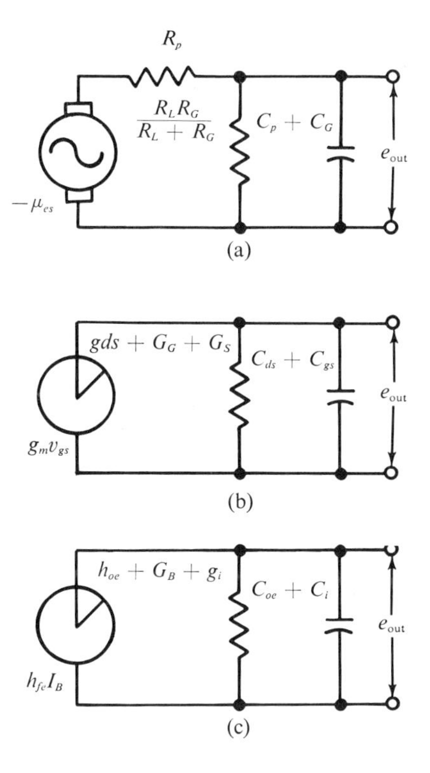

Figure 7-6. Simplified high-frequency equivalent circuits, (a) Vacuum-tube amplifier; (b) Field-effect transistor amplifier; (c) Junction-transistor amplifier.

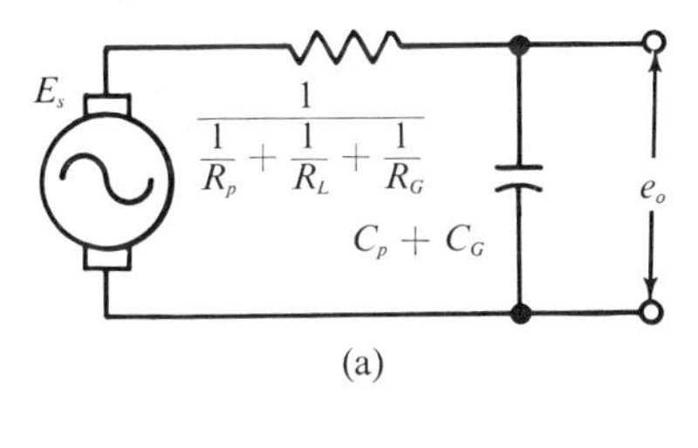

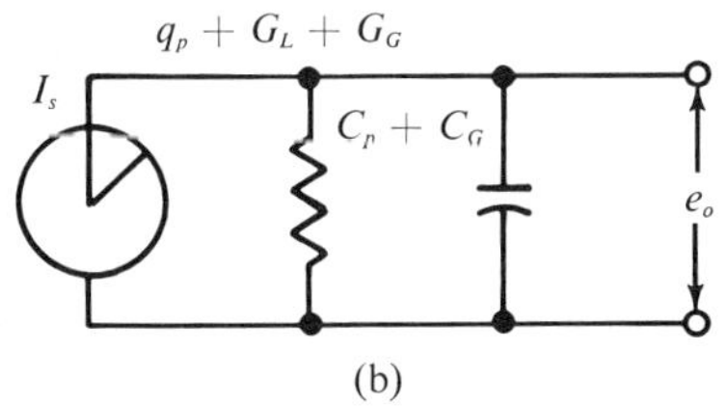

Figure 7-7. Equivalent circuit of an *RC*-coupled amplifier at high frequencies. (a) Thevenin's theorem; (b) Norton's theorem; (These two circuits may represent the FET, vacuum tube, or junction transistor).

value of which may be calculated by the student as an exercise. We will call this equivalent source voltage E_s, as indicated in Fig. 7-7. Now, this equivalent source voltage E_s is connected in series (by Thevenin's theorem) with a resistance R which has a value equal to r_p and $(R_L R_G)/(R_L + R_G)$ connected in parallel. Thus, our final equivalent voltage circuit depicted in Fig. 7-7(a) comprises a voltage generator (E_s) connected to a simple RC series circuit.

An amplifier may also be evaluated by Norton's equivalent circuit method in which the energy source is a current generator (I_{sc}), and the internal impedance is placed in parallel with the generator and the load. The equivalent Thevenin's circuit in Fig. 7-7(a) is shown as an equivalent Norton's circuit in Fig. 7-7(b). Note that the resultant current and voltage across the load are independent of the method of circuit analysis.

These equivalent circuits are extremely useful analyses, because we can now derive a universal RC frequency-response chart for a one-stage RC-coupled amplifier at high audio frequencies. Hence, the charts in Figs. 7-8 and 7-9 may be used for either method with the percentage of output being voltage in Thevenin's method and current in Norton's method.

Example 5. Determine the percentage of maximum output current for the junction-transistor equivalent circuit in Fig. 7-6(c) when $C = 1{,}000$ pf, $R_{\text{int}} = 2$ kΩ, and $f = 100$ kHz.

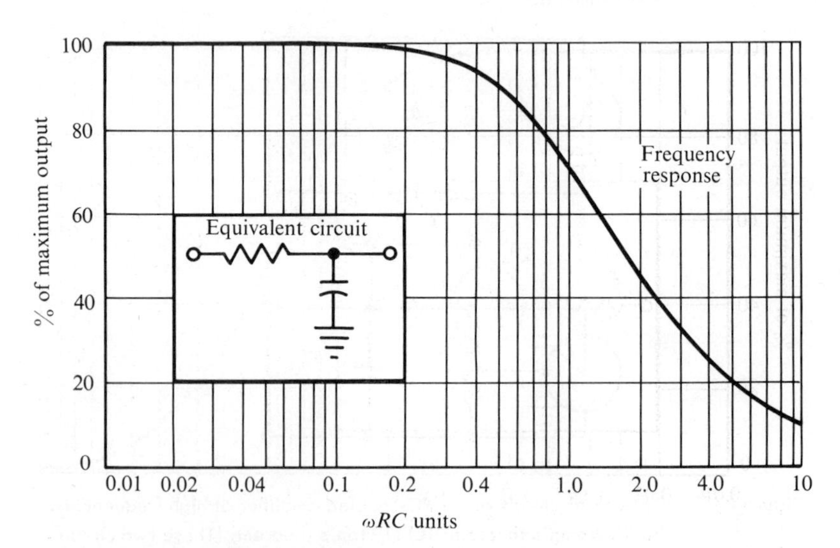

Figure 7-8. **Universal RC frequency-response chart for one-stage amplifier at high frequencies.**

$$\omega RC = 2\pi \times 10^5 \times 2 \times 10^3 \times 10^{-9}$$
$$= 1.26$$

From the chart in Fig. 7–8, the per cent of maximum output is

$$I_O = 0.61\ I_{max} \qquad \text{(answer)}$$

The student may reason from the previous worked-out example that the percentage of maximum output voltage is a function of ωRC. He can then see that this is true from the universal frequency-response curve plots shown in Fig. 7–8. We conclude that any *RC*-coupled amplifier stage has the same form of high-frequency response.

Example 6. The equivalent high frequency circuit of a single-stage FET amplifier is composed of a 50-kΩ series resistor and a 200-pf shunt capacitor. What is the percentage of maximum output voltage at 20 kHz?

$$\omega RC = 2\pi \times 2 \times 10^4 \times 5 \times 10^4 \times 2 \times 10^{-10}$$
$$= 1.256$$

From the chart in Fig. 7–8, we find that the per cent of maximum voltage is

$$E = 0.62\ E_{max} \qquad \text{(answer)}$$

Next, suppose that we have two identical *RC*-coupled amplifier

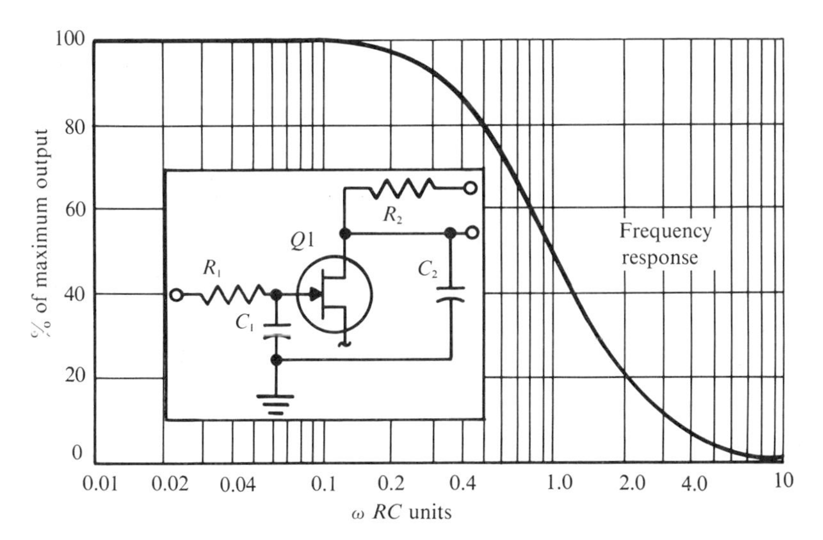

Figure 7–9. Universal *RC* frequency-response chart for symmetrical two-stage amplifiers at high frequencies.

stages that are isolated by a *tube*, or an FET, connected in cascade. Since there is isolation between the two *RC* circuits, we can evidently

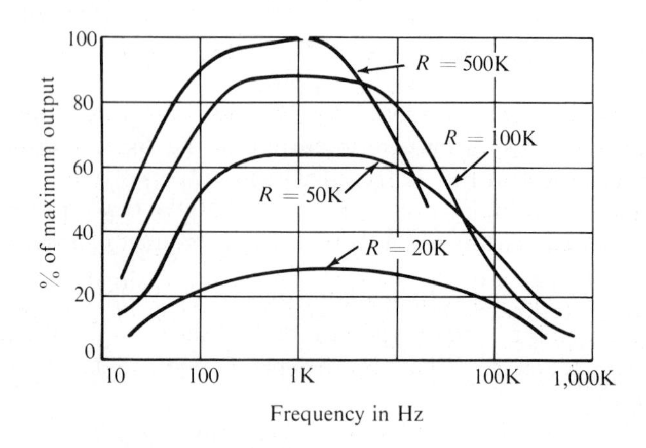

Figure 7-10. **Comparative gain and bandwidth values for an *RC*-coupled stage versus value of plate-load resistor.**

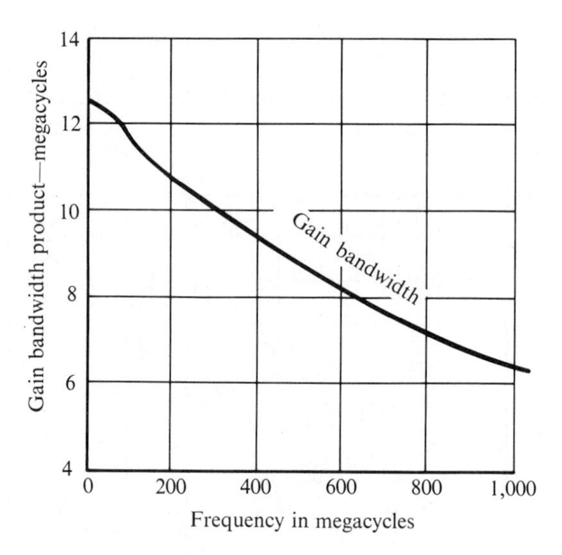

Source: J. F.Pierce, *Transistor Circuits: Theory and Design* (Charles E. Merrill Publishing Co., 1963), p. 265.

Figure 7-11. The gain-bandwidth product for a single-stage feedback amplifier.

derive a universal chart for the two-stage amplifier by simply squaring the percentage values given in the single-stage chart. Thus, Fig. 7–9 shows a universal *RC* frequency-response chart for any symmetrical two-stage amplifier at high audio frequencies. In summary, we have arrived at generalized analyses of amplifier frequency responses at low-, midband-, and high-audio frequencies.

We are now in a good position to recognize the fact that gain and bandwidth are contradictory parameters in any *RC*-coupled amplifier design. As shown in Fig. 7–10, the gain of an *RC* stage increases rapidly with initial increase in the value of the plate-load resistance. Simultaneously with this increase in gain, the bandwidth of the stage decreases. The gain of the stage increases more slowly for very large values of plate-load resistance, because the effect of the dynamic plate resistance becomes progressively masked. Of course, the bandwidth of the stage continues to decrease as the plate-load resistance is increased.

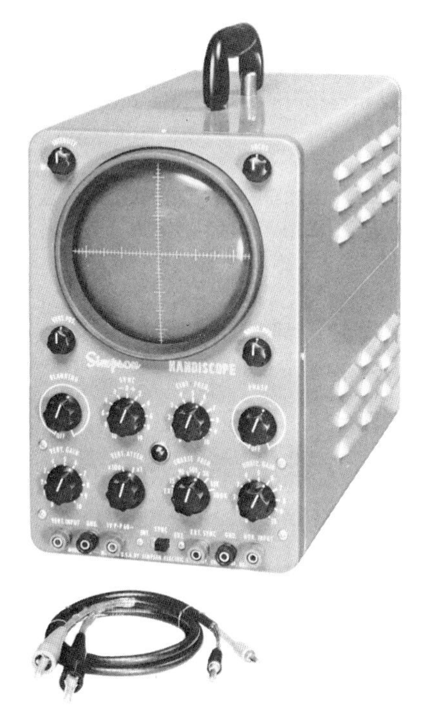

Courtesy of Simpson Electric Co.

Figure 7–12. A service-type oscilloscope that utilizes cascaded *RC*-coupled amplifier stages.

If increased gain is obtained by cascading identical *RC*-coupled stages, we perceive that the bandwidth will be reduced. The bandwidth reduction can be evaluated from the universal *RC* frequency-response charts previously illustrated. Gain increases faster than bandwidth decreases when identical *RC*-coupled stages are cascaded. Suppose that we have a stage with a voltage gain of A, and a bandwidth of 200 kHz. If we connect two stages in cascade, the gain of the amplifier becomes A^2, and the bandwidth is reduced to 193 kHz. Thus, the gain has been increased by 100 per cent, while the bandwidth has been reduced by 2.5 per cent. Figure 7–11 on page 218 illustrates the relationship of the gain-bandwidth product to frequency. Cascaded *RC*-coupled amplifiers are commonly utilized in service-type oscilloscopes, such as the one illustrated in Fig. 7–12 on the preceding page.

7–5. Basic Algebraic Analyses

Let us proceed at this point to consolidate our new knowledge by review and inspection of the basic algebraic analyses that apply to simple *RC*-coupled amplifiers. The following paragraphs discuss voltage gain formulas for triode audio amplifiers or FET audio amplifiers. We start with the least complex situation, in which the amplifier is operated in the midband frequency range, and consider the constant-voltage generator analysis. With reference to Fig. 7–13, the total resis-

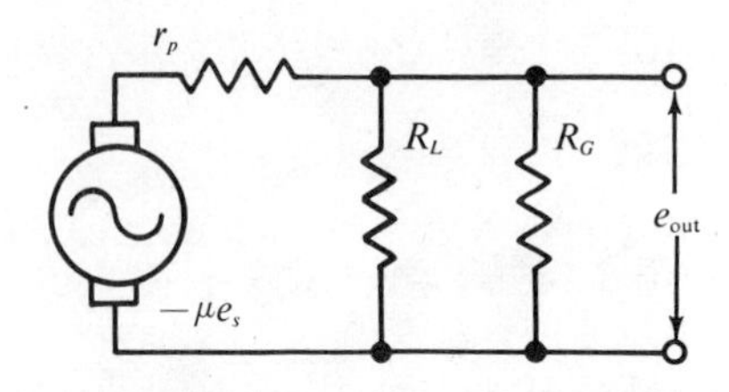

Figure 7–13. Equivalent circuit at midband frequencies.

tance "seen" by the generator is formulated

$$R_T = r_p + \frac{R_L R_G}{R_L + R_G} \tag{7–12}$$

If R_G should have a very high resistance in comparison with R_L, we could neglect R_G as was done in Formula 7–11. However, when R_G has a value in the same order of magnitude as R_L, it cannot be neglected without incurring objectionable error. Let us formulate the value of plate-current flow from Formula 7–12.

$$i_p = \frac{-\mu e_s}{R_T} = \frac{-\mu e_s}{r_p + \dfrac{R_L R_G}{R_L + R_G}} \tag{7-13}$$

or,

$$i_p = \frac{-\mu e_s(R_L + R_G)}{R_L R_G + r_p(R_L + R_G)} \tag{7-14}$$

The output voltage is equal to the product of i_p and the effective resistance of R_L and R_G in parallel. Therefore

$$e_o = \frac{-i_p R_L R_G}{R_L + R_G} = \frac{-\mu e_s(R_L + R_G)R_L R_G}{[R_L R_G + r_p(R_L + R_G)][R_L + R_G]} \tag{7-15}$$

or,

$$e_o = \frac{-\mu e_s R_L R_G}{R_L R_G + r_p(R_L + R_G)} \tag{7-16}$$

Voltage gain, symbolized by A_V, equals e_o/e_s. Accordingly

$$A_V = \frac{-\mu R_L R_G}{R_L R_G + r_p(R_L + R_G)} \tag{7-17}$$

Example 7. A vacuum-tube *RC*-coupled amplifier has the following values: $R_L = 30$ kΩ, $R_G = 100$ kΩ, $r_p = 20$ kΩ, and $\mu = 20$. What is the voltage gain at mid-frequency?

$$A_V = \frac{-20 \times 3 \times 10^4 \times 10^5}{3 \times 10^4 \times 10^5 + 2 \times 10^4(3 \times 10^4 + 1 \times 10^5)}$$

$A_V \simeq -17$ (answer)

As previously explained, the minus sign denotes that the plate-output voltage is shifted 180° in phase with respect to the grid-input voltage.

Recall that if R_G should be much greater in value than R_L, we draw the equivalent circuit shown in Fig. 7-14, and the foregoing

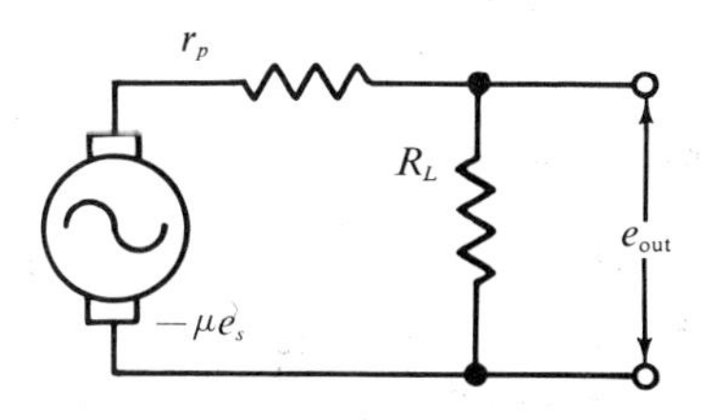

Figure 7-14. Equivalent circuit when $R_G \gg R_L$.

formulas become simplified, as follows.

$$R_T = r_p + R_L \tag{7–18}$$

$$i_p = \frac{-\mu e_s}{r_p + R_L} \tag{7–19}$$

$$e_o = \frac{-\mu e_s R_L}{r_p + R_L} \tag{7–20}$$

$$A_V = \frac{-\mu R_L}{r_d + R_L} \tag{7–21}$$

Example 8. What is the voltage gain of an FET amplifier when $R_D =$ 33 kΩ, $r_{ds} =$ 30 kΩ, $R_G =$ 100 MΩ, and $\mu = -30$.

$$A_V = \frac{-30 \times 3.3 \times 10^4}{3 \times 10^4 + 3.3 \times 10^4}$$
$$\simeq -15.9 \qquad \text{(answer)}$$

The voltage gain at low frequencies, such as 15 to 200 Hz, entails recognition of the reactance in the coupling capacitor. Therefore, we draw our low-frequency equivalent circuit as shown in Fig. 7–15, and

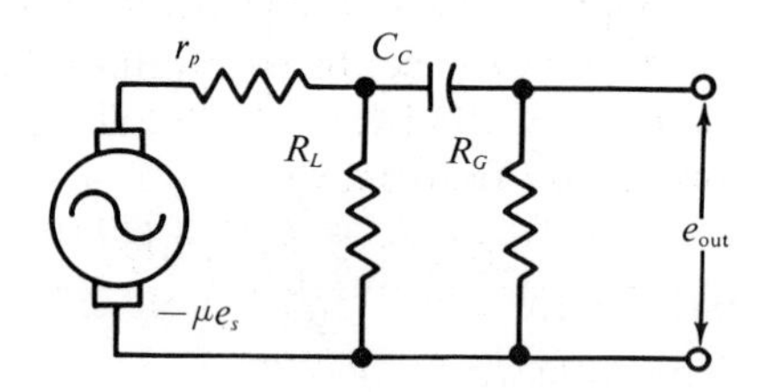

Figure 7–15. Equivalent circuit of low frequencies.

then we take the reactance of coupling capacitor C_C into account. Our voltage-gain formula becomes

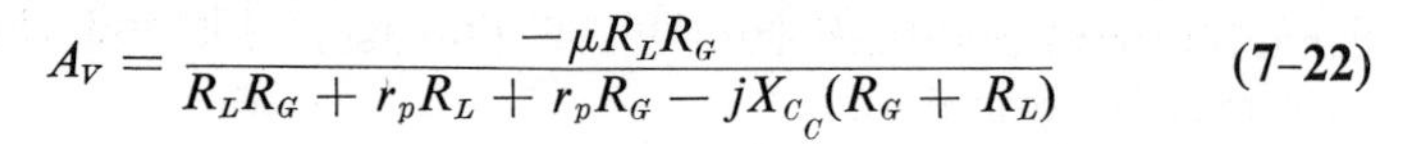

$$A_V = \frac{-\mu R_L R_G}{R_L R_G + r_p R_L + r_p R_G - jX_{C_C}(R_G + R_L)} \tag{7–22}$$

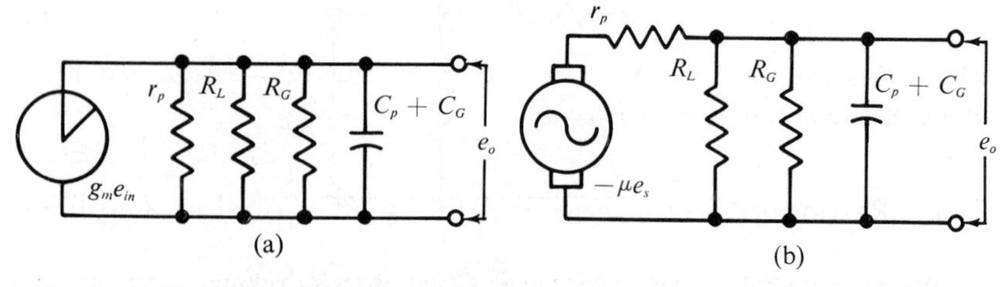

Figure 7–16. Equivalent circuits of vacuum tubes and field-effect transistor amplifiers at high frequencies. (a) Equivalent voltage circuit; (b) Equivalent current circuit.

The voltage gain at high frequencies, such as 10 kHz and higher, entails consideration of the total stray capacitance in shunt to the resistance of the stage, and we draw our equivalent circuit as depicted in Fig. 7-16(a). Then, when we take the reactance of C into account, our voltage-gain formula becomes

$$A_V = \frac{-\mu R_L R_G}{R_L R_G + r_p(R_L + R_G) + j\omega C r_p R_L R_G} \tag{7-23}$$

In view of the comparative complexity of the above formulas, it is obviously advantageous to calculate the gain of *RC*-coupled amplifiers with the aid of the universal charts that we previously developed. While the charts do not eliminate calculations completely, they reduce the necessary labor of calculation to a bare minimum.

The calculations for gain of vacuum-tube and FET amplifiers can sometimes be simplified by using an equivalent current circuit as in Fig. 7-16(b). The amplifying device appears as a current generator ($g_m e_{in}$) in parallel with the internal device impedance (r_p or r_{ds}), the device load resistor (R_L or R_D), the load resistor (R_G), and the reactance of capacitor C. When the values of the capacitive reactance are large in relation to the resistor values, the voltage gain may be formulated

$$A_V \simeq -g_m R_T \tag{7-24}$$

where R_T is r_{int}, R_L and R_G in parallel.

The voltage gain at low frequencies, which entails the reactance of the coupling capacitor, may be formulated

$$A_{V\text{low}} = -g_m Z \frac{R_G}{R_G + \dfrac{1}{j\omega C}} \tag{7-25}$$

where Z is the total impedance of the network at the particular low frequency.

Again, the voltage gain at high frequencies entails the total stray capacitance in parallel with the resistance of the stage. The voltage gain may be formulated

$$A_{V\text{high}} = -g_m Z \tag{7-26}$$

where Z is the total parallel impedance of the resistances and resistance of the shunt capacitance.

7-6. Resonant-Circuit Analogy to an RC-Coupled Amplifier

We may be surprised to discover that a low-Q resonant circuit can be formulated to describe the frequency-response characteristic of an *RC*-coupled amplifier. That is, a suitable parallel-resonant *LCR* circuit

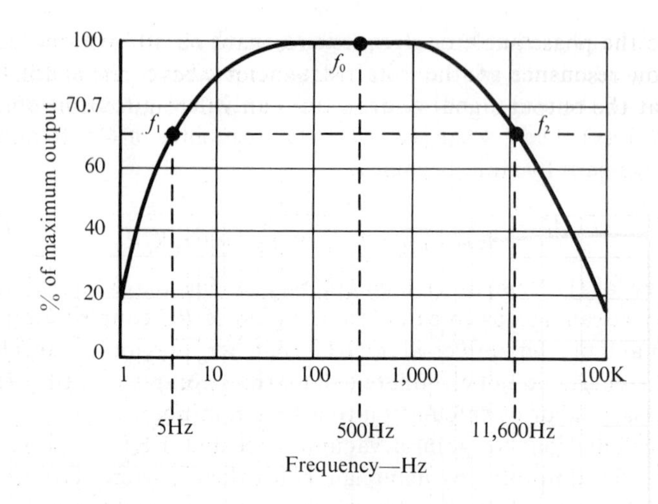

Figure 7–17. Frequency response of an *RC*-coupled amplifier.

is equivalent to a chosen *RC*-coupled amplifier circuit insofar as its frequency response is concerned. If we plot the percentage of output voltage versus frequency for an *RC*-coupled amplifier, we obtain a curve such as shown in Fig. 7-17. The resemblance of this curve to the frequency characteristic of a low-Q parallel-resonant circuit is evident. This method of analysis is best adapted to comparatively narrow-band amplifiers, and becomes unsatisfactory for wide-band amplifiers.

First let us ask what the Q value of the response characteristic in Fig. 7–17 may be. Recall that the Q value is related approximately to the bandwidth as follows.

$$BW = f_2 - f_1 \tag{7–27}$$

$$Q \simeq \frac{f_0}{f_2 - f_1} \simeq \frac{f_o}{BW}$$

where BW denotes bandwidth, Q denotes X_L/R, f_0 denotes the resonant frequency, and f_2 and f_1 denote the frequencies at the half-power or -3 db points.

Thus, with reference to Fig. 7–17, the bandwidth of the amplifier is 11,600 Hz (approximately), f_0 is 500 Hz, and the Q of the equivalent parallel-resonant circuit is approximately 0.043. Of course, this is a very low Q value by ordinary standards. A narrow-band audio amplifier such as used in military speech work might have an equivalent Q value of 0.5. In any case, the equivalent Q of an *RC*-coupled amplifier is a small value.

Since the phase angle of a parallel-resonant circuit leads at frequencies below resonance and lags at frequencies above resonance, it follows that the output signal voltage from an *RC*-coupled amplifier will

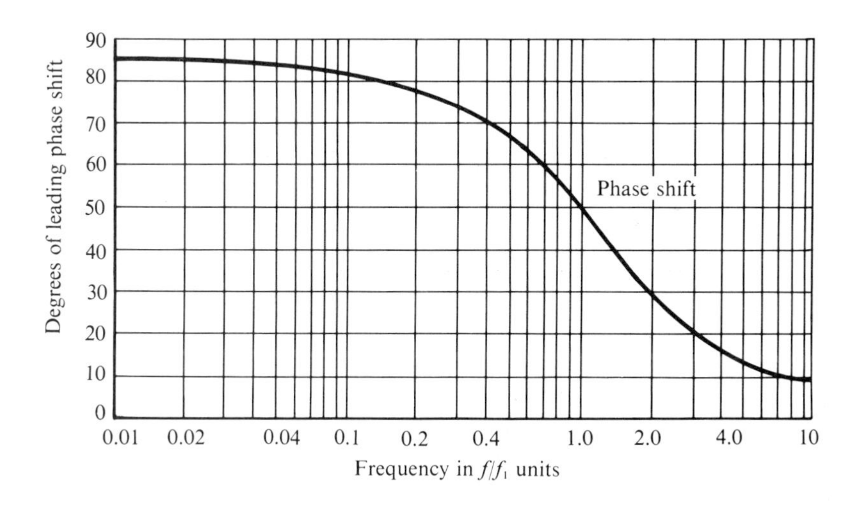

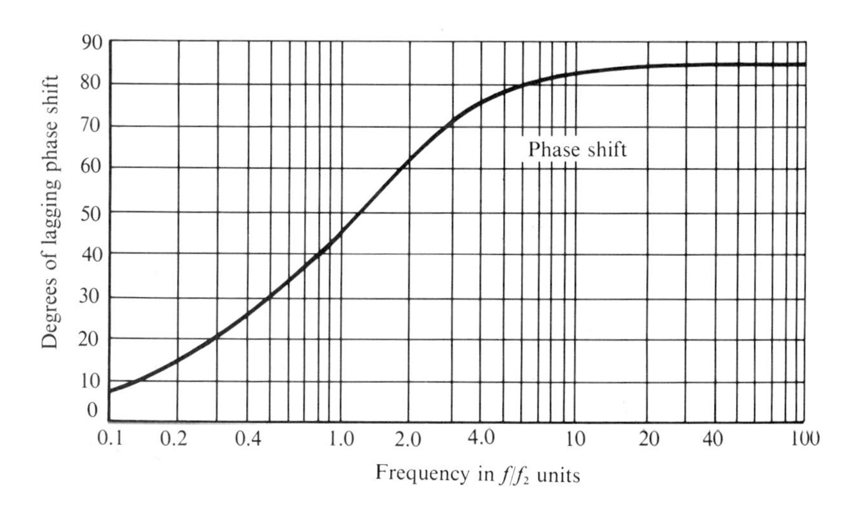

Figure 7–18. (a) Universal phase-shift chart for *RC*-coupled amplifier at low frequencies; (b) Universal phase-shift chart for *RC*-coupled amplifier at high frequencies.

lead the input signal voltage at low frequencies, and lag at high frequencies. It is instructive to use this knowledge of equivalent circuits to derive a universal phase-shift chart for RC-coupled amplifiers, as shown in Fig. 7–14. Note that f_1 and f_2 refer to the frequency of test

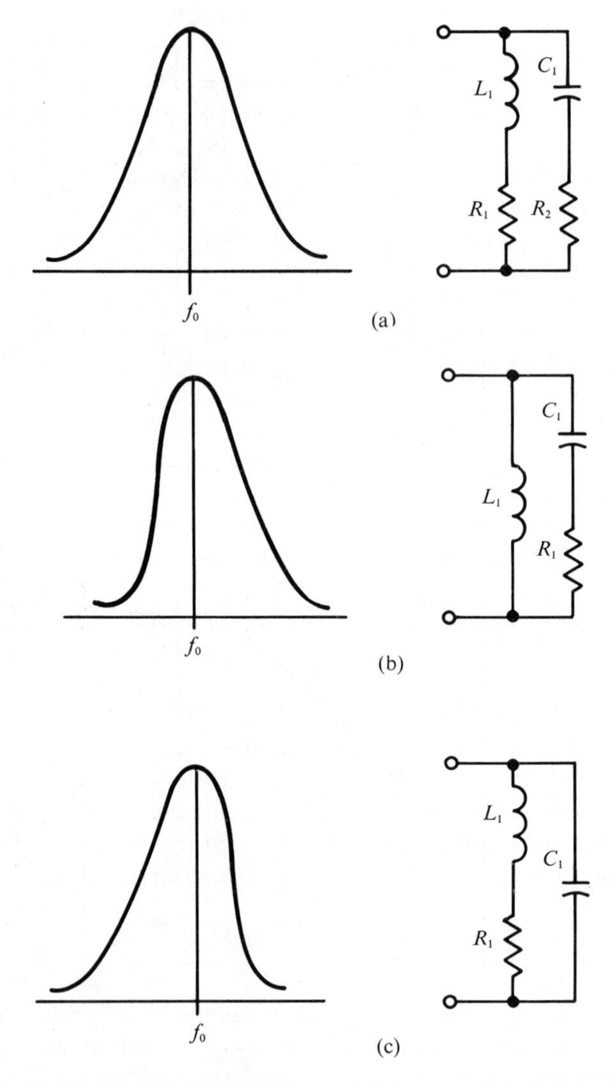

Figure 7–19. Symmetry or asymmetry of frequency-response curve depends on resistance proportions in *L* and C branches. (a) Equal resistances; (b) Resistance in capacitance branch; (c) Resistance in inductive branch.

to the half-power point at low frequencies and to the half-power point at high frequencies, respectively. Refer to Fig. 7-17 for an indication of f_1 and f_2.

At the frequencies of f_1 and f_2, the circuit capacitive reactance is equal to the equivalent circuit resistance. Examination of Figs. 7-15 and 7-16 will show that at f_1 and f_2 the circuit phase angle is $+45°$ and $-45°$, respectively.

Figure 7-18(a) shows that the phase angle leads almost 90° at very low test frequencies and leads by only a small amount as the midband frequency is approached. Of course, at f_0, the phase shift of the amplifier is zero. Figure 7-18(b) shows that the phase angle lags by a small amount as the midband frequency is passed and lags by almost 90° at very high test frequencies. Beginning students sometimes suppose that an *RC*-coupled amplifier has zero phase shift through the entire midband region from f_1 to f_2. We perceive that this is not so. The phase shift is exactly zero only at f_0. In other words, whenever we have a circuit with reactive components, the condition of zero phase shift exists at only one frequency, although the phase shift might be very small through the midband region of the circuit's frequency response.

The frequency-response curve of an audio amplifier may be symmetrical at low- and high-frequency ends, but it is more common for the low-frequency end to rise comparatively rapidly and for the high-frequency end to fall off gradually. In any case, an *LCR* parallel-resonant curve can be fitted to the curve of an *RC*-coupled amplifier. The bandwidth, of course, is determined by the amount of resistance in the *LCR* circuit. Moreover, the symmetry or asymmetry of the curve depends upon the resistance proportions in the *L* and *C* branches, as shown in Fig. 7-19.

If the *L* and *C* branches have approximately equal values of resistance, the frequency-response curve is approximately symmetrical. On the other hand, if there is more resistance in the *C* branch, the curve falls off gradually at high frequencies. Conversely, if there is more resistance in the *L* branch, the curve rises slowly at low frequencies.

SUMMARY

1. A coupling capacitor in an *RC*-coupled amplifier passes the signal voltage from the output of one stage to the input of the next stage and provides d-c voltage isolation between the two stages.

2. The value of the coupling capacitor and of the load resistor determines the low-frequency response of an *RC*-coupled amplifier.

3. The values of the shunt capacitance and all associated resistors

in the output of an amplifier determine the high-frequency cutoff of an *RC*-coupled amplifier.

4. An audio amplifier may be analyzed as a voltage source ($-\mu e_{in}$) in series with the internal resistance of the device (r_{int}) and the load impedance. The voltage gain is therefore always less the μ of the device.

5. An audio amplifier may also be analyzed as a current source ($g_m i_{in}$) in parallel with the internal resistance of the device (r_{int}) and the load impedance.

6. Higher values of coupling and bypass capacitors must be employed in transistor amplifiers than in amplifiers using field-effect or vacuum devices, due to the inherent low input and output resistances of the junction transistor.

7. The emitter resistor in a junction-transistor circuit performs a temperature-stability function rather than a bias function compared with cathode and source resistors.

8. In a junction-transistor *RC*-coupled amplifier, the emitter of the transistor is forward-biased with respect to base, and the collector is reverse-biased with respect to base.

9. The percentage of maximum output voltage of a single *RC*-coupled stage may be determined by the chart in Fig. 7–4 when the ratio of R/X_C is known, where R is the equivalent circuit resistance and X_C is the reactance of the coupling capacitor.

10. A symmetrical two-stage amplifier is one in which two identical stages are connected in cascade.

11. When amplifier stages are connected in cascade to obtain greater gain, the bandwidth decreases.

12. Mid-frequency of an amplifier is that range of frequencies where the reactance of the coupling capacitors is sufficiently small to appear as a short circuit, and the reactance of the shunt capacitance is sufficiently large enough to appear as an open circuit.

13. The voltage gain of a multi-stage amplifier is the product of the individual gains of each stage.

14. The phase angle of an amplifier will lead the input signal at low frequencies and will lag the input signal at high frequencies.

Questions

1. What are the functions of the interstage coupling capacitor in a two-stage amplifier?
2. What factors limit the high-frequency cutoff of an *RC*-coupled amplifier?
3. Compare the elements of a field-effect transistor to those of a triode vacuum tube.
4. What is the function of R_G in the grid of a vacuum tube and R_G in the gate of an FET?
5. Compare the source resistor of an FET to the cathode resistor of a vacuum tube, and explain their purpose.
6. What is the major limiting factor of amplification in an amplifier at high frequencies?
7. What are the two most noticeable differences between an audio amplifier using vacuum tubes and one using junction transistors?
8. Why is the input impedance to a common-emitter transistor amplifier much less than the output impedance?
9. What is the purpose of the emitter resistor R_E in a transistor amplifier?
10. From the chart in Fig. 7-4, find the percentage of maximum output voltage when $\omega C = R$, when $\omega C = 2\ R$, and when $\omega C = 0.5\ R$.
11. What is the relationship between gain and bandwidth in an amplifier?
12. The individual voltage gain of each stage of a two-stage *RC*-coupled amplifier is 22. What is the overall voltage gain (assuming no losses in the coupling circuit)?
13. Draw an equivalent mid-frequency circuit, an equivalent low-frequency circuit, and an equivalent high-frequency circuit of an amplifier.
14. Explain each of the equivalent circuits in Question 13.
15. What is the relationship between bandwidth and the load resistance of an amplifier?
16. Determine the voltage gain of an FET amplifier when $\mu = 35$, $R_D = 20\ \text{k}\Omega$, $r_{ds} = 50\ \text{k}\Omega$, and $R_G = 1\ \text{M}\Omega$.
17. Compare a low-*Q* resonant circuit to an *RC*-coupled amplifier.
18. Why is the equivalent *Q* value of an *RC*-coupled amplifier a small value?
19. What is the phase angle between the input and the output signal of an amplifier at the cutoff frequencies f_1 and f_2?
20. From the charts in Figs. 7-18 and 7-19, determine the phase shift of the output signal when $f/f_1 = 2$ and $f/f_2 = 2$.
21. Compare the high- and low-frequency response curves.

PROBLEMS

1. From the chart in Fig. 7-4, find the percentage of maximum output voltage when $\omega C = 2\ R$, when $\omega C = 0.3\ R$, and when $\omega C = 4\ R$.
2. Suppose the amplifier *T*1 in Fig. 7-4(b) has a maximum output of 10 v; what is the output voltage at 20 Hz if the equivalent resistance of the circuit is 1 MΩ?
3. Determine the frequency at which the base voltage of the transistor (*T*2) in Fig. 7-1(c) is 70 per cent of the maximum voltage, when the equivalent circuit resistance is 1 kΩ and the coupling capacitance is 5 μf.
4. What value of coupling capacitor C_C is necessary to allow the grid voltage in Fig. 7-1(a) to be at 70 percent of maximum at a frequency of 100 Hz, when the equivalent circuit *R* is 80 kΩ?
5. The equivalent high-frequency circuit of a single-stage FET amplifier is composed of a 1-MΩ series resistor and a 2-nf shunt capacitor. If the maximum output is 20 v, what is the output voltage at 1 MHz?
6. The equivalent high-frequency circuit of a junction transistor can be represented by a 1.59-kΩ parallel resistor and a 2,000-pf shunt capacitor. At what frequency is the output current 70 per cent of the maximum current?
7. An amplifier is to have an output current of 70 per cent maximum at a frequency of 100 kHz. What value of equivalent-shunt resistance must be employed if the shunt capacitance is 1.59 nf ? (Refer to Fig. 7-9.)

CHAPTER 8

RC-Coupled Amplifier

8–1. Introduction

Some of the capabilities and limitations of *RC*-coupled amplifiers were noted in the preceding chapter. At this point, it is advisable for us to consider these factors in somewhat greater detail. Typical *RC*-coupled triode field-effect transistors (FET) and vacuum-tube stages are shown in Fig. 8–1. We are aware of the fact that the value of the gate- or grid-leak resistor R_{G1} should not exceed a maximum rated value which is specified by the manufacturer. If this resistance value is exceeded, unstable amplifier operation often results, and there is also the possibility of device damage. To understand this limitation in amplifier operation, we must recognize that the symbol for a vacuum tube or field-effect transistor is a very abbreviated technical "shorthand," which implies a number of physical principles.

We know that the gate-leak resistor R_{G1} in Fig. 8–1 (a) applies the bias voltage developed across R_S to the gate. If the amplifier stage is

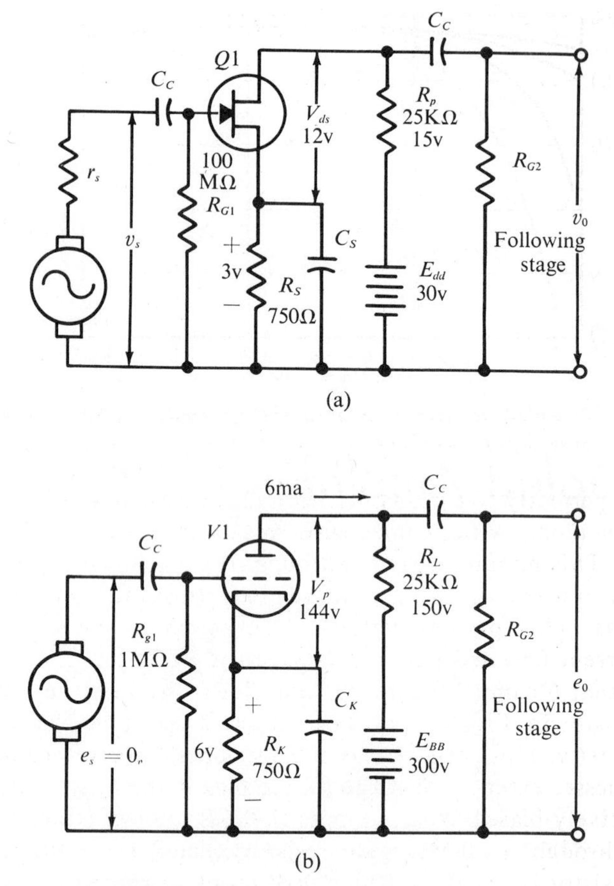

Figure 8–1. **Triode amplifier stage. (a) Field-effect transistor; (b) Vacuum tube.**

not overdriven (operates in Class *A*), the d-c voltage on the gate is constant, since no gate current flows to produce a voltage drop across R_{G1}. It is evident that the value of R_{G1} should be sufficiently large that the time constant $R_{G1}C_C$ is large, thereby providing good low-frequency response. Observe the low-frequency response curves shown in Fig. 8–2. As the value of the gate resistance is progressively increased, the low-frequency response is also extended from a to b, to c, to d.

Maximum gate resistance values from 1 to 100 MΩ are typically specified by FET manufacturers for various types of field-effect transistors. An upper limit is specified because of the presence of minority carriers in the PN gate-to-source junction. No semiconductor material

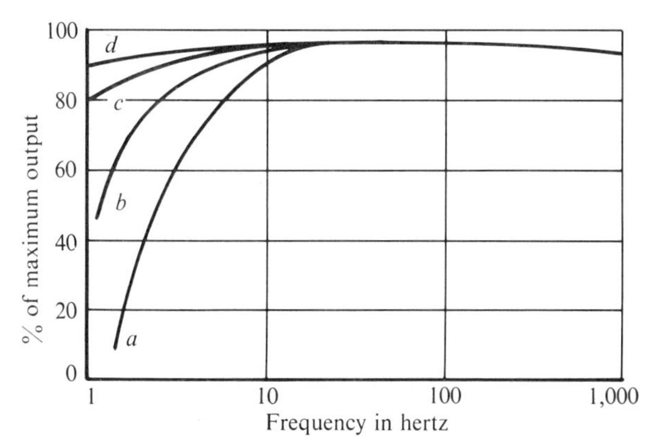

Figure 8-2. Extended low-frequency response is provided by higher values of grid-leak resistance.

can be made perfectly free of impurities, and the reverse gate-to-source junction bias along with temperature results in a very small gate-current flow. This minute current through the gate resistor R_G is in the direction to make the gate positive and reduce the reverse gate-to-source bias. This positive voltage drop across R_G could increase the drain-current flow and possibly damage the FET.

Suppose that 6 namp of gate current flows through the 100-MΩ gate-leak resistor in Fig. 8–1 (a). The voltage drop across the resistor is $6 \times 10^{-1} = 0.6$ v. This positive bias of 0.6 v opposes the source bias to a greater or lesser extent, and if the source bias is small, the gate will become positively biased. Next, suppose that 6.0 namp of minority current flows through a 10 MΩ gate resistor. Then, the voltage drop across the resistor is $6 \times 10^{-2} = 0.06$ v. Reduction of the source voltage is much less and the possibility of damage to the FET is practically eliminated.

Maximum grid-leak resistance values from 1 to 10 MΩ are typically specified by vacuum-tube manufacturers for various types of triodes. An upper limit is specified because of the presence of residual gas molecules in a vacuum tube. No tube can be pumped to a perfect vacuum, and electron current flow in a tube tends to release a few gas molecules from the metal electrodes. Although a "getter" is provided in the tube (a highly reactive chemical substance which tends to capture gas molecules), a perfect vacuum cannot be produced.

Residual gas particles become ionized by bombardment from the cathode. These positive ions are attracted to the negative grid and capture electrons from the grid metal. In turn, a small grid current flows. This grid current flows through R_{G1} in Fig. 8–1, and its direction of flow is evidently in a direction that makes the grid less negative. If the voltage drop across R_{G1} should be excessive, a very large plate current would flow. Excessive plate-current flow will damage or destroy a tube; thus, all vacuum tubes are rated for maximum plate-current flow.

When marginal values of grid-leak resistance are used in an *RC*-coupled amplifier, operation becomes unstable. Gas current values are not constant, but tend to vary with the peak plate-current flow. If a small grid-drive signal is applied, the tube remains stable; however, if a large grid-drive signal is applied, gas current flow increases, and the coupling capacitor tends to maintain the resulting positive grid bias. Unstable operation results, and the signal output is distorted. A critical point may be passed if the grid-signal drive is excessive, at which grid-current flow "snowballs" and the tube becomes damaged. This is called grid-current blocking action.

The alert student will perceive from Figs. 8–1 and 8–2 that low-frequency response can be extended by increasing the value of the coupling capacitor. It might be supposed that the low-frequency re-

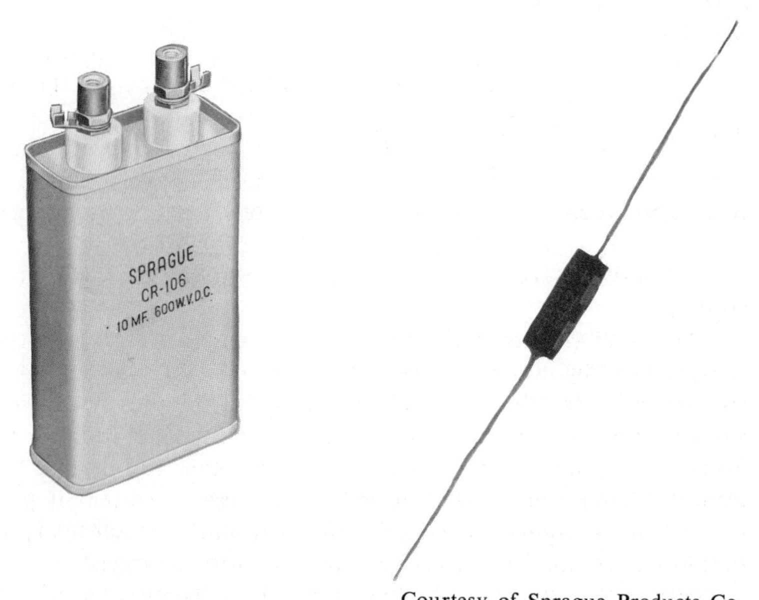

Courtesy of Sprague Products Co.

Figure 8–3. (a) A 10-μf coupling capacitor; (b) A 10-μf, 50 wv d-c tantalum capacitor.

sponse could be extended indefinitely merely by utilizing a suitably large value of coupling capacitance; however, we again encounter a practical limitation in our effort to extend the low-frequency response by this means. First, a very large coupling capacitor is physically very large. Consider the 10-μf, 600-wv capacitor for vacuum-tube circuits and the 10-μf, 50-wv capacitor for transistor and field-effect transistor circuits illustrated in Fig. 8–3. The 10 μf, 600-wv capacitor has a surface area of 55 sq in. In turn, a physically large capacitor has a high value of stray capacitance to the chassis and components of the amplifier. Thus, the high-frequency response of the amplifier becomes impaired.

Second, it is difficult to maintain a very high value of leakage resistance in a capacitor that has a very large capacitance value. All capacitors tend to become more "leaky" after extended operation. It is evident that leakage resistance in a coupling capacitor permits current flow from the plate or drain power supply through the grid-leak or gate resistance. The result is the same as gas-current flow insofar as tube operation is concerned, and the same as minority gate current flow insofar as FET operation is concerned. Thus, it is generally considered that coupling capacitors in *RC*-coupled FET or vacuum-tube amplifiers should be limited to a maximum value of 0.25 μf. In junction-transistor amplifiers, coupling capacitors of 10 to 20 μf are common, because of the low input resistance of these devices. We will find that the low-frequency response can nevertheless be extended by other circuit means to be discussed subsequently.

8–2. Source- or Cathode-Bias Circuit

With reference to Fig. 8–1, let us consider the circuit action of the cathode- or source-bypass capacitors, C_K and C_S. We will use the FET stage for convenience, as the action is the same in a vacuum-tube or a junction-transistor stage. It is instructive to consider the stage operation when C_S is omitted as in Fig. 8–4. We observe that the drain current flows through R_S and that part of the output signal voltage drops across R_s. The phase relations are such that the source signal drop subtracts from the gate signal. This is just another way of saying that the total gate-signal is reduced. In turn, the maximum available output signal is not obtained from the amplifier. We say that *negative feedback* is present when the source-bias resistor is unbypassed.

Again, with reference to Fig. 8–1 (a), it is evident that if C_S has a very large capacitance, there is no signal-voltage drop across the

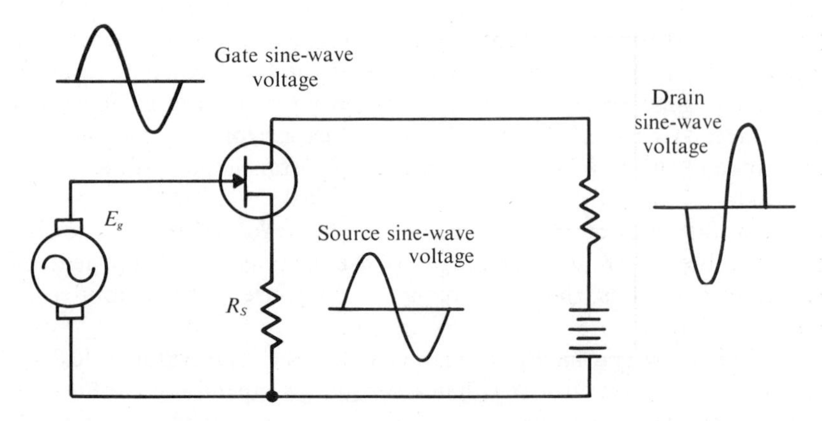

Figure 8–4. **The source signal opposes the gate signal.**

source-bias circuit. Hence, no negative feedback is present, and we obtain the maximum available gain from the stage. Note that the reactance of C_S must be chosen to be a small fraction of the resistance value of R_S, at the lowest frequency of operation. A rule of thumb for selecting C_S is that $X_C = 0.1\ R_S$ at the lowest frequency we wish to pass.

Example 1. Let us assume that we wish to obtain the maximum available gain of the stage at 60 Hz and that R_S has a value of 750Ω.

$$X_C = 0.1R_s$$
$$= 75\ \Omega \text{ at } 60 \text{ Hz}$$

and

$$C_S = \frac{0.159}{60 \times 75}$$
$$\simeq 35\mu\text{f} \qquad \text{(answer)}$$

Therefore, C_S must have a value of at least 35 μf. An electrolytic capacitor will be utilized. The comparatively low leakage resistance of an electrolytic capacitor is of no consequence, because the capacitor is shunted by a low resistance of 750 Ω.

It is interesting to note that we can often obtain improved high-frequency response at the expense of gain by *partial bypassing* of the source emitter or cathode resistor. For example, suppose that we wish to improve the high-frequency response of a stage when a drain-load resistance of 0.5 MΩ is employed. We observe from Fig. 8–5 (a) that when the source-bias resistor is completely bypassed, the frequency

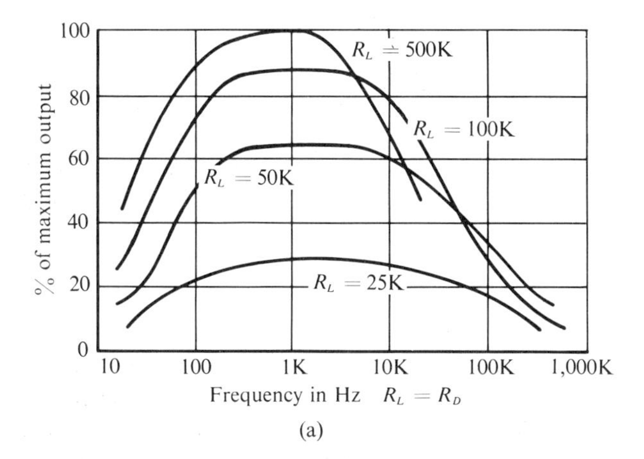

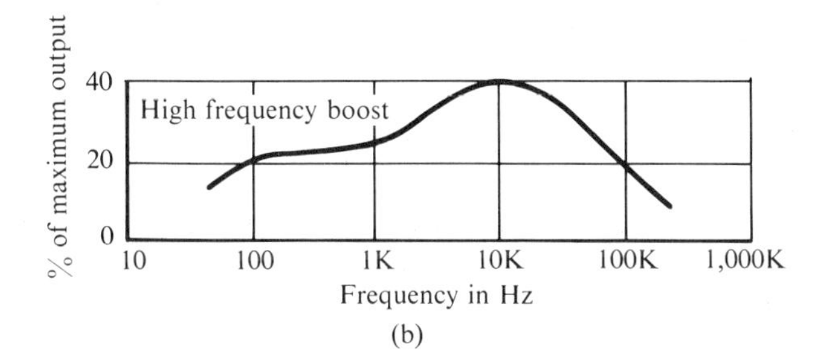

Figure 8–5. **(a) Stage response with completely bypassed source resistor; (b) High-frequency boost by partially bypassed source resistor.**

response starts to drop off soon after 1 kHz. Now, suppose that we choose a 0.2-μf source-bypass capacitor and connect it across a 750-Ω source resistor. We perceive that this capacitor begins to bypass appreciable signal around the source-bias resistor at frequencies higher than 1 kHz. Therefore, the stage gain increases past 1 kHz, and the high-frequency response is improved [Fig. 8–5 (b)].

Of course, there is a limit to the extension of high-frequency response that can be obtained by sacrifice of maximum available gain. At 10 kHz, for example, the bypassing action of a 0.2-μf capacitor is almost complete, and we can obtain practically no extension of high-frequency response at frequencies higher than 10 kHz. Within its limits, however, the technique of trading available gain for extended high-frequency response is a very useful expedient in practical design work.

8–3. Drain- or Plate-Load Resistance

Even if an amplifier is only required to operate over a very limited frequency range, there are nevertheless practical limits imposed on the highest value of d-c load resistance that may be used. These limitations are first, the maximum convenient or allowable supply voltage; and second, a desire for a linear output signal and a power gain. To demonstrate these limitations, we will evaluate a vacuum tube amplifier.

With reference to Fig. 8–1 (b), we observe that when a 25-kΩ plate-load resistor is utilized, there is a d-c drop of 150 v across the resistor. In turn, the plate-cathode d-c voltage is 144 v. If we should increase the plate-load resistance to 0.5 MΩ, the plate-cathode voltage would become objectionably low unless a very high value of plate-supply voltage were used. In other words, it is impractical to use a plate-supply voltage of several thousand volts in conventional amplifiers.

Let us see why it is impractical to obtain a plate-cathode voltage of 150 v if we use a 0.5-MΩ plate-load resistor. The plate-current flow is 6 ma. If the plate-load resistance is 25,000 Ω as in Fig. 8–1 (b), the power dissipation in the resistance is 0.9 w, and this is not an unreasonable value. On the other hand, if we use a plate-load resistance of 0.5 MΩ, the power dissipation in the resistance at 6 ma current flow becomes 18 w, which is an objectionably high value from a practical viewpoint. Moreover, there would be a 3,000-v drop across the plate-load resistor, and the power supply must provide 3,150 v to obtain a plate-cathode voltage of 144 v. This is an absurd value from a practical standpoint.

It is practical to use a plate-supply voltage of about 300 v. If a plate-load resistance of 25,000 Ω is chosen for a 6J5 triode, the output waveform will be reasonably undistorted, as seen from Fig. 8–6. In other words, the load-line intervals from $E_g = 0$ to $E_g = -2$, from $E_g = -2$ to $E_g = -4$, from $E_g = -4$ to $E_g = -6$, and so on, are reasonably uniform. On the other hand, if a plate-load resistance of 0.5 MΩ is used, we are then operating on the rapidly-curved portions of the characteristics in the vicinity of plate-current cutoff. These curvatures vary considerably from one tube to another; they even vary for the same tube as it ages. Accordingly, the output waveform will be appreciably distorted, and the distortion will vary in an unpredictable manner. For this reason, it is impractical to use extremely high values of plate-load resistance, even if the plate-supply voltage is chosen at a suitable value.

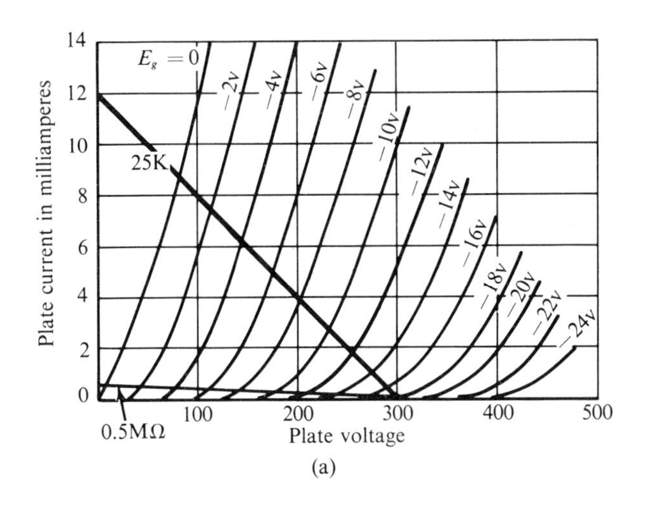

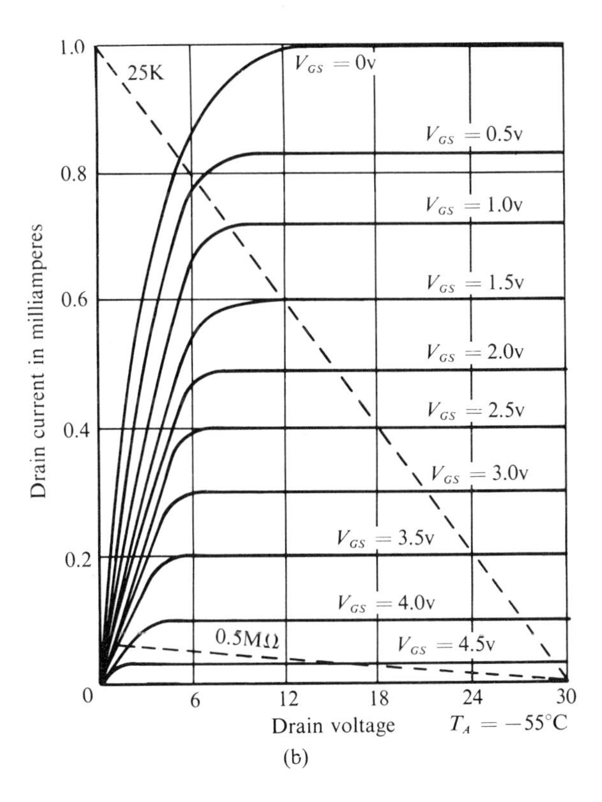

Figure 8-6. **Output characteristic curves. (a) Vacuum tube; (b) Field-effect transistor.**

This same reasoning applies to the selection of a drain resistor for the FET amplifier shown in Fig. 8-1 (a). The two load lines, 25 kΩ and 0.5 MΩ, are drawn on the drain characteristics in Fig. 8-6 (b). We may also note that selection of an extremely large value of plate or drain resistance limits the linear output-current swing and, therefore, the output power.

8-4. Biasing Junction-Transistor Circuits

Operating differences between junction transistors, vacuum tubes, and field-effect transistors justify a review of the characteristics of the junction transistor at this point.

1. In a Class-A amplifier, the input to the junction transistor is forward-biased. Therefore, the base always draws current and usually offers an input impedance of less than a thousand ohms to a signal voltage.
2. The output impedance of a transistor is usually a lower value than that of an FET or vacuum-tube amplifier, usually 2 to 20 kΩ.
3. Junction transistor characteristics vary greatly with a change of temperature; we will return to this subject in Chapter 9.
4. The input and output circuits of a junction transistor are not isolated from each other; any change in the input circuit reflects into the output, and any change in the output circuit reflects into the input circuit.

The relationship of the reflected output admittance on the input impedance can be observed in Formula 8-1 for the input impedance (Z_i).

$$Z_i = h_{ie} - \frac{h_{re}h_{fe}}{h_{oe} + G_L} \tag{8-1}$$

where h_{oe} is the transistor's output admittance, h_{re} is the transistor's reverse-voltage feedback ratio, h_{fe} is the transistor's forward-current gain, h_{ie} is the transistor's input resistance, Z_i is the transistor's input impedance.

Any change of the load admittance G_L ($G_L = 1/R_L$) results in a proportional change in the input impedance. In other words, an increase in the load admittance increases the input impedance, and a decrease in the load admittance decreases the input impedance.

Example 2. Suppose the equivalent transistor circuit in Fig. 8-7 has a load admittance (G_L) of 200 μmho ($R_L = 5$ kΩ) and uses a transistor with the small signal parameters of $h_{ie} = 8$ kΩ, $h_{oe} = 40$

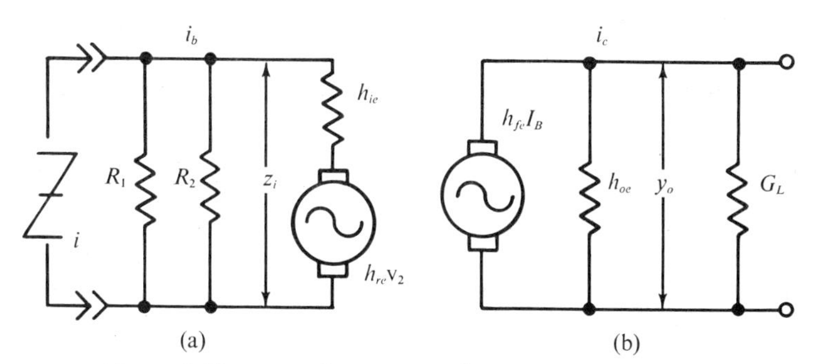

Figure 8–7. Equivalent circuit of a junction transistor.

μmho, $h_{fe} = 200$, and $h_{re} = 5 \times 10^{-4}$. What is the input impedance (z_i)?

$$\begin{aligned} z_i &= h_{ie} - \frac{h_{re}h_{fe}}{h_{oe} + G_L} \\ &= 8 \times 10^3 - \frac{5 \times 10^{-4} \times 2 \times 10^2}{4 \times 10^{-5} + 20 \times 10^{-5}} \\ &= 8 \times 10^3 - 0.425 \times 10^{-3} \\ &= 7.58\ \text{k}\Omega \qquad \text{(answer)} \end{aligned}$$

Now, let us assume that the load resistance is decreased to 1 kΩ ($G_L = 1{,}000$ μmho) and calculate the new input impedance to demonstrate the effects of the load being reflected into the input circuit.

Example 3.

$$\begin{aligned} z_i &= 8 \times 10^3 - \frac{10 \times 10^{-2}}{4 \times 10^{-5} + 100 \times 10^{-5}} \\ &= 8 \times 10^3 - 0.093 \times 10^3 \\ &\simeq 7.91\ \text{k}\Omega \qquad \text{(answer)} \end{aligned}$$

We will note that the input impedance (z_i) is approximately equal to h_{ie} and that the lower the value of R_L, the closer the input impedance approaches h_{ie}. In practical transistor amplifiers ($R_L < 10$ kΩ), we may assume z_i to equal h_{ie}.

The relationship of the reflected input impedance (z_i) to the output circuit can be observed in Formula 8–2 for the output admittance (y_0).

$$y_0 = h_{oe} - \frac{h_{re}h_{fe}}{h_{ie} + Z_i} \qquad \textbf{(8–2)}$$

where $Z_i = z_i$ and all input resistors in parallel (Fig. 8–7).

Any change of the input circuit impedance (Z_i) has an inverse effect on the output admittance, or, the output impedance changes. To illustrate this effect, let us calculate the output admittance for Example 2.

Example 4. Suppose the amplifier in Example 2 has values of $R_1 = 100\ \text{k}\Omega$ and $R_2 = 10\ \text{k}\Omega$ for an equivalent input impendance, Z_i of R_i in parallel with R_1 in parallel with R_2 to equal a total input impedance Z_i of approximately 4.2 kΩ. What is the value of output admittance y_0?

$$
\begin{aligned}
y_0 &= h_{oe} - \frac{h_{re}h_{fe}}{h_{re} + Z_i} \\
&= 40 \times 10^{-6} - \frac{5 \times 10^{-4} \times 2 \times 10^2}{8 \times 10^3 + 4.2 \times 10^3} \\
&= 40 \times 10^{-6} - 7.9 \times 10^{-6} \\
&\simeq 32.1\ \mu\text{mho} \qquad \text{(answer)}
\end{aligned}
$$

Example 5. Suppose the input impedance (Z_i) in Example 4 is reduced by a decrease in R_1 or R_2 to an equivalent value of 1 kΩ. Then,

$$
\begin{aligned}
y_0 &= 40 - \frac{10 \times 10^{-2}}{9 \times 10^{-3}} \\
&= 40^{-6} - 11.1 \times 10^{-6} \\
&= 29\ \mu\text{mho} \qquad \text{(answer)}
\end{aligned}
$$

We will note that the larger the value of the input impedance, the closer the output admittance approaches h_{oe}; and the smaller the value of y_0 becomes (larger r_0).

As you will recall, establishment of an operating point in a junction transistor is considered to be by base current rather than by base voltage. This means that calculations for determining the bias of a junction transistor are somewhat more complicated than those for the FET and the vacuum tube.

Examination of Fig. 8–8 shows that the base current flows through the forward-biased emitter-base diode and develops a voltage (V_{EB}) equal to the voltage barrier. We will consider V_{EB} to be a constant of about 0.3 v for germanium transistors and about 0.6 v for silicon transistors at an ambient temperature of 25° C. Note that these approximations are well within the tolerances of our circuit components.

Example 6. Assume that the circuit in Fig. 8–8 (a) employs a germanium transistor. What is the base current? From Fig. 8–8 (c),

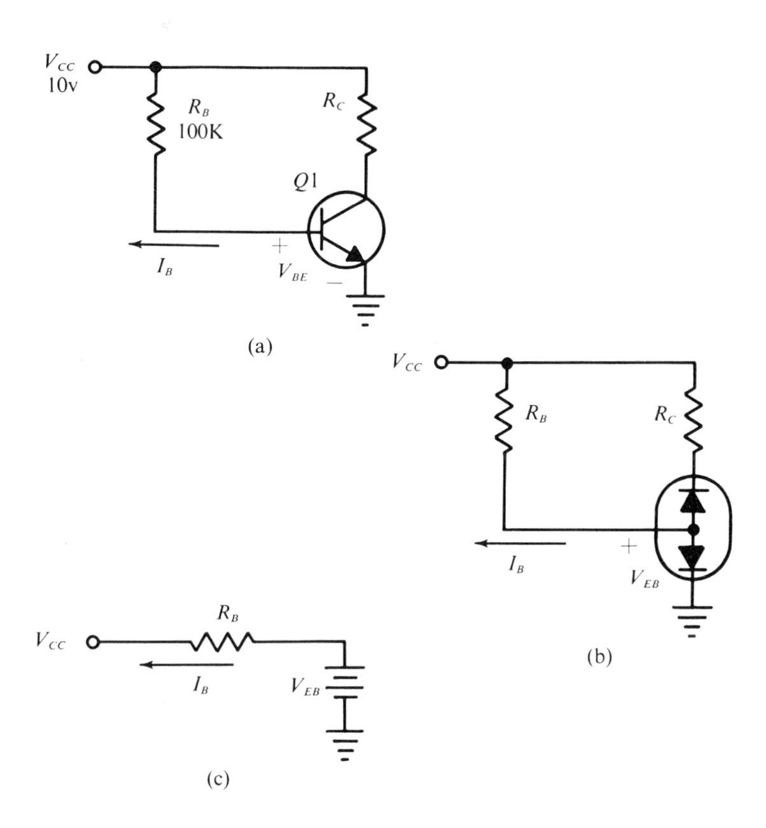

Figure 8–8. (a) Transistor with unstabilized bias circuit; (b) Equivalent circuit consisting of two back-to-back diodes; (c) Equivalent circuit for base-currrent approximation.

$$I_B = \frac{V_{CC} - V_{EB}}{R_B}$$

$$\simeq \frac{9.7}{100\text{ k}\Omega} = 97\ \mu\text{a} \qquad \text{(answer)}$$

Observe that employment of a silicon transistor would decrease the base current slightly to 94 μa.

A more practical bias circuit is depicted in Fig. 8–9. One method of calculating the bias current (I_B) or the circuit value would be to employ Kirchhoff's voltage law with two loop equations.

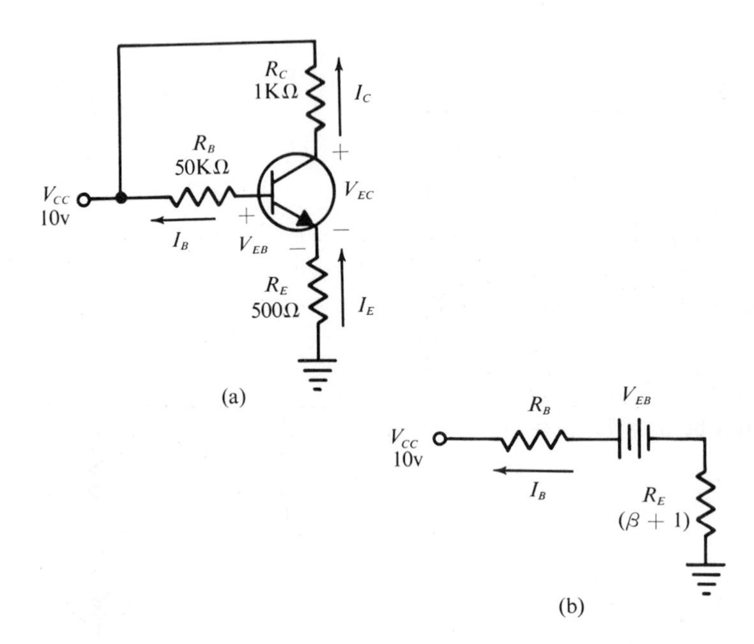

Figure 8–9. **Transistor bias circuits. (a) Circuit employing emitter stabilization; (b) Equivalent circuit for base-current calculation.**

$$0 = -I_E R_E - V_{EC} - I_C R_C + V_{CC} \tag{8–3}$$
$$0 = -I_E R_E - V_{EB} - I_B R_B + V_{CC} \tag{8–4}$$

Examination of the two equations will show that we have five unknowns: I_E, I_C, I_B, V_{CE}, and V_{EB}. Assuming a value for V_{EB} still leaves four unknowns. Obviously, we must eliminate some of these unknowns. Let us rearrange Equation 8–4 by substituting some known relationships.

$$V_{CC} - V_{EB} = I_E R_E + I_B R_B \tag{8–5}$$

Substituting $I_C + I_B$ for I_E in Equation 8–5 gives

$$V_{CC} - V_{EB} = I_C R_E + I_B R_E + I_B R_B \tag{8–6}$$

Again, substituting βI_B for I_C gives

$$V_{CC} - V_{EB} = \beta I_B R_E + I_B R_C + I_B R_B \tag{8–7}$$

Factoring gives

$$V_{CC} - V_{EB} = R_E (\beta + 1) I_B + R_B I_B \tag{8–8}$$

Then,

$$V_{CC} - V_{EB} = [R_E(\beta + 1) + R_B] I_B$$

and finally,

$$I_B = \frac{V_{CC} - V_{EB}}{R_E(\beta + 1) + R_B} \qquad \textbf{(8–9)}$$

Formula 8–9 is obvious if we compare the equivalent circuit in Fig. 8–9 (b) to the formula. In other words, the resistor R_E appears as $(\beta + 1)\ R_E$ to the base circuit. This is because the voltage developed across R_E by I_E is $(\beta + 1)R_E I_B$ as we observed from Equation 8–8.

Example 7. Assuming the transistor in Fig. 8–9 is a germanium type with a beta of 79, what is the base current?

$$I_B = \frac{10 - 0.3}{500\,(79 + 1) + 50 \times 10^3}$$

$$= \frac{9.7}{90\text{ k}\Omega} \simeq 113\ \mu\text{a} \qquad \text{(answer)}$$

The transistor beta value has a large control over the base current.

Example 8. Assume the transistor in Example 7 is replaced by a germanium transistor with a beta of 19. This results in a base current of

$$I_B = \frac{9.7}{10\text{ k}\Omega + 50\text{ k}\Omega}$$

$$= 164\ \mu\text{a} \qquad \text{(answer)}$$

We should also note that the value of R_E has a large control over base current as it appears as $(\beta + 1)\ R_E$ to the base circuit. We shall discuss this point further in Chapter 9 under the subject of Temperature Stability.

Example 9. Suppose only the emitter resistor in Example 8 is increased to 1,000 Ω. What is the effect to the base current?

$$I_B \simeq \frac{9.7}{(19 + 1)\ 1\text{ k}\Omega + 50\text{ k}\Omega}$$

$$\simeq \frac{97}{20\text{ k}\Omega + 50\text{ k}\Omega}$$

$$\simeq 140\ \mu\text{a} \qquad \text{(answer)}$$

Increasing R_E from 500 Ω to 1,000 Ω in Example 8 (beta = 19) had the effect of decreasing I_B by 14.5 per cent. On the other hand, in

Example 8 (beta = 79) a similar increase in R_E produces a decrease in I_B of 34 per cent.

The most common bias arrangement for a junction transistor is depicted in Fig. 8–10. Resistors R_1 and R_2 form a bleeder. Current I_2

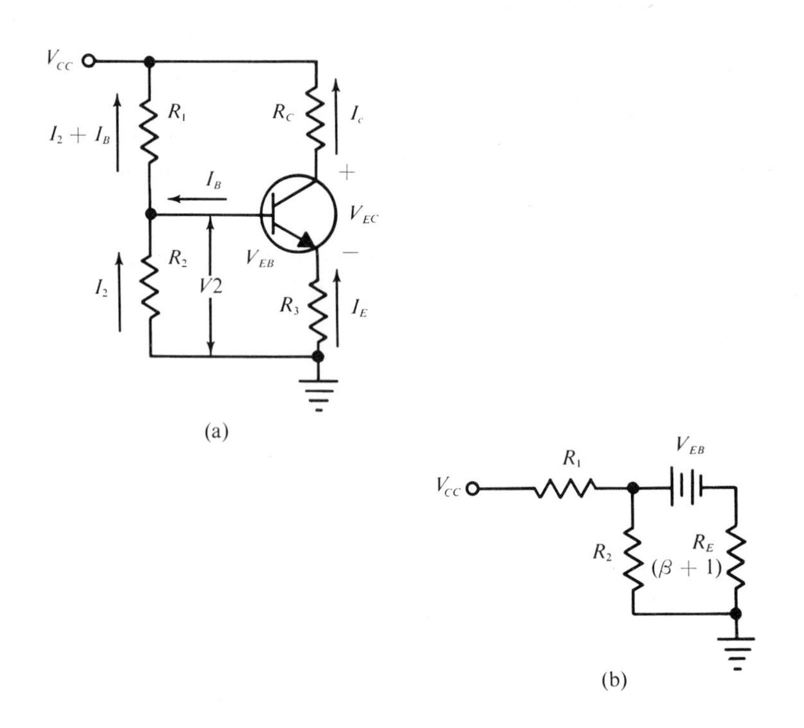

Figure 8–10. Transistor biasing circuits. (a) Bleeder biasing circuit; (b) Equivalent circuit for calculation of base current.

is usually quite large in relation to I_B so that the base-to-ground voltage V_2 is determined primarily by the ratio of R_1 and R_2. When $I_2 \gg I_B$,

$$V_2 \simeq \frac{R_2}{R_1 + R_2} V_{CC} \tag{8–10}$$

and

$$V_E \simeq V_2 - V_{EB} \tag{8–11}$$

then,

$$I_E \simeq \frac{V_E}{R_E} \tag{8–12}$$

and

$$\simeq I_B \; \frac{I_E}{\beta + 1} \text{ or } \frac{V_E}{(\beta + 1)\, R_E} \tag{8-13}$$

Example 10. Determine the base current for the circuit in Fig. 8–10 when $R_1 = 68\text{ k}\Omega$, $R_2 = 4.7\text{ k}\Omega$, $R_E = 1\text{ k}\Omega$, beta $= 99$, $V_{CC} = 20\text{ v}$, and the transistor is a germanium type.

$$R_E\,(\beta + 1) = (99 + 1)\; 1\text{ k}\Omega$$
$$\simeq 100\text{ k}\Omega \qquad \text{(answer)}$$

Note that the value of the emitter resistance that the base current sees is over 20 times the value of R_2. In other words, I_2 is over twenty times I_B. Therefore, we may approximate V_2 by assuming that R_E is open.

$$V_2 \simeq \frac{R_2}{R_1 + R_2}\, V_{CC}$$
$$V_2 \simeq \frac{4.7\text{ k}\Omega}{4.7\text{ k}\Omega \times 68\text{ k}\Omega} \times 20$$
$$\simeq 1.3\text{ v}$$

then,

$$V_E \simeq V_2 - V_{EB}$$
$$V_E \simeq 1.3 - 0.3 \simeq 1\text{v}$$
$$I_E \simeq \frac{V_E}{R_E}$$
$$\simeq \frac{1}{1\text{ k}\Omega} \simeq 1\text{ ma}$$
$$I_B \simeq \frac{I_E}{\beta + 1}$$
$$\simeq \frac{1\text{ma}}{100} \simeq 10\;\mu\text{a} \qquad \text{(answer)}$$

or,

$$I_B \simeq \frac{V_E}{(\beta + 1)\, R_E}$$
$$\simeq 10\;\mu\text{a} \qquad \text{(answer)}$$

In a well designed amplifier (stabilized), the value of R_2 should be *less than one-tenth* R_E $(\beta + 1)$. However, we will have occasion to analyze poorly designed circuits which are *not within this criterion, in which case the bias circuit in* Fig. 8–10(b) *must be analyzed by Kirchhoff's law* or by Thevenin's theorem and by the use of the equivalent circuits in Fig. 8–11.

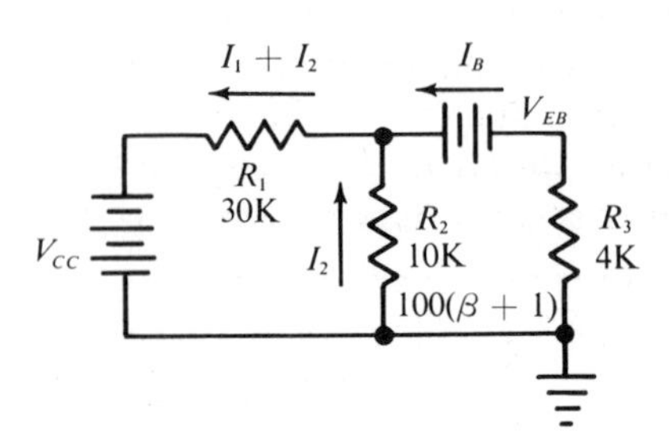

Figure 8–11. **Equivalent bias circuit for a poorly designed amplifier, $R_E\ (\beta + 1) < 10\ R_2$.**

Example 11. Suppose the values of the components in the amplifier in Fig. 8–10 (a) are $R_1 = 30\ \text{k}\Omega$, $R_2 = 10\ \text{k}\Omega$, $R_E = 100$; and the transistor is a germanium type with a beta of 39.

We see that $R_E\ (\beta + 1)$ is 4 kΩ and does not satisfy the design criteria for a short-cut analysis of bias. Then, writing the two loop equations from Fig. 8–11 gives

$$\text{I.}\quad 19.7 = 34\ \text{k}\Omega\ I_B + 30\ \text{k}\Omega\ I_2$$
$$\text{II.}\quad 20 = 30\ \text{k}\Omega\ I_B + 40\ \text{k}\Omega\ I_2$$

Multiplying Equation I by 40 and Equation II by −30, then adding, gives:

$$\begin{aligned} 78.8 &= 136\ \text{k}\Omega\ I_B \times 120\ \text{k}\Omega\ I_2 \\ -\ 60.0 &= -90\ \text{k}\Omega\ I_B - 120\ \text{k}\Omega\ I_2 \\ \hline 18.8 &= 46\ \text{k}\Omega\ I_B \end{aligned}$$

and finally,

$$I_B = \frac{18.8}{46\ \text{k}\Omega} \simeq 410\ \mu\text{a} \qquad \text{(answer)}$$

Observe that calculation by the approximation method $[R_E(\beta + 1) > 10\ R_2]$ would give an erroneous calculation of $I_B \simeq 1.17$ ma.

In the analysis of junction transistor biasing the temperature and therefore the voltage barrier (V_{EB}) are constant for either the silicon or the germanium transistors. However, the voltage barrier changes inversely with temperature (2 mv per °C for germanium and 2.5 mv per °C for silicon) and must be considered when the transistor is expected to operate above or below the ambient temperature of 25°C. For example, the value of V_{EB} for a germanium transistor operating at 65°C is approximately 0.22 v and for a silicon transistor operating at this temperature V_{EB} is approximately 0.5 v.

In our analysis of an FET or a vacuum-tube amplifier, the practical maximum value of the drain or source resistance was limited by the power supply voltage and required power output of the amplifier. That analogy, of course, applies to the junction transistor amplifier. To demonstrate the effects of power output versus load resistance, suppose the amplifier stage in Fig. 8–12 is to be biased Class *A*.

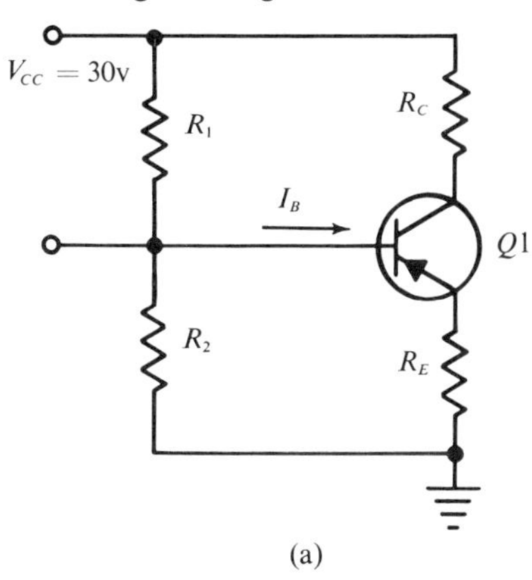

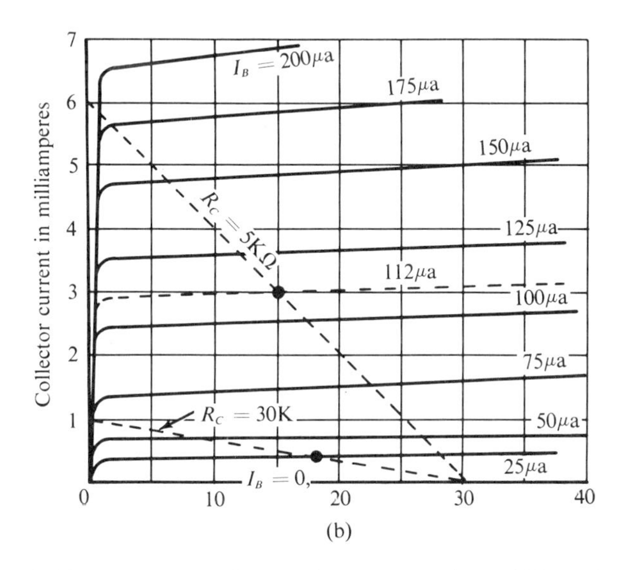

Figure 8–12. Comparison of Class-A operation with a 5-kΩ load to Class-A operation with a 30-kΩ load. The maximum output power for 5 kΩ is 22.5 mw and for 30 kΩ is 3.75 mw.

In the first case, the collector resistor (R_C) is to be 5 kΩ, and the second case R_C is to be 30 kΩ.

The output power is formulated as in any circuit.

$$p_o = \frac{I_{pp}}{2\sqrt{2}} \times \frac{V_{pp}}{2\sqrt{2}} \quad \textbf{(8–14)}$$

$$p_o = \frac{I_{pp}\,V_{pp}}{8} \quad \textbf{(8–15)}$$

The load line in Fig. 8–12 (b) for a 5-kΩ collector resistor gives a maximum output power.

$$P_o \simeq \frac{5.8 \text{ ma} \times 30}{8}$$
$$\simeq 21.7 \text{ mw}$$

The 30-kΩ collector resistor gives a maximum output power of

$$P_o \simeq \frac{1 \times 30}{8}$$
$$\simeq 3.74 \text{ mw}$$

Obviously, the small value of collector resistance can give a larger output power with a much larger input power. The 30-kΩ collector could be used in the first stage of a preamplifier, and the 5-kΩ collector could be used in the second stage of such an amplifier.

The maximum value of the collector resistor in a junction-transistor amplifier is also limited by the collector leakage current, I_{CEO}. As R_C is increased, $I_{CEO} \times R_C$ approaches V_{CC} and signal changes cannot appear across R_C.

8–5. Square-Wave and Pulse Response

We are now familiar with the frequency response of an *RC*-coupled amplifier, but frequency response alone does not give complete information about an amplifier's square-wave or pulse response. We can calculate the square-wave response of an amplifier from its frequency characteristic and its phase characteristic. However, this is an involved calculation, and it is far more practical to make use of universal *RC* time-constant charts. We know that the low-frequency square-wave response of a stage or of cascaded stages will depend upon the time constant of the coupling capacitor(s) and the grid, gate, or base resistance(s). Thus, for a single-stage amplifier, the low-frequency square-wave response is given by the familiar universal *RC* time-constant chart depicted in Fig. 8–13. Again, we know that the high-frequency square-wave response of a single-stage amplifier depends upon the time constant of the plate-, drain-, or collector-output

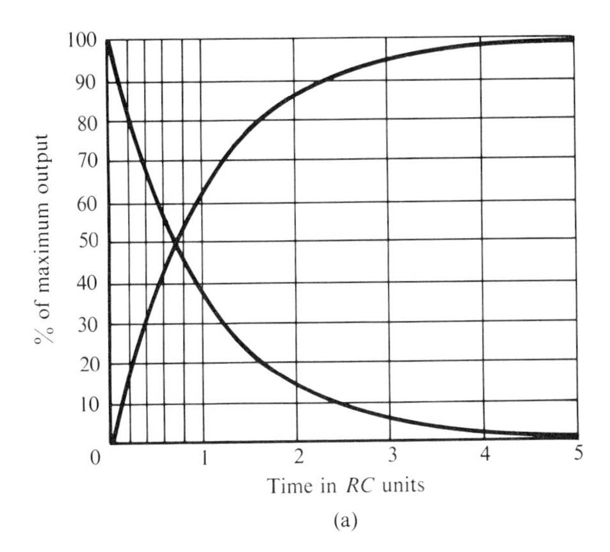

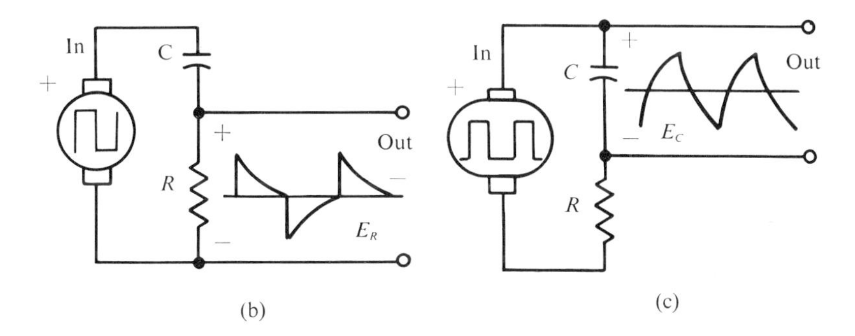

Figure 8–13. (a) Universal *RC* time-constant chart; (b) Typical low-frequency square-wave response; (c) Typical high-frequency square-wave response.

circuit. Hence, for a single-stage amplifier, the high-frequency response is also given by the familiar universal *RC* time-constant chart. At medium square-wave frequencies, the square-wave output is virtually undistorted.

Next, to visualize the square-wave response of a two-stage *RC*-coupled amplifier, we observe that at low frequencies, we have two differentiating circuits connected in cascade and isolated by the control device between them, as with FET's depicted in Fig. 8–14 (a).

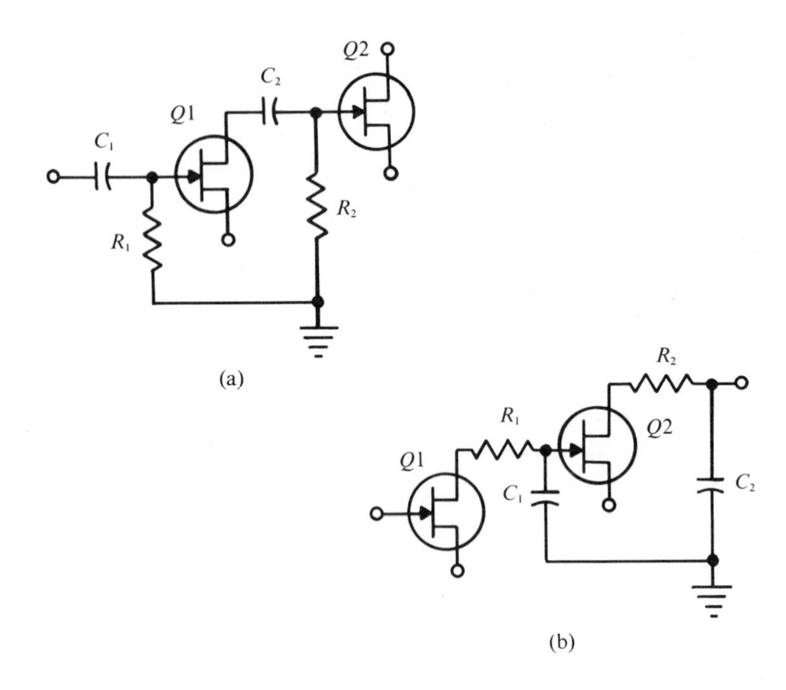

Figure 8–14. (a) Equivalent circuit at low square-wave frequencies; (b) Equivalent circuit at high square-wave frequencies.

Similarly, at high frequencies, we observe that we have two integrating circuits connected in cascade and isolated by the control device between them, as in Fig. 8–14 (b). In turn, the square-wave output is modified; the first *RC* section is energized by a square wave, but the second *RC* section is energized by an exponential wave.

The end result is seen in Fig. 8–15. We assume equal *R* and *C* values. At low square-wave frequencies, the output waveform decays more rapidly than for a single-stage amplifier. We observe that the output waveform also falls to a minimum value in approximately two time constants, and then rises again to more than 10 per cent amplitude. In other words, the square-wave response *undershoots* in a two-stage *RC*-coupled amplifier. This undershoot results from the fact that the second *RC* section is energized by an exponential waveform. At high square-wave frequencies, the output waveform rises more slowly than for a single stage. We observe that the waveform has a double curvature, with a point of inflection at about 0.5*RC*. Again, this point

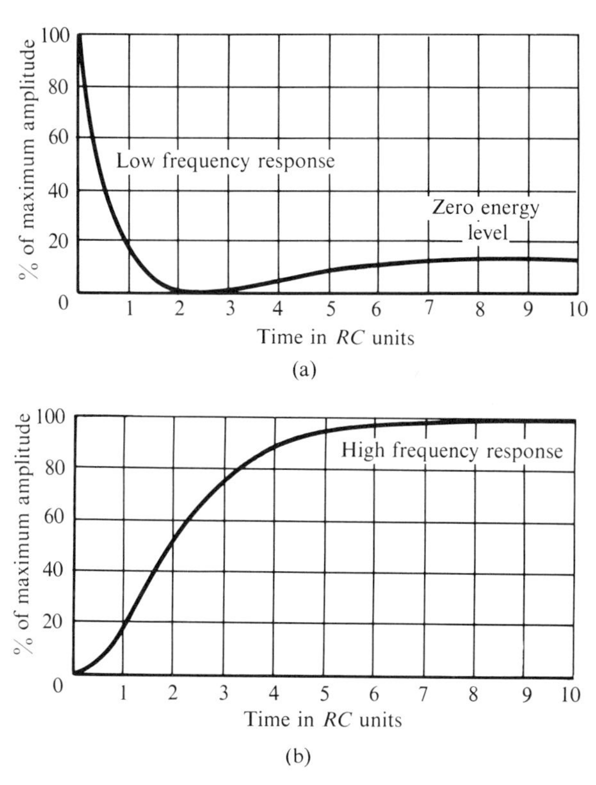

Figure 8–15. (a) Universal *RC* time-constant chart for symmetrical two-stage *RC*-coupled amplifier (low-frequency square-wave response); (b) Universal *RC* time-constant chart for symmetrical two-stage *RC*-coupled amplifier (high-frequency square-wave response).

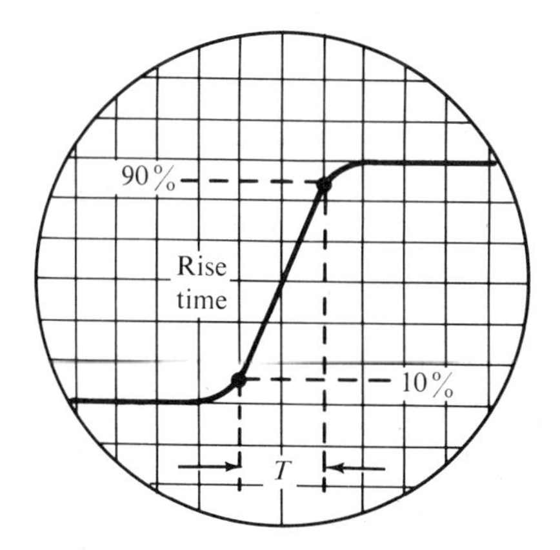

Figure 8–16. Measurement of rise time.

of inflection results from the fact that the second *RC* section is driven by an exponential waveform. At medium square-wave frequencies, the square-wave output is virtually undistorted.

Rise time of the square-wave output from an amplifier is measured as shown in Fig. 8–16. That is, rise time is the elapsed time T from the 10 per cent to the 90 per cent point of maximum amplitude along the leading edge of the square wave. Rise time is most conveniently measured with a triggered-sweep scope that has a calibrated time base. The rise time of an amplifier is related to its frequency response as follows.

$$T_{\text{rise}} = \frac{1}{3\text{f}} \qquad \textbf{(8–16)}$$

where T_{rise} denotes rise time, and f_{CO} denotes the upper frequency at which the amplifier response is 3 db down.

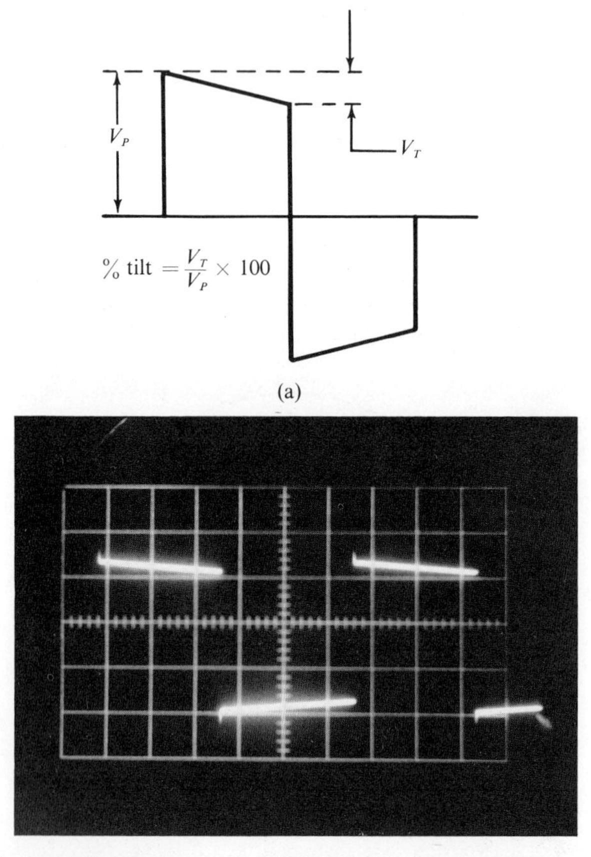

Figure 8–17. (a) Definition of square-wave tilt; (b) Photograph of a tilted square wave.

Example 12. Determine the upper cutoff frequency of an amplifier in which the rise time of an amplified pulse is 1 μsec.

$$f_{co} \simeq \frac{1}{3T_{\text{rise}}}$$
$$\simeq \frac{1}{3 \times 10^{-6}}$$
$$\simeq 333 \text{ kHz} \qquad \text{(answer)}$$

At low square-wave frequencies, *tilt* produced by phase shift is a dominant feature of square-wave reproduction. Tilt is defined as shown in Fig. 8–17. The relation of percentage tilt to the -3 db low-frequency response point (f_1) of an amplifier is formulated

$$f_1 = \frac{f\phi}{3} \tag{8–17}$$

where $\phi =$ the per cent tilt $= \dfrac{V_T}{V_P} \times 100$

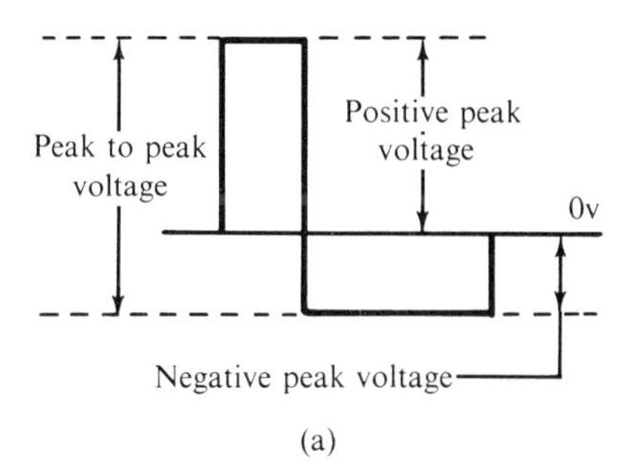

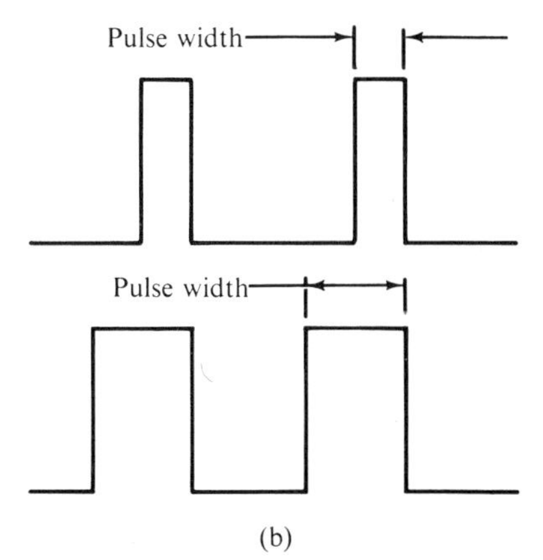

Figure 8–18. (a) Voltage specifications in a pulse waveform; (b) Pulse waveforms with the same repetition rate but with different pulse widths.

$$\phi = \frac{V_T}{V_P} \times 100 \qquad \textbf{(8–18)}$$

Example 13. Suppose the square wave in Fig. 8–17 is a 100-Hz output from an amplifier with $V_T = 2$ v and $V_P = 22$ v. What is the low-cutoff frequency?

$$\phi = \frac{2}{22} \times 100 = 9.1 \text{ per cent}$$

$$f_1 = \frac{100 \times 9.1}{3}$$

$$\simeq 300 \text{ Hz} \qquad \text{(answer)}$$

In practice, the square-wave frequency is reduced until the tilt of the reproduced square waveform is easily measured on the scope screen; thus, a 15 per cent tilt value might be measured. In turn, this value will be substituted into Formula 8–17 to calculate the approximate frequency at which the amplifier response is 3 db down at the low-frequency end.

A pulse is an unsymmetrical square wave. Figure 8–18 shows the voltage specifications in a pulse waveform. When a pulse passes through an *RC* coupling circuit, the waveform necessarily has an average value of zero. In other words, the positive-peak excursion in Fig. 8–18 (a) has exactly the same area as the negative-peak excursion. It follows that the positive and negative peak voltages of a pulse are always unequal. An average value of zero means that the product of positive-peak voltage and the duration of the positive excursion is equal to the product of negative peak voltage and the duration of the negative excursion. To put it another way, the number of coulombs in the positive and negative excursions is the same.

When the pulse width is considerable, the leading edge of the reproduced pulse from an *RC* coupling circuit has full amplitude, as shown in Fig. 8–19. The percentage tilt in the top of the reproduced pulse depends on the time constant of the circuit, and is the same as for a square wave that has the same duration for its positive excursion. Of course, the negative excursion of the reproduced pulse is also differentiated over a longer elapsed time. As can be seen from Fig. 8–19 (d), the reproduced waveform positions itself with respect to the 0-v axis so that the area of the positive excursion is equal to the area of the negative excursion.

When the pulse width is narrow, the output waveform (across the capacitor) does not have time to rise to the full amplitude of the applied pulse voltage, as depicted in Fig. 8–20. Note that the time constant under discussion here is the time constant of the output circuit [Fig. 8–21 (a)]. In other words, we are concerned solely with

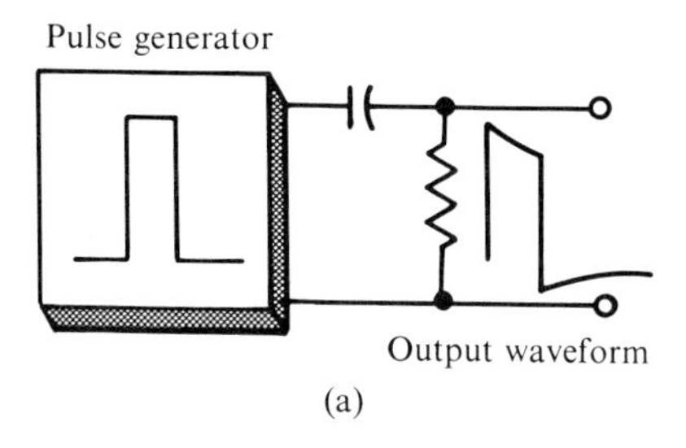

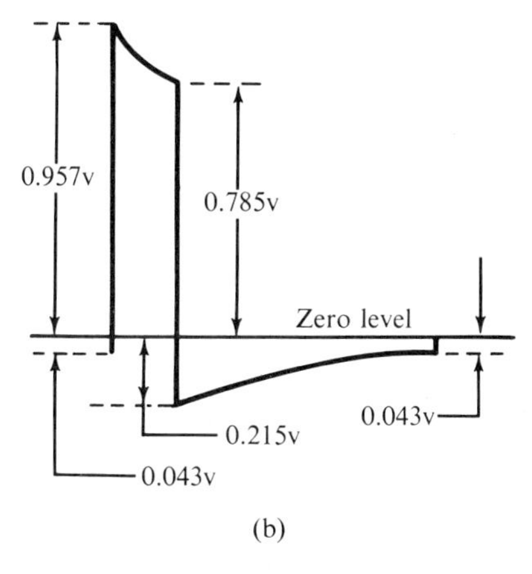

Figure 8-19. Pulse voltage applied to *RC*-coupling circuit. (a) Applied waveform; (b) Output waveform in detail.

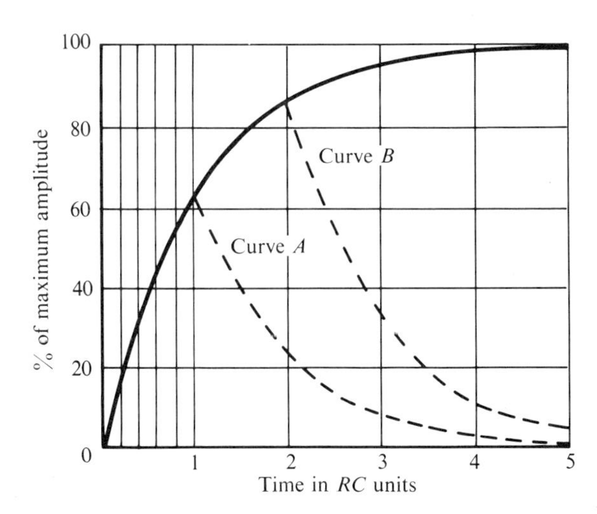

Figure 8-20. Dotted lines *A* and *B* show the reproduced waveform when the pulse width is equal to one time constant, and to two time constants.

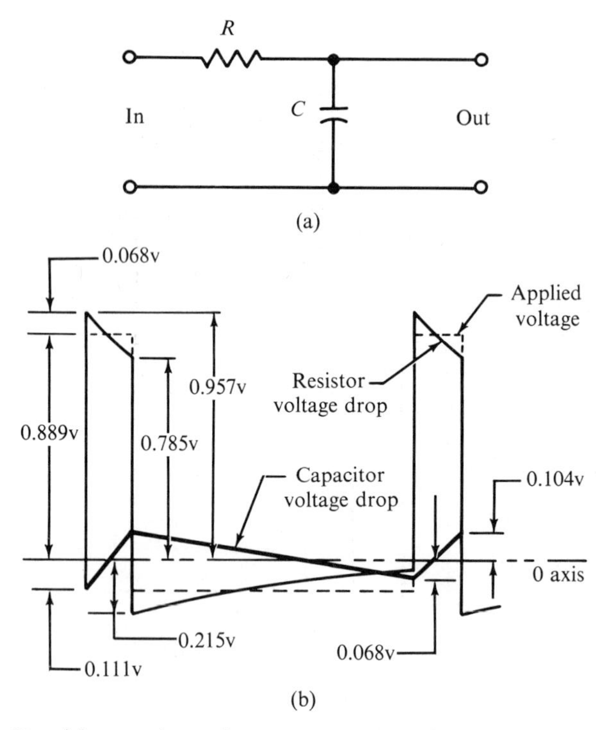

Figure 8-21. (a) Equivalent plate-output circuit; (b) Detail of an output pulse waveform (diagrammatic).

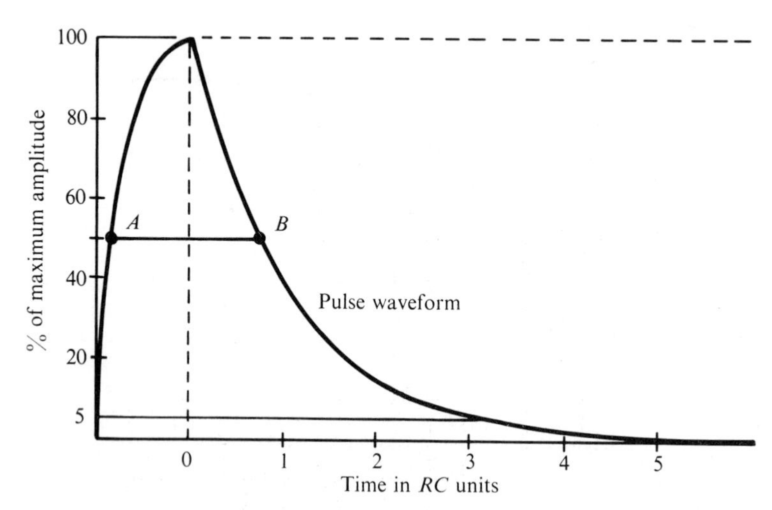

Figure 8-22. Pulse width is measured between the 50 per cent amplitude points between points *A* and *B*.

the rise of the reproduced pulse in Fig. 8–20, while we were concerned solely with the decay of the reproduced pulse in Fig. 8–19 (b). Thus, a wide pulse exhibits the transient response of the coupling circuit in an *RC* stage, while a narrow pulse exhibits the transient response of the output circuit. The waveform of a reproduced narrow pulse is detailed in Fig. 8–21 (b). At medium pulse widths and moderate repetition rates, the output pulse waveform is virtually undistorted.

A differentiated pulse has a width that is measured between its 50 per cent of maximum amplitude points, as shown in Fig. 8–22. We use

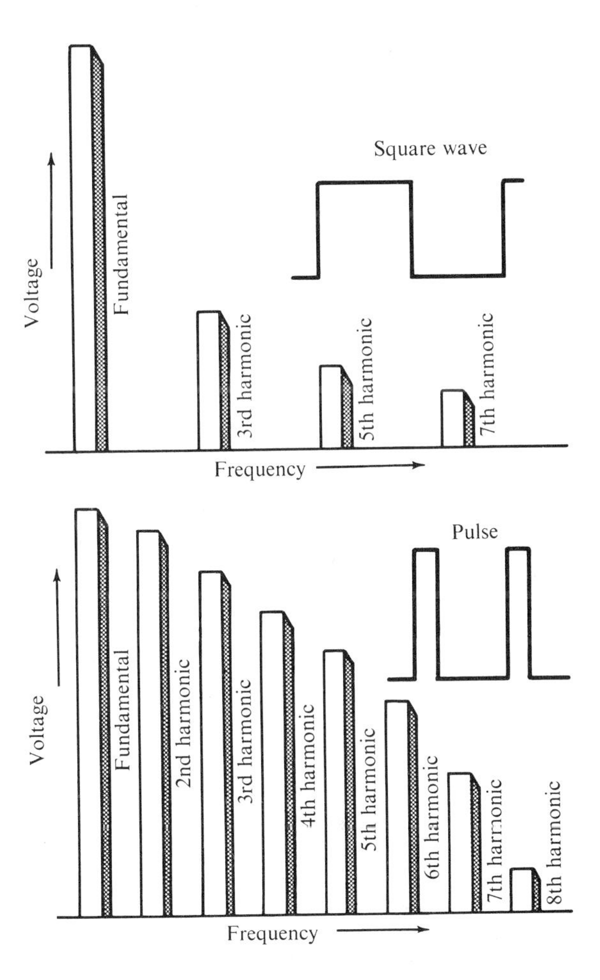

Figure 8–23. Comparative harmonic amplitudes for a square wave and a pulse.

a triggered-sweep scope with a calibrated time base and measure the elapsed time in microseconds or milliseconds from point A to point B. A pulse waveform has twice as many harmonic frequencies as a square wave, because a square wave has only odd harmonics; on the other hand, a pulse has both even and odd harmonics, as shown in Fig. 8–23. The harmonic amplitudes are also comparatively large in a pulse waveform. We will find that although the same basic information concerning transient response is provided by both square waves and pulses, low-frequency information is often presented to better advantage by square waves and high-frequency information by pulses.

Thus far, we have assumed for purposes of simplicity that the square and pulse waveforms used to test the transient response of an amplifier are ideal. That is, we assumed that the leading edge of the applied square wave or pulse had zero rise time. In some cases, this assumption is justified in practice; in other cases, it is not. No square wave or pulse waveform has a zero rise time. Although the rise time might be very fast, it can nevertheless be measured with a high-quality oscilloscope. When the rise time of an amplifier is in the same order of magnitude as the rise time of the applied square wave or pulse, we must take the rise time of the applied waveform into account. The rise time of a reproduced waveform is formulated

$$T_r = \sqrt{T_a^2 + T_o^2} \qquad \textbf{(8–19)}$$

where T_r denotes the rise time of the reproduced waveform, T_a denotes the rise time of the applied waveform, and T_0 denotes the actual rise time of the amplifier under test.

Example 14. What is the rise time of the amplifier if the measured rise time is 250 nsec, and the pulse rise time is 40 nsec (assume the oscilloscope is perfect)?

$$\begin{aligned} T_o^2 &= T_r^2 - T_a^2 \\ &= (250)^2 - (40)^2 \\ T_o^2 &\simeq 60{,}900 \\ T_o &= 246 \text{ nsec} \qquad \text{(answer)} \end{aligned}$$

It is evident from Formula 8–19 that if the amplifier has a slow rise time in comparison with the rise time of the applied wave form, the rise of the reproduced waveform will be virtually equal to the amplifier rise time. Suppose, however, that the amplifier rise time should be the same as the rise time of the applied waveform. Then, the rise time of the reproduced waveform will be approximately 40 per cent greater than the rise time of the amplifier.

Care must be taken to insure that the rise time of the input waveform and the oscilloscope are better than that of the device under test. The rise time of an oscilloscope with a plug-in is formulated

$$T_r = \sqrt{T_{\text{scope}}^2 + T_{\text{plug-}in}^2} \qquad \textbf{(8-20)}$$

Example 15. Suppose the rise time of an oscilloscope is 10 nsec, and the rise time of the plug-in is 14 nsec. What is the rise time of the unit?

$$T_r = \sqrt{(10)^2 + (14)^2}$$
$$= 17.2 \text{ nsec} \qquad \text{(answer)}$$

8-6. Amplifier Classifications

Thus far, we have defined amplifier operation in Classes *A*, *B*, and *C*. In practice, these classes are also subclassified. If a tube is biased so that it operates in an intermediate manner to Classes *A* and *B*, it is called a Class-*AB* amplifier. We know that a Class-*A* stage operates over the complete input cycle without plate-current cutoff. Class-*AB* stages operate at higher bias than in Class *A*, but at lower bias than in Class *B*. Class-*B* tubes, as we recall, are biased almost to plate-current cutoff. Class-*A* amplifiers are designated Class A_1 or Class A_2. In Class A_1, the tube draws no grid current at the positive peak of drive. On the other hand, in Class A_2, the tube draws more or less grid current over the peak of positive drive. Similarly, a Class AB_1 amplifier circuit draws no grid current, while an AB_2 amplifier circuit draws grid current on positive peaks of drive. Class-*B* and Class-*C* stages always draw grid current when driven to maximum rated output.

8-7. Limiting Frequency Response of Cascaded Stages

We have seen that the overall frequency-response curve of an amplifier changes as more stages are connected in cascade. If an *infinite* number of stages were connected in cascade, a *definite form of frequency-response curve* would be obtained, which has great significance in electronics technology. We will find that when several stages are connected in cascade, this limiting frequency-response curve is approached rather rapidly. For example, an oscilloscope that has a half-dozen stages connected in cascade exhibits a frequency-response curve that approximates the limit of an infinite number of cascaded stages. This curve is called by many names, among which are the "bell" curve, the normal curve of error, the probability density

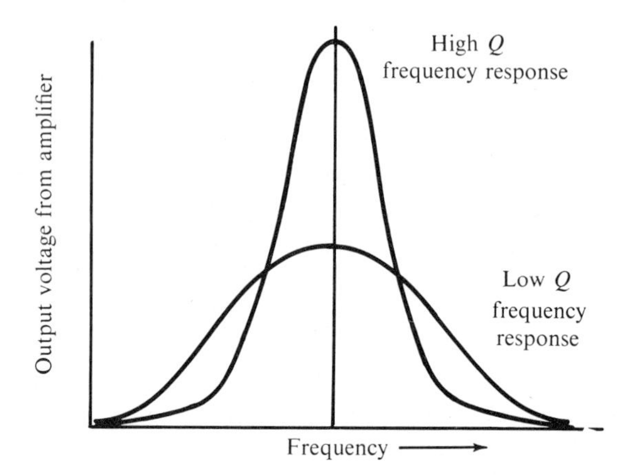

Figure 8–24. Gaussian frequency-response characteristics or bell curve.

function, or simply the frequency curve. Two examples of this curve are illustrated in Fig. 8–24. Its equation is written

$$E_o = A\epsilon^{-Bf^2} \quad \textbf{(8–21)}$$

where E_o is the value of the output signal voltage, A is an amplifier, ϵ is the base of natural logarithms, B is another amplifier parameter, and f is the frequency at which E_o is measured.

When A and B have various values in Equation 8–21, the aspect of the "bell" curve changes, as shown for two typical examples in Fig. 8–24. Nevertheless, the curve always has the same basic characteristics, because A and B are merely constants in the exponential function 8–21. Of particular interest in electronics technology is the manner in which the "bell" curve falls off following the cutoff frequency. That is, the "bell curve" has a particularly important *cutoff characteristic.* This is called a *Gaussian cutoff characteristic* and is named after the mathematician who discovered this curve. Any *RC* amplifier with an appreciable number of stages connected in cascade will have a cutoff characteristic that approximates a Gaussian cutoff characteristic.

The reason for development of the Gaussian cutoff characteristic is that each amplifier stage has a frequency-response curve that is in accord with the universal *RC* frequency-response curve. The output from each stage is multiplied by the response of the following stage. As more stages are connected in cascade, the product approaches Equation 8–21 as a limit. The Gaussian cutoff is of basic importance in electronics technology because it provides optimum response to square-wave or pulse signals when the amplifier's frequency response is limited. Let us consider the fundamental principles that are involved.

We know that an ideal pulse or square waveform contains an infinite number of harmonic frequencies. Of course, ideal pulses and square waves cannot be generated in practice. On the other hand, it is readily possible to generate square waves and pulses that more nearly approach the ideal than it is to design amplifiers that approach the ideal. Therefore, in most practical situations we will conduct our test work with pulses or square waves that are virtually ideal, although the amplifiers in oscilloscopes depart significantly from the ideal. This is just another way of saying that the pulse and square-wave waveforms that we employ usually have more harmonics than can be passed by the vertical amplifier in an oscilloscope.

Since the vertical amplifier in an oscilloscope usually fails to pass all the harmonics of the applied square-wave or pulse signal, the question immediately arises concerning what cutoff characteristic should be chosen for the amplifier to provide optimum waveform reproduction. Minimum distortion for a given bandwidth is obtained when the amplifier has a Gaussian cutoff characteristic. Hence, designers of quality (laboratory-type) oscilloscopes devise vertical-amplifier circuits that have the desired Gaussian cutoff characteristics. If simple *RC* coupling is used throughout, this characteristic is obtained automatically. We will find, however, that *RC* coupling is often supplemented by extended high-frequency response. Methods of obtaining a desired Gaussian cutoff characteristic when an *RC* amplifier is elaborated in this manner will be explained subsequently.

8-8. High-Frequency Compensation

We have learned that an *RC*-coupled amplifier with a number of stages has a high-frequency cutoff characteristic that approximates a Gaussian curve, as shown in Fig. 8-24. Recall that this curve is formulated

$$E_o = A\epsilon^{-Bf^2} \tag{8-22}$$

A Gaussian cutoff characteristic provides optimum transient response for square-wave or pulse waveforms. When series and/or shunt peaking is utilized to extend the high-frequency cutoff of an *RC*-coupled amplifier, the cutoff characteristic no longer approximates

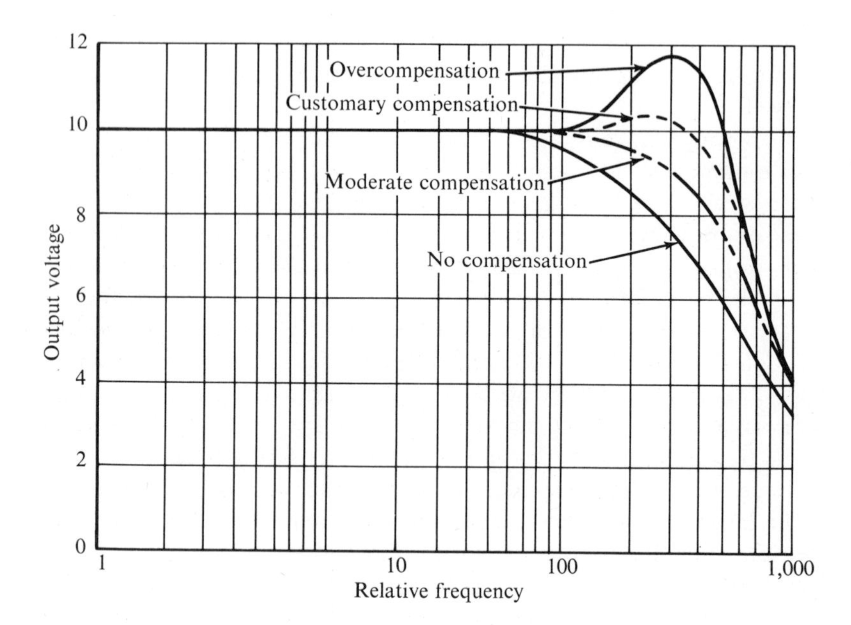

Figure 8–25. **High frequency response curve showing the effects of peaking coils.**

a Gaussian curve. This fact can be seen in Fig. 8–25. When customary high-frequency compensation is obtained by means of peaking coils, the amplifier response rises slightly at the high-frequency end and then drops off rather abruptly. We observe that this curve departs considerably from the curve for no compensation, which roughly corresponds to a Gaussian cutoff characteristic.

Therefore, vertical amplifiers in high-quality (laboratory-type) oscilloscopes do not employ simple series-shunt peaking-coil compensation. Instead, specially designed amplifiers are used that provide a good approximation to a Gaussian cutoff characteristic. An example is Fig. 8–26. This arrangement is called a distributed amplifier. It comprises low-pass filter networks in the grid and plate circuits. Plates are coupled by inductors, as are grids; the capacitive components in the low-pass filter networks are the plate and grid interelectrode capacitances. A distributed amplifier provides very uniform (flat) frequency response out to the cutoff point, which may be designed for a very high value. Moreover, the cutoff characteristic is essentially Gaussian, which provides optimum transient response.

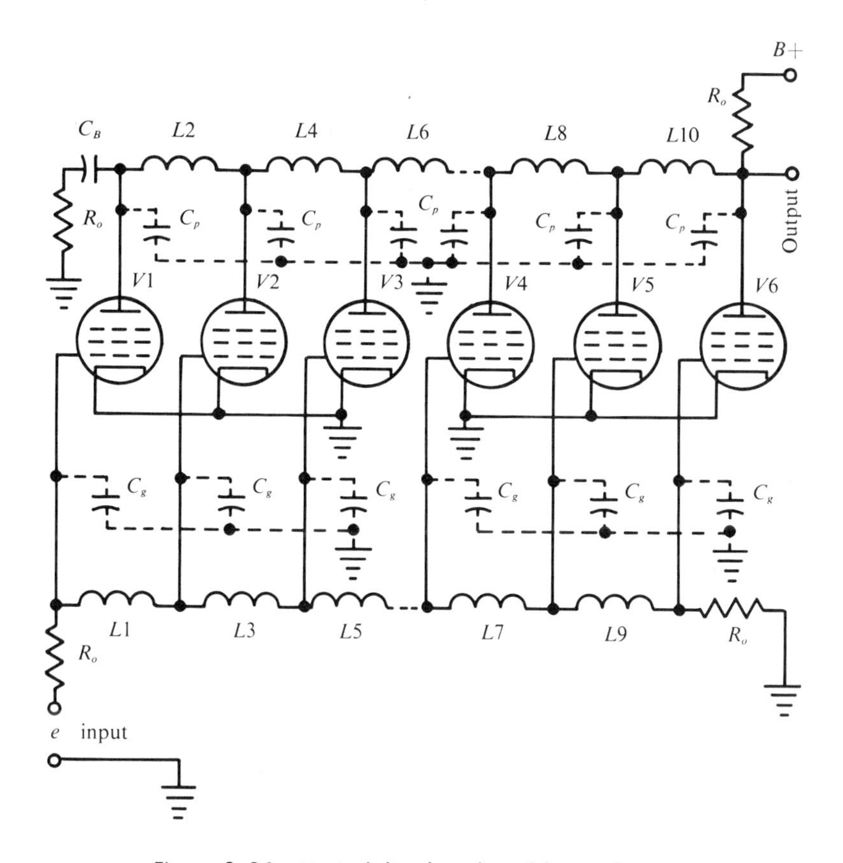

Figure 8-26. Typical distributed-amplifier configuration.

SUMMARY

1. Increasing the value of the grid-leak resistor of a vacuum tube amplifier and the gate resistor of an FET amplifier generally increases the low frequency response of the amplifier.

2. The maximum value of grid-leak resistor is limited by grid-current caused by residual gas molecules left in the tube.

3. The maximum value of the gate resistor is limited by gate current through the reverse-biased gate junction by minority carriers.

4. Values of resistance of 1 to 100 MΩ are recommended by manufacturers as gate resistors for field-effect transistor Class-*A* amplifiers.

5. Values of grid-leak resistances from 1 to 10 MΩ are recommended for vacuum tube Class-*A* amplifiers.

6. The low-frequency response of an amplifier can be extended by increasing the value of the coupling capacitor or the bypass capacitor.

7. Limitations on the value of coupling capacitors are physical size, leakage current, and stray capacitance.

8. The maximum value of a collector load resistor for a transistor is limited by supply voltage, power output, and collector leakage current values.

9. The operating point of the collector current in a transistor is established by the base current through the forward-biased base to emitter junction.

10. The internal input impedance characteristics of a junction transistor are much lower in value than those for vacuum tubes or field-effect transistors. These characteristics make circuit analysis for the junction transistors somewhat more difficult than for FET's or vacuum tubes.

11. Changes in the output circuit in a junction transistor reflects into the input; a decrease in the load admittance (G_L) increases the input impedance (z_i).

12. Changes in the input circuit in a junction transistor reflects into the output; an increase in the input impedance (Z_i) produces an increase in the output admittance (decrease in output impedance r_0).

13. Transistor bias configurations are selected to establish the operating point of collector current and to stabilize the collector current against changes due to temperature change.

14. The frequency response of an amplifier may be determined by observing its distorting effects on a square wave signal.

15. Pulse rise time is the time elapsed between 10 per cent and 90 per cent of the pulse rise.

16. Pulse-fall time is the time elapsed between 90 per cent and 10 per cent of the pulse decay.

17. The upper and lower cutoff frequencies of an amplifier may be determined by examination of an amplified square wave on an oscilloscope and application of the measured values to a formula.

18. Care must be taken when measuring rise time of a pulse that the rise time of the oscilloscope is faster than that of the pulse.

19. A wide pulse exhibits the transient response of the coupling circuit in a *RC* stage, and a narrow pulse exhibits the transient response of the output circuit (plate, drain, or collector).

20. The high-frequency response of an amplifier depends upon the output impedance of the device, the impedance of the total a-c load, the internal device capacitance, and the stray circuit capacitance.

21. The width of a pulse is measured between its 50 per cent of maximum amplitude points.

Questions:

1. Give three limitations of the maximum value for the collector resistor in a Class-*A* transistor amplifier.
2. What are the limitations on the maximum value for the grid-leak resistor of a vacuum-tube amplifier and the gate resistor of an FET amplifier?
3. What are the effects on the low frequency response of an amplifier when the value of the coupling capacitor is increased?
4. What factors limit the maximum value of the coupling capacitor in an amplifier?
5. Discuss the employment of a partially bypassed source resistor as a method to extend the low-frequency range.
6. Discuss each of the three limitations on the maximum value of the collector resistor in a Class-*A* transistor amplifier.
7. Discuss the differences between biasing a junction transistor and a field-effect transistor.
8. Discuss the problems involved in circuit analysis employing junction transistors.
9. Discuss the effects of a change of input impedance on the output admittance of a transistor amplifier.
10. Discuss the effects of a change in output admittance on the input impedance of a transistor amplifier.
11. Why does an amplifier distort a square-wave signal which it amplifies?
12. What is the visual effect of low-frequency distortion in an amplifier?
13. What is the visual effect of high-frequency distortion in an amplifier?
14. Discuss the measurement of the rise time of a pulse.
15. Discuss the measurement of the fall time of a pulse.
16. What problems are involved when measuring the rise time of a pulse on an oscilloscope?
17. Compare a pulse to a square waveform.
18. What is the effect on a pulse which is coupled through an *RC*-coupling circuit?
19. What are the noticeable differences in an amplified wide pulse and an amplified narrow pulse displayed on an oscilloscope screen?
20. Draw a pulse and indicate the rise time, fall time, and pulse width.
21. Determine the rise time of an amplifier in which the rise time of the input waveform is 50 nsec and the measured rise time is 160 nsec (assume the oscilloscope rise time is 0).

22. Determine the measured rise time of the output pulse in Question 21 when an oscilloscope with a rise time of 22 nsec is employed.

23. Discuss the classifications of amplifiers, such as Class *AB*, Class AB_2, etc.

PROBLEMS

1. Calculate the value of bypass capacitors to allow a low-frequency response of 100 Hz when the value of R_E is 470 Ω.

2. Calculate the value of base current in the circuit in Fig. 8-10 when the circuit values are $R_E = 680\ \Omega$, $R_1 = 82\ k\Omega$, $R_2 = 6.8\ k\Omega$ and a silicon transistor with a beta of 99 is used with $V_{CC} = 20$ V.

3. In reference to Fig 8-10, the circuit uses a germanium transistor with a beta of 49 and has values of $R_E = 1\ k\Omega$, $R_2 = 4.7\ k\Omega$. What is the value of R_1 which will establish a base current of 50 μa at a temperature of 75° C when $V_{CC} = 20$ v?

4. The circuit in Fig. 8-10 uses a germanium transistor with a beta of 110 and circuit values of $R_1 = 10\ k\Omega$, $R_2 = 4.7\ k\Omega$. What value of R_E will allow a base current of 25 μa at 65° C?

5. Suppose the beta of the transistor in Problem 3 decreases to 19. What is the new value of base current?

6. If the transistor in the circuit of Problem 3 is replaced by one with a beta of 100, what is the new base current?

7. If the transistor in the circuit in Problem 5 is replaced with a silicon type which has a beta of 50, what is the new base current?

8. Calculate the values of input impedance (Z_i) and output admittance (y_o) for the circuit in Fig. 8-10 (a) when $R_1 = 68\ k\Omega$, $R_2 = 4.7\ k\Omega$, $R_c = 2.2\ k\Omega$, and the transistor is a 2N3926 type which has the parameters $h_{ie} = 6\ k\Omega$, $h_{oe} = 40\ \mu$mho, $h_{fe} = 100$, and $h_{re} = 8 \times 10^{-4}$.

9. Determine the new output admittance for the transistor in Problem 8 when the value of R_1 is reduced to 22 kΩ.

10. Determine the upper-cutoff frequency of an amplifier in which an amplified pulse has a rise time of 2.5 μsec.

11. Determine the upper-cutoff frequency of an amplifier in which an amplified 1-kHz pulse has a rise time of 250 nsec.

12. What is the low-cutoff frequency of the amplifier in Problem 11 if the measured tilt is 2 per cent?

13. A 2-kHz pulse observed on an oscilloscope after being amplified is found to have a rise time of 50 nsec. What is the upper-cutoff frequency?

14. Suppose the square waveform in Problem 13 is measured to have a value of $V_p = 20$ v and a value V_T of 19.5 v. What is the per cent of tilt, and what is the lower-cutoff frequency of the amplifier?

15. Determine the new value of base current for the transistor in Examples 6, 7 and 8: (a) at 0° C, (b) at 65° C.

CHAPTER 9

RC-Coupled Amplifier Difficulties

9–1. Introduction

Unexpected operating difficulties may occur in *RC*-coupled amplifier systems. A common difficulty in a three-stage amplifier concerns system oscillation, termed *motorboating*. This name derives from the "putt-putt" sound output of a motorboating amplifier. System oscillation can occur at any frequency within the operating range of the amplifier; however, the oscillating frequency is usually quite low, for reasons that will be explained. We will use a junction transistor amplifier for our analysis with the understanding that it applies for field-effect transistor and vacuum-tube amplifiers. Figure 9–1 shows the basis of positive feedback in a three-stage *RC*-coupled amplifier which leads to the instability condition called motorboating. For simplicity, a battery is depicted as the collector power supply.

Observe that the signal voltage undergoes phase inversion at each stage; thus, v_1 becomes inverted at v_2, v_2 becomes inverted at v_3, and v_3

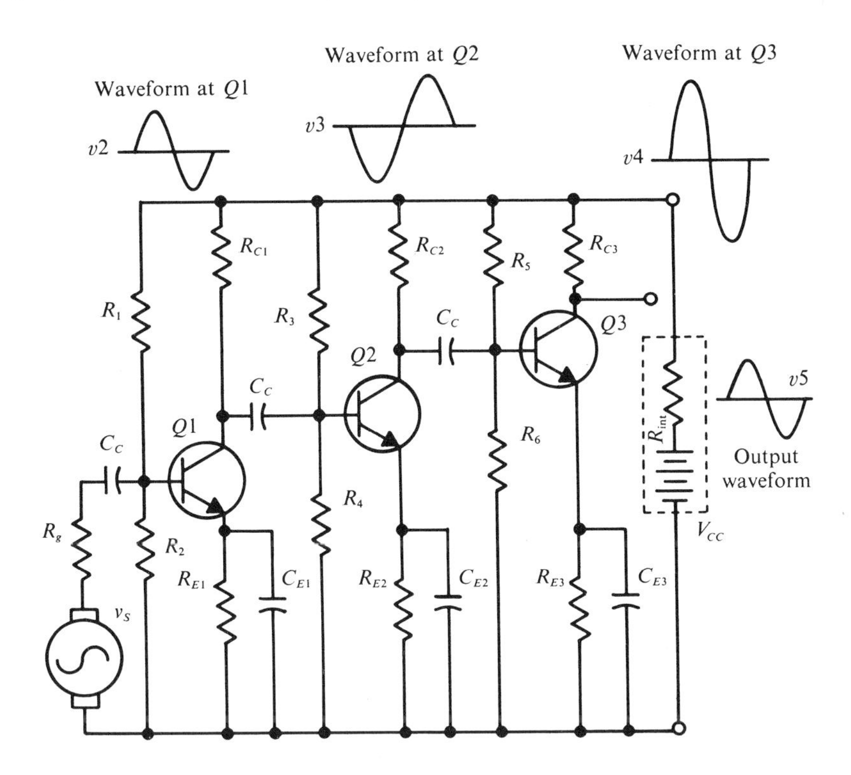

Figure 9–1. Signal voltage (v_s) is fed back in phase with v_2 via R_{int} of the collector supply (V_{CC}).

becomes inverted at v_4. Note also that V_{CC} is the common d-c source for $Q1$, $Q2$, and $Q3$. The collector currents for the three transistors flow via R_{C1}, R_{C2}, and R_{C3} through V_{CC} to the common emitter line or bus, often called B-bus. Next, we perceive that battery V_{CC} necessarily has internal resistance, R_{int}. Evidently, the collector currents for the three transistors flow through R_{int} and signal voltages similar to v_2, v_3, and v_4 are dropped across R_{int}. Each of these drops has the same phase as the collector-signal currents, and each is attenuated in proportion to the ratio $R_{\text{int}}/(R_{\text{int}} + R_C)$. We assume here for simplicity that $R_C = R_{C1} = R_{C2} = R_{C3}$.

Since v_4 has the greatest amplitude, the resultant phase of the drop v_5 across R_{int} is the same as that of v_4. We recognize that v_5 has an amplitude equal to $R_{\text{int}}(i_{c3} + i_{c1} - i_{c2})$. Insofar as system operation is concerned, it is as if an a-c generator with an output equal to v_5 were connected in series with R_{C1}. In turn, a substantial fraction of v_5 adds to v_2 at C_{C2}, and the amplitudes of v_3 and v_4 are thereby increased. Note carefully that if v_5 should have the same amplitude, or greater amplitude, than v_2, the $Q2$-$Q3$ system *then generates its own input at sufficient amplitude to maintain self-oscillation.* In other words, we can short-circuit v_1, but the waveforms v_3, v_4, and v_5 are maintained by system feedback.

This is called *positive feedback*, because it adds to the amplitude of an applied signal voltage. We recall from the previous chapter that negative feedback subtracts from the amplitude of an applied signal voltage. Self-oscillation, or motorboating caused by positive feedback, can be eliminated by several expedients. The most obvious method is to reduce the impedance of the collector power supply by means of a large bypass capacitor, as depicted in Fig. 9–2(a). This approach is

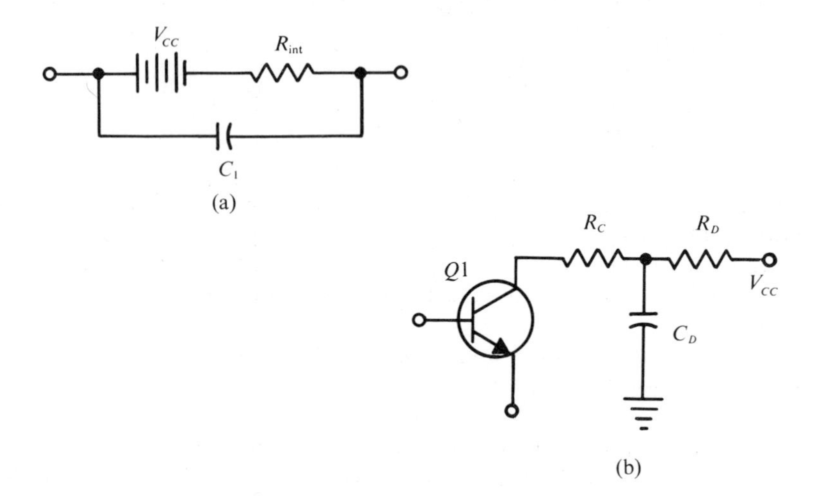

Figure 9–2. (a) Feedback reduced by power-supply bypassing; (b) Feedback reduced by circuit decoupling.

satisfactory, provided that the value of R_{int} is small and that the amplification of $Q2$ and $Q3$ is not very great. In the majority of practical situations, we find that the addition of C_1 in Fig. 9–2(a) merely reduces the frequency of motorboating and that a prohibitively large capacitance value would be required to eliminate self-oscillation.

We can easily understand the need for an extremely large bypassing capacitance in Fig. 9–2(a) when we observe that the value of R_{int} is comparatively small. Accordingly, the reactance of C_1 must be still smaller at the lowest frequency that can be amplified by the system. It is evident that the bypassing action of C_1 must be supplemented by other circuit means in most cases to eliminate motorboating in a three-stage RC-coupled amplifier. Instead of attempting to minimize the internal impedance of the collector power supply, it is economically advisable to elaborate the collector-load circuits of the transistor amplifier to minimize the signal-current flow through the power supply.

This is commonly accomplished by insertion of *decoupling circuits* in the collector-load circuits, as depicted in Fig. 9–2(b). R_C is the familiar collector-load resistor which is connected to the power supply via decoupling resistor R_D and decoupling capacitor C_D. Since the value of R_D may be chosen many times the value of R_{int} without incurring an excessive d-c voltage drop, it follows that a small value of C_D in Fig. 9–2(b) has the same effectiveness in reducing positive feedback as a large value of C_1 in Fig. 9–2(a). Such decoupling circuits are used in the majority of RC-coupled amplifiers.

9–2. Stabilization of the Operating Point of Junction Transistors

When junction transistors are used in amplifiers, one of the basic operating problems concerns maintenance of proper biasing conditions despite variations in temperature. The electrical properties of the transistor depend to a great extent upon the temperature of its junctions. In turn, the crystal temperature depends upon its internal thermal losses, the outside ambient temperature, and the thermal conductivity. Especially troublesome is the change of collector current resulting from a change of temperature, as shown in Fig. 9–3 on page 274.

The effect of an increase in temperature on the operating point is illustrated in Fig. 9–4. The collector current is established at 2.5 ma (point A) with a 100-μa base current. An increase in temperature moves all the current curves upward (dotted lines) as the leakage current I_{CEO} ($I_B = 0$) increases. The 100-μa base current then results in a collector current of 2.8 ma. The 0.3-ma collector current increase is the increased leakage current (I_{CEO}) due to the increase in temperature. Hence, the collector current is formulated

$$I_C = \beta I_B + I_{CEO} \tag{9–1}$$

The increased leakage current (I_{CEO}) increases the collector-base junction temperature, resulting in a further increase in temperature, and thereby increased leakage current, and so on. This regenerative action will produce *thermal runaway* unless bias stabilization is employed. We observe in Fig. 9–4 that the operating point of 2.5 ma could be held as the temperature increased, if the base current was

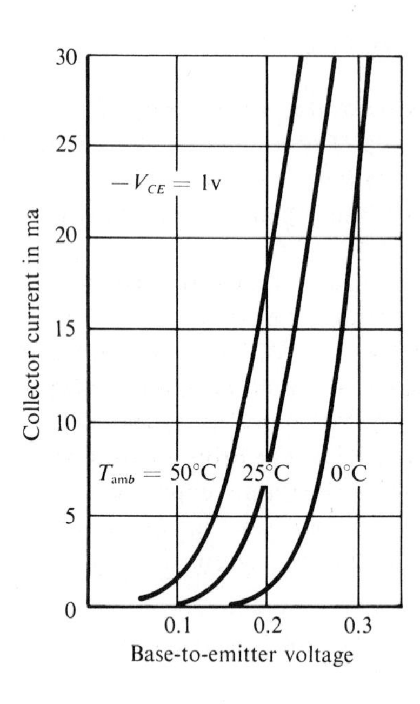

Figure 9-3. Illustration of collector current as a function of base voltage for various ambient temperatures.

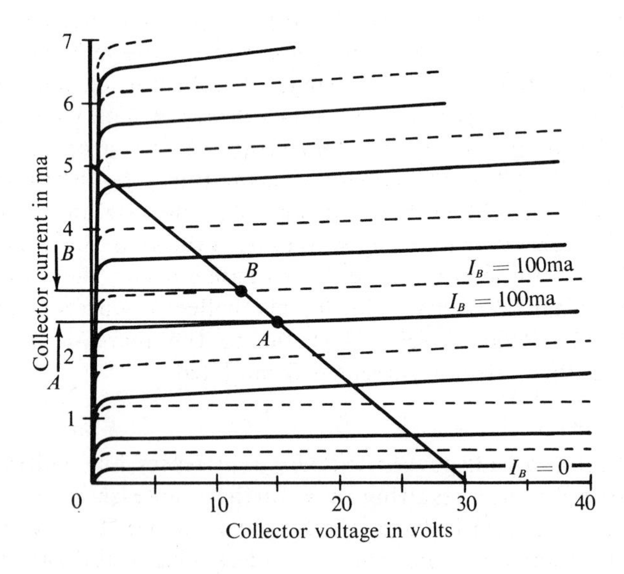

Figure 9-4. Illustration of the shift of operating point with an increase of temperature.

decreased simultaneously as the temperature increased. In other words, *we can minimize the effects of a temperature rise on the collector current by decreasing base current.*

At this point, it might be wise to review the definitions of several terms and phenomena.

1. Leakage current, I_{CEO}, is defined as the collector to emitter current with the base open ($I_B = 0$).

2. Leakage current, I_{CO}, is that current which flows in a back-biased diode at a given temperature and is equal to I_{CBO}, the leakage current of a common-base transistor amplifier.

3. Leakage current, I_{CO}, is *multiplied* in a common-emitter circuit by the d-c beta of the transistor.

$$I_{CEO} \simeq \beta_{\text{d-c}} I_{CO} \tag{9-2}$$

4. The stability factor (S) of a transistor amplifier is the ratio of the corresponding changes in the collector current I_C to the leakage current I_{CO} for a change of temperature.

$$S = \frac{\Delta I_C}{\Delta I_{CO}} \tag{9-3}$$

Formula 9-3 states that the stability factor (S) is the ratio of a change of collector current in a common-emitter amplifier,* for a given temperature change, to the change of leakage current which would occur in a back-biased diode for the same temperaure change. Obviously, a small value of stability factor is desirable. A factor of 5 or less is usually considered to provide good stability.

Three of the simpler types of bias circuits are shown in Fig. 9-5. The bias arrangement in Fig. 9-5(a) offers no bias stability, while the circuit shown in Fig. 9-5(b) offers moderate stability. Stability in Fig. 9-5(b) occurs as an increase in temperature causes an increased collector current resulting in an increased voltage drop across R_C and a lower voltage drop across R_B. Hence, current through R_B decreases, thereby decreasing I_C. This bias arrangement results in a decreased gain as the a-c signal is also coupled back to the base circuit by degeneration. We may overcome the degeneration by splitting the base resistor and bypassing one part, as depicted in Fig. 9-6. However, this results in a decreased output impedance and a signal loss as one-half of the base resistor shunts the collector resistor. The stability factor is formulated

$$S = \frac{R_C + R_B}{R_C + R_B(1 - \alpha)} \tag{9-4}$$

*The stability factor of a common-base amplifier is 1.

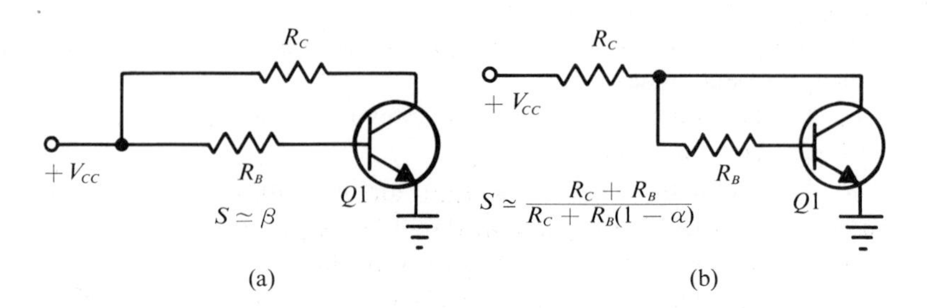

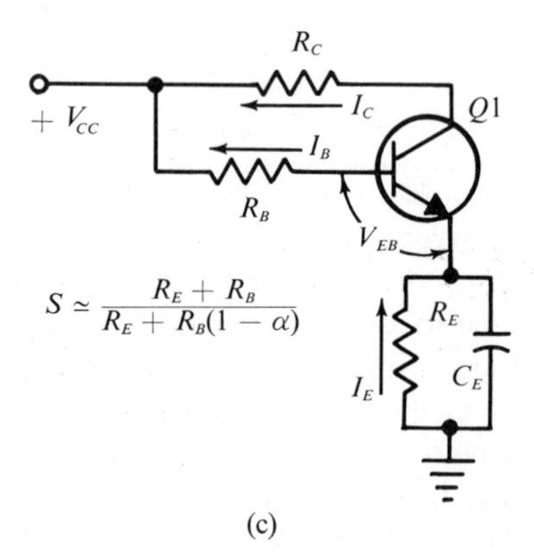

Figure 9–5. (a) Simple unstabilized bias circuit; (b) Connection of R_2 to the collector provides moderate bias stability; (c) Emitter resistor R_E provides considerable increase in bias stability.

The bias arrangement in Fig. 9–5(c) offers greater stability than the previous method without affecting stage gain or output impedance. Stability is the result of the negative d-c feedback voltage developed across R_E. As temperature increases, emitter current increases, resulting in an increase in emitter voltage (V_E). This increase in emitter voltage results in a decreased voltage drop across R_B, resulting in a decreased base current I_B and a decreased collector current I_C.

Finally, the decrease in base current results in a decreased collector

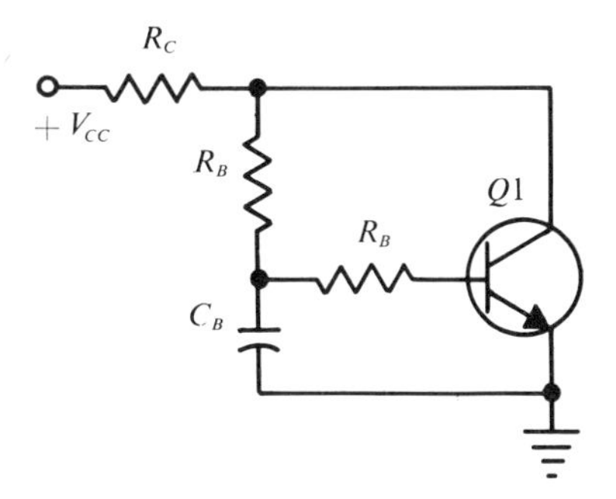

Figure 9–6. **Bypass capacitor C_B prevents signal feedback to the base.**

current, and the operating point is stabilized. The stability factor is formulated

$$S = \frac{R_E + R_B}{R_E + R_B(1 - \alpha)} \qquad \textbf{(9–5)}$$

Example 1. Suppose the circuit in Fig. 9–5(c) has the values $R_E =$ 1 kΩ, $R_B = 33$ kΩ, and beta $= 49$. What is the stability factor?

$$\alpha = \frac{\beta}{\beta + 1} = 0.98$$

$$S = \frac{1 + 33}{1 + 33 \times 0.02}$$

$$S = \frac{34}{1.66} \simeq 20.5 \qquad \text{(answer)}$$

The stability factor is rather high, even though R_B is small and R_E is rather large. In a well-stabilized circuit ($S < 5$), we may calculate the stability factor with the short formula

$$S = 1 + \frac{R_B}{R_E} \qquad \textbf{(9–6)}$$

Applications of this formula to Example 1 would result in the answer of $S = 34$, an error of 70 per cent.

In many practical situations, more bias stability (smaller value of S) is required than is provided by the foregoing configurations. Accordingly, a resistor (R_2) is usually connected from the base to ground to form a bleeder circuit, as illustrated in Fig. 9–6. The value of R_2 is selected to be small in relation to R_1 and $R_E(\beta + 1)$. The ratio of R_2 to R_E is determined by the required stability factor.

$$S \approx 1 + \frac{R_2}{R_E} \tag{9–6}$$

The correct ratio for a stability of less than 5 is $R_2/R_E = 4$. This low value of R_2 establishes a constant value of d-c voltage at the base of the transistor so that any change of collector current results in an increase in the voltage across the emitter resistor. The increase of emitter voltage decreases base current and finally results in a decrease in collector current.

Example 2. Suppose the value of the stability factor is to be 9. What is the ratio of R_2 to R_E?

$$S = 9 = 1 + \frac{R_2}{R_E}$$

$$\frac{R_2}{R_E} = 8 \qquad \text{(answer)}$$

Obviously, R_2 must be less than 8 times R_E. Note that in Example 2 transistor characteristics do not enter into our calculations. The stick-bias configuration results in a much smaller effective value of $R_B(R'_B)$; and as we may observe from Formulas 9–3 and 9–4, this decreases the stability factor. The effective value of $R_B(R'_B)$ is the parallel combination value of R_1 and R_2.

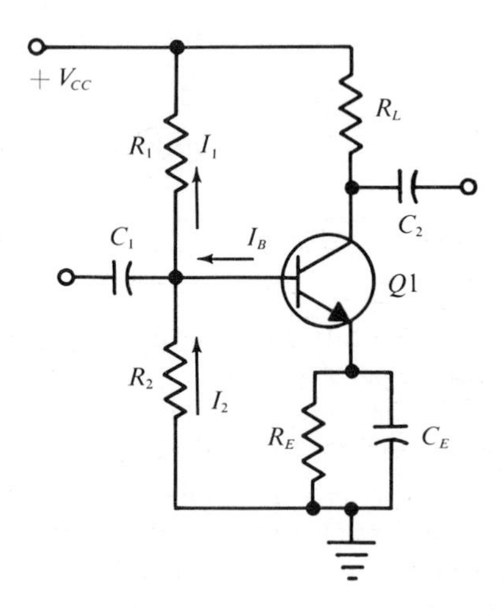

Figure 9–7. Complete *RC*-coupled stage with bias stabilization.

Example 3. Suppose the circuit in Fig. 9–7 has the values $R_1 =$ 68 kΩ, $R_2 = 3.3$ kΩ, $R_E = 1$ kΩ and beta $= 49$ ($\alpha = 0.98$). What is the stability factor? $R'_B = R_1$ in parallel with $R_2 \simeq 3.16$ kΩ

$$S = \frac{1 + 3.16}{1 + 3.16(1 - 0.98)}$$

$$S \simeq \frac{4.16}{1.06}$$

$$\simeq 4 \qquad \text{(answer)}$$

We will now compare this calculation to the calculation using the short Formula 9–4.

Example 4.

$$S = 1 + \frac{R'_B}{R_E}$$

$$\simeq 1 + \frac{3.14}{1}$$

$$\simeq 4.14 \qquad \text{(answer)}$$

Obviously, our error in Example 4 is well within the tolerance of circuit components and transistor parameters. We are then justified in employment of the short formula.

Recall that a transistor has maximum ratings as visualized in Fig. 9–8. Collector dissipation is equal to the product of collector voltage and collector current and decreases with an increased temperature.

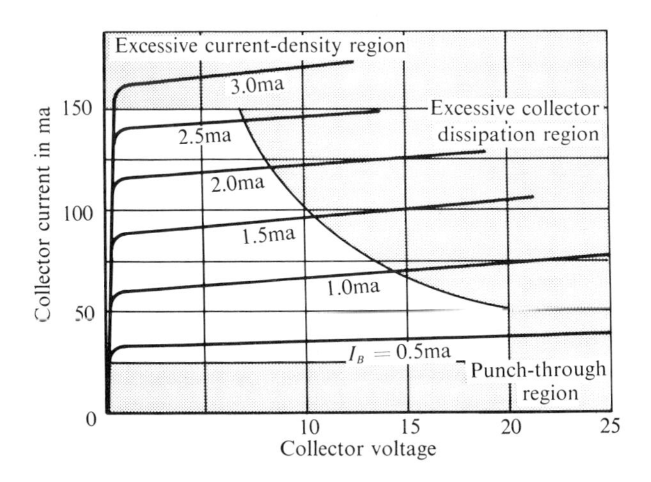

Figure 9–8. Forbidden regions of operation are shown as shaded areas.

The transistor can also be damaged, even if its maximum collector dissipation is not exceeded, when the path of operation extends into the excessive current-density region. Again, if the collector voltage is excessive and the punch-through region is entered, the transistor will be damaged even if the maximum collector dissipation is not exceeded. Punch through occurs as the depletion region of the back-biased collector-base junction reaches farther into the base region. The base becomes thinner until the collector and emitter regions touch, resulting in a rupture of the base and a collector-emitter short.

9–3. Low-Frequency Boost or Compensation

Although bypassing of the power supply and decoupling of the collector-load circuits usually suffice to reduce positive feedback below the level at which low-frequency oscillation becomes self-sustained, it is obvious that positive feedback cannot be completely eliminated in practice unless a separate power supply is used for each control device. Of course, it is not economical to duplicate power supplies in an amplifier. Therefore, a three-stage *RC*-coupled amplifier commonly has residual positive feedback present at low frequencies.

This residual positive feedback at low frequencies tends to overcome the attenuating action of the coupling capacitors and input

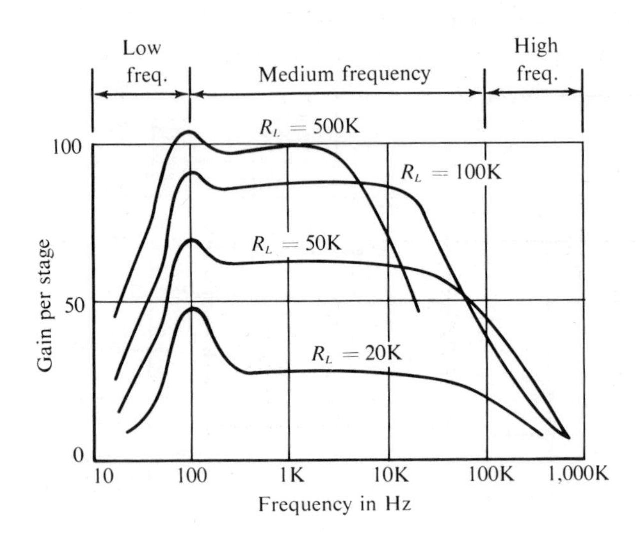

Figure 9–9. **Excessive low-frequency compensation distorts the frequency-response curve.**

biasing resistors in the amplifier. Hence, the low-frequency response of the amplifier is extended by the residual positive feedback. This circuit action is called *regeneration*, *low-frequency boost*, or *low-frequency compensation*. In many cases, the values of R_D and C_D in Fig. 9–2(b) are chosen to obtain maximum extension of the amplifier's low-frequency response. It is evident that excessive low-frequency compensation will distort the amplifier's frequency-response curve and produce objectionable peaking at low frequencies as shown in Fig. 9–9.

9–4. Hum in *RC*-Coupled Amplifiers

Power-supply hum may impose difficulty in amplifier operation using any type of control device, particularly if the amplifier gain is high. We will employ FET amplifiers in our analysis. Since hum voltage is amplified in successive stages, it is evident that more filtering is required for the early stages than for the final stages. Note that decoupling circuits can assist considerably in hum reduction. Thus, with reference to Fig. 9–2(b), the value of C might be chosen much larger than is required to prevent regeneration, in order to obtain adequate hum reduction in an early stage. Hum voltage is introduced via the load resistor of a stage, as shown in Fig. 9–10. Note that if coupling

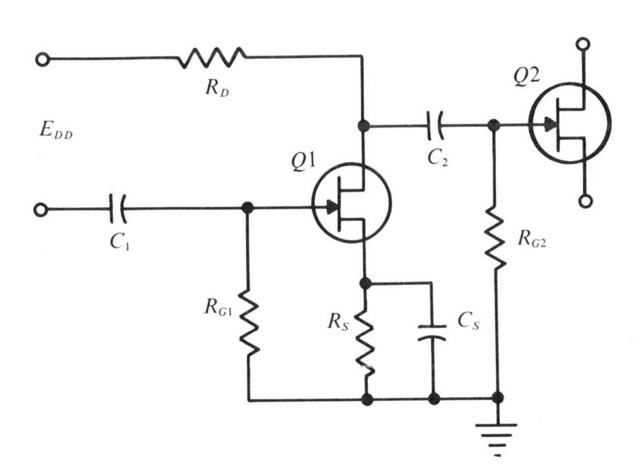

Figure 9–10. Hum voltage v_h is applied to the gate of Q2 via drain resistor R_D.

capacitor C has a comparatively small value, 60-Hz or 120-Hz hum voltage will be proportionally attenuated at the gate of Q_2. This expedient is used to some extent in small radio receivers. However, it is not a very satisfactory method of hum reduction, because low-frequency response is sacrificed.

With reference to Fig. 9–10, let us suppose the value of coupling capacitor C is sufficiently large to provide full gain down to low audio frequencies, such as 20 Hz. Then, the hum voltage applied to the gate of $Q2$ is formulated

$$v_{hG} = \frac{v_h r_{ds}}{R + r_{ds}} \tag{9–7}$$

where

$$R = \frac{R_D R_{G2}}{R_D + R_{G2}} \tag{9–8}$$

and v_h is the value of hum voltage from the power supply, r_{ds} is the dynamic drain resistance of $Q1$, R is formulated by 9–8, v_{hG} is the value of hum voltage applied to the gate of $Q2$, R_D is the drain-load resistance for $Q1$, and R_{G2} is gate resistance for $Q2$.

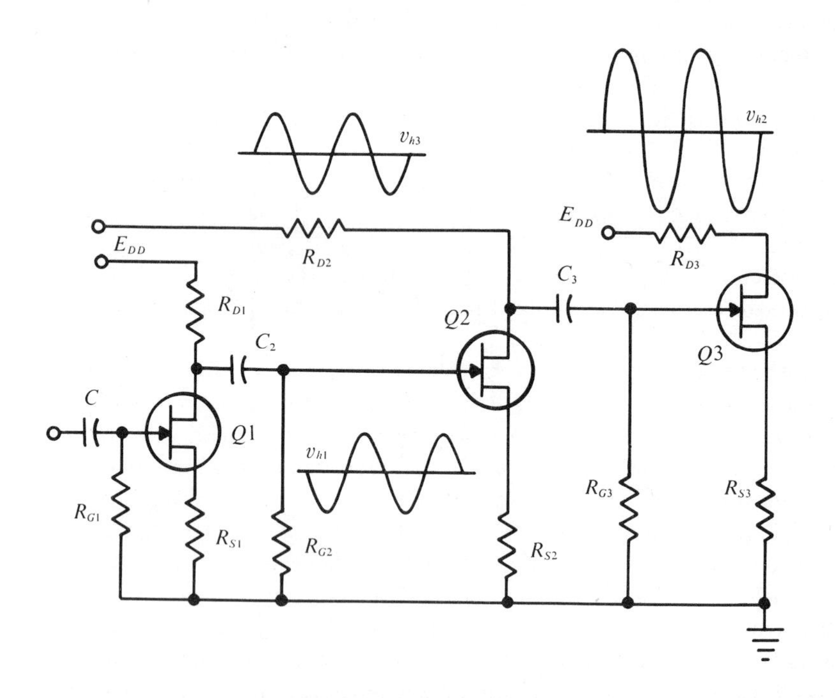

Figure 9-11. Plan of hum cancellation.

Of course, hum voltage is amplified by successive stages, as depicted in Fig. 9–11. Moreover, the phase of the hum voltage is inverted from the gate to the drain of each field-effect transistor. This fact permits a circuit arrangement that provides hum cancellation. Let us see how this works. Observe that hum voltage v_h is amplified and inverted by $Q2$, and hum voltage v_{h1} appears at the drain of $Q2$. Next, v_{h1} is amplified and inverted by $Q2$, and hum voltage v_{h2} appears at the drain of $Q3$. Observe carefully that if a hum voltage v_{h3} were applied to R_{D2}, cancellation takes place between v_{h1} and v_{h3}, reducing v_{h3} to zero.

Since the required value of v_{h3} is greater than that of v_h in order to obtain cancellation, it will not do to connect R_{D2} at the output of the power supply. However, we know that a multisection filter has a progressively reduced hum level from the first to the last section. Therefore, we can connect R_{D2} to a suitable intermediate section of the power-supply filter, and the hum voltages will cancel. Hence, this is a useful method of hum reduction. It is not possible to obtain complete hum cancellation by this means, because the ripple waveform is not quite the same at various sections of a power-supply filter. It is desirable to obtain as much hum reduction as possible in the first stage of an amplifier, because any hum signal from the first stage is progressively amplified and becomes more difficult to eliminate.

The gate or grid circuits in an *RC*-coupled amplifier are often high-impedance circuits. In turn, objectionable hum can be coupled from a-c leads into the input circuits via stray capacitance. Hum voltage from electrostatic coupling can be minimized by keeping a-c leads well separated from gate or grid-circuit leads. Heater leads in vacuum tubes are commonly twisted to minimize hum. If further reduction of electrostatic coupling is required in vacuum tubes, the heater circuit can be wired with shielded leads.

Another common source of objectionable hum in high-gain *RC*-coupled vacuum tube amplifiers stems from heater-cathode *leakage* or heater *emission* in tubes used in the early stages. Cathodes commonly operate above ground potential in order to obtain cathode bias. If there is leakage resistance R_l between heater and cathode, as depicted in Fig. 9–12, it is evident that a 60-Hz hum voltage will be applied to the cathode. This difficulty is easily minimized or eliminated by selection of tubes that have negligible heater-cathode leakage.

Electrostatic coupling of hum voltage into high-impedance grid circuits from nearby heater terminals via stray capacitance is shown in Fig. 9–13. It is helpful to ground a socket terminal between heater and grid terminals, if this is possible. For example, terminals 1 and 6

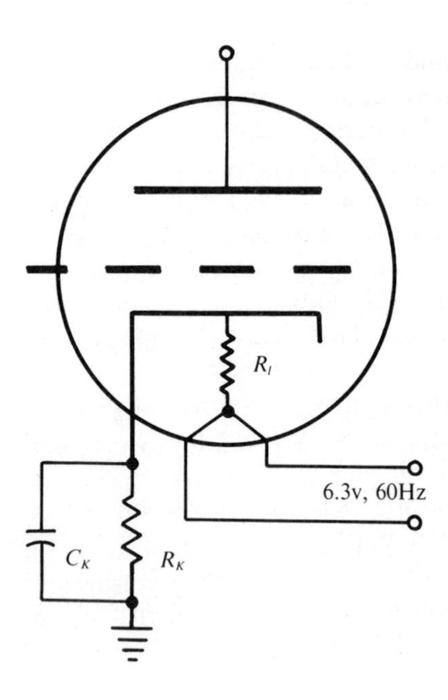

Figure 9-12. **Heater-cathode leakge R_L energizes the cathode with 60-Hz voltage.**

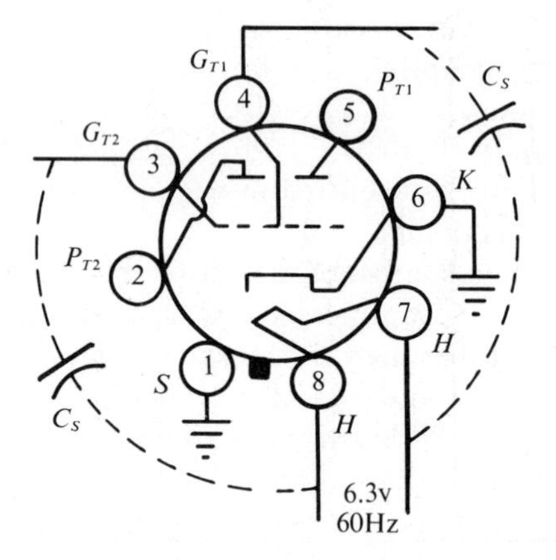

Figure 9-13. **Hum is minimized by grounding the socket terminals between heater and grid terminals.**

are shown grounded. Preamplifiers are most affected by slight hum-injection voltages. Later stages do not require elaborate hum-reduction measures. Preamplifiers present fewer difficulties if the grid circuits can be designed with comparatively low internal impedance.

Heater emission is an elusive source of hum in a vacuum-tube preamplifier, if its presence is unsuspected. Electron emission from heater to cathode occurs in tubes that have traces of chemical impurities in the heater insulation material or that are contaminated by cathode emissive substance. This small 60-Hz electron flow from heater to cathode poses no practical problem if the cathode is grounded. However, if the cathode operates above ground potential, tube selection becomes necessary, or a suitable heater-biasing and/or hum-bucking circuit is required, as will be explained later in the chapter.

Cathode-to-heater emission is another elusive source of hum in a vacuum-tube preamplifier. However, this is a less common source of residual hum than heater-to-cathode emission. Tube selection will eliminate difficulties due to cathode-to-heater emission, or a heater-biasing and/or hum-bucking circuit may be designed into the preamplifier.

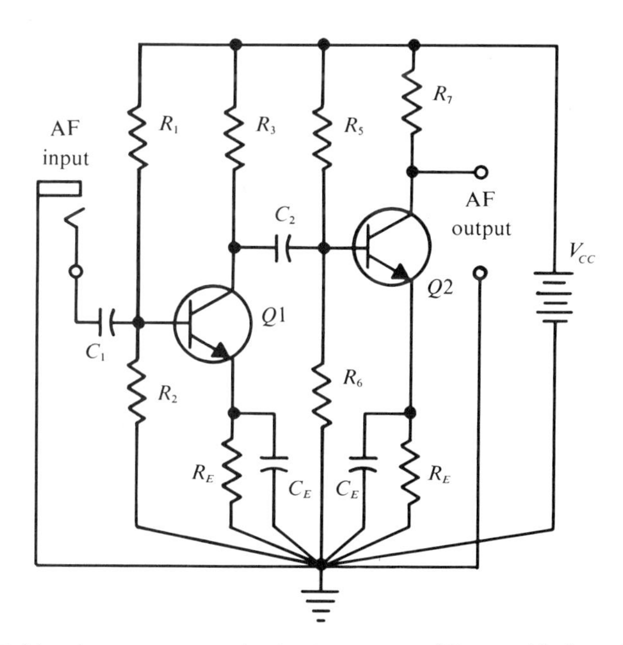

Figure 9–14. A common ground point in a preamplifier avoids hum due to circulating ground currents.

A heater has a small residual magnetic field, and some heaters have stronger magnetic fields than others. One type of heater will produce a minimum magnetic field when one particular end of the heater is grounded, and another type of heater produces a minimum magnetic field when the heater circuit "floats" with respect to ground. In any case, the residual magnetic field from the heater in a preamplifier tube may modulate the electron cloud in the tube sufficiently to produce audible hum. Obviously, a preamplifier tube should be located far from power transformers and filter inductors.

Since a preamplifier stage is extremely responsive to residual hum sources, circulating ground currents must also be minimized or eliminated. Circulating ground currents stem from components grounded to separated points on the preamplifier chassis. Although the resistance of the chassis metal is quite small, there is nevertheless a slight *IR* drop between separated ground points. In turn, 60-Hz hum voltage can be coupled into the input circuit via circulating ground currents. Therefore, it is good practice to choose a single ground point in a preamplifier chassis, as depicted in the transistor circuit, Fig. 9–14. All ground connections are then made at this common point.

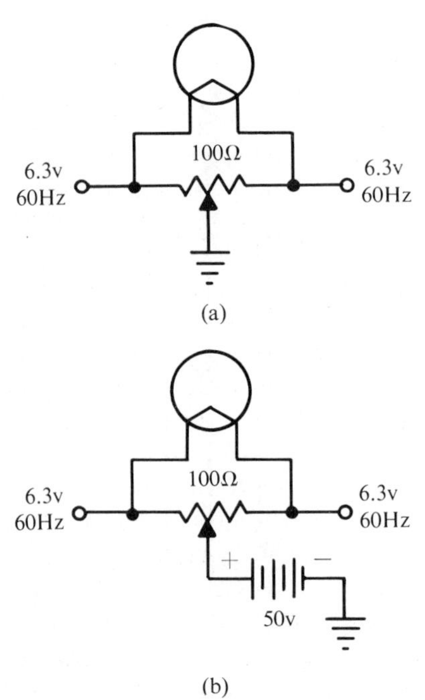

Figure 9–15. (a) Hum-bucking circuit; (b) Hum-bucking circuit with positive heater bias.

In summary, objectionable hum from a vacuum-tube preamplifier stage can stem from more than one source. After all the recognized principles of good practice have been observed, it then becomes necessary to minimize or eliminate residual hum by biasing the heater in suitable polarity and/or employing a hum-bucking circuit as shown in Fig. 9-15. When heater emission is not a significant factor, and the hum stems chiefly from capacitive coupling or magnetic fields from the heater, the hum-bucking arrangement depicted in Fig. 9-15(a) is usually satisfactory. The potentiometer is adjusted for minimum hum output. If a mid-point adjustment should be optimum, a center tap on the heater winding may be grounded, and the potentiometer omitted.

When heater emission is also a significant factor in preamplifier hum, a 50-v positive bias is commonly applied to the heater circuit as shown in Fig. 9-15(b). This positive bias prevents escape of electrons emitted from the heater. As noted previously, cathode-to-heater emission is a less-frequent trouble source than heater-to-cathode emission. However, if cathode-to-heater emission should be a significant factor, the bias arrangement shown in Fig. 9-15(b) will increase the hum level; in such case, it becomes necessary to reverse the polarity of the heater bias voltage.

9-5. Microphony in *RC*-Coupled Amplifiers

In vacuum-tube *RC*-coupled amplifiers, microphony is almost always due to mechanical vibration of a tube, particularly in the preamplifier stage. If a tube is microphonic, a ringing sound is heard from the speaker when the tube is tapped lightly. Evidently, if substantial radiated sound is absorbed by a microphonic tube, the amplifier may supply its own input and howl (oscillate) uncontrollably. Tube selection usually suffices to eliminate microphony. However, if a speaker is mounted in the same cabinet with a high-gain amplifier, microphony may become a stubborn problem at high-level output. In such case, "hangover" ringing following loud tones is eliminated to best advantage by utilizing a speaker in a separate cabinet.

When it is impractical to isolate the speaker in order to eliminate "hangover," it becomes necessary to provide acoustic damping for the preamplifier tube, and possibly the second-stage tube. Common means of acoustic damping include a heavy lead shield for the tube and use of a floating-spring socket. A lead shield lowers the mechanical resonant frequency of the tube to a very low value and provides a high inertia. A floating-spring socket minimizes the mechanical coupling between chassis and tube; in turn, chassis vibration is largely suppressed at the tube itself.

Microphony and acoustic feedback can obviously result from a pickup or other input device which is located in a strong acoustic field. Therefore, it is advisable to replace the input device with a short-circuit when checking an amplifier for microphony. Then, if tests establish that the amplifier is not objectionably microphonic, the remedy for acoustic feedback will be to remove the input device from strong acoustic fields. In public-address systems, the input device is a conventional microphone; and in this situation, it is often possible to use a directional microphone that is oriented away from the loudspeaker.

9–6. Parasitic Oscillation

Parastic oscillation can be compared with motorboating, in that it is a spurious oscillating condition in an *RC*-coupled amplifier. However, motorboating occurs at low audio frequencies, whereas parasitic oscillation occurs at very high frequencies—typically 100 MHz. Various operating difficulties result from parasitic oscillation, such as reduced amplification, pseudo-static noise, and short control device life. We will employ a vacuum tube for an analysis of parasitic oscillation with the understanding that the analysis applies for junction transistors and field-effect transistors.

Parasitic oscillation is most likely to be encountered in an output stage that uses a high-gain device, and this type of oscillation is usually triggered by a high-amplitude signal that drives the tube momentarily into grid-current flow. However, parasitic oscillation is also occasionally an operating difficulty in earlier stages.

Connecting leads have inductance, and if grid and plate leads are not widely separated, stray capacitance can form a high-frequency feedback situation as seen in Fig. 9–15(a). Sometimes the tube oscillates parasitically only on the peaks of strong drive signals. However, if positive feedback is considerable, the parasitic oscillation may persist in the absence of drive signal. When the level of parasitic oscillation is high, the resulting high self-bias due to grid-current flow will make it impossible to obtain an audible output from the amplifier. It is good practice to route the grid and plate leads as far apart as practical to minimize the possibility of parasitic oscillation.

However, proper lead dress alone does not always suffice to eliminate parasitic oscillation. In such case, it is common practice to insert a grid-stopping resistance called a "parasitic suppressor," of approximately 50,000 Ω in series with the grid circuit at the socket terminal, as shown in Fig. 9–16(b). The grid-stopping resistor is so-called be-

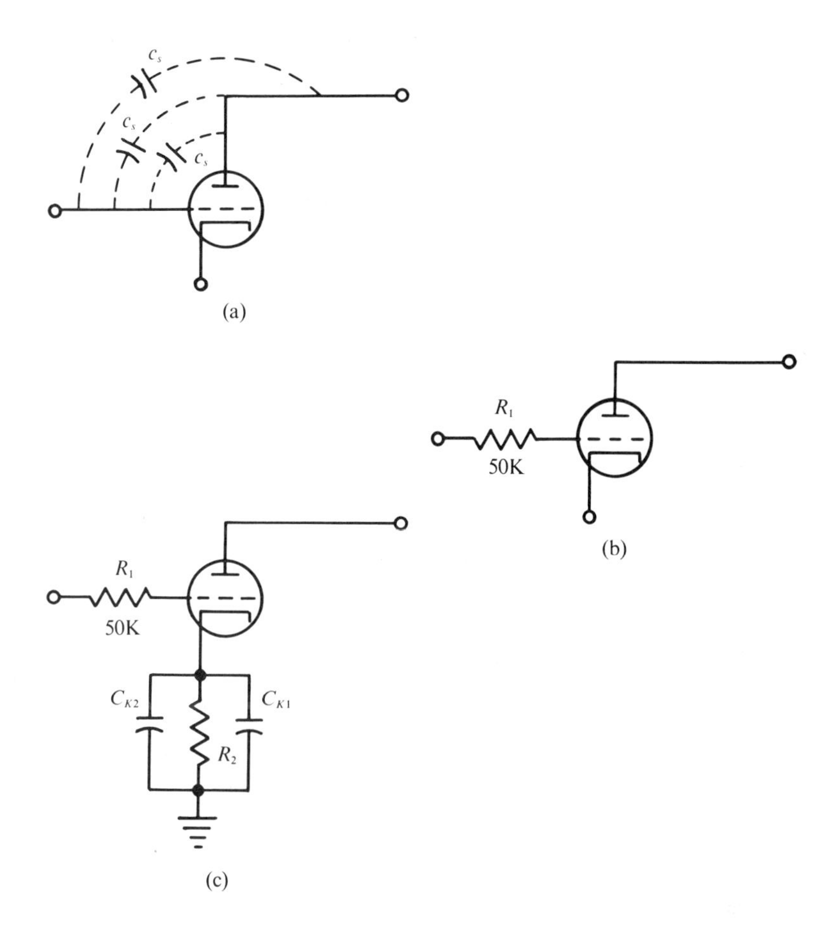

Figure 9-16. (a) Parasitic oscillation can be caused by coupling of grid and plate leads; (b) Grid-stopping resistor; (c) Auxiliary cathode bypassing.

cause it stops parasitic oscillation in most cases. This resistance lowers the Q of the high-frequency resonant circuit formed by grid-lead inductance and stray capacitance. If a grid-stopping resistance does not entirely eliminate parasitic oscillation, a small bypass capacitor may also be connected from the plate terminal to ground; however, this method tends to attenuate the high-frequency response.

Electrolytic capacitors, generally used to bypass the cathode bias resistor, may not appear capacitive at high frequencies; instead, an electrolytic capacitor may appear as a small inductance at 100 MHz. In turn, a significant inductance is present in the cathode circuit at a parasitic oscillation frequency. We will find that positive feedback exists in a stage when the cathode and plate circuits have certain reactive relations. Therefore, as shown in Fig. 9–16(c), a small fixed capacitor C_{K2} is shunted across the electrolytic capacitor C_{K1} in stubborn situations. The small capacitor has a low capacitive reactance to ground at very high frequencies and effectively places the cathode at r-f ground potential.

9–7. D-C versus A-C Load Lines

Beginning students sometimes do not recognize that an a-c load line may have significantly different slope from a d-c load line. In turn, incorrect conclusions may be drawn concerning amplifier operation. Let us consider a practical example. Figure 9–17(a) depicts an *RC*-coupled amplifier circuit in which the d-c load of *Q*1 is the sum of the 9,000-Ω collector resistor and the 1,000-Ω emitter resistor. The a-c load is the parallel combination of R_C, R_1, R_2, and r_{i2}. With an 8.47-kΩ r_i value, the a-c load is approximately 2.86 kΩ. For the sake of simplicity, we will assume that C_C is practically a short-circuit for a-c at the frequency of amplifier operation. It is apparent that the d-c load line is determined by 10,000 Ω, since the coupling capacitor blocks d-c current flow.

If the base bias of *Q*1 is 75 μa, the operating point for the first stage falls at point *P* of Fig. 9–17(b). Insofar as the gain is concerned, we cannot obtain a correct answer if we assume that the a-c collector load is 10,000 Ω. Instead, we must draw the actual a-c load line for 2,800 Ω through *P*, as shown in the diagram. Note carefully that the d-c load line is *used only to locate the operating point P.* The slope of the a-c load line is $1/R'_L$ or $\Delta I/\Delta E$. The path of a-c signal operation is followed along the a-c load line.

Suppose that the base is driven by a 25-μa peak signal. Then, the collector-current flow and collector voltage are denoted by point P_1. On the opposite half cycle, the base is driven by a —25-μa signal, and the collector current flow and collector voltage are denoted by point P_2. We perceive that the voltage amplification of the stage is somewhat less than, and the current amplification is somewhat more than if operation were along the d-c load line.

If R_C should have a higher value, or if the a-c load should have a lower value, the separation of a-c and d-c load lines would be increased accordingly. In most *RC*-coupled junction transistor amplifier configurations, the a-c and d-c load lines have widely different slopes and great error is incurred by assuming that signal operation is along the

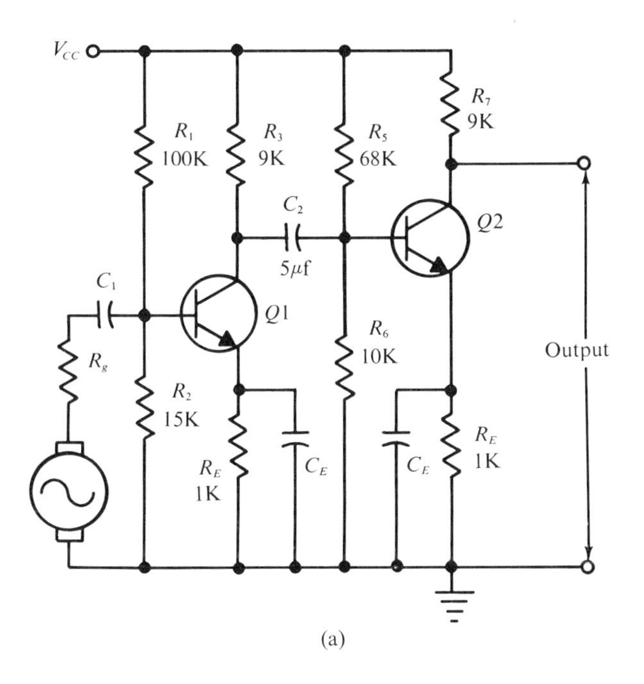

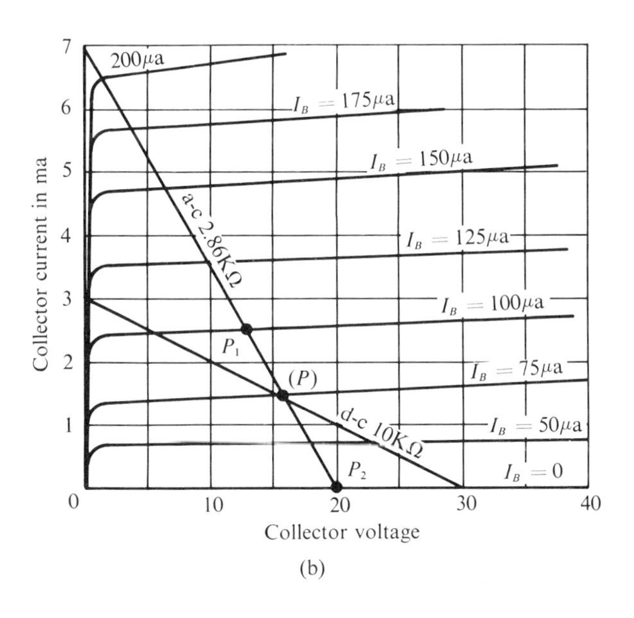

Figure 9-17. (a) *RC*-coupled amplifier; (b) Plots of a-c and d-c load lines on the characteristic curves.

d-c load line. On the other hand, when we proceed to study FET or vacuum-tube amplifiers, we will find that no great error is incurred by assuming that signal operation is along the d-c load line.

Finally, let us consider the change that occurs in the a-c load line of an amplifier when the operating frequency is comparatively low. Using the vacuum-tube amplifier in Fig. 9–18(a) for analysis, the results are that C_C in Fig. 9–18(a) has substantial capacitive reactance.

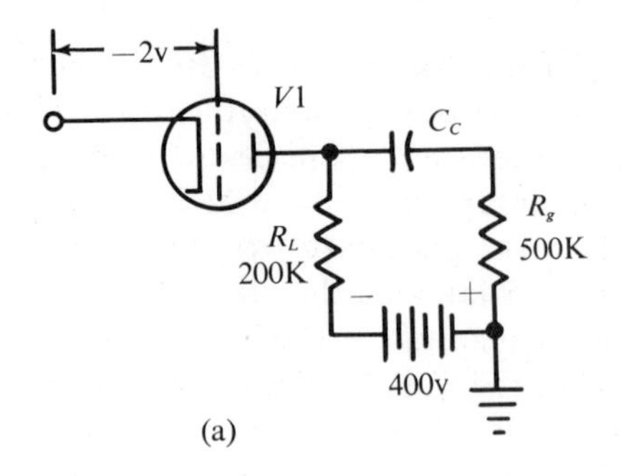

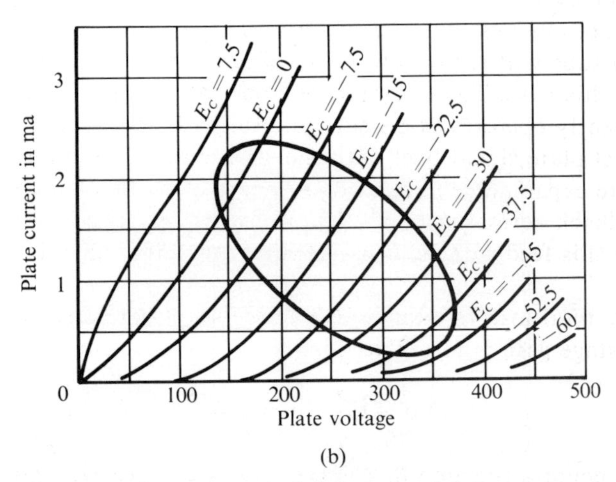

Figure 9–18. Amplifier circuit at high frequency. (a) A reactive plate load; (b) Corresponding elliptical a-c load "line".

In this region of operation, the effective a-c plate load is no longer resistive, but becomes reactive. In other words, the plate works into a capacitive impedance. It can be shown that the a-c load line then opens up into an ellipse, depicted in Fig. 9–18(b). Stage operation from one grid potential to the next is then followed around the ellipse.

The same considerations apply to the change that occurs in the a-c load line when the operating frequency is comparatively high, and the shunt capacitance in the plate circuit draws a substantial fraction of the a-c signal current. As before, the load line opens up into an ellipse. With this brief introduction to reactive loads, we will reserve details of amplifier operation for subsequent discussion.

9-8. Pentode Audio Amplifiers

RC-coupled amplifiers often utilize pentode vacuum tubes instead of triodes. Comparative two-stage amplifiers are shown in Fig. 9-19. More components are required in a pentode amplifier circuit. However, a pentode has the advantage that its grid-input capacitance is lower than for a triode. In turn, plate-load resistance R_3 in Fig. 9-19(b) is shunted by less total capacitance than is R_3 in Fig. 9-19(a). Because of this reduced shunt capacitance, the plate-load resistance may have a higher value in a pentode stage, with consequently higher voltage gain at the same bandwidth. A pentode has comparatively low grid-input capacitance because of the electrostatic shielding action provided by the screen grid.

It is not obvious that a screen grid between the plate and control grid of a tube will reduce the value of grid-input capacitance. Let us see why this is so. The capacitance from control grid to plate in a triode evidently operates as a small coupling capacitor that couples the high-level plate signal back into the grid circuit. In other words, the grid-plate capacitance in a triode provides a residual feedback circuit; this feedback action is called the *Miller Effect* in a triode. We will discover this feedback is equivalent to an added grid-input capacitance.

If the total load resistance in a triode-amplifier stage is denoted by R_L, the stage gain is formulated

$$\text{Voltage gain} = \frac{-\mu R_L}{R_L + r_p} \qquad \textbf{(9-9)}$$

where μ denotes the amplification factor of the triode, and r_p denotes its dynamic plate-resistance value. Thus, if the input signal voltage is equal to e_s, the signal voltage at the plate is equal to $-Ae_s$, where A is the voltage-gain figure given by Formula 9-9. In turn, the difference between the grid-signal voltage and the plate-signal voltage is formulated:

$$e_s - (Ae_s) = e_s(A + 1) \qquad \textbf{(9-10)}$$

For any capacitor, Coulomb's law states that $Q = CE$; hence, the

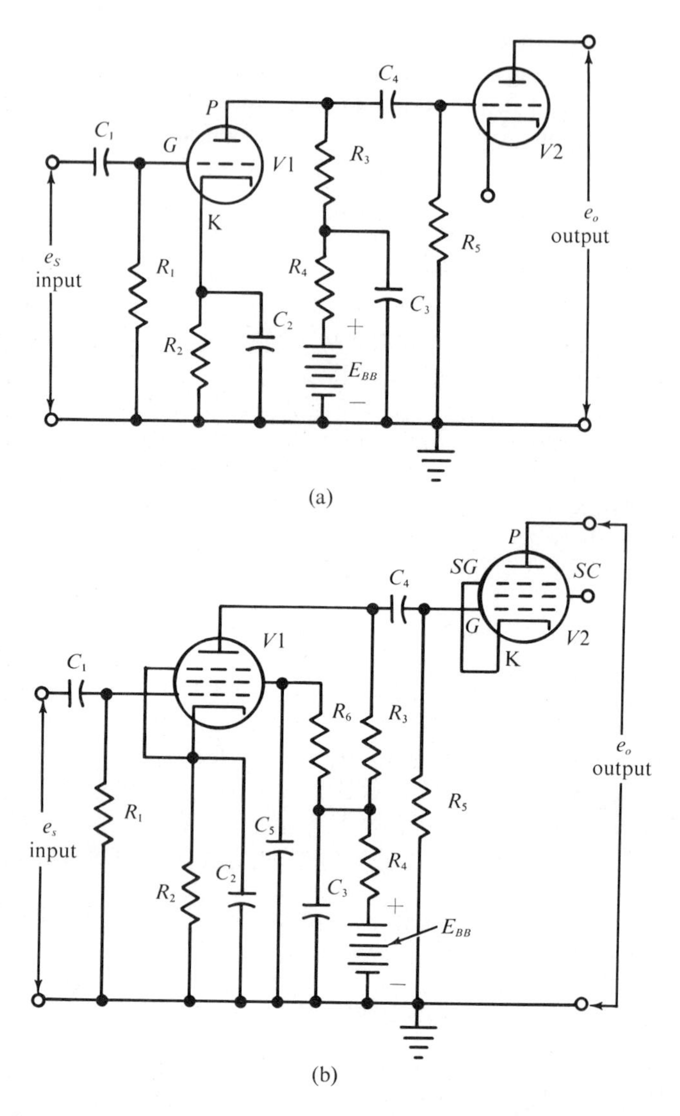

Figure 9-19. Comparative triode and pentode RC-coupled amplifier configurations. (a) Triode amplifier circuit; (b) Pentode amplifier circuit.

charge on the grid that results from the grid-cathode capacitance is formulated

$$q_1 = C_{gk}e_s \tag{9–11}$$

where G_{gk} denotes the grid-cathode capacitance value. Similarly, the charge on the grid that results from grid-plate capacitance is formulated

$$q_2 = C_{gp}e_s(A + 1) \tag{9-12}$$

The total charge on the grid that results from interelectrode capacitance is equal to the sum of q_1 and q_2.

$$q_1 + q_2 = e_s[C_{gk} + C_{gp}(A + 1)] \tag{9-13}$$

Finally, again invoking Coulomb's law, the total effective grid-input capacitance is formulated

$$C_{in} = C_{gk} + C_{gp}(A + 1) \tag{9-14}$$

Although the beginning student might suppose that this increased grid capacitance is negligible, note that Formula 9-14 states that the value of C_{gp} is multiplied approximately by the stage gain. Thus, if we are concerned with high stage gain, the Miller Effect can be quite troublesome. For example, let us suppose that a triode is used that has a C_{gk} value of 1.8 pf, and a C_{gp} value of 1.8 pf, and that the stage gain is 65. Perhaps to our surprise, Formula 9-14 shows that the effective grid-input capacitance is more than 120 pf. This is a very significant value of capacitance in shunt with the plate-load resistance, and the high-frequency response is considerably reduced.

To avoid this reduction in high-frequency response, pentodes can be utilized instead of triodes, as shown in Fig. 9-19(b). The screen grid prevents feedback of signal voltage from plate to grid by means of its electrostatic shielding action. Note that the screen-dropping resistor R_6 is bypassed by C_5. Bypassing is essential, because the screen would otherwise operate as a plate. That is, if the screen resistor were unbypassed, the screen grid would develop an out-of-phase signal voltage that would produce Miller effect and also reduce the stage gain. Therefore, it is essential to maintain the screen grid at a-c ground potential. We recall that a suppressor grid is placed in a pentode to nullify the effect of secondary electron emission, thereby preventing the screen from collecting secondary electrons.

9-9. Constant-Current Gain Calculation

Pentodes have a very high value of dynamic plate resistance, as implied by the slight slope of the plate characteristic curves shown in Fig. 9-20. Since a pentode has a very high plate resistance, it can be regarded as a constant-current source. In turn, the formula for voltage gain of a stage is simplified. Figure 9-21 shows a comparison of the constant-voltage equivalent circuit for a triode with the constant-cur-

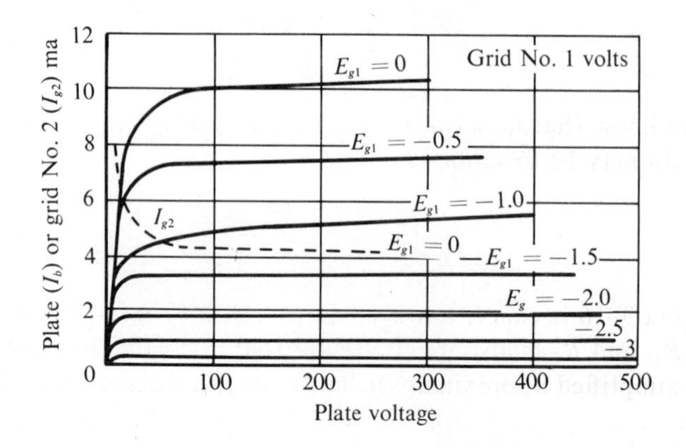

Figure 9–20. **A pentode has a very high dynamic plate resistance.**

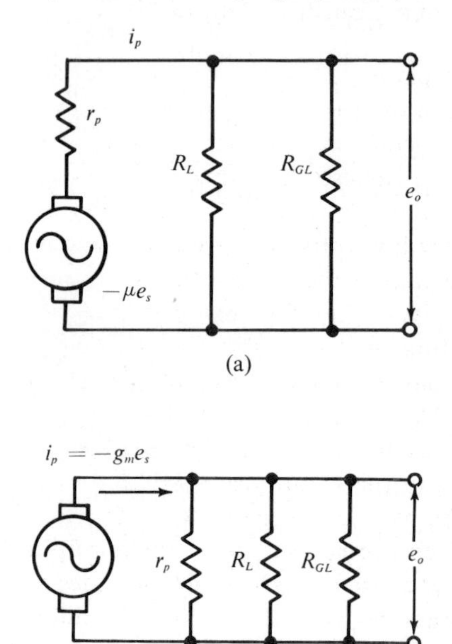

Figure 9–21. **(a) Constant-voltage equivalent circuit at midband frequencies; (b) Constant-current equivalent circuit at midband frequencies.**

rent equivalent circuit for a pentode. With reference to Fig. 9–20(b), the output signal voltage e_o is evidently formulated

$$e_o = \frac{i_p}{\dfrac{1}{r_p} + \dfrac{1}{R_L} + \dfrac{1}{R_{GL}}} \tag{9–15}$$

We know that $i_p = (\mu e_s)/r_p$, and that $g_m = \mu/r_p$. In turn, the voltage gain may be formulated

$$A_v = \frac{g_m}{\dfrac{1}{r_p} + \dfrac{1}{R_L} + \dfrac{1}{R_{GL}}} \tag{9–16}$$

Formula 9–16 is exact; however, in practice, r_p is usually much greater than R_L, and R_G is also much greater than R_L. Therefore, we ordinarily use a simplified approximate formula for the voltage gain of a pentode stage.

$$A_v = g_m R_L \tag{9–17}$$

9–10. Impedance-Coupled Amplifier

Although the plate-load resistor in a pentode stage has a significantly larger value than in a triode stage for the same band-width, a pentode resistance-coupled amplifier must nevertheless be elaborated to obtain uniform response out to high frequencies such as 4 MHz. Of course, we could obtain a reasonably uniform response out to 4 MHz by utilizing a very low value of plate resistance with a pentode. However, the voltage gain of the stage would be so low that this approach would be impractical. Therefore, conventional video amplifiers employ impedance coupling, as depicted in Fig. 9–22.

A basic impedance-coupled stage of the shunt-peaked type is depicted in Fig. 9–22(a). At comparatively low frequencies, inductor L has very low reactance and can be regarded as a short-circuit. However, at high frequencies, in the vicinity of 4 MHz, the inductor has high reactance and may be parallel-resonant with C_S. Since the impedance of the plate-load circuit remains almost constant up to 4 MHz, if the value of L is properly chosen this circuit provides acceptable video amplification. In practice, we will find that R_L might have an optimum value of 1,700 Ω, and that L might have an optimum value of 100 μh for reasonably uniform response out to 4 MHz. Thus, if V should have a transconductance of 9,000 μmho, the stage gain would be 18.

Another basic impedance-coupled stage of the series-peaked type is diagrammed in Fig. 9–21(b). This configuration employs series resonance to maintain good high-frequency response. At low frequencies, L has negligible reactance and can be regarded as a short-circuit. On the other hand, at high frequencies, the reactance of L becomes sub-

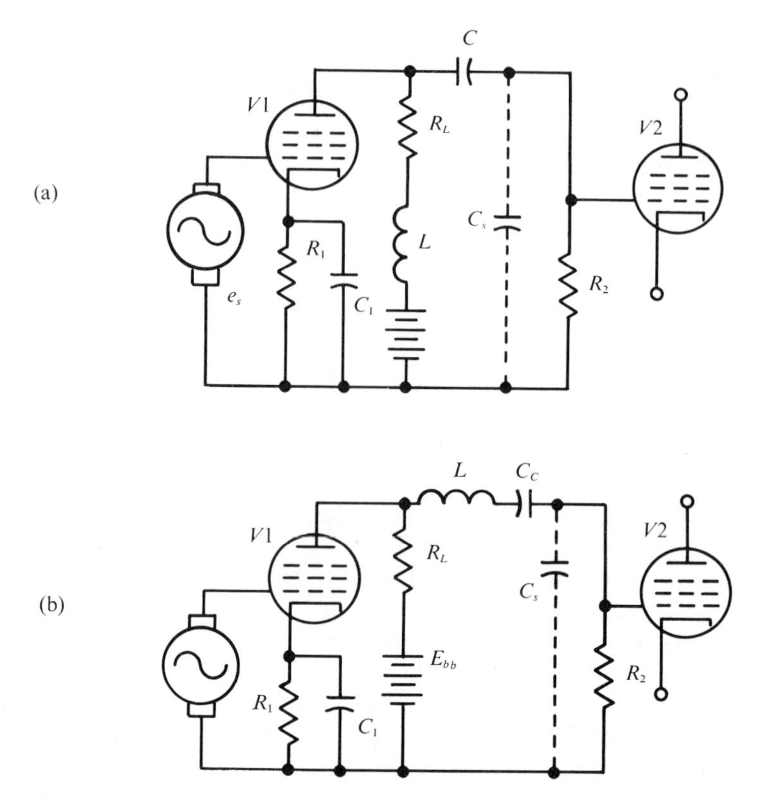

Figure 9–22. (a) Impedance-coupled amplifier stage, shunt-peaked; (b) Impedance-coupled amplifier stage, series-peaked.

stantial, and a typical configuration is designed to resonate X_L and X_{Cs} at 4 MHz. Since this is a series-resonant circuit, a signal voltage magnification of Q times appears across C_S at resonance. Note that inductor L in the series-peaked configuration isolates the plate-output capacitance of the first tube from the grid-input capacitance of the second tube. If we use approximately the same value for L as in Fig. 9–22(a), the stage characteristics are similar.

The maximum available gain from the series-peaked arrangement in Fig. 9–22(b) depends on the ratio of plate capacitance to grid capacitance. It can be shown that optimum gain results when the plate capacitance is half the grid capacitance of the following tube. Under this condition, R_L can have a maximum value of about 2,500 Ω, and the series-peaked stage provides 50 per cent more gain than the shunt-peaked stage.

It might be anticipated that still more gain could be obtained from a video-amplifier stage by utilizing a combination of series and shunt peaking. This is actually the case, and a video amplifier stage such as shown in Fig. 9–23 provides the maximum gain obtainable from an

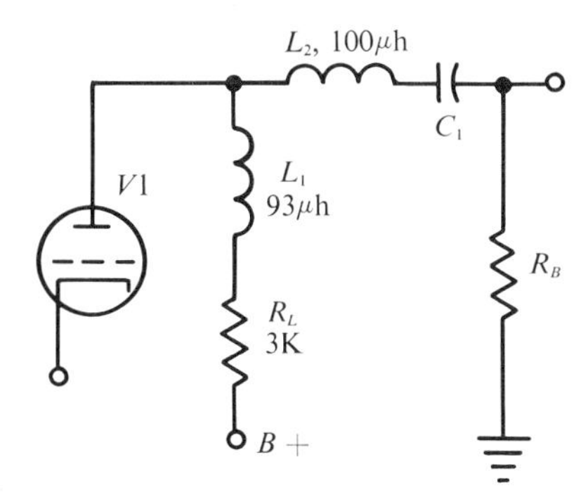

Figure 9–23. Series- and shunt-peaked impedance-coupled stage with uniform response to 4 MHz.

impedance-coupled amplifier. With the values indicated, virtually flat frequency response is obtained up to 4 MHz, and the stage gain is about 80 per cent greater than for a shunt-peaked stage. As in the case of a series-peaked stage, the maximum gain is not obtained unless the plate capacitance is about half of the grid capacitance in the following tube.

The series-parallel circuit depicted in Fig. 9–23 is complex, but it can be readily analyzed by drawing its equivalent circuit diagrammed in Fig. 9–24. We perceive that the tube can be replaced by an equivalent generator with an internal resistance R that replaces r_p and R_L. L_1 and L_2 form a conventional low-pass filter in combination with C_p and C_g. We recall that a low-pass filter has a flat frequency response

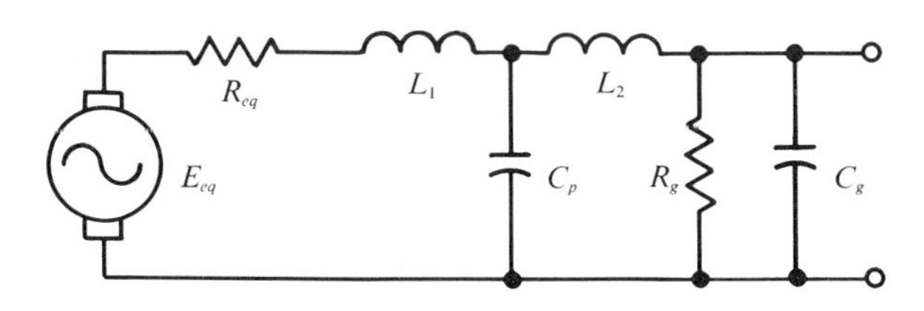

Figure 9–24. Equivalent low-pass filter circuit for the configuration of Figure 9–23.

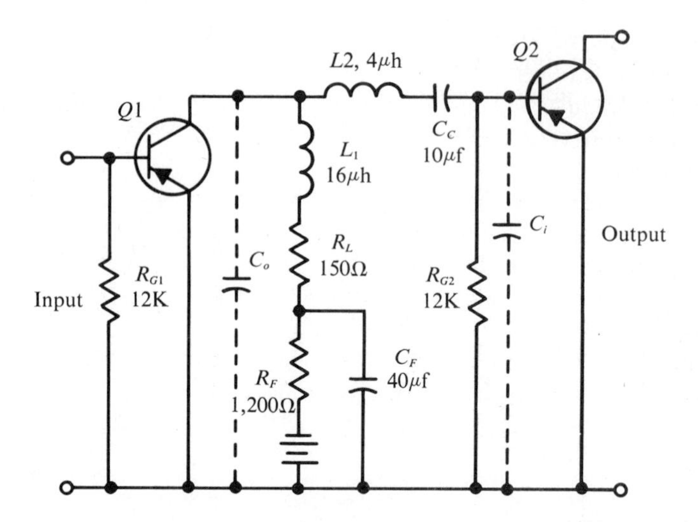

Figure 9–25. Typical 4-MHz transistor video amplifier.

up to its high-frequency cutoff point. Accordingly, it is evident that the configuration in Fig. 9–23 must have a similar frequency response, provided that the inductive and capacitive values are suitably related to the resistance values.

Figure 9–25 shows a typical transistor video-amplifier stage that has a flat response out to 4 MHz. Observe that the collector load resistor has a low value and the peaking coils have low inductances compared with the vacuum-tube counterpart. The reason for these low values stems from the high junction capacitances of a transistor, compared with the interelectrode capacitances of a vacuum tube. Of course, the transistor video amplifier has the same basic equivalent coupling circuit as a tube amplifier (Fig. 9–24). We merely replace C_p and C_g with the junction capacitances of the transistor.

Any impedance-coupled *RC* amplifier may employ low-frequency compensation to obtain extended low-frequency response. A low-frequency compensation circuit is also called a low-frequency boost circuit. It is used to provide increased amplification at low frequencies; thus, if low frequencies are attenuated by a grid-coupling capacitor, we can use a low-frequency compensation circuit to bring these low-frequency signal components back to normal amplitude. Low-frequency compensation can be obtained by utilizing a suitable plate-load or collector-load impedance instead of a plate-load or collector-load resistor in an amplifier stage. This plate-load or collector-load impedance commonly comprises an *RC* configuration that develops an increased impedance value as the signal frequency is decreased. Since the stage gain increases when the plate-load impedance is increased, low-frequency compensation is obtained. For example, R_F and C_F in Fig. 9-25 comprise a low-frequency compensation circuit. Coupling capacitor C_C has a comparatively large value of 10 μf; however, the coupling capacitor works into a comparatively low-impedance base circuit. In turn, a low-frequency compensation circuit is required to obtain good low-frequency response. Note in passing that extended low-frequency and high-frequency response can be obtained (at the expense of gain) by use of negative-feedback circuits. However, this topic must be reserved for subsequent detailed discussion.

Typical frequency-response curves for a shunt-peaked stage are shown in Fig. 9-26 for conditions of no compensation, correct compensation, and over-compensation. When no peaking coil is used, the

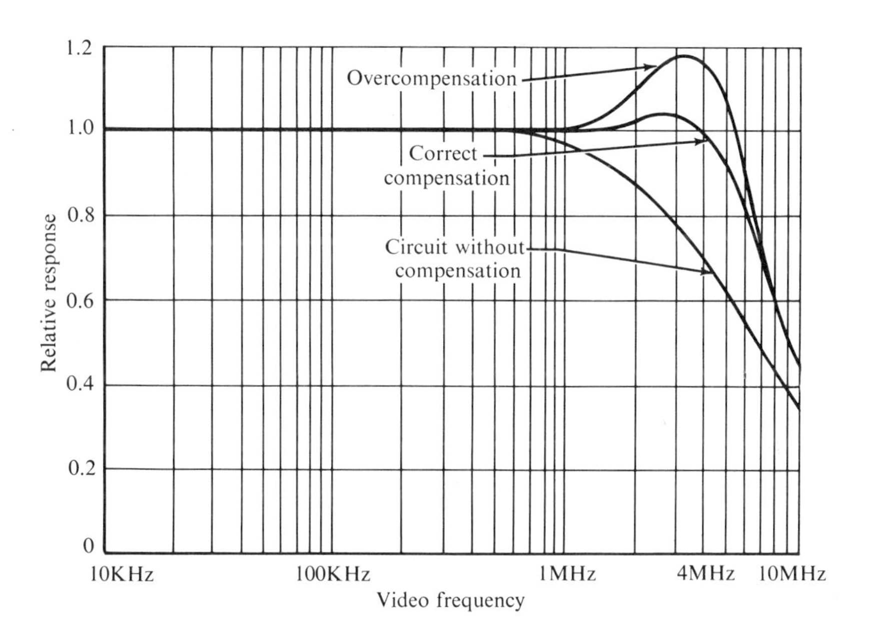

Figure 9-26. Frequency-response curves for a stage with and without shunt peaking.

frequency response starts to drop off at 800 kHz. With correct peaking inductance, frequency response starts to drop off at 4 MHz. This is an enormous extension of high-frequency response. Note that if the stage is over-compensated, an objectionable high-frequency rise occurs from 800 kHz to 6 MHz. Over-compensation is undesirable, because the transient response of the amplifier is thereby seriously impaired.

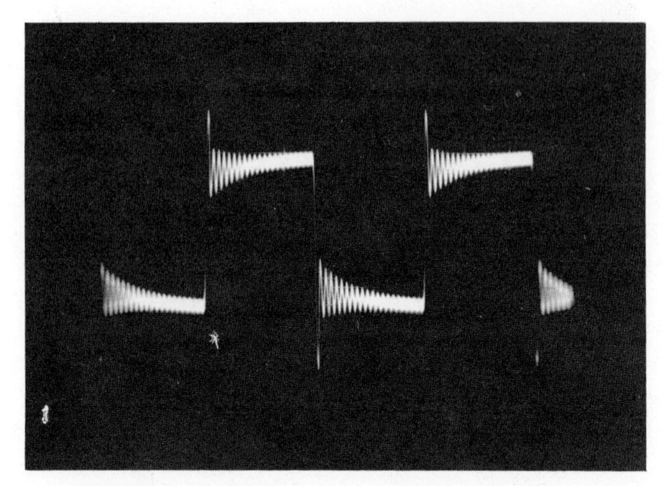

Figure 9-27. Square-wave overshoot and ringing in overcompensated video amplifiers.

Transient response is checked with a square-wave or pulse generator. Most video amplifiers tend to overshoot and ring noticeably. However, if over-compensated, the amount of overshoot and ringing becomes unacceptably large, as illustrated in Fig. 9-27. In a television receiver, objectional overshoot and ringing are seen as picture distortion. Edge reproduction of image is broken up into successive vertical black and white lines. Ringing denotes a shock-excited oscillation resulting from the response of *L*, *C*, and *R* components to a suddenly-applied voltage. A peaking coil has a self-resonant frequency, and if its effective resistance value is less than the critical value, the coil will be shock-excited into oscillation by a square-wave or pulse voltage. This is a damped sine waveform of voltage; its rate of decay depends upon the value of effective resistance associated with the peaking coil.

Observe the simple *LCR* configuration depicted in Fig. 9-28(a). A switch is provided to apply the battery voltage *E* to the circuit suddenly. If *R* has a large value, the current flow has a waveform similar to the curve shown in Fig. 9-28(b). On the other hand, if *R* has a small value, the current waveform is oscillatory, as seen in Fig. 9-28(b). We perceive that if *R* has a large value, *C* will charge slowly, and the current will rise to a peak value and subsequently decay to zero as *C* becomes fully charged. Inductor *L* develops an opposition to rapid charge of *C*, due to counter emf. We state that current flow in Fig. 9-28(b) is aperiodic, because the current flows in one direction only and does not reverse its direction of flow periodically.

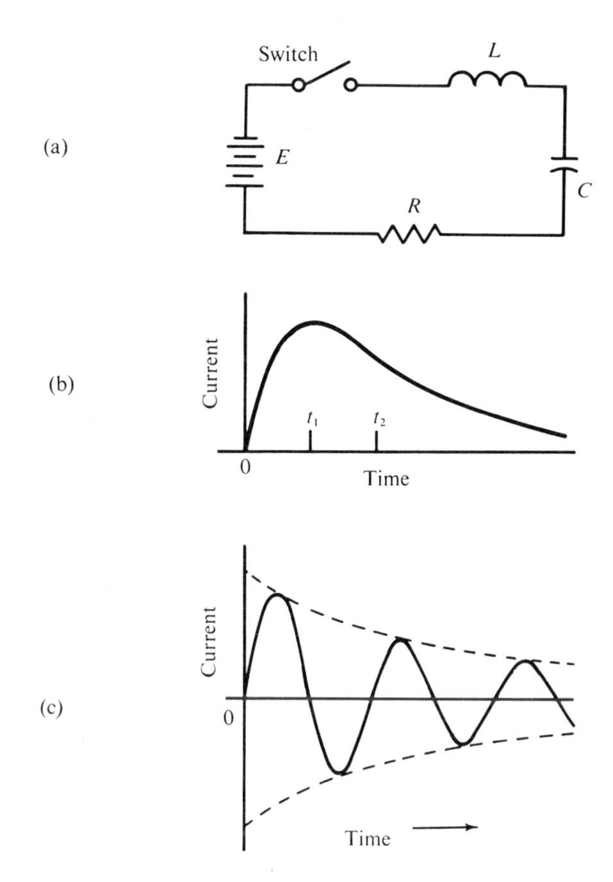

Figure 9–28. (a) *LCR* circuit with switch for sudden application of voltage; (b) Non-oscillatory response; (c) Oscillatory response.

As the value of R is reduced in Fig. 9–28(a), a critical value is reached, below which aperiodic current flow changes into an oscillatory flow as depicted in Fig. 9–28(c). This critical value of resistance is formulated

$$R - 2\sqrt{\frac{L}{C}} \tag{9–18}$$

where R denotes ohms, L denotes henrys, and C denotes farads.

When R has its critical value, we state that the circuit is critically damped. If R has less than the critical value, we say that the circuit is under-damped; or, if R has more than the critical value, we define the circuit as over-damped. The current waveform in an under-damped circuit is called a damped sine wave (Fig. 9–28c). It is a sine wave that decays in accordance with an exponential envelope. Thomson's formula states the frequency of the damped sine wave

$$f = \frac{1}{2\pi}\sqrt{\frac{1}{LC} - \frac{R^2}{4L^2}} \qquad \textbf{(9–19)}$$

The formula for the ringing waveform is written

$$i = \frac{E}{\omega L}\epsilon^{-\alpha t}\sin \omega t \qquad \textbf{(9–20)}$$

where i denotes the instantaneous current value, E denotes the applied voltage, ω denotes $2\pi f$, f denotes the frequency stated by Thomson's formula, L denotes the inductance value in henrys, ϵ denotes the base of natural logarithms (2.718), α denotes $R/2L$, and t denotes time in seconds.

When the value of R in Fig. 9–28(a) is slightly less than the critical value, the oscillatory waveform decays very rapidly, and it might make only one positive excursion and one negative excursion. On the other hand, when the value of R is much less than the critical value, the oscillatory waveform decays slowly and many positive and negative excursions are completed before the transient interval is completed.

SUMMARY

1. "Motorboating" is a common form of oscillation that occurs in a low-frequency amplifier, in which some of the output signal is coupled through the power supply to the input of the amplifier in the form of positive feedback.
2. Positive feedback can be eliminated by such methods as decoupling of the power supply, collector and plate decoupling, and the use of negative feedback.
3. The electrical characteristics of a junction transistor depend very much upon the crystal temperature. The crystal temperature, in turn, depends upon its internal losses and upon the outside factors, such as ambient temperature and thermal conductivity.
4. The most noticeable effect of an increase in temperature on a junction transistor is the increase in collector current.
5. The collector current of a junction transistor in the common-emitter configuration is equal to beta times the base current, plus the leakage current.
6. The stability factor (S) of a transistor amplifier can be thought of as a *figure of merit:* the smaller the stability factor, the less the effects of a temperature change.

7. A well-designed common-emitter amplifier should have a stability factor of 5 or less.

8. A transistor may be stabilized against the effects of collector current runaway due to a temperature change by the use of negative feedback to the d-c base current I_B.

9. Shunt feedback, used to stabilize a transistor against temperature effects, reduces the output gain and output impedance.

10. The emitter resistor employed in series stabilization must be bypassed to prevent a loss of amplifier gain.

11. The short formula will yield a close approximation of the stability factor when the factor is low and the transistor beta is high.

12. Thermal runaway is a "snowballing" action that may cause destruction of a transistor.

13. A transistor has maximum ratings of current, voltage, temperature, and power.

14. Excessive low-frequency compensation in an amplifier distorts the frequency response curve.

15. Amplifier noise, such as hum, is amplified in successive stages. Therefore, filtering is more important for the early stages than for the final stages.

16. *Hum* in high impedance circuits can be coupled from a-c leads into the input circuit via stray capacitances.

17. A common source of hum in vacuum-tube amplifiers stems from heater cathode leakage or heater emission in the early stages.

18. Ground loops in an amplifier may be eliminated by choosing a single ground point on the chassis in the stages with low-signal levels.

19. Microphony in vacuum tubes is almost always due to mechanical vibrations.

20. Acoustical feedback in an amplifier results from the pickup being located in a strong acoustical field.

21. Parasitic oscillations are the result of internal oscillations within an amplifying device; however, such oscillations may occur through shunt capacitance between the input and output leads.

22. The d-c load line is used to determine the operating point only.

23. The slope of the a-c load is determined by the a-c load and is located by drawing a line at this slope through the operating point.

24. When the a-c load has substantial capacitance, the load line is an ellipse.

Questions

1. Explain the term "motorboating" as applied to *RC*-coupled amplifiers and discuss the remedy for this occurrence.
2. What outside factors have the greatest effect upon transistor parameters?
3. What is the relationship of the leakage current in a common-emitter amplifier to that in a common-collector amplifier?
4. Explain the term I_{CEO}.
5. Discuss the stability factor of a common-emitter amplifier.
6. What is the maximum value of stability desired in a well-designed common-emitter amplifier?
7. What is the value of the stability factor in a common-base amplifier?
8. Discuss the basic concept of stabilization of the collector current against a change of temperature.
9. In reference to the circuit in Fig. 9-5(b), discuss the method of bias stabilization.
10. In reference to the circuit in Fig. 9-5(c), discuss the method to stabilize the collector current.
11. When may we use the short formula to calculate the stability factor of a transistor amplifier?
12. Explain the term "thermal runaway."
13. With reference to Fig. 9-6, what are the effects of adding capacitor C_B to the circuit?
14. Discuss the limitations placed on a given transistor such as is depicted by the output characteristics in Fig. 9-8.
15. In reference to Fig. 9-9, what causes the peaks on the low end of the bandpass curves?
16. Where is the most important point to apply corrective measures to eliminate noise, such as hum, in an amplifier?
17. Discuss the plan of hum cancellation illustrated in Fig. 9-11.
18. How can hum from electrostatic coupling be minimized?
19. Discuss methods used to reduce heater-cathode leakage hum in vacuum tubes.
20. Discuss methods used to reduce heater-emission hum in vacuum tubes.
21. What are circulating ground loops, and what is the remedy for such loops?
22. Discuss the methods employed to reduce microphonics in a vacuum tube.

23. What causes acoustic feedback in a public address system, and what is the remedy for this problem?

24. Explain the causes of parasitic oscillations. What are the remedies for these oscillations?

25. What is a grid-stopping resistor? Can this type of resistor be used in junction transistor and field-effect transistor amplifiers?

26. In reference to Fig. 9-17(b): (a) What is the slope of the d-c load line? (b) What is the slope of the a-c load line?

27. In reference to Fig. 9-17(b): (a) What are the quantities which may be determined by the d-c load line? (b) What are the quantities which may be determined by the a-c load line?

28. What are the effects on the a-c load line of an amplifier when the load is capacitive?

PROBLEMS

1. In reference to the circuit shown in Fig. 9-5(a), what is the stability factor when the circuit values are $R_B = 100$ kΩ, $R_C = 2$ kΩ, and beta $= 99$?

2. In reference to the bias circuit shown in Fig. 9-5(b), what is the stability factor when circuit values are $R_B = 100$ kΩ, $R_C = 2$ kΩ, and beta $= 99$?

3. In reference to the circuit shown in Fig. 9-5(c), what is the stability factor when the circuit values are $R_B = 80$ kΩ, $R_E = 1$ kΩ, $R_C = 2$ kΩ, and beta $= 99$?

4. In reference to the circuit shown in Fig. 9-7, what is the stability factor when the circuit values are $R_1 = 68$ kΩ, $R_2 = 4.7$ kΩ, $R_E = 1$ kΩ, and beta $= 99$?

5. In reference to the circuit shown in Fig. 9-7, what is the approximate value of R_E to give a stability factor of 5 when the circuit values are $R_1 = 47$ kΩ and $R_2 = 3.3$ kΩ?

6. Suppose the value of R_E in Problem 5 is 560 Ω and the value of R_1 is 47 kΩ. What value of R_2 will give a stability factor of approximately 4?

7. In reference to the circuit shown in Fig. 9-10, suppose the hum voltage (E_h) from the power supply is 3 mv. What is the hum voltage applied to the gate of $T2$ when the circuit values are $r_{ds} = 50$ kΩ, $R_D = 20$ kΩ, and $R_{G2} = 10$ MΩ?

8. Calculate the base current for the circuit in Problem 1 for a silicon transistor at 25°C with $V_{CC} = 20$ v.

9. Calculate the base current for the circuit in Problem 3 for a silicon transistor at 25°C with $V_{CC} = 20$ v.

CHAPTER 10

Power Amplifiers

10–1. Introduction

We recall that two basic classifications of audio amplifiers comprise voltage amplifiers and power amplifiers. The same classification applies to video amplifiers. Note that the primary function of a power amplifier is to deliver signal power to a load; any increase in signal voltage that might occur is of minor consideration. Because most power amplifiers require a relatively high signal-voltage input, a power amplifier must usually be preceded by one or more voltage amplifiers to increase the source signal voltage to a suitable level for driving the power stage.

Various types of devices may be used as audio power amplifiers, including junction-power transistors, power field-effect transistors, and vacuum tubes. Types of vacuum tubes include triodes (or multi-grid tubes operated as triodes), beam-power tetrodes, and pentodes. These devices may be operated singly as Class-*A* amplifiers, in parallel pairs, or as push-pull pairs in stages that are biased for Class-*A*, Class-*AB*, or Class-*B* operation. *Class-A operation*, commonly utilized in an amplifier stage with a single tube, denotes that the plate current is not cut off over any portion of the cycle. In other words, the bias voltage and input signal amplitude are such that plate current flows at all times. Another class of operation is called *limiting Class-A push-pull operation*; it is the condition in which one tube just reaches plate-current cutoff when the other reaches zero bias. This is a borderline case between Class-*A* and Class-AB_1, as explained below. The subscript 1 indicates that grid current does not flow during any part of the input cycle.

Class-AB operation indicates an overbiased operating condition and is used only in push-pull amplifiers that cancel even harmonics. A Class-AB_1 amplifier is an arrangement in which the grid bias and a-c grid voltages permit plate current in a particular tube to flow for appreciably more than half, but less than the entire, input cycle. The meaning of the subscript 1 is the same in either *A* or *B* classifications. We will find that Class-A_1 triode tubes may be operated from power supplies with comparatively poor regulation without serious loss of output power, due to the comparatively small rise of current at the maximum signal point. Class-AB_1 triode tubes require good power-supply regulation.

Class-B operation denotes push-pull tubes that are biased almost to the point of plate-current cutoff. We have noted that the subscript 1, used with the *A* and *AB* classifications, denotes that no grid current flows at any point of the operating cvcle. A subscript 2 denotes that grid current flows for at least a part of the operating cycle. In Class-*B* operation, the subscript 2 is usually omitted because grid-current flow is the normal characteristic. Of course, if a comparatively small drive signal is applied to a Class-*B* tube, grid current does not flow.

In general, audio power amplifier tubes have low amplification factors, low dynamic resistance, and high current flow. We will find that a higher amplification factor increases the power output of a stage (with a given power input), but a higher internal resistance (plate or collector) decreases the power output. To obtain low plate resistance, the space between plate and cathode is made smaller in a power tube than in a voltage-amplifier tube. A power tube has a comparatively large plate area, and the cathode is designed for high emission. Since the control grid must permit a comparatively large number of electrons to flow to the plate, the grid wires have wide spacing; this results in a reduced amplification factor. Pentodes designed for power amplification have higher amplification factors than power triodes; however, power pentodes have higher plate resistance.

We will cover the limitations of power transistors in a later chapter.

Power amplifiers have multitudes of applications in communications equipment, regulated power supplies, industrial control equipment, and electronic instrumentation. All radio transmitters employ power amplifiers, such as the amateur transmitter illustrated in Fig. 10–1. In a radio receiver, audio power is required to operate the loudspeaker; therefore, the last audio stage is operated as a power amplifier. It follows that the load resistance for a power amplifier seldom exists as a physical resistor, but instead, as a load such as a loudspeaker.

It is instructive to begin our analysis with consideration of a triode power amplifier stage as depicted in Fig, 10–2, with the understanding that the concepts apply to field-effect-transistor and junction-transistor power amplifiers. The power output of the amplifier configuration shown in Fig. 10–2 (b) may be calculated from the $I_b - E_b$ curves of Fig. 10–2 (a). Note that the plate voltage swings between

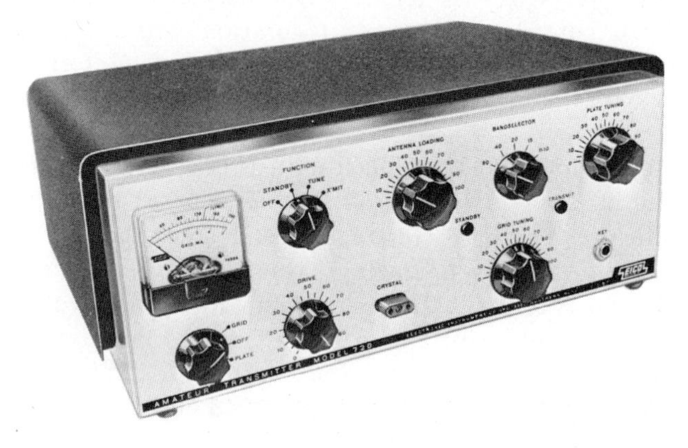

Courtesy of EICO, Electronic Instrument Co. Inc.

Figure 10–1. Power amplifiers are used in radio transmitters.

150 v and 350 v as the plate current swings from 50 ma to 10 ma. Thus, the peak a-c plate current is (50–10)/2 = 20 ma. The maximum a-c signal voltage is (350–150)/2 = 100 v. In turn, the power output from the stage is formulated

$$P_o = \frac{E_{max} I_{max}}{2} = \frac{100 \times 0.02}{2} = 1\text{w} \tag{10–1}$$

We can also calculate the power output from the equivalent circuit depicted in Fig. 10–2 (c); the power output, of course, is the power in R_L. We know that the voltage output from an amplifier stage is pro-

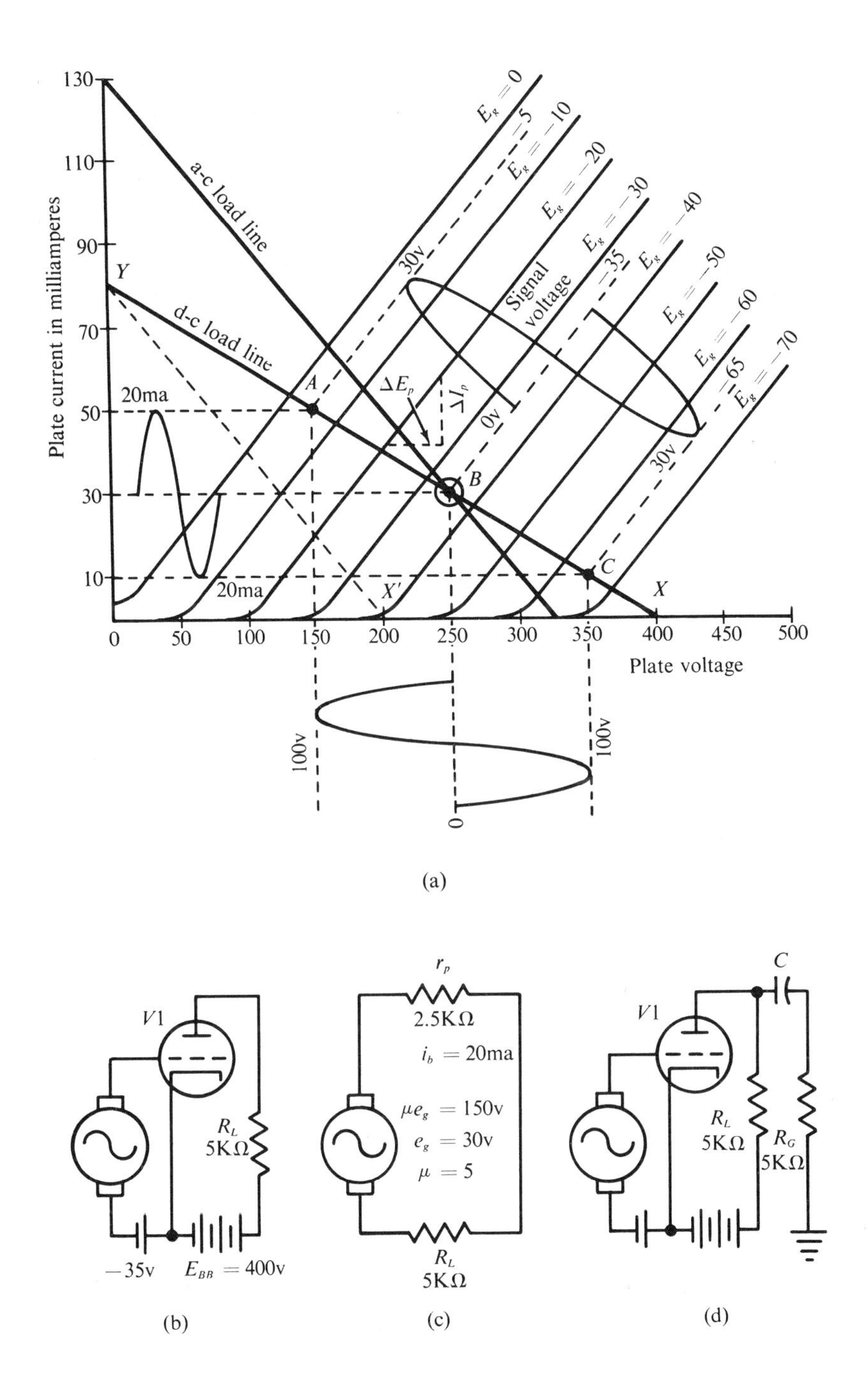

Figure 10-2. Load line and circuits for a triode power amplifier.

portional to μe_g, and that the power output is proportional to $(\mu e_g)^2$. Because μe_g operates in series with r_p and R_L in the equivalent circuit, the a-c component of plate current is formulated

$$i_b = \frac{\mu e_g}{R_L + r_p} \qquad \textbf{(10–2)}$$

In turn, the output voltage appears across R_L as $i_b R_L$. Let us define the following symbols.

e_o = instantaneous values of output voltage
E_o = rms values of output voltage
i_b = instantaneous values of plate current
i_b = rms values of plate current

It is evident that we may write

$$e_o = i_p R_L = \frac{\mu e_g R_L}{r_p + R_L} \qquad \textbf{(10–3)}$$

The output power in watts, when I_b and E_o denote effective (rms) values, becomes

$$P_o = I_b E_o = \frac{R_L(\mu e_G)^2}{(r_p + R_L)^2} \qquad \textbf{(10–4)}$$

Example 1. In Fig. 10–2 (c), R_L has a value of 5,000 Ω, r_p equals 2,500 Ω, E_g equals 21.2 rms v, and the amplification factor is 5. Hence, the power output may be formulated

$$P_o = \frac{(5 \times 21.2)^2 \times 5 \times 10^3}{(2.5 \times 10^3 + 5 \times 10^3)^2} = 1\text{w} \qquad \text{(answer)}$$

Because maximum power transfer occurs when $R_L = r_p$, the power output for this condition will be formulated

$$P_o = \frac{(\mu E_g)^2 r_p}{(2r_p)^2} = \frac{(\mu E_g)^2}{4r_p} \qquad \textbf{(10–5)}$$

Example 2. In the example of Fig. 10–2 (c), if R_L is changed to 2,500 Ω, maximum power output will be obtained, and its value will be

$$P_o = \frac{(5 \times 21.2)^2}{4 \times 2.5 \times 10^3} = 1.12 \qquad \text{(answer)}$$

Any other value of R_L with a given e_g will produce less output power.

Next, observe the pentode power-amplifier stage depicted in Fig. 10–3. Except for the screen circuit and the output transformer, the configuration is the same as the triode stage that we have analyzed.

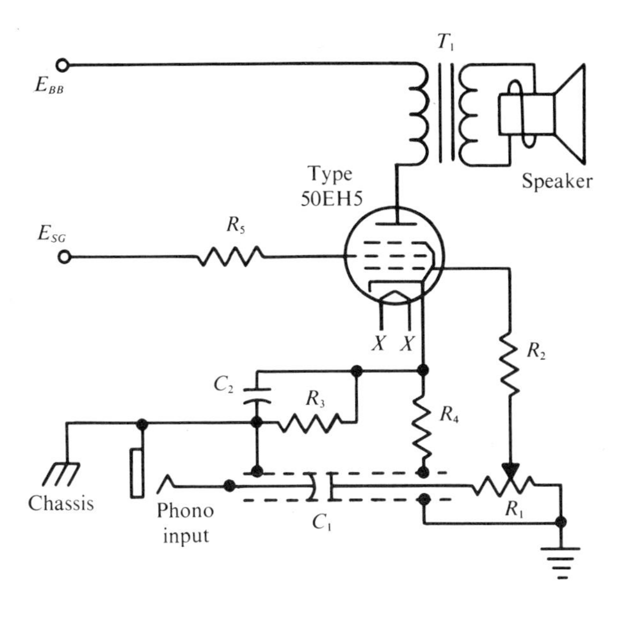

Figure 10-3. Typical pentode power amplifier with a rated output of 1w.

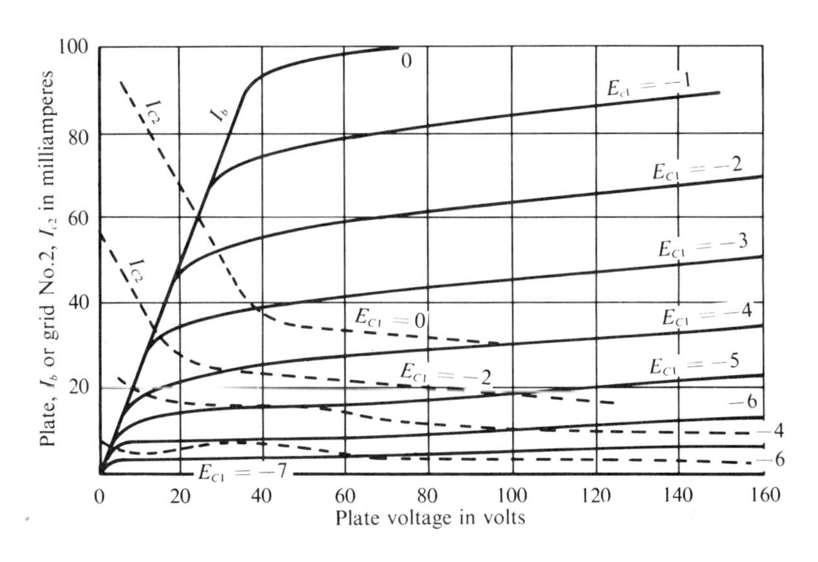

By permission Radio Corporation of America

Figure 10-4. Plate characteristics for the 50EH5 power pentode.

The primary of the output transformer presents a 3,000 Ω plate load. Therefore, we can draw a 3,000-Ω load line on the characteristics shown in Fig. 10–4 and analyze the stage operation as previously explained. Since the pentode can be regarded as a constant-current source, our calculations can be simplified somewhat by using the transconductance value instead of the amplification factor of the tube. The transconductance of a 50EH5 is 14,600 μmho.

Note that the output transformer serves as an impedance-matching device. Because the dynamic plate resistance of a 50EH5 tube is 11,000 Ω, and since the impedance of the loudspeaker voice coil might be as low as 4 Ω, the output transformer has a step-down turns ratio to provide the correct ratio of primary voltage and current to secondary voltage and current. Impedance matching by means of a transformer is depicted in Fig. 10–5. We recall that the output voltage of a trans-

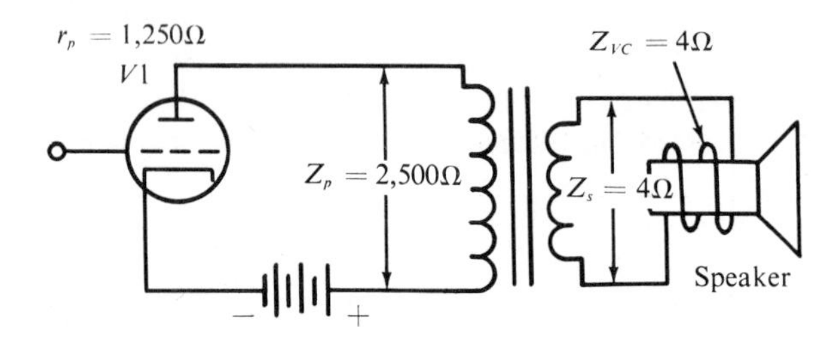

Figure 10–5. The output transformer is an impedance-matching device.

former varies directly as the turns ratio and that the output current varies inversely as the turns ratio. Accordingly,

$$\frac{E_p}{E_s} = \frac{N_p}{N_s} \tag{10–6}$$

and,

$$\frac{I_s}{I_p} = \frac{N_p}{N_s} \tag{10–7}$$

If the two left members and the two right members of the preceding formulas are multiplied together, the impedance relations may be determined.

$$\frac{E_pI_s}{E_sI_p} = \frac{N_p^2}{N_s^2} \tag{10–8}$$

The primary impedance of a matching transformer is defined as the ratio of rated primary voltage to rated primary current. Similarly, the secondary impedance is defined as the ratio of rated secondary voltage to current. If Z_p is substituted for E_p/I_p and Z_s is substituted for I_s/E_s in Formula 10–8, we obtain the formula

$$\frac{Z_p}{Z_s} = \left(\frac{N_p}{N_s}\right)^2 \tag{10–9}$$

If the loudspeaker voice coil should present a pure resistance, with no reactance, we could denote Z_p by R_p, and denote Z_s by R_s. A voice coil "looks like" a pure resistance to an a-c signal at only one frequency of operation, in the majority of cases. Let us consider a practical example of impedance matching. We will find the turns ratio needed for the transformer depicted in Fig. 10–5. To minimize distortion, we will discover subsequently that R_L should be chosen greater than r_p,

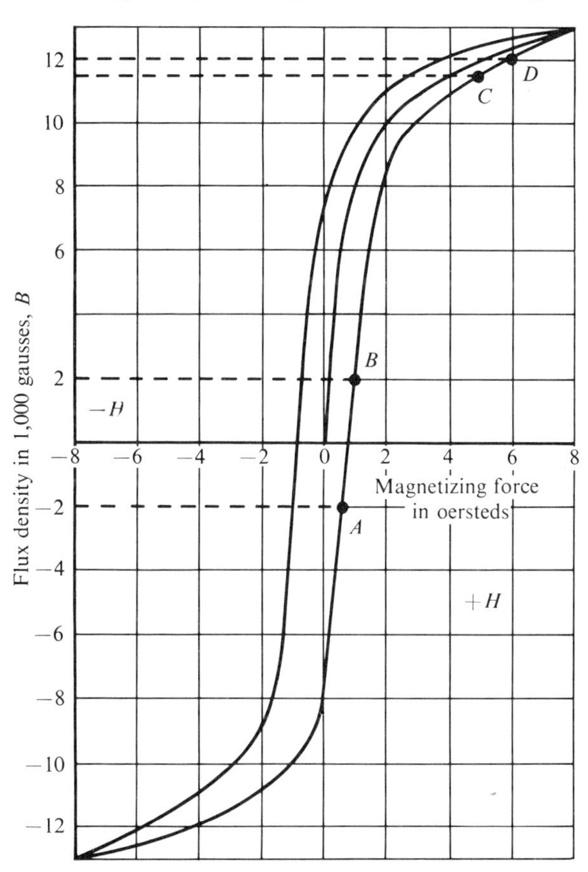

Figure 10–6. The change in flux density with respect to magnetizing force is much greater from *A* to *B* than from *C* to *D*.

although this choice does not provide maximum power transfer. In turn, because the plate resisance in this example is 1,250 Ω, we will choose a value of 2,500 Ω for R_L. In accordance with customary practice, R_L is denoted by Z_p in Fig. 10–5.

Of course, the power that is fed to the 4-Ω voice coil in Fig. 10-6 depends on the turns ratio of the output transformer. We have chosen an operating condition in which 4 Ω is to be transformed into 2,500 Ω. The turns ratio required is calculated by taking the square root of both members of Formula 10–9 and substituting our chosen values.

Exampe 3.

$$\frac{N_p}{N_s} = \frac{\sqrt{2{,}500}}{\sqrt{4}} = \frac{50}{2} = 25 \qquad \text{(answer)}$$

Just as vacuum tubes have maximum plate dissipation ratings, so do output transformers have maximum power ratings. The amount of power that can be transformed is determined by the current and voltage ratings of the transformer windings and the allowable losses. We see from Fig. 10–5 that the primary current may contain a d-c component in addition to the a-c signal component. A d-c component limits the incremental inductance. This fact is shown in Fig. 10–6. Note that the same magnetizing force is expended from A to B as from C to D. However, the resulting change in flux density is much less from C to D than from A to B. Operation between C and D corresponds to a large d-c component. Operation from A to B corresponds to zero d-c component. Hence, an output transformer is rated for maximum d-c flow through the primary.

The formula for the induced voltage of a transformer winding in rms volts is written

$$E = 4.44\, fNBA\, 10^{-8} \qquad \textbf{(10–10)}$$

where f denotes the operating frequency in Hertz, N denotes the number of turns in the winding, B denotes the flux density in the core, and A denotes the cross-sectional area of the core. In any transformer, the induced voltage is proportional to the product of the frequency and the flux density. At low frequencies, the flux density is high, and distortion is greater because the path of operation on the BH curve is longer, or the path of operation is more nonlinear than for small flux densities. A certain amount of core saturation is often tolerated at low frequencies to avoid using a massive core that would greatly increase the cost of the transformer. Thus, the maximum permissible flux density is a compromise with respect to tolerable low-frequency distortion.

We will find that an output transformer causes a reduction in power output at both the low end and the high end of the audio range, as depicted in Fig. 10–7. Reduced output at low frequencies stems from the shunting action of the transformer primary inductance

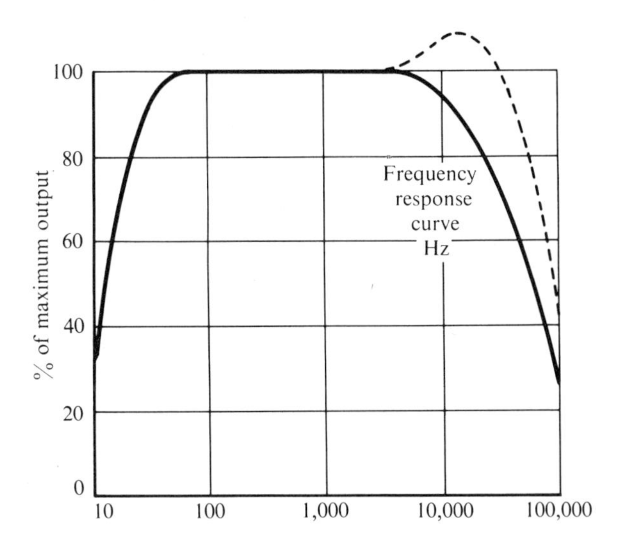

Figure 10–7. Typical frequency response for an audio output transformer.

on the tube, as indicated in Fig. 10–8 (a). Note the following definitions of symbols.

μ : amplification factor of tube
E_g : rms input voltage
r_p : dynamic plate resistance
R_1 : primary winding resistance
L_1 : primary leakage inductance
L_p : incremental primary inductance
N : primary-to-secondary turns ratio
L_2 : secondary leakage inductance
R_2 : secondary winding resistance
R_L : load resistance
E_L : rms voltage developed across the load
k : coefficient of coupling

Midband gain is essentially independent of frequency, as indicated by the absence of reactance in the equivalent circuit of Fig. 10–8 (b). Over the midband region, the reactance of the primary inductance is large enough so that its shunting action can be neglected. Also, the leakage reactances are small at low and midband frequencies and can

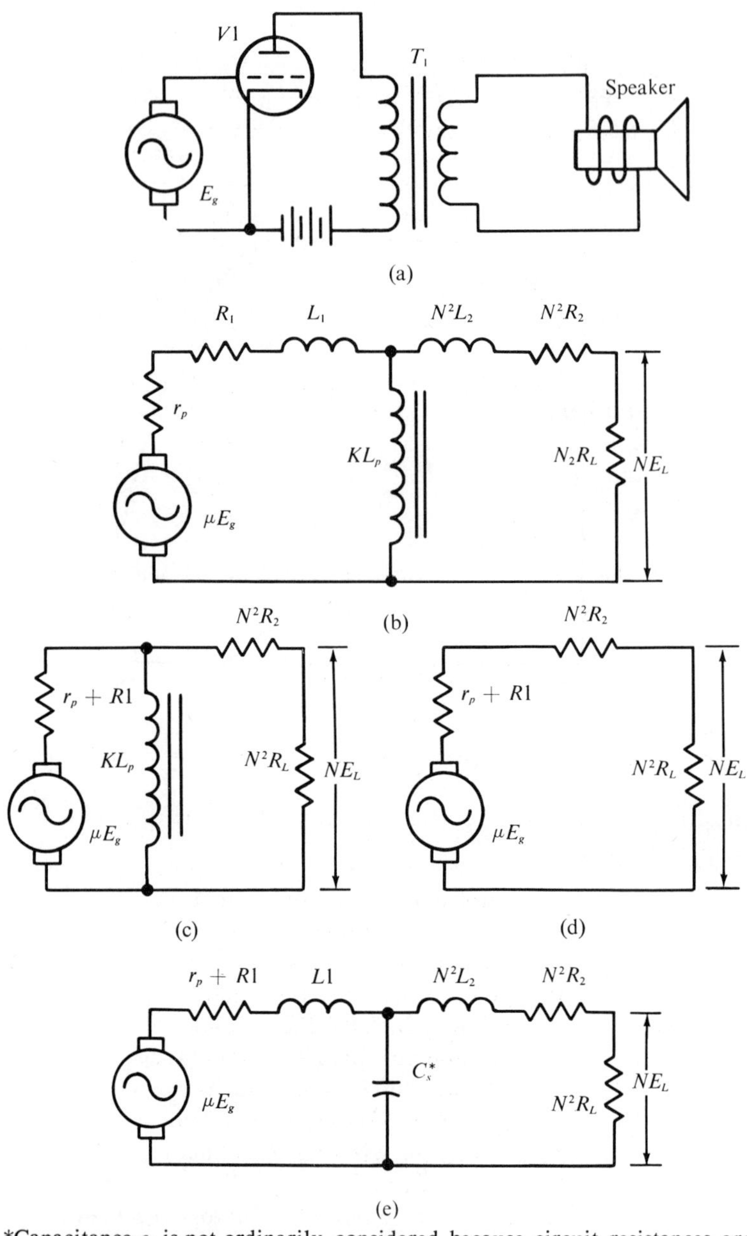

*Capacitance c_s is not ordinarily considered because circuit resistances are usually small.

Figure 10-8. Equivalent circuits for an audio output transformer. (a) Actual circuit; (b) Equivalent circuit reduced to unity turns ratio; (c) Low frequency equivalent circuit; (d) Middle frequency equivalent circuit; (e) High frequency equivalent circuit.

be neglected. Reduced output at high frequencies [Fig. 10–8 (c)] results from voltage drop or loss across the leakage reactances inasmuch as leakage reactance becomes substantial at high frequencies. Thus, there is an IX drop across each leakage reactance. Furthermore, the distributed capacitances of the windings have a comparatively low reactance at high frequencies and tend to shunt signal current from the load R_L.

In order to extend the flat portion of the frequency-response curve into the low-frequency region, the output transformer must have a high value of primary inductance. Again, in order to extend the flat portion into the high-frequency region, the leakage-inductance values L_1 and L_2 should be minimized. Thus, output-transformer design is a complex problem. With a given tube and transformer, an increase in value of R_L improves the high-frequency response without affecting the low-frequency response substantially. The signal voltage E_L across the load at various frequencies is most conveniently determined from a chart, such as shown in Fig. 10–7. Otherwise, we can calculate E_L from the appropriate equivalent circuit in Fig. 10–8. For example, at midband, we may write

$$E_L = \frac{\mu E_g N R_L}{(r_p + R_1) + N^2(R_2 + R_L)} \qquad \textbf{(10–11)}$$

Finally, let us note that the effect of c_s in Fig. 10–8 (c) may produce resonant action in combination with the windings in such manner that a resonant rise of voltage is found across the secondary terminals at high frequencies. Proper control of resonance in an interstage transformer helps to maintain output at high frequencies.

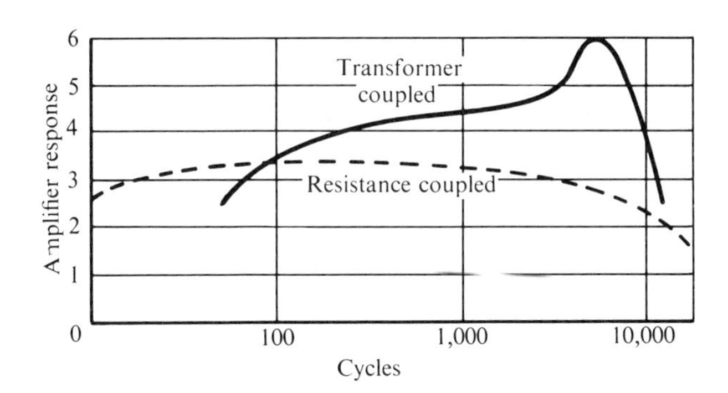

Figure 10–9. Reactance in audio-transformer windings often results in a peaked high-frequency response in voltage amplifiers (but not in power amplifiers).

However, excessive resonant response produces an objectionable high-frequency peak, depicted in Fig. 10–9. This type of frequency distortion causes the sound output to have a shrill and "tinny" timbre.

10–2. Parallel Pairs

When more signal power output is required than can be provided by a particular tube, a larger tube with higher ratings may be used. Another approach is to connect two tubes in parallel, as shown in Fig. 10–10. In parallel operation, the plate-load resistance becomes one-half the value used with one tube; output signal distortion is about the same as for one tube ; the grid-signal voltage is the same as

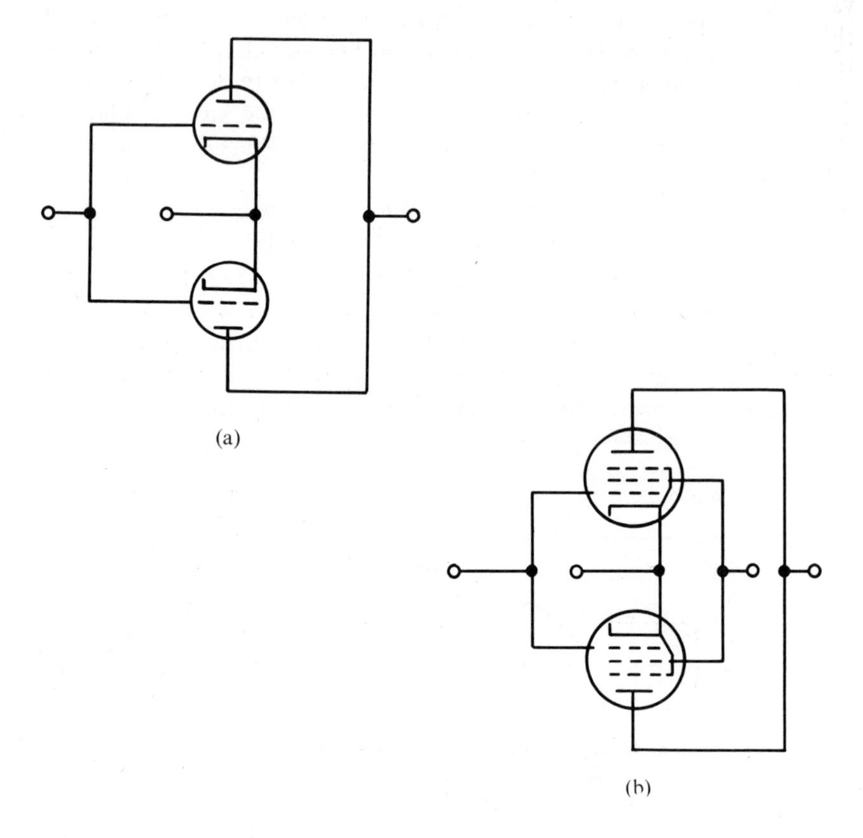

Figure 10–10. (a) Parallel connection of triodes; (b) Parallel connection of pentodes.

for one tube; plate current flow is doubled; the dynamic plate resistance becomes one-half the value for one tube. As one would anticipate, the grid-input capacitance is doubled and the plate-output capacitance is doubled when tubes are operated in parallel. Parasitic oscillation is often troublesome, and stopping resistances are frequently required in the control-grid and screen-grid circuits, as shown in Fig. 10–11.

Beam-power tubes may also be connected in parallel. Let us briefly review the characteristics of beam-power tubes. This type of tube is capable of providing comparatively high levels of signal-output power

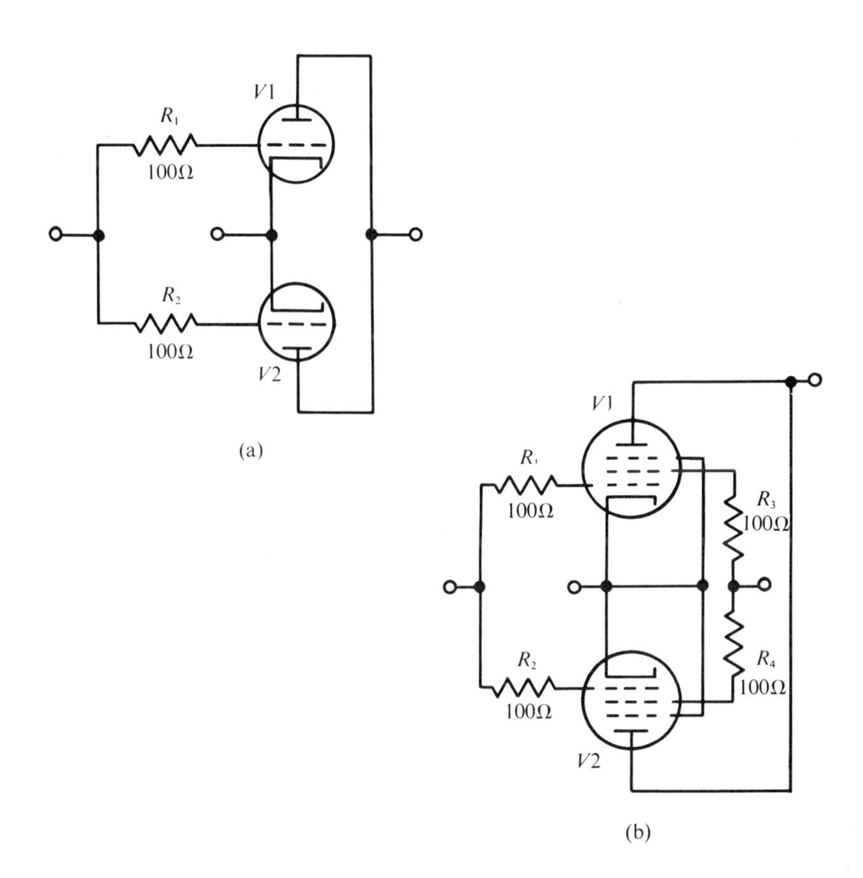

Figure 10–11. (a) Control-grid stopping resistors for parallel-connected triodes; (b) Control-grid and screen-grid stopping resistors for parallel-connected pentodes.

because of concentration of the plate-currrent electrons into "beams" or "sheets" of moving charges. In triodes, for example, the plate-current electrons move in a predetermined direction but are not confined into a "beam." However, in a beam-power tetrode, beam-forming plates are employed as depicted in Fig. 10–12. It should be

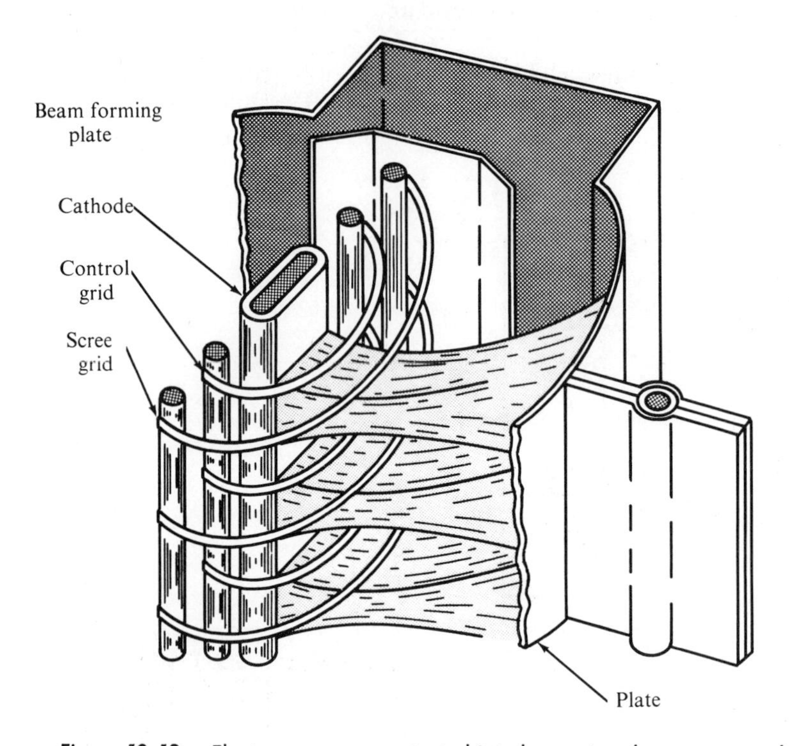

Figure 10–12. Electrons are concentrated into beams in a beam-power tube.

stressed, also, that the control-grid and screen-grid wires have the same pitch in a beam-power tetrode; the wires of each grid are physically in line relative to the paths of the plate-current electrons, as shown in Fig. 10–13. Because of this arrangement of control grid and screen grid in a beam-power tetrode, the screen intercepts fewer electrons than in a pentode. Also, the beam-forming plates prevent interception of electrons by the grid-supporting wires. Since the screen current is less in a beam-power tube, more electrons arrive at the plate and the plate current is greater. We perceive that the plate and control grid are electrically isolated in a beam-power tube, much as in a pentode. However, the plate resistance is comparatively low, and a substantial amount of signal power can be developed at relatively low distortion.

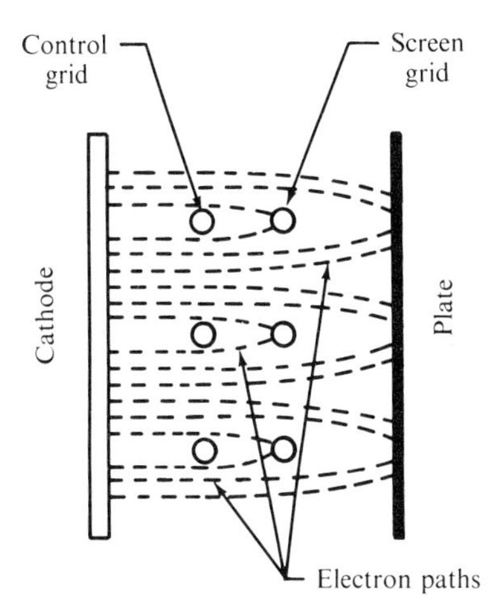

Figure 10-13. Electron paths in a beam-power tube.

The beam electrodes in a beam-power tube are internally connected to the cathode. Because of the electron-beam concentration that is produced, a comparatively large space charge exists in the space between the screen and the plate. It is as if a *virtual cathode* were operating between screen and plate, as depicted in Fig. 10-14. This

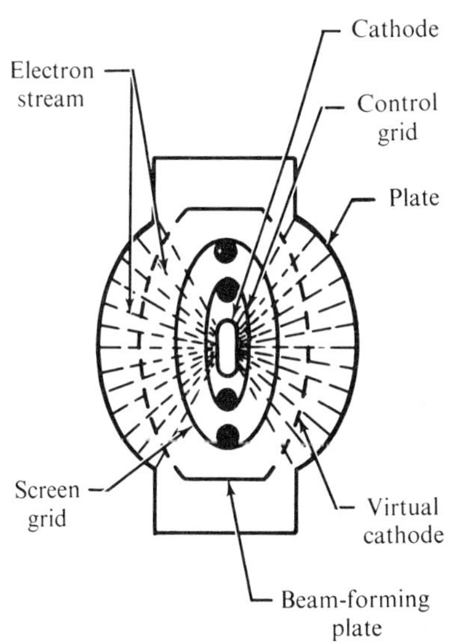

Figure 10-14. A concentrated space charge, or virtual cathode, is produced by the beam-forming plates.

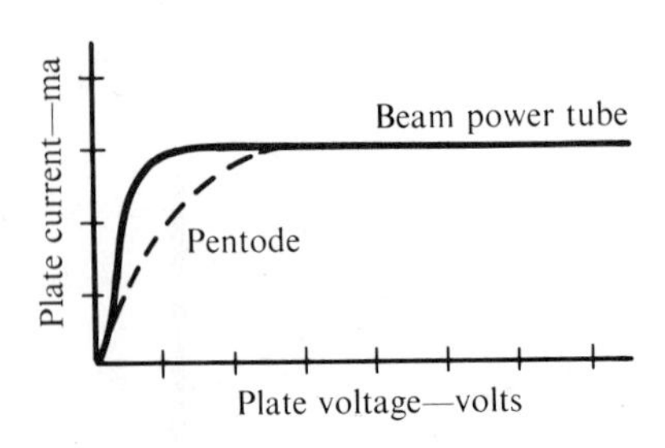

Figure 10–15. Characteristic curves for a pentode and a beam-power tube.

virtual cathode suppresses flow of secondary electrons from plate to screen much as if a suppressor grid were present, and moreover, suppression is more effective in a beam-power tube. Figure 10–15 shows the plate-current plate-voltage characteristics for a beam-power tube and a pentode. Improved suppression in a beam-power tube permits greater signal-power output and lower values of d-c plate voltage than in a pentode.

Beam-power tubes may be connected in parallel, just as triodes and pentodes. The same problems of parastic oscillation are encountered, and stopping resistors are usually required at the control-grid terminals of the sockets. If parasitic oscillation is not completely suppressed, stopping resistors are also required at the screen-grid terminals of the sockets. This approach will be completely satisfactory unless the control-grid and screen-grid lead wires are quite long. If unusually long leads must be used in an amplifier, stopping resistors may be supplemented by small high-frequency choke coils. Choke-coil inductances must be chosen by experiment; their basic function is to shift the high-frequency resonance values so that oscillation cannot take place.

Many of the disadvantages of parallel-tube operation can be avoided by elaborating the circuitry somewhat to obtain push-pull operation of two devices. A push-pull amplifier consists of two tubes (or transistors) whose grid and plate signals (or base and collector signals) are 180° out of phase. In other words, the grid signal of one tube is 180° out of phase with the grid signal of the other tube; in turn, the plate signal of one tube is 180° out of phase with the plate signal of the other tube. The two tubes may be operated in Class *A*, Class *B*, or Class *AB*. Push-pull amplifier circuits are often used in audio-frequency power amplifiers because less distortion, greater power output, and higher plate efficiency are obtained in push-pull operation.

Classes of amplifiers are often compared with one another in terms of *plate efficiency*. Plate efficiency is defined as the ratio of a-c signal power developed in the plate load to the d-c power that is supplied to the plate. We recall that the a-c signal power is equal to the product of measured values of a-c signal voltage and a-c current expressed in rms values. The d-c power supplied to the plate is equal to the product of d-c plate voltage and d-c plate current. In Class-*A* amplifiers, the plate efficiency is about 20 per cent or less, whether a tube is operated singly or in parallel with another tube. This low efficiency is due to the high average value of d-c plate current and its consequent high plate dissipation.

On the other hand, a push-pull stage can be operated in Class *AB* with little distortion, and the plate efficiency may approach 55 per cent. A push-pull stage can also be operated in Class *B* without objectionable distortion, and the plate efficiency may approach 65 per cent. Power-supply drain is also much less on the average, because a Class-*B* stage draws negligible plate (or collector) current when no input signal is present. When an input signal is applied, the average plate current then rises in accordance with the peak voltage of the input signal. Figure 10-16 compares the operating point for a Class-*A* and Class-*B* transistor amplifier.

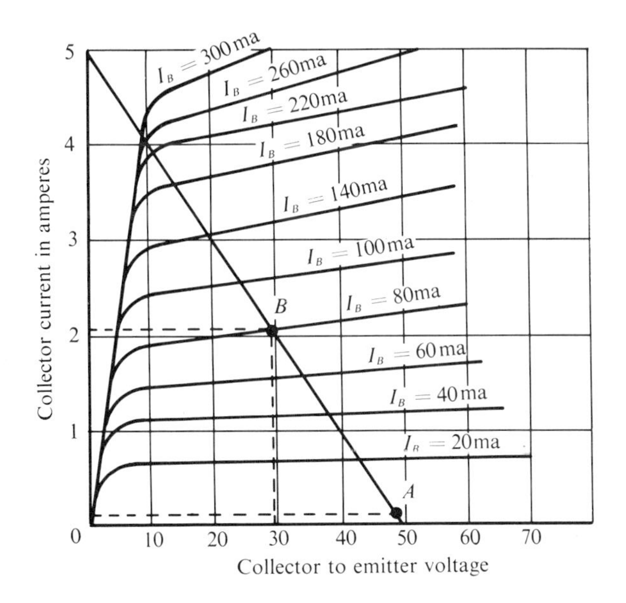

Figure 10-16. Amplifier operating point: Point *A*, Class-*A* operation; Point *B*, Class-*B* operation.

10–3. Analysis of Push-Pull Amplifiers

It is instructive to start our analysis with consideration of triode push-pull amplifier operation. A push-pull triode amplifier circuit is shown in Fig. 10–17. The upper and lower sections of the circuit are

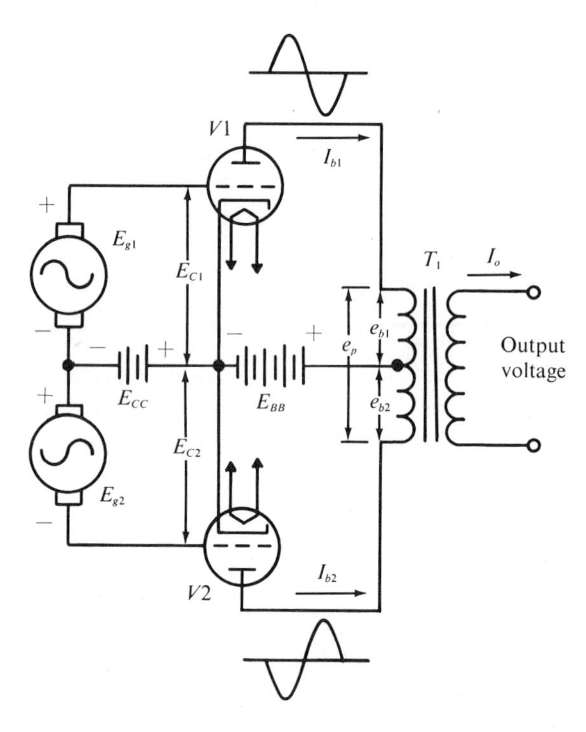

Figure 10–17. Basic push-pull triode amplfier circuit.

similar. Triodes $V1$ and $V2$ are the same tube type and therefore have similar characteristics. The two grid signals e_{g1} and e_{g2} have the same amplitude and frequency but are 180° out of phase with each other. A single bias battery and a single plate-supply battery are used to supply both tubes with proper d-c operating potentials. Transformer T_1 acts as the plate load for both tubes. The primary of transformer T_1 is center-tapped (point A) so that the output voltages e_{b1} and e_{b2} are equal in magnitude. Plate currents i_{b1} and i_{b2} are also equal in magnitude.

Of course, a push-pull signal source (e_{g1} and e_{g2}) as depicted in Fig. 10–17 is not ordinarily available to drive the grids of $V1$ and $V2$. Therefore, a *phase inverter* of some type will be employed to change a single-ended input signal into a double-ended (push-pull) output signal. Let us consider some basic types of phase inverters. In operation, all transformers produce across the secondary winding an emf that is opposed to the change in flux that induces the emf. In turn, the instantaneous polarity of the output depends on how the load is connected to the secondary.

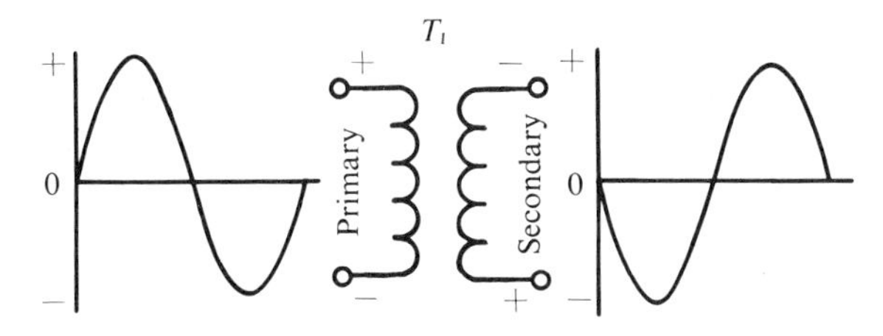

Figure 10–18. Phase inversion by transformer action.

Figure 10–18 depicts phase inversion of sine waves and square waves by a transformer. We will see that a center-tapped secondary will further provide a double-ended output from a single-ended input. A center-tapped secondary is used in Class-*B* push-pull circuits to obtain signal voltages of opposite instantaneous polarity to the grids of the tubes, as shown in Fig. 10–19. If, at a given instant, the

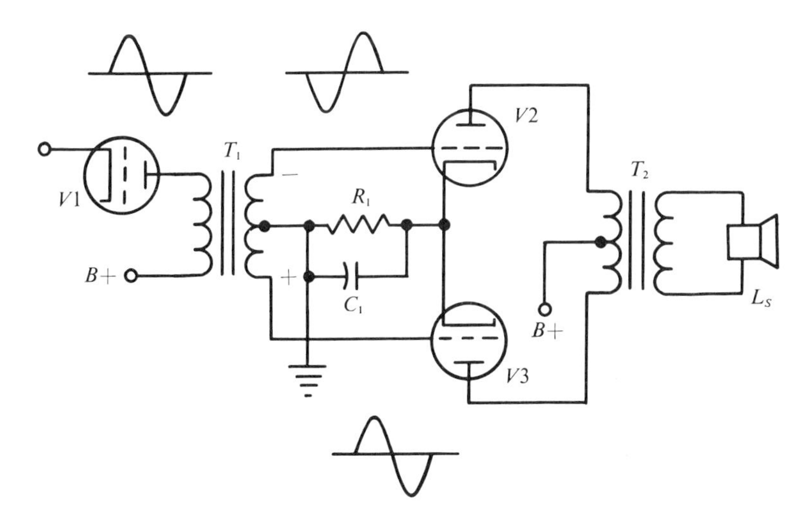

Figure 10–19. Transformer *T* provides double-ended drive from a single-ended source.

polarity of point X is negative-going with respect to the grounded center tap, the polarity of point Y will be positive-going with respect to the center tap. Thus, a negative-going potential is applied between the grid and ground of $V2$, and at the same time, a positive-going potential is applied between the grid and ground of $V3$.

Figure 10–19 depicts one type of the difficulty encountered in designing low-distortion transformers with reasonably low losses. For example, the loss from leakage reactance is greater for high frequencies than it is for low frequencies. Again, the shunting effect of distributed capacitances poses a difficult design problem. Hysteresis losses increase with frequency. Hence, transformers are not used unless there is no other practical means of obtaining phase inversion. We will consider electronic phase inverters subsequently. For our present purposes, the transformer phase inverter serves adequately for analysis of push-pull amplifier operation and simplifies the circuitry to be considered.

With reference to Fig. 10–17, symbols for various voltages and currents are similar to those employed in a single-ended triode amplifier circuit. When no input signal is applied, it is evident that $i_o = 0$ in the secondary of the output transformer. Steady d-c plate-current flow through the primary induces no voltage in the secondary. The d-c plate current of $V1$ flows from cathode to plate, through the upper half of the primary winding (from point B to A), and thence back to the cathode. The d-c plate current of $V2$ flows from cathode to plate, through the lower half of the primary winding (from point C to A), and thence back to the cathode.

Observe that points B and C in Fig. 10–17 are equally negative with respect to point A since the magnetizing force is zero, and d-c saturation of the core cannot occur. For this reason, a much smaller output transformer can be used in a push-pull amplifier than in a single-ended amplifier. Note in Fig. 10–19 that the d-c plate-current flow of $V1$ tends to produce core saturation in transformer T_1. Thus, the design of this transformer is more difficult than the design of the following output transformer (T_2).

Again, with reference to Fig. 10–17, when sine-wave signals e_{g1} and e_{g2} are applied to the respective grids of the tubes, sine-wave plate currents i_{b1} and i_{b2} flow in the primary winding of the transformer. Current i_{b1} is 180° out of phase with i_{b2} since the two grid signals are 180° out of phase with each other. During the positive swing of i_{b1}, point B on the primary becomes more negative with relation to point A. At the same time, the fall in i_{b2} causes point C to become less negative with respect to point A by an equal amount.

We perceive that the voltage across the entire primary in Fig. 10–

18, e_p, is twice the value of either e_{b1} or e_{b2}. In other words, e_p is equal to e_{b1} plus e_{b2}. A half cycle later, all the polarities reverse. Here, again, the voltage e_p across the primary is equal to e_{b1} plus e_{b2}. The relationship of e_p in terms of e_{b1} and e_{b2} holds true for all instantaneous values of plate current. The output signal from transformer T is single-ended, and is fed to a utilization device, such as a loudspeaker.

The dynamic characteristic for two tubes operating in push-pull is constructed from the individual dynamic characteristics of both

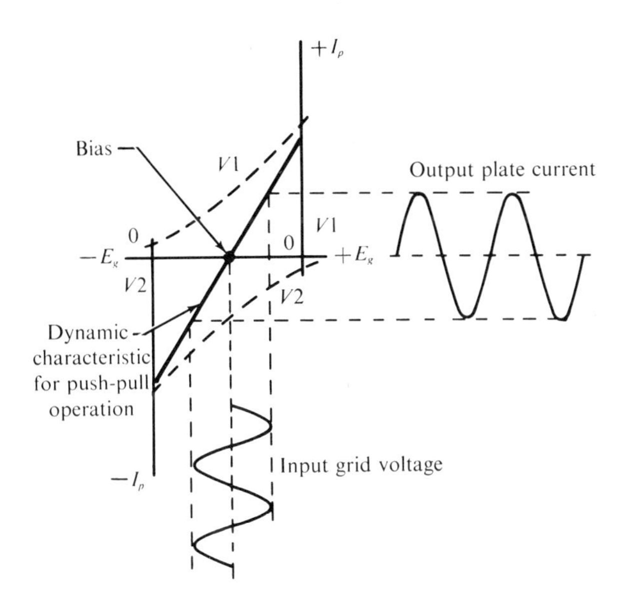

Figure 10–20. Dynamic characteristic of two tubes in push-pull Class-*A* operation.

tubes. Figure 10–20 shows a dynamic characteristic for two tubes operating in Class-*A* push-pull. It is obtained in the following manner. The dotted curve marked $V1$ is the dynamic characteristic of one tube, and $V2$ is the dynamic characteristic of the other tube. These dynamic characteristics are identical, and because the plate currents of the two tubes are 180° out of phase, the characteristics are placed 180° out of phase with a common horizontal axis. Then, the characteristics are lined up so that the bias voltage of one tube meets the same bias voltage of the other tube.

We will suppose that $E_{CC} = -5$v for the tubes represented in Fig. 10–20. This same voltage value occurs at the same place on both grid-voltage axes. The resultant dynamic characteristic is obtained by algebraically adding the instantaneous values of plate current for various values of grid-signal voltage. Of course, in single-tube Class-*A* operation, little distortion occurs in the output plate-current waveform, because the grid signal operates along the most linear portion of the dynamic characteristic. In Class-A push-pull operation, the distortion is even less, because the dynamic characteristic is even more linear. This is another advantage of push-pull operation.

By projecting various points of the input grid signal to the push-pull characteristic (shown by dashed lines in Fig. 10–20), the output plate-current waveform is obtained. Note that a greater grid-voltage swing is possible in push-pull operation without causing appreciable distortion. The reason for this is that the dynamic characteristic is linear for a greater amount of voltage variation. It follows that if a greater grid signal can be employed in push-pull operation, then a correspondingly greater power output can be obtained. The efficiency of a Class-*A* push-pull amplifier may approach 30 per cent, whereas

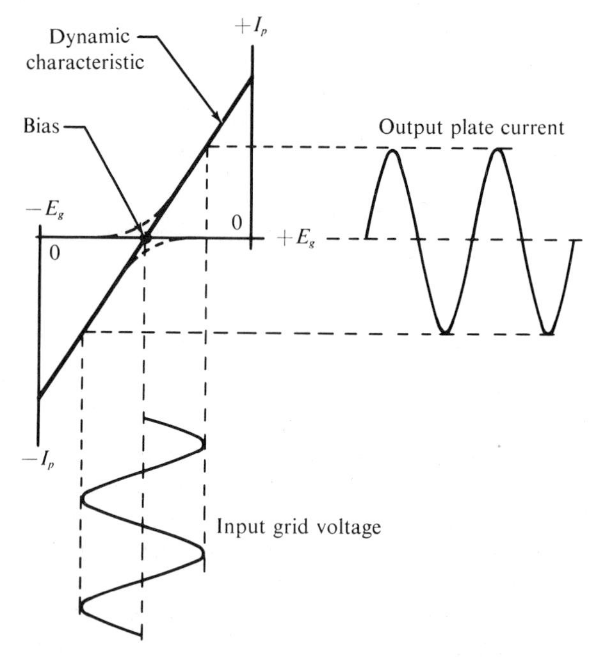

Figure 10–21. Dynamic characteristic for two tubes operating in push-pull Class *B*.

in Class-*A* single-tube operation, an efficiency of 20 per cent is about the limit.

As previously noted, push-pull amplifiers are commonly operated in Class *AB*. The resultant dynamic characteristic is constructed in the same general manner as in Fig. 10–21. Since the higher value of grid bias provides a longer characteristic in Class-*AB* operation than in Class *A*, it is evident that a greater grid swing is permissible. Hence, greater power output and higher plate efficiency are obtained. The plate efficiency may approach 55 per cent.

Push-pull amplifiers are advantageously operated in Class B. The resultant characteristic for Class-*B* operation is shown in Fig. 10–21. We recall that the bias voltage is nearer to plate-current cutoff than in Class-*AB* operation. Here again, the output waveform is not seriously distorted (see Fig. 10–22). Note that the bias is not quite at the plate-current cutoff value. If complete cutoff bias were employed, objectionable crossover distortion would occur, as depicted in Fig. 10–23. In other words, complete cutoff bias causes the resultant dynamic characteristic to have an "S" shape, which produces severe

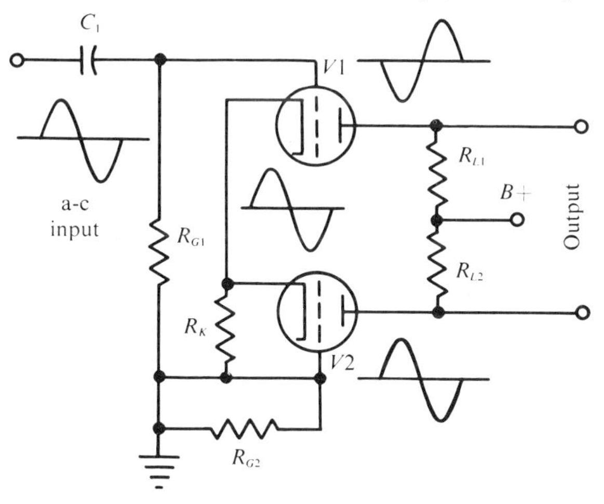

Figure 10–22. Push-pull Class-*B* amplifier circuit.

nonlinear distortion of the output plate-current waveform. The plate efficiency of a Class-*B* push-pull amplifier, operated as shown in Fig. 10–23, is about 60 per cent to 65 per cent. Table 10–1 provides a resume of the operating points, distortion, power output, and plate efficiency of the classes of amplifiers that we have studied.

Power-type transistors, such as illustrated in Fig. 10–24, are utilized in Class-*B* push-pull audio-output stages. The circuit for a

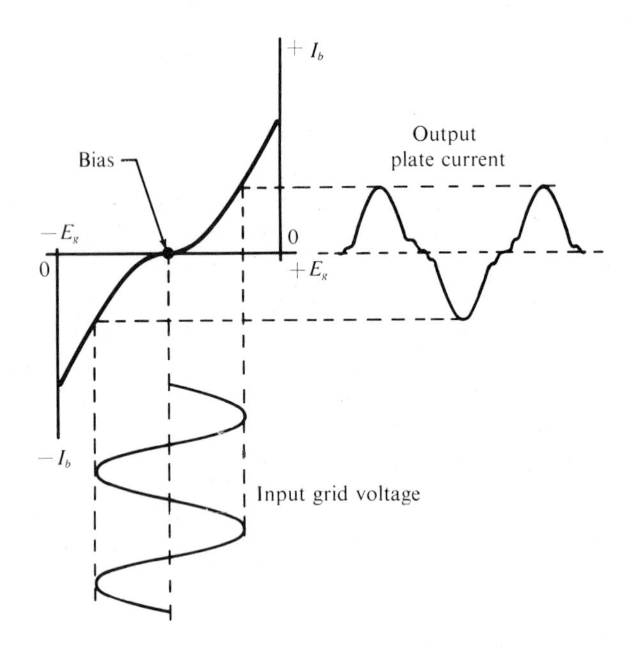

Figure 10-23. Operation of a push-pull Class-*B* amplifier biased to plate-current cutoff showing the effects of crossover distortion on the output signal.

typical push-pull Class-*B* output stage is diagrammed in Fig. 10–25. Note that the voltage divider R_1-R_2 provides a slight forward bias of about 0.14 v on the transistors. This is about the minimum bias that can be used without encountering crossover distortion, as depicted in Fig. 10–26 (b) for transistors. Crossover distortion in a transistor is caused by the voltage barrier between the emitter base junction. The resistors (R_E) in the emitter leads (Fig. 10–25) stabilize the transistors so that they will not go into thermal runaway as long as the ambient temperature is less than 55°C. Typical collector characteristics for a power transistor with a-c and d-c load lines are seen in Fig. 10–26 (a).

It can be shown that maximum signal-output power without clipping in a transistor push-pull stage is formulated

$$P_{o\ \max} = \frac{I_{\max} V_{CE}}{2} \tag{10-12}$$

where V_{CE} denotes the collector-to-emitter voltage with no signal input and is usually equal to V_{CC}.

Example 4. Suppose the Class-*B* amplifier in Fig. 10–25 has a maximum collector current of 2 amp and a maximum collector voltage of 20 v. What is the maximum output power?

$$P_{o\,\max} = \frac{2 \times 10}{2}$$
$$\simeq 20\text{w} \qquad \text{(answer)}$$

The load resistance presented by the output transformer is formulated

$$R_C = \frac{V_{CE}}{I_{\max}} \tag{10–13}$$

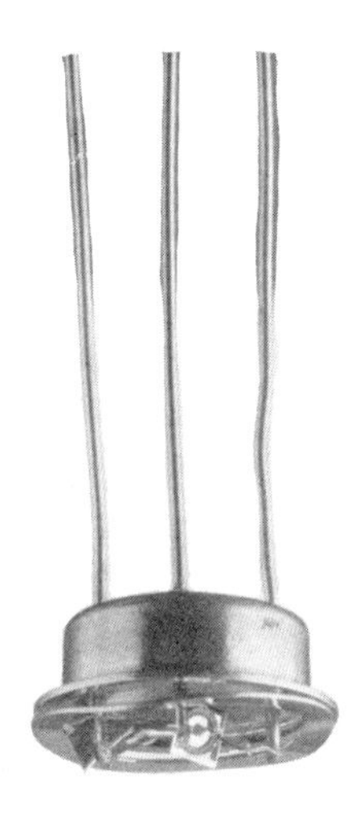

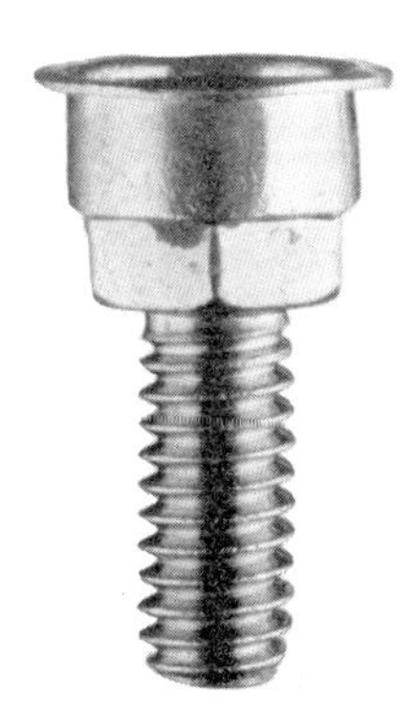

Courtesy of Texas Instruments Co.

Figure 10–24. Construction of a typical power transistor.

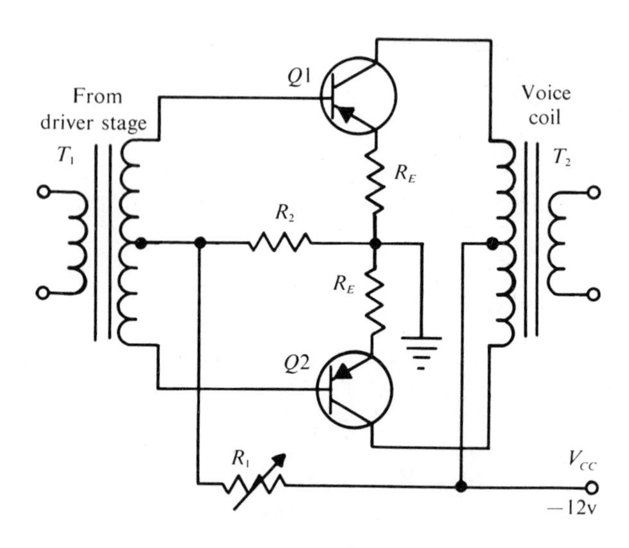

Figure 10–25. A Class-*B* push-pull audio-output configuration.

TABLE 10–1

Vacuum Tube Amplifier Characteristics

Class	*Location of operating point on dynamic characteristic*	*Relative distortion*	*Relative power output*	*Approximate percentage of plate efficiency*
A *single-tube*	On linear portion	Low	Low	Under 20%
A *push-pull*		Very low	Moderate	20 to 30%
AB *single-tube*	Between linear portion and plate-current cut-off	Moderate	Moderate	40%
AB *push-pull*		Low	High	50 to 55%
B *single-tube*	At vicinity of cut-off	High	High	40 to 60%
B *push-pull*		Low	Very high	60 to 65%
C *single-tube*	About $1^1/_2$ to 4 times plate-current cut-off	Very high	Very high	60 to 80%

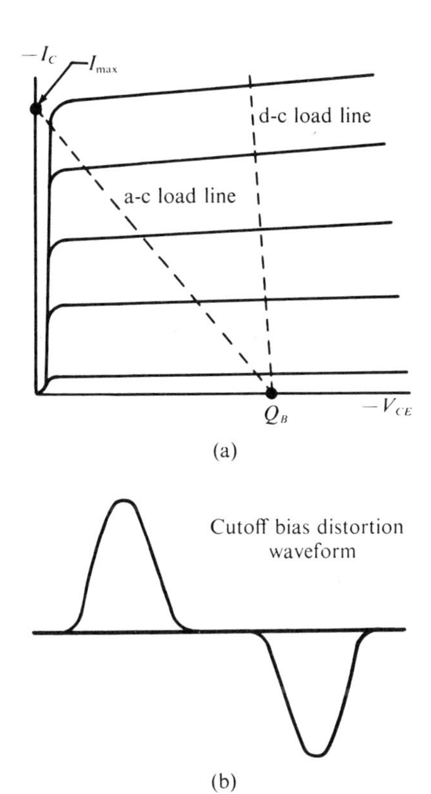

Figure 10–26. (a) Load line drawn on a family of collector characteristics; (b) Cross-over distortion due to cutoff bias in a transistor.

Example 5. The load impedance seen by the transistor in Example 4 is

$$R_C = \frac{20}{2} = 10\ \Omega \qquad \text{(answer)}$$

The collector-to-collector impedance is equal to four times the load resistance per collector, and the maximum output power may be stated in the form

$$P_o = \frac{2V_{CE}^2}{R_{C-C}} \tag{10–14}$$

Example 6. The primary impedance of the transformer in Problem 2 is figured at the top of page 336.

$$R_{C-C} = 4R_C$$
$$= 4 \times 10 = 40\Omega \qquad \text{(answer)}$$

Example 7. The maximum output power of the amplifier illustrated in Example 4 may be checked by

$$P_o = \frac{2(20)^2}{40}$$
$$= 20 \text{ w} \qquad \text{(answer)}$$

Thus, for a specified output power and collector voltage, the collector-to-collector load resistance can be determined. For output power values in the range of 50 mw to 850 mw, the load impedance is so low that it is essentially a short-circuit compared with the output impedance of the transistors, Hence, no attempt is made in practice to match the output impedance of transistors in power-output stages. The *power gain* of the push-pull configuration depicted in Fig. 10–25 is formulated

$$\text{Power gain } (A_P) = \frac{P_{\text{out}}}{P_{\text{in}}} = \frac{I_o^2 R_C}{I_{\text{in}}^2 R_{\text{in}}} \qquad \textbf{(10–15)}$$

Example 8. Suppose the amplifier in Fig. 10–25 has the following values: $I_o = 2$ amp, $R_c = 20\ \Omega$, $I_{\text{in}} = 20$ ma, and $R_{\text{in}} = 100\ \Omega$. What is the power gain?

$$A_P = \frac{(2)^2 20}{(.02)^2 100}$$
$$= 2{,}000 \qquad \text{(answer)}$$

Of course, I_o/I_{in} is the current gain, beta, from a practical standpoint, because the load resistance has a small value. In turn, the power gain of the amplifier may be written

$$\text{Power gain} = \beta^2 \frac{R_C}{R_b}, \text{ approximately} \qquad \textbf{(10–16)}$$

where R_C denotes the collector load resistance, R_B denotes the base input resistance, and beta is the common-emitter current gain of the transistor.

Example 9. Suppose a push-pull amplifier has a base-input resistance of 20 Ω and an output-collector load resistance of 12 Ω. What is the power gain if the transistor has a beta of 75?

$$A_P = (75)^2 \frac{10}{20}$$
$$\simeq 2{,}800 \qquad \text{(answer)}$$

Since the load resistance value is determined by the required maximum undistorted power output, the power gain can be written in terms of the maximum output power by combining Formulas 10–14 and 10–15.

$$\text{Power gain} = \frac{2\beta^2 V_{ce}^2}{R_b P_{\text{out}}} \qquad \textbf{(10–17)}$$

We have noted several limitations on the operation of the transistor. From observation of the output characteristics of a typical transistor in Fig. 10–27, we can conclude that there are many limita-

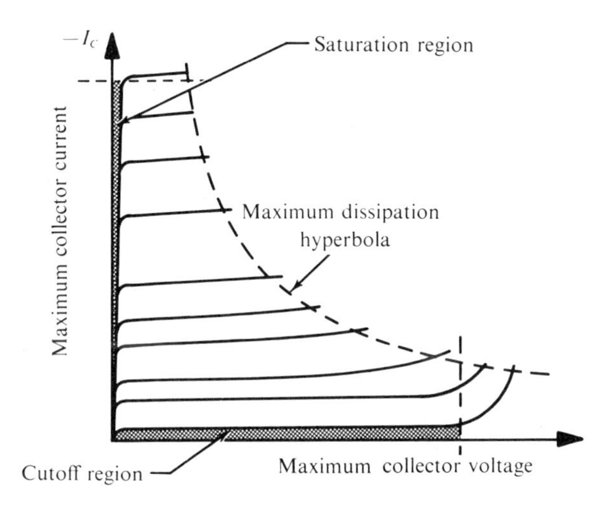

Figure 10–27. **Transistor output characteristic showing maximum current, minimum voltage, and maximum power dissipation hyperbola.**

tions, such as the choice of an operating point and signal excitation. We shall now repeat these limitations.

1. The maximum collector dissipation represented by the hyperbola depends upon the ambient temperature and the resistance to the removal of heat.

2. Maximum collector current, above which high distortion and high collector temperature occurs.

3. Maximum collector voltage, above which breakdown of the reverse-biased collector junction occurs.

4. The cutoff region, which limits output signal at low collector currents.

5. The saturation region, which limits the output signal at low-collector voltage values.

We noted in passing that a transistor amplifier has a maximum operating temperature. In fact, when a transistor is used at high junction temperatures (high power dissipation and/or high ambient temperature), it is possible for regenerative heating to occur at the collector base junction, which will result in thermal runaway and possible destruction of the transistor. In any circuit, the junction temperature (T_j) is determined by the ambient temperature (T_A) and the resistance to heat flow (Φ).

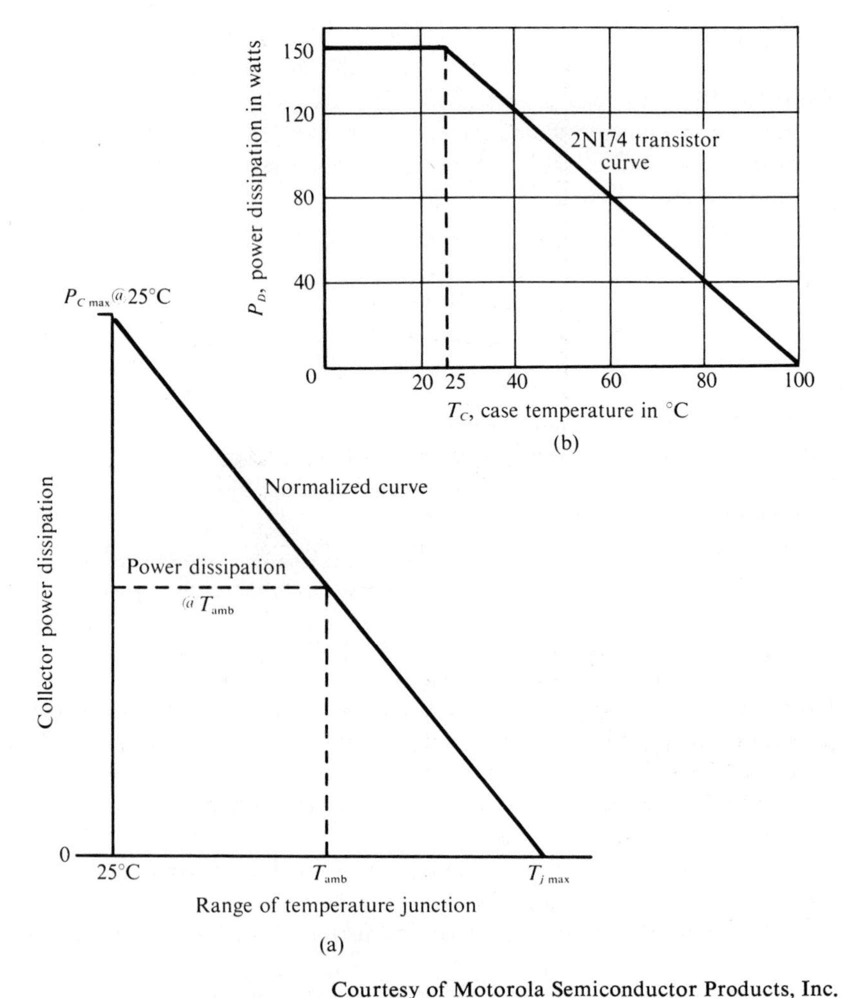

Courtesy of Motorola Semiconductor Products, Inc.

Figure 10–28. **Power-temperature derating curve for a transistor. (a) Normalized curve; (b) Curve of a 2N174 transistor.**

The maximum power dissipation is normally given by the manufacturer. The maximum power dissipation at any temperature may be derived from Ohm's law of thermodynamics and the formula

$$P_C = \frac{T_{j\,max} - T_{amb}}{R_T} \tag{10-18}$$

where T_j is the maximum recommended collector temperature, T_{amb} is the ambient temperature about the transistor, and R_T is the thermal resistance in °C/watt.

Note : Manufacturers' data give Φ in watt/°C or R_T in °C/watt $(R_T = \frac{1}{\Phi}.)$

Example 10. Suppose the transistor represented in Fig. 10–28 has a maximum junction temperature (T_j) of 85°C, a derating factor (Φ) of 0.5w/°C, and is operating at 50°C. What is the maximum allowable collector dissipation?

$$R_T = \frac{1}{\Phi} = \frac{1}{0.5} = 2\text{°C/watts}$$

$$P_C = \frac{85 - 25}{2}$$

$$= 30 \text{ w} \qquad \text{(answer)}$$

Allowable collector dissipation is a linear inverse function of temperature as illustrated in Fig. 10–28. In other words, the allowable dissipation at 25°C decreases to zero as the ambient temperature approaches the maximum collector temperature. The allowable collector dissipation at any temperature may be determined by the formula for similar triangles, as illustrated in Fig. 10–28 (a).

$$\frac{P_{C\,amb}}{P_C\,25\text{°C}} = \frac{T_{j\,max} - T_{amb}}{T_{j\,max} - 25\text{°C}} \tag{10-19}$$

Example 11. The maximum collector dissipation of the 2N174 germanium transistor in Fig. 10–28(b) is 150 w at 25°C, and the maximum allowable collector temperature is 100°C. What is the allowable collector dissipation at 65°C?

$$P_C\,65° = \left(\frac{100 - 65}{100 - 25}\right)150$$

$$= \frac{35}{75} \times 150$$

$$\simeq 70.2 \text{ w} \qquad \text{(answer)}$$

The collector dissipation of a transistor in a well designed Class-*B* push-pull amplifier is approximately 0.25 times the output power.

10–4. Phase Inversion

We have seen why transformers are not ideal phase-inversion components. Accordingly, let us consider typical vacuum-tube and transistor phase inverters that may be used in combination with push-pull output stages. One of the simplest forms of single-tube phase inverters is depicted in Fig. 10–29. Resistors R_2 and R_3 have the same value;

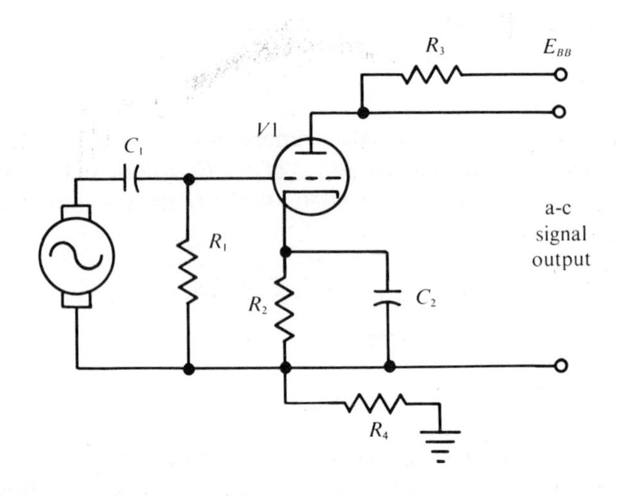

Figure 10–29. Single-tube paraphase inverter circuit that provides maximum voltage gain.

therefore, the voltage drop across them is the same, since the same plate current flows through both resistors. The instantaneous signal polarities, however, are exactly opposite; at the instant that a positive-going signal is applied to the grid, point X becomes less positive with respect to ground and point Y becomes more positive. These signals, with the indicated polarities, are impressed across load resistors R_4 and R_5 through the blocking capacitors C_3 and C_4. C_2 is the plate-supply bypass capacitor.

Since the configuration depicted in Fig. 10–29 does not provide voltage gain, a somewhat elaborated circuit shown in Fig. 10–30 is often used. Maximum available gain is provided by this circuit because the cathode resistor is bypassed. However, the necessity for provision of an input that "floats," or operates above ground potential, may make the higher-gain arrangement impractical.

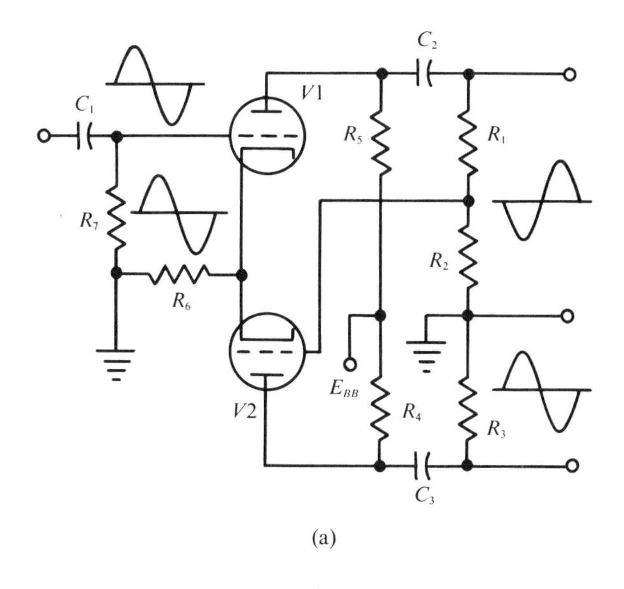

(a)

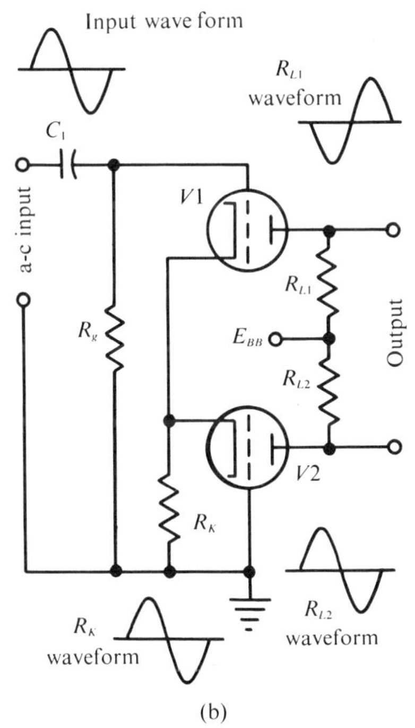

(b)

Figure 10-30. Two-stage paraphase amplifiers. (a) Vacuum tube; (b) Transistor.

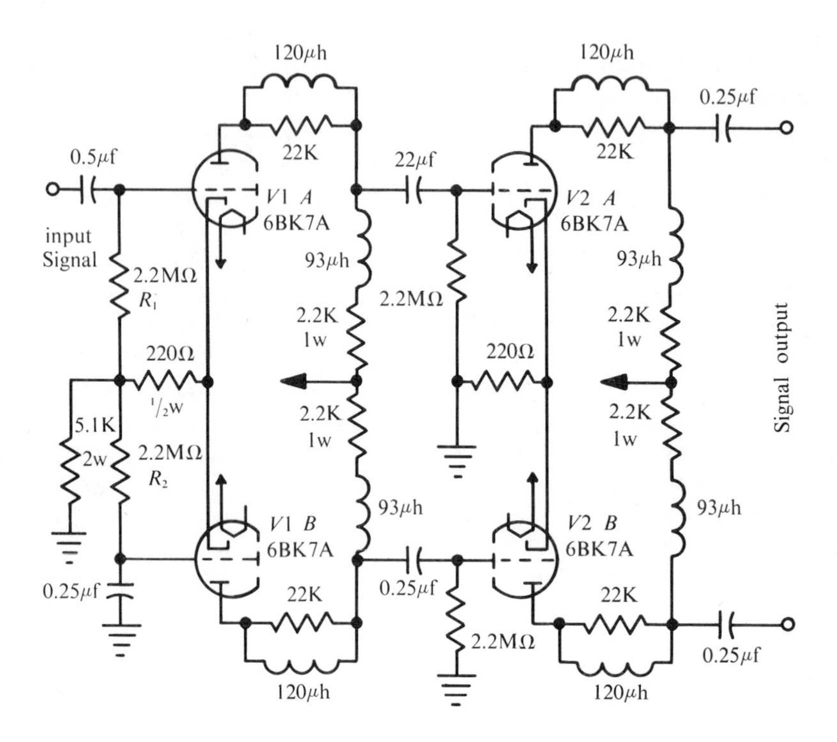

Figure 10–31. A two-stage cathode-coupled inverter-amplifier with plate-load impedances.

When the signal does not extend into the video-frequency range, a two-tube paraphase inverter-amplifier is often utilized, as diagrammed in Fig. 10–31. One tube (*V*1) operates as a conventional amplifier, and the other tube inverts the first signal. In combination, the two tubes thus provide two equal amplified output signals that have opposite polarity. This arrangement has approximately twice the voltage gain of a single-tube paraphase inverter such as depicted in Fig. 10–30. Although no advantage is obtained on a tube-for-tube basis, the two-tube configuration in Fig. 10–31 can be driven from a source that has one side grounded.

In Fig. 10–30, *V*1 amplifies the input signal and impresses the amplified output across the voltage divider consisting of R_1 and R_2. Resistor R_2 has a carefully-chosen value such that the signal drop across it has the same amplitude as the signal on the grid of *V*1. The signal drop across R_2 is fed to the grid of *V*2 and amplified in the plate circuit of *V*2. Since the effective plate loads for *V*1 and *V*2 are made equal, the two output signals have equal amplitudes, and are, of course, in opposite phase. This arrangement is not useful at very high frequencies, because the stray capacitances in the two plate-load circuits are unequal and cause unequal amplifications and phase shifts. An even worse phase-shift and frequency-response situation is encountered at very low frequencies. However, at audio frequencies, for example, the two-tube paraphase inverter is quite satisfactory.

Oscilloscopes exemplify electronic equipments which require a double-ended signal output that ranges from very low frequencies to very high frequencies, such as 4 MHz or more. Hence, phase inverters are utilized that provide balanced performance at high frequencies. This requirement entails phase inversion in low-impedance circuits so that stray capacitances can be neglected, even at high operating frequencies. The cathode-coupled phase inverter-amplifier diagrammed in Fig. 10–31 is widely used in this application. Each tube operates as a combined inverter and amplifier. Although the voltage gain of the configuration is typically the same as for one tube in a conventional amplifier circuit, push-pull output is obtained at high operating frequencies.

The student will perceive that when a positive-going signal is applied to the grid of $V1$ in Fig. 10–31, a positive-going signal appears across R_K. In turn, a positive-going signal is applied to the cathode of $V2$, or $V2$ is cathode-driven. Since $V1$ is grid-driven and $V2$ is cathode-driven, the amplified signal drop across R_{L1} is opposite in phase to the amplified signal drop across R_{L2}. The push-pull output signal will obviously be balanced in half-cycle amplitudes if $R_{L1} = R_{L2}$. Observe that phase inversion takes place in the low-impedance common-cathode circuit. Moreover, stray capacitances are balanced in this configuration, and balanced performance can be obtained at unusually high operating frequencies.

As in a conventional amplifier, high-frequency response will be attenuated unless the plate-load resistors have suitably low values. In other words, the higher the frequency range that we desire, the lower must be the values of the plate-load resistors. Of course, this entails a reduction in the maximum available voltage gain, just as in any RC-coupled amplifier. Therefore, for high-frequency application such as in oscilloscope amplifiers, the plate-load circuits are impedance-coupled, as depicted in Fig. 10–31. This is a two-stage arrangement that provides greater voltage amplification than a single stage.

Note that the input stage employs a common-cathode resistance of 5,320 Ω. This comparatively large value is used to obtain a wide dynamic range of operation. Since proper bias is dropped across 220 Ω, the grid-leak resistances for *V1A* and *V1B* are returned to the junction between the 220-Ω and 5,100-Ω cathode resistors. Operation at frequencies up to 4 MHz requires that the plate-load resistors have values of 2,200 Ω each. Full amplification at this high frequency also requires the use of series and shunt peaking coils, as shown. The 22,000-Ω damping resistors in shunt with the 120-μh peaking coils lower the *Q* of the coils and provide optimum transient response.

Since phase inversion is accomplished by *V1A* and *V1B*, the common cathode resistor for *V2A* and *V2B* has a value of 220 Ω, which is required for bias voltage. The student may perceive that inasmuch as a push-pull signal is applied to the grids of *V2A* and *V2B*, no increase in gain would be obtained if the common-cathode resistor were bypassed. The voltage gain of the amplifier is approximately the same as if two of the triodes were used in a two-stage impedance-coupled configuration. In other words, voltage gain that could be obtained by connecting the tubes in another configuration is foregone in order to obtain balanced push-pull output at frequencies up to 4 MHz.

The 10-w transistor amplifier in Fig. 10–32 (a) employs many of the techniques we have studied. Transistor *Q1* is a common-emitter amplifier. Transistor *Q3* is a paraphase amplifier operating in conjunction with *Q2* to form a Class-*B* amplifier. *Q2* and *Q3* are direct coupled with *Q4* and *Q5* in a Darlington connection for maximum current gain. The output is in the form of a bridge circuit with the 1,500-μf capacitor allowing current to flow into and out of the speaker as the collector of *Q5* rises and falls with the signal voltage. The bias resistor *R2* is used to adjust the d-c level of *Q5* to one-half the value of E_{CC}. Negative feedback is coupled from the output to the input via R_{12} to increase the input impedance and the bandwidth, and to decrease internal distortion. The overall result of the direct-coupled amplifier (no transformer) is an amplifier with low distortion, good bandwidth, and about 1-Ω output impedance for good speaker damping.

Another name for the Darlington configuration is a "super-alpha" arrangement. A high-input impedance preamplifier employing this circuit is depicted in Fig. 10–32 (b). By connecting two common-collector stages in cascade, the a-c input impedance is made nearly equal to β^2 multiplied by the load comprising R_5 in parallel with R_6. The input impedance of this configuration at room temperature is greater than 2 MΩ. In spite of this high a-c input impedance, the d-c resistance in series with the base of *Q1* (R_1 plus the parallel resistance of R_2 and R_3) is considerably lower. This low d-c resistance in series with the base is essential for temperature stabilization.

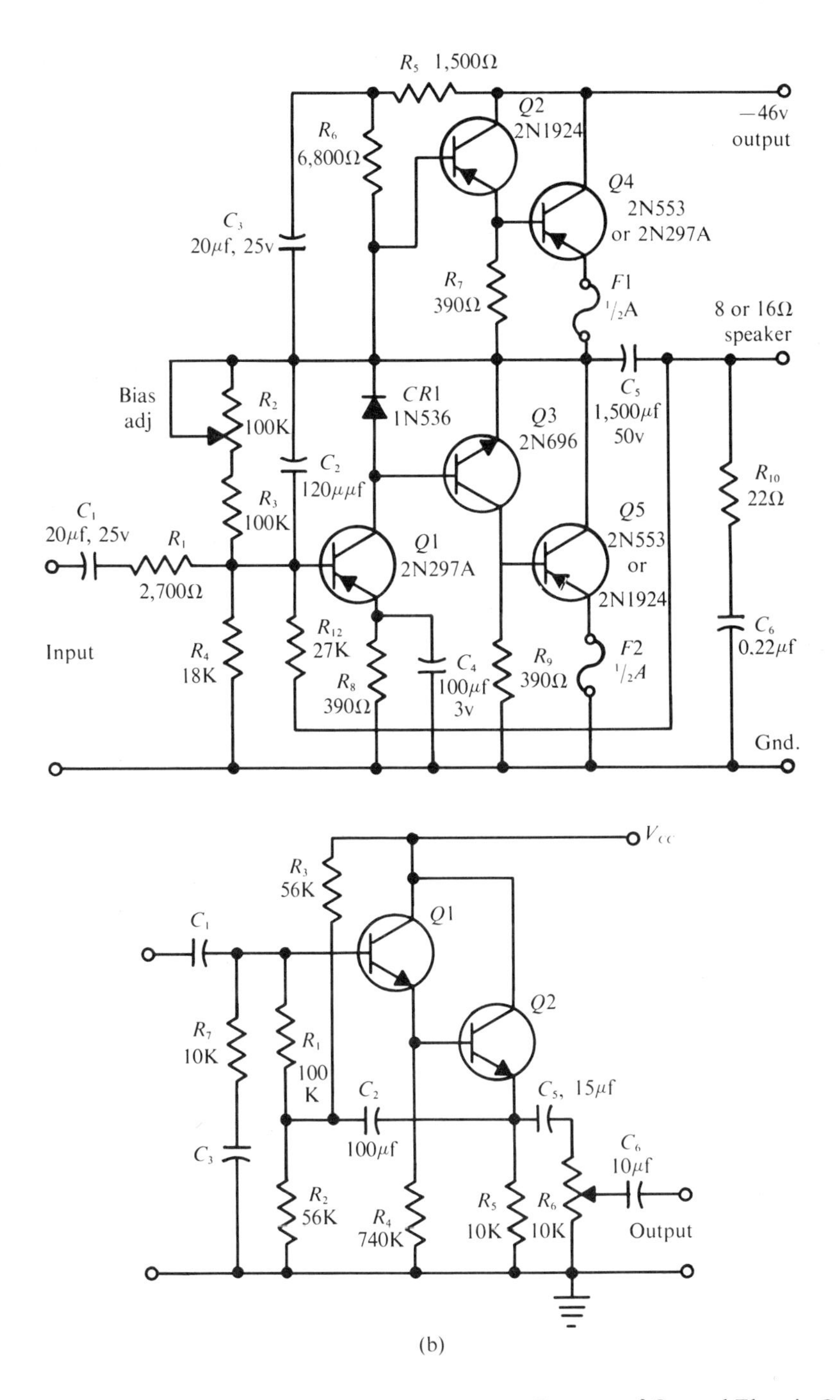

Courtesy of General Electric Co.

Figure 10-32. (a) Ten-watt audio amplifier; (b) Basic Darlington or "Super-alpha" configuration.

The high ratio of a-c input impedance to d-c input resistance is made possible by the use of capacitor C_2 connected between the emitter of $Q2$ and the junction of R_1 and R_2. This capacitor couples the output voltage back to the junction. Since the output voltage is nearly the same as the input voltage, the difference voltage across R_1 is very small and the effective value of R_1 is increased by several orders of magnitude. The upper limit of R_1 is set by temperature considerations. R_4 adds sufficient current to $Q1$ to maintain a high β value and good frequency response. This circuit has good performance over a temperature range from $-25°C$ to $+55°C$. Its frequency response is flat within 0.5 db from 10 Hz to 250 kHz. So, if germanium transistors are used, the temperature range is somewhat limited, but has good performance within the restricted temperature range.

10–5. Magnetic Amplifiers

A magnetic amplifier consists of saturable reactors, resistors, rectifiers, and transformers, connected in such manner that a comparatively low-level signal controls large amounts of power. Thus, a magnetic amplifier is basically a power amplifier. We sometimes find a saturable reactor described as a magnetic amplifier, but this is an incorrect description. Although the saturable reactor is the key component in a magnetic amplifier, the saturable reactor requires various associated components to develop amplifier action. To understand magnetic-amplifier action, we must start with the nonlinear characteristic of a magnetic circuit that utilizes an iron core. Let us briefly review this topic.

Figure 10–33 (a) shows the linear characteristic of an air magnetic circuit. An iron magnetic circuit [Fig. 10–34 (b)] can be regarded as linear only over a restricted range of magnetizing force. Different types of ferromagnetic cores have different *BH* curves, as shown in Fig. 10–33 (c). Permeability characteristics for ferromagnetic cores are nonlinear and can be seen in Fig. 10–33 (d). We know that a change in permeability (μ) will change the inductance of a coil; when a core is saturated, its permeability is very low. The characteristics for an ideal ferromagnetic core material are as follows:

1. Minimum hysteresis and eddy-current losses (see Fig. 10–34). Note that the area enclosed by the hysteresis loop is proportional to the core loss.

2. High saturation flux density; that is, when the core has become saturated, it has a maximum number of flux lines. This denotes the power capacity of the core. Note that a large power capacity is not necessarily the same as high permeability.

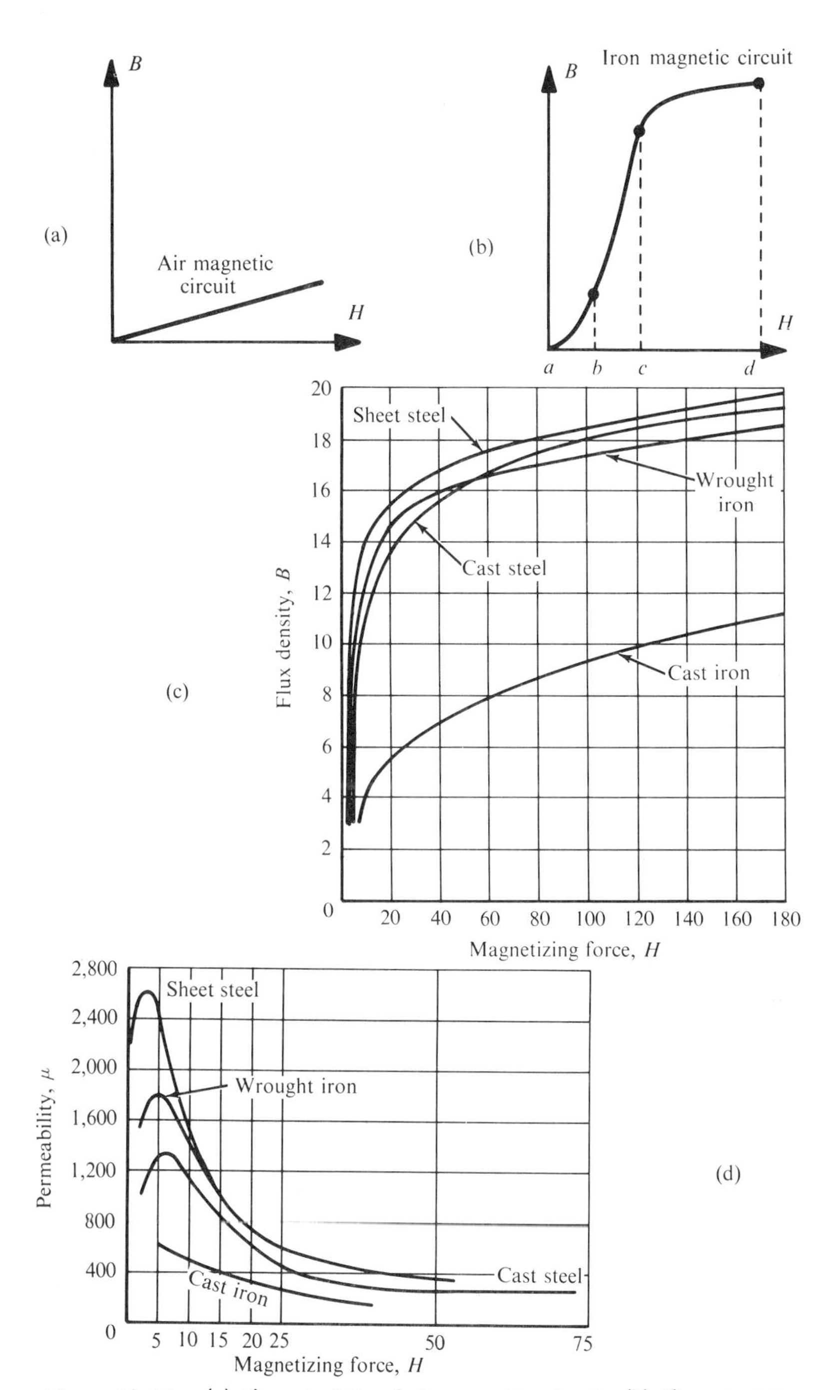

Figure 10-33. (a) Characteristic of air magnetic circuit; (b) Characteristic of iron magnetic circuit; (c) *BH* curves for common ferromagnetic cores; (d) Permeability curves for common ferromagnetic cores.

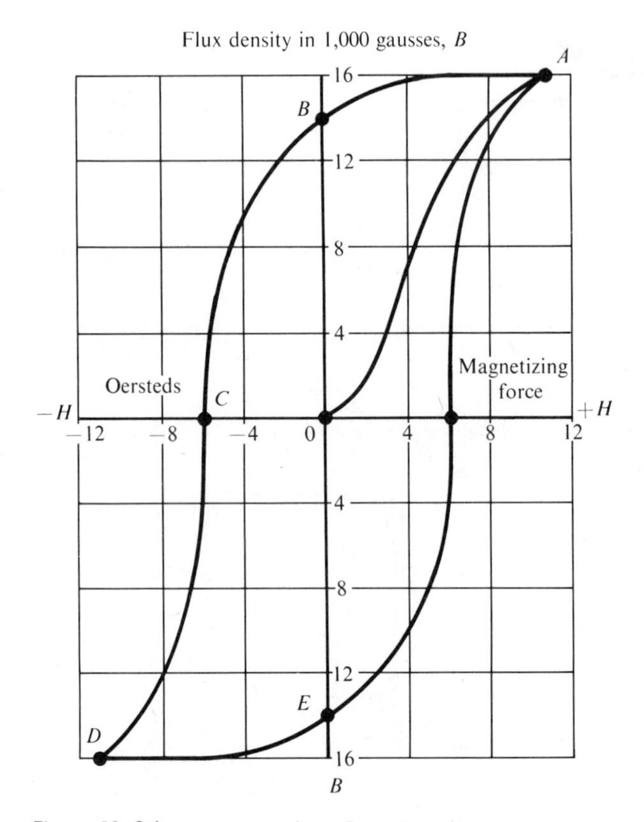

Figure 10–34. Hysteresis loop for a ferromagnetic core.

3. High permeability provides a narrow hysteresis loop with steep sides; in turn, a small magnetizing force produces core saturation.

4. Core materials should be stable and relatively unaffected by temperature changes, mechanical strain or shock, and previous magnetization.

In analyzing the action of a saturable reactor, it is helpful to start with the control circuit. Figure 10–35 (a) shows a ferromagnetic core with a control winding N_c connected in series with variable resistor R, switch S, and the control-voltage source $E_{d\text{-}c}$. When the control voltage is applied, the amount of magnetizing force is adjusted by means of the rheostat. As the current flow is increased, the flux density increases, and the core permeability is low at the start. However, as the control current (magnetizing force) is increased into region B in Fig. 10–35 (b), the flux density increases very rapidly.

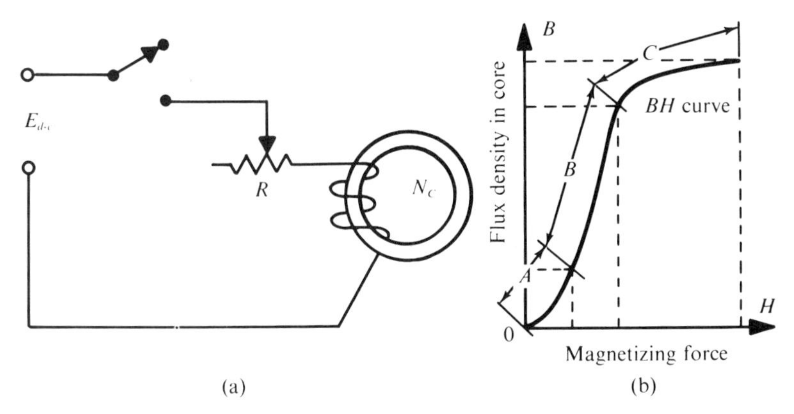

Figure 10–35. (a) Saturable reactor core and control circuit; (b) *BH* curve for reactor core.

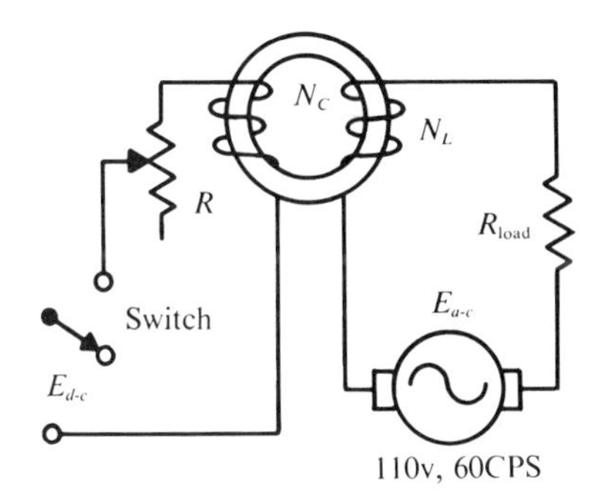

Figure 10–36. Saturable reactor circuit.

Then, as the control current enters region *C*, the core is being driven into saturation and the permeability again becomes very low. Refer to Fig. 10–33 (d).

Next, if a load winding is added to the core as shown in Fig. 10–36, we have a complete saturable reactor. The load circuit comprises the load winding, load resistor, and a-c voltage source connected in series. We perceive that the a-c voltage across the load resistor will depend on the inductive reactance of the load winding and the value of the load resistor. Of course, the permeability depends on the control current that flows. If the control current is large, the core is driven into saturation, permeability falls to a low value, the inductive reactance falls to a low value, the load current becomes high, and most of the a-c source voltage then drops across the load resistor.

The basic reactor circuit shown in Fig. 10–36 is inefficient, due to a-c current flowing through the load winding. This produces two detrimental effects.

1. Voltage is induced from the load winding to the control winding by transformer action. Since the control winding has low impedance, large currents will flow in the control circuit and dissipate a large amount of power.

2. During one-half cycle, the load-winding current flow will produce a flux that opposes the flux produced by the control winding. In turn, power is required from the control winding to return the core flux to its normal operating point.

This problem of opposing flux can be eliminated by placing a *rectifier* in series with the load winding. Then, the load current becomes undirectional, and if the rectifier is connected in proper polarity, the flux fields never oppose each other. The addition of a rectifier elaborates the saturable reactor into a magnetic amplifier. Details of circuit action are explained below. Note in passing that another method of reducing the limitations previously mentioned consists in using a nonpolarized three-legged core, as shown in Fig. 10–37.

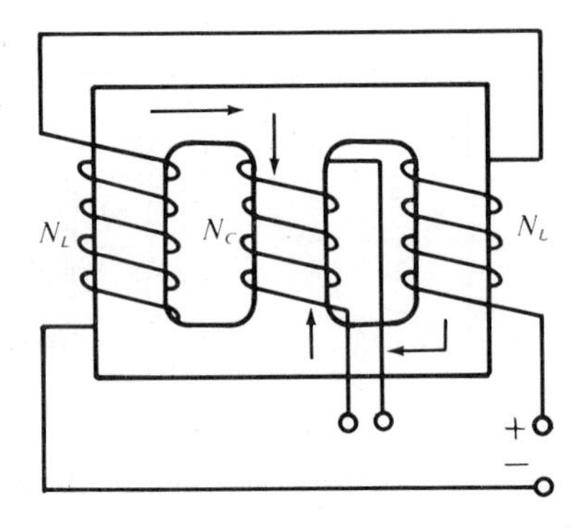

Figure 10–37. Nonpolarized three-legged core.

We perceive that one load winding is wound around one outside leg, and the other load winding is wound around the opposite leg; the two are connected so that the two flux fields cancel in the center leg.

A schematic representation of the arrangement shown in Fig. 10–37 is depicted in Fig. 10–38. We observe that the flux fields produced by the two load windings are in opposition to each other and induce no voltage in the control winding. When the load voltage reverses, the field produced by each load winding reverses, and the two flux fields

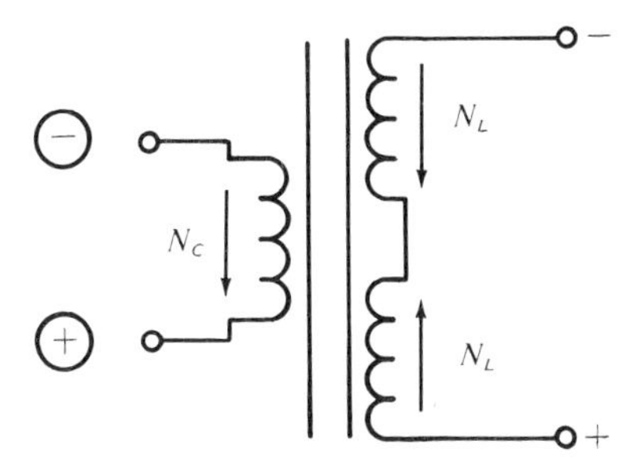

Figure 10-38. Nonpolarized three-legged winding.

again cancel out in the center leg. Note that one flux field aids the control flux, and the other flux field opposes the control flux. Therefore, a control voltage of either polarity can be used, and this is the basis of the term "nonpolarized" three-legged core. In some cases, it is necessary to use a three-legged core such as shown in Fig. 10-39. Here, the load windings are connected so that the load fluxes add in the center leg.

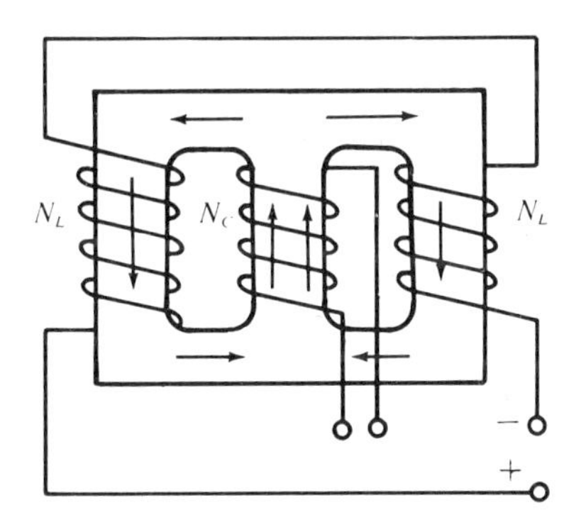

Figure 10-39. Polarized three-legged core.

Figure 10-40 depicts a schematic representation of the arrangement shown in Fig. 10-39. Flux fields produced by the two load windings aid each other in the center leg. In turn, the load-winding fields may aid the control field and thereby help to saturate the core; or the load-winding fields may oppose the control field and thereby reduce the flux density. We see that the polarity of the applied voltages for either the control coil, or for the load coils, will determine the flux density. Therefore, this arrangement is said to be *polarized.*

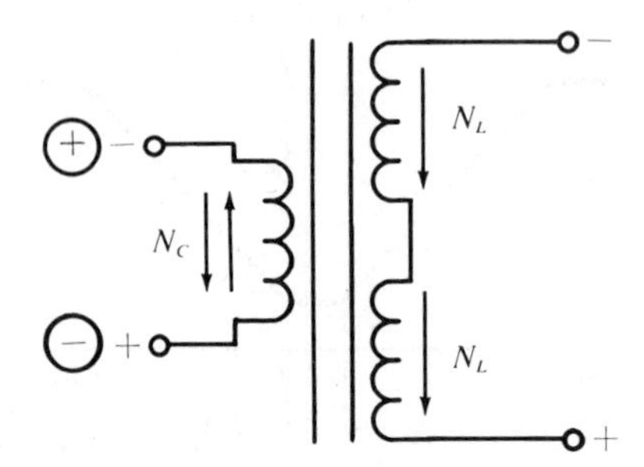

Figure 10-40. Polarized three-legged core winding.

We recall that the addition of a rectifier (Fig. 10-41) elaborates the reactor circuit into a magnetic amplifier. In this circuit, the saturable reactor is used in combination with a semiconductor rectifier CR_1 to produce a controllable d-c voltage across the load resistance R_L. The reactor has a three-legged core, and since the two load coils produce opposing fields, the reactor is nonpolarized. As has been explained, the voltage across R_L depends on the level of the d-c control voltage.

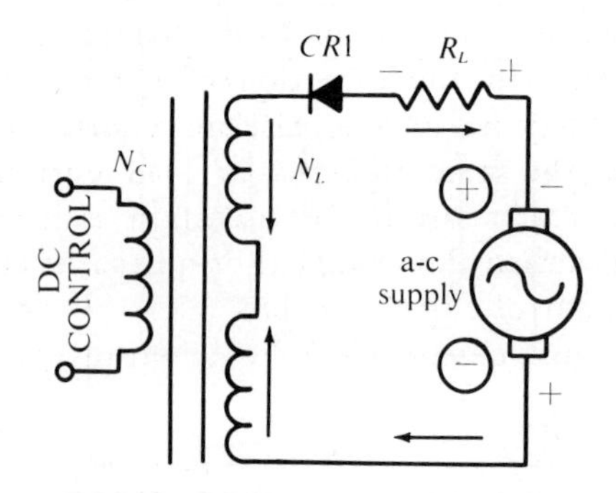

Figure 10-41. Basic nonpolarized magnetic-amplifier circuit.

Let us consider the operation of the magnetic amplifier in Fig. 10-41 with a control winding (N_c) voltage of zero. When the top of the a-c supply is positive, current will flow through the load windings (N_L), *CR*1, R_L, and back to the source. The absence of a d-c voltage to saturate the core results in high permeability, which means that the inductive reactance is high. Current flow through the circuit is small, and the voltage drop across R_L is low. Next, when a d-c voltage is applied across the control winding, d-c current flows and the core is magnetized. If this d-c current is of sufficient magnitude, the core

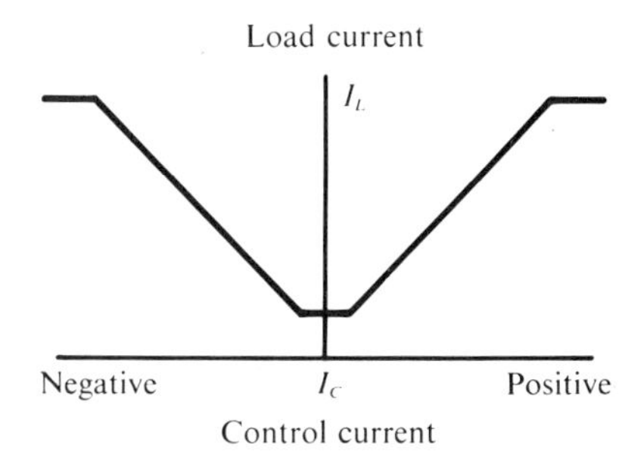

Figure 10–42. Dynamic characteristic curve for a nonpolarized magnetic amplifier.

saturates and its permeability is greatly reduced. Inductive reactance in the load windings becomes very small, and load current increases accordingly. In turn, a large voltage is dropped across R_L.

With suitable parameters in the magnetic-amplifier circuit, a small control voltage can "valve" a large amount of power in the load circuit. A dynamic characteristic curve can be plotted to show the control action. Figure 10–42 is a plot of load current versus control current. Observe that an increase in control current (up to the point of saturation) causes an increase in load current in a nonpolarized magnetic amplifier, with the direction of control-current flow being inconsequential. That is, an increase in control current in either direction produces the same increase in load current. Note that the load current never reaches zero, but a certain minimum value. This minimum value is called the no-signal or quiescent current value, and occurs when the control current is zero.

The term no-signal current may sound strange at first, since the

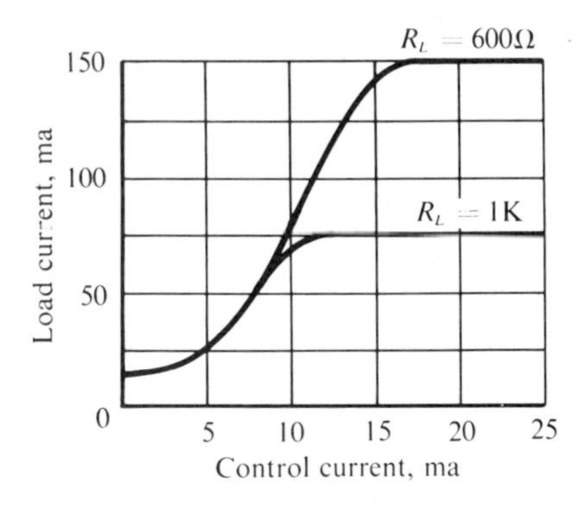

Figure 10–43. Characteristic curves for different values of load resistance.

control current is d-c. However, we perceive that the control signal is analogous to an input signal applied between the control grid and cathode of a vacuum tube. The load current in a magnetic amplifier is affected by the value of load resistance R_L. Figure 10–43 shows an output characteristic curve when load current I_L is plotted against control current I_C, for two different values of load resistance. This curve illustrates the fact that the load current is much greater at saturation in a magnetic amplifier that has a small load resistance. This curve also exemplifies the amplification (current amplification) provided by a magnetic amplifier.

SUMMARY

1. The beam-power tetrode tube is capable of providing comparatively high levels of signal-output power.
2. The arrangement of the control grid and the screen grid in conjunction with the beam-forming plates increases the number of electrons reaching the plate.
3. The production of a large space charge between the screen and plate of a beam-power tube forms a virtual cathode which suppresses the flow of secondary electrons.
4. Plate efficiency is defined as the ratio of a-c signal power developed at the plate load to the d-c power that is supplied to the plate circuit.
5. The advantages of a push-pull Class-*B* amplifier circuit are greater efficiency, cancellation of second harmonic distortion, and a very small standby power consumption.
6. The disadvantages of push-pull amplifiers are the necessity of an output transformer, the necessity of input-phase inversion, the necessity of matched control devices, and possible crossover distortion.
7. A transformer is a simple phase inverter which has a relatively high distortion level and a relatively low efficiency when used in an amplifier.
8. Single-ended transformers are more difficult to design than push-pull transformers because the d-c plate current through a single-ended transformer tends to saturate the core whereas the magnetic fields produced by each plate current in a push-pull transformer cancel.
9. Greater grid-voltage swing is possible in Class-*A* push-pull operation without appreciable distortion.
10. Crossover distortion occurs in vacuum-tube and transistor

Class-*B* amplifiers biased at cutoff. This type of distortion is the result of extreme nonlinearity of the amplifying device near cutoff.

11. Crossover distortion in a transistor is more extreme than in a vacuum tube.

12. The derating factor of a transistor is given in watts/°C and is the amount of power the collector dissipation must be derated for each degree centigrade rise in temperature above 25°C.

13. The single-stage paraphase amplifier provides two output signals out of phase to each other with no gain.

14. A two-stage paraphase amplifier develops twice the gain of a conventional single-stage paraphase amplifier (Fig. 11–18) and does not require a floating input as the high gain single-stage paraphase amplifier in Fig. 11–19 does.

15. The cathode-coupled inverter amplifier has the advantage of being capable of operating at relatively high frequencies.

16. The collector dissipation of each transistor in a well designed Class-*B* push-pull amplifier is approximately 0.25 times the output power.

$$P_{\text{diss}} \simeq 0.25\, P_{\text{diss}}$$
$$\simeq 0.25 \left| \frac{2}{\pi}(I_{c\,\text{peak}} - I_{CEO}) \right| V_{CC}$$
$$P_{\text{diss}} \simeq 0.159\, I_{c\,\text{peak}} V_{CC} \qquad \textbf{(10–20)}$$

Questions

1. Why is suppression of secondary electrons more effective in a beam-power tube?
2. Why can a beam-power tube supply greater signal power output with lower values of plate supply voltage than a pentode?
3. What are the effects on a circuit when two tubes are connected in parallel to obtain a greater output power?
4. Define the term "plate efficiency" as applied to an amplifier circuit.
5. What are the advantages of a Class-*B* push-pull amplifier over a Class-*A* amplifier?
6. What are the disadvantages of a Class-*B* push-pull amplifier in comparison with a Class-*A* amplifier?
7. What is the function of a phase inverter in a push-pull amplifier?
8. What are the disadvantages of using a transformer in an amplifier as a phase inverter?
9. Why does d-c saturation not occur in the output transformer of the amplifier in Fig. 11-8?
10. With reference to Fig. 11-10, compare the single-ended transformer in the plate circuit of *V*1 to the push-pull transformer in the output circuit.
11. Compare the linearity of a Class-*A* push-pull amplifier with that of a Class-*A* single-stage amplifier.
12. How is greater efficiency obtained in an amplifier with Class-*AB* operation than with Class-*A* operation?
13. What causes crossover distortion in Class-B amplifiers, and what is the remedy for this type of distortion?
14. What causes crossover distortion in a transistor Class-*B* push-pull amplifier?
15. List five limitations of transistors and give an explanation for each limitation.
16. What is the relationship between the load impedance of a Class-*B* push-pull amplifier and the maximum output power?
17. Why is the primary impedance of the transformer used in a Class-*B* push-pull amplifier four times the impedance of one-half the transformer (R_C)?
18. What is the disadvantage of the single-stage paraphase amplifier?
19. What are the advantages and disadvantages of a two-tube paraphase amplifier?

20. Why can the cathode-coupled paraphase amplifier operate at higher frequencies than other types of paraphase amplifiers?

21. Draw the circuit in Fig. 11-21, and trace a single cycle of audio through each stage.

22. What is the key component in a magnetic amplifier?

23. What are the characteristics of an ideal ferromagnetic core material?

24. What causes the hysteresis loop in a ferromagnetic core?

25. With reference to Fig. 10-36, what undesirable effects are produced by the a-c current flowing through the control winding?

26. With reference to Fig. 10-39, what are the advantages of the polarized three-legged core in a magnetic amplifier?

PROBLEMS

1. A Class-*B* push-pull amplifier uses type 2N176 transistors. The maximum collector current in the amplifier is 4 amp, and the maximum collector voltage is equal to V_{CC}, which is 20 v. What is the maximum possible output power of the circuit?

2. What is the maximum output power in Problem 1 if the collector bias (V_{CC}) is reduced to 10 v?

3. What is the value of the load impedance seen by each transistor in Problem 1?

4. Determine the primary impedance of the transformer used in Problem 1.

5. What is the power gain of the amplifier represented in Problem 1 when $I_o = 3$ amp, $I_{in} = 60$ ma, and the input resistance is 300 Ω?

6. A Class-*B* push-pull amplifier uses a 2N376A transistor with the following circuit values: $V_{ce} = 25$ v, $I_C = 4$amp, $R_L = 2.2\ \Omega$, $R_{in} = 4\ \Omega$, and beta $= 45$. Determine the circuit power output, power gain, load resistance, and the primary impedance of the output transformer.

7. The maximum collector power dissipation of the 2N376A transistor is 90 w at 25°C and the maximum junction temperature is 100°C. Will the transistor in Problem 6 operate within a safe power dissipation level at 50°C?

8. A 2N1365 power transistor has the ratings 120 w at 25°C and maximum junction temperature of 100°C. What is maximum output power if two of these transistors are used in a Class-*B* push-pull amplifier at an ambient temperature of 60°C in a circuit in which the total output power is four times the collector dissipation of each transistor?

9. A Class-*B* push-pull amplifier uses two types of 2N739 silicon transistors with the characteristics $\beta = 40$, $T_{j\,max} = 200°C$, and $V_{ce\,max} = 300$ v. The circuit has the values of $V_{ce} = 100$ v, $I_o = 0.1$ amp, $R_c = 100\ \Omega$, and $R_b = 100\ \Omega$. What is the power gain and power output of the circuit ?

10. What is the derating factor in watts/°C of the 2N3739 silicon transistor if $T_{j\,max} = 200°C$ and $P_{diss\ 25°C} = 20$ w ?

11. Suppose the collector current in the 2N3739 transistor in Problem 10 is 1 amp at an ambient temperature of 100°C. What is the maximum allowable collector voltage under these conditions?

CHAPTER 11

Transducers

11–1. Introduction

Transducers are defined in electronics technology as devices that change electrical energy into some other form of energy, or vice versa. Since there are many forms of energy, there are many types of transducers utilized in combination with amplifiers and recording or radiating systems. Perhaps the most familiar transducers are exemplified by microphones and receivers employed in telephone handsets. The loudspeakers used in radio and television receivers are also a familiar type of transducer. A microphone is often called a telephone transmitter. Many types of microphones are in common use in various areas of electronics technology. Only the more basic types of microphones and other transducers can be discussed in an introductory text.

Figure 11–1 depicts the plan of a simple telephone transmitter. We will find that the battery is an important source of energy, as well as the sound energy that strikes the microphone diaphragm. Diaphragm

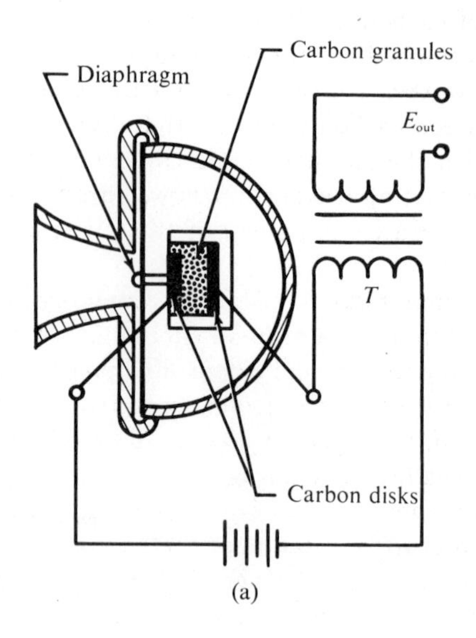

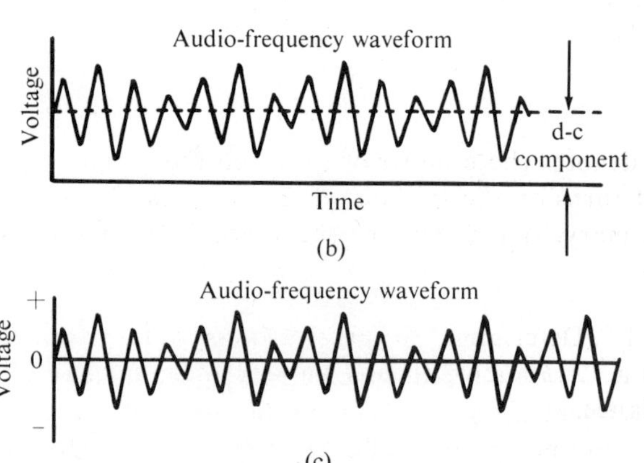

Figure 11-1. (a) Plan of a simple carbon microphone; (b) Current flow in primary winding; (c) Current flow in secondary winding.

vibration operates to valve the electrical energy supplied by the battery; the device construction is such that the electrical energy output is considerably greater than the sound energy input. That is, this type of microphone operates as an amplifier. In Fig. 11-1, the battery causes a constant flow of direct current through the microphone and the primary of the matching transformer. This matching transformer is also called an induction coil.

Behind the thin diaphragm of the microphone in Fig. 11-1 is

mounted a cup of carbon granules. An electrode, consisting of a small carbon disk, is attached to the diaphragm. When the diaphragm is struck by sound waves, it is set into vibration, and this vibration varies the pressure between carbon granules, thereby varying their resistance in accordance with the amplitude and frequency of the applied sound wave. Hence, a pulsating d-c current flows through the primary of the transformer, and the a-c component of this current is induced into the secondary winding.

Note that the matching transformer depicted in Fig. 11-1 might provide either a step-up or a step-down in output voltage. The winding ratio of primary to secondary depends upon the impedance of the circuit or device to which the secondary delivers electrical energy. In a microphone system, it is desired to obtain maximum power transfer. Therefore, the transformer winding ratio is chosen to match the resistance of the microphone to the resistance (or impedance) of the load. A carbon-granule microphone has an average impedance of about 75 Ω. If the microphone drives an amplifier that has an input impedance of 600 Ω, the transformer is then designed to match 75 Ω to 600 Ω.* We recall that winding ratios are related to impedance ratios according to the formula

$$\frac{N_p^2}{N_s^2} = \frac{Z_p}{Z_s} \tag{11-1}$$

where N_p denotes the number of turns on the primary, N_s denotes the number of turns on the secondary, Z_p denotes the source impedance for the primary, and Z_s denotes the reflected load impedance for the secondary.

Example 1. Determine the turns ratio of a transformer designed to match a 75-Ω microphone to an amplifier that has a 600-Ω input impedance.

$$\frac{N_p}{N_s} = \sqrt{\frac{Z_p}{Z_s}}$$

$$\frac{N_p}{N_s} = \sqrt{\frac{75}{600}}$$

$$\simeq 1:2.85 \qquad \text{(answer)}$$

Again, if the microphone drives a loudspeaker that has a voice-coil impedance of 5 Ω, the transformer will be designed to match 75 Ω to 5 Ω. In this case, matching transformer provides a voltage step-down. In any case, a matched system provides maximum power transfer at an

*The impedance of audio devices is taken at a reference frequency of 1 KHz.

efficiency of 50 per cent. In other words, half of the power available in the source is transferred to the load. The amplifier following a microphone is commonly called a microphone amplifier, as seen in Fig. 11-2.

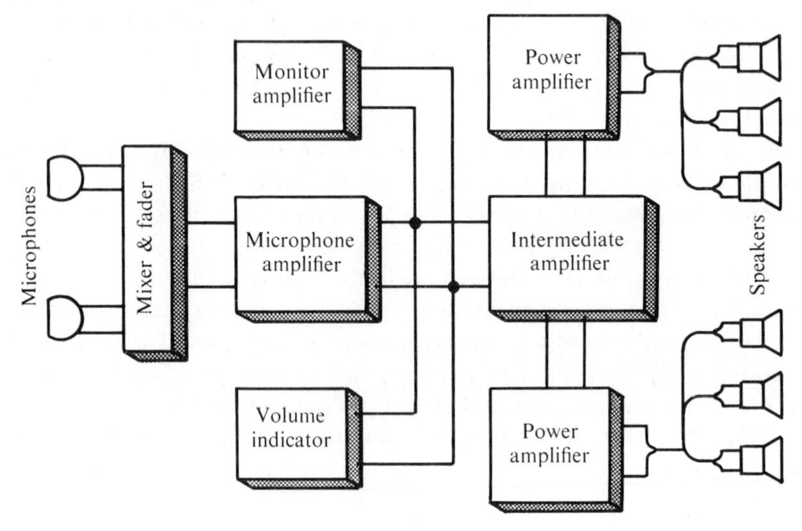

Figure 11-2. Block diagram of a public-address system.

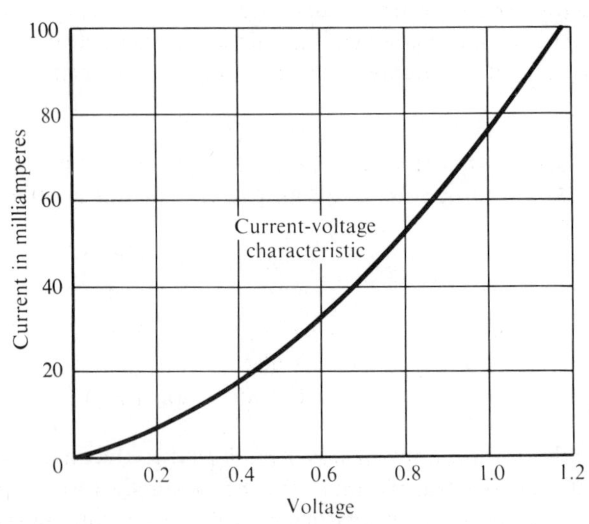

Figure 11-3. Current-voltage characteristic for a single-button carbon microphone.

11-2. Double-Button Carbon Microphones

When carbon granules are compressed in a microphone, the resistance of the microphone button decreases. Microscopic arcs provide conduction from one granule to the next. Carbon has a nonlinear resistance characteristic, as seen in Fig. 11-3; this is the voltage-current graph for a single-button carbon microphone. Because the resistance characteristic is nonlinear, audio waveforms are somewhat distorted. That is, the peaks of the waveform have a greater amplitude than they should with respect to the initial portion of the waveform. This produces noticeable even-harmonic distortion.

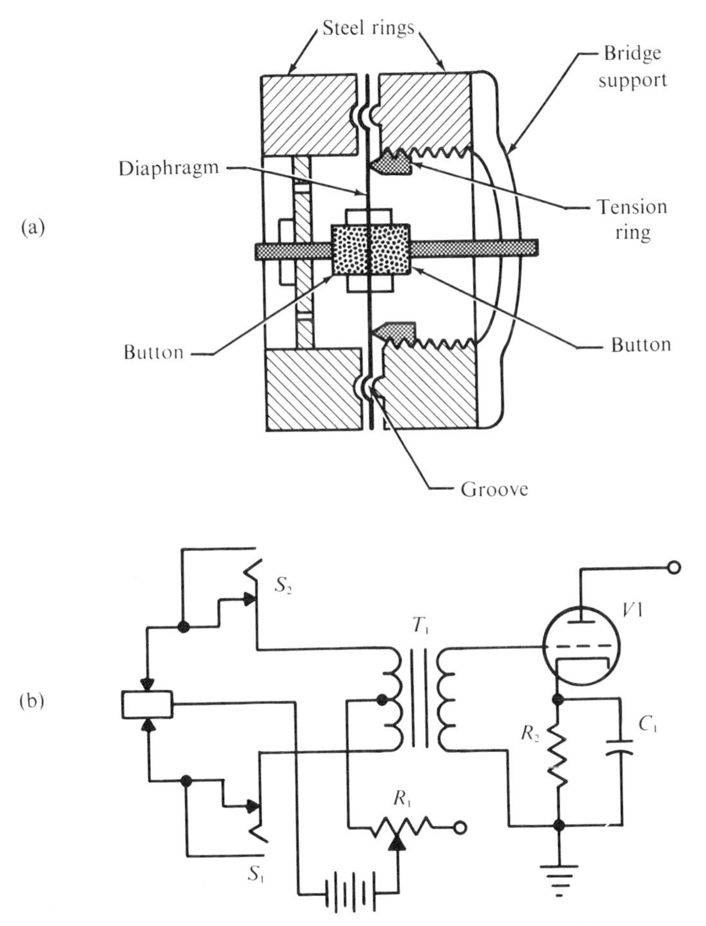

Figure 11-4. (a) Double-button carbon microphone; (b) Input circuit.

To minimize this type of distortion in a carbon microphone, the double-button construction can be utilized, as shown in Fig. 11–4(a). A double-button microphone has carbon-granule cups on both sides of the diaphragm. Thereby, it provides push-pull operation which tends to cancel out even-harmonic distortion. The matching transformer has a center-tapped primary, as shown in Fig. 11–4(b). The jacks provide for plugging a milliammeter into the circuit; the meter indicates the average d-c current flow which is adjusted by means of the rheostat. Double-button microphones commonly have a total resistance of several hundred ohms. On the other hand, the grid-input impedance of the amplifier tube is extremely high. Accordingly, exact impedance matching is not practical; the matching transformer is commonly designed to match several hundred ohms to 500,000 Ω.

Operation of a double-button microphone is depicted in Fig. 11 5.

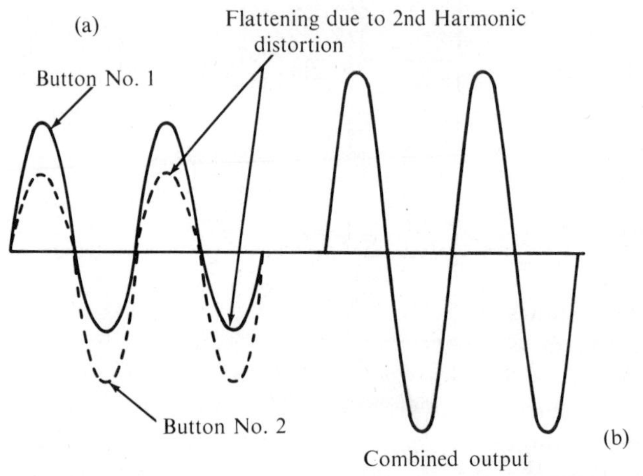

Figure 11–5. (a) Outputs from individual buttons; (b) Push-pull output.

The output waveform from each button is distorted, principally by generation of a second harmonic. However, the push-pull matching transformer combines the pair of waveforms as shown, providing cancellation of the second harmonic. Note that push-pull action cancels even harmonics only; it enhances odd harmonics. However, odd-harmonic distortion from a double-button microphone is very small compared with even-harmonic distortion. The maximum rated d-c current for a carbon microphone should not be exceeded, because the tiny arcs between carbon granules develop heat and some of the granules may be fused. In such case, the sensitivity of the microphone is reduced, and the granules are said to be "packed." This difficulty can be overcome by reducing the d-c current flow and tapping the diaphragm lightly to unpack the granules.

A high quality single-button carbon microphone will provide about 0.25 a-c v across 75 Ω. On the other hand, a high-quality double-button carbon microphone provides about 0.05 a-c v across 200 Ω. Rated d-c current flow is typically 25 ma for each button.

Example 2. Suppose the output of the microphone in Fig. 11–4 is 50 mv across the 200-Ω primary impedance. What is the voltage developed across the 500,000-Ω grid resistor in the secondary of the transformer?

$$\frac{E_s}{E_p} = \frac{N_s}{N_p} = \sqrt{\frac{Z_s}{Z_p}} \tag{11–2}$$

then,

$$E_s = E_p \sqrt{\frac{E_s}{E_p}}$$

$$= 50 \sqrt{\frac{50{,}000}{200}}$$

$$\simeq 790 \text{ mv} \qquad \text{(answer)}$$

11–3. Other Types of Microphones

Figure 11–6 shows the input circuit for a piezoelectric or crystal type of microphone coupled into an amplifier employing a field-effect transistor. A pair of Rochelle-salt crystals (potassium-sodium tartrate) are cemented together; electroplated metal films provide contact to opposite sides of the unit. In one type of construction, sound waves strike one of these metal films, causing the crystal unit to vibrate. In turn, the piezoelectric effect of the crystal lattice generates an audio output voltage. Another type of construction employs a diaphragm that is mechanically coupled to the crystal unit. This arrangement provides a greater output voltage, although it has poorer frequency response than the crystal unit alone.

In some microphones, several crystal units are connected in series, parallel, or series-parallel to obtain higher output voltage, lower internal impedance, or both. Note that a crystal microphone is basically a generator and does not utilize a local battery. If substantial d-c voltage were accidentally applied to a crystal microphone, the unit would be damaged. Although the output level of various crystal microphones will vary considerably, depending upon the design, an output of 0.02 a-c v is typical. Since a crystal microphone has a very high internal im-

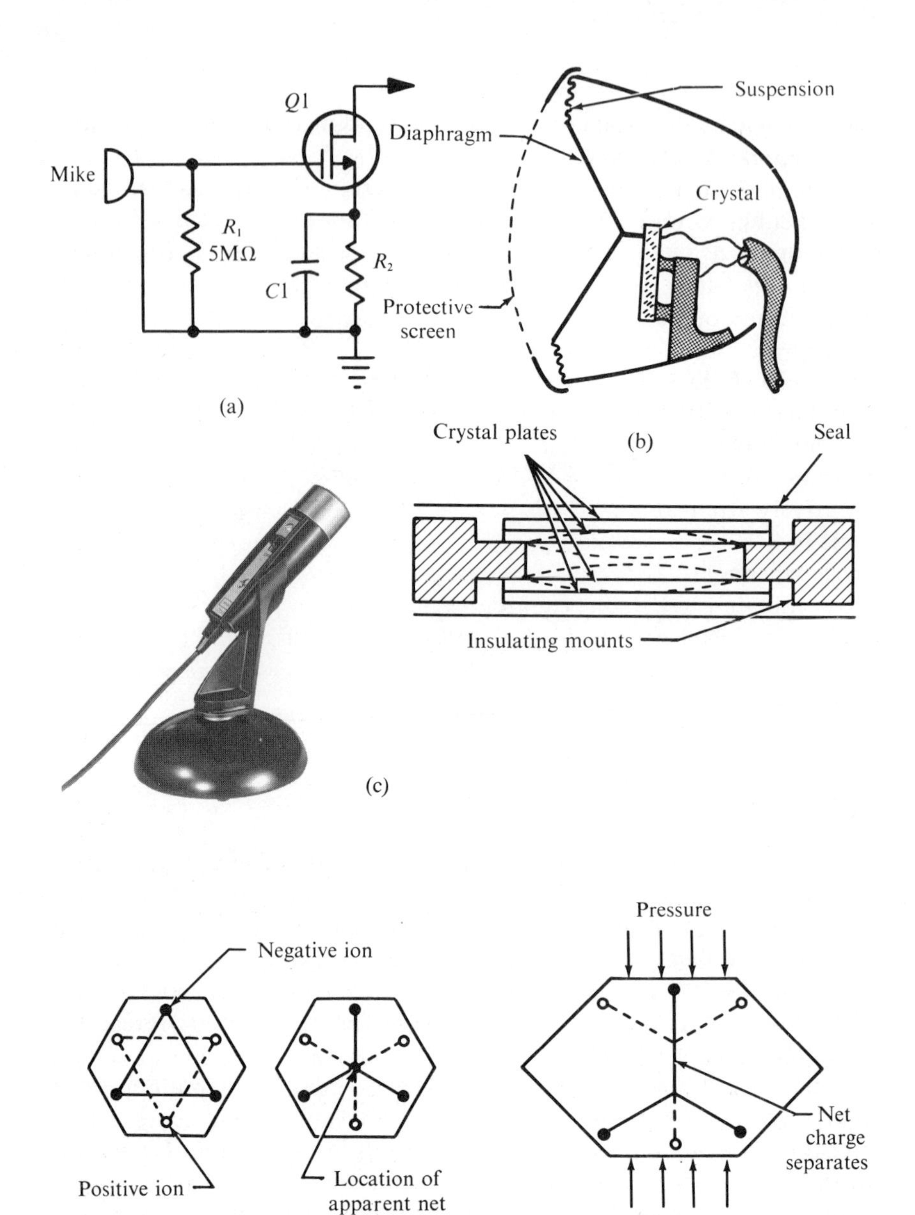

Courtesy of Shure Brothers, Inc.

Figure 11-6. (a) Crystal microphone circuit; (b) Plan of crystal microphone; (c) Appearance of a crystal microphone; (d) Generation of voltage by a Rochelle-salt crystal.

pedance (typically 250,000 Ω), the audio-output voltage that it provides is attenuated by the length of cable used to connect the microphone to the microphone amplifier. Shielded cable must be used to avoid stray-field pickup by the high-impedance circuit.

A shielded coaxial cable for a crystal microphone cannot be terminated in its characteristic resistance, because the impedance match between microphone and cable would be extremely poor. We recall that all practical coaxial cables have a fairly low value of characteristic resistance. Therefore, the output end of the cable must be left open, and the cable "looks like" a capacitor to the microphone over the audio-frequency range. The cable is equivalent to a fixed capacitor connected in shunt to the microphone. The higher audio frequencies are comparatively attenuated at the input to the microphone amplifier. The level of 0.02 a-c v previously noted is typical at the midband audio range when a 6-ft coaxial output cable is used.

To obtain a satisfactorily uniform frequency response, various circuit designs may be employed. For example, a matching transformer may be inserted between the crystal microphone and a properly terminated output cable. Typical transformers provide a flat frequency response from about 100 to 5,000 Hz with attenuated output past these limits. If the crystal microphone works directly into a cable, the value of grid-leak or gate resistance for the amplifier microphone is chosen as high as possible, such as 5 MΩ. Lower values of grid-leak resistance cause more attenuation of the signal, although high-frequency response is improved. Again, the microphone amplifier may be designed with frequency-compensating *RC* circuitry to equalize the response of the crystal-microphone and cable arrangement.

Figure 11-6(b) provides a visualization of voltage generation by a Rochelle-salt crystal when subjected to mechanical pressure. The crystal lattice is a structure of ions, which is the basis of the piezoelectric effect. When pressure is applied to the crystal, a displacement of charge centers occurs in the lattice. The crystal is made up of equal numbers of positive and negative ions arranged in equilateral triangles. In turn, the average or net charge in each group of three like charges is in the geometric center of this triangle. We observe that the centers of net negative and net positive charges coincide when no pressure is applied to the crystal.

On the other hand, when pressure is applied, the ions in the lattice are forced to move. This movement results when each triangle of charges is compressed in a vertical direction [Fig. 11-6(b)] and the triangle is spread out in a lateral direction. As the shape of each triangle changes, the location of the apparent net charges also changes.

The net positive charge moves upward and the net negative charge moves downward. Because of this charge separation, a voltage appears within the crystal. The individual charge separations produce a combined voltage difference that appears on the surfaces of the crystal. In this illustration, the top surface of the crystal exhibits a positive potential, and the bottom surface exhibits a negative potential.

This type of crystal develops a voltage difference between the surfaces to which the pressure is applied. Other types of crystals will develop a voltage at right angles to the stressed surfaces. Still other

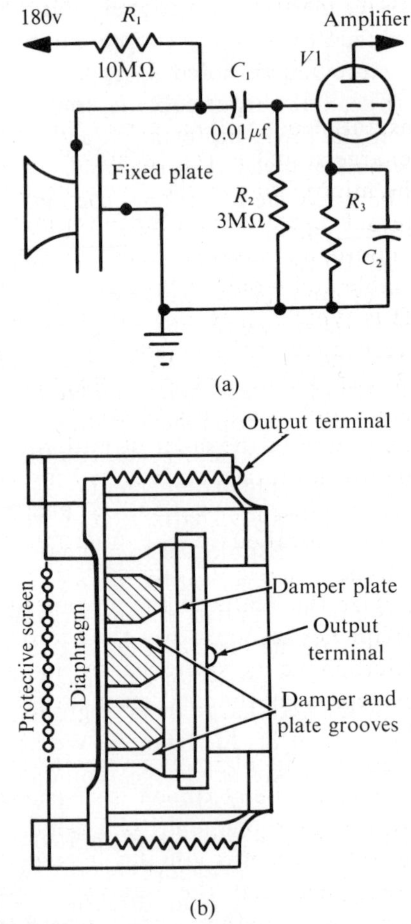

Figure 11-7. (a) Typical capacitor-microphone input circuit; (b) Cross-sectional view.

types of ionic crystals have no externally evident piezoelectric effect. Quartz crystals are similar in many respects to Rochelle-salt crystals. However, the Rochelle crystal is more efficient for use in microphones. Other transducers, in which comparatively large pressures are employed, make use of quartz crystals. A quartz crystal is more rugged than a Rochelle-salt crystal.

Figure 11-7 depicts the plan of a capacitor microphone and a circuit employing a crystal microphone. It comprises an air-dielectric capacitor, one plate of which is a diaphragm. Plate separation is typically 0.001 in. $B+$ voltage is applied across the plates; in turn, vibrations of the diaphragm vary the capacitance value, and the voltage drop across the series resistor varies according to the formula

$$e = E_{BD} - \frac{Q}{C_t} \tag{11-3}$$

where e is the instantaneous voltage across the microphone resistor, Q is the coulomb charge stored in the capacitor, and C_t is the capacitance value of the microphone at the given instant.

Since the internal reactance of a capacitor microphone is very high, the value of the microphone resistor must also be very high in order to obtain appreciable audio-output voltage. As depicted in Fig. 11-7, a value of 10 MΩ is typical. It is also advantageous to make the value of the grid-leak resistance in the preamplifier as high as possible; however, a value of 3 MΩ is about the tolerable maximum in view of residual grid current which can cause erratic operation. The audio voltage applied to the grid is in the order of millivolts.

Because of the low-level output, it is not practical to use a cable between a capacitor microphone and its preamplifier; instead the preamp is commonly installed in the microphone housing or stand. Residual hum voltage is a significant design consideration, and it is customary to energize the heater of the preamp tube from a d-c source. To obtain good frequency response over the audio range, the diaphragm of a capacitor microphone is stretched by its mounting; a stress is provided that develops a mechanical resonant frequency in the range from 5 to 10 kHz.

Figure 11-8 depicts the plan of a velocity or ribbon-type microphone. A thin corrugated metal ribbon is suspended in a strong magnetic field. Thus, the microphone utilizes generator action; a voltage is induced in the ribbon when it vibrates in response to incident sound waves. We will recognize that the internal impedance of a velocity microphone is very low; therefore, a microphone transformer is generally used to step up the microphone impedance and match it to the

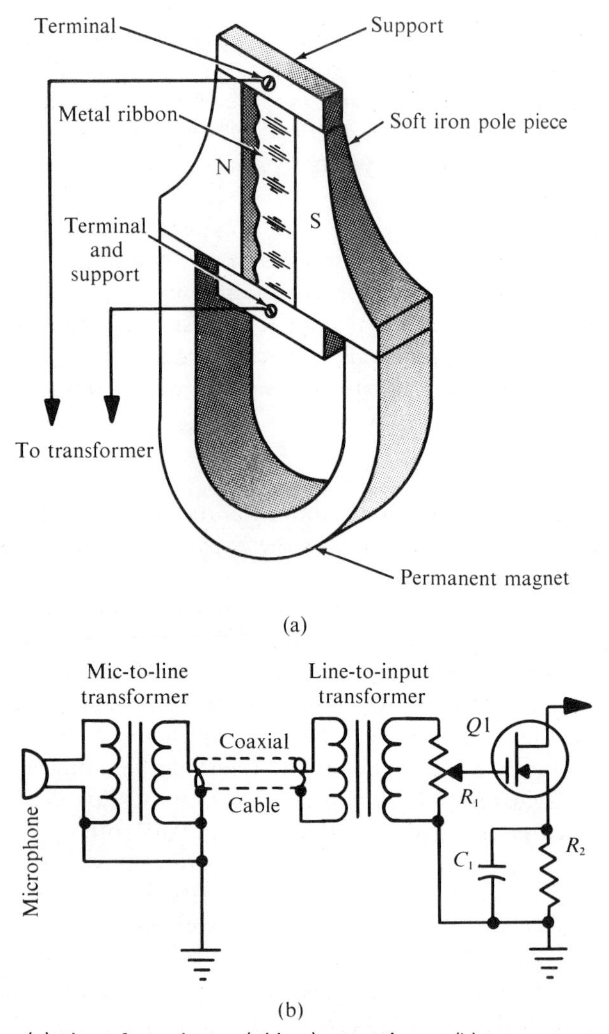

Figure 11-8. (a) Plan of a velocity (ribbon) microphone; (b) Input circuit.

characteristic impedance of the cable. In turn, a line transformer is used to match the cable impedance to the high impedance of the input circuit in the amplifier. In Fig. 11-8(b), an audio signal amplitude of approximately 0.04 v a-c is applied to the gate. To obtain good frequency response over the audio range, the mass of the ribbon in the microphone is chosen to provide a resonant frequency below the range of speech and music waveforms.

A dynamic microphone arrangement is depicted in Fig. 11-9. It can be compared with a ribbon-type microphone and also with a small

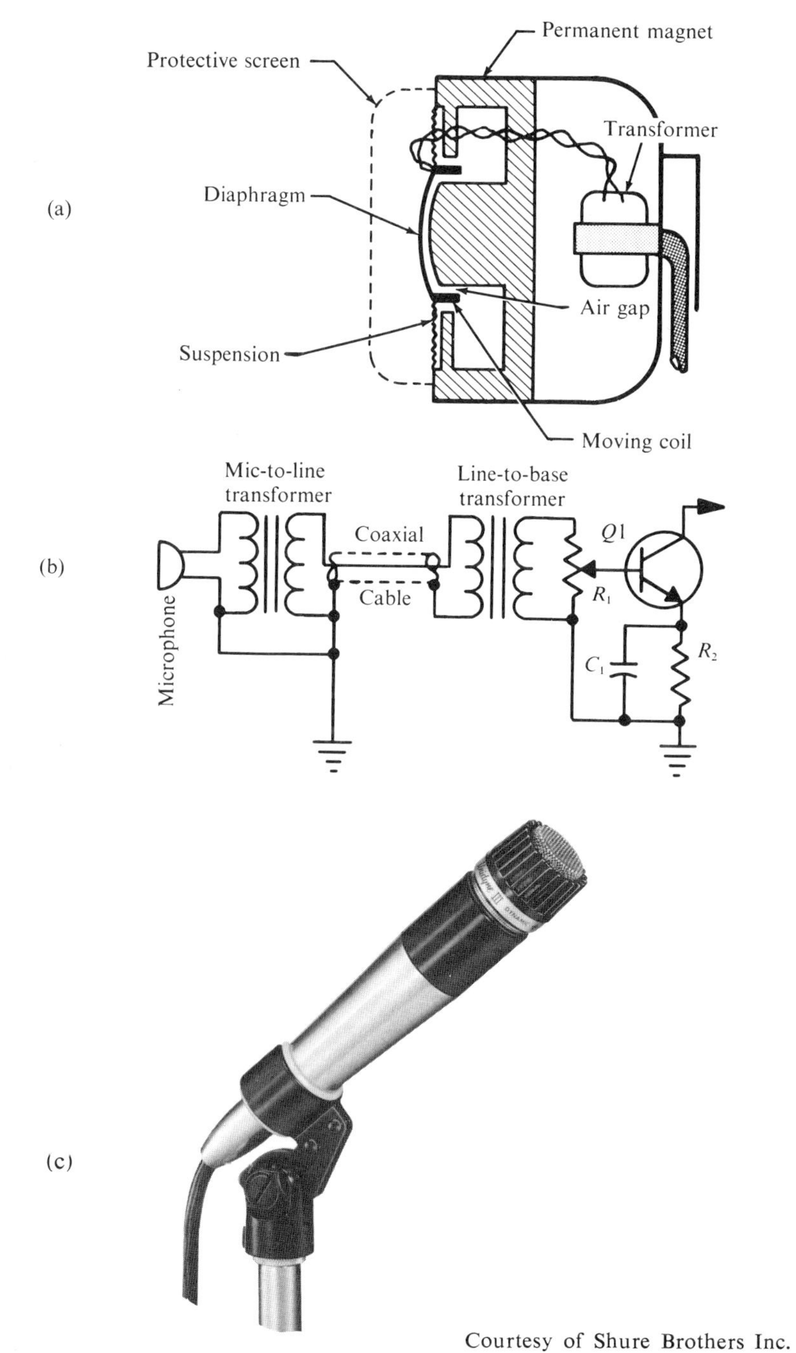

Courtesy of Shure Brothers Inc.

Figure 11-9. (a) Plan of a dynamic microphone; (b) Input circuit; (c) Appearance of a dynamic microphone.

loudspeaker. We perceive that generator action is employed. The damper chambers behind the diaphragm are designed to provide a satisfactorily uniform response over the audio-frequency range. However, a ribbon-type microphone will usually provide somewhat better fidelity. A dynamic microphone has the advantage of greater audio-output voltage. In the first analysis, with other things being equal, a moving coil with 50 turns will provide 50 times the output voltage that a ribbon provides. It is evident that a dynamic microphone has a low value of internal impedance, although its impedance is substantially greater than that of a ribbon microphone.

11–4. Loudspeakers and Earphones

An audio amplifier is generally used to energize a pair of earphones, or a loudspeaker. Hence, we will briefly review the operation of these devices. The earliest form of telephone receiver was very simple and utilized a permanent bar magnet, as seen in Fig. 11–10(a). A soft-iron pole piece at the end of the permanent magnet operates as the core for an electromagnet. Efficiency was improved by use of a horseshoe permanent magnet, with two soft-iron pole pieces and two

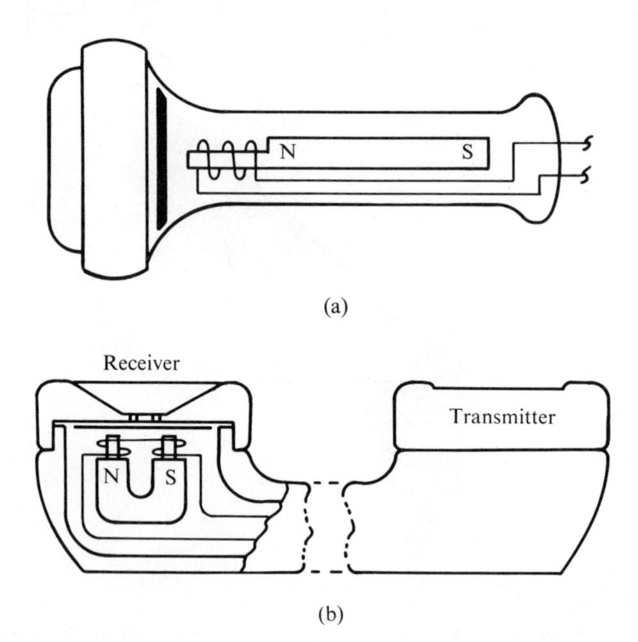

Figure 11-10. (a) Original form of telephone receiver; (b) Improved form.

electromagnets, as seen in Fig. 11-10(b). Audio-frequency current is passed through the coil windings, and in turn the thin iron diaphragm is vibrated and radiates sound waves into the surrounding air.

The permanent magnet is important, not only because it increases

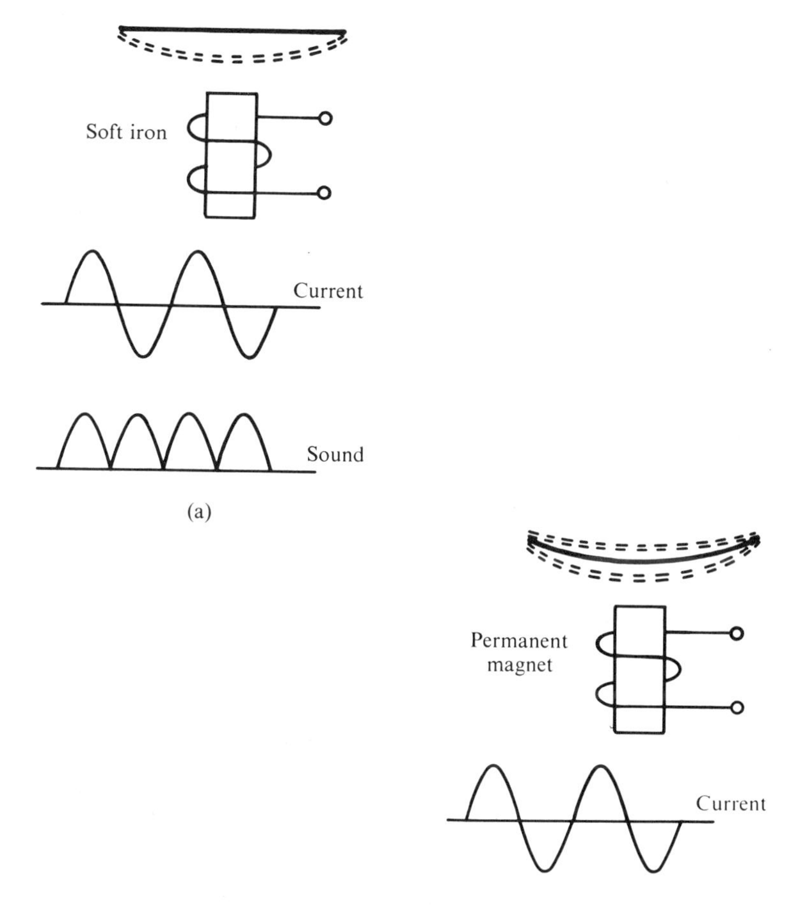

Figure 11-11. (a) Double-frequency distortion produced in absence of a permanent-magnet field; (b) Elimination of double-frequency distortion by presence of a permanent-magnet field.

the amplitude of diaphragm vibration when audio-frequency currents flow through the windings, but also because it permits the diaphragm to follow both the positive and negative half-cycles of the audio frequency waveform. In other words, as shown in Fig. 11-11, the permanent magnet prevents the diaphragm from vibrating at twice the frequency of the audio signal. Insofar as the electromagnet itself is concerned, the diaphragm will be attracted to the core regardless of the direction of current flow. If an a-c current is passed through an electromagnet, attraction is exerted on each half cycle, or, the attractive force has twice the frequency of the a-c current. On the other hand, if a permanent magnetic field is provided, the electromagnet merely increases or decreases the flux field of the permanent magnet, and the diaphragm then vibrates at the same frequency as the applied a-c current.

We will recognize that a telephone receiver will "work backwards," and generate an audio-frequency signal if we speak into the receiver. This generator action (the reverse of the former motor action) causes an audio-frequency voltage to be induced in the electromagnet windings. Accordingly, a telephone receiver also operates as a telephone transmitter.

Various types of telephone receivers have been used as loudspeakers in the past. Their chief disadvantage is that the diaphragm has a limited range of vibration, and the unit tends to "rattle" and distort strong audio signals. If the diaphragm is moved back farther from the pole pieces, the unit does not have good sensitivity. The *dynamic*

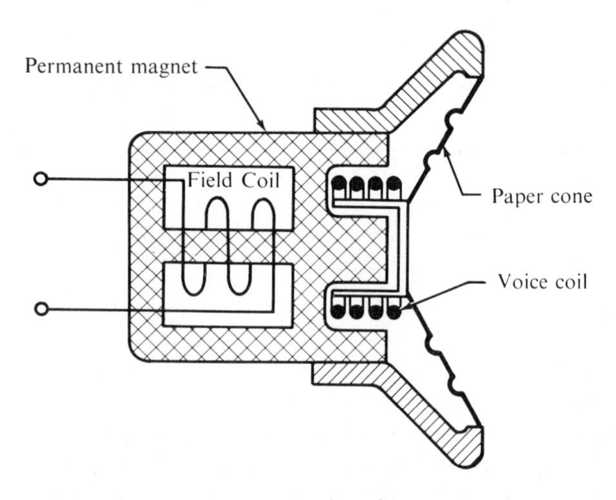

Figure 11-12. Dynamic loudspeaker construction.

speaker depicted in Fig. 11-12 eliminates this disadvantage; it operates by motor action. Magnetic flux produced by voice currents reacts with the strong field flux to move the voice coil and its attached cone. The field is established by a strong permanent magnet. Older designs employed a field coil that was often also used as an inductor in the power-supply filter section of the receiver. However, permanent field magnets have supplanted electromagnets in loudspeaker design. One disadvantage of an electromagnet field is that the supply voltage must be pure d-c to avoid residual hum. Although various expedients can be and have been widely used in the past to cancel residual hum, the *PM* construction is preferred because it leads to manufacturing economies.

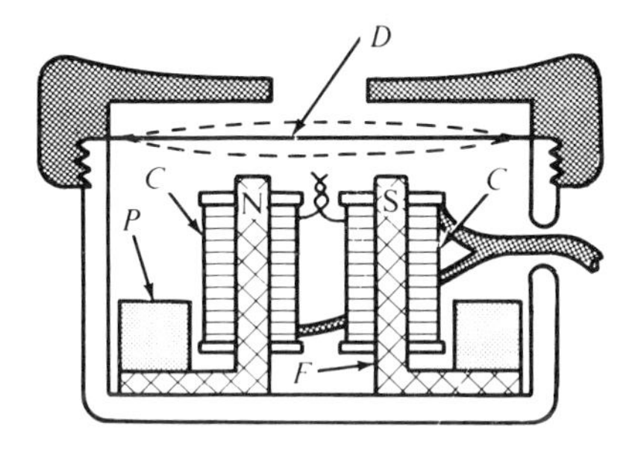

Figure 11-13. Construction of an earphone.

Figure 11-13 illustrates the construction of a conventional earphone. A horseshoe-type permanent magnet *P* provides magnetomotive force to the pair of soft-iron pole pieces *E* and *F*. The magnetic circuit is completed through an air gap to diaphragm *D*. Each pole piece is energized by a coil comprising approximately 5,500 turns of No. 40 gauge copper wire. Thus, the complete electromagnetic circuit comprises about 11,000 turns. Two earphones are customarily used on a headband. Together, the pair of earphones has a series resistance of approximately 3,000 Ω. At 1 kHz, the a-c resistance of the headset is about 4,000 Ω, and the reactance is approximately 11,000 Ω. Thus, the impedance of a conventional headset at 1 kHz is in the order of 12,500 Ω.

The foregoing a-c resistance and reactance values are not constant, but increase with the value of a-c current flow through the headset. For example, the values noted above are typical for a current flow of 1 μa. At a current flow of 1 ma, the a-c resistance doubles to 8,000 Ω, and the reactance increases to about 14,000 Ω. Distributed capacitance is high, and the coil windings form low-Q resonant circuits. The chief resonant action, however, is due to the motional impedance of the diaphragm; most earphones exhibit a pronounced mechanical resonance in the vicinity of 1 kHz. However, when an earphone is pressed against the listener's ear, this mechanical resonance peak is largely damped. Suitable design of air chambers in an earphone provides reasonably uniform frequency response over the audio-frequency range when the receiver is pressed against the listener's ear. The efficiency of a headset is very poor, and is in the order of 1 per cent. In other words, only 1 per cent of the available electrical power is applied to the eardrum as acoustical power.

11-5. Phasing of Loudspeakers

When more than one loudspeaker is used, as in Fig. 11-2 or Fig. 11-14(b), it is essential that the loudspeakers be operated in phase with each other. That is, if the cone of one speaker moves inward at the same time that the cone of the other speaker moves outward, more or less cancellation of sound output will occur. Therefore, a test of each speaker is made at the time of installation to make certain that the system operates in phase. The simplest way to check speaker phasing is to connect a dry cell across the voice-coil terminals and observe whether the cone is moved in or out. Then, the next speaker is checked in the same manner observing the same battery polarity. For example, the first speaker cone might move in when a positive d-c voltage is applied to one of the voice-coil terminals. This positive terminal is noted. The second speaker is checked, and we note the positive terminal for inward movement of the cone. Finally, the voice coils are connected in series-aiding (positive to negative) for proper operation. Of course, if the voice coils were to be connected in parallel, the positive voice-coil terminals would be connected together, and the negative terminals would be connected together.

It is just as essential that speakers that have built-in output transformers be properly phased. In such case, we commonly connect the primaries of the output transformers in parallel. If the transformers have good low-frequency response, we can use a battery test, as previously explained, to identify the primary terminals that are to be connected together. However, difficulty may be encountered in case the cone moves so quickly that we are not sure whether it moves in or out when the battery voltage is applied. However, if a d-c voltmeter is connected across the voice-coil terminals of each speaker in turn, a certain indication is obtained. The pointer "kicks" up or down on the scale when the battery voltage is applied to the primary of the output transformer.

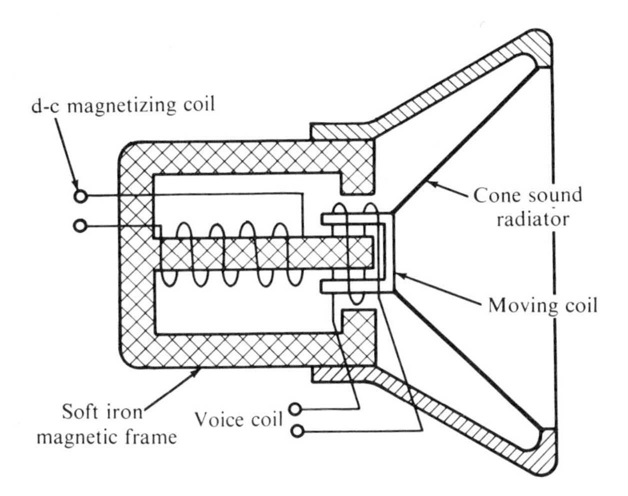

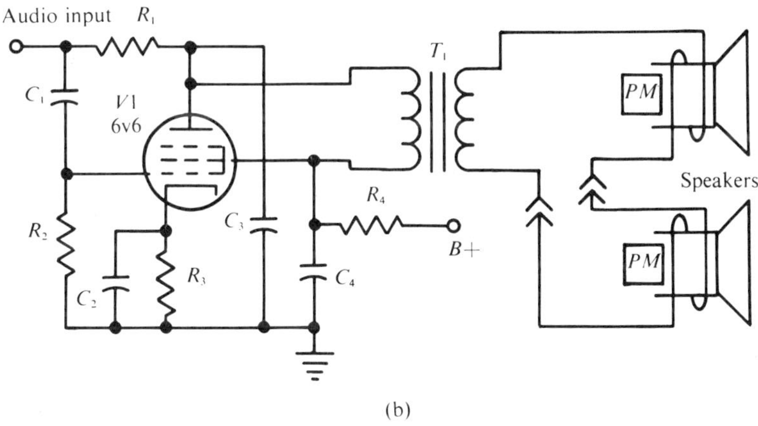

(b)

Figure 11-14. (a) Plan of a dynamic loudspeaker with an electromagnetic field; (b) A pair of permanent-magnet dynamic speakers matched to the plate resistance of an audio-output tube.

We simply note the positive transformer terminal that provides an upscale "kick," for example, for each speaker. In turn, we connect the positive primary terminals together and connect the negative primary terminals together for parallel operation from an audio amplifier. The Electronic Industries Association (EIA) recommends public-address amplifiers that provide a constant-voltage 70.7-v output. The output transformers have their primaries connected in parallel across the 70.7-v output line. Impedance matching, in the conventional sense, is neglected in this constant-voltage system. The output transformers commonly have a primary winding rated in watts, and a secondary winding rated in ohms. We select a watt rating to match the power rating of the speaker, and we select a resistance rating to match the voice-coil impedance rating.

The acoustics of a loudspeaker, or of an array of speakers in a public-address system, is an extensive topic that cannot be covered here. We will merely note that speakers are mounted in baffles (sound boards) that tend to isolate sound radiation from the rear surface of the cone. This rear radiation is 180° out of phase initially with the front radiation. Special speaker cabinets are often utilized to bring the rear radiation more or less in phase with the front radiation so that rear radiation is not lost. In this way, operating efficiency is improved.

When two speakers are mounted in the same cabinet, one speaker is often designed to handle low audio frequencies and is called a "woofer." The other speaker is designed to handle high audio frequencies and is called a "tweeter." Low-frequency speakers have much larger cones than high-frequency speakers. Filter networks are often employed between dual speakers and the audio amplifier to separate the low- and high-frequency audio ranges. These are called crossover networks. You will have an opportunity to learn about these topics in your advanced courses.

11–6. The 70.7-volt Speaker-Matching System

The Electronic Industries Association (EIA) has established a constant-voltage system of audio power distribution called the 70.7-volt System. It provides a simple way to connect method matching a number of speakers to an amplifier. This system employs 70.7-v matching transformers that have their primary terminals marked in watts, and secondary terminals marked in ohms. To use this system, we proceed according to the following steps.

1. Determine the number of watts required by each speaker.
2. Add these values to find the total power demand, and select a 70.7-v amplifier with a rated power output at least equal to this demand.
3. Select 70.7-v transformers with appropriate primary wattage ratings.
4. Connect the primary terminals in parallel with the 70.7-v line from the amplifier output. A primary mismatch up to 25 per cent is permissible.
5. Connect each secondary to its speaker, using the matching ohms tap.

6. If the matching transformers should be rated in impedance values, the primary wattage may be calculated as follows.

$$Z_p = \frac{70.7^2}{p} \qquad \textbf{(11–4)}$$

where Z_p is the rated primary impedance, and p is the number of watts required by the speaker.

From Formula 11–4 we will observe the following.

1 w corresponds to 5,000 Ω Z_p
2 w corresponds to 2,500 Ω Z_p
5 w corresponds to 1,000 Ω Z_p
10 w corresponds to 500 Ω Z_p

An example is shown in Fig. 11–15. The 6-w speaker has a 4-Ω voice coil, and the paralleled 10-w speakers have 8-Ω voice coils. In

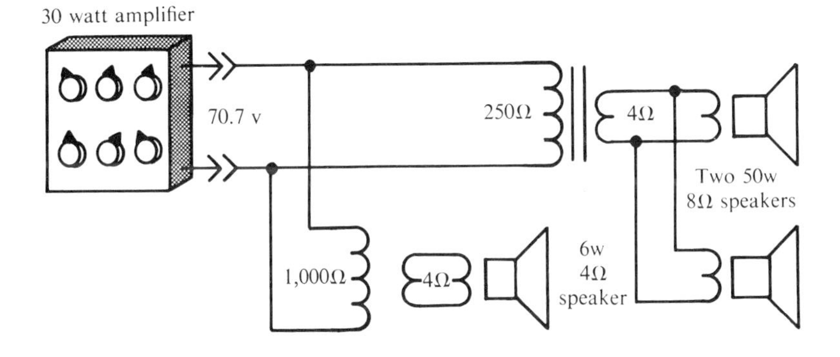

Figure 11–15. Illustration of a speaker matching network.

turn, the total power demand is 26 w, and the amplifier would be rated for about 30 w. From Formula 11–4, the primary impedance for the 6-w transformer will be 833 Ω; a 1,000-Ω impedance could be used. For 20-w speaker combination, the transformer impedance will be 250 Ω.

SUMMARY

1. Transducers are devices that change electrical energy into some other form of energy.

2. The transformer that is used to match the impedance of a microphone to that of a preamplifier is usually a step-up transformer.

3. The resistance characteristic of a carbon granule microphone is nonlinear.

4. Carbon has a nonlinear resistance characteristic which may produce even harmonic distortion in an audio signal. This type of distortion is minimized by using two carbon cups in a double-button microphone to cancel out even harmonics.

5. A quality single-button carbon microphone will provide about 250 mv a-c across 75 Ω, and a quality double-button carbon microphone will provide about 50 mv a-c across 200 Ω.

6. A crystal microphone must be coupled to the preamplifier by a shielded cable to prevent the pickup of stray noise signals.

7. Crystal microphones have a poor output level at high audio frequencies which can be compensated in the microphone amplifier.

8. A preamplifier is usually installed in the housing of a capacitor pickup microphone because of the low level output voltage from this type of microphone.

9. The internal impedance of a velocity- or ribbon-type microphone is very low and must be matched to the amplifier by a transformer.

10. A dynamic microphone can be compared to the ribbon-type microphone or to a speaker, in that generator action is employed.

11. The permanent magnet in an earphone or speaker prevents a doubling of the audio frequency and increases the output sound level.

12. When more than one loudspeaker is used at the output of an amplifier, it is essential that the loudspeakers be operated in phase with each other.

13. The purpose of the permanent magnet in an earphone is to prevent the diaphragm from oscillating at twice the signal frequency and to increase the amplitude of diaphragm vibrations.

14. A telephone receiver speaker may be used to produce audio-frequency voltages by speaking into the receiver.

15. The chief disadvantage of using a telephone receiver as a speaker is that the diaphragm has a limited range of vibration.

16. An electrodynamic speaker with a field coil has disadvantages over a *PM* speaker entailing higher cost and power supply hum.

Questions

1. What is the function of a microphone as a transducer?
2. What is the function of a speaker as a transducer?
3. Discuss the purpose of a diaphragm in a carbon microphone.
4. What is the average impedance of a carbon microphone?
5. Explain the operating principles of a carbon granule microphone.
6. What is the maximum efficiency of a matching transformer?
7. What is the cause of even-harmonic distortion in a carbon microphone?
8. How is harmonic distortion minimized in carbon microphones?
9. What limits the maximum value of d-c current in a carbon microphone?
10. Compare the single- and double-button carbon microphones.
11. Compare a crystal microphone to a carbon microphone.
12. Compare the capacitor microphone to both the carbon and the crystal microphones.
13. Why is an amplifier stage usually placed in the housing of a capacitor-type microphone?
14. What is the purpose of applying stress to the diaphragm of a capacitor microphone?
15. What is the operating principle of a dynamic microphone?
16. Compare the ribbon-type microphone to the dynamic microphone.
17. A 75-Ω microphone is coupled through an input transformer to a 1,000-Ω transistor amplifier. What is the transformer turns ratio?
18. What is the purpose of the permanent magnet in an earphone?
19. Discuss methods which can be employed to assure that loudspeakers are operating in phase.
20. What is the purpose of the permanent magnet in a speaker?
21. Explain the principle of operation of a crystal telephone transmitter.
22. What is the chief disadvantage of using a telephone receiver as a speaker?
23. What are the disadvantages of an electrodyamic speaker with a field coil in comparison to the *PM* speaker?
24. What is the effective impedance of a conventional earphone at 400 Hz?
25. What is the order of efficiency of a headset?

CHAPTER 12

Distortion Analysis

12–1. Introduction

Previous discussion has introduced us to the basic characteristics of nonlinear harmonic distortion, frequency distortion, and phase distortion. As would be anticipated, various other types of distortion are also encountered in vacuum-tube and transistor amplifier circuits and systems. We will start with an analysis of nonlinear distortion, since this is one of the simplest and most common forms of distortion. All vacuum tubes and transistors have transfer characteristics that are more or less nonlinear; for example, the $e_g - i_b$ characteristic of a triode is nonlinear, as depicted in Fig. 12–1(a). Similarly, the dynamic transfer characteristic of a transistor is nonlinear, as shown in Fig. 12–1(b). As the value of the plate-load resistance is increased, the resultant transfer characteristic for a triode vacuum tube becomes more nearly linear; however, complete linearity can only be approached.

Figure 12–2 provides a visualization of amplitude distortion caused

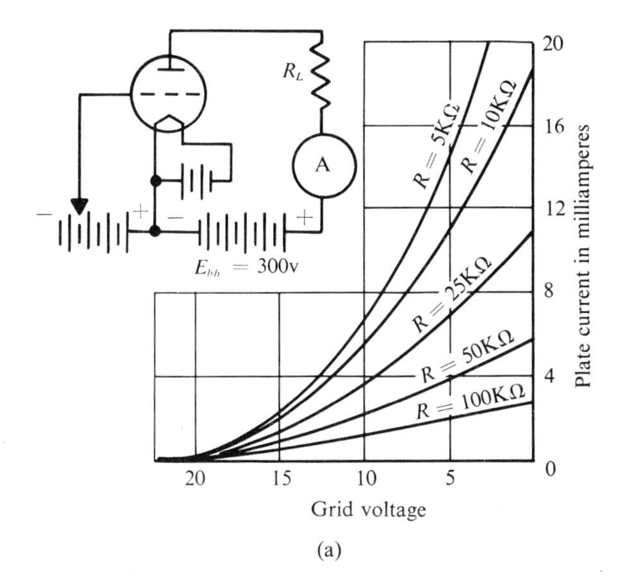

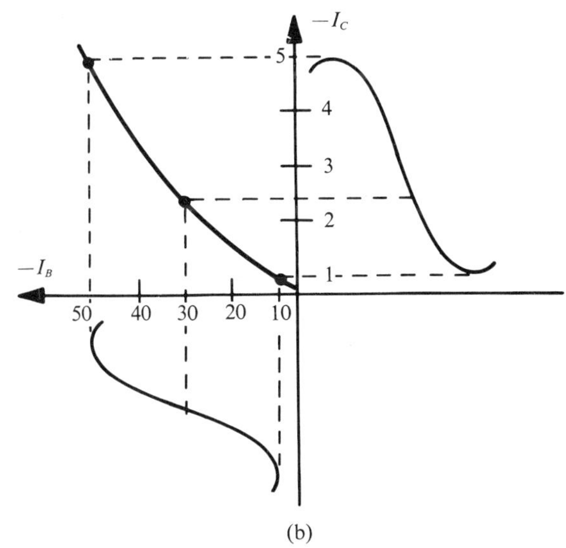

Figure 12-1. (a) Dynamic transfer characteristic curves for a triode with various plate-load resistances; (b) Dynamic transfer characteristic for a transistor in the common emitter configuration.

by a nonlinear transfer characteristic, as would be displayed on an oscilloscope screen. We will find that when a sine wave is distorted in this manner, even harmonics are generated in the output waveform, as

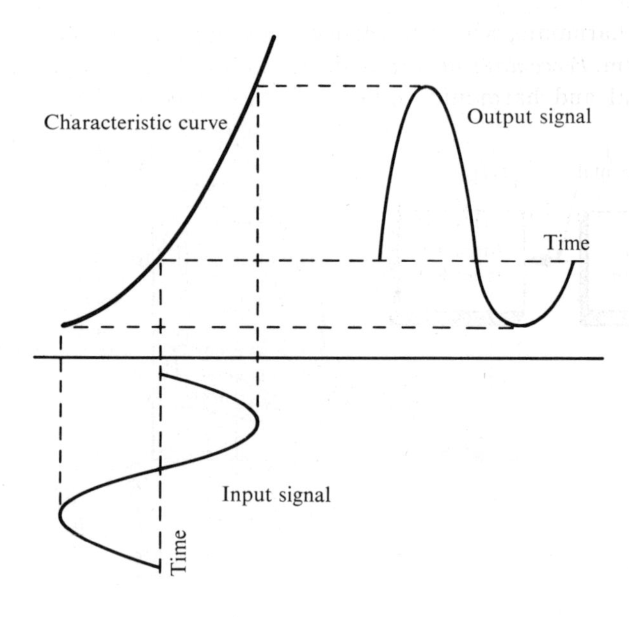

Figure 12-2. Visualization of distortion caused by a nonlinear transfer characteristic.

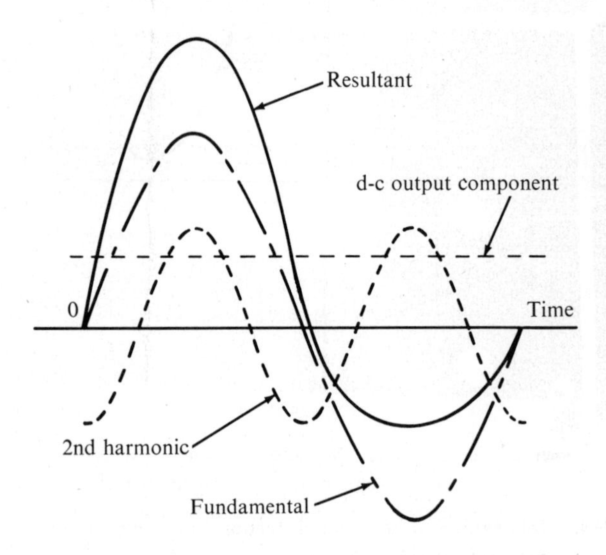

Figure 12-3. Generation of a second harmonic by nonlinear distortion.

depicted in Fig. 12–3. The most prominent even harmonic is the second harmonic, which has twice the frequency of the input signal waveform. *Percentage distortion* denotes the relative amplitudes of fundamental and harmonic voltages. For example, the second harmonic

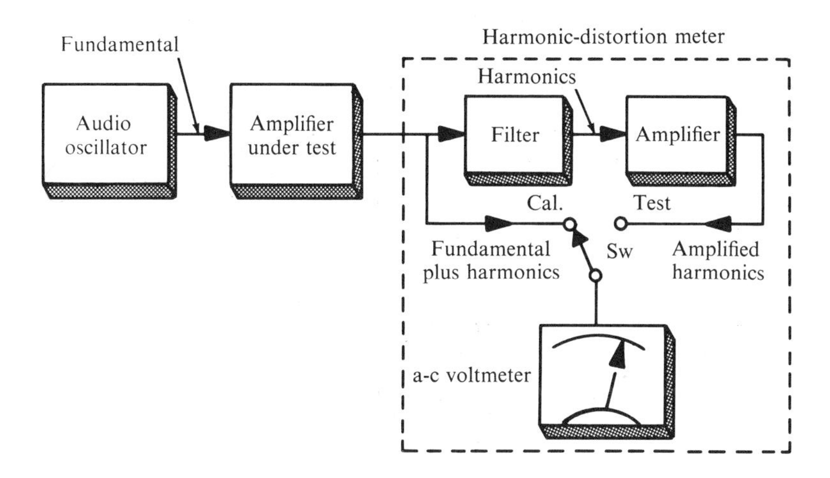

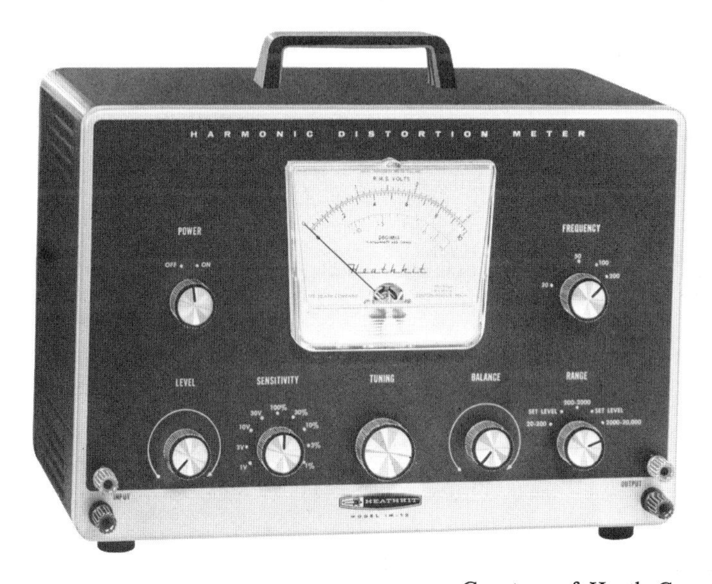

Courtesy of Heath Company

Figure 12–4. (a) Method of harmonic-distortion measurement; (b) View of a typical harmonic-distortion meter.

in Fig. 12–3 has 31 per cent of the fundamental amplitude. Therefore, we say that there is 31 per cent second-harmonic distortion present.

Although there is no industry standard that defines "high-fidelity" system performance, it is generally agreed that harmonic distortion should not exceed 1 per cent. Harmonic distortion is commonly measured as shown in Fig. 12–4. An audio oscillator with a low-distortion

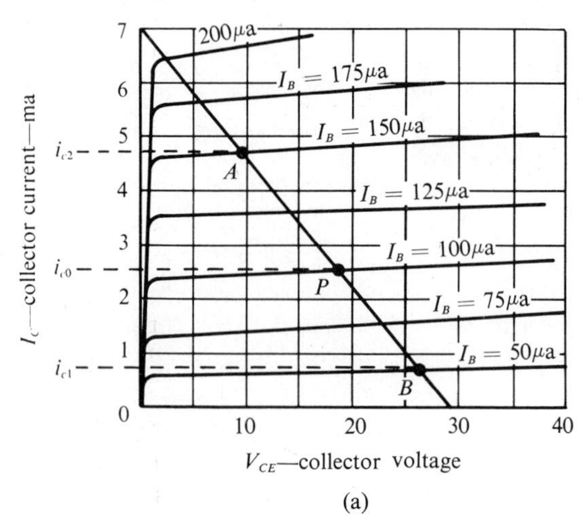

(a)

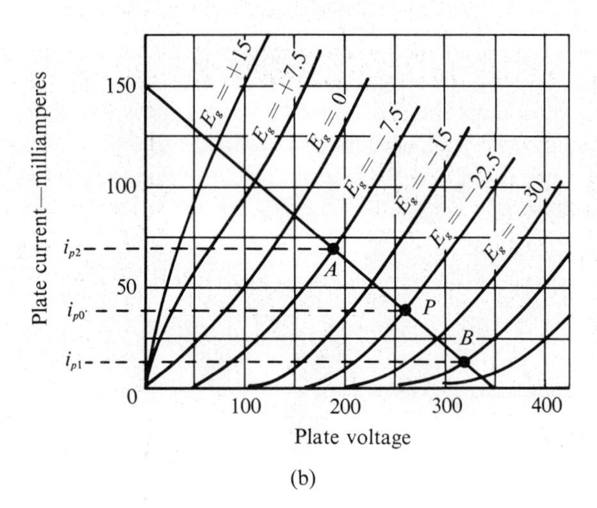

(b)

Figure 12–5. Illustration of harmonic distortion. (a) From the $V_C - I_C$ characteristics of a junction transistor; (b) From the $E_p - I_b$ characteristics of a vacuum tube.

sine-wave output is used as a signal source. The audio-oscillator signal is passed through an amplifier, which is connected in turn to a harmonic-distortion meter as depicted in Fig. 12–4(a). First, the signal output from the amplifier is indicated by the a-c voltmeter; the operator adjusts the meter circuit for an indication of 100 per cent on the scale. Thus, a reference level is established. Next, the output from the amplifier is passed through a filter that eliminates the fundamental component; in turn, any harmonics that may be present energize the meter, and their amplitude is indicated as percentage harmonic distortion on the scale.

The method of harmonic-distortion measurement shown in Fig. 12–4(a) does not denote the relative amplitudes of fundamental and harmonic voltages; instead it denotes the relative amplitudes of total output and harmonic voltages. A more elaborate arrangement is required to measure the relative amplitudes of fundamental and harmonic voltages. However, the simple arrangement depicted in Fig. 12–4 provides reasonably accurate measurements and is in wide use by shops that make high-fidelity amplifier and system tests. In summary, the simple method of distortion measurement compares the rms value of the second harmonic in Fig. 12–3 with the rms value of the resultant.

Generation of harmonic distortion can be observed and analyzed from the path of operation along the load line on a chart of collector-family characteristics, as shown in Fig. 12–5(a). The operating point is depicted at P. If the base signal swings 50 μa from P in either direction, the resulting collector-current flow varies from i_{c1} to i_{c2}. Because the transistor is nonlinear, the collector-current change from i_{co} to i_{c2} is greater than the collector-current change from i_{c0} to i_{c1}. This inequality in peak-current values corresponds to the amplitude distortion

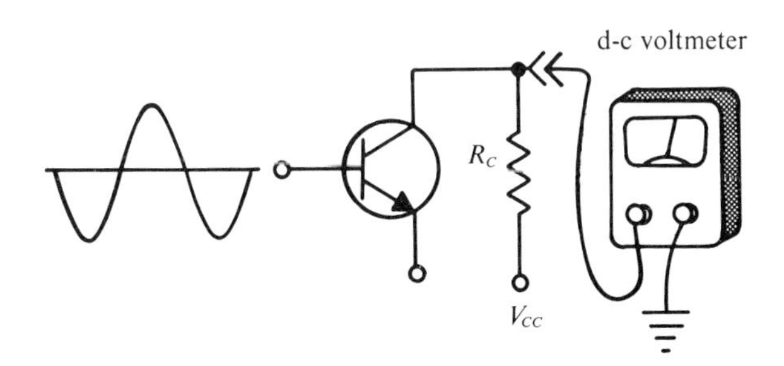

Figure 12–6. The d-c voltmeter measures any shift in collector potential when an a-c signal is applied.

shown in Fig. 12–3. Thus, even harmonics are generated in the collector-circuit. Evidently, the greater the base current swing in Fig. 12–5, the greater the amplitude (harmonic) distortion. The same considerations apply to operation on a plate family of tube characteristics, as illustrated in Fig. 12–5(b). Therefore, amplifiers are rated for harmonic distortion at maximum signal output. We shall return to power-output versus harmonic-distortion relations subsequently.

It is evident from Fig. 12–3 that this type of amplitude distortion is accompanied by a d-c component. In other words, the amplifier device operates in part as a rectifier, because it is a nonlinear device. This d-c component can be measured as a collector-potential shift, as shown in Fig. 12–6. We measure the collector voltage first with no a-c signal applied to the base. Then, a chosen value of a-c signal drive is applied; if the collector voltage changes in value, we know that amplitude distortion of the type depicted in Fig. 12–3 is occurring. This test is most informative when the base-signal drive is gradually advanced from zero to the maximum rated output of the stage, because the stage operating characteristics change with the signal level.

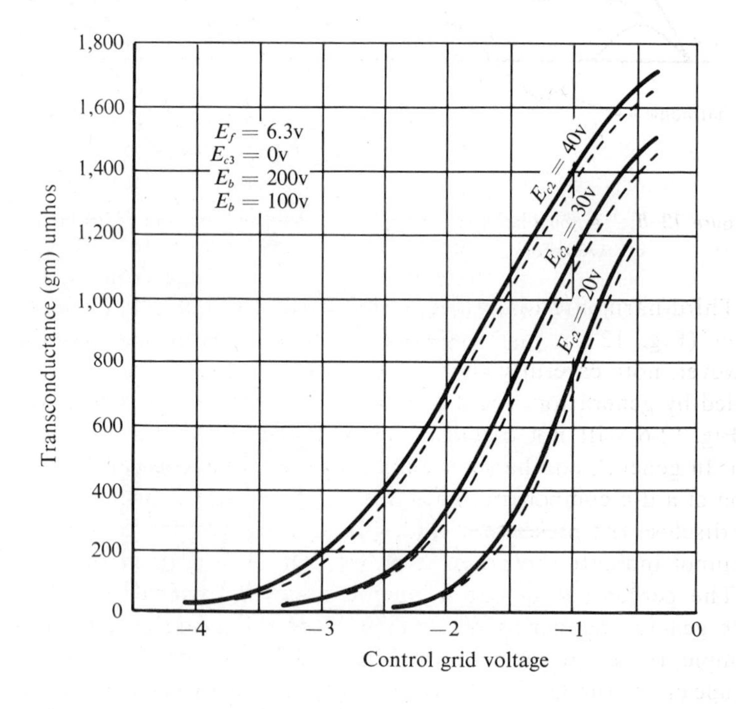

Figure 12–7. The transconductance characteristic has opposite curvatures at either end.

The alert student will perceive that other types of amplitude distortion can occur in vacuum-tube and transistor amplifiers. For example, Fig. 12–7 depicts the transconductance characteristics of a pentode. Compare these curves with the transistor curves shown in Fig. 12–1(b). The characteristics have opposite curvatures at either end. If the tube is driven strongly, so that operation extends into both curved intervals, it is apparent that both the top and the bottom of the output waveform will be flattened. This is an example of *odd-harmonic* distortion; if a third harmonic is added to the fundamental, as shown in Fig. 12–8, the basis of top-and-bottom flattening is evident. Although other odd harmonics are generated by symmetrical flattening (symmetrical compression) of a sine waveform, the third harmonic is most prominent, as set forth by Fourier series for nonsinusoidal waveforms.

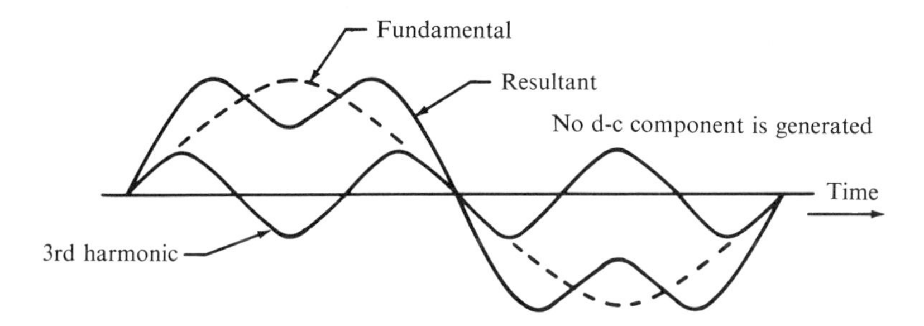

Figure 12–8. A third harmonic produces symmetrical flattened tops in a sine waveform.

Third-harmonic distortion is indicated by a harmonic distortion meter (Fig. 12–4), just as second-harmonic distortion is indicated. However, note carefully that third-harmonic distortion is *not* accompanied by generation of a d-c component; therefore, the test depicted in Fig. 12-6 will not disclose the presence of third-harmonic distortion. In general, odd-harmonic distortion is not accompanied by generation of a d-c component. Thus, although the test shown in Fig. 12–6 can disclose the presence of 2nd, 4th, and 6th harmonics, for example, it cannot indicate the presence of 3rd, 5th, or 7th harmonics.

The percentage of odd-harmonic distortion is calculated in the same general manner as percentage of even-harmonic distortion. For example, the third harmonic in Fig. 12–8 has 39 per cent of the amplitude of the fundamental. Accordingly, the third-harmonic distortion is 39 per cent. As noted previously, if the simple harmonic-distortion measuring arrangement of Fig. 12–4 is employed, the meter will indicate the percentage of third-harmonic distortion with respect to the resultant waveform. In summary, if only odd-harmonic distortion occurs, the voltmeter test depicted in Fig. 12–6 would lead to the false conclusion that the stage is not distorting the signal; however, a harmonic-distortion meter will indicate the presence of the odd-harmonic distortion.

On the basis of this introductory discussion, it might be assumed that even-harmonic distortion is always accompanied by generation of a d-c component. However, this assumption would be incorrect in several cases. For example, let us consider the second-harmonic distortion depicted in Fig. 12–9. Observe that the negative half cycle of the resultant is a mirror image of the positive half cycle. Since the area

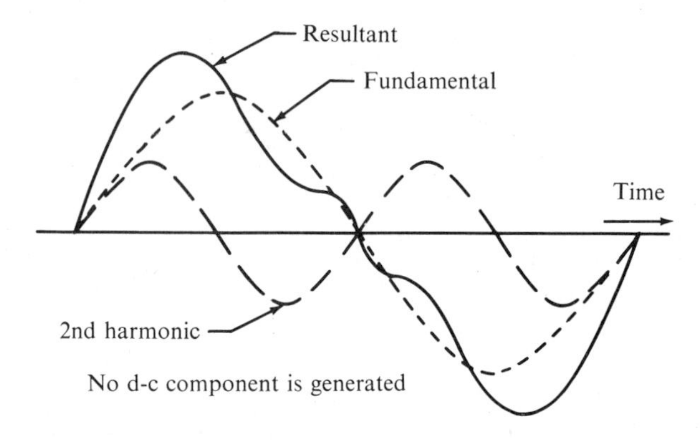

Figure 12–9. This form of second-harmonic distortion is not accompanied by generation of a d-c component.

under the positive half cycle of the resultant is equal to the area under the negative half cycle, it is apparent that no d-c component is generated. Therefore, the test depicted in Fig. 12–6 cannot disclose the presence of this second-harmonic distortion. On the other hand, a harmonic distortion meter will indicate the percentage distortion that is present.

Compare the phase relations of fundamental and second harmonic in Figs. 12–9 and 12–3. We observe that the phase difference is zero in Fig. 12–9, but that the phase difference is 30° with respect to the fundamental in Fig. 12–3. Accordingly, we draw the important conclusion that the magnitude of the d-c component generated by even-harmonic distortion is dependent upon the harmonic phase relation with the fundamental. By way of contrast, odd-harmonic distortion cannot generate a d-c component, regardless of phase relations because of symmetry relations. Thus, the test depicted in Fig. 12–6 is useful, although it is not conclusive. A harmonic-distortion meter provides the only definitive test for amplitude distortion.

An amplifier often generates both even and odd harmonics. For example, Fig. 12–10 illustrates the resultant of a fundamental, second harmonic, and third harmonic. The second harmonic is somewhat out of phase with the fundamental. In turn, the positive half cycle of the

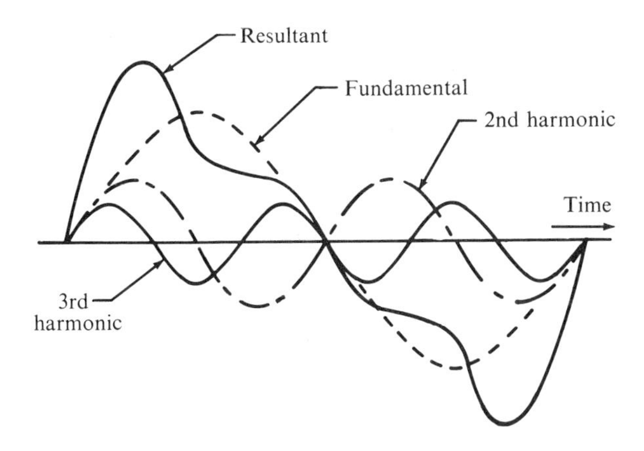

Figure 12–10. Typical resultant of both second- and third-harmonic distortion.

resultant is not quite a mirror image of the negative half cycle. Accordingly, a small d-c component is generated by the second harmonic. Of course, the third harmonic cannot generate a d-c component. Let us investigate the percentage distortion that is present in the resultant. The second harmonic alone produces approximately 41 per cent distortion. The third harmonic alone produces approximately 18 per cent distortion. The *total harmonic distortion* is defined as the square root of the sum of the squares of the individual distortions. Thus,

$$\text{Total distortion} = \sqrt{H_2^2 + H_3^2} \qquad \textbf{(12-1)}$$

where H_2 represents the value of second-harmonic distortion, and H_3 represents the value of third-harmonic distortion.

We conclude, accordingly, that the total percentage distortion is approximately 50 per cent in Fig. 12–10. If we had added the individual percentage distortions arithmetically, we would have obtained a figure of about 59 per cent. This figure, of course, neglects the phase relations among the harmonics and the fundamental. To obtain an average value in view of arbitrary phase relations encountered in practice, we define total distortion in accordance with Formula 12–1.

Harmonic distortion meters are designed to indicate a total distortion that is in reasonable accord with Formula 12–1.

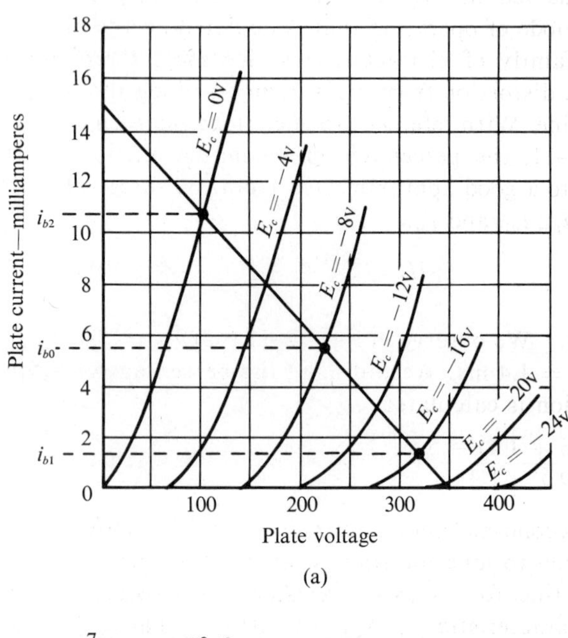

(a)

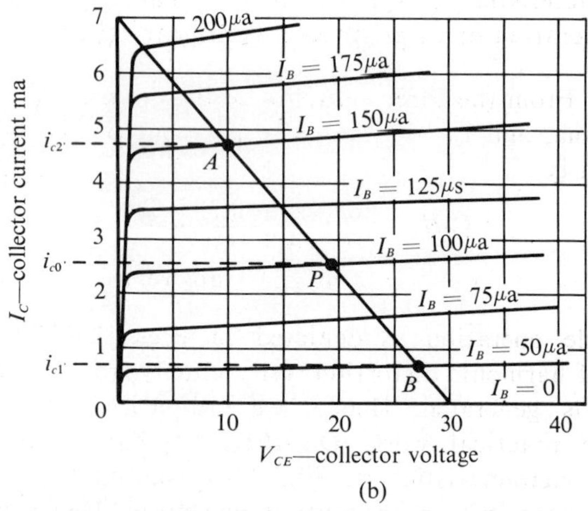

(b)

Figure 12–11. Determination of second-harmonic distortion from triode load line. (a) Triode amplifier; (b) Transistor amplifier.

12-2. Distortion Calculation from Plate Characteristics

We observed in Fig. 12-5 that even-harmonic distortion results from the mode of operation that is depicted on the plate family or the collector family of characteristics. We may, therefore, evaluate the percentage distortion from measurements along the path of operation on a load line. With reference to the triode plate characteristics shown in Fig. 12-11, the percentage of second-harmonic distortion may be evaluated to a good approximation from the three values of plate-current flow, i_{p0}, i_{p1}, and j_{p2}.

$$\%H_2 = \frac{0.5(i_{p2} + i_{p1}) - i_{p0}}{i_{p2} - i_{p1}} \times 100 \qquad \textbf{(12-2)}$$

Example 1. We observe in Fig. 12-11 that $i_{p2} = 10.6$ ma, $i_{p0} = 5.4$ ma, and $i_{p1} = 1.4$ ma. Accordingly, the percentage of second-harmonic distortion is calculated

$$\frac{0.5(10.6 + 1.4) - 5.4}{10.6 - 1.4} \times 100 = 6.6\%, \text{ approximately} \qquad \text{(answer)}$$

This second-harmonic analysis may be applied in exactly the same manner to junction-transistor or field-effect transistor amplifiers. The load line for a 4.28 kΩ resistor is constructed on the transistor output characteristics in Fig. 12-11(b). A change of 50 μa above and below the operating point produces a nonlinear collector change.

Example 2. From the three values of collector current: $i_{c0} = 2.6$ ma, $i_{c2} = 4.8$ ma, and $i_{c1} = 0.7$ ma, the per cent of second harmonic distortion is

$$\%H_2 = \frac{0.5(4.8 + 0.7) - 2.6}{4.8 - 0.7}$$

$$= 3.66\% \qquad \text{(answer)}$$

For triode operation as depicted in Figs. 12-11(a) and (b), the percentage of harmonic distortion stems chiefly from the second harmonic that is generated. Hence, we ordinarily neglect the third harmonic in practical work. On the other hand, a pentode has a nonlinearity characteristic (see Fig. 12-7) that contributes an appreciable third harmonic in the output waveform. Hence, we will next investigate the calculation of second-harmonic and third-harmonic distortion from measurements along the path of operation on a pentode load line.

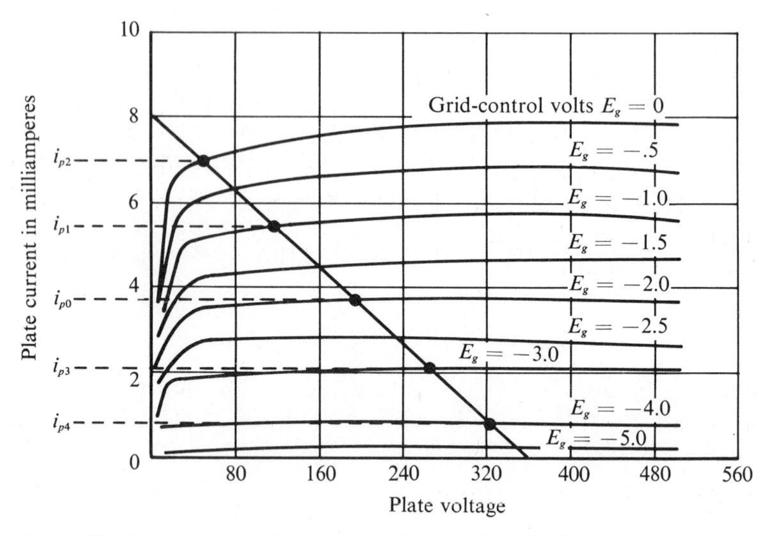

Figure 12-12. Determination of second-harmonic and third-harmonic distortion from a pentode load line.

With reference to Fig. 12-12, a 45,000-Ω resistive load line has been drawn on the plate family of characteristics. The grid bias is −2 v, and the grid signal has a peak amplitude of 2 v. Thus, the plate current varies from i_{p2} to i_{p4}. To calculate the percentage of harmonic distortion, we also determine the plate-current values i_{p1} and i_{p3} that correspond to half-of-peak grid voltages −1 and −3 v. It may be shown by Fourier-series analysis that the percentage of second-harmonic distortion is formulated

$$\%H_2 = \frac{i_{p2} + i_{p4} - i_{p1} - i_{p3}}{i_{p2} - i_{p4} + i_{p1} - i_{p3}} \times 100, \text{ approximately} \qquad \textbf{(12-3)}$$

Also, it may be shown that the percentage of third-harmonic distortion is formulated

$$\%H_3 = \frac{i_{p2} - i_{p4} - 2i_{p1} + 2i_{p3}}{2i_{p2} - 2i_{p4} - i_{p1} + i_{p3}} \times 100, \text{ approximately} \qquad \textbf{(12-4)}$$

Example 3. We observe from Fig. 12-12 that $i_{p1} = 5.4$ ma, $i_{p2} = 6.9$ ma, $i_{p3} = 2.1$ma, and $i_{p4} = 0.9$ ma. Accordingly, the percentage of second-harmonic distortion is calculated

$$\%H_2 = \frac{6.9 + 0.9 - 5.4 - 2.1}{6.9 - 0.9 + 5.4 - 2.1} \times 100$$

$$= 1.1\%, \text{ approximately} \qquad \text{(answer)}$$

Example 4. The percentage of third-harmonic distortion is calculated

$$\%H_3 = \frac{6.9 - 0.9 - 10.8 + 4.2}{13.8 - 1.8 + 5.4 - 2.1} \times 100$$
$$= 4\%, \text{ approximately} \qquad \text{(answer)}$$

The total harmonic distortion in Fig. 12–12 is equal to the square root of the sum of the squares of second-harmonic and third-harmonic percentage distortions.

Example 5. The total harmonic distortion represented in Fig. 12–12 is

$$\%H_T = \sqrt{1.11^2 + 4^2} = 4.2\%, \text{ approximately} \qquad \text{(answer)}$$

We should note here that these analyses apply in exactly the same manner on the plate characteristics, the collector characteristics, and the drain characteristics of vacuum tubes, junction transistors, and field-effect transistors, respectively.

12–3. Intermodulation Distortion

Harmonic distortion is a very useful measure of amplifier performance. However, a certain amplifier might have a low percentage of harmonic distortion and nevertheless produce a sound output that has an objectionable tone quality. When this occurs, the poor tone quality is attributed to *intermodulation distortion.* Analysis of the amplifier operation in such cases will show that the nonlinear characteristic is curved in a manner that a *pair* of tones is amplified with the generation of appreciable *sum-and-difference* frequencies that have an inharmonic relation to the fundamentals. To clarify this amplifier action, we may recall that whenever two frequencies pass through a nonlinear device, each tone is modulated upon the other.

Figure 12–13 illustrates the amplitude modulation of a high-frequency sine wave by a low-frequency sine wave. It may be shown that the modulated waveform depicted in Fig. 12–13 contains the unmodulated sine-wave frequency and a pair of additional signals with frequencies that are equal to the sum and difference of the unmodulated frequency and the modulating frequency. These three frequencies are visualized in Fig. 12–14. Suppose that the unmodulated frequency (which may be called the carrier) has a frequency f_0, and that the modulating voltage has a frequency f_1. Then, a pair of *sideband frequencies* are generated that have values of $f_0 + f_1$, and $f_0 - f_1$.

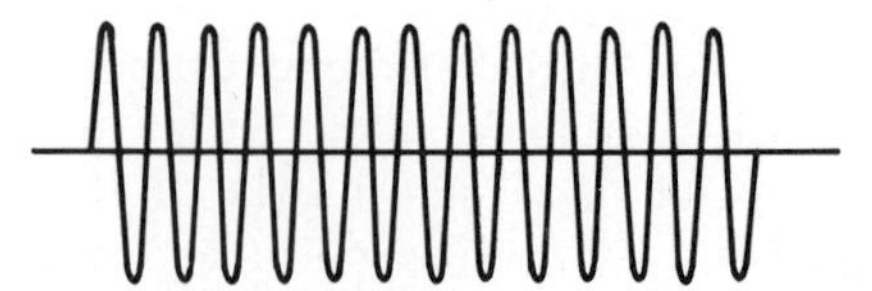

High frequency unmodulated voltage

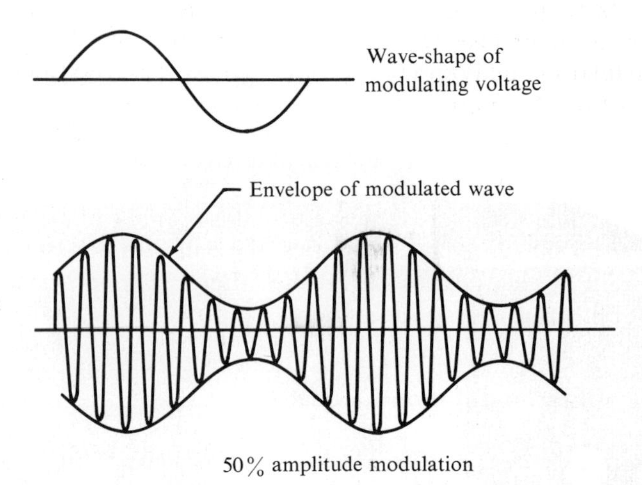

Figure 12–13. **Amplitude modulation of a high-frequency sine wave by a low-frequency sine wave.**

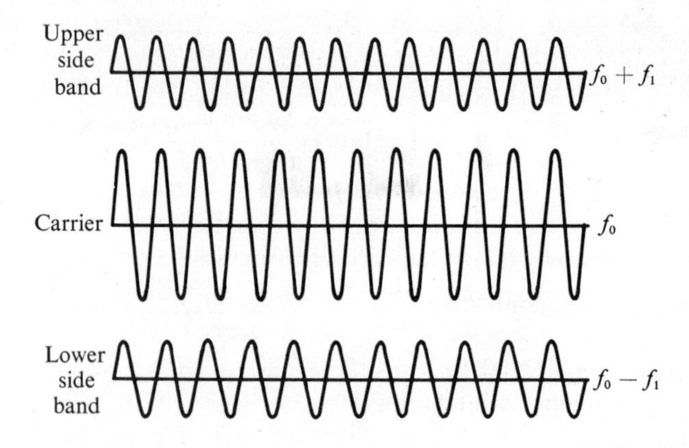

Figure 12–14. **Generation of sum-and-difference, or sideband frequencies.**

Intermodulation tests are commonly made with a pair of applied frequencies that have values of 60 Hz and 6 kHz. If the amplifier is nonlinear, sideband frequencies will appear in the output from the amplifier; these sideband frequencies will have values of 6,060 Hz and 5,940 Hz. We call these sum-and-difference signals *intermodulation frequencies*. Since they have an inharmonic relation to the applied 60 Hz and 6 kHz frequencies, intermodulation frequencies are more objectionable to the listener than are harmonic frequencies. For this reason, intermodulation analyzers (often called audio analyzers) such as illustrated in Fig. 12-15 are widely used to check amplifier performance.

Courtesy of Heath Company

Figure 12-15. Typical intermodulation analyzer.

An intcrmodulation analyzer contains high- and low-frequency sine-wave sources; hence, it is sometimes called a two-tone generator. The analyzer also contains filters, whereby the sum-and-difference signals generated by the amplifier under test can be separated from the fundamentals. An a-c meter indicates the percentage of sum-and-difference signal amplitudes with respect to the fundamental amplitudes. To establish a standard reference for percentage intermodulation measurements, the low-frequency signal and high-frequency signal are

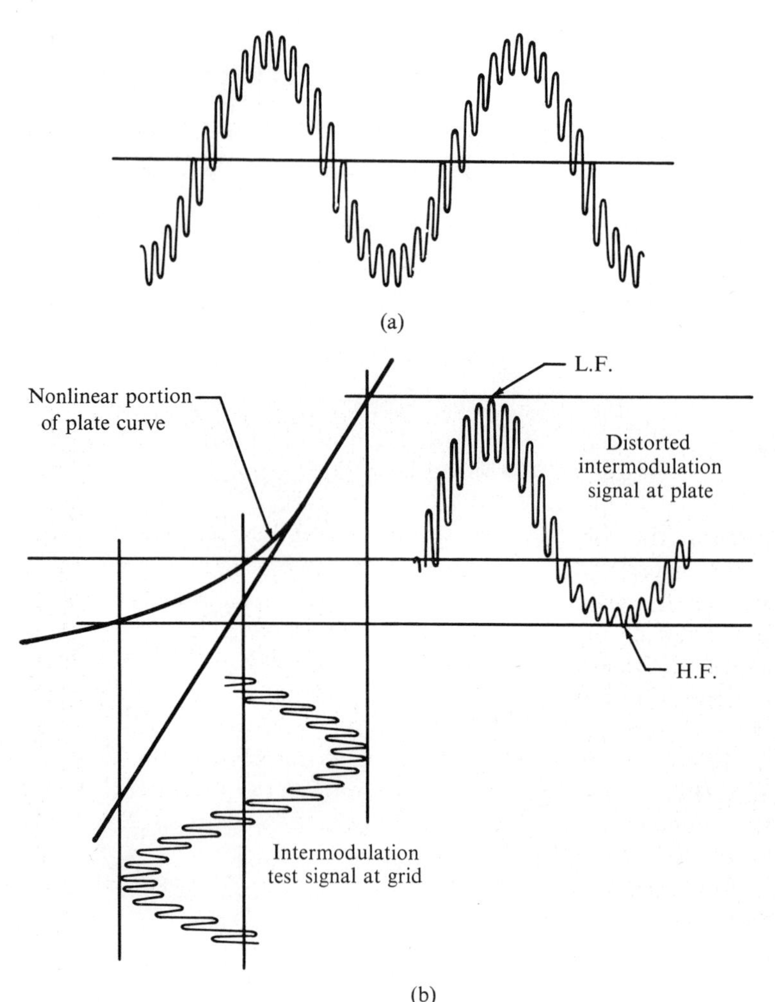

Figure 12–16. (a) Standard intermodulation test-signal waveform; (b) Nonlinear amplifier characteristic generates intermodulation frequencies.

adjusted to the relative amplitudes depicted in Fig. 12–16(a). When this two-tone signal is passed through a nonlinear amplifier, the output waveform becomes distorted as shown in Fig. 12–16(b).

This distorted output waveform contains the high-frequency sum-and-difference signals that impair the fidelity of an amplifier. Accordingly, these intermodulation frequencies are filtered from the other signal components by the analyzer, and their relative amplitude is

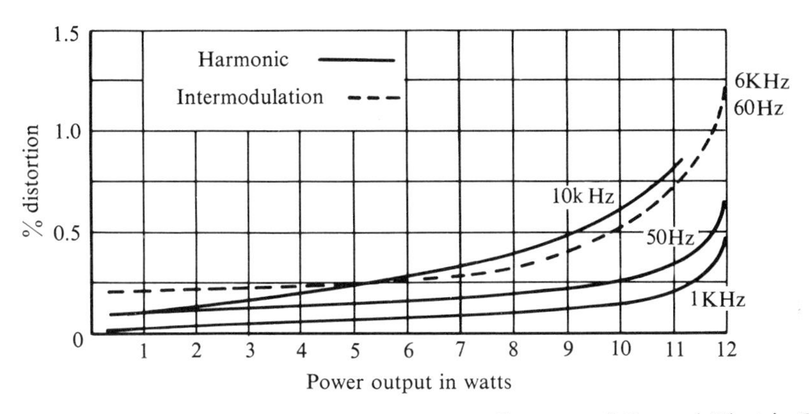

Courtesy of General Electric Co.

Figure 12–17. Comparison of harmonic and intermodulation distortion percentages for a typical high-fidelity amplifier.

indicated by the meter in terms of percentage intermodulation distortion. The meter indicates the root-mean-square value of the intermodulation frequencies. There is no definite relation between the percentages of harmonic distortion and intermodulation distortion for an amplifier. For example, Fig. 12–17 shows the performance of one high-fidelity transistor audio amplifier. The following characteristics will be observed in this particular example.

1. Both harmonic distortion and intermodulation distortion increase as the power output from the amplifier is increased.
2. Harmonic distortion is greater at 50 Hz than at 1 kHz, but increases to a maximum at 10 kHz.
3. Intermodulation distortion is greater than harmonic distortion at 50 Hz, but is approximately the same at 10 kHz.
4. Harmonic and intermodulation distortion values cross over on a 10 kHz harmonic-distortion test when the amplifier power output is about 5 w.

It is considered good practice in high-fidelity technology to measure *both* percentage harmonic distortion and percentage intermodulation distortion at *maximum rated power output*. Both harmonic and intermodulation measurements will be less than 1 per cent for a properly operating high-fidelity amplifier. Because the sound volume is very large at maximum power output, it is customary practice to disconnect the speaker from the amplifier, and to connect a power resistor of correct value in place of the speaker. In turn, the power output is measured with an a-c voltmeter connected across the resistor.

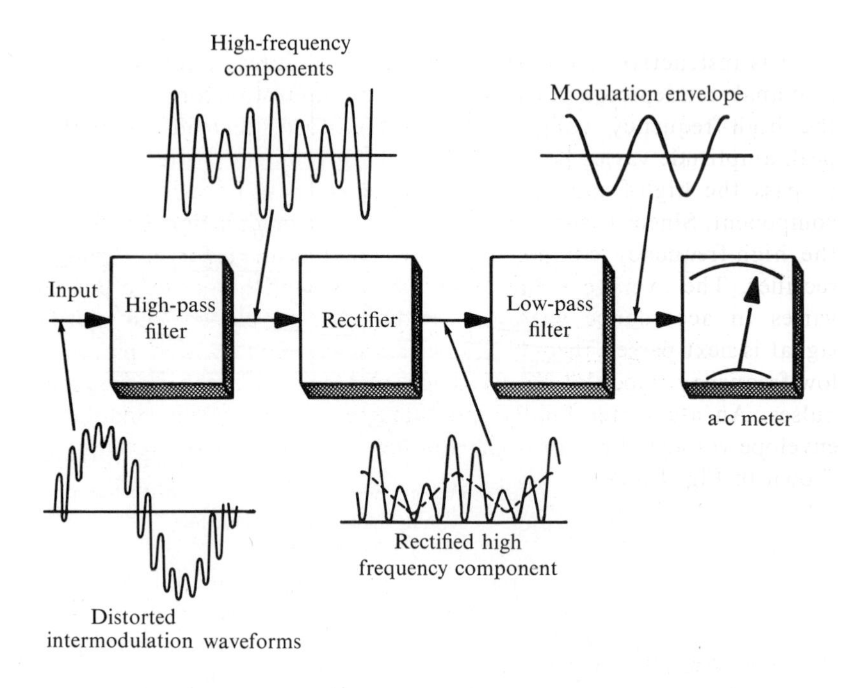

Figure 12-18. **Plan of an intermodulation distortion analyzer.**

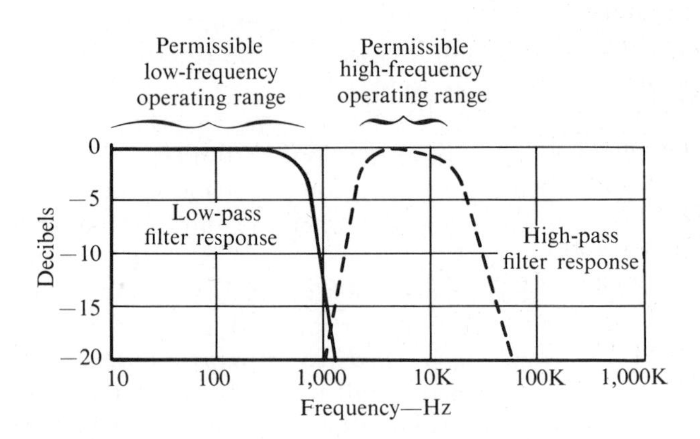

Figure 12-19. **Filter characteristics in a typical intermodulation distortion analyzer.**

It is instructive to observe the plan of an intermodulation distortion analyzer depicted in Fig. 12–18. If the input waveform is distorted, the high-frequency component is not uniform; instead, its peak-to-peak amplitude varies periodically. A high-frequency filter is utilized to pass the high-frequency components and reject the low-frequency component. Since it is desired to recover the modulation envelope of the high-frequency components, the signal is next passed through a rectifier. The average value of the signal is no longer zero but now varies in accordance with the modulation envelope. The rectified signal is next passed through a low-frequency filter which passes the low-frequency modulation envelope and rejects the high-frequency pulses. An a-c meter finally indicates the value of the modulation-envelope voltage. Typical high- and low-pass filter characteristics are shown in Fig. 12–19.

SUMMARY

1. All vacuum tubes and transistors have transfer characteristics that are somewhat nonlinear.
2. The percentage distortion denotes the relative amplitudes of fundamental and harmonic voltages in a waveform.
3. Harmonic distortion of an amplifier or amplifying system may be determined by passing a low distortion sine wave through the amplifier or system, thereafter removing the fundamental frequency, and then comparing the amplitude of the remaining harmonics to the amplitude of the fundamental.
4. Presence of second harmonic distortion may be determined by observation of the d-c voltage measurement on the output load resistor as a signal is applied to an amplifier stage.
5. Odd-order harmonic distortion alone does not shift the d-c level of an output voltage.
6. The magnitude of the d-c component generated by even-harmonic distortion is dependent upon the harmonic phase relation with the fundamental.
7. The total distortion present in a waveform is the square root of the sum of the squares of the odd- and even-harmonic distortion.
8. Intermodulation distortion is the result of modulation effects between pairs of tones amplified in a nonlinear device.
9. Graphical distortion analysis may be applied to transistor, vacuum tube or field-effect transistor output characteristic curves.
10. When two frequencies are modulated together, side-band fre-

quencies are generated which have values of the sum ($f_0 + f_1$) and the difference ($f_0 - f_1$) in relation to the two original frequencies.

11. Intermodulation tests are commonly made with an audio analyzer using a pair of applied frequencies that have a value of 60 Hz and 6 kHz.

12. An intermodulation meter filters out the sum and difference signals generated by the amplifier under test; it then compares their amplitude to the amplitudes of the original test signals in terms of percentage of harmonic distortion.

13. Harmonic and intermodulation distortion should both be measured at maximum rated power output into a resistor of the proper value.

14. Odd harmonics are generated by symmetrical compression of a sine wave, while even harmonics are generated by inequality in the minimum and maximum peaks of a sine wave.

Questions:

1. Define the term percentage of distortion as referred to a sine-wave amplifier.
2. Discuss the internal operation of a distortion meter used to measure second harmonic distortion.
3. Discuss the visual differences between odd-order harmonic distortion and even-order harmonic distortion.
4. Discuss a method for determining the presence of even-harmonic distortion in an amplifier using a d-c voltmeter.
5. Discuss a method for determining the presence of odd-harmonic distortion in an amplifier using an oscilloscope.
6. Which of the methods discussed in either Question 4 or 5 can be used to detect odd-harmonic distortion in an amplifier?
7. Can an amplifier generate both odd- and even-harmonic distortion? Explain.
8. To what type of distortion is poor tone quality attributed?
9. Explain the modulation effect.
10. Three frequencies, f_1, f_2, and f_3, are amplified through a nonlinear device. What are the output frequencies?
11. Explain the internal operation of an intermodulation distortion analyzer.
12. What is the maximum allowable harmonic and intermodulation distortion for high-fidelity operation of an amplifier?
13. Why is the speaker replaced by a load resistor when making distortion measurements?
14. Compare the operations of a high-pass and a low-pass filter.

PROBLEMS

1. Calculate the percentage of second-harmonic distortion developed along the load line in the collector characteristic curve in Fig. 12-20 for a base current of ± 100 μa at about a 200-μa operating point.
2. Determine the percentage of the second-order harmonic distortion along the load line on the characteristic curves in Fig. 12-21 (page 404) when the static base bias is 125 μa and ΔI_B is 100 μa.
3. Determine the values of second harmonic distortion developed along the load line in Fig. 12-21 for a variation of 50 μa either side of the following five operating points: $I_B = 175$ μa, $I_B = 150$ μa, $I_B = 125$ μa, $I_B = 100$ μa, and $I_B = 75$ μa.

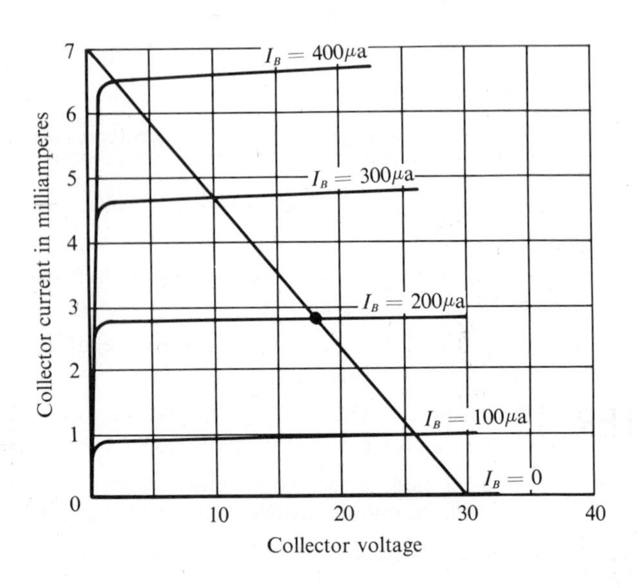

Figure 12-20. **Collector characteristic curve.**

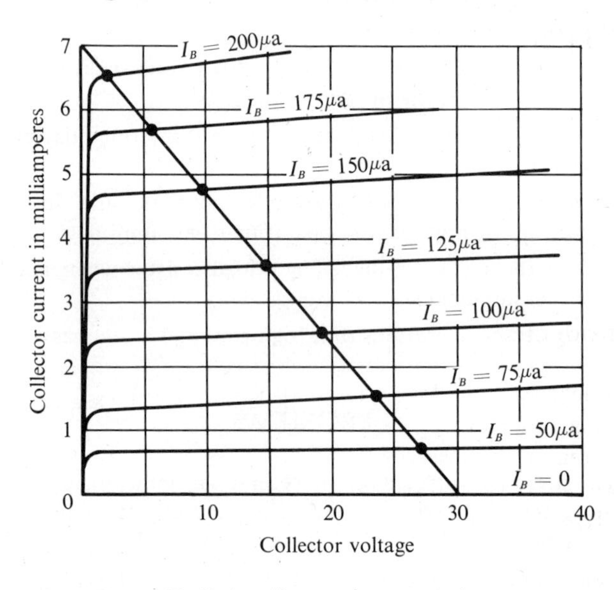

Figure 12-21. **Collector characteristic curve.**

CHAPTER 13

Principles of Negative Feedback

13–1. Introduction

We have learned that amplitude distortion is inherent in all amplifiers, because all amplifying devices have characteristics that are nonlinear to a greater or lesser extent. In high-fidelity amplifiers, for example, we seek to reduce amplitude distortion to a practical minimum. The designer is quite willing to "trade" gain for improved fidelity. As will be explained, circuits that develop *negative feedback* permit any desired compromise between amplifier gain and percentage harmonic distortion (or intermodulation distortion).

Previous discussion has also pointed out that limited frequency response, or nonuniform amplification at various frequencies (called frequency distortion), may be a problem in simple amplifier arrangements. Here again, elaboration of an amplifier configuration to include negative feedback permits us to obtain reduced frequency distortion at the expense of gain. The bandwidth of an amplifier can be extended, and its frequency response made uniform by means of negative feedback. In high-fidelity systems, the bargain is obviously a good one. The present state of the art could not exist if circuit designers did not have the powerful tool of negative feedback with which to work.

We recognize that any amplifier has a certain phase characteristic, and that good transient (square-wave) response requires a linear phase relation versus frequency. Recall that a curved or nonlinear phase characteristic can cause tilt in a reproduced square wave; other types of transient distortion are also encountered when a phase characteristic is very nonlinear. As will be explained, negative feedback can linearize the phase characteristic of an amplifier; again, this is at the cost of gain. If we desire to obtain good transient response in an oscilloscope, for example, we will utilize a considerable amount of negative feedback in the vertical-amplifier system.

Still another benefit of negative feedback is realized in amplifier *stabilization.* In other words, transistors and tubes have characteristics that change with age. The emission, amplification factor, mutual conductance, and dynamic plate resistance of a tube change in value as a tube ages. A transistor gradually develops greater leakage current, reduced alpha (or beta), and changed input and output impedances in extended service. Electronic instruments such as VTVM's and TVM's

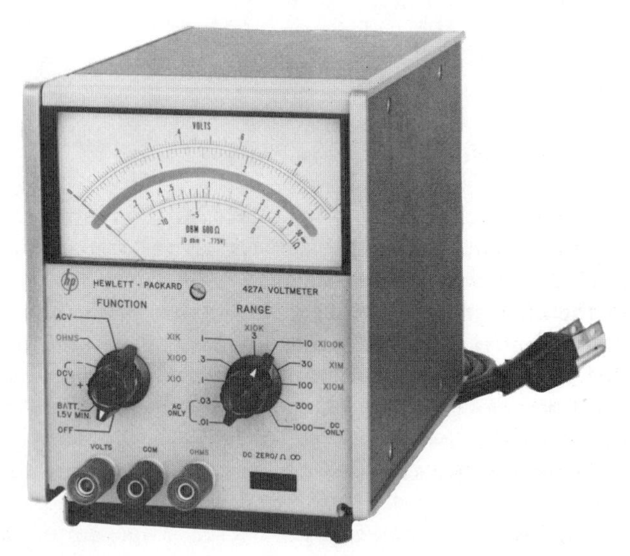

Courtesy of Hewlett Packard, Inc.

Figure 13–1. Typical low-range a-c transistor voltmeter (TVM).

(see Fig. 13–1) must maintain accurate calibration in spite of component drift. Accordingly, a large amount of negative feedback is utilized in their amplifier configurations.

We have become familiar also with tolerance problems. When a transistor or vacuum tube ages sufficiently that its performance is unsatisfactory in spite of ample negative feedback, we must, of course, replace the component. A new transistor or vacuum tube, for example, might have some parameter tolerances as great as ± 30 per cent. Fortunately, negative feedback minimizes the necessity for replacement control device selection and for meter recalibration. Let us see how the foregoing benefits are obtained in practice.

13–2. Feedback in Amplifiers

Whenever a portion of the amplified output is fed back into the input circuit of an amplifier, changed performance characteristics result. We recall that in-phase feedback via a common impedance in the power supply can cause motorboating in a three-stage amplifier. This was an example of positive or regenerative feedback. Again, we recall that the grid-plate, collector to base, or drain to gate interelectrode capacitance in a triode device permits feedback of signal energy from the output element to the control element and results in greatly increased input capacitance. This is also an example of positive or regenerative feedback.

However, if a portion of the amplified output is fed back in opposing phase to the input circuit of an amplifier, we state that *negative*, *degenerative*, or *inverse* feedback is present. These are equivalent terms that mean the same thing. To simplify our discussion of negative feedback insofar as possible, we will return to consideration of circuit action entailing an unbypassed source resistor, as shown in Fig. 13–2. We will use a field-effect transistor as a control device with the understanding that our analysis applies for vacuum tubes and junction transistors. The configuration depicted in Fig. 13–2(a) is called a *source-follower* stage. Note that the gate-input signal is applied between gate and ground, and that the output signal is taken from across the unbypassed source resistor. Since the output signal has the same phase as the input signal, we say that output "follows" the input.

In this preliminary example, we have chosen a value of source resistance which serves both as a bias source and as a load resistor. Observe in Fig. 13–2(b) that the operating point falls at *P*. Let us see

(a)

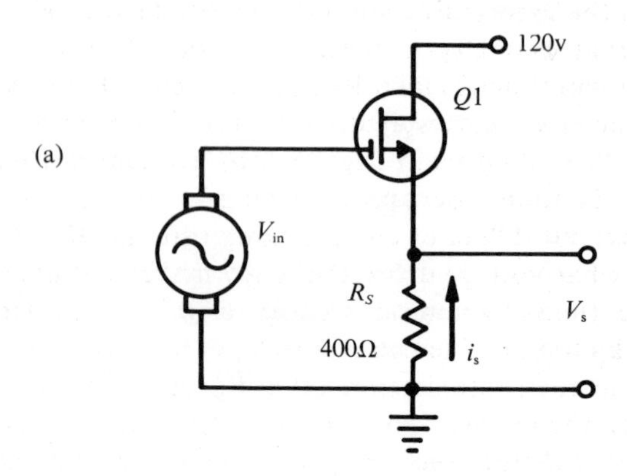

(b)

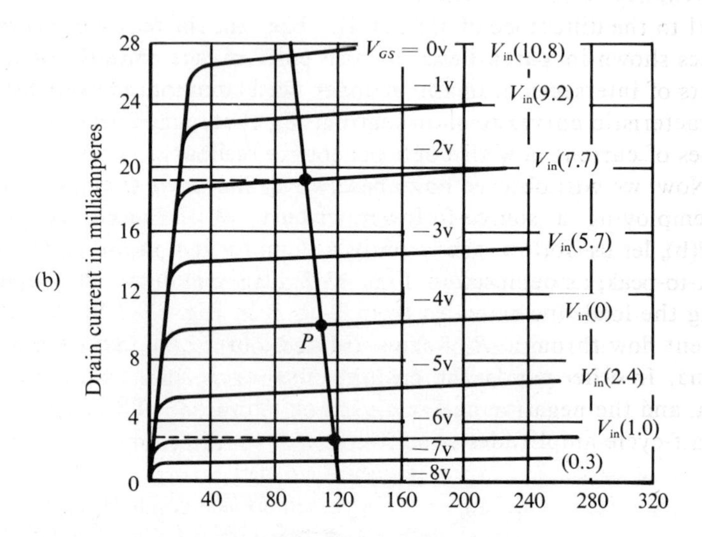

Figure 13–2. **Comparison of distortion in a source-follower and a common-source amplifier. (a) Source follower circuit; (b) Output characteristic curves.**

why this is so. If $V_{in} = 0$ in Fig. 13–2(a), it is evident that the gate-source potential is equal to the *IR* drop across the source resistor. The load line is drawn in the customary manner; that is, $\tan\theta = 1/R_L = 1/0.4\text{K}$ or $\tan\theta = \Delta I/\Delta V = 25/(120 - 110) = 25/10$. Observe that the source-current flow i_{so} is 10 ma at point *P* in Fig. 13–2(b). Since $400 \times 0.01 = 4$ v, it is obvious that the operating point must fall at *P* when the E_{in} terminals are short-circuited ($E_{in} = 0$). This fact has been denoted by marking (0) at the end of the −4 v V_G curve. In other words, −4 denotes the gate-source voltage, and (0) denotes V_{in}.

Next, suppose that we apply a certain amount of positive drive voltage to the gate which will make V_G equal to zero. That is, there is a particular value of V_{in} corresponding to the intersection of the load line with the $V_G = 0$ curve in Fig. 13–2(b). Evidently, this particular value of V_{in} is found by applying Ohm's law to the source resistor. Or, we ask: "If 27 ma of current flows through 400 Ω at the point in question, what voltage drop results (27 ma from drain current curve)?" Of course, the voltage drop is equal to 10.8 v, approximately. Therefore, we mark (10.8) at the end of the $V_G = 0$ curve.

It is clear that at each intersection of a V_G curve with the load line in Fig. 13–2(b), the current flow through the source resistor is given by projecting the intersection point to the current axis. V_S is obviously equal to $i_S R_S$. In turn, the corresponding value of V_{in} is equal to the difference of V_G and V_S. The student may verify the V_{in} values shown in parentheses at the ends of the curves for various points of intersection. In this manner, we elaborate the plate-family of characteristic curves to show steps of V_{in} that correspond to various values of current flow through the source resistor.

Now, we will observe how the *linearity* has been greatly improved by employing a source-follower circuit. With reference to Fig. 13–2(b), let us follow the circuit action for an input signal of 8 v peak-to-peak, as depicted in Fig. 13–3. It is evident that operation along the load line extended from P_1 to P_2 in Fig. 13–2(b). In turn, the current flow through R_k varies from 10 to 19 ma, and from 10 to 2.8 ma. In other words, the positive half cycle has an amplitude of 9 ma, and the negative half cycle has an amplitude of 7.2 ma. The ratio of half-cycle amplitudes is 1.25 to 1.

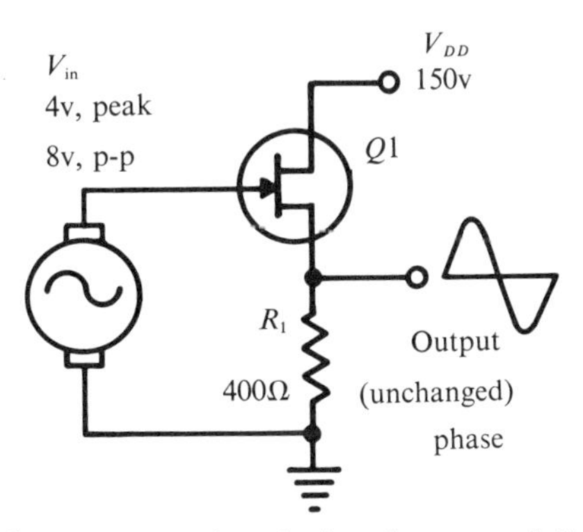

Figure 13–3. Input signal applied to the source-follower stage.

Of course, the output waveform is distorted accordingly. However, we will observe next that the distortion would be much greater if a source-follower stage were not used. For purposes of comparison, let us place the load resistor in the drain circuit of the FET, as depicted

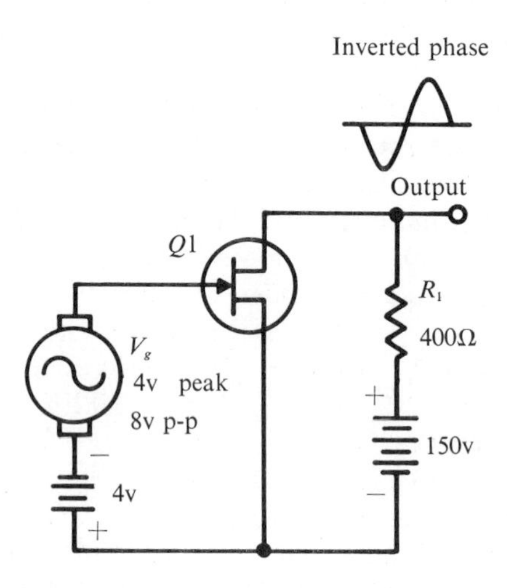

Figure 13–4. Load resistor placed in the drain circuit.

in Fig. 13–4. Now, let us refer to Fig. 13–2(b) and drive the drain-load stage over approximately the same current swing as before. Operation then extends along the load line from $V_T = -4.5$ to $V_G = -8$ v. We see that the positive half cycle of current flow has an amplitude of 2.5 ma, and the negative half cycle has an amplitude of 10 ma. The ratio of half cycle amplitudes is 4.

Therefore, the source follower provided a great improvement in output amplitude linearity. Let us again note the ratios of positive-peak to negative-peak output voltages. For the source follower,

$$\frac{+V_p}{-V_p} = 1.25 \tag{13–1}$$

For the drain-loaded stage,

$$\frac{+V_p}{-V_p} = 4 \tag{13–2}$$

In other words, the source-follower stage provided a reduction in the dissymmetry of the output waveform. Of course, this reduction in amplitude distortion was obtained at the cost of voltage gain. Roughly speaking, the source-follower stage must be driven 3.3 times as hard as the drain-load stage to obtain the same output voltage; this fact follows directly from inspection of Fig. 13–2(b).

However, this is only one part of the follower circuit analysis. In addition to the great reduction in amplitude distortion provided by a follower stage, the input capacitance to the control device is greatly reduced. That is, we will find that the Miller effect is much less in a follower-type circuit. In turn, a follower type circuit can be satisfactorily driven from a very high impedance source. This is very desirable in many practical applications. We observe further that although the input impedance to a follower circuit is very high, the output impedance is quite low. Otherwise stated, the follower circuit is an electronic impedance transformer. This is also a desirable feature in various practical applications.

13–3. Algebraic Analysis of Negative Feedback

We have made a graphical analysis of a source-follower stage in order to visualize the action of negative feedback. Now, we are in a good position to proceed with algebraic analyses of negative-feedback situations. Figure 13–5 shows a simple block diagram of a negative-feedback amplifier. We will assume midband operation, in which there

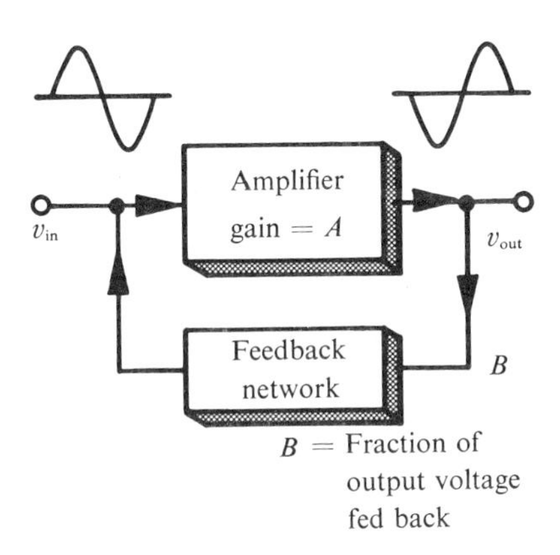

Figure 13–5. Basic negative-feedback system.

is no phase shift through the amplifier; we will also assume that we are considering a single drain-loaded stage in which V_{out} has opposite phase to V_{in}. The feedback network is simply a voltage divider that permits us to feed a chosen amount of output signal back into the input circuit in opposing phase.

If the stage gain is A times, then the net output voltage in Fig. 13–5 is $v_{out} = A_v(V_{in} - BV_{out})$, where B is the fraction of the output voltage that is fed back into the input circuit. To put it another way, if we feed a voltage Bv_{out} back into the input circuit, the net input voltage is obviously formulated

$$v_{in}\,(\text{total}) = V_{in} + Bv_{out} \qquad \textbf{(13–3)}$$

Both the input voltage V_{in} and the feedback voltage BV_{out} are amplified through the amplifier, or the net output voltage is formulated

$$v_{out} = A(v_{in} - Bv_{out}) \qquad \textbf{(13–4)}$$

Then we rearrange Formula 13–4 and write

$$\frac{v_{out}}{v_{in}} = \frac{A_v}{1 - A_v B} \qquad \textbf{(13–5)}$$

Since the denominator in Formula 13–5 is greater than unity (B is negative), it is apparent that negative feedback decreases the gain of a stage.

Example 1. Suppose that $A_v = 15$, and that 1/4 of the output signal voltage is fed back into the input circuit. Then, the stage gain is calculated

$$\text{Stage gain} = (A_v f)\,\frac{15}{1 + 15/4} = 3.15 \text{ times} \qquad \text{(answer)}$$

Otherwise stated, 25 per cent negative feedback reduced the stage gain in Example 1 to 21 per cent of its original value. If we were utilizing a source-follower stage, it is evident that 100 per cent negative feedback would then be present, or $B = 1$. In turn, if $A_v = 15$, the "gain" of the stage is equal to 0.94, in accordance with Formula 13–5. We perceive that the gain of a source follower may approach 1 as the mu of the device is increased; however, the output signal voltage must always be less than the input signal voltage. In any negative-feedback amplifier, the voltage gain may be made to approach $1/B$, but cannot quite attain this value, as stated implicity by Formula 13–5.

Note in passing that Formula 13–5 assumes that the gain of the device does not vary appreciably over the range of signal swing. This assumption is essentially true for small values of drive voltage (small-signal operation). On the other hand, the assumption is not true for larger-signal operation. Therefore, when we are concerned with large-signal operation, it is advisable to employ graphical analysis, as we did in Fig. 13–2.

Next, we may ask what the exact output impedance of a source follower or cathode follower will be for small-signal operation. Reasoning from our knowledge of tube operation, it is evident that the equivalent circuit is drawn as shown in Fig. 13–6. The symbols r_{ds} and

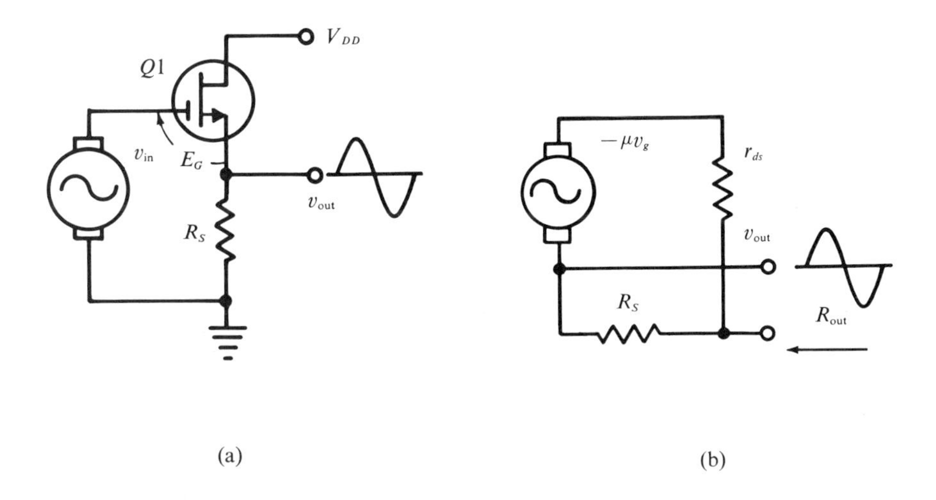

Figure 13–6. (a) Source-follower stage; (b) Equivalent circuit.

r_b represent the dynamic drain or plate resistance of the field-effect transistor and vacuum tube. Essentially, r_{ds} and R_S are connected in parallel; in addition, the effective total resistance depends upon the E/I value of generator $-\mu v_G$. Algebraic solution of the equivalent circuit is formulated

$$r_{out} = \frac{r_{ds} R_S}{r_{ds} + R_S(\mu + 1)} \tag{13–6}$$

Example 2. Suppose that $R_S = 400\ \Omega$, $r_{ds} = 6{,}700\ \Omega$, and $\mu = 20$. Then, the output resistance r_{out} in Fig. 13–6(b) is equal to

$$r_{out} = \frac{6{,}700 \times 400}{6{,}700 + 400(21)} \simeq 180\ \Omega \qquad \text{(answer)}$$

We recognize that the output resistance of source follower is substantially less than the value of the source resistor because of the parallel relations in its equivalent circuit. Note that if a pentode or MOSFET is used in a cathode-follower or source-follower stage, r_p or r_{ds} is very large compared with R_S, and we may write the approximate formula

$$r_{\text{out}} = \frac{R_S}{1 + g_m R_S}, \text{ approximately} \qquad \textbf{(13–7)}$$

Example 3. A MOSFET or pentode vacuum tube with a transconductance of 9,000 μmho operates in a source- or cathode-follower circuit with a source resistor of 1,000 Ω. What is r_{out}?

$$r_{\text{out}} = \frac{1{,}000}{1 + (9 \times 10^{-3} \times 1{,}000)} \simeq 100\ \Omega \qquad \text{(answer)}$$

The great reduction in effective output resistance stems from the large value of transconductance provided by the device. When we design a source- or cathode-follower stage, we are interested chiefly in its output resistance (stated by Formula 13–7), in its voltage "gain," and in the effective input capacitance. In the case of a triode device, which is the most usual situation, the voltage "gain" follows from the equivalent circuit.

$$A_v = \frac{e_{\text{out}}}{e_{\text{in}}} = \frac{\mu R_S}{r_{ds} + R_S(\mu + 1)} \qquad \textbf{(13–8)}$$

Example 4. Determine the voltage gain of a cathode- or source-follower stage if $\mu = 15$, $R_S = 400\ \Omega$, and $r_{ds} = 6{,}700\ \Omega$.

$$A_v = \frac{15 \times 400}{6{,}700 + 400(10)} \simeq 0.45 \qquad \text{(answer)}$$

The signal voltage has been reduced to less than half, but we obtain a very low output impedance. Moreover, the follower stage can be satisfactorily driven from a very high-impedance source, because the input capacitance has been greatly reduced. It can be shown that the input capacitance for a follower stage is formulated

$$c_{\text{in}} = c_{gs} + c_{gd}(1 - A_v) \qquad \textbf{(13–9)}$$

where A_v denotes the voltage gain of the stage and is positive for a follower stage.

Recall that if the source resistor were connected into the amplifier circuit as a drain-load resistor, the Miller effect would greatly increase the input capacitance. In such a case, we write the formula for the input capacitance

$$c_{\text{in}} = c_{gs} + c_{gd}(1 + A_v) \qquad \textbf{(13–10)}$$

where A_v denotes the voltage gain of the stage.

To illustrate the reduction of input capacitance provided by a stage follower, let us consider two examples.

Example 5. Consider an FET-source follower amplifier with $r_{ds} =$ 40 kΩ, $c_{gs} = c_{gd} = 3.4$ pf, and a g_m of 500 μmho working into a source load of 100 Ω. What is the voltage gain and input capacitance?

$$\begin{aligned}
\mu &= g_m r_{ds} \\
&\simeq 500 \times 10^{-6} \times 40 \times 10^3 \\
&\simeq 20 \\
A_v &= \frac{\mu R_S}{r_{ds} + R_S(\mu + 1)} \\
&\simeq \frac{20 \times 100}{40 \times 10^3 + 100(20 + 1)} \\
&\simeq 0.47 \qquad \text{(answer)} \\
c_{\text{in}} &= c_{gs} + c_{gd}(1 - A) \\
&= 3.4 + 3.4(1 - 0.47) \\
&\simeq 5.2 \text{ pf} \qquad \text{(answer)}
\end{aligned}$$

We calculated that the gate-input capacitance of the source follower is 5.2 pf. Now, we will calculate as a comparison the input capacitance of a common-source amplifier.

Example 6. Consider the FET in Example 5 in a common-source amplifier in which the drain-load (R_D) is 20 kΩ.

$$\begin{aligned}
A_v &= -g_m R_D \\
A_v &= -500 \times 10^{-6} \times 40 \times 10^3 \\
&\simeq 20 \\
c_{\text{in}} &\simeq c_{gs} + c_{gd}(1 - A_v) \\
&\simeq 3.4 + 3.4(1 + 20) \\
&\simeq 74.8 \text{ pf} \qquad \text{(answer)}
\end{aligned}$$

Thus, the input capacitance of a drain-loaded, plate-loaded, or collector-loaded stage is many times greater than the input capacitance of a follower stage.

When the input signal is at the base and the output signal is taken from the emitter of the junction transistor, the circuit is called a common-collector or emitter-follower amplifier. The characteristics of an emitter-follower (high-input resistance, low-output resistance, low gain, and low distortion) are similar to those for the cathode and source follower. However, as in other amplifier configurations, the input resistance of the junction-transistor amplifier is much less than that for a cathode- or source-follower amplifier. For example, the common-emitter amplifier in Fig. 13–7(a) has an input resistance of about 2 kΩ. At the same bias point, the follower circuit in Fig. 13–7(b)

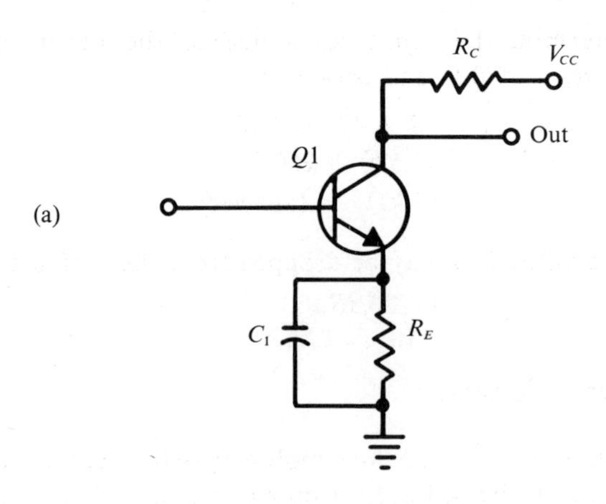

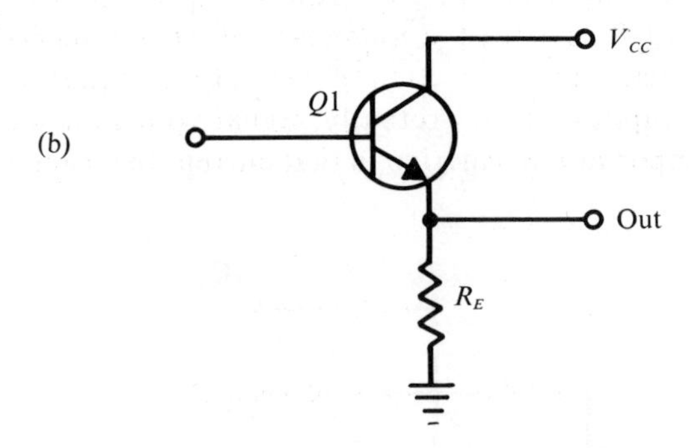

Figure 13–7. (a) The input impeaance of the common-emitter stage is approximately 2 kΩ; (b) The input impedance is approximately 2 kΩ plus $(\beta + 1)R_E$.

has an input resistance of 2 kΩ plus $(\beta + 1)R_E$. In other words, the input resistance is formulated

$$r_{\text{in}} = R_E(\beta + 1) \tag{13–11}$$

where r_i is the internal resistance of the transistor.

$$r_{\text{in}} \simeq r_e + \frac{r_{bb}}{\beta}$$

where $r_e = 26\,\Omega/I_c$ ma, and r'_{bb} = base spreading resistance, typically about 250 Ω. Formula 13–11 may be simplified to

$$r_{\text{in}} \simeq \beta\left(R_E + \frac{250}{\beta}\right) \qquad \textbf{(13–12)}$$

Example 7. Determine the input resistance to the circuit in Fig. 13–7(b) when $R_E = 470\ \Omega$ and beta = 99.

$$\begin{aligned} r_{\text{in}} &\simeq 99 \times 470 + 250 \\ &\simeq 46{,}530 + 250 \\ &\simeq 47\ \text{k}\Omega \qquad \text{(answer)} \end{aligned}$$

Obviously, the value 250 may be dropped from the formula to give

$$r_{\text{in}} \simeq \beta R_E \qquad \textbf{(13–13)}$$

13–4. Feedback Networks

Thus far, we have considered the simplest type of negative-feedback circuit, exemplified by the follower circuit. This is called a *current-feedback* or *series-feedback* arrangement, because the feedback voltage is proportional to the load current. Another general arrangement depicted in Fig. 13–8 is called a *voltage-feedback* or *shunt-feedback* arrangement, because a given fraction of the output signal voltage is fed back into the input circuit. Actually, signal voltage is fed back in both cases. The important distinction is that current feedback increases

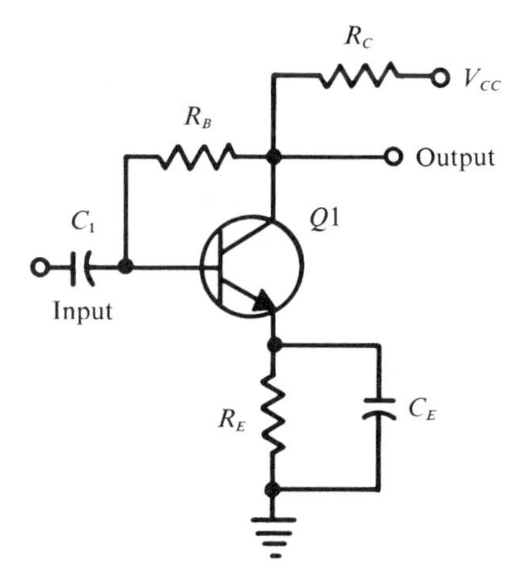

Figure 13–8. A degenerative amplifier stage employing voltage feedback.

the input and the output impedances of an amplifier, while voltage feedback decreases the input and output impedances of an amplifier. In multi-stage amplifiers, negative feedback may be obtained in various ways. The feedback circuit, or "loop," may include one or two stages (rarely more than two stages). *Compound* feedback denotes two sources of feedback voltage, such as an unbypassed cathode resistor and a voltage-divider circuit from the output to the input of the amplifier.

An example of a voltage-feedback network for a transistor stage is shown in Fig. 13–8. Feedback is obtained via the feedback resistor R_B. A part of the output voltage at the collector is fed back through R_B to the base circuit to oppose base current.

The effects of shunt feedback on the amplifier are to decrease the output and input resistance, decrease distortion, increase bandwidth, and decrease gain. The gain (A_f) with shunt feedback is formulated

$$A_f = \frac{A}{1 - BA} \tag{13–14}$$

where B is $-(R_C/R_B)$.

Example 8. The current gain of the circuit in Fig. 13–8 without feedback is 99. What is the gain when the circuit is connected with $R_B = 100\ \text{k}\Omega$ and $R_C = 2\ \text{k}\Omega$?

$$B = -\frac{2}{100} = (-0.02)$$

$$A_f = \frac{99}{1 - (-0.02 \times 99)}$$

$$\simeq 33 \qquad \text{(answer)}$$

Current feedback is provided by the unbypassed emitter resistor in Fig. 13–9. It is apparent that i_c produces a signal voltage drop across R_E that opposes the input signal voltage. The gain of stage is formulated

$$A_{if} = \frac{A_i}{1 - BA_i} \tag{13–15}$$

where $B = -(R_E/R_C)$.

Example 9. The current gain of the amplifier stage in Fig. 13–9 is 50 with R_E bypassed. What is the current gain with R_E unbypassed?

$$B = -\frac{100}{2{,}000} = -0.05$$

$$A_f = \frac{50}{1 - 50 \times 0.05}$$

$$\simeq 14.3 \qquad \text{(answer)}$$

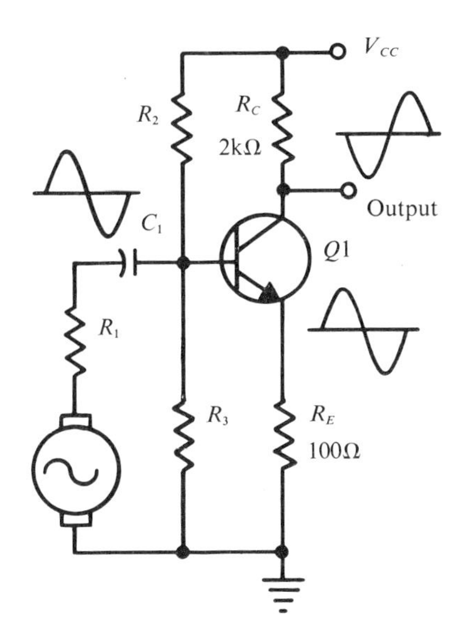

Figure 13–9. A degenerative amplifier stage employing current feedback.

Feedback over two stages is less straightforward, because any phase shift present in one stage adds to the phase shift present in the other stage. If the total phase shift approaches 180°, we can perceive that our intended negative feedback will then become positive feedback with consequent self-oscillation of the amplifier. To avoid this difficulty, the values of reactive components in the amplifier circuit and in the feedback network must be chosen so that the overall phase shift does not approach 180° at any frequency within the operating range of the amplifier. This may entail a trial-and-error design approach, if multigrid tubes are used in circuits that include a number of reactive components. Note in this regard that stray capacitances may be significant at high operating frequencies.

Figure 13–10 depicts a two-stage *RC*-coupled amplifier employing compound negative feedback. The unbypassed portion of R_E in the emitter of $Q2$ develops degeneration at the emitter of $Q2$ by current feedback and at the base of $Q1$ through the series resistor R_E to the base of $Q1$. Note that the feedback signal is in phase with the base signal of $Q2$ and out of phase with the base signal of $Q1$. Negative feedback, such as that applied between $Q2$ and $Q1$, is called *current sampling* and *voltage injection.*

The *RC*-coupled amplifier depicted in Fig. 13–11 employs compound negative feedback. A part of the output signal is picked up by voltage

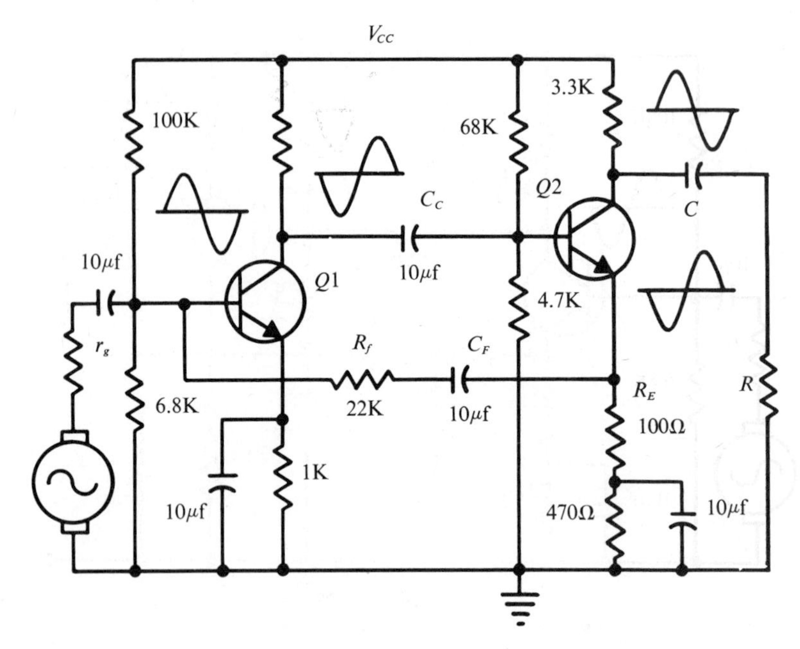

Figure 13–10. A two-stage *RC* amplifier employing compound feedback; R_E gives series feedback to Q2, and feedback to Q1 by current sampling and voltage injection.

sampling and coupled back to the emitter of $Q1$ as current injection. The feedback signal is in phase with the input signal to the base of $Q1$.

Stable operation of the circuits depicted in Figs. 13–10 and 13–11 entails suitable relations between the values of coupling capacitors, emitter bypass capacitors, feedback capacitors, and stray capacitances in the amplifier circuit and in the feedback networks. Thus, although the amplifier circuits appear quite simple, there is more than meets the eye. For example, a change in feedback-component layout might change the stray capacitance relations sufficiently to cause high-frequency oscillation.

Coupling transformers are often a significant source of frequency distortion. Therefore, the frequency response of an amplifier can usually be improved by utilizing negative feedback around the transformers, as shown in Fig. 13–12. The feedback loops are indicated in the diagram. In some cases, selected types of transformers may be required due to phase-shift characteristics at low and high frequencies. In other words, a total phase shift that approaches the positive-feedback value must be avoided, or the amplifier will break into self-

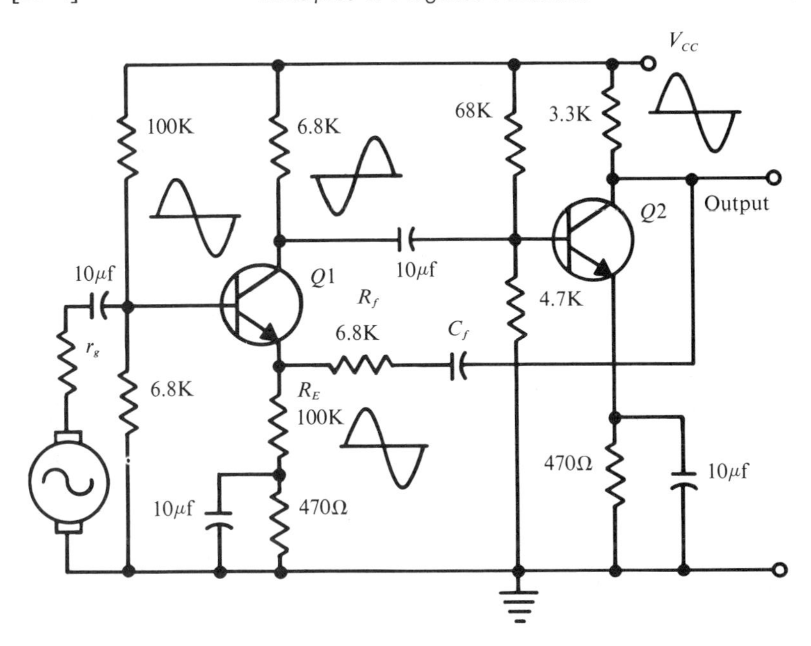

Figure 13–11. A two-stage *RC* amplifier employing feedback by voltage sampling from the collector of Q2 and current injection at the emitter of Q1.

oscillation. It would be impractical to feed back over three stages, because the phase-shift problem then becomes formidable.

Observe in Figs. 13–10 and 13–11 that any power-supply hum that enters the collector-load circuits appears subsequently in the output circuit. In turn, the hum voltage is fed back in inverted phase via C_f to the emitter circuit of $Q1$. Thus, the hum level is reduced by negative-feedback action. This is another benefit provided by negative feedback. We recall that any tube or transistor generates more or less random noise voltage that may be audible as a "rushing" noise in the amplifier output. It is clear that negative feedback is also effective in reducing the noise-voltage output.

Negative current feedback, as illustrated in Fig. 13–9, increases the input resistance of the amplifier according to the formula below.

$$r_i \simeq h_{ie} + R_E(\beta + 1) \qquad \textbf{(13–16)}$$

Example 10. Suppose the input resistance to a common-emitter amplifier without feedback is 2 kΩ (h_{ie}). What is the input resistance if the beta is 99 and a 400-Ω emitter resistor is unbypassed?

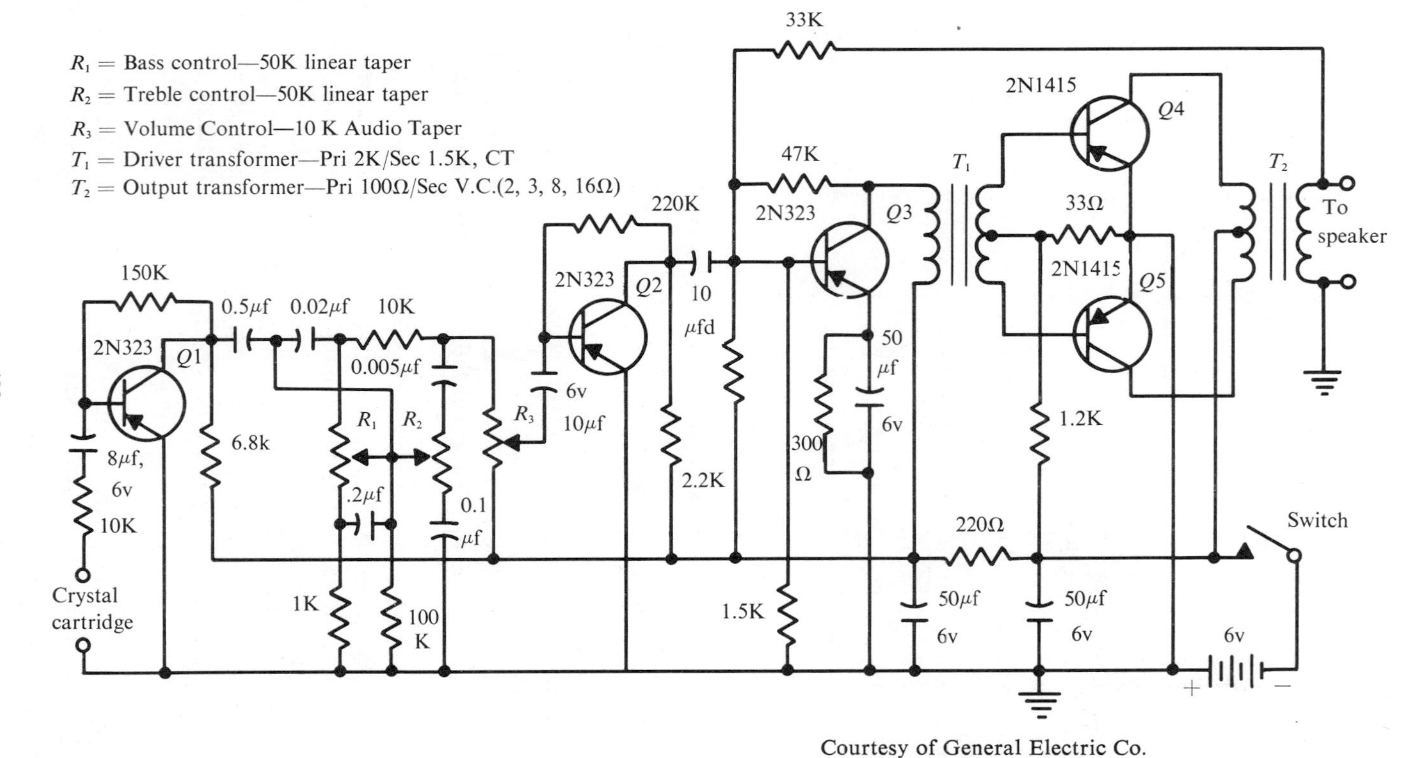

Courtesy of General Electric Co.

Figure 13-12. A phonograph amplifier with negative valtage feedback.

$$r_i \simeq 2{,}000 + 400(100)$$
$$r_i \simeq 42\ \text{k}\Omega \qquad \text{(answer)}$$

The unbypassed emitter resistor also increases the output resistance of the amplifier and causes the gain of the amplifier to be practically independent of the device gain. For example, suppose the circuit in Fig. 13-9 has the emitter resistor bypassed and uses a transistor with a beta of 30. The circuit gain would be approximately 30. If the transistor were replaced with one with a beta of 120, the circuit gain would then be approximately 120. Obviously, this change of gain is undesirable.

On the other hand, if the emitter resistor were unbypassed to give a feedback factor of -0.2, what is the effect to the unit change on the circuit gain? In the first case with a beta of 30, the circuit gain is

$$A_{if} = \frac{30}{1 - (-0.2 \times 30)}$$
$$= 4.28$$

In the second case with a beta of 120, the circuit gain is

$$A_{if} = \frac{120}{1 - (-0.2 \times 120)}$$
$$= 4.8$$

A change of beta of 400 per cent produces a circuit gain change of only 12 per cent.

The phonograph amplifier depicted in Fig. 13-12 is a good example of a practical amplifier employing compound negative feedback. Voltage or shunt feedback is employed from the collector to the base of $Q1$, $Q2$, and $Q3$, and voltage feedback is coupled from the output transformer to the base of $Q3$.

With reference to Fig. 13-9, let us suppose that the line voltage fluctuates with the result that the $B+$ voltage tends to drift. Since d-c current feedback takes place across R_E, gain variation and bias shift are reduced. A somewhat more elaborate d-c current-feedback configuration is seen in Fig. 13-13. This will be recognized as a direct-coupled version of the two-device paraphase inverter-amplifier that was discussed previously. A large amount of negative feedback is provided via the cathode resistors. Phase inversion for push-pull output is obtained by means of the 33-kΩ common-cathode resistor. Correct operating points for the tubes are obtained by using a power supply that has its positive and negative terminals above and below ground potential.

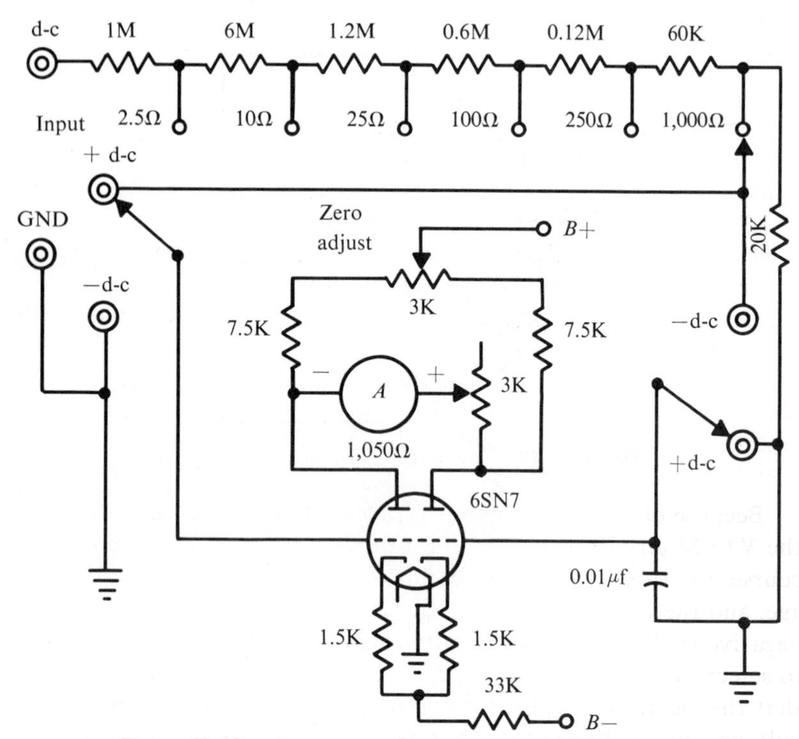

Figure 13-13. A vacuum-tube voltmeter configuration.

In the configuration of Fig. 13-13, the grids operate at ground potential. Observe that this is basically a balanced-bridge circuit, two arms of which are provided by the plate resistances of the triodes. (See Fig. 13-14). If we apply a positive voltage at the input, the grid of the first triode is driven in a positive direction; phase inversion causes the grid of the second triode to be driven an equal amount in the negative direction. Hence, the plate resistance of the first triode decreases, and the plate resistance of the second triode increases. The electronic bridge is thereby unbalanced, and the meter indicates the unbalance-current flow.

Note that the zero-adjust control in Fig. 13-13 permits the pointer to be brought exactly to zero on the meter scale when no input voltage is applied. That is, the zero-adjust control is set for exact balance of the bridge. The 3-kΩ rheostat is a calibrating resistor that is set to make the meter indicate correct full-scale voltage when this value of input voltage is applied. Note that a polarity-reversing switch is provided, whereby the input voltage may be applied to either the grid of the first triode or to the grid of the second triode. Thereby, d-c voltages of either positive or negative polarity may be measured.

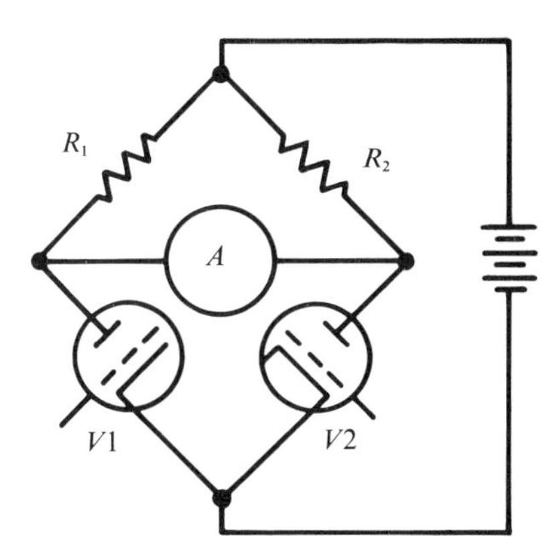

Figure 13-14. Simplified electronic bridge circuit.

Because of the large amount of negative feedback that is used in the VTVM circuit, meter calibration is very stable. Eventually, of course, the tube characteristics will change quite substantially due to age, and the instrument calibration will be impaired in spite of the negative-feedback action. The tubes must then be replaced. It is desirable to select tubes that have negligible grid-current flow. It is apparent that the electronic bridge cannot distinguish between an applied input voltage and a voltage due to contact potential, grid emission, or gas current. New tubes are customarily aged for forty-eight hours in the VTVM with the power turned on before the meter calibration is checked. This procedure insures that the "getter" on the tubes has "cleaned up" residual gas. Then, if the calibration needs to be "touched up," the 3 kΩ calibrating resistor in Fig. 13-13 is adjusted as required.

SUMMARY

1. When applied to an amplifier, negative feedback (degeneration) reduces gain, increases bandwidth, and reduces distortion.
2. An unbypassed emitter resistor in a transistor amplifier causes increased input and output resistance, increased bandwidth response, and decreased distortion.
3. The follower stage is used to match a high impedance source of energy to a low impedance load.
4. The follower-amplifier stage has good high-frequency response due to the reduced Miller capacitance effect.
5. An unbypassed resistor in the emitter of a common-emitter amplifier increases the input and output resistance of the amplifier.

6. Negative feedback in an amplifier decreases the effects of aging and replacement of the control device upon the circuit.

7. Phase shift within an amplifier changes as the frequency changes; therefore, a multistage amplifier may oscillate at high or low frequencies.

8. Feedback over three stages of amplification is seldom used due to the problem of oscillation.

9. Negative feedback is employed in vacuum-tube and transistor voltmeters to compensate for drift due to aging of the components and the control devices.

Questions

1. What are the advantages of using negative feedback in an amplifier?
2. What are the effects of using an unbypassed emitter resistor in a transistor amplifier?
3. What is the Miller-capacitance effect in a triode amplifying device?
4. What is the primary application of follower circuits?
5. Why is the voltage gain less than 1 in a follower circuit?
6. What effects does an unbypassed emitter resistor in a circuit (such as Fig. 13-9) have on (a) circuit gain; (b) distortion; (c) input resistance and output resistance; and (d) bandwidth?
7. By what factor is the input resistance of a transistor amplifier increased when the emitter resistor is unbypassed?
8. What are the effects on the input and output resistance of an amplifier when voltage feedback is employed?
9. Could feedback be used in amplifiers to match impedances? Explain.
10. What are the effects of an unbypassed source resistor in a common-source amplifier?
11. With reference to Fig. 13-10, discuss the feedback employed and its effects on the amplifier.
12. Compare the method of establishing feedback in the circuit in Fig. 13-11 to the method employed in Fig. 13-10.
13. What is meant by current sampling as applied to feedback?
14. What is meant by voltage sampling as applied to feedback?
15. Discuss the effects of negative feedback by current and voltage injections on an amplifier.
16. In reference to Fig. 13-12, discuss each type of negative feedback employed and the effects of each feedback component of the circuit.
17. Why is negative feedback employed in transistor and vacuum-tube voltmeters?

PROBLEMS

1. An amplifier without feedback has a voltage gain of 22. What is the voltage gain when 1/5 of the output is fed back into the input?
2. What feedback factor in Problem 1 would reduce the voltage gain to 2?
3. Determine the output resistance of a vacuum-tube cathode-follower circuit which has the values $R_K = 200\ \Omega$, $r_p = 7{,}000\ \Omega$, and $\mu = 19$.

4. A field-effect transistor source-follower stage has the values $R_S = 100\ \Omega$ and $g_m = 8{,}000\ \mu$mho. What is the approximate output impedance?

5. What value of cathode resistance is necessary to produce an output resistance of $50\ \Omega$ for a cathode follower which has a value of $g_m = 10{,}000\ \mu$mho?

6. If the value of g_m in Problem 5 reduces to $6{,}000\ \mu$mho, what is the new value of output resistance of the circuit?

7. Calculate the input capacitance value for a triode-grounded cathode amplifier when $c_{gk} = 6.8$ pf, $c_{gp} = 7$ pf, and the voltage gain of the circuit is -18.

8. What is the input capacitance of the amplifier in Problem 7 if the circuit is connected as a cathode follower with a gain of 0.6?

9. What is the input capacitance of a common-emitter amplifier in which the voltage gain is 25 and the value of $c_{be} = 10$ pf and the value of $c_{bc} = 8$ pf?

10. What is the value of the input capacitance of the transistor in Problem 9 when the circuit is connected as an emitter follower and the voltage gain is 0.72?

11. What is the value of the input capacitance of a field-effect transistor connected as a grounded source when the capacitances are $C_{gd} = C_{gs} = 3.5$ pf and the voltage gain is 22?

12. What is the value of the input capacitance of the field-effect transistor in Problem 11 connected as a source follower when the gain is 0.9?

CHAPTER 14

Sine-Wave Oscillators

14–1. Introduction

We have seen that an amplifier may break into self-sustained oscillation if appreciable positive (in-phase) feedback is present from output to input. That is, the amplifier supplies its own input. An oscillator circuit consists basically of an amplifying component, such as a vacuum tube, transistor, or tunnel diode, connected into a configuration that provides substantial positive feedback. In turn, the circuit becomes a waveform generator. There are many types of waveform generators utilized in electronics technology. This chapter is concerned with analysis of the prototype sine-wave oscillator. Sine-wave oscillators find very extensive application, as in radio receivers [see Fig. 14–1(a)] and in radio transmitters. A variable frequency oscillator (VFO) for a transmitter is illustrated in Fig. 14–1(b).

As would be anticipated, the basic function of an oscillator is to

Courtesy of Knight Electronics

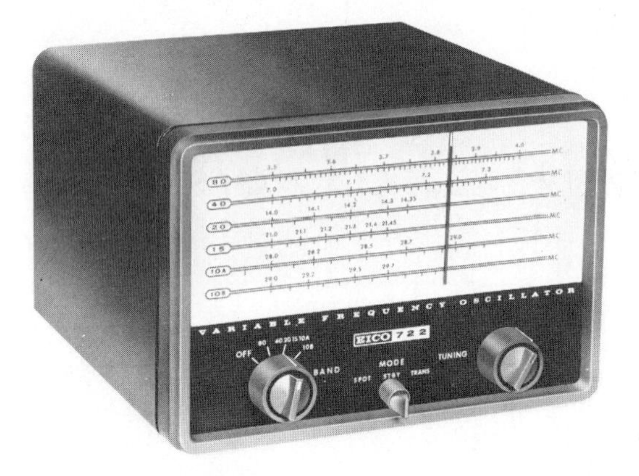

Courtesy of EICO, Electronic Instrument Co., Inc.

Figure 14-1. (a) Sinusoidal oscillators are used in radio tuners; (b) A variable-frequency oscillator.

generate a waveform of specified frequency and amplitude. The frequency of oscillation, in particular, is usually critical, and engineers are concerned with various techniques of frequency stabilization. Some of these techniques will be noted in the latter part of this chapter. Low-frequency sine-wave oscillators commonly employ *RC* networks; on the other hand, high-frequency oscillators generally utilize *LC* resonant circuits. We will find that *LC* oscillators operate on the basis of resonant interchange of energy between a capacitor and an inductor with a tube or semiconductor device supplying pulses of energy in proper phase and magnitude to maintain oscillation.

All understanding of this basic principle is provided by the response of the pulsed *LC* circuit depicted in Fig. 14-2(a). When the leading edge of the pulse voltage energizes the *LC* circuit, the circuit rings transiently, much as a bell rings when it is struck. Unless energizing pulses are supplied at frequent intervals, the ringing waveform

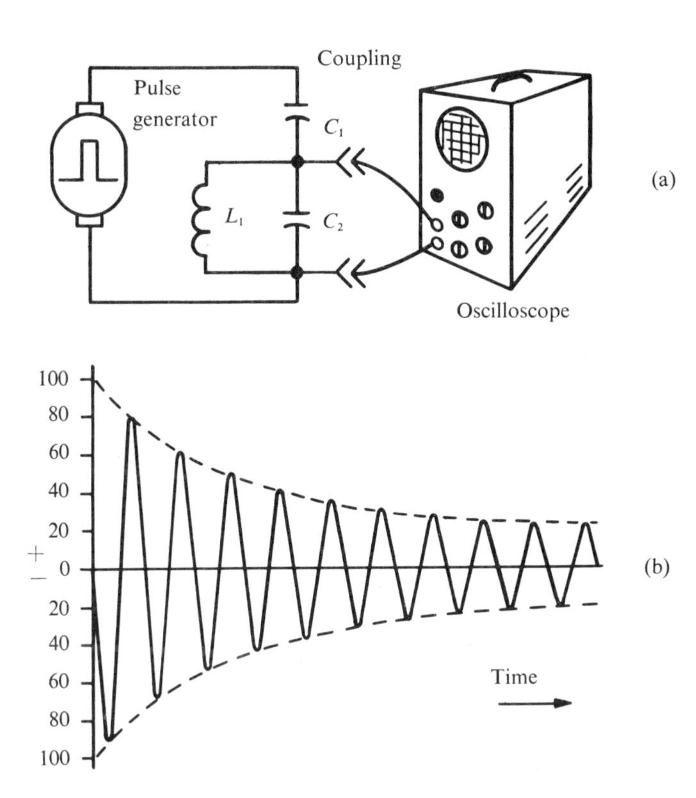

Figure 14-2. (a) Pulsed LC circuit; (b) Typical ringing waveform.

decays exponentially [Fig. 14-2(b)]. However, in a conventional *LC* oscillator circuit, an energizing pulse is provided at the peak of each oscillatory cycle, and a sustained oscillatory voltage is developed across the *LC* resonant circuit. Now, with our understanding of amplifier action, we are in a good position to analyze oscillator circuits.

An oscillator is basically an energy converter that changes d-c electrical energy from the plate power supply into a-c electrical energy. Rotating machines are practical energy converters only at low frequencies. A basic oscillator block diagram is shown in Fig. 14-3. There are two necessary conditions for production of sustained oscillation. First, the feedback voltage from the plate circuit must be in phase with the original excitation voltage at the grid; in other words, feedback must be positive, or regenerative. Second, sufficient energy must be fed back to the grid to compensate for circuit losses. This is just another way of saying that an oscillator must feed back

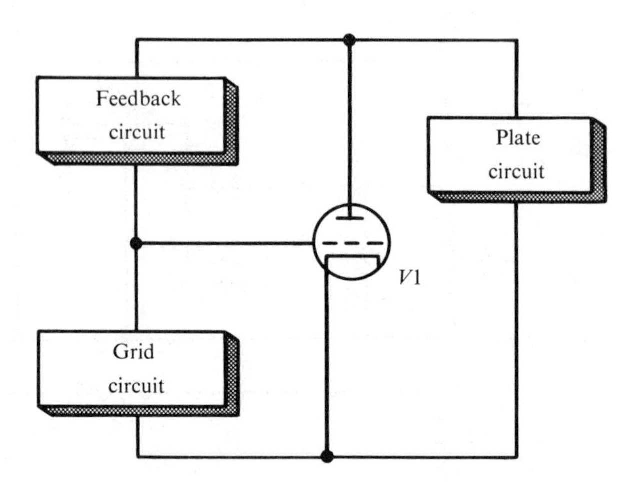

Figure 14–3. **Basic plan of an oscillator circuit.**

an amount of energy that is at least equal to the original excitation energy. The output a-c energy may range from a small fraction of a watt (as in a radio receiver), to many thousands of watts, as in a radio broadcast transmitter or induction heater.

Positive feedback may be accomplished by inductive, capacitive, or resistive coupling between the grid and plate circuits of a tube, or between the collector and base circuits of a transistor. Various circuits have been developed to provide positive feedback in proper phase and suitable amplitude. Each of these circuits has certain characteristics that make its use advantageous in particular circumstances. If appropriate values of inductance and capacitance are utilized, tuned-circuit oscillators may be designed to generate sine-wave voltages at frequencies from the audio range to very high radio frequencies.

The lower frequency limit of *LC* oscillators is imposed by the physical size and cost of the inductors that are required. Hence, audio oscillators commonly use suitable *RC* circuitry. The upper frequency limit imposed on an *LC* oscillator is determined by the values of distributed inductance and capacitance of the circuit components, leads, interelectrode capacitances of tubes, and junction capacitances of semiconductor devices. Of itself, a vacuum tube or transistor is not an oscillator. As depicted in Fig. 14–2, oscillation actually takes place in a tuned circuit, a part of which is composed of interelectrode capacitances or junction capacitances, and the distributed capacitances and inductances of the circuit.

A vacuum tube, transistor, or tunnel-diode oscillator functions

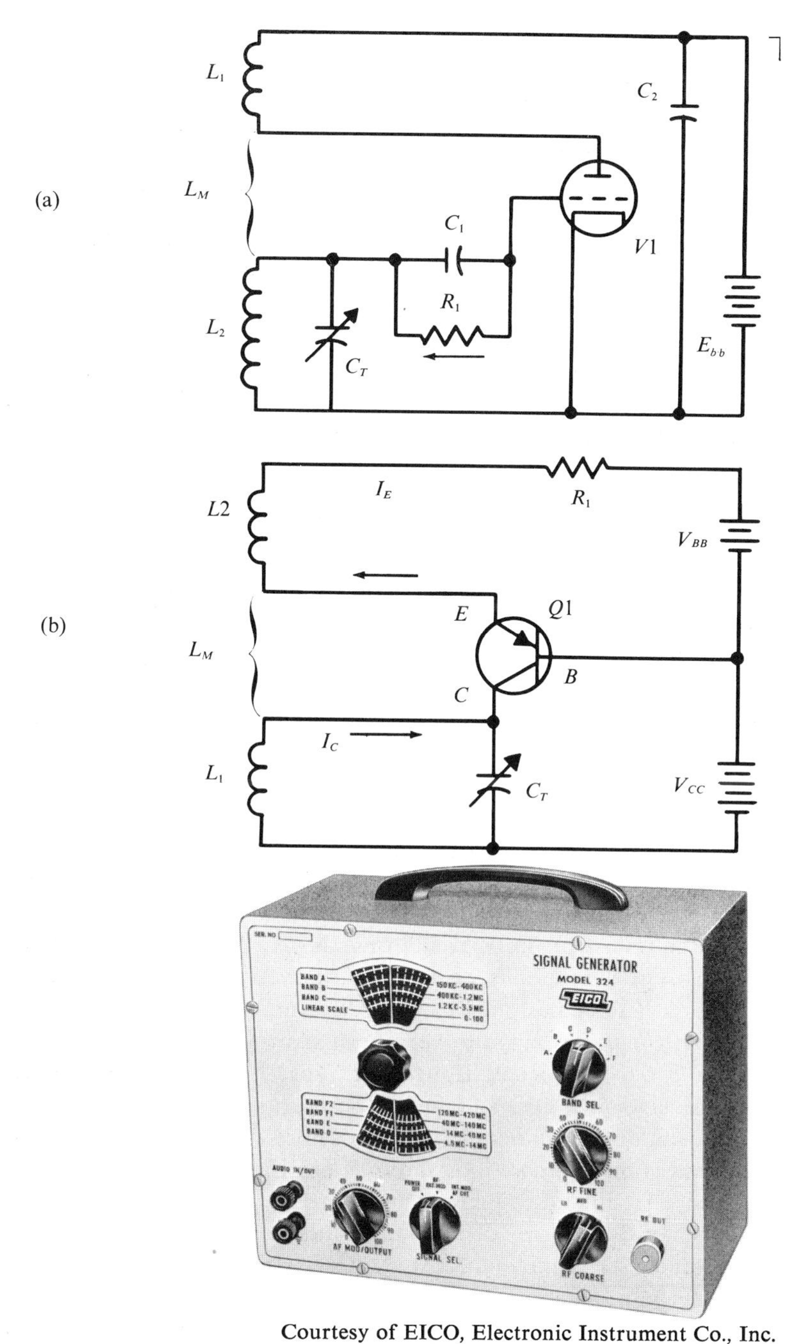

Courtesy of EICO, Electronic Instrument Co., Inc.

Figure 14-4. Tickler feedback oscillator circuits. (a) Triode vacuum tube; (b) Triode transistor; (c) A signal generator utilizes a frequency-calibrated sine-wave oscillator.

primarily as an electrical valve that amplifies its own output. It automatically delivers to the grid (or base) circuit a suitable amount of energy in aiding phase (at correct intervals of time) to sustain oscillation of the *LC* tuned circuit. One of the simplest types of oscillator circuit employs tickler feedback, as shown in Fig. 14–4(a). A signal generator that utilizes an oscillator with tickler feedback is illustrated in Fig. 14–4(c). Feedback voltage in proper phase is transferred from the plate to the grid circuit by mutual inductive coupling between the oscillator tank coil L_2 and the tickler feedback coil L_1. The amount of feedback voltage is determined by the amount of flux from L_1 that links L_2. Thus, the feedback voltage is varied by moving L_1 with respect to L_2.

The frequency-determining portion of the oscillator circuit in Fig. 14–4(a) is the tank circuit L_2-C_T. In operation, the coil and tuning capacitor interchange energy at the resonant-frequency rate, and the

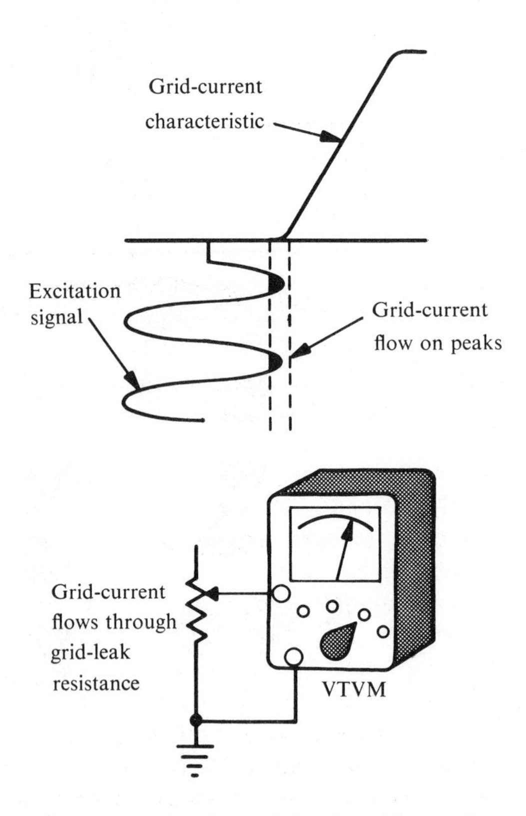

Figure 14–5. **Principle of signal-developed bias voltage.**

excitation voltage developed across C_T is applied to the triode grid in series with the grid-bias section R-C_1. Grid current flowing through R as depicted in Fig. 14–5 establishes a negative bias voltage for the grid. Capacitor C_1 in Fig. 14–4(a) charges up to the peak voltage of the applied grid signal and decays more or less through R between peak intervals. A comparatively steady grid-bias voltage is accordingly developed during oscillator operation. One test for proper operation is a measurement of d-c voltage across the grid-leak resistor, as depicted in Fig. 14–5. To avoid disturbance of oscillator operation, a high-resistance voltmeter such as a VTVM should be used. The tickler-feedback oscillator in Fig. 14–4(b) employs a transistor in the common-base configuration. Resistor R_1 is a bias resistor.

14–2. Series-Fed Hartley Oscillator

We will find that a configuration called the series-fed Hartley oscillator is widely used in various electronic instruments and systems. Its most useful frequency range is from 100 Hz to 30 MHz. Figure 14–6 shows the basic circuit for a series-fed Hartley oscillator, with the curves of grid voltage, plate voltage, and plate current. This type of oscillator commonly operates in Class *C*. Its efficiency may approach 90 per cent. Note that the plate circuit of the configuration is drawn with heavy lines. Inductor L_1 is a part of the tuned circuit comprising L_1, L_2, and C_T. L_1 also serves to couple energy from the plate circuit back into the tuned grid circuit via mutual inductance between L_1 and L_2. Capacitor C_1 acts in combination with resistor R_G to provide a suitable value of grid-bias voltage. Capacitor C_2 and the radio-frequency choke coil (rfc) improve circuit efficiency by keeping the a-c component of plate current from flowing through the power supply.

Note that the *B*-minus terminal of the power supply in Fig. 14–6 is connected to the resonant tank coil L_1. Accordingly, the tuned circuit contains a d-c component of plate current in addition to the a-c signal component. Because the power supply is connected in series with the triode plate and tank coil L_1, this is termed a series-fed arrangement. A Hartley oscillator operates both with mutual inductive coupling between L_1 and L_2, and with the coupling provided by capacitor C_T. The following analysis of circuit action is based simply on the mutual inductive coupling.

1. When power is applied to the triode, cathode emission takes place, and the d-c plate voltage causes a rise in plate-current flow.
2. This increase in plate current flow through L_1 in Fig. 14–6

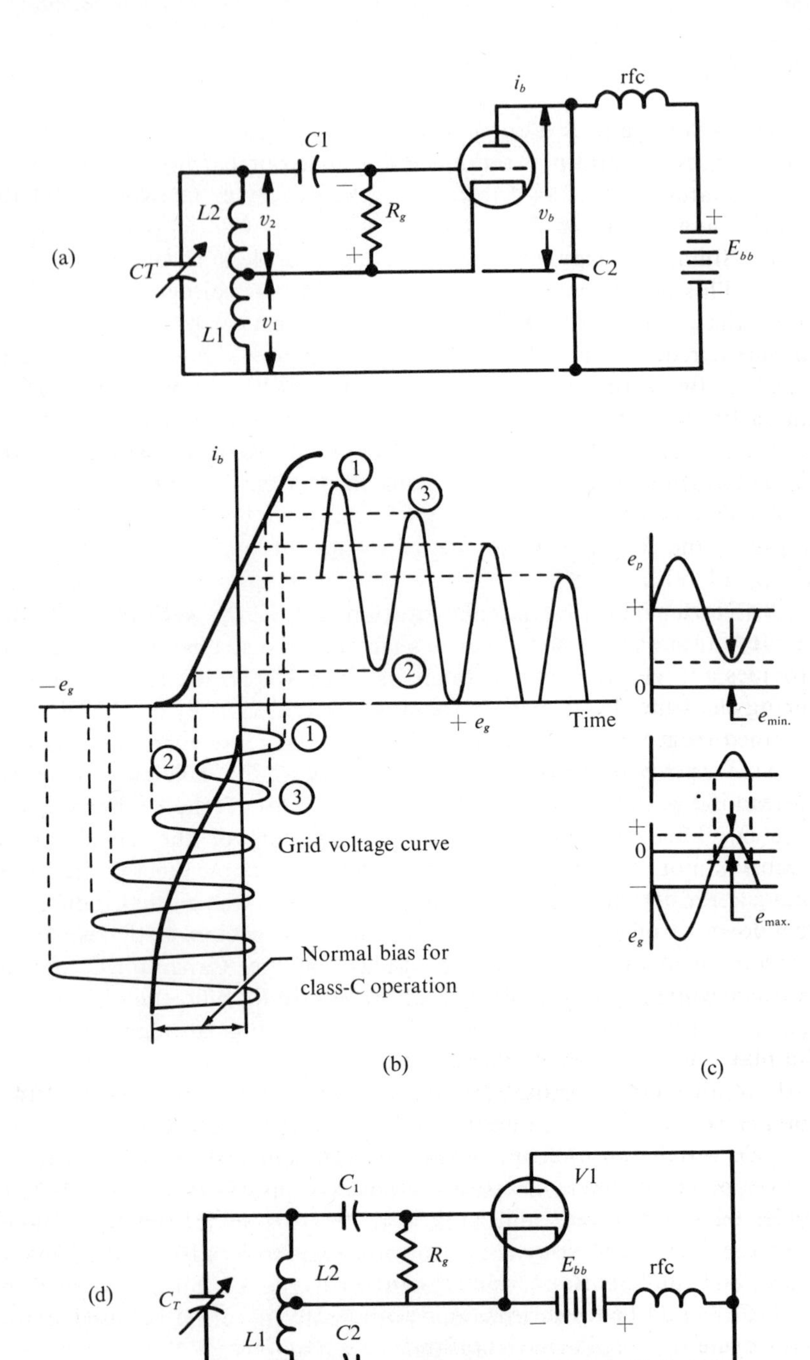

Figure 14-6. Hartley oscillator circuits and operating waveforms.

induces a voltage in L_2 that drives the grid positive; in turn, the plate current rises to point (1) in Fig. 14–6(b).

3. Plate-current flow through L_1 stops rising at point (1), and the voltage induced in L_2 falls to zero.

4. Simultaneously, the plate-current flow starts to fall.

5. This decrease in plate-current flow through L_1 induces a voltage in L_2 that swings the grid in a negative direction, and the plate current is reduced further.

6. Plate-current flow through L_1 stops falling at point (2), and the mutually induced voltage in L_2 again drops to zero.

7. In the absence of negative-going voltage from L_2, the plate current again starts to rise, and the cycle repeats (point 3).

8. On each subsequent cycle of operation, bias voltage builds up across C_1 and R_G until it reaches the normal steady-state value shown in Fig. 14–6(b).

9. Normal bias voltage corresponds to Class-*C* operation [Fig. 14–6(c)]. Interchange of energy between the tank coil and the capacitor provides a "flywheel effect" that maintains the oscillatory waveform during the time that the plate current is zero and no energy is being obtained from the power supply.

Next, we will observe that the Hartley oscillator continues to operate if L_M, the mutual inductance between L_1 and L_2, is zero [Fig. 14–6(a)]. The coupling effect of tank capacitor C_T provides positive feedback. For example, if the plate current is increasing, C_T will change to a higher voltage and the accompanying charge current will flow down through L_2. Decay in the charge current is associated with a self-induced voltage e_2 in L_2, and has positive polarity at the grid. In other words, this self-induced voltage is in the same direction as the charge current and tends to sustain the charge current. Conversely, if the plate current is decreasing, tank capacitor C_T will discharge. The path of discharge current is up through L_2. In turn, self-induced voltage (e_2) in L_2 now drives the grid negative with respect to the cathode which causes a further decrease in plate current.

We perceive that C_T couples a self-induced voltage on both half cycles back to the grid, and that this self-induced voltage is in phase with the initial excitation voltage. Thus, positive feedback is provided, and the condition for oscillation is sustained. The time constant of $R_G C_1$ should be fairly long compared with the period of one oscillator cycle. If the time constant is too short, the grid bias voltage will be insufficient, and if it is too long, the oscillator will cease operation and block periodically because of excessive bias voltage. Self-bias (signal-developed) bias voltage is advantageous because the oscillator circuit is self-starting when operating voltages are applied to the tube.

To extract oscillatory energy from an oscillator circuit, a load must be added. Loading of an oscillator may be accomplished by placing a relatively high resistance in shunt with the tank circuit, or placing a low resistance in series with the tank. In either event, the effective Q value of the tank is reduced, and the amplitude of generated signal voltage is decreased. A load may be inductively coupled to the tank; in such case, it is best practice to locate the coupling point near *r-f* ground potential to decrease the loading effect. In any case, oscillator loading tends to detune the tank circuit. Hence, isolating or buffer amplifier stages are often inserted between oscillator and load to minimize frequency shift with load variation.

An oscillator operates close to the resonant frequency of the tank circuit if its effective Q value is high. Thus, the oscillating frequency is given to a good approximation by the familiar resonant-frequency formula

$$f_o = \frac{1}{2\pi\sqrt{LC}} \qquad \textbf{(14–1)}$$

where L denotes the inductance of the tank coil and C denotes the shunt capacitance. (Observe that the tank circuit depicted in Fig. 14–6 is a series-parallel network that must be reduced to an equivalent parallel circuit for resonant-frequency calculation.)

In general practice, a more convenient expression of Formula 14–1 may be derived as follows.

$$\begin{aligned} L &= L_{\mu h} \times 10^{-6} \\ C &= C_{pf} \times 10^{-12} \\ \frac{1}{2\pi} &= 0.159 \\ f &= f_{MHz} \times 10^{6} \end{aligned} \qquad \textbf{(14–2)}$$

Example 1. Suppose that the tank inductance is 100 μh, and the shunt tank capacitance is 100 $\mu\mu$f. Then, we calculate the oscillator frequency as follows.

$$f_{MHz} = \frac{0.159 \times 10^{-6}}{\sqrt{L_{\mu h} \times 10^{-6} \times C_{pf} \times 10^{-12}}}$$

or,

$$f_{MHz} = \frac{159}{\sqrt{L_{\mu h} C_{pf}}}$$

$$f_o = \frac{159}{\sqrt{100 \times 100}} = 1.59 \text{ MHz} \qquad \text{(answer)}$$

The shunt-fed (parallel-fed) Hartley oscillator circuit depicted in Fig. 14–6(d) differs from the series-fed version in that direct current does not flow through the tank circuit. The power supply is connected in shunt with the triode plate and the portion of the tank circuit that includes L_1. We observe that the d-c component of plate current is isolated from L_1 by blocking capacitor C_2, and that the a-c component is isolated from the power supply by the *r-f* choke coil. In high-power oscillators, such as used in radio transmitters and industrial-electronic equipment, shunt feed is advantageous because the tank coil is less dangerous as a source of shock to operating personnel. Although *r-f* burns can result, the operator is protected from more serious injury by the high d-c voltage from the power supply.

Operation of Hartley oscillators employing either vacuum tubes or field-effect transistors is identical. However, we must classify certain differences in the operation of the junction-transistor Hartley oscillator. Careful comparison of the transistor circuit in Fig. 14-7(a) to the vacuum-tube circuit in Fig. 14-6(a) reveals resistor R_B in transistor circuit. This resistor establishes forward-bias to bring the transistor

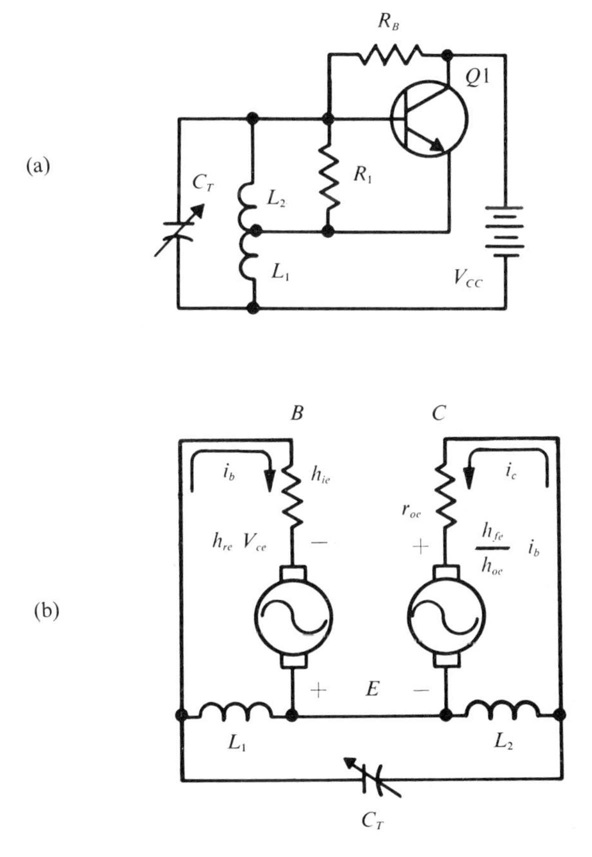

Figure 14-7. Transistor Hartley oscillator circuits. (a) Amplifier circuit; (b) Equivalent circuit.

into conduction. *Note that junction transistors must be forward-biasea to conduct; whereas, field-effect transistors and vacuum tubes will conduct with a zero-volt bias.*

The equivalent circuit of the Hartley oscillator in Fig. 14–7(b) reveals that the frequency of the transistor Hartley oscillator is affected by the parameters of the transistor. The frequency of oscillation is formulated

$$f_o = \frac{1}{2\pi\sqrt{C(L_1 + L_2) - L_1 L_2 h_{oe} h_{ie}}} \qquad \textbf{(14–3)}$$

In practical application, the product $L_1 L_2 h_{oe} h_{ie}$ is small and may be dropped from Formula 14–3.

Example 2. Suppose the values of inductors in the oscillator in Fig. 14–7(a) are $L_1 = 100\ \mu\text{h}$ and L_2 is $10\ \mu\text{h}$. What is the oscillating frequency when $C_T = 100\ \mu\text{f}$?

$$f = \frac{0.159}{\sqrt{1 \times 10^{-10}(110 \times 10^{-6})}}$$

$$f = 1.51\ \text{MHz} \qquad \text{(answer)}$$

As in any oscillator, the amplifying device must have a minimum gain factor to overcome the circuit losses. The conditions for sustained oscillations in a Hartley oscillator are

$$\frac{h_{fe}}{\Delta h} \simeq \frac{L_2}{L_1} \qquad \textbf{(14–4)}$$

where $\Delta h = h_{ie} h_{oe} - h_{fe} h_{re}$.

These h parameters are given for most types of transistors in a manufacturer's data sheet.

h_{ie} = input resistance
h_{oe} = output admittance
h_{fe} = forward-current transfer
h_{re} = reverse-voltage transfer

Example 3. Suppose the transistor used in the circuit in Example 2 is type 2N508 with parameters $h_{fe} = 110$, $h_{ie} = 2.8\ \text{k}\Omega$, $h_{re} = 7.5 \times 10^{-4}$, and $h_{oe} = 43\ \mu\text{mho}$. What is the required ratio of L_2 to L_1 for sustained oscillations?

$$\frac{L_2}{L_1} \simeq \frac{110}{2.8 \times 10^3 \times 4.3 \times 10^{-6} - 1.1 \times 10^2 \times 7.5 \times 10^{-4}}$$

$$= \frac{110}{-70.2 \times 10^{-3}}$$

$$\frac{L_2}{L_1} \simeq \frac{15.7}{1} \qquad \text{(answer)}$$

Note: The negative sign in the denominator is ignored.

There are situations such as depicted in Fig. 14-8 where there is a possibility of oscillation at two or more frequencies. In case the amount of positive feedback is the same at each frequency, the circuit

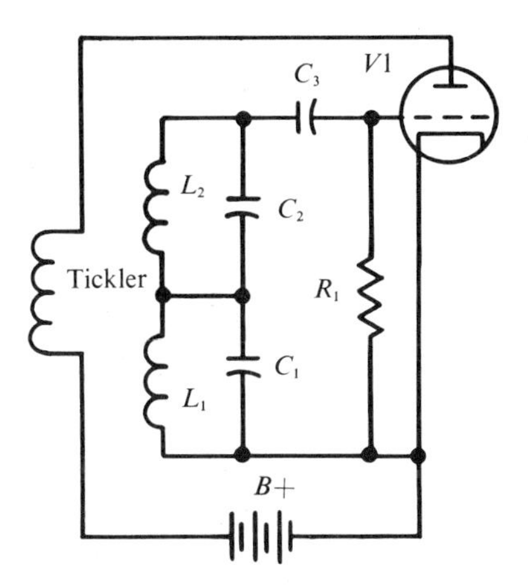

Figure 14-8. A tank resonant at two frequencies oscillates at the higher-Q frequency.

will oscillate at the frequency that is associated with the higher Q value. The frequency associated with the lower Q value will be suppressed. Thus, if a tank coil is poorly designed and has a large amount of leakage reactance, the oscillator may operate at a considerably higher frequency than anticipated. Again, if there is a possibility of oscillation at two different frequencies, and both resonant circuits have practically the same Q value, the oscillating frequency may "jump" up or down intermittently. This type of circuit action is called "mode-hopping" in ultra-high frequency oscillators.

We will find that there are certain types of oscillators that may "jump" in output frequency as the output loading varies. Again, this is an example of an oscillator that has two possible frequencies of oscillation. Load variation changes the Q of one of the associated resonant circuits which results in a frequency jump.

14-3. Colpitts Oscillator

Another widely used oscillator circuit, the Colpitts configuration, is depicted in Fig. 14-9(a). It is somewhat similar to the Hartley oscillator, with the exception that a split-tank capacitor is employed as part of the feedback circuit instead of a split-tank inductor. The

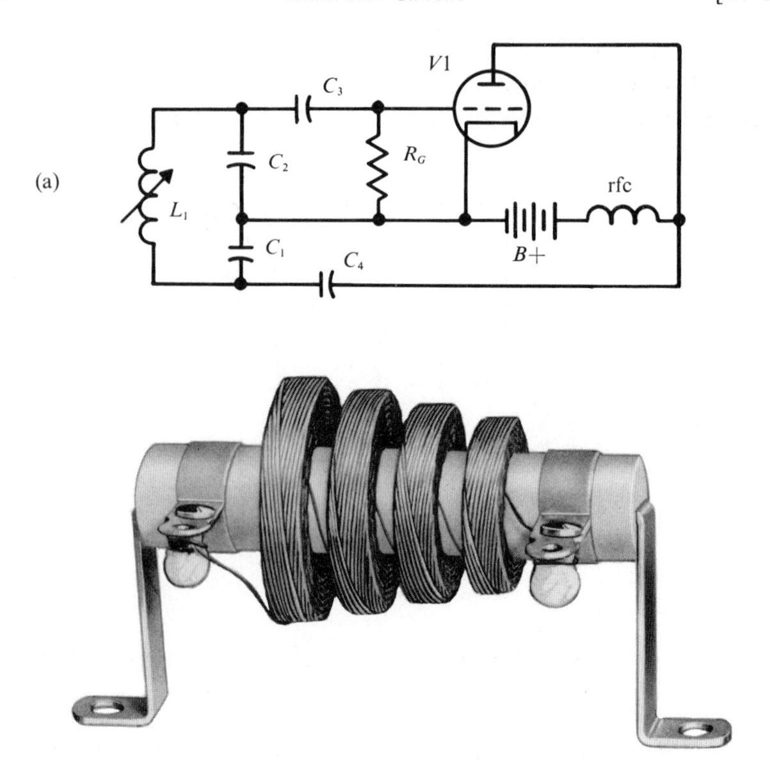

Courtesy of J. W. Miller Co.

Figure 14–9. (a) Basic Colpitts oscillator circuit; (b) A transmitting radio frequency choke.

frequency of oscillation is determined by the values of L_1, C_1, and C_2; grid excitation voltage appears across C_2 instead of across L_2 as in the shunt-fed Hartley oscillator. C_3 and C_4 perform the same function in the Colpitts circuit as C_1 and C_2 in the shunt-fed Hartley circuit. As illustrated in Fig. 14–9(b) *R-F* chokes used in oscillators are commonly split up into several series sections. This design provides a reasonably high value of impedance over an appreciable frequency range. The Colpitts oscillator has a useful frequency range from 100 Hz to 1,000 MHz.

A simplified vector analysis of the Hartley and the Colpitts oscillator circuits is seen in Fig. 14–10. In the Hartley configuration [Fig. 14–10(a)], C_2 couples the triode plate to the lower end of L_1, placing e_b across L_1 and across the series combination of L_2 and C_T. The L_2-C_T branch is predominantly capacitive—(X_C is greater than X_L). Current

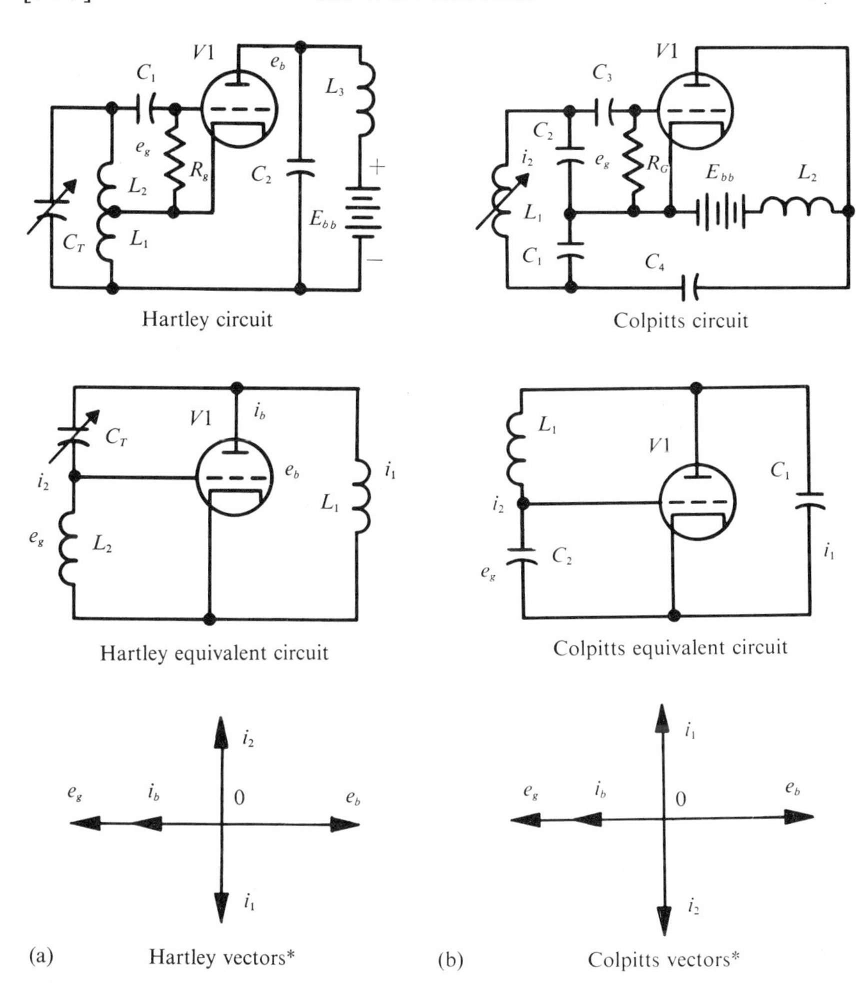

*Tank circuits are assumed to have very high Q values.

Figure 14–10. Vector diagrams for the Hartley and Colpitts oscillator circuits.

flow i_2 in this branch leads e_b by 90°. Grid-excitation voltage e_g is developed by L_2, and this voltage leads current i_2 by 90°. Thus, the voltage e_b across the plate load impedance is 180° out of phase with the grid-excitation voltage e_g.

Next, in the Colpitts oscillator configuration [Fig. 14–10(b)], C_4 couples the plate to the lower end of C_1, thus placing e_b across the C_1 and the series combination of C_2 and L. The a-c component of current

i_1 in C_1 leads e_b by 90°. Note that the C_2-L branch is predominantly inductive because X_L is greater than X_C in this branch; current i_2 lags voltage e_b across the L-C_2 branch by 90°. Again, the voltage e_b across the plate load impedance is 180° out of phase with e_g.

A transistor Colpitts oscillator is depicted in Fig. 14–11(a). Again, the significant difference between using the vacuum tube and the

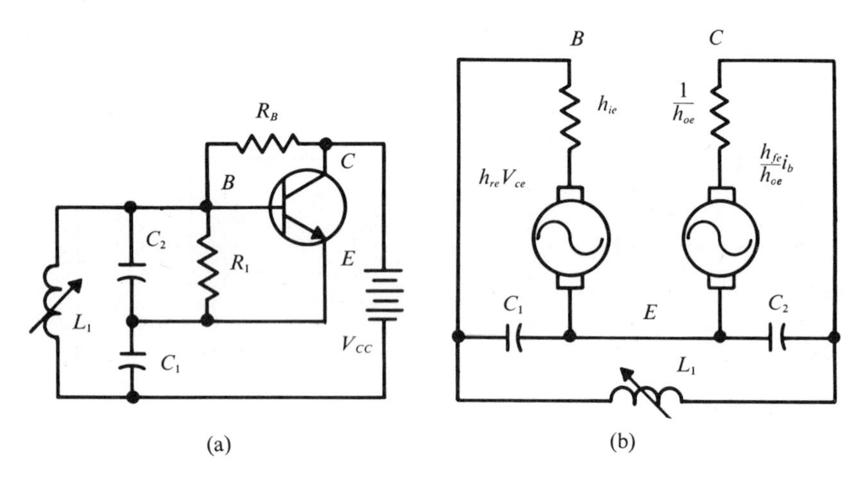

Figure 14–11. A transistor Colpitts oscillator. (a) Circuit diagram; (b) Equivalent circuit.

field-effect transistor or the junction transistor as an amplifying device is that the latter must be forward-biased. From the equivalent circuit in Fig. 14–11(b), the frequency of oscillations may be formulated

$$f^2 = \left(\frac{h_{22}}{C_1 C_2 h_{11}} + \frac{1}{LC_1} + \frac{1}{LC_2}\right)\frac{1}{4\pi^2} \qquad \textbf{(14–5)}$$

In a practical circuit, the first term on the right side of the equation is very small and may be eliminated. The result is that the frequency of oscillations, for either the vacuum tube or the transistor circuit, is determined by the resonant circuit with capacitors C_1 and C_2 connected in series.

$$f_o \simeq \frac{1}{2\sqrt{LC_T}} \qquad \textbf{(14–6)}$$

where

$$C_T = \frac{C_1 C_2}{C_1 + C_2}$$

and when

$$C_2 \gg C_1, C_T \simeq C_2$$

Example 4. Suppose a Colpitts oscillator circuit has the following tank circuit values: $L = 100$ h, $C_1 = 500$ pf, and $C_2 = 50$ pf. What is the oscillating frequency?

$$C_T \simeq C_2$$

$$\therefore \quad f_o = \frac{1}{2\pi\sqrt{1 \times 10^{-4} \times 50 \times 10^{-12}}}$$

$$= 2.25 \text{ MHz} \qquad \text{(answer)}$$

14-4. Ultra-Audion Oscillator

Another oscillator, commonly utilized at ultrahigh frequencies, was invented by Dr. DeForest, and named the ultra-audion circuit by him. Its useful frequency range extends to 1,000 MHz. A basic circuit arrangement is depicted in Fig. 14-12(a). There is a basic similarity between the ultra-audion oscillator and the Colpitts oscillator. We observe that the grid-to-cathdoe and plate-to-cathode interelectrode capacitances indicated in Fig. 14-12(a) provide this similarity.

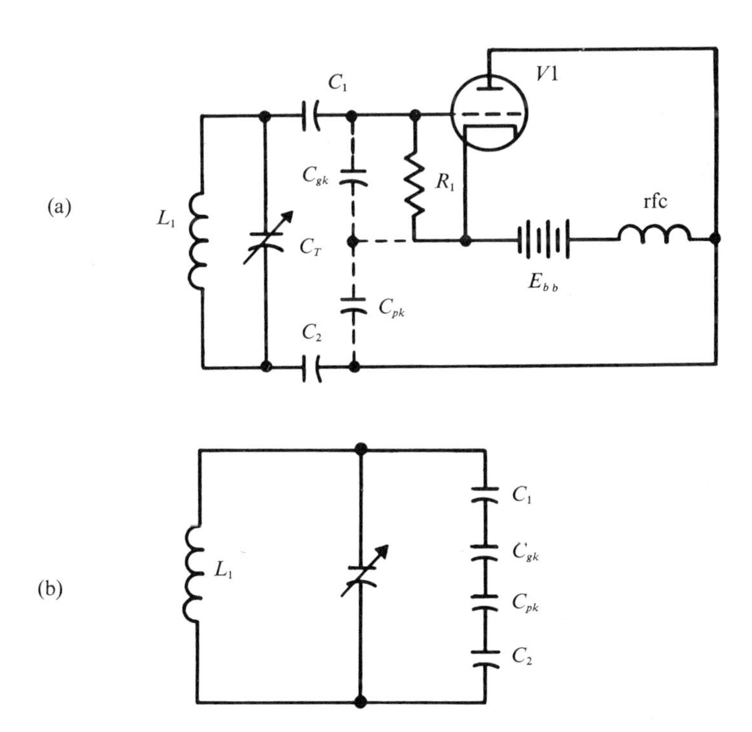

Figure 14-12. (a) Ultra-audion oscillator circuit; (b) Equivalent tuned circuit.

Parallel d-c plate supply is necessarily utilized; the *r-f* choke prevents the a-c component of plate current from entering the power supply. Blocking capacitor C_2 is inserted to provide a low-reactance path for *r-f* current flow, while preventing the tank coil from assuming supply potential.

The voltage drop across C_{gk} in Fig. 14–10(a) is appreciable at the operating frequency, and this voltage drop provides grid excitation. Grid-bias voltage is developed by flow of grid current through R. We observe in Fig. 14–12(b) that the total tank capacitance comprises C_T in parallel with the series combination of C_1, C_{gk}, C_{pk}, and C_2. Capacitors C_1 and C_2 have comparatively large values so that little reactance is imposed to *r-f* current flow. Therefore, at high frequencies the oscillating frequency is approximately

$$f_o = \frac{1}{2\pi\sqrt{LC}}$$

where

$$C = C_T + \frac{C_{gk}\,C_{pk}}{C_{gk} + C_{pk}} \tag{14–7}$$

Since C_{gk} and C_{pk} have excessively high reactance at low radio frequencies, the circuit does not oscillate unless external capacitances are connected to supplement the interelectrode capacitances. When this is done, the circuit becomes a conventional Colpitts configuration.

14–5. Tuned-Plate Tuned-Grid Oscillator

Another basic type of oscillator utilizes a tuned circuit in both the input circuit and the output circuit, as shown in Fig. 14–13. This oscillator conventionally uses a vacuum tube as a control device and is called a tuned-plate, tuned-grid (TPTG) oscillator. However, we will employ a field-effect transistor to illustrate that a transistor may assume the function of a vacuum tube. When the FET is employed as a control device, the circuit is called a tuned-drain, tuned-gate (TDTG) oscillator. We will observe that positive feedback takes place via gate-drain interelectrode capacitance. Hence, the TDTG oscillator is less useful at low frequencies of operation than some of the oscillator configurations discussed previously. However, the TDTG oscillator finds wide application from 200 kHz to 30 MHz.

In the equivalent circuit for a TDTG oscillator [Fig. 14–13(b)], both (parallel) tanks operate slightly below resonance and appear as highly inductive components L_1 and L_2. Since the drain-gate capacitance C_{dg} is small, the reactance X_{cgd} is correspondingly large; X_{cgd}

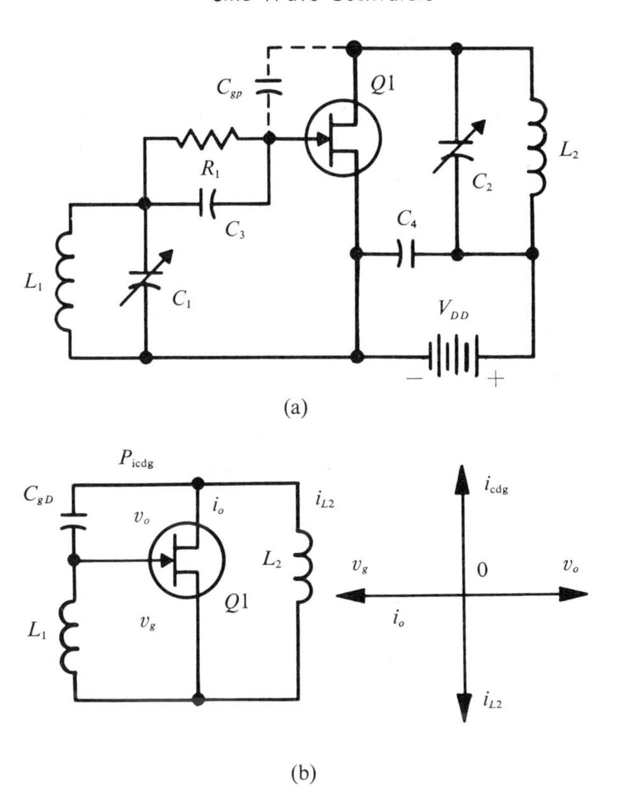

Figure 14–13. Tuned-plate tuned-grid oscillator. (a) Employing a field-effect transistor; (b) Equivalent circuit with vector diagram.

is greater than $X_{L'1}$, and, in turn, the left hand branch is capacitive. The right-hand branch is inductive, and, at the oscillating frequency,

$$X_C - X_{L'1} = X_{L'2} \tag{14–8}$$

Drain voltage v_d [Fig. 14–13(c)] is the reference vector; e_d appears across both branches. Current i_{L2} lags e_d by 90°, because it flows through the inductive circuit L'_2. Current i_{cgd} leads v_d by 90°, because it flows through the left-hand branch which contains more capacitive reactance than inductive reactance. Gate voltage v_g leads i_{cdg} by 90° because it drops across the inductive portion L'_1 of the left-hand branch. Note that i_d is in phase with v_g and 180° out of phase with v_d; this is the relation that is necessary to sustain oscillation. Note that the tank having the highest Q value in Fig. 14–11(a) determines the oscillator frequency. Thus, if the drain tank is more heavily loaded than the grid tank, the oscillator frequency will be determined by the gate tank, inasmuch as the gate tank will then have the higher Q value.

14–6. Push-Pull Oscillators

To obtain a sine-wave power output that is larger than can be provided by a control device of a certain type, an additional control device may be included in the oscillator circuit to operate in push-pull. Note that parallel operation of oscillator control devices is

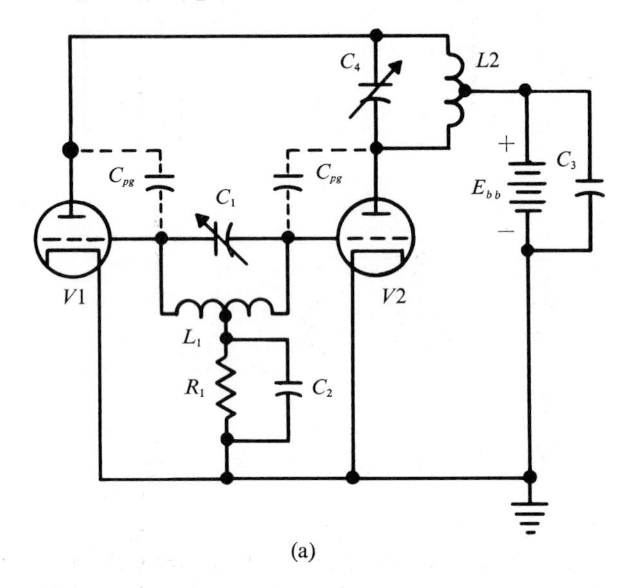

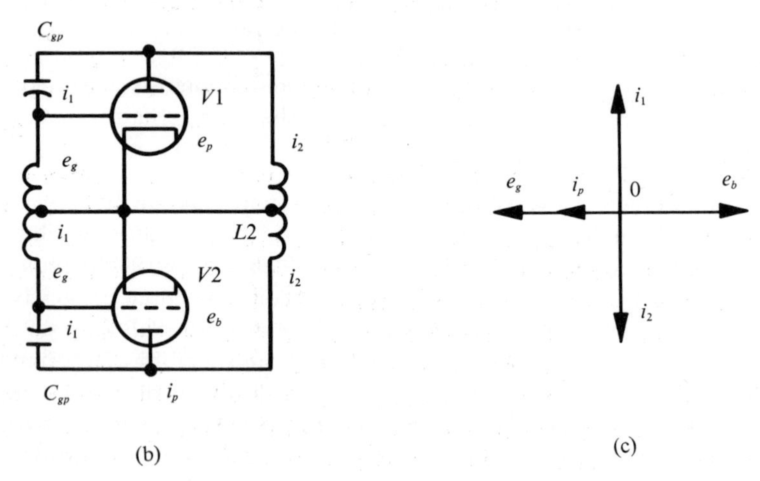

Note: Diagram assumes that the Q value of the tank is very high.

Figure 14–14. Basic push-pull oscillator circuit.

difficult because of their tendency to develop parasitic oscillation. As in the case of push-pull amplifiers, the even-harmonic content is reduced in the output waveform of a push-pull oscillator. Also, the frequency stability of a push-pull oscillator is greater than that of a single-ended oscillator. Interelectrode capacitances are effectively connected in series, which reduces the capacitance in shunt to the tank. This is an advantage in very high frequency oscillators to obtain an extended high-frequency range.

A basic push-pull oscillator circuit employing vacuum tubes is shown in Fig. 14–14. It utilizes the interelectrode capacitances of each tube to feed back sufficient signal voltage to the grid tank for maintenance of oscillation. When the oscillator is first turned on, it is improbable that the two tubes will be operating under exactly the same conditions. In turn, one tube will conduct more current than the other. Since there is a net current flow through the tank, a small a-c voltage appears across the tank terminals at the resonant frequency; this a-c voltage rapidly builds up to a level determined by the value of signal-developed bias in relation to the plate resistance and amplification factor of the tubes.

We perceive that a push-pull oscillator is effectively a pair of TPTG oscillators connected in push-pull. Thus, the sequence of events in V_1 is similar to the action previously described for the TPTG oscillator in Fig. 14–13. Like push-pull amplifiers, grid voltages in a push-pull oscillator are 180° out of phase. It is important that the center taps on L_1 and L_2 in Fig. 14–12 be located at the exact center of the coil windings in order to avoid distortion in the output waveform. In normal operation, even harmonics are cancelled by a push-pull oscillator.

14–7. Electron-Coupled Oscillators

Three basic factors that affect the frequency stability of an oscillator are (1) a change in value of L or C; (2) a change in load; (3) variation in the power-supply voltage. An electron-coupled oscillator (ECO) circuit provides light loading to the utilization circuit, especially when the plate tank is used as a frequency multiplier. We will find that power-supply voltage variations tend to cancel in the screen and plate circuits of an electron-coupled oscillator. Because of light loading and its self-regulating feature, the ECO has comparatively good frequency stability.

With reference to Fig. 14–15, the electron-coupled oscillator

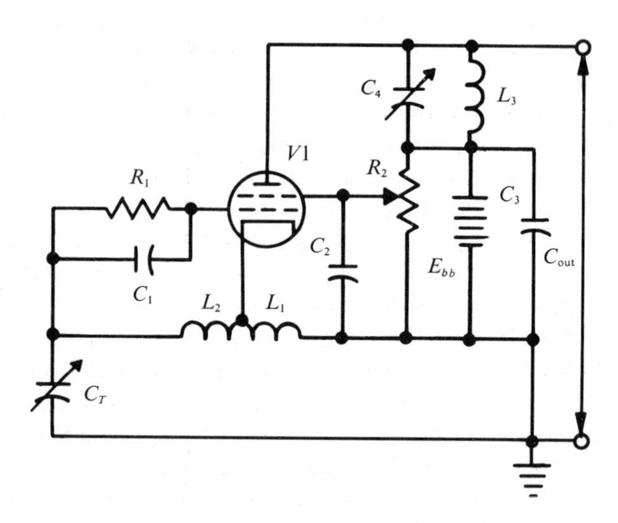

Figure 14-15. An electron-coupled oscillator circuit.

circuit combines the functions of an oscillator and of a power amplifier. Note that the control grid, tank circuit, (C_T, L_1, and L_2), cathode and screen grid form a series-fed Hartley oscillator with the screen grid functioning as a plate. Capacitor C_2 places the screen at zero *r-f* potential, and, like C_3, bypasses the power supply. The output tuned circuit, C_4-L_3, is in the plate branch. We observe that the electron stream is the only coupling medium between the grid tank and the plate tank; this fact is the basis for the name electron-coupled oscillator. Insofar as capacitive coupling is concerned, the two tank circuits are isolated by the screen grid, which is at *r-f* ground potential.

In turn, this type of oscillator is comparatively stable. Load variations have little effect on the oscillation frequency. An increase in screen voltage decreases the oscillation frequency, while an increase in plate voltage increases the oscillation frequency. If the screen and plate voltages are obtained from the same power supply, and the supply voltage increases, the tendency of the plate to increase the frequency and of the screen to decrease the frequency contributes to frequency stabilization. Conversely, a decrease in supply voltage causes opposing frequency shifts. Potentiometer R_2 is adjusted for optimum screen voltage, after which no further adjustment should be necessary. The frequency stabilization is optimum when the ratio of plate-to-screen voltage is approximately 3 to 1.

14-8. Negative-Resistance Oscillator

Negative resistance is defined as a voltage-current relation in which an increase in voltage is associated with a decrease in current, and vice versa. Analysis will show that negative-resistance action is present in any oscillator, such as those that have been discussed. However, it is customary to consider that negative-resistance oscillators are those that exploit secondary emission in triodes or tunneling action in semiconductors. One of the oldest negative-resistance oscillators is called the dynatron. It utilizes a screen-grid tube and depends for its operation on secondary emission from the plate. The dynatron does not employ positive feedback of the control grid; instead, it employs interaction between screen grid and plate.

In Fig. 14-14, plate current starts to flow when the tube is turned on. At point B [Fig. 14-14(b)], the plate current i_b starts to decrease because of secondary emission at the plate. The decrease in i_b from points B to C is accompanied by a decrease in the voltage drop across the plate-load impedance (parallel LC tank) and an increase in plate voltage e_b. At point C, secondary emission ceases, and i_p stops decreasing and starts to increase. The increasing voltage across the plate-load impedance is accompanied by a decrease in e_b. At point B, i_b stops rising and starts to fall. The voltage across the plate-load impedance again starts to fall as e_b rises, and the cycle of operation repeats.

The vectors depicted in Fig. 14-16(c) indicate the phase relations between the a-c components of plate voltage e_b, plate current i_b, inductor current i_L, and capacitor current i_C. As would be anticipated, the circuit oscillates at approximately the resonant frequency of the tank circuit. Because the plate-load impedance is in series with the tetrode plate and E_b, the plate voltage will decrease as the voltage e_t and current in the plate-load impedance increase. In turn, e_b and i_b are 180° out of phase. Current i_L in the inductive branch lags e_t by almost 90° (the losses in the tank are associated with the coil). Current i_C leads e_t by 90° (losses in the capacitive branch are negligible). The vector sum of i_L and i_C is the total input current flowing to and from the terminals of the plate-load impedance. This is the a-c component of the plate current i_b.

A tunnel diode has a voltage-current characteristic that can be compared with that of the dynatron, as seen in Fig. 14-17. The tunnel diode with its negative resistance does not take, but rather gives, energy to the circuit. As with the vacuum tube or transistor, this energy is supplied by the d-c power source.

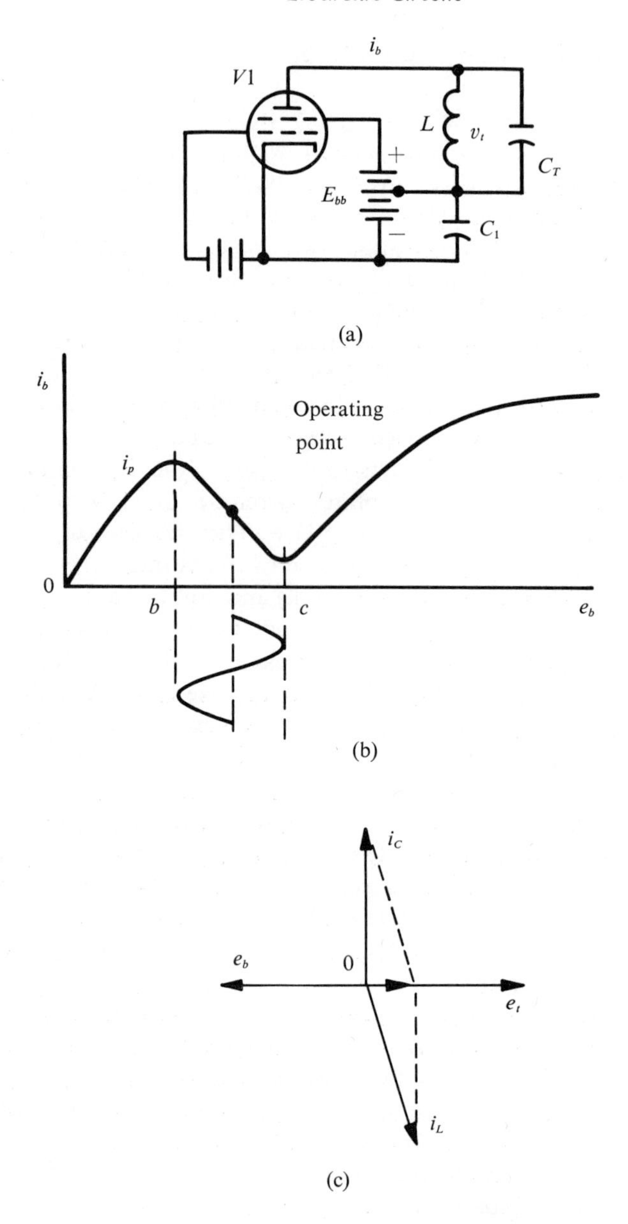

Figure 14-16. **Dynatron oscillator.**

The necessary conditions for the tunnel diode to operate as an oscillator is that its negative resistance must be less than the equivalent parallel resistance of the circuit and satisfy the conditions

$$r_n < \frac{L}{R_S C} \tag{14–9}$$

where r_n is the negative resistance of the diode, L is the total circuit inductance, C is the total circuit capacity, and R_S is the equivalent resistance across the diode of all the circuit losses as well as the resistance of the power supply.

The necessary condition for the tunnel diode to oscillate at high frequencies is that the self-resonant frequency of the equivalent circuit (Fig. 14–18) be greater than the resonant frequency of the LC circuit. The self-resonant frequency of the tunnel diode is formulated

$$f_o = \frac{1}{2\pi\, r_n C}\sqrt{\frac{r_n^2 C}{L_s} - 1} \tag{14–10}$$

Example 5. Determine the self-resonant frequency of a type 1N652 tunnel diode.

$$-r_n = 20\ \Omega$$
$$L_s = 8\ \text{ph}$$
$$C = 40\ \text{pf}$$
$$R_s = 1\ \Omega$$

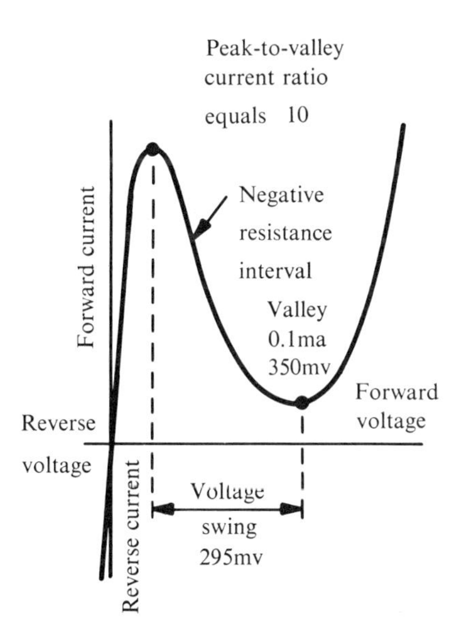

Figure 14–17. Voltage-current characteristic of a tunnel diode.

$$f_o = \frac{1}{2\pi \times 20 \times 40 \times 10^{-12}} \sqrt{\frac{(20)^2 \times 40 \times 10^{-12}}{8 \times 10^{-9}} - 1}$$

$$\simeq 200 \text{ MHz} \qquad \text{(answer)}$$

Except for the fact that the voltage-current characteristic is attributed to tunneling action, operation of the tunnel-diode oscillator in Fig. 14–18 is basically the same as that of the dynatron oscillator. A tunnel-diode circuit can oscillate in the GHz range. As the voltage across the diode is increased from 55 mv to 350 mv, the current flow through the diode lessens. On the other hand, as the voltage across the diode is decreased from 350 mv to 55 mv, the current flow through the diode increases. This negative-resistance phenomenon is a dynamic characteristic of which the conductance value is the slope of a tangent drawn through the operating point.

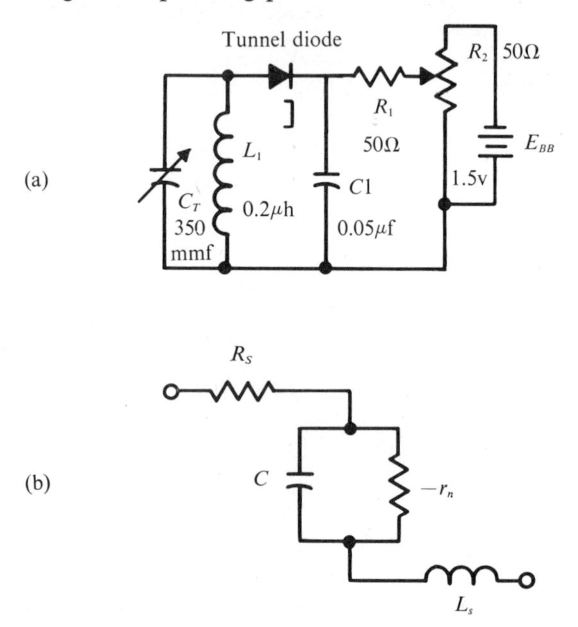

Figure 14–18. (a) Tunnel-diode oscillator circuit; (b) Equivalent circuit of a tunnel diode.

SUMMARY

1. An oscillator circuit consists basically of an amplifying device connected in conjunction with a frequency-selective network that supplies positive feedback.

2. Low-frequency sine-wave oscillators employ *RC* networks as frequency-selective components.

3. High-frequency oscillators generally utilize *LC* resonant circuits as frequency-selective components.

4. The two conditions necessary for production of sustained oscillations in an oscillator circuit are positive feedback and sufficient gain to overcome circuit and load losses.

5. Self-bias is advantageous in an oscillator circuit because the circuit is self-starting when operating voltages are applied to the circuit.

6. Loading an oscillator reduces the amplitude of the generated signal voltage and tends to detune the tank circuit.

7. A buffer-stage isolates the oscillator from the load to minimize loading effects.

8. There may be a possibility of oscillation at two frequencies in a poorly designed oscillator, and this may cause the oscillator to jump from one frequency to the other intermittently.

9. The Colpitts oscillator employs a split-tank capacitor instead of a split inductor as part of the feedback circuit.

10. The useful range of the ultra-audion oscillator circuit extends to 1 GHz.

11. The ultra-audion oscillator utilizes the internal capacitance of the control device in parallel with a variable capacitor to control the oscillating frequency.

12. The positive feedback in a TPTG oscillator takes place through the interelectrode capacitance of the control device.

13. The plate-tank circuit of a TPTG oscillator is more heavily loaded than the grid tank. The result is a much lower plate tank *Q*, allowing the grid tank to control the oscillating frequency.

14. Control devices may be operated in parallel to develop greater output power.

15. Parallel operation of control devices in an oscillator is difficult because of a tendency for the devices to develop parasitic oscillations.

16. Three basic factors which affect the frequency stability of an oscillator are a change in the value of *L* or *C*, a change in the load, and a variation in the power-supply voltage.

17. The electron-coupled oscillator has comparatively good frequency stability due to the light loading of the resonant tank and the self-regulating action against power supply variations.

18. The dynatron oscillator employs interaction between the screen grid and the plate to develop regeneration.

19. A tunnel diode is a negative-resistance device which obtains energy from the d-c supply and returns it to the tank circuit to produce oscillations.

20. A tunnel-diode oscillator can be constructed to oscillate in the GHz range.

21. The requirements for a tunnel-diode circuit to oscillate are that the negative resistance of the diode be less than the circuit shunt resistance and that the resonant frequency of the circuit be less than the self-resonant frequency of the diode.

Questions

1. Explain two parts of a basic oscillator.
2. What are the basic functions of an oscillator?
3. Why do low-frequency oscillators usually employ *RC* circuits for frequency selection?
4. Why do high-frequency oscillators usually employ *LC* circuits for frequency selection?
5. What are two requirements for a circuit to oscillate?
6. How may regeneration be developed in a transistor or vacuum tube?
7. Why do the Hartley and the Colpitts oscillators have very high values of efficiency?
8. What is the most useful frequency range of the Hartley oscillator?
9. What is the function of an *r-f* choke in an oscillator circuit?
10. Explain the operation of a transistor Hartley oscillator through one cycle.
11. What are the effects of loading an oscillator circuit?
12. Why is a buffer stage used between an oscillator and the load?
13. The term "mode hopping" refers to ultrahigh frequencies. What is the cause of this action?
14. What are the basic differences between the Colpitts and the Hartley oscillator circuit?
15. A student constructs the Hartley oscillator in Fig. 14-7 as a laboratory experiment, and it does not oscillate upon application of the supply voltage. What are the four possible troubles with the circuit?
16. What is the useful operating frequency range of the Colpitts oscillator?
17. What is the useful operating frequency range of the ultra-audion oscillator?
18. With reference to the oscillator in Fig. 14-12(a), what is the function of the choke (rfc)? What is the function of capacitors C_1 and C_2?
19. How is grid bias developed in the oscillator depicted in Fig. 14-12(a)?
20. What is the useful frequency range of a TPTG oscillator?
21. What circuit components control the frequency in the circuit in Fig. 14-13?
22. What are the advantages of using an FET rather than a vacuum tube in the oscillator represented in Fig. 14-13?
23. What are the advantages of a push-pull oscillator?

24. To what basic oscillator can the push-pull oscillator in Figure 14–14 be compared?

25. Explain the operation of the oscillator in Fig. 14–14.

26. Why must the tap on coil L_1L_2 in Fig. 14–4(b) be located in the exact center of the coil?

27. What are the advantages of electron-coupled oscillators (ECO)?

28. What is the proper screen-to-plate voltage ratio for optimum frequency stability?

29. Explain the principle of operation of the dynatron oscillator.

30. Discuss one cycle of operation of the dynatron oscillator.

31. What are the circuit requirements for a tunnel diode to operate as an oscillator?

PROBLEMS

1. A Hartley oscillator employs a vacuum tube as an amplifier with a tank inductance of 200 μh and a shunt tank capacitance of 160 pf. What is the oscillating frequency?

2. A Hartley vacuum-tube oscillator circuit is to oscillate at 20 MHz. What value of tank capacitance must be employed if the tank inductance is to be 28 μh?

3. A Hartley field-effect transistor oscillator circuit uses a 100 μh tank inductor and a 22 pf / 200 pf-variable capacitor. What is the frequency range of the oscillator?

4. A junction transistor Hartley oscillator has a tank circuit with a 300-pf capacitor and an equivalent inductance of 80 μh. What is the operating frequency?

5. What is the ratio of L_1 and L_2 in Problem 4 if a 2N1097 transistor is used in which $h_{ie} = 1.4\ \text{k}\Omega$, $h_{oe} = 30\ \mu$mho, $h_{re} = 4.5 \times 10^{-4}$ and $h_{fe} = 44$?

6. The Colpitts oscillator in Fig. 14–9 employs the values $C_1 = 200$ pf, $C_2 = 40$ pf, and $L = 200\ \mu$h. What is the oscillating frequency?

7. An ultra-audion oscillator has the following circuit values: $L = 10\ \mu$h, $C_T = 22$ pf, $C_1 = C_2 = 300$ pf, $C_{gk} = 15$ pf, and C_{pk} 12 pf. What is the oscillating frequency of the circuit?

8. Determine the self-resonant frequency of a type 1N652 tunnel diode with the internal values $L_s = 8$ ph, $-r_n = 40\ \Omega$, $C = 40$ pf, and $R_s = 1\ \Omega$.

9. Will the tuned-diode circuit in Problem 8 oscillate in a circuit in which $L = 100$ ph and $C = 47$ pf?

CHAPTER 15

Piezoelectric and RC Sine-Wave Oscillators

15–1. Introduction

When the operating frequency of an electronic system must be held within close limits, a crystal-controlled oscillator is ordinarily used. This is an oscillator in which the mechanical resonant frequency of a specially cut crystal controls the operating frequency. Figure 15–1 illustrates a citizens-band transceiver that utilizes quartz-crystal oscillators. Crystal-controlled oscillators are a standard means for maintaining the frequencies of radio transmitting stations within assigned limits. Highly accurate time intervals are utilized in radar systems, and a crystal-controlled oscillator is generally used as a basic circuit to generate a highly accurate frequency.

Crystals exhibit a characteristic called the *piezoelectric effect*, by which mechanical stresses produce electric charges, and conversely, electric charges produce mechanical stresses. That is, if an a-c voltage

Courtesy of EICO, Electronic Instrument Co., Inc.

Figure 15–1. Piezoelectric oscillators are used in CB transceivers.

is applied to a crystal, it will vibrate at the frequency of the applied voltage. The amplitude of vibration of a particular crystal, however, is much larger at a certain frequency than at all other frequencies. This is called the *resonant frequency* of the crystal. It is a function of the mechanical dimensions and physical properties of the crystal lattice. We will find that inductance, capacitance, and resistance have mechanical analogies; in turn, it is not surprising that mechanical systems have resonant frequencies.

The piezoelectric effect is exhibited to different degrees by crystals of various substances, such as quartz, tourmaline, and Rochelle salt. Of these three, Rochelle salt exhibits the greatest piezoelectric effect. However, quartz crystals are used for crystal control in most oscillators because quartz has great mechanical strength. A natural quartz crystal has a hexagonal cross-section and pointed ends, as shown in Fig. 15–2. Sections cut from the crystal have electrical properties associated with mechanical stresses that are expressed in terms of three sets of axes. The axis joining the points at the ends of the crystal is called the *optical* or *Z axis*. Note that mechanical stresses along this particular axis produce no piezoelectric effect.

When a quartz crystal is cut in a plane perpendicular to its optical axis, a hexagonal cross-section is seen, as depicted in Fig. 15–2(b). The *X* axes drawn through the corners of the hexagon are called the *electrical axes*, and the *Y* axes drawn perpendicular to the faces of the hexagon are called the *mechanical axes*. Designation of the axes as electrical or mechanical aids in defining the type of cut. For example, in Fig. 15–2(c), the *X*-cut crystal is shown cut perpendicular to one of the *X* axes. During one-half cycle of the a-c voltage impressed across

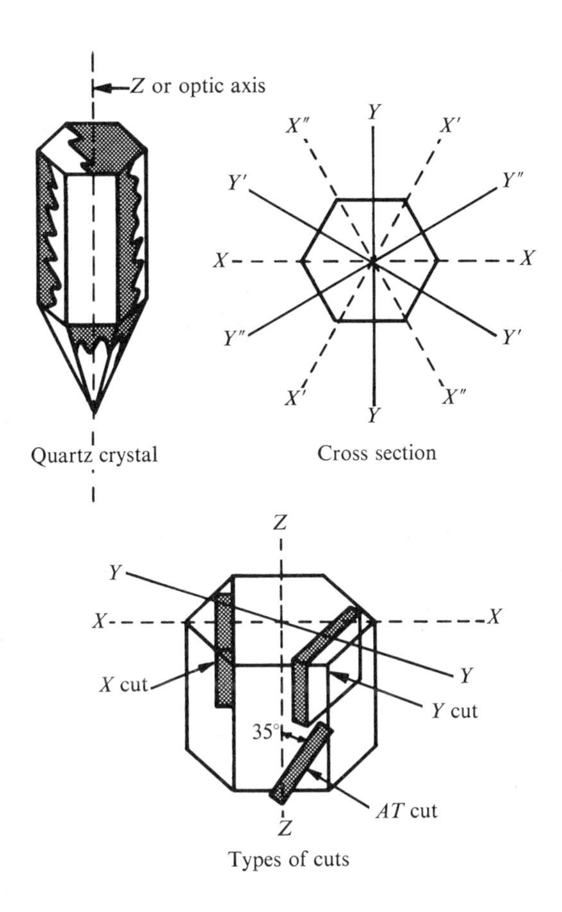

Figure 15–2. Quartz-crystal outline and terminology.

the flat surfaces of this *X*-cut crystal, the crystal will expand along the axis perpendicular to its flat surfaces, and during the other half cycle it will contract. Similarly, the *Y*-cut crystal is cut perpendicular to one of the *Y* axes. The *AT* cut is made at approximately a 35° angle with the *Z* axis. Other types of cuts are called *BT*, *CT*, *DT*, and *GT* cuts; each has distinctive characteristics. They are cut with the face of the crystal at an angle to the *Z* axis.

Crystals used in oscillators are thin sheets that are cut from a natural crystal and then ground to suitable thickness to obtain the desired resonant frequency. For any given crystal cut, the thinner the crystal, the higher is the resonant frequency. The purpose of the special cuts is to increase the stability of the crystal by reducing its coefficient of temperature drift. Temperature drift denotes the change in resonant frequency of the crystal with changes in temperature. The *temperature coefficient* is the relation between the frequency change and the temperature change. If the frequency increases with a temperature rise, the crystal has a positive temperature coefficient. If temperature variation has little or no effect on resonant frequency, the crystal has a zero temperature coefficient. The temperature coefficient is usually expressed as the number of Hertz changes per megaHertz per degree centigrade.

An *X* cut has a negative temperature coefficient. A *Y* cut has a temperature coefficient that is both negative and positive, depending upon the amount of temperature change. An *AT* cut has a very low or practically zero temperature coefficient. The *GT* cut has a zero temperature coefficient over a wide temperature range. A summary of various crystal-cut characteristics is given in Table 15–1. Transmitters that require a very high frequency stability, such as broadcast transmitters, employ temperature-controlled ovens to maintain the crystal temperature within the proper range (see Fig. 15–3). These ovens are thermostatically controlled containers in which the crystals are placed.

The type of cut also determines the activity of the crystal. Some crystals can vibrate at more than one frequency, and will thus operate at harmonic frequencies. These crystals are said to have multiple peaks. Crystals that do not have uniform thickness may have two or more resonant frequencies. Usually, one resonant frequency is more pronounced than the others, and these other frequencies are called *spurious frequencies.* Occasionally, such a crystal oscillates at two frequencies at the same time. In other cases, the crystal jumps in frequency as the plate-tank capacitor is varied past a certain point. The amount of a-c current that can be safely passed through a crystal ranges from 50 to 200 ma. If a crystal carries too much current, it will crack. Crystal temperature increases due to a-c current flow; it is possible to scorch or burn a crystal.

Courtesy of Bliley Electric Company

Figure 15–3. Quartz-crystal holder and a crystal oven.

TABLE 15-1
Characteristics of Various Crystal Cuts

Type Cut	*Frequency Range*	*Temperature Coefficient*	*Remarks and/or Characteristics*
X	100 kHz to 1100 kHz	Negative values −10 to −25 parts per million per degrees C	Complicated spectrum. Low activity. Obsolete cut.
Y	Up to 10 MHz	−20 to +100 parts per million per degrees C	Discontinuities in frequency with temperature changes. Obsolete cut.
AT	500 kc to 10 MHz	Zero, 40° to 50°C	High activity, extensively used.
BT	500 kc to 10 MHz	Zero, 20° to 30°C	High activity, extensively used.
CT	50 kc to 500 kHz	Zero, 20° to 30°C	Extensively used cuts. Avoids use of large quartz plates for low frequencies. These two cuts are directly related to *AT* & *BT* cuts; however, they are physically smaller for these low frequencies.
DT	50 kc to 500 kHz	Zero, 20° to 40°C	
ET	100 kc to 1 MHz	Zero, 70° to 80°C	Harmonic types. Crystal produces output voltages at 3rd, 4th, 5th, 6th, and 7th harmonics of fundamental vibration.
FT	100 kc to 1 MHz	Zero, 70° to 80°C	
GT	100 kHz	Zero, 10° to 100°C	Excellent secondary standard of frequency. Excellent temperature-versus-frequency characteristics. Operates best at 100 kHz.

15-2. Crystals as Circuit Elements

Crystals are mounted in holders, as previously noted. A holder provides mechanical support for the crystal, with electrodes for application of voltage. A holder is designed to permit maximum freedom of crystal vibration. Typically, the crystal rests on one electrode, with a small air gap between the top electrode and the crystal. This arrangement may be used with *AT* and *BT* crystal cuts, as well as in cases where the crystal is not subjected to mechanical shock. Note that crystal holders have a significant effect on the resonant frequency of the crystal. Some variable-gap holders, for example, can shift the resonant frequency of a 14-MHz crystal as much as 20 to 50 kHz when the air gap is varied.

It is instructive to note the example of mechanical resonance depicted in Fig. 15–4. A mass is attached to a flat spring which is secured

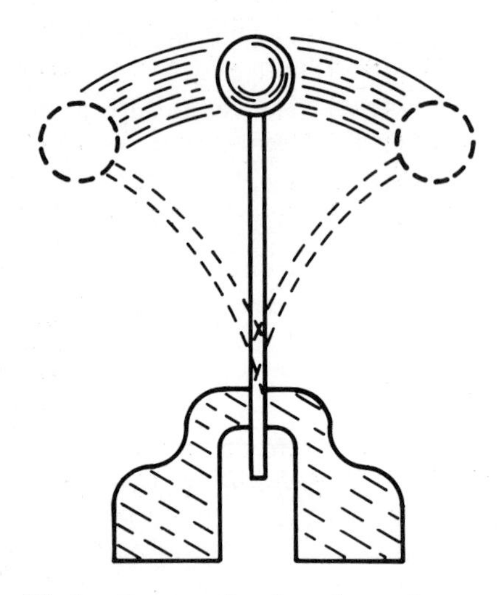

Figure 15–4. An example of mechanical resonance.

in a support. If the mass is drawn to one side and then released, it will vibrate. The mass is analogous to inductance, the spring to capacitance, and the total friction to resistance. The frequency of oscillation is formulated

$$f = \frac{1}{2\pi\sqrt{\dfrac{\text{mass}}{\text{stiffness}}}} \tag{15–1}$$

In Formula 15–1, we observe that mass has the significance of inductance, and that the reciprocal of stiffness has the significance of capacitance. The reciprocal of stiffness is called compliance. The frequency stability of a crystal-controlled oscillator depends on the Q of the crystal and its temperature coefficient. Q values of quartz crystals are very high, often 100 times greater than LC tanks. Exact Q values depend upon the cut, type of holder, and accuracy of grinding. Commercial crystals range in Q value from 5,000 to 30,000, while some laboratory experimental crystals have Q values as high as 400,000. The following data show the size and electrical characteristics of a typical quartz crystal and its holder.

Dimensions:

Thickness . . . . 0.25 in.
Width . . . . 1.3 in.
Length . . . . 1.08 in.

Resonant Frequency: 430 kHz

Equivalent Electrical Parameters:

Inductance . . . . 3.3 h
Capacitance . . . . 0.042 $\mu\mu$f
Holder Capacitance 5.8 $\mu\mu$f
Q 23,000

When placed in an electrical circuit, a crystal can respond like a very high Q series-resonant circuit. The electrical characteristics associated with a vibrating crystal can be represented by an equivalent circuit comprising resistance, inductance, and capacitance, as shown in Fig. 15–5. When the crystal is not oscillating, it is represented

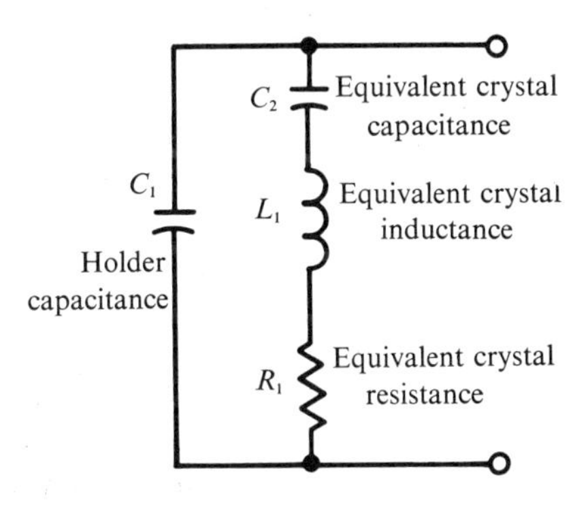

Figure 15–5. Equivalent circuit for a crystal.

simply by C_1. The frequency stamped on a crystal holder usually denotes the parallel-resonant frequency of the crystal and its holder as applied to an oscillator circuit. This does not mean that an ordinary inductor and capacitor could be substituted for the crystal to give the same electrical characteristics. That is, the Q value of the crystal circuit is many times greater than that of any ordinary LC combination.

15–3. Crystal-Controlled Oscillators

Since a vibrating quartz crystal is equivalent to a resonant circuit, it can be used in place of the usual tuned circuit as a frequency-controlling element in a vacuum-tube or transistor oscillator. The circuits depicted in Fig. 15–6 show common crystal-oscillator configurations. The vacuum-tube oscillator is similar to the TPTG configuration that we studied previously, with the crystal replacing the tuned-grid circuit. Feedback takes place through the plate-to-grid interelectrode capacitance. Oscillation occurs at the resonant frequency of the crystal, and the plate circuit is tuned slightly higher (inductive side).

In the transistor oscillator circuit (Fig. 15–6)(b), the equivalent circuit and impedance versus frequency characteristic indicate that two modes of operation exist. The lower mode is called the series-resonant mode, and the upper mode is called the parallel-resonant mode. Note that the series resistance R_s of the crystal operates together with the equivalent inductance to determine the impedance of the equivalent circuit at mode resonance. Different modes have different impedances. Best frequency stability is associated with lowest equivalent-circuit resistance. The depicted arrangement is designed to operate a crystal in its third harmonic mode.

In Fig. 15–6(a), a resistor is connected from grid to ground through a choke to allow electrons which have been attracted to the grid on the positive swing to return to the cathode ground point. The flow of d-c current through this resistor develops grid-bias voltage. Note that the capacitor usually associated with a grid-leak resistor is unnecessary in this circuit, because the crystal and holder serve in its stead. A meter is generally used to tune a crystal oscillator. In Fig. 15–6(a), the meter is connected in series with the plate circuit; d-c plate current flows from the cathode to the plate, and then through the meter. Any change in plate current is accordingly indicated by the meter (a d-c milliammeter). When the oscillator is generating its greatest amount of *r-f* current, the grid draws maximum current and grid bias is maximum. This holds the plate current to a minimum, and the meter indicates a minimum value. To tune the circuit, the minimum current point is determined, and the tank capacitance is then reduced to increase the plate-current flow about 4 ma.

With reference to Fig. 15–7, we observe that there is a pronounced drop in plate current when a crystal oscillator circuit goes into oscillation. Although maximum output occurs at point *A*, the operating point should be placed between points *B* and *C*. The reason for this reduction in output is that operation is unstable at point *A*. In other words, if the crystal is operated at point *A*, erratic or intermittent

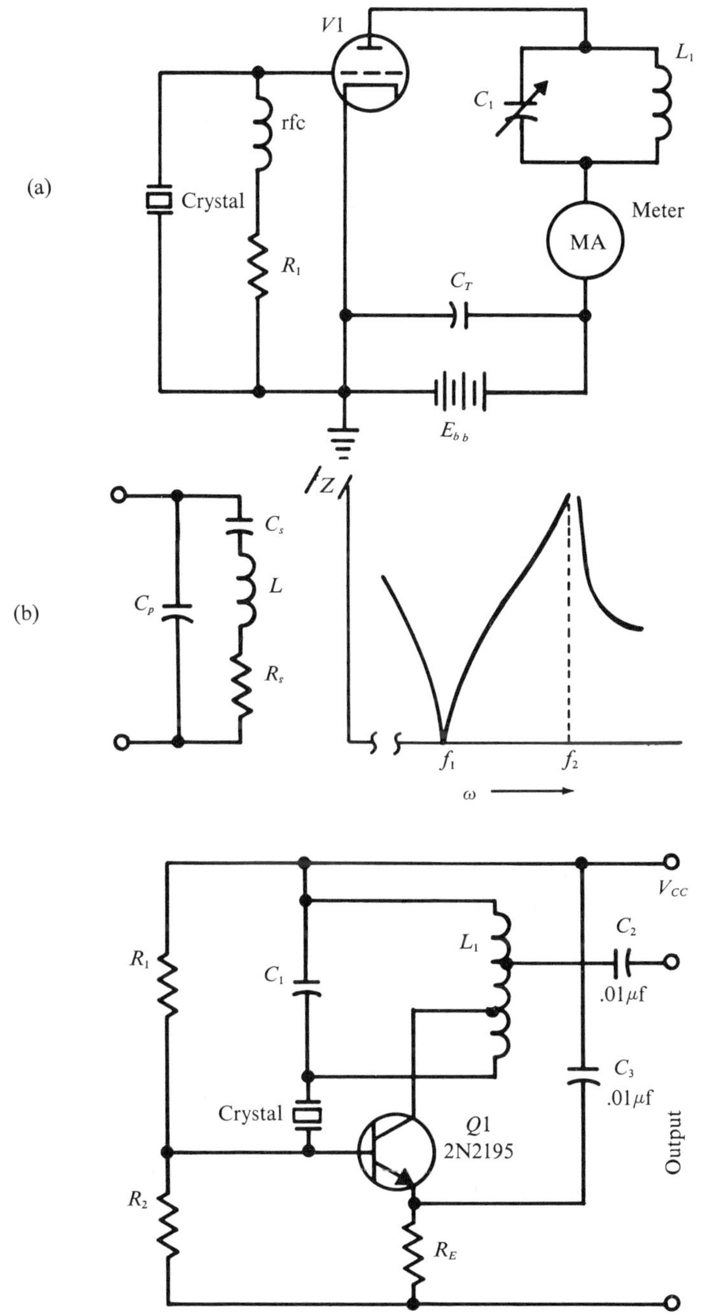

Figure 15-6. (a) Vacuum-tube crystal-oscillator circuit; (b) Transistor crystal-oscillator circuit with equivalent circuit and impedance versus frequency characteristics.

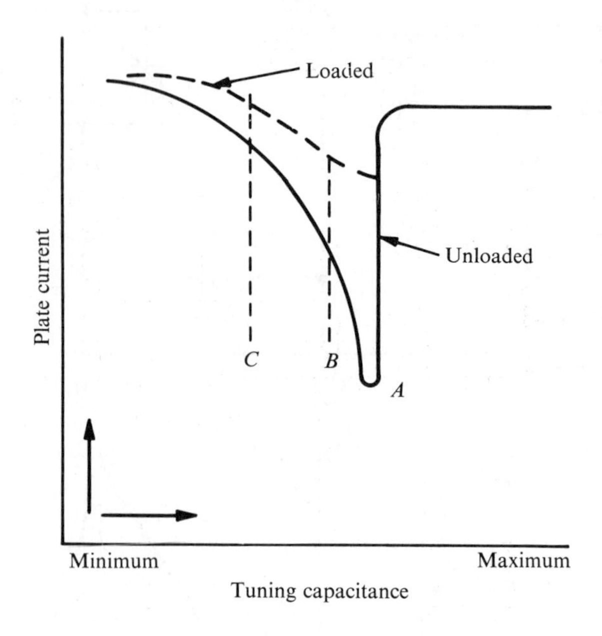

Figure 15-7. **Plate current versus tank-circuit tuning.**

output can be anticipated. Note how abruptly the response drops to the nonoscillatory state to the right of point *A*. The power output that can be obtained from a crystal oscillator is limited. If too high an excitation voltage is applied to a crystal, it will shatter. Hence, a tetrode or pentode crystal oscillator circuit is usually preferred to a triode configuration; multigrid tubes have a comparatively high power sensitivity, and greater output is provided by a small excitation voltage. The small grid-to-plate capacitance of a tetrode or pentode presents a high impedance to *r-f* current flow and limits the grid excitation to the small required value. In fact, it is sometimes necessary to add a small amount of capacitance between grid and plate of a pentode.

Figure 15-8 shows a typical pentode crystal-oscillator circuit. To obtain sufficient positive feedback, a small capacitor (C_4) may need to be added. Grid-leak resistance bias is supplemented by cathode bias in this configuration. Cathode-bias voltage is developed by resistor R, and d-c current flowing from cathode to plate develops a voltage drop across the resistor that makes the end connected to cathode positive with respect to ground. Since the lower end of the cathode resistor is effectively connected to the grid by R_1 and the choke, any voltage drop developed across R_1 appears between the grid and cathode. This makes the grid negative with respect to the cathode by an amount equal to

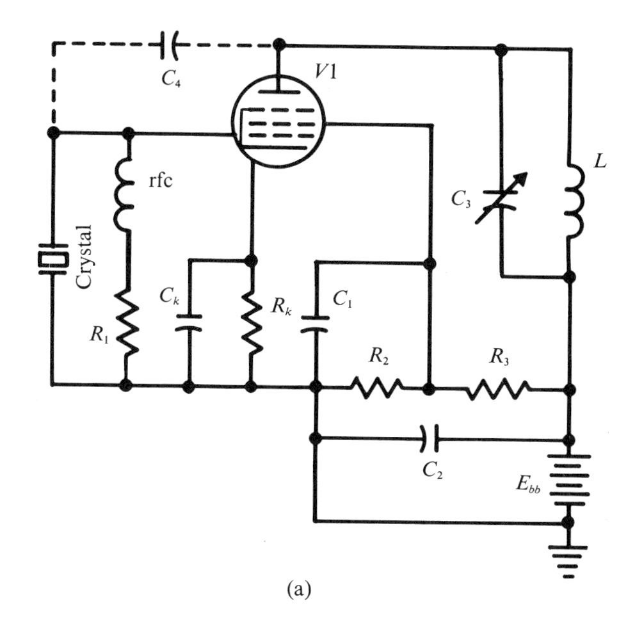

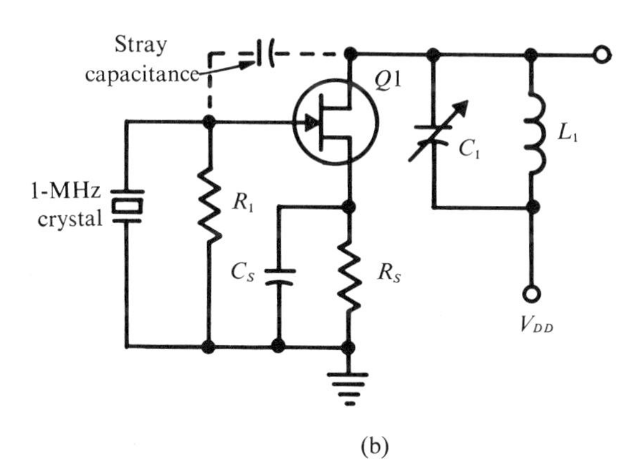

Figure 15–8. (a) A pentode crystal-oscillator circuit; (b) A field-effect transistor crystal oscillator circuit.

the voltage drop across R_1 and R_k. Capacitor C_k in Fig. 15–8 bypasses the *r-f* component of plate current around the cathode resistor. We sometimes refer to the cathode resistor R_k as a minimum bias resistor, because bias voltage is developed across R_k for any plate-current flow. Thus, this source of bias voltage is independent of oscillation and prevents excessive tube current if oscillation stops.

The FET is preferred to the vacuum tube or junction transistor as a control device in a crystal oscillator, as it has the advantages of small size, ruggedness, and high input impedance. The high-input impedance of the FET (1,000 MΩ or greater) and the 100-MΩ gate resistor does not affect the Q of the crystal. In the modified Miller circuit in Fig. 15–8(b), the values of L and C are tuned to a frequency slightly higher than the parallel resonant frequency of the crystal, as the drain tank must appear slightly inductive to start oscillations. The drain-tank capacitor is made variable to allow for stray capacitance and for the tolerance of other components.

15–4. RC Sine-Wave Oscillators

RC sine-wave oscillators are widely used in electronics technology (see Fig. 15–9). An oscillator in which a frequency-selective *Wien-bridge* circuit is utilized as the *RC* feedback network is called a Wien-bridge oscillator. One circuit that is extensively employed for this type of oscillator is shown in Fig. 15–10. In Fig. 15–10(a), the phase-shifting configuration is depicted as a bridge at the left of the diagram. In Fig. 15–10(b), the circuit has been redrawn to indicate the feedback paths more clearly. Triode $V1$ is one of the oscillator tubes. Triode $V2$ operates as an amplifier and phase inverter. Thus, even without the bridge circuit, the system would oscillate, because any signal that is fed to the grid of $V2$ is amplified and inverted by both $V1$ and $V2$. The signal voltage fed back to the grid of $V1$ is in phase with the initial signal voltage, so that oscillation is established. However, without the bridge section, the system is not frequency-selective.

The bridge section functions to eliminate feedback voltages of all frequencies except the single desired output frequency. The frequency of operation depends on the settings of variable capacitors C_1 and C_2, and the values of R_1 and R_2 in Fig. 15–10. Normally, $C_1 = C_2$, and $R_1 = R_2$. We will observe that the bridge permits a voltage of one frequency only to be effective in the circuit, because of degeneration and phase shift provided by the reactive components. Oscillation can take place only at the frequency f_o, which permits the voltage across R_2 (the input signal to $V1$) to be in phase with the output voltage from $V2$, and for which the positive feedback voltage exceeds the negative feedback voltage. Voltages of any other frequency cause a phase shift between the output of $V2$ and the input of $V1$ and are attenuated by the large degeneration of the bridge circuit, so that the positive feedback is then inadequate to maintain oscillation at any frequency other than f_o.

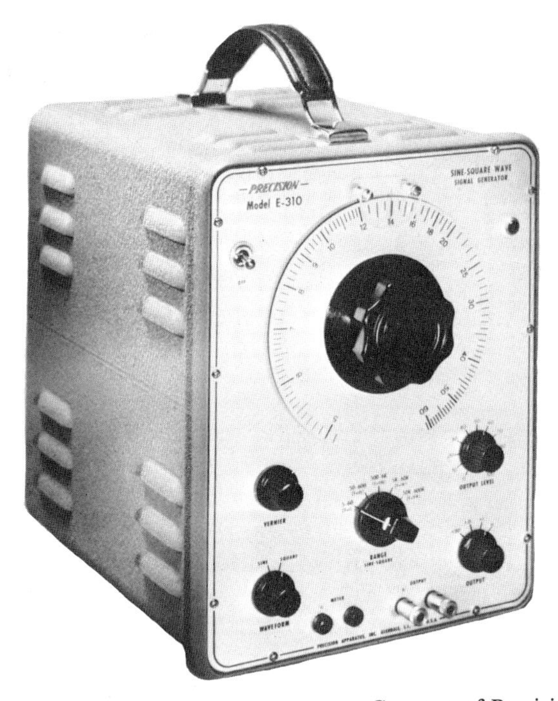

Courtesy of Precision Apparatus,
Division of B and K Dynascan Corp.

Figure 15-9. RC oscillators are used in audio-frequency sine-wave generators.

We will observe that a negative feedback voltage is provided via the voltage divider R_3 and the lamp LP_1 in Fig. 15-10. Because there is no phase shift across this divider, and because the resistances are independent of frequency, the amplitude of the negative feedback voltage is also independent of frequency. Positive feedback voltage is provided via voltage divider $R_1C_1 - R_2C_2$. If the frequency is very high, the reactance of the capacitors is almost zero. In this case, R_2 is shunted by a very low reactance, making the $V2$ grid voltage almost zero. On the other hand, if the frequency is reduced toward zero, the current that can flow through either C_2 or R_2 is reduced to virtually zero by the very high reactance of C_1. Therefore, the grid voltage of $V2$ falls to almost zero. At some intermediate frequency, the positive feedback voltage is maximum, as shown in Fig. 15-10(c). The curve is rather flat in the vicinity of f_o, but the phase shift that occurs in the positive-feedback circuit aids in permitting only a single frequency to be generated.

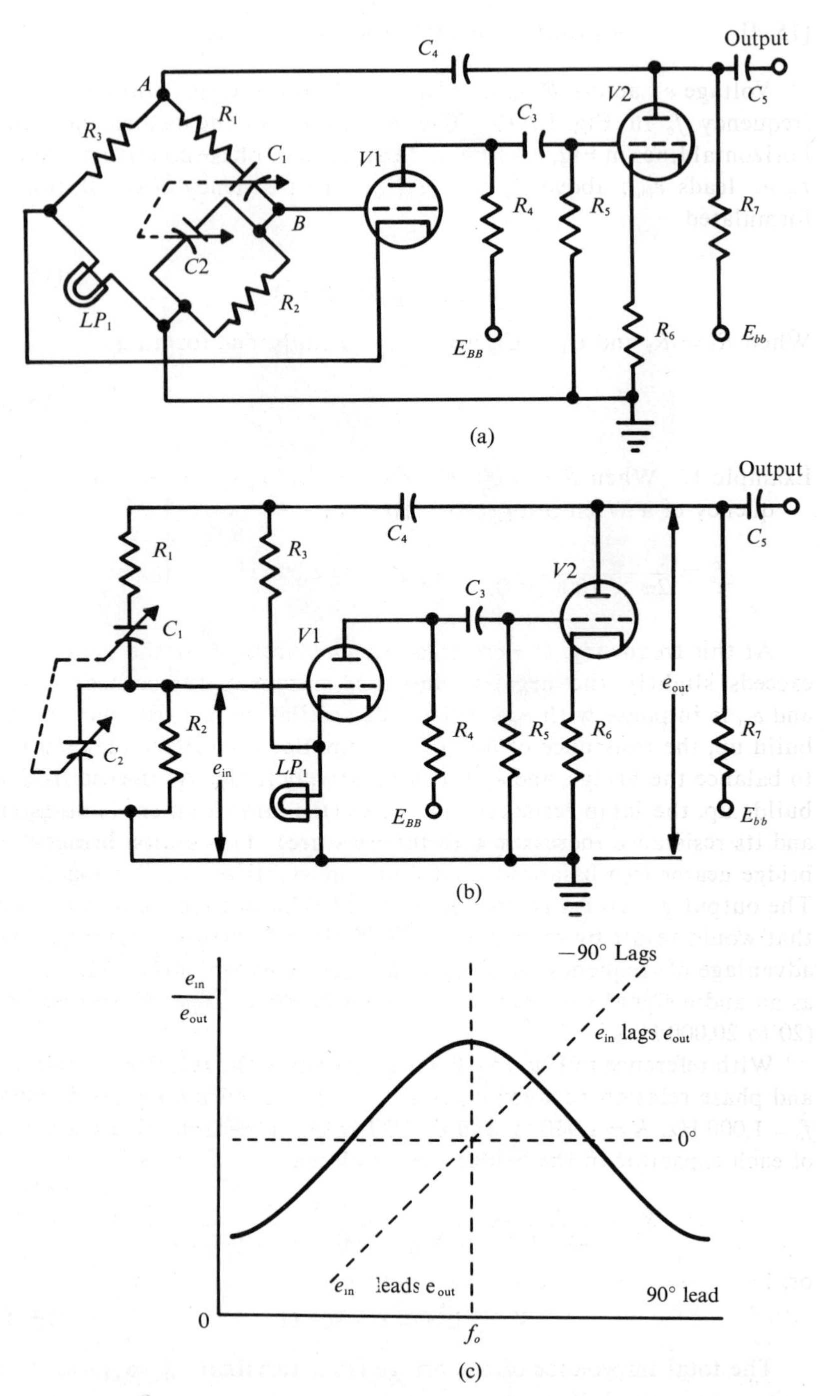

Figure 15-10. **Wien-bridge oscillator circuit and frequency characteristic.**

Voltage e_{in} across R_2 is in phase with the output voltage e_{out} at frequency f_o in Fig. 15–10. The intersection of the dashed slant and horizontal lines in Fig. 15–10(c) indicates the in-phase condition. Below f_o, e_{in} leads e_{out}; above f_o, e_{in} lags e_{out}. The frequency of oscillation is formulated

$$f_o = \frac{1}{2\pi\sqrt{(R_1C_1)(R_2C_2)}} \tag{15–2}$$

When $R_1 = R_2$ and $C_1 = C_2$, we have the simplified formula

$$f_0 = \frac{1}{2\pi RC} \tag{15–3}$$

Example 1. When $R = 1{,}000\ \Omega$ and $C = 0.159\ \mu\text{f}$, the oscillating frequency of a Wien-bridge oscillator is

$$f_o = \frac{1}{2\pi \times 1000 \times 0.159 \times 10^{-6}} = 1{,}000\ \text{Hz} \qquad \text{(answer)}$$

At this frequency, the positive feedback voltage on the grid of $V1$ exceeds slightly the negative feedback voltage at the cathode of $V1$, and e_{in} is in phase with e_{out}. When the oscillation is just starting to build up, the resistance of lamp LP_1 is smaller than the value required to balance the bridge, and e_{in} is comparatively large. As the oscillation builds up, the lamp resistance increases (the lamp filament is tungsten and its resistance increases with temperature). This action brings the bridge nearer to a balanced condition and stabilizes oscillation at f_o. The output waveform is sinusoidal, and LP_1 helps to prevent distortion that would result by overdrive of $V1$. A Wien-bridge oscillator has the advantage of frequency stability and true sine-wave output when used as an audio signal generator over a considerable range of frequencies (20 to 20,000 Hz).

With reference to Fig. 15–10, let us calculate the relative magnitude and phase relation between e_{in} and e_{out} for the following conditions: $f_o = 1{,}000$ Hz, $R = 1{,}000\ \Omega$, and $C = 0.159\ \mu\text{f}$. The capacitive reactance of each capacitor in the bridge is formulated

$$X_C = \frac{1}{2\pi f_o C} = \frac{1}{6.28 \times 1{,}000 \times 0.159 \times 10^{-6}}$$

or,

$$X_C = 1{,}000 \angle -90^\circ\ \Omega \tag{15–4}$$

The total impedance of the bridge from terminal A to ground is written as in the following equations.

$$Z_t = R - jX_C + \left[\frac{R(-X_C)}{R - jX_C}\right]$$

$$Z_t = 1{,}000 - j1{,}000 + \left[\frac{(1{,}000\angle 0^\circ)(1{,}000\angle -90^\circ)}{1{,}414\angle -45^\circ}\right]$$

or,

$$Z_t = 2{,}100\angle 45^\circ\ \Omega \tag{15–5}$$

We may next assume some value of output voltage at terminal A of the bridge, such as $2.1\angle 0^\circ$ v. Then, the current at terminal A is evidently

$$i_t = \frac{e}{Z_t} = \frac{2.1\angle 0^\circ}{2{,}100\angle -45^\circ} = 0.001\angle +45^\circ \tag{15–6}$$

In turn, the voltage e_{in} across the parallel impedance Z_b comprising R_2C_2 is calculated

$$e_{in} = i_t Z_b = (0.001\angle +45^\circ)(707\angle -45^\circ) = 0.707\angle 0^\circ \text{ v} \tag{15–7}$$

Therefore, we perceive that e_{in} is in phase with e_{out} at $f_o = 1{,}000$ Hz, for the circuit values depicted in Fig. 15–10.

In the second part of our analysis, we will proceed to calculate the lead of e_{in} with respect to e_{out} when $f = 500$ Hz. As before, we first calculate the capacitive reactance of C when the frequency is changed to 500 Hz.

$$X_C = \frac{1}{2\pi fC} = 2{,}000\angle -90^\circ \tag{15–8}$$

Next, the total impedance of the bridge from terminal A to ground at 500 Hz is calculated

$$Z_t = R - jX_C + \left[\frac{(R)(-jX_C)}{R - jX_C}\right] = 3{,}000\angle -53^\circ\ \Omega \tag{15–9}$$

We may assume the same output voltage as before, and calculate the total current at terminal A of the bridge.

$$i_t = \frac{e}{Z_t} = 0.0007\angle +53^\circ \text{ amp} \tag{15–10}$$

In turn, the input voltage e_{in} across the parallel impedance Z_B comprising R_2C_2 is calculated

$$e_{in} = i_t Z_B = 0.615\angle +26.5^\circ \text{ v} \tag{15–11}$$

We conclude accordingly that e_{in} is reduced and leads e_{out} at $f = 500$ Hz.

In the third part of our analysis, we will repeat the foregoing calculations to find the lag of e_{in} with respect to e_{out} at $f = 2{,}000$ Hz.

$$X_C = 500 \angle -90^\circ \ \Omega \tag{15-12}$$
$$Z_t = 1{,}495 \angle -36.7^\circ \ \Omega \tag{15-13}$$
$$i_t = 0.0014 \angle +36.7^\circ \tag{15-14}$$
$$e_{\text{in}} = 0.620 \angle -26.8^\circ \text{ v} \tag{15-15}$$

We conclude that e_{in} is again reduced, and now lags e_{out} at $f = 2{,}000$ Hz.

Another widely used *RC* sine-wave oscillator is the phase-shift configuration depicted in Fig. 15–11. A simple amplifier circuit is shown in Fig. 15–11(a); when this amplifier is combined with the phase-shift network of Fig. 15–11(b), oscillation occurs at a frequency that provides a 360° total phase shift. Observe that 180° of phase shift is furnished by the common-emitter amplifier, and another 180° is furnished by the *RC* network. A 5-kΩ potentiometer is provided in the oscillator circuit [Fig. 15–11(c)] to adjust the oscillating frequency; a variation from about 200 to 400 Hz is available. The basic frequency of oscillation is approximately formulated

$$f_o = \frac{1}{2\pi\sqrt{6R^2C^2 + 4RR_LC^2}} \text{ Hz, approximately} \tag{15-16}$$

Example 2. Determine the oscillating frequency of a phase-shift oscillator when $R_L = 4.7$ kΩ, $R = 6.8$ kΩ, and $C = 0.01$ μf.

$$f_o = \frac{0.159}{\sqrt{6 \times (6.8 \times 10^3)^2 \times (10^{-8})^2 + 4 \times 6.8 \times 10^3 \times 4.7 \times 10^3 \times (10^{-8})^2}}$$
$$f_o = 7.88 \text{ kHz} \qquad \text{(answer)}$$

The addition of an emitter follower to the junction transistor phase-shift oscillator in Fig. 15–11(b) reduces the effects of the low-input impedance of the common-emitter stage on the phase-shift circuits. Further improvement of the phase-shift oscillator may be accomplished by employing an FET as an amplifier, as depicted in Fig. 15–12. The operating frequency of the FET phase-shift oscillator is also determined by Formula 15–16.

Any oscillator must employ an amplifying device with sufficient gain to overcome the internal losses of the circuit and the losses to the load. For oscillations to occur in the phase-shift oscillator in Fig. 15–11(a), the beta of the transistor must be greater than

$$h_{fe} \geq 23 + \frac{30R}{R_L} + \frac{4R_L}{R} \tag{15-17}$$

Example 3. Observe the circuit in Fig. 15–11(c) on page 476.

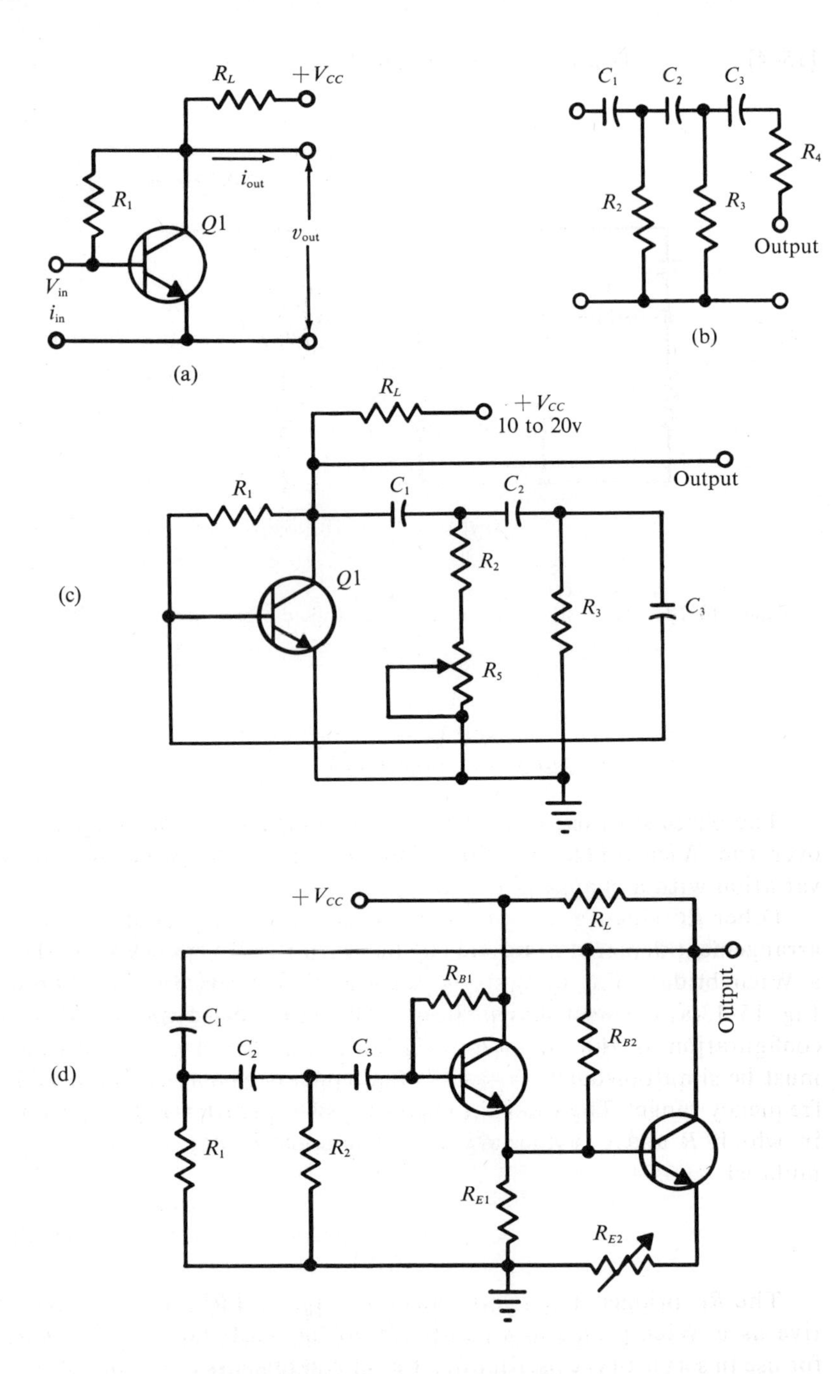

Figure 15-11. Phase-shift oscillator. (a) Basic amplifier circuit; (b) Phase-shift circuit; (c) Complete junction transistor phase-shift oscillator; (d) Improved junction transistor phase-shift oscillator.

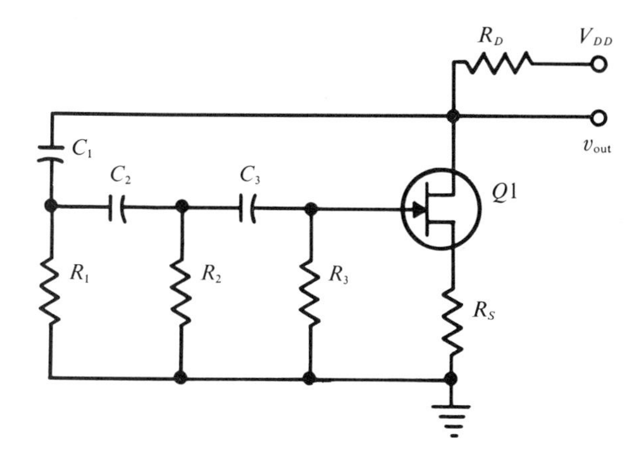

Figure 15–12. Phase-shift oscillator using a field-effect transistor as an amplifier.

$$h_{fe} \geq 23 + \frac{30 \times 4.7\text{ K}}{3.3\text{ K}} + \frac{4 \times 3.3\text{ K}}{4.7\text{ K}}$$

$$h_{fe} \geq 68.4 \qquad \text{(answer)}$$

The phase-shift oscillator has the advantage of fewer components over the Wien-bridge oscillator but the disadvantage of amplitude variation with a change of frequency.

Other *RC* sine-wave oscillators use configurations similar to the arrangement depicted in Fig. 15–10, but with a null network other than a Wien bridge. For example, a parallel-T *RC* network as shown in Fig. 15–13(a) is sometimes utilized. The chief disadvantage of this configuration is the comparatively large number of parameters that must be simultaneously varied to obtain operation over an appreciable frequency range. The resonant frequency of a parallel-T *RC* network in which *R* and *C* values are related as noted in Fig. 15–13(a) is formulated

$$f_o = \frac{1}{4\pi RC} \tag{15–18}$$

The *RC* bridged-T network shown in Fig. 15–13(b) is not as selective as a Wien-bridge or a parallel-T *RC* network, but is satisfactory for use in service-type oscillators. Fewer components are required, and operation over a large frequency range can be obtained by means of a two-section variable capacitor and a switch to vary the values of R_1

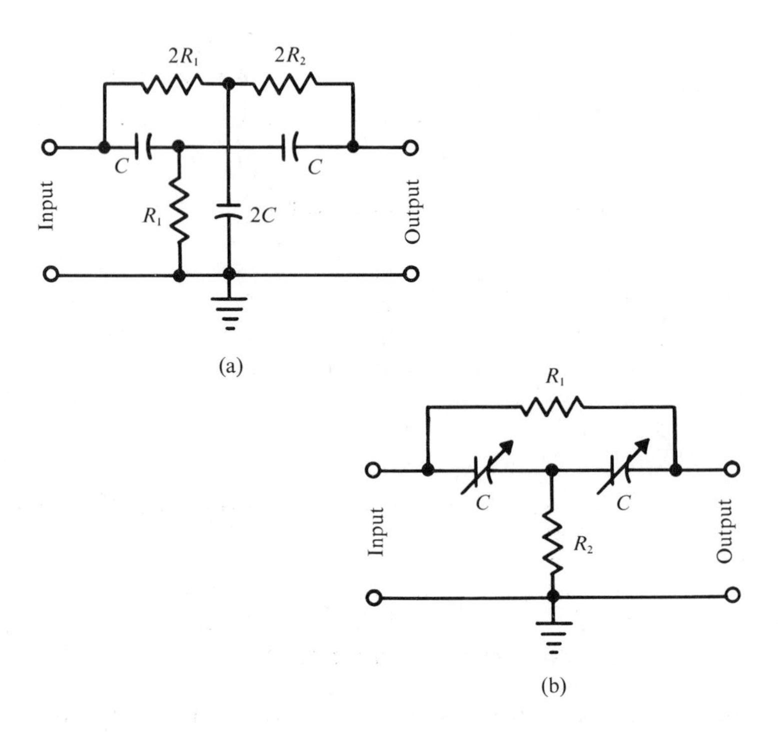

Figure 15–13. (a) Parallel-T RC (twin T) network; (b) RC bridged-T network.

and R_2 in comparatively large steps. The resonant frequency of an *RC* bridged-T network is formulated

$$f_o = \frac{1}{2\pi C_1 \sqrt{R_1 R_2}} \quad \textbf{(15–19)}$$

Improved bridged-T characteristics can be obtained by using inductors instead of resistors. However, large values of inductance are required for low-frequency operation; practical design problems arise because of the extensive shielding that is required to minimize stray-field interference. Hence, laboratory type oscillators for audio frequencies generally employ Wien bridges instead of *RC* or *LC* bridged-T networks.

SUMMARY

1. A crystal-controlled oscillator is ordinarily used when the frequency of an electronic device must be held to close limits.

2. A crystal exhibits a characteristic called the piezoelectric effect by which mechanical stress produces electric charges and electrical charges produce mechanical stresses.
3. There is an indirect relation between the thickness of a crystal and the resonant frequency of the crystal.
4. The temperature coefficient of a crystal is the relation between changes of the resonant frequency of the crystal and changes in temperature.
5. The temperature coefficient of a crystal is expressed as the number of Hertz change per megahertz, per one degree-centigrade change in temperature. The resonant frequencies other than the most pronounced are called spurious frequencies.
6. Too much current through a crystal will cause the crystal to crack or to scorch.
7. Crystal holders have a significant effect on the resonant frequency of the crystal.
8. The frequency stability of a crystal depends upon the temperature coefficient of the crystal and the Q of the crystal.
9. Crystal-controlled oscillators may be designed to operate on a harmonic of the crystal fundamental frequency.
10. The plate current of a vacuum-tube oscillator or the collector current of a transistor oscillator shows a pronounced decrease as the circuit goes into oscillation.
11. Multigrid tubes are usually preferred over triode tubes as crystal oscillators because of their higher gain.
12. As a control device, FET crystal oscillators have the advantages of both vacuum tubes and transistors.
13. The Wien-bridge oscillator has good frequency and amplitude stability over a considerable range of audio frequencies.
14. The Wien-bridge oscillator will oscillate only at a frequency which has no phase shift from the output of the second amplifier to the input of the first amplifier.
15. The phase-shift oscillator effectively has two frequency selecting methods: the bridge *RC* circuits and the negative feedback versus the positive feedback voltage.

Questions

1. Discuss the characteristics of the piezoelectric effects in a crystal.
2. Draw an equivalent-electrical circuit of a crystal used to stabilize an oscillator, and explain how the crystal controls an oscillator.
3. What is the relationship between the thickness of a crystal and the resonant frequency of the crystal?
4. What are the characteristics of *X*, *Y*, *AT*, and *GT* cuts of crystals?
5. Discuss the term "temperature coefficient" as related to a crystal.
6. What is meant when a crystal is said to have multiple peaks?
7. What is the cause of spurious frequencies?
8. What are the possible harmful effects of an excessive a-c current being carried by a crystal?
9. What is the function of a crystal holder?
10. Discuss the operation of a crystal oven.
11. In a crystal oscillator, what is the necessary electrical relationship between the plate-tank and the crystal-resonant frequencies?
12. Explain the relationship between plate current and oscillating frequency of the circuit shown in Fig. 15-7.
13. Why is the pentode vacuum tube preferred over the triode vacuum tube as a crystal oscillator?
14. What are the functions of the cathode resistor in the oscillator shown in Fig. 15-8?
15. Explain the operation of the Wien-bridge oscillator.
16. The frequency is controlled by two methods in a Wien-bridge oscillator. Explain each method.
17. How does negative feedback aid in keeping the output voltage level of a Wien-bridge oscillator constant with a change in load?
18. Would a phase-shift oscillator employing four identical *RC* components oscillate?
19. What is the advantage of using an FET rather than a junction transistor as an amplifier in a phase-shift oscillator?
20. Why are Wien-bridge rather than *LC* bridged-T networks used in laboratory oscillators?

PROBLEMS

1. Determine the values of R_1 and R_2 which would cause a Wien-bridge oscillator to oscillate at 15.9 kHz when $C_1 = C_2 = 0.001\ \mu f$. (Let $R_1 = R_2$).

2. What is the oscillating frequency of the circuit depicted in Fig. 15-10 when $R_1 = R_2 = 15\ \text{k}\Omega$ and $C_1 = C_2 = 0.01\ \mu\text{f}$?

3. Suppose a voltage of 10 v at the operating frequency is applied to the top of the positive feedback network in Problem 2. What is the voltage at the grid of the vacuum tube?

4. Determine the frequency of the phase-shift oscillator depicted in Fig. 5-11 when $R = 10\ \text{k}\Omega$, $R_L = 5.6\ \text{k}\Omega$, and $C = 0.001\ \mu\text{f}$.

5. Determine the value of the three feedback capacitors necessary to cause the phase-shift oscillator circuit in Fig. 5-11 to oscillate at 10 kHz when $R = 6.8\ \text{k}\Omega$ and $R_L = 4.7\ \text{k}\Omega$.

6. What is the minimum value of beta which will cause the circuit in problem 4 to oscillate?

7. What is the minimum necessary beta for the circuit in Problem 5?

8. Prove that changing the *RC* components in the feedback circuit in Fig. 5-11(b) will not change the ratio of the output to input voltage for the network.

9. Assuming no loading effect from R_L or r_i for the feedback network in Fig. 5-11(b), derive a formula for the oscillating frequency.

10. Prove that the voltage out of a 4-section *RC* feedback circuit (45° each) for a phase-shift oscillator is twice as much as would be developed out of a 3-section *RC* feedback circuit (60° each).

11. Determine the frequency which will have a 180° phase shift across the parallel-T *RC* network in Fig. 15-13(a) when $R = 10\ \text{k}\Omega$ and $C = 0.001\ \mu\text{f}$.

12. What values of R will produce a 180° phase shift of a 15-kHz signal through the parallel-T *RC* network when $C = 0.01\ \mu\text{f}$?

13. What value of C will produce an 180° phase shift of a 400-Hz signal through the parallel-T *RC* network when $R = 47\ \text{k}\Omega$?

14. Determine the frequency at which the *RC* bridged-T network will produce an 180° phase shift when $R_1 = 10\ \text{k}\Omega$, $R_2 = 4.7\ \text{k}\Omega$, and $C_1 = 0.1\ \mu\text{f}$.

15. Determine the value of C_1 necessary for the *RC* bridged-T circuit to produce an 180° phase shift on a 20 kHz signal when $R_1 = 15\ \text{k}\Omega$ and $R_2 = 2.2\ \text{k}\Omega$.

CHAPTER 16

Tuned Amplifiers

16–1. Introduction

Bandpass amplifiers find wide application in electronics technology. A bandpass amplifier is commonly termed a *tuned amplifier*, because bandpass response is usually obtained by means of *LC* resonant circuits. Let us consider a typical example of a tuned amplifier. The familiar radio broadcast band extends from 550 kHz to 1,500 kHz. Each broadcast transmitter that operates in this band occupies a range of approximately 10 kHz. To avoid interference, it is necessary that a radio receiver have an adjustable (tunable) bandpass response of 20 kHz; this 20-kHz "slice" of the frequency spectrum must be tunable from 550 kHz to 1,500 kHz in order to select any transmission within the broadcast band. Hence, radio receivers utilize tunable bandpass amplifiers.

A basic field-effect transistor (or vacuum-tube) tuned amplifier stage is diagrammed in Fig. 16–1. We observe that the drain-load

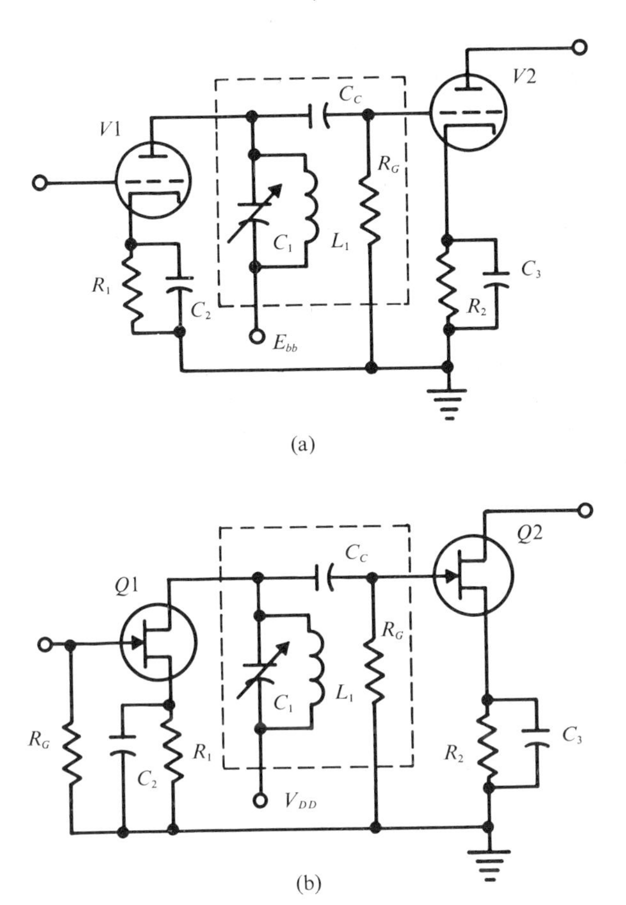

Figure 16–1. Basic single tuned amplifier stage. (a) Employing a triode vacuum tube; (b) Employing field-effect transistors.

circuit is impedance-coupled to the following FET. This impedance comprises L_1, the a-c resistance of L_1 (not shown), the dynamic drain resistance of T_1, R_g, C_1, the drain capacitance of T_1, the gate-input capacitance of T_2, and the stray circuit capacitances. In turn, these parameters can be represented by an equivalent parallel-resonant circuit that has a certain Q value. The frequency response of the stage is the same as the frequency response of this equivalent parallel-resonant circuit. Of course, the bandwidth of the stage is given by the approximate formula

$$BW = \frac{f_o}{Q} \text{ (approximately)} \qquad \textbf{(16–1)}$$

where bandwidth BW is in Hertz, f_o is the resonant frequency in Hertz, and Q is equal to the inductive reactance of L at f_o divided by the equivalent resistance in series with L.

The impedance of the drain-load circuit is formulated

$$Z = \omega LQ \text{ (approximately)} \tag{16–2}$$

Stage gain is calculated in the same basic manner as for a resistive plate load.

$$A_v = \frac{\mu Z}{r_{ds} + Z} \tag{16–3}$$

where A_v is the number of times that the gate-input signal is amplified, mu is the amplification factor of the FET, Z is the drain-load impedance in ohms (actually a pure resistance at f_o), and r_{ds} is the dynamic drain resistance of the FET.

Let us consider the typical example depicted in Fig. 16–2. The resonant frequency of the stage is equal, for practical purposes, to

$$f_o = \frac{1}{2\pi\sqrt{LC}} \tag{16–4}$$

Example 1. The resonant frequency of the stage in Fig. 16–2 is

$$f_o = \frac{1}{2\pi\sqrt{169 \times 10^{12} \times 15 \times 10^{-5}}}$$

$$f_o \simeq 1 \text{ MHz} \qquad \text{(answer)}$$

The circuit Q is approximately

$$Q = \frac{X_L}{R}$$

$$= \frac{6.28 \times 10^6 \times 1.5 \times 10^{-4}}{7.6}$$

$$\simeq 124 \qquad \text{(answer)}$$

The bandwidth at the half-power points (0.707 voltage) is formulated by Eq. 16–1.

$$BW = \frac{f_o}{Q}$$

$$= \frac{1 \times 10^6}{1.24 \times 10^2}$$

$$\simeq 8.2 \text{ kHz} \qquad \text{(answer)}$$

The impedance at resonance in accordance with Formula 16–2 is

$$Z = \omega LQ$$

$$\simeq 6.28 \times 10^6 \times 150 \times 10^{-6} \times 124$$

$$\simeq 117 \text{ k}\Omega$$

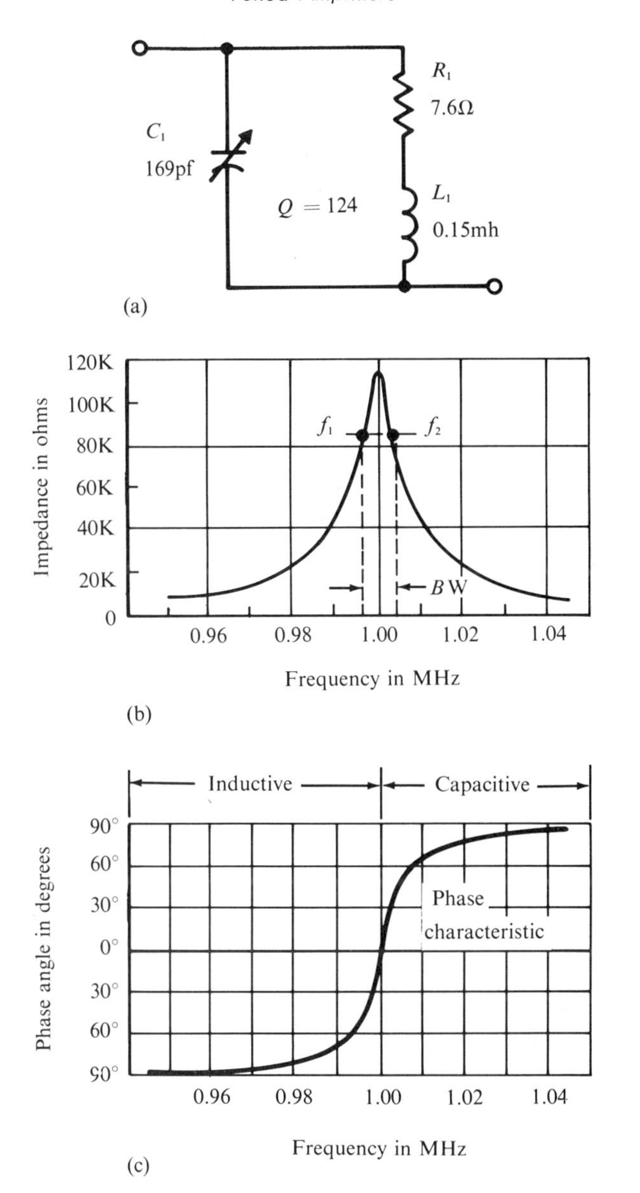

Figure 16–2. **(a) Equivalent plate load impedance; (b) Frequency response; (c) Phase characteristic.**

We recall that the circuit will "look" inductive at frequencies below resonance and will "look" capacitive at frequencies above resonance, as depicted in Fig. 16–2(c).

There are both advantages and disadvantages associated with the tuned amplifier shown in Fig. 16–1. Its chief advantage is simplicity and economy of construction. One of its disadvantages is that the L/C ratio of the drain-load impedance varies greatly over the range from 550 kHz to 1,500 kHz. Formula 16–2 shows that the drain-load impedance rises to a maximum value at the high-frequency end of the band; hence, the stage gain falls at low frequencies, in accordance with Formula 16–3. It would be desirable, of course, to obtain the maximum possible gain over the entire band. Another disadvantage is that the Q of the drain-load impedance depends on the L/C ratio (among other factors), and the stage is least selective (has its greatest bandwidth) at low frequencies.

Still another disadvantage is due to positive feedback from drain to gate via interelectrode capacitance in the configuration of Fig. 16–1. If the drain-load impedance of T_2 is similar to that of T_1, we evidently have a tuned-drain tuned-gate oscillator similar to the tuned-plate tuned-grid (vacuum-tube) configuration. Again, if the gate circuit of T_1 comprises a tuned circuit similar to the drain-load circuit (as is necessarily the case in practical situations), a tuned-drain tuned-gate configuration results. Hence, a simple tuned-amplifier stage employing a three-element control device oscillates and is useless for radio reception unless suitable means are taken to suppress the effect of positive feedback.

In the past, various expedients and circuit elaborations have been employed to suppress oscillation in a triode tuned amplifier. The simplest method entailed connection of a high resistance in series with the grid lead to the tube. In other words, a sufficient amount of positive resistance will cancel out the negative resistance developed by interelectrode feedback. However, this series resistance incurs a signal drop and reduces the available amplification. This method of oscillation control was called the "losser" method. Although the "losser" resistor was a distinct disadvantage, some amplification could still be obtained, and this method was widely used before better techniques were developed.

The first important advance was the introduction of neutralizing circuits. This is a circuit arrangement that feeds a signal back to the grid that is equal in amplitude and opposite in phase to the Miller feedback signal. Figure 16–3 depicts a basic configuration. We observe that the collector tank is center-tapped, and that an out-of-phase signal voltage is present at its lower terminal. When neutralizing capacitor C_N is adjusted to the same capacitance value as the collector-base interelectrode capacitance, the Miller feedback is exactly cancelled,

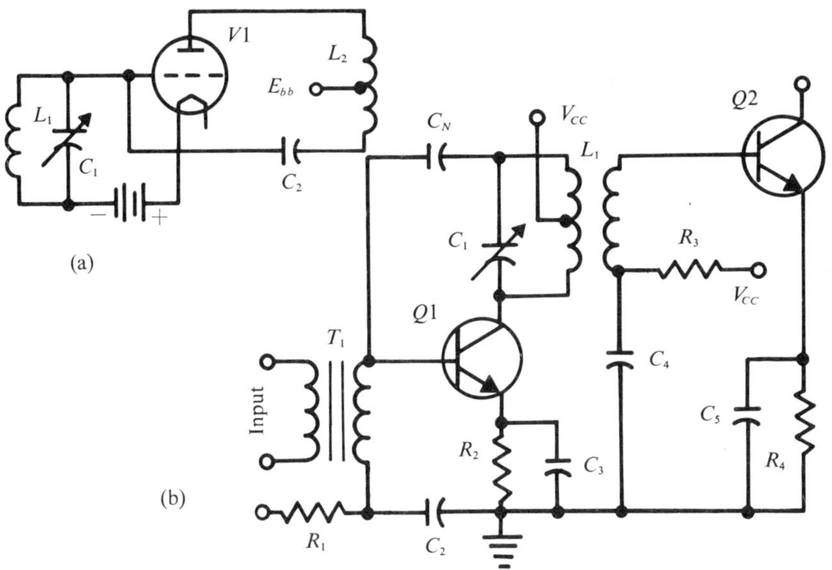

Figure 16-3. Single-tuned amplifier. (a) Employing a vacuum tube; (b) Employing a junction transistor with neutralizing capacitor.

and the tuned amplifier stage does not oscillate. This arrangement, or an equivalent configuration, is still widely utilized today in tuned power amplifiers utilizing vacuum triodes or junction transistors operating at a fixed frequency.

There are, however, disadvantages to the configuration of Fig. 16-3 when a tuned amplifier must operate over a wide frequency range, as in a radio receiver. Circuit parameters vary with frequency in such manner that Miller feedback is greater than the neutralizing feedback at high frequencies when the circuit is adjusted for exact cancellation at low frequencies. In this case, the stage oscillates when tuned to a high frequency. Again, suppose that C_N in Fig. 16-3 is adjusted for exact cancellation at the high frequency end of the band. Then, the stage is overneutralized at low frequencies, and gain is reduced. In practice, it is necessary to accept reduced gain at low frequencies. Neutralizing arrangements provide higher overall gain than "losser" methods, and are accordingly in very wide use for radio reception in tuned amplifiers employing vacuum tabes, junction transistors or field-effect transistors.

With the advent of the screen-grid tube, high-gain tuned vacuum tube amplifiers became possible. Disadvantages of both "losser" methods and overneutralization at low frequencies were overcome. Moreover, the pentode is inherently a higher-gain tube than a triode.

A basic pentode tuned amplifier is diagrammed in Fig. 16–4. Of course, the Miller effect is negligible, and circuit operation is not impaired by plate-to-grid signal feedback. The plate-load circuit is analyzed

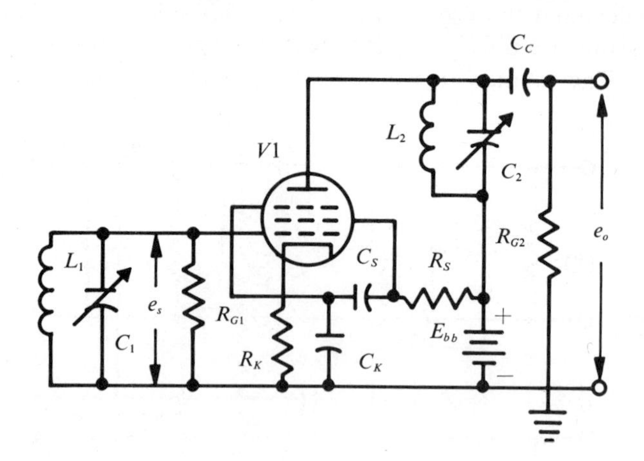

Figure 16–4. Basic pentode tuned amplifier.

in the manner previously explained for the collector-load circuit, and the stage gain is formulated

$$A_v = g_m Z_L \tag{16–5}$$

where Z_L is the impedance at resonance of the equivalent parallel *LCR* tank.

We know that Z_L and Q vary with frequency and with the L/C ratio; in turn, optimum performance is obtained at the high-frequency end of the band. Hence, the basic pentode tuned amplifier is not without its disadvantages. Nevertheless, its characteristics are greatly superior to those of a triode tuned amplifier, and the configuration of Fig. 16–4 is in extensive use. This arrangement is commonly used as a preamplifier in radio receivers, both to provide signal amplification and initial selectivity (discrimination against unwanted signal frequencies). Following the preamplifier, more sophisticated circuitry is employed to provide additional gain and selectivity.

16–2. Junction Transistors at High Frequencies

The principles of tuned-amplifier circuits apply for vacuum-tube, field-effect transistor, and junction-transistor circuits. As we have found in low-frequency amplifiers, some circuit modification is necessary to compensate for the junction transistor's low input and output impedances. We will evaluate these necessary modifications as they occur in the tuned-amplifier theory.

To understand the necessity for special consideration of the junction transistor, we must compare the equivalent circuits depicted in Fig. 16–5. The vacuum triode [Fig. 16–5(a)] and the field-effect tran-

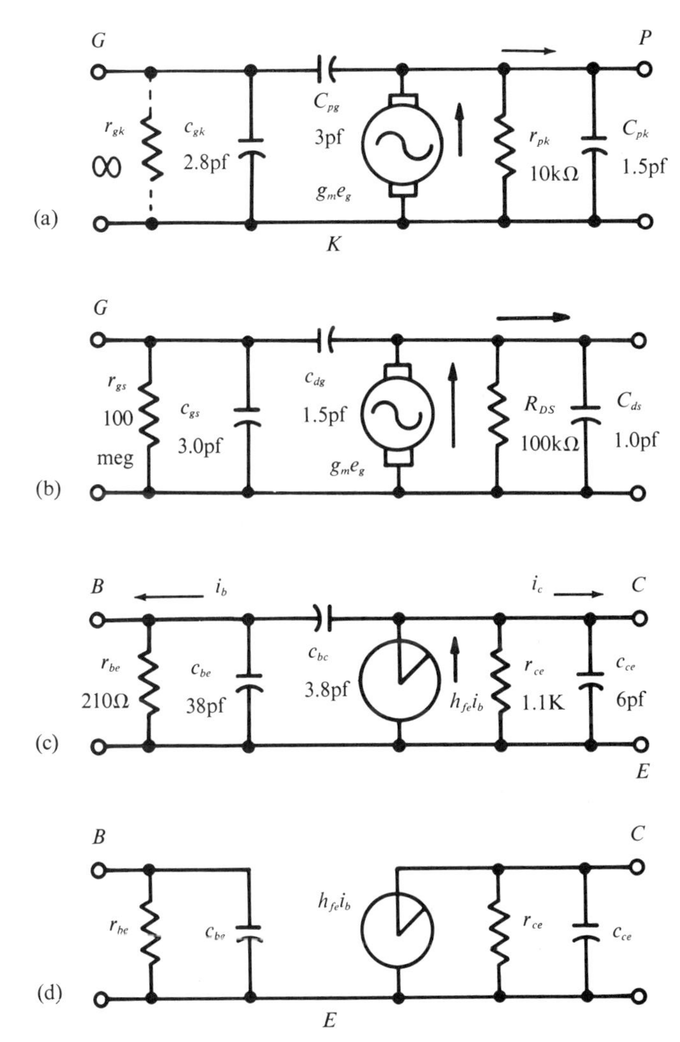

Figure 16–5. Equivalent circuits with average internal values. (a) Vacuum tube; (b) Field-effect transistor; (c) Junction transistor; (d) Junction transistor after neutralization.

sistor (Fig. 16–5(b)) compare very closely; the exception is the output impedance of the FET which compares with that of a vacuum pentode. On the other hand, the equivalent resistance values of the silicon junction transistor (Fig. 16–5(c)), are much smaller than either the FET or the vacuum tube.

We shall now note the physical significance of all the elements in the junction-transistor model in order to establish their relationship to one another and their significance in a high-frequency amplifier. The input impedance (Z_i) at high frequencies is primarily composed of r_{be}, *the base to emitter resistance*, and c_{be}, *the base-to-emitter capacitance*, in parallel. These two components are the electrical equivalent of the forward-biased depletion layer between the base and the emitter.

The output impedance is comprised of r_{ce}, *the collector-to-emitter resistance*, and c_{ce}, *the collector-to-emitter capacitance.* These two components are the equivalent of the changes in output voltage which result from changes of output current and changes in the thickness of the collector-to-base depletion layer.

The generator $h_{fe} i_b$ represents the changes of output current in relation to the transistor input current. The forward-current transfer ratio (h_{fe}) decreases at high frequencies, and for the transistor depicted in Fig. 16–5(c), is 30 to 75 at 40 MHz. As with the vacuum-tube and the field-effect transistor, the internal capacitance (c_{cb}) between the output and the input of the transistor results in positive feedback which can result in oscillation of a tuned amplifier. The effect of this internal capacitance can be cancelled in transistors (Fig. 16–6) just as in vacuum tubes (Fig. 16–3). Obviously, the formulas for calculation of feedback are the same for either the FET, vacuum tube, or junction transistor.

Resistors R_N, R'_N, and C_N form a negative-feedback network to cancel the positive feedback voltage developed across r_{be} and b_{be}. Calculation of these feedback components is made with the assumption that the internal capacitances and resistances of the device will remain constant with frequency.

The values of the feedback components are formulated

$$C_N \simeq \frac{1}{n} c_{bc} \tag{16–6}$$

where n is the turns ratio of the transformer $L_1 L_2$.

$$R_N \simeq n\, r_{bb} \frac{c_{be}}{c_{bc}} \tag{16–7}$$

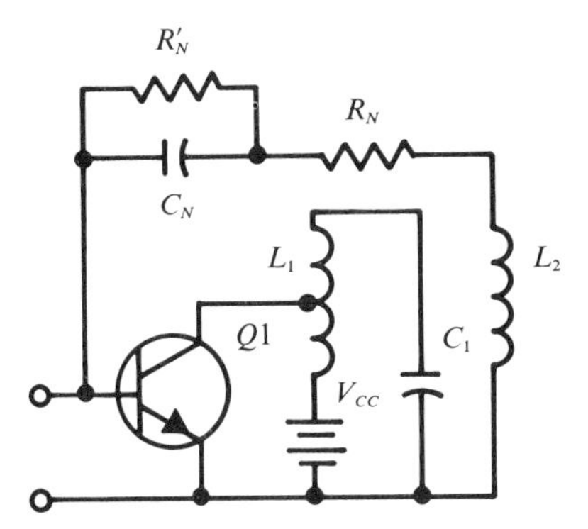

Figure 16-6. Neutralized transistor stage showing only the a-c components; resistor R_N and capacitor C_N form the neutralization feedback network.

where r_{bb} is the resistance between the base lead and the base junction (approximately 100 Ω).

For the field-effect transistor and the vacuum tube, the term r_{bb} is dropped. The value of the compensating resistor (R'_N) is calculated by the formula

$$R'_N \simeq n\, r_i \tag{16-8}$$

Example 2. Calculate the value of the neutralizing components necessary to use the 2N2221 transistor from Fig. 16-5(c) in the circuit for Fig. 16-6. Assume a turns ratio (n) of 3:1.

$$C_N \simeq 1/3 \times 3.8 \times 10^{-9}$$
$$\simeq 1.6 \text{ pf} \qquad \text{(answer)}$$
$$R_N \simeq 3 \times 100 \frac{38}{3.8}$$
$$\simeq 3 \text{ k}\Omega \qquad \text{(answer)}$$
$$R'_N \simeq 3 \times 210$$
$$\simeq 630\ \Omega \qquad \text{(answer)}$$

As a comparison of the relative values of feedback components in a junction transistor to those in a vacuum tube or a field-effect transistor, let us calculate the component values necessary for neutralization of the FET in Fig. 16-5(b).

Example 3.

$$C_N \simeq 3 \times 1.5 \times 10^{-12}$$
$$\simeq 4.5 \text{ pf} \qquad \text{(answer)}$$

$$R_N \simeq 3 \times \frac{3}{1}$$

$$\simeq 9\,\Omega \qquad \text{(answer)}$$

$$R'_N \simeq 3 \times 100 \times 10^6$$

$$\simeq 300\ \text{M}\Omega \qquad \text{(answer)}$$

Obviously, the neturalization circuit for a vacuum-tube or field-effect transistor circuit can be simplified to the circuit shown in Fig. 16-6. The series resistor R_N can be omitted due to its small value, and the shunt resistor R'_N can be omitted due to its large value. Likewise, the entire neutralizing network can sometimes be omitted in a junction-transistor tuned amplifier due to the extremely low input and output impedances.

Next, let us consider the maximum frequency band that can be covered practically by a tuned amplifier. Recall that the equivalent capacitance in the tank circuit comprises the distributed capacitance of the coil, interelectrode capacitances, and the stray capacitances of wiring and components. Therefore, the minimum capacitance that can be placed in shunt with the coil has a definite limit. This minimum capacitance value determines the highest frequency to which the tank can be tuned. By increasing the value of C_2 in Fig. 16-4, the tank can be resonated to progressively lower frequencies. However, there is a practical limit to the upper value of C_2; the L/C ratio is progressively decreased, and the impedance value decreases in turn. Stage gain is decreased, and it is accepted as a practical rule in design that a ratio of 3-to-1 in frequency is the tolerable maximum. In other words, if the resonant frequency is 1,500 kHz when C is set to minimum capacitance, then the lowest practical frequency is 500 kHz when C_2 is set to maximum capacitance.

16-3. Cascade-Tuned Amplifiers

When tuned-amplifier stages are connected in cascade, and each stage is tuned to the same frequency, as in a conventional tuned *RF* amplifier, the output from the preceding stage is multiplied by the gain of the following stage. In turn, the overall frequency-response curve becomes narrower, or the overall bandwidth is reduced. Its progressive decrease in bandwidth is visualized in Fig. 16-7. This is called a synchronously-tuned amplifier. Bandwidths are shown with respect to the response of the first stage. In the first stage, the bandwidth extends between the 0.707 of maximum amplitude points. In

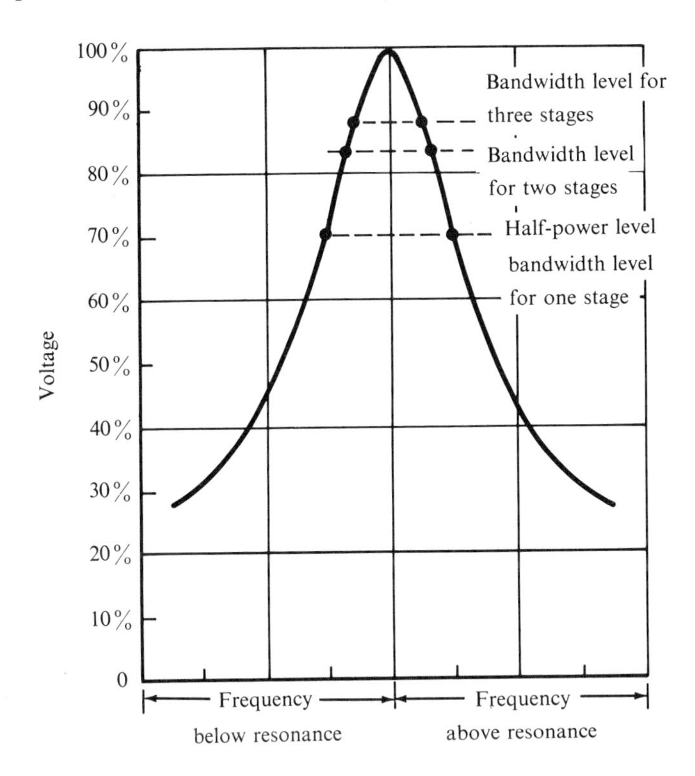

Figure 16-7. **Bandwidth reduction for cascaded tuned-amplifier stages.**

the second stage, the effect of gain multiplication is to move the bandwidth interval up to the 0.84 of maximum amplitude points on the reference curve. In the third stage, the bandwidth is further reduced, and extends between the 0.89 of maximum amplitude points on the reference curve.

Therefore, when a synchronously tuned amplifier is employed, the tank-circuit Q values must be lower than in a single tuned-amplifier stage, if the same bandwidth is to be retained. The individual stage gain is reduced to some extent, but this is a minor disadvantage of a synchronously tuned amplifier. It may be overcome by use of more elaborate tank-circuit configurations, as discussed subsequently.

It is also possible to increase the bandwidth of a cascaded tuned amplifier without reducing the Q values of the individual tank circuits by *stagger tuning*. With reference to Fig. 16-8, L_1C_1 is tuned to a lower frequency than L_2C_2. In turn, increased bandwidth and essen-

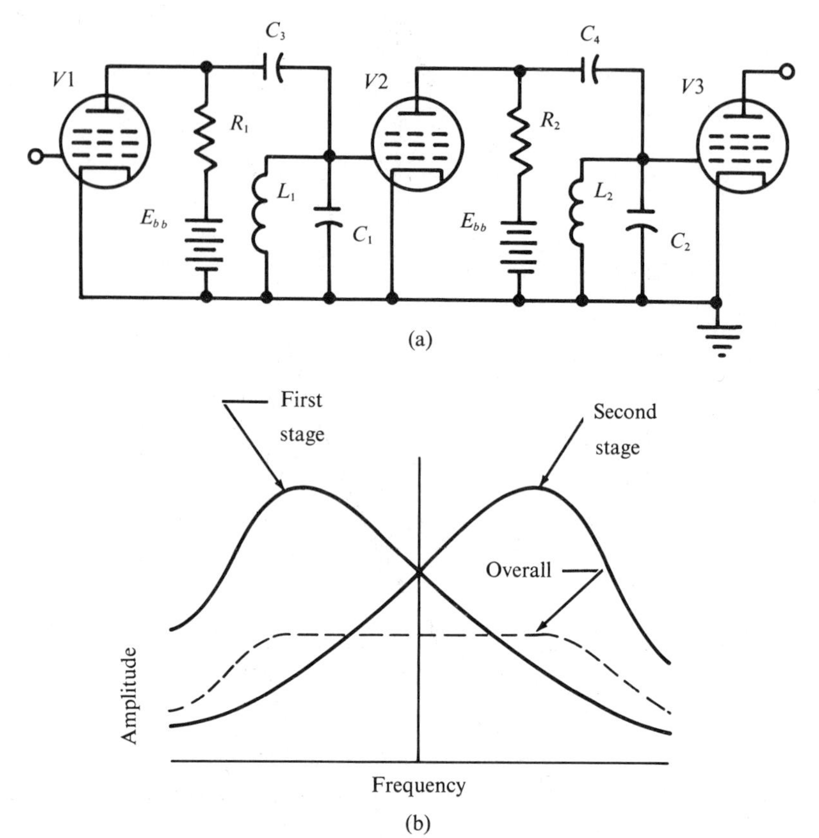

Figure 16–8. **(a) Two-stage stagger-tuned amplifier circuit; (b) Individual stage responses and overall response.**

tially flat-topped frequency response is obtained. Of course, the gain of the two-stage amplifier is considerably less than if it were synchronously tuned. In television receivers, three or more stagger-tuned stages may be cascaded. It can be shown that the shape of the overall response curve in a stagger-tuned amplifier remains the same if the gain of one or more of the devices declines. Only the overall gain of the amplifier is affected by a slump in g_m or h_{fe}. A stagger-tuned amplifier provides an economical means of obtaining wide-band flat-topped response with reasonably steep sides on the overall frequency-response curve.

As an example of a transistor tuned amplifier at medium frequencies (455 kHz), we will discuss the *i-f* amplifier circuit in Fig. 16–9. The transistors are an NPN silicon type, and their beta cutoff is 1

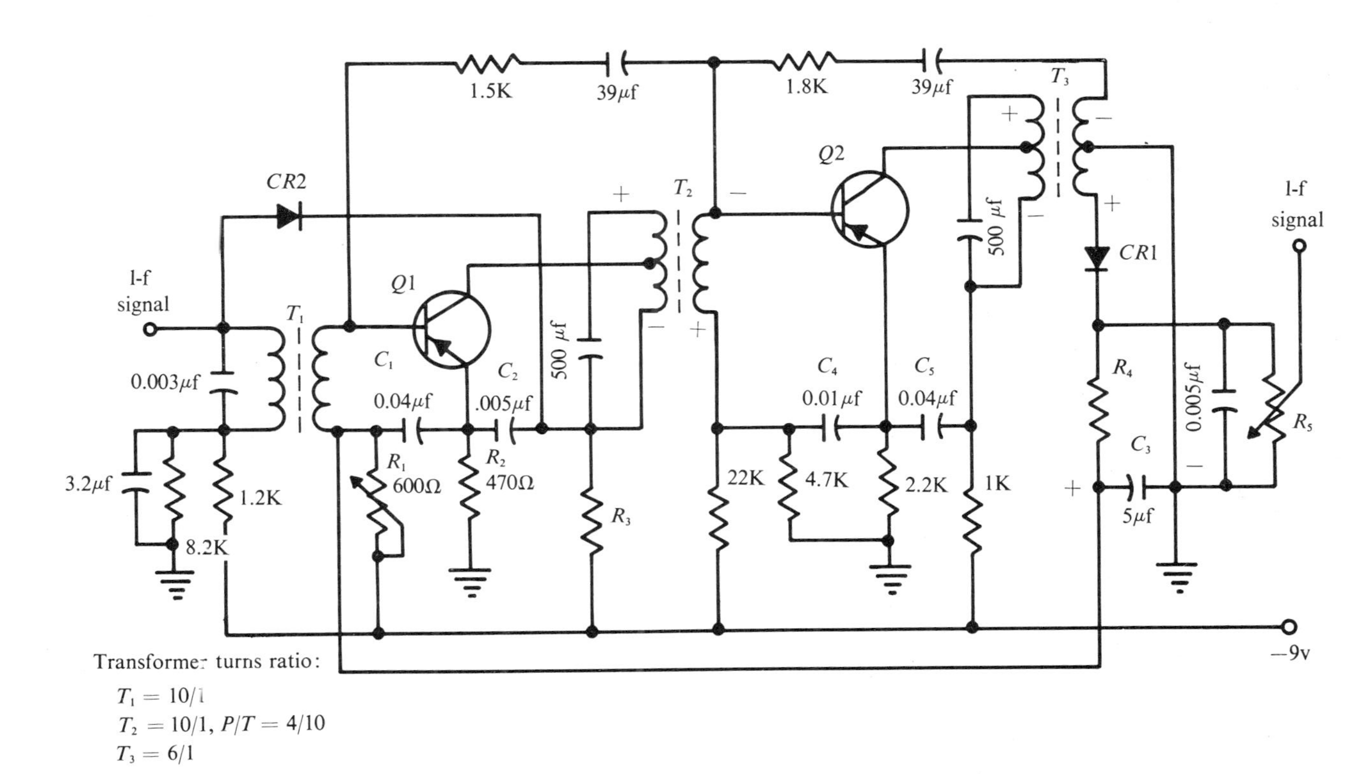

Figure 16-9. Illustration of a 455-kHz *i-f* amplifier with an AM detector.

MHz. The operating point of each transistor is set by a voltage divider and an emitter resistor which aids in temperature stabilization. The first stage is set to about 3 μa base current which is changed by the AVC (automatic volume control) voltage feedback through R_4. The second stage has about 10 μa base current to allow a larger signal swing at the collector. The tuned-collector circuits are transformer-coupled to the next stage for optimum impedance matching and, at the same time, to develop a 180° phase-shift neutralization. Notice that the collector is tapped into the tuned-tank circuit (at 1/5) to decrease the loading effect of the transistor's internal capacitance and resistance. This technique provides a much higher effective tank circuit Q and, hence, a narrower overall bandwidth.

Example 4. As an exercise, let us calculate the values of neutralizing components in each stage when $r_{bb} = 110\ \Omega$, $n = N_5/N_p = 1/5$, $c_{be} = 800$ pf, $c_{bc} = 7$ pf, and $r_{be} = 100\ \text{k}\Omega$.

$$C_N \simeq \frac{1}{n} c_{bc} \tag{16–6}$$

$$C_N \simeq 5 \times 7 \text{ pf} \simeq 35 \text{ pf} \qquad \text{(answer)}$$

$$R_N \simeq n r_{bb} \frac{c_{be}}{c_{bc}} \tag{16–7}$$

$$R_N \simeq \frac{1}{5} 110 \times \frac{800 \times 10^{-12}}{10 \times 10^{-12}} \simeq 1.76\ \text{k}\Omega \qquad \text{(answer)}$$

$$R'_N \simeq n\, r_{be} \tag{16–8}$$

$$R'_N \simeq \frac{1}{5} 100\ \text{k}\Omega \simeq 20\ \text{k}\Omega \qquad \text{(answer)}$$

We will use variable 10-to-47 pf capacitors for neutralization, as the calculated values are only approximate. Capacitors C_1, C_2, C_3, C_4, and C_5 are bypass capacitors to the *i-f* frequency. Diode $CR1$, resistor R_5, and capacitor C_6 demodulate the amplitude-modulated *i-f* frequency to develop an audio output signal.

Another interesting tuned *i-f* transistor is depicted in Fig. 16–10. Both transistors are silicon type ZN1605 operating in the common-emitter configuration at a frequency of 10.7 MHz. The first stage uses double tuning for maximum gain. The second stage uses only single tuning. The amplifier is broadly tuned, because the low input impedance of T_2 is connected directly across the tuned secondary tank. This heavy damping eliminates the need of neutralization. Note that the unloaded Q of L_1 is approximately 100; whereas, the loaded Q is approximately 30.

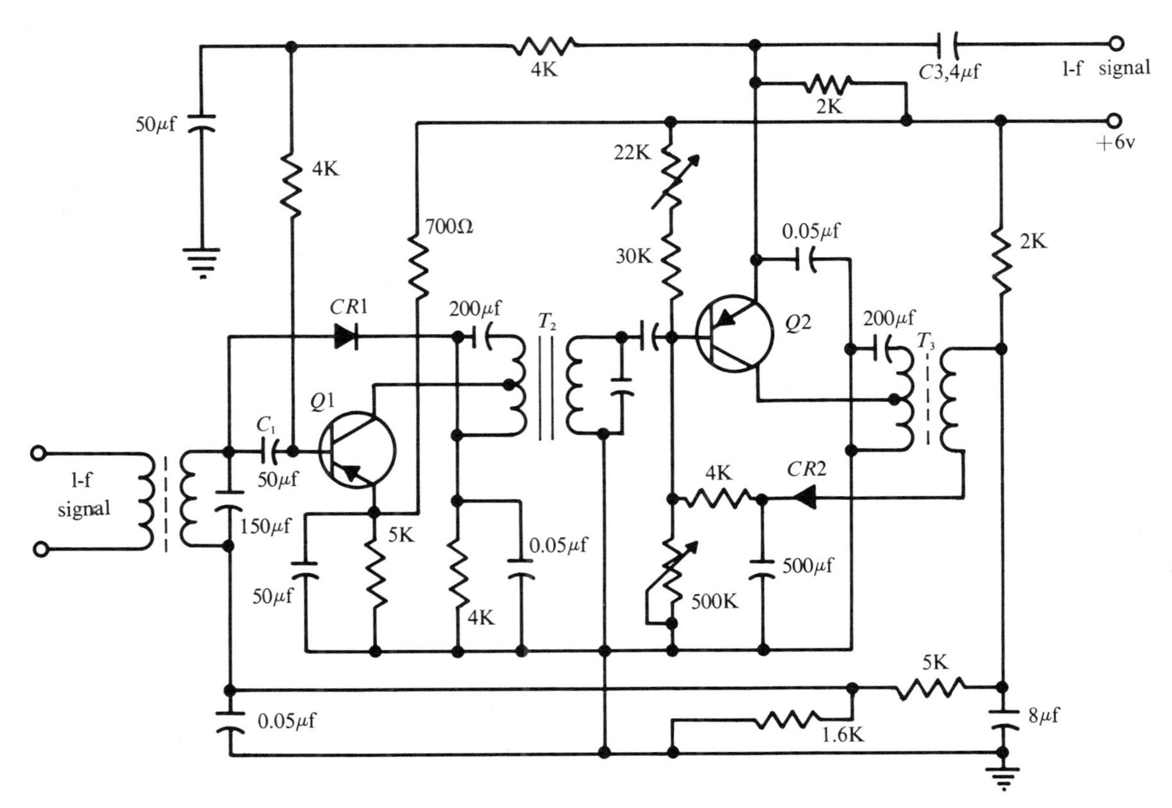

Figure 16–10. A two-stage *i-f* amplifier of a portable radio receiver.

Transistor $Q2$ operates as a tuned amplifier in the common-emitter configuration for the 10.7-MHz signal and as an RC-coupled amplifier in the common-collector configuration for the demodulated audio signal. The signal from the secondary of T_2 is demodulated by $CR1$. The audio frequency is then coupled back to the base of $Q2$, amplified, and taken from the emitter.

16–4. Amplifiers with Tuned-Transformer Coupling

As noted earlier, higher gain at a given bandwidth can be obtained from a tuned amplifier by using double-tuned transformer coupling, as shown in Fig. 16–11. The frequency response of the tuned stage is the same as the frequency response of the transformer. Both primary and secondary have the same Q values, and are tuned to the same center frequency in conventional applications. Bandwidth of the circuit is dependent chiefly upon the spacing (coupling) of primary and secondary. As the spacing between the coils is reduced, the energy transfer increases to a maximum as the bandwidth increases. At very close spacing, a double-humped response is obtained with still greater bandwidth.

To understand this circuit action, let us start with inductance considerations; with reference to Fig. 16–12(a), primary and secondary P and S are magnetically coupled. This is just another way of saying that the primary and secondary have mutual inductance. We may draw the equivalent circuit depicted in Fig. 16–12(b), which shows the circuit significance of this mutual inductance. The value of the mutual inductance can be measured by first connecting the coils in series-aiding, as shown in Fig. 16–12(c). The mutual inductance is common to both primary and secondary. Accordingly, the total measured inductance value is formulated

$$L_{\text{total}} = L_p + L_s + 2L_m \qquad \textbf{(16–9)}$$

Next, we may connect the coils in series-opposing, as shown in Fig. 16–12(d). Now, the total measured inductance value is formulated

$$L_{\text{total}} = L_p + L_s - 2L_m \qquad \textbf{(16–10)}$$

Then, we can calculate the value of mutual inductance for the given coil spacing by subtracting Formula 16–10 from Formula 16–9.

$$\begin{array}{r} L_{\text{total}}\ (\text{aiding}) = L_p + L_s + 2L_m \\ L_{\text{total}}\ (\text{opposing}) = L_p + L_s + 2L_m \\ \hline L_{\text{total(aiding)}} - L_{\text{total(opposing)}} = 4L_m \end{array}$$

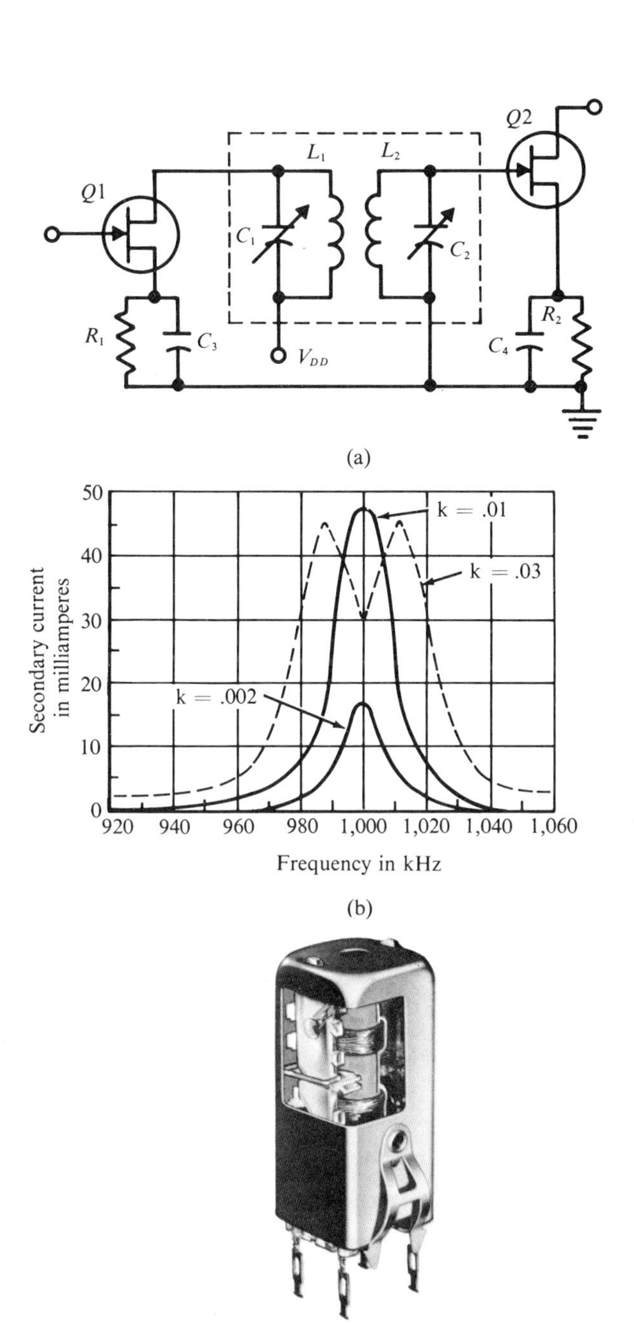

Courtesy of J. W. Miller Co.

Figure 16-11. (a) Amplifier stage with double-tuned transformer coupling; (b) Frequency response of transformer secondary; (c) Cutaway view of a double-tuned transformer.

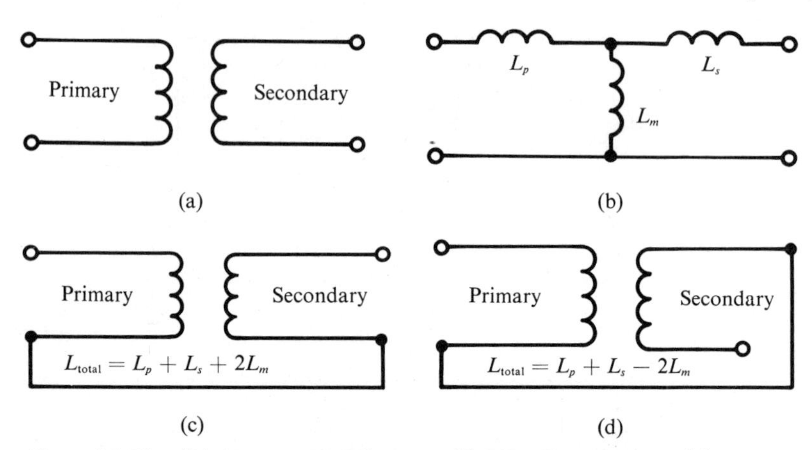

Figure 16-12. (a) Air-core transformer; (b) Equivalent circuit; (c) Series-aiding connection; (d) Series-opposing connection.

or,

$$L_m = \frac{L_{\text{aiding}} - L_{\text{opposing}}}{4} \tag{16-11}$$

Let us consider the theoretical derivation of the *coefficient of coupling*. If the coupling between the primary and secondary is complete, and there is no leakage flux, the theoretically maximum value of mutual inductance exists. This maximum value is formulated

$$L_{M(\max)} = \sqrt{L_p L_s} \tag{16-12}$$

Evidently, if $L_p = L_s$ (as is usually the case), $L_{M(\max)}$ is equal to L_p or to L_s. When we discuss tuned transformers, it is convenient to describe the amount of coupling in terms of the *coefficient of coupling*, which is formulated

$$k = \frac{L_m}{\sqrt{L_p L_s}} \tag{16-13}$$

where L_m is the *mutual* inductance.

Evidently, if $L_p = L_s$, Formula 16-13 is written

$$k = \frac{L_m}{L_p} = \frac{L_m}{L_s} \tag{16-14}$$

Example 5. A high-frequency transformer is tested for mutual inductance and coefficient of coupling. Using Fig. 16-12, $L_p = 50\ \mu$h, $L_s = 20\ \mu$h, $L_{\text{aiding}} = 133.2\ \mu$h, and $L_{\text{opposing}} = 6.8\ \mu$h. What are the values of the mutual inductance L_m and the coefficient of coupling k?

$$L_m = \frac{L_A - L_O}{4}$$

$$= \frac{126.4}{4} = 31.6\ \mu\text{h} \qquad \text{(answer)}$$

$$k = \frac{L_m}{\sqrt{L_p L_s}}$$

$$= \frac{31.6}{\sqrt{50 \times 20}} = 0.1 \qquad \text{(answer)}$$

In other words, L_M is the maximum possible value of L_m, and we usually work with values of L_m that give values of k in the order of 1 per cent, as noted in Fig. 16–11(b). When $k = 1$ per cent in this example, we say that *critical coupling* is present. This term denotes the maximum possible ouptut and also the point at which the response curve still has a single peak. If coupling exceeds the critical value, the single peak separates into a double-humped curve. We may note in passing that tuned transformers are often operated with greater than critical coupling, and a resulting double-humped frequency-response curve.

The coefficient of critical coupling is related to the Q values of primary and secondary by the formula

$$k_{\text{(critical)}} = \frac{1}{\sqrt{Q_p Q_s}} \qquad \textbf{(16–15)}$$

If both primary and secondary have the same Q value, we write

$$k_{\text{(critical)}} = \frac{1}{Q_p} = \frac{1}{Q_s} \qquad \textbf{(16–16)}$$

Example 6. Determine the coefficient value for critical coupling of the transformer in the circuit of Fig. 16–11 if $Q_p = Q_s = 100$.

$$k = \frac{1}{\sqrt{Q_s Q_p}}$$

$$= \frac{1}{100} = 0.01 \qquad \text{(answer)}$$

In other words, when the Q value is low, we must employ tighter coupling to obtain critical coupling, and vice versa. Figure 16–13 provides a visual summary of frequency-response curve variation as k is varied with Q held constant, and as Q is varied with k held constant. With critical coupling, the Q value of the tuned transformer is given approximately by the formula

$$Q_{\text{(trans.)}} = \sqrt{Q_p Q_s} \qquad \textbf{(16–17)}$$

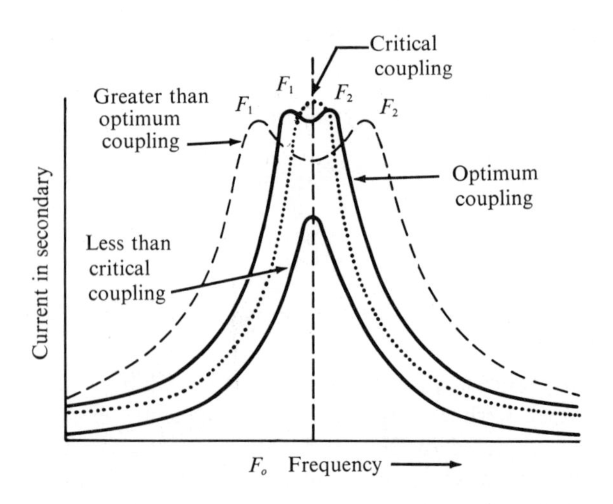

(a)

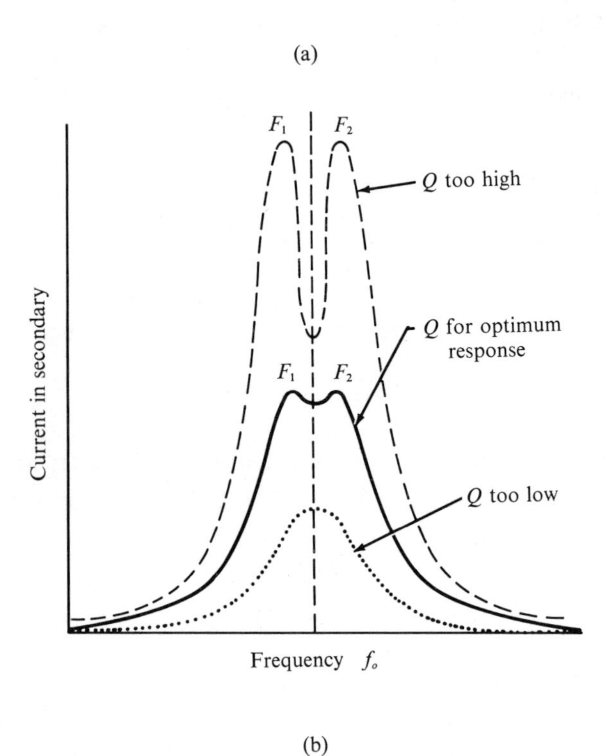

(b)

Figure 16-13. (a) Response curves for various coefficients of coupling, with primary and secondary Q values held constant; (b) Response curves for various values of Q, with the coefficient of coupling held constant.

If both primary and secondary have the same Q value, we write

$$Q_{(\text{trans.})} = Q_p = Q_s \tag{16–18}$$

Transformer bandwidth at critical coupling is given by the formula

$$BW = \frac{\sqrt{2} f_o}{Q_{(\text{trans.})}}, \text{ approximately} \tag{16–19}$$

Example 7. In reference to Fig. 16–11 and Example 6, if the transformer Q is 100, the bandwidth is then

$$\begin{aligned} BW &\simeq \frac{\sqrt{2} f_o}{Q_{(\text{trans.})}} \\ &\simeq \frac{\sqrt{2} \times 10^6}{100} \\ &\simeq 14.1 \text{ kHz} \qquad \text{(answer)} \end{aligned}$$

The Q value in Example 7 was approximately 100, and the bandwidth was calculated to be 14,100 Hz, in accordance with Formula 16–19. Inspection of the curve for $k = 0.01$ shows that the calculation is in reasonably good agreement with the measured response. Approximate formulas are also available for bandwidth calculation in the overcoupled situations, and these may be referred to in electronics design handbooks. As one might anticipate, the bandwidth decreases when tuned-transformer stages are cascaded. Again, approximate formulas for cascade amplifiers may be found in handbooks.

16–5. Transient Response of Tuned-Transformer Stages

The transient response of a tuned-transformer stage is the same as the transient response of the transformer. This response is perhaps unexpected. With reference to Fig. 16–14(a), it might be supposed that the transformer would have a simple ringing pattern. However, analysis of the equivalent circuit depicted in Fig. 16–14(b) reveals that two ringing frequencies are present. So this is basically a bridge circuit; if L_p and L_s are tuned to the same frequency (f_o), we perceive that there is no voltage drop across L_m with respect to current flow around the complete circuit. This situation corresponds to one of the ringing frequencies in which L_p and L_s are connected in series with C_p and C_s.

Next, another current path is provided around half the circuit comprising L_p, C_p, and L_m in shunt with L_s and C_s. An identical current path (insofar as the resonant frequency is concerned) is provided around the other half of the circuit comprising L_s, C_s, and L_m in shunt with L_p and C_p. Evidently, this ringing frequency is different from

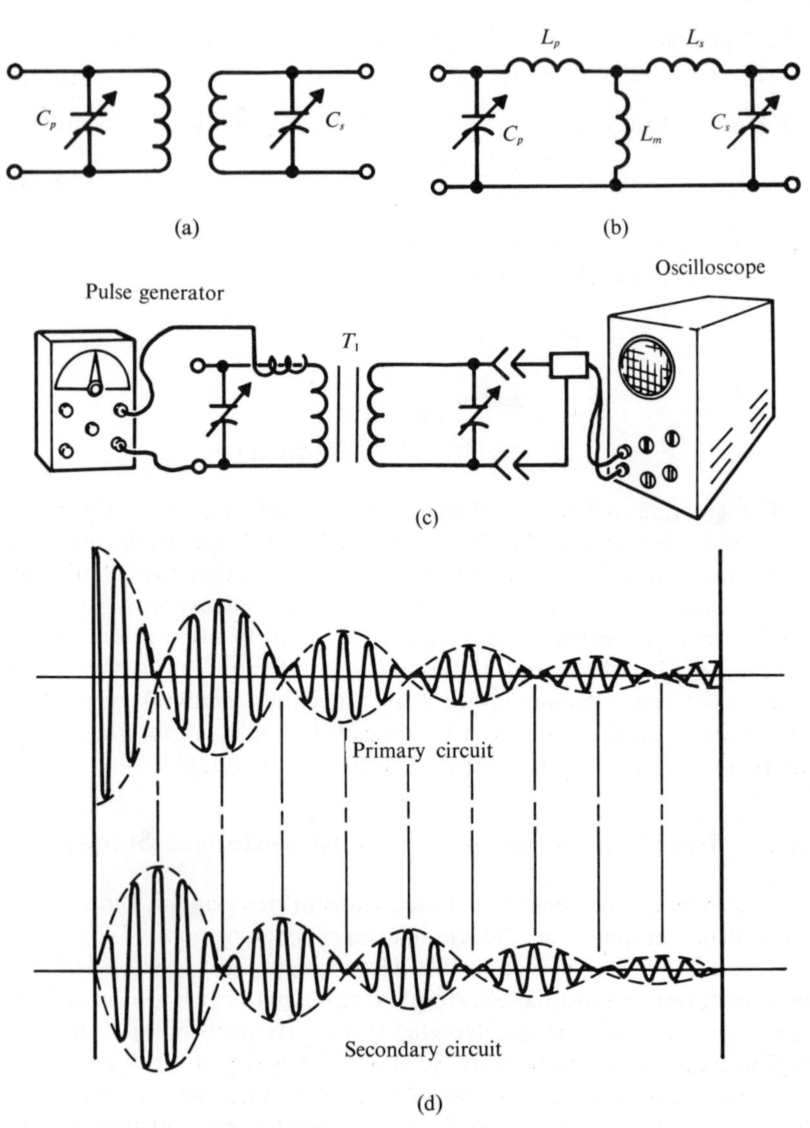

Figure 16–14. (a) Double-tuned transformer; (b) Equivalent circuit; (c) Test setup; (d) Ringing waveforms.

that associated with current flow around the complete circuit. Thus, two ringing frequencies are present. A ringing test can be made as shown in Fig. 16–14(c). An energizing pulse is provided by a pulse generator coupled to the primary by a "gimmick." A gimmick is simply a small coupling capacitance provided by a turn or two of insulated wire. An oscilloscope is connected to the secondary terminals to display the ringing waveform.

Since two ringing frequencies are present, they *beat* together and successively add and subtract as they progress in and out of phase with each other. (See Fig. 16–15) Note the secondary ringing pattern illus-

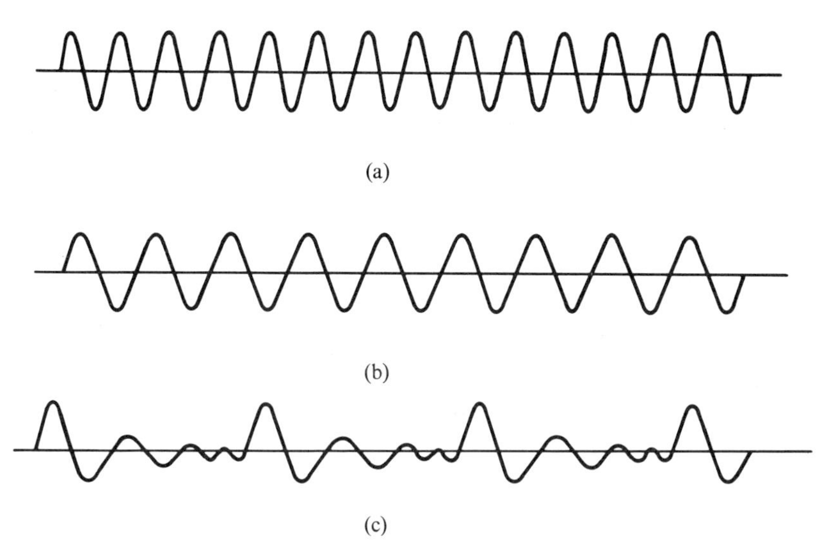

Figure 16–15. (a) First frequency; (b) Second frequency; (c) Beat frequency.

trated in Fig. 16–14(d). If the oscilloscope were connected across the primary terminals, we would observe the primary ringing pattern shown in Fig. 16–14(d). Observe that the ringing pattern of the primary goes through its maximum amplitude at the time that the secondary ringing pattern goes through zero amplitude, and vice versa. In other words, oscillatory energy is coupled back and forth from secondary to primary, and from primary to secondary. The amplitudes of the ringing waveforms gradually decay because of I^2R losses in the a-c resistances of the primary and secondary windings.

All of the characteristics of the double-tuned transformer are implicit in the ringing waveforms, and detailed analyses are given in advanced electronics texts. Since the detailed analysis is somewhat involved, it is not developed in this introductory presentation of the topic. We will merely note that the ringing frequency is equal to f_o, and that it is the average of the two beating frequencies. The beating frequencies are approximately equal to F_1 and F_2 depicted in Fig. 16–13. If we increase the coupling coefficient, the beat intervals in Fig. 16–14(d) become shorter; each envelope excursion then contains a fewer number of cycles. If the coupling coefficient is greatly decreased, the ringing waveform will decay to zero before the first envelope excursion is traversed.

16–6. Summary of Basic Formulas Containing a Q-Value Term

It is advisable at this point in our studies to summarize the basic formulas that contain a Q-value term. We will consider series and parallel *RLC* circuits which utilize ideal resistive, inductive, and capacitive components. The Q values are formulated at the resonant frequency of the circuit. In a series-resonant circuit [Fig. 16–16(a)],

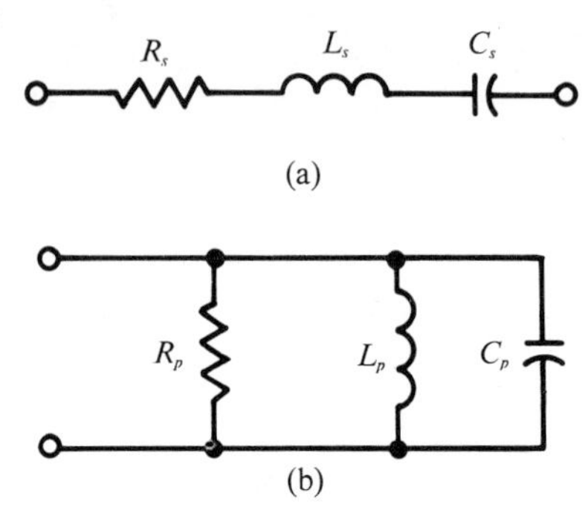

Figure 16–16. (a) Series *RCL* circuit; (b) Parallel *RCL* circuit.

the inductive and capacitive reactances cancel at resonance; nevertheless, the inductor and the capacitor have certain reactances at resonance, and these reactance values are equal. The *total* reactance is zero, and only the series resistance is "seen" from the input terminals. A voltage rise of approximately Q times appears across the inductor and across the capacitor.

In a parallel-resonant circuit such as Fig. 16–16(b), the inductive and capacitive reactances draw equal and opposite currents that make the impedance of the parallel L and C components infinite. Thus, the *total* reactance is infinite, and only the parallel resistance is "seen" from the input terminals. Nevertheless, the inductor and capacitor have certain reactances at resonance, and these reactance values are equal. With reference to Fig. 16–16, we may write

$$Q = \frac{\omega L_s}{R_s} = \frac{1}{\omega C_s R_s} = \frac{R_p}{\omega L_p} = R_p \omega C_p = \frac{\sqrt{\frac{L_s}{C_s}}}{R_s} = \frac{R_p}{\sqrt{\frac{L_p}{C_p}}} \tag{16–20}$$

If the resonant frequency is held constant, and the value of L is increased, then the value of C must be reduced accordingly. In a series-resonant circuit, a higher L/C ratio results in a higher Q value; however, in a parallel-resonant circuit, a higher L/C ratio results in a lower Q value. These facts follow directly from the relations expressed in Formula 16–20. Other basic formulas that are very useful in *RCL* circuit analysis are shown in the following table.

General formulas	*Formulas for Q greater than* 10	*Formulas for Q less than* 0.1
$R_s = \frac{R_p}{1 + Q^2}$	$R_s \simeq \frac{R_p}{Q_2}$	$R_s \simeq R_p$
$X_s = X_p \frac{Q^2}{1 + Q^2}$	$X_s \simeq X_p$	$X_s \simeq X_p Q^2$
$L_s = L_p \frac{Q^2}{1 + Q^2}$	$L_s \simeq L_p$	$L_s \simeq L_p Q^2$
$C_s = C_p \frac{1 + Q^2}{Q^2}$	$C_s \simeq C_p$	$C_s \simeq \frac{C_p}{Q^2}$
$R_p = R_s(1 + Q^2)$	$R_p \simeq R_s Q^2$	$R_p \simeq R_s$
$X_p = X_s \frac{1 + Q^2}{Q^2}$	$X_p \simeq X_s$	$X_p \simeq \frac{X_s}{Q^2}$
$L_p = L_s \frac{1 + Q^2}{Q^2}$	$L_p \simeq L_s$	$L_p \simeq \frac{L_s}{Q^2}$
$C_p = C_s \frac{Q^2}{1 + Q^2}$	$C_p \simeq C_s$	$C_p \simeq C_s Q^2$

In a series-resonant circuit with zero series resistance, a voltage rise of Q times appears across the inductor and across the capacitor.

$$V_L = V_C = QV_s \tag{16–21}$$

On the other hand, if the series resistance is finite and sufficiently large to make the circuit Q comparatively small, we may write the following useful approximation.

$$V_L = V_C \simeq V_s \sqrt{Q^2 + 1} \tag{16–22}$$

The bandwidth of either a series-resonant or a parallel-resonant circuit that has fairly high Q is stated by the approximate formula

$$BW \simeq \frac{f_o}{f_2 - f_1} \tag{16–23}$$

The Q value of either a series or parallel *RCL* circuit may be approximated by varying the value of C and noting the capacitance values of f_o, f_2, and f_1. In turn, the Q value is given approximately by the following formula.

$$Q = \frac{2C_o}{C_2 - C_1} \quad \textbf{(16–24)}$$

Although only the exceptional student can be expected to memorize Formulas 16–20 through 16–24, all students should make an effort to memorize as many of the formulas as possible. These formulas are of basic importance in circuit analysis, and it is both inconvenient and time consuming to make repeated reference to these pages.

SUMMARY

1. Tuned amplifiers are commonly called bandpass amplifiers.
2. Tuned amplifiers may use vacuum tubes, junction transistors, or field-effect transistors as active devices.
3. The bandwidth of a tuned amplifier is determined by the resonant frequency of the tuned circuit, the inductance, and the effective resistance of the tuned circuit.
4. A neutralizing circuit feeds back to the input a signal that is equal in amplitude and opposite in phase to the signal feedback by the internal capacitance of the device.
5. Employment of a resistor in series with the grid of a triode tuned amplifier decreases the circuit gain and prevents oscillation of the stage.
6. Using a pentode vacuum tube in a tuned amplifier eliminates the need of neutralization in the stage.
7. As a tank circuit is tuned lower in frequency, the L/C ratio and the tank circuit impedance are decreased.
8. Cascading amplifier stages results in increased gain and decreased bandwidth.
9. Neutralization of a tuned junction-transistor amplifier is sometimes unnecessary because the low values of input and output impedance "damp" any oscillations.
10. The neutralization resistors, R_N and R'_N, are usually unnecessary in a tuned vacuum-tube or tuned field-effect transistor amplifier.
11. The tuning limits of a resonant LC circuit are limited by the stray capacitance of the circuit and the L/C ratio of the inductor and capacitor.
12. When tuned amplifiers are connected in cascade, the overall gain increases and the overall bandwidth decreases.
13. A stagger-tuned amplifier provides a wide-band flat-topped frequency response with reasonably steep sides.
14. The coefficient of coupling of the transformer in a tuned amplifier determines the bandwidth of the stage.

15. The value of the coupling coefficient for critical coupling varies inversely as the square root of the product of primary and secondary Q values.

Questions

1. What factor affects the bandwidth of a tuned amplifier?
2. What factors affect the impedance of a tuned amplifier?
3. With reference to Fig. 16-1, what are the advantages and disadvantages of the circuit?
4. What factors tend to produce oscillation in a circuit such as the one in Fig. 16-1?
5. Explain the "losser" method of suppressing oscillation in an amplifier.
6. Explain the Miller effect in an amplifying device.
7. Discuss neutralization of an amplifying device.
8. With reference to the circuit shown in Fig. 16-3, what are the disadvantages of this neturalization method?
9. Why doesn't the pentode vacuum tube require neutralization?
10. What are the disadvantages of using a pentode vacuum tubc in a tuned amplifier?
11. What determines the highest frequency to which a given tuned amplifier may be tuned?
12. What are the effects of a lower L/C ratio on a tuned amplifier?
13. What is the reason for the complex neutralization circuit in the transistor amplifier depicted in Fig. 16-5?
14. Explain why the complex neturalization circuit in Fig. 16-5 may be simplified when a field-effect transistor or vacuum tube is employed.
15. Why are the input and output impedances of a junction transistor extremely low at high frequencies?
16. Why does the performance of a tuned LC circuit improve at the higher-frequency end of the band?
17. What is the accepted frequency tuning ratio of an LC tank circuit when a variable capacitor is employed?
18. What are the effects of connecting tuned amplifiers in cascade?
19. What are the advantages of stagger tuning?
20. Why are the transistors in Fig. 16-10 tapped into the low end of the transformer of the tuned circuit?
21. Why is it unnecessary to employ neutralization in the tuned amplifiers in Fig. 16-11?
22. With reference to the tuned amplifier in Fig. 16-11: (a) What is the base current of transistor $Q1$? (b) What is the base current of transistor $Q2$? (c) In what configuration is transistor $Q2$?

23. Explain critical coupling as applied to a tuned transformer.
24. A single-stage tuned amplifier has a 60-kilohertz bandpass; what is the bandpass when three of these stages are connected in cascade?
25. What are the characteristics of a field-effect transistor tuned amplifier in which the transformer is critically coupled?
26. What is the effect of overcoupling in a tuned transformer-coupled amplifier?
27. What is the relationship of circuit Q to critical coupling in a transformer-coupled amplifier?
28. What is the relationship of circuit Q and the bandwidth in a tuned amplifier?
29. Discuss the ringing pattern of the transformer represented in Fig. 16-14.
30. What is a "gimmick"?
31. Why are two ringing frequencies observed in the test setup depicted in Fig. 16-14?

PROBLEMS

1. With reference to the circuit depicted in Fig. 16-2(a), determine the resonant frequency, the Q, the impedance, and the bandwidth when $C =$ 220 pf.
2. What value of series R would be necessary in Problem 1 to increase the bandwidth to 20 kHz?
3. With reference to the circuit in Fig. 16-2(a), calculate the circuit impedance and bandwidth for values of C of 33 pf and 200 pf. Discuss the reason for the change in impedance in these two examples.
4. With reference to Fig. 16-5, determine the values of R_N, R'_N and C_N to neutralize a 2N2904 transistor with the parameter values of $h_{fe} = 20$, $c_{ce} = 2$ pf, $c_{cb} = 8$ pf, $c_{be} = 9.5$ pf, $r_{be} = 550\ \Omega$, and $r_{ce} = 2.2\ \text{k}\Omega$.
5. Suppose the circuit in Problem 4 is to operate at 200 MHz using an MF1164 transistor with values of $c_{cb} = 2.5$ pf, $c_{oe} = 1.0$ pf, $c_{be} = 3$ pf, $r_{be} = 180\ \Omega$, and $r_{ce} = 1.5\ \text{k}\Omega$. What are the values of the neutralizing components?
6. Suppose the tuned circuit in Fig. 16-3 is to use a field-effect transistor with the following parameter values: $g_m = 1{,}000\ \mu$mho, $r_{ds} = 250\ \text{k}\Omega$, $r_{gs} = 10\ \text{M}\Omega$, $c_{ds} = 1.5$ pf, $c_{gs} = 4$ pf, and $c_{dg} = 3.2$ pf. What are the values of the neutralizing components?
7. Determine the values of the neutralizing components for the circuit in Problem 6 when a triode vacuum tube is used in which the parameters

are mu $= 20$, $c_{pk} = 5$ pf, $c_{pg} = 9.2$ pf, $c_{gk} = 7.5$ pf, $r_p = 20\ \text{k}\Omega$, $N = 5$, and $r_i = \infty$.

8. Determine the mutual inductance and coefficient of coupling of a transformer in which $L_s = L_p = 100\ \mu\text{h}$, $L_A = 230\ \mu\text{h}$, and $L_O = 170\ \mu\text{h}$.

9. What is the new value of mutual inductance in Problem 8 if the coefficient of coupling is increased to 0.2? decreased to 0.05?

10. The tuned transformer in the *i-f* amplifier in Fig. 16–10 has a primary Q of 20 and a secondary Q of 15. What is the coefficient of coupling value for critical coupling?

11. What is the bandwidth of the circuit in Problem 10 when $f_o = 455$ kHz?

12. With reference to the 10.7 MHz amplifier in Fig. 16–11, the values of circuit Q are $Q_p = 30$ and $Q_s = 20$. What is the coefficient value for critical coupling?

13. What is the bandwidth of one stage of the circuit in Problem 12? What is the bandwidth of two identical stages?

CHAPTER 17

Amplitude Modulation and Demodulation

17–1. Introduction

Modulation is the process whereby information is inserted into a sine waveform called the *carrier*. There is a very large number of modulation methods, of which only the most basic method called amplitude modulation, is discussed in this chapter. In amplitude modulation, the amplitude of a sine-wave voltage or current (carrier) is made to vary with time according to the voltage or current variations of another signal called the *modulating* signal. In general, the carrier has a higher frequency than the modulating signal.

Demodulation, also called *detection*, is the process by which the modulating signal is recovered from the modulated carrier at the receiver. There is a very large number of demodulation methods; we will limit our introductory survey to AM (amplitude-modulation) detectors. We will also restrict our discussion to the audio-frequency range of modulation and demodulation. Audio frequencies are generally considered to extend to 15 or 20 kHz, as depicted in the audio-frequency spectrum shown in Fig. 17–1.

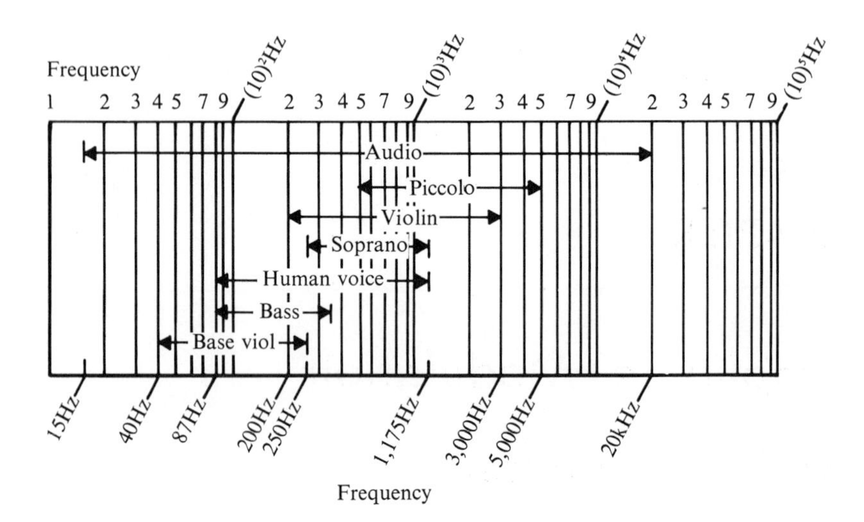

Figure 17–1. Audio-frequency spectrum.

The human voice range extends from about 87 Hz to 1,175 Hz, a violin has a range from about 200 Hz to 3 kHz, and the bass viol range extends from about 40 Hz to 250 Hz. Pure tones of the piccolo extend to approximately 5 Hz. However, combinations of sound frequencies entail harmonics that extend to 20 kHz. These combinations of frequencies give to speech or music the identifying characteristics that distinguish one person's voice from another's, and one musical instrument's tone from another's. It is not practical, for the following reasons, to transmit electromagnetic waves at audio frequencies—that is, without the use of a modulated radio-frequency carrier.

1. The range of audio-frequency electromagnetic waves is very limited, chiefly because of the poor radiation efficiency of antennas having a reasonable size.

2. All transmitters that employed audio-frequency electromagnetic radiation would necessarily operate in the same frequency range; therefore, it would be impossible to separate various transmissions at the receiver.

3. Even if it were mechanically practical to construct an antenna to radiate in the audio-frequency range with reasonable efficiency, such an antenna would radiate poorly at high frequencies if tuned to low frequencies, and vice versa.

4. Inductors and capacitors would have to be excessively large to resonate at low audio frequencies, and adequate resonant-circuit bandwidth would be difficult or impossible to realize with any reasonable percentage of operating efficiency.

These basic problems are overcome by use of a modulated *r-f* carrier at the transmitter and by demodulation at the receiver. Efficiency is reasonably high, even with the simplest methods of modulation. In effect, the entire audio-frequency spectrum is inserted into a small "slice" of the radio-frequency spectrum. Interference between two or more transmissions is still possible and may occur occasionally. However, interference is practically no problem in properly designed systems.

17–2. Amplitude Modulation

Amplitude modulation is defined as variation in amplitude of an *r-f* carrier wave at an audio-frequency rate. In other words, the initial *r-f* sine wave is made to increase and decrease in voltage according to the instantaneous amplitudes of an audio (sound) signal. Evidently, if the audio signal has a high frequency, the amplitude of the modulated carrier will vary more rapidly than if the audio signal has a low frequency. Loud audio-frequency tones cause the amplitude of the modulated wave to increase and decrease to a considerable extent; on the other hand, weak audio-frequency tones cause the amplitude of the modulated wave to vary but slightly.

Figure 17–2 shows that a single-tone modulated *r-f* waveform comprises three separate frequencies. These three frequencies can be derived by trigonometric-formula analysis of the modulation process.

$$e = E_c(1 + m \sin 2\pi f_m t) \sin 2\pi f_c t \qquad \textbf{(17–1)}$$

where e denotes the instantaneous voltage of the modulated wave, E_c denotes the voltage of the carrier wave, f_m denotes the modulating signal frequency, and f_c denotes the carrier frequency, and m is the modulation factor.

In other words, the instantaneous amplitude of the modulated wave is formulated as the sine-wave carrier plus the product of the sine-wave carrier and the sine-wave modulating signal. If the student should have difficulty in recognizing that Formula 17–1 describes the waveform depicted in Fig. 17–2, it will be helpful to "work backward" from the next formula and Fig. 17–3. The student may show that Formula 17–1 can be expanded by familiar trigonometric formulas to state (assuming that $m = 1$):

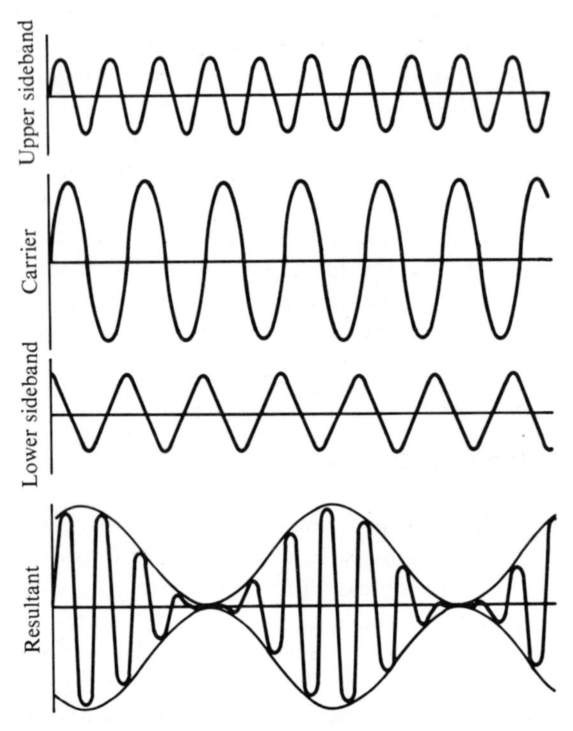

Figure 17–2. How the carrier and two sidebands combine to form a 100 per cent modulated waveform.

$$e = E_c \sin 2\pi f_c t + 0.5E_c \cos 2\pi(f_c - f_m)t - 0.5E_c \cos 2\pi(f_c + f_m)t \tag{17–2}$$

Observe that Formula 17–2 states that e is equal to the sum of three radio frequencies: *the carrier frequency, the difference of the carrier frequency and the modulating frequency, and the sum of the carrier and the modulating frequency.* The pair of sum-and-difference frequencies are called the two sideband frequencies. We perceive that the carrier contains no information; all the information is contained in the sideband frequencies. It is evident that in the case of complete sinusoidal modulation, the sideband power is equal to 50 per cent of the carrier power. Therefore, since the carrier contains no information, it might seem to be a waste of power to transmit the carrier. However, the presence of the carrier is necessary to provide the audio information at the receiver. Let us see why this is so.

Figure 17–3(a) shows the beat pattern of the carrier and both sideband frequencies. We observe that the *envelope* (outline) of the modu-

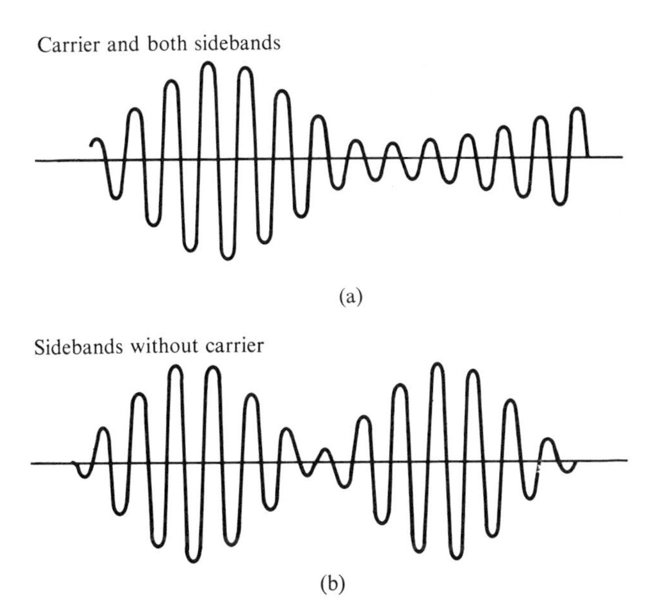

Figure 17–3. (a) Beat pattern of the carrier and both sideband frequencies; (b) Beat pattern of the two sideband frequencies alone.

lated waveform is a sine wave that represents the modulating signal. In other words, the envelope rises and falls in exact accordance with the audio signal that was used to modulate the carrier. On the other hand, Fig. 17–3(b) shows the beat pattern that results if we suppress the carrier wave from the modulated wave; now, the envelope rises and falls twice as fast as the audio signal that was used to modulate the carrier. That is, we have introduced serious frequency distortion by eliminating the carrier from the modulated wave. If a speech waveform were transmitted by this method, the receiver output would be unintelligible. Hence, although the carrier contains no information, the carrier is nevertheless an essential component of the modulated waveform.

Further analysis of Formula 17–2 reveals that the *bandwidth* of the modulated wave in the *r-f* spectrum is double the frequency of the modulating signal, inasmuch as both sum and difference frequencies are generated in the amplitude-modulation process. We must not suppose that the sideband frequencies given by Formula 17–2 are merely mathematical fictions; if we use suitable filters, each of the sideband frequencies (as well as the carrier component) can be separated from the modulated waveform and displayed on an oscilloscope screen.

Thus far, we have considered a completely (100 per cent) amplitude-modulated waveform. Next, if the audio-modulating voltage is only half as great, we obtain the 50 per cent modulated waveform depicted in Fig. 17–4(b). Again, if the modulating voltage is reduced to zero, we have only the carrier remaining, as shown in Fig. 17–4(c). Calculation of the percentage modulation in the general situation is made as seen in Fig. 17–4(d). The formula is written

$$\text{Percentage modulation} = \frac{B - A}{B + A} \times 100 \qquad \textbf{(17–3)}$$

where B and A are the crest and trough amplitudes, as depicted in Fig. 17–4(d).

Example 1. Suppose the crest voltage of the modulation envelope in Fig. 17–4(d) is 20 v, and the trough voltage is 13 v. What is the percentage of modulation?

$$PM = \frac{20 - 13}{20 + 13} \times 100$$
$$= 21.2\% \qquad \text{(answer)}$$

In the case of 100 per cent modulation, we perceive that the carrier rises to double its initial amplitude and falls to zero. The sidebands contain a power equal to one-half of the carrier power at 100 per cent modulation. In other words, a completely modulated wave contains 50 per cent more power than the unmodulated carrier. If the carrier is modulated less than 100 per cent, the sideband power is proportionally less. Speech and music waveforms contain a rather small average-power value; therefore, amplitude modulation entails a rather large proportion of carrier power that does not contribute directly to information transfer. We will learn subsequently that other forms of modulation provide greater utilization of modulating information; however, amplitude modulation is comparatively simple and economical of bandwidth, and so it is widely used.

Of course, amplitude modulation cannot be "improved" by use of more than 100 per cent modulation. When a carrier is overmodulated, there are intervals during which "gaps" occur in the transmission, as shown in Fig. 17–5. These gaps are called "sideband splatter," because the audio signal is distorted and sounds as if it were mixed with a splattering tone. Sideband splatter also entails abrupt stopping and starting of the modulation envelope along the zero axis, and these discontinuities comprise high transient frequencies. These high-frequency transients greatly increase the normal bandwidth of the

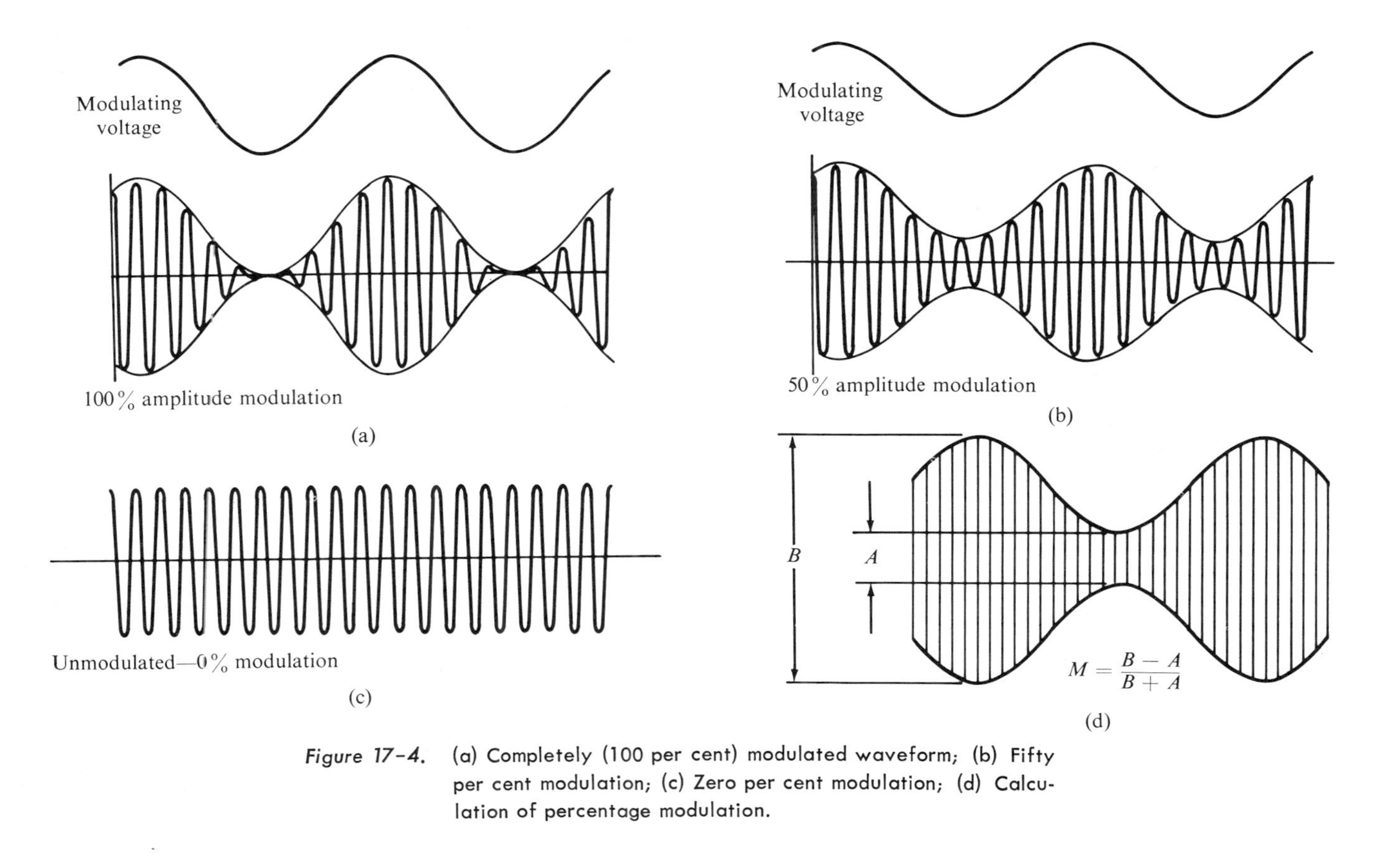

Figure 17-4. (a) Completely (100 per cent) modulated waveform; (b) Fifty per cent modulation; (c) Zero per cent modulation; (d) Calculation of percentage modulation.

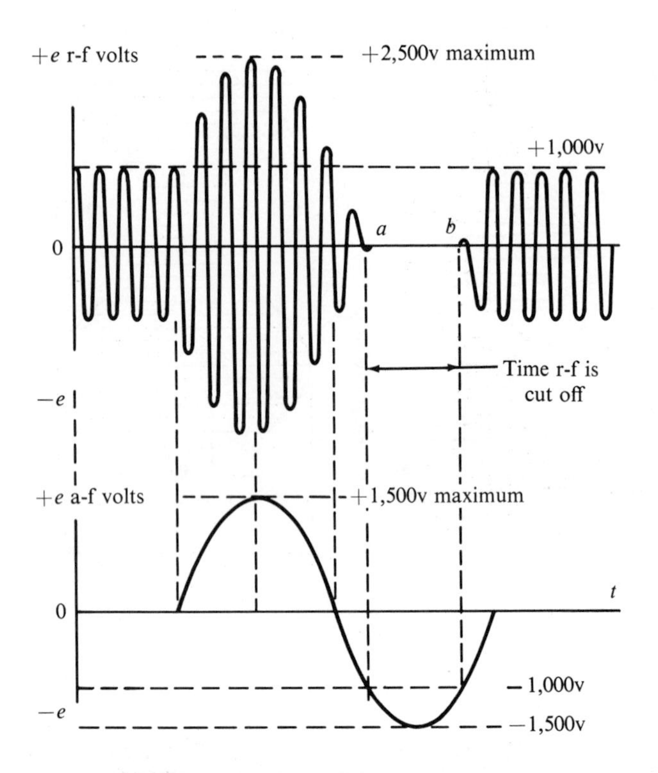

Figure 17–5. **Result of amplitude modulation in excess of 100 per cent (overmodulation).**

transmission and cause interference with other transmissions that operate on nearby carrier frequencies.

17–3. Simple Radiotelephone System

A block diagram of an amplitude-modulated radiotelephone transmitter is shown in Fig. 17–6. The top row of blocks indicate the *r-f* section and below are blocks that indicate the audio-frequency section; at the bottom of the diagram, the power supply is indicated that provides all d-c operating potentials to the transmitter. We perceive that the *r-f* section generates the high-frequency carrier that is fed to and radiated by the antenna. In the *a-f* section, a speech amplifier is energized by a few millivolts of audio signal from the microphone. In turn, the audio signal is amplified to several volts at the driver stage. This stage comprises power amplifiers that provide a com-

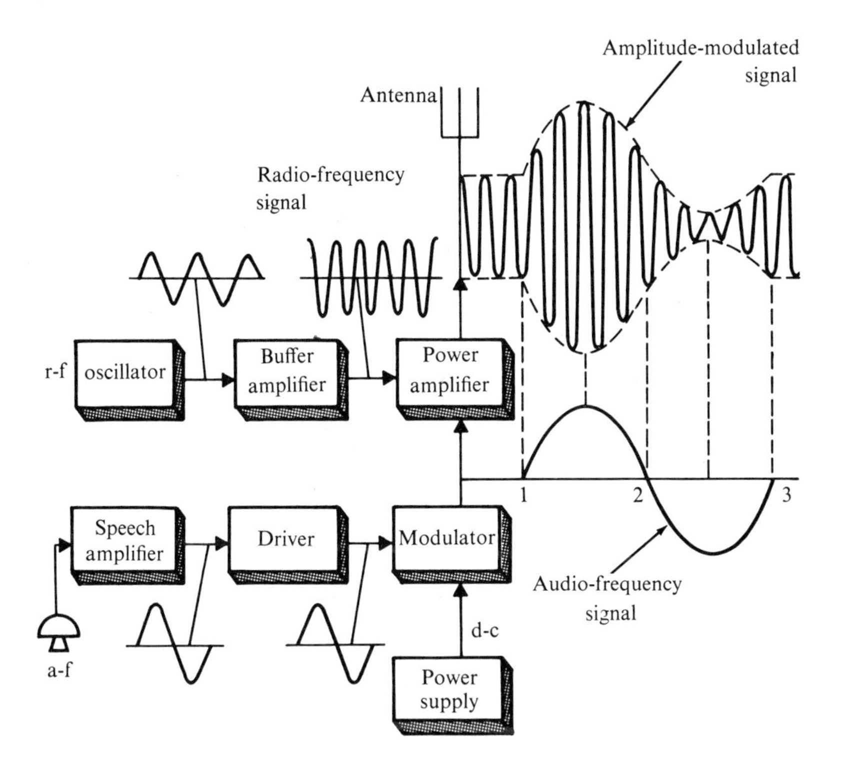

Figure 17–6. Block diagram of a simple AM radiotelephone transmitter.

paratively large voltage and substantial current to the modulator input.

Output from the modulator (an audio signal with substantial power) is fed to the *r-f* power amplifier through a circuit that causes the audio signal to alternately add to and subtract from the plate power of the *r-f* power amplifier. The amplitude of the *r-f* carrier rises and falls in accordance with the amplitude of the audio signal. One basic method of amplitude modulation is called *high-level linear modulation*, as depicted in Fig. 17–7. This is termed a high-level modulation system because the audio signal is fed into the plate circuit or the collector circuit of the *r-f* power amplifier and modulates the final amplifier. The *r-f* power amplifier circuit in Fig. 17–7(a) is tuned in this example to a frequency of 1 MHz. Plate-current flow is 100 ma, and the plate-supply voltage is 1,000 v. A grid-input driving signal is provided by the *r-f* oscillator and buffer amplifier; the buffer amplifier is used to isolate the oscillator from the power stage, thereby maintaining a stable oscillator frequency.

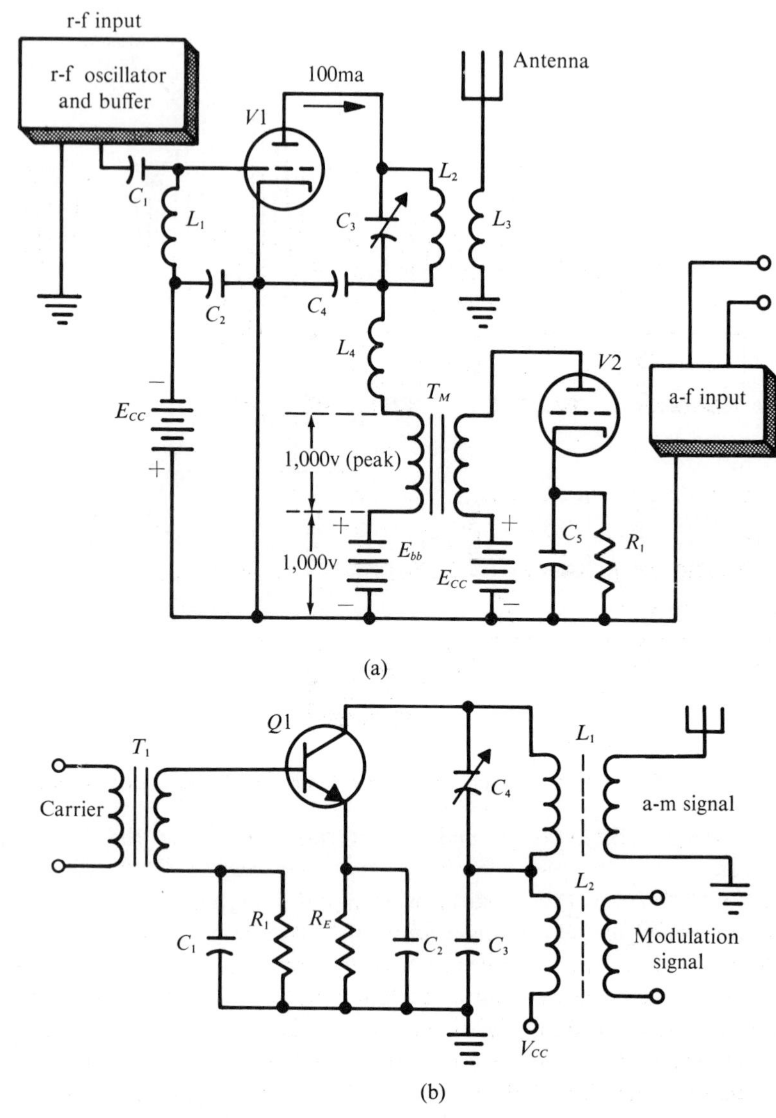

Figure 17–7. Large signal or linear modulation. (a) Vacuum tube output; (b) Transistor output.

An audio signal of approximately 1,000 peak v that has a frequency of 1 kHz in this example is provided by the *a-f* section shown as a block at the right of Fig. 17–7. This audio signal appears as the output of the modulation transformer T_M. Note that this audio signal operates in series with the *r-f* voltage across the tank circuit L_2C_3 and the plate power supply E_{BB}. To simplify our analysis, this example considers modulation of the 1 MHz *r-f* output voltage by a single *a-f* signal with a sine waveform. The 1-MHz *r-f* carrier wave combines with the *a-f* signal to form sidebands, as stated by Formulas 17–1 and 17–2. In this example, the frequency of the *r-f* voltage developed across the plate-tank circuit is 1,000 times the audio frequency. Thus, the time for one complete audio cycle is long enough to include 1,000 cycles of *r-f* energy in the *LC* tank circuit.

Before an audio-modulating signal is applied in Figure 17–7, the *r-f* signal at the grid of V_1 causes the triode to conduct periodically (Class-*C* operation provides high efficiency). During each conducting period, capacitor *C* charges. When the grid swings below cutoff, the triode stops conducting and the capacitor discharges through the coil. In turn, the *LC* circuit starts to "ring" with each pulse of current flow. It is this exchange of energy back and forth between the coil and capacitor that accounts for the a-c voltage developed across the tank. In this example, the peak *r-f* voltage across the tank capacitor *C*, which receives its charge from the 1,000-v plate supply, is equal to 1,000 v minus incidental circuit losses.

Because the tank circuit operates as a shock-excited oscillator in response to the pulses of plate-current flow, the plate voltage varies between 1,000 $\pm$ 1,000 v, or between 2,000 and 0 v in Fig. 17–7. The plate voltage is above the plate-supply voltage during that part of the cycle when the grid voltage is below cutoff and the triode is not conducting. On the other hand, the plate voltage is below the plate-supply voltage during the time that the grid is above cutoff and the triode is conducting. Energy to supply the tank-circuit losses comes, of course, from the plate supply, and it is the flywheel effect in the tank circuit that accounts for the sine waveform of *r-f* voltage across, and current through, the tank.

Now, let us assume that the *a-f* voltage with its sine waveform is introduced by modulation transformer T_M, in series with the *r-f* tank and the plate supply. Consider that portion of the *a-f* cycle in which the voltage is gradually rising sinusoidally, so that its polarity aids the plate voltage. For 250 cycles of radio frequency, or one-quarter of an audio-frequency cycle, the total voltage available for charging capacitor *C* in Fig. 17–7 is increasing from 1,000 v to a maximum of 2,000 v. Assume that the capacitor charges to 2,000 v; the plate voltage at the end of 250 *r-f* cycles attains a maximum value of 2,000 v. Thus, with 100 per cent modulation, and on the positive audio peaks, the value of the peak-to-peak voltage across capacitor *C* is approximately double the peak-to-peak value it would have in the absence of modulation, or the peak-to-peak voltage is four times the value of the plate-supply voltage.

The tank voltage in Fig. 17–7(a) starts to decrease between the 250th *r-f* cycle and the 500th *r-f* cycle as the *a-f* voltage falls from a maximum of 1,000 v to 0 v during this period. From the 500th to the 750th *r-f* cycle, the polarity of the *a-f* voltage is reversed, and it reaches a maximum negative value at the 750th *r-f* cycle; the tank capacitor charges up to the difference voltage between the 1,000-v supply value and the instantaneous *a-f* voltage value. At the instant that the *a-f* voltage attains its maximum negative value (−1,000 v at the 750th *r-f* cycle), the tank capacitor is not charging at all. Note that the tank voltage is momentarily zero, and the *r-f* output energy also falls to zero at this instant. Then, the *a-f* voltage starts to progress from its maximum negative value toward zero, and the opposition that it presents to the $B+$ voltage is lessened. The plate-tank capacitor starts charging again, and the *r-f* tank current again increases during the subsequent 500 *r-f* cycles of operation.

The antenna is connected to one terminal of the tank circuit in Fig. 17–7(a), and the other tank-circuit terminal is grounded. In other words, the antenna coupling coil is a part of the complete tank circuit. We will find that an antenna-ground circuit "looks like" a resonant *LCR* circuit. Since the lines of force in the antenna system spread out extensively into surrounding space, the tank coil is coupled to an extensive field thus established in space. It can be shown that the phase relation of this field changes progressively as we proceed away from the antenna, and that time is required for field energy to travel out and return to the antenna conductors. Therefore, part of the returning energy is opposed by a changed conductor voltage, and this part of the returning energy cannot re-enter the antenna conductors. Instead, it is forced to detach itself and to travel out through space as a radiation field.

Current flow into the antenna-ground system rises and falls in accordance with tank-voltage variation. An *r-f* ammeter connected in series with the antenna lead in Fig. 17–7 will indicate the rms value of current flow in the antenna system. Let us consider this effective (rms) current indication corresponding to an AM signal in somewhat greater detail. Before any *a-f* signal is applied, the tank circuit is oscillating due to shock-excitation by Class-*C* operation of $V1$. In other words, plate-current flow is 100 ma and is composed of a rapid succession of pulses. Power supplied to the tank circuit, of course, comes from the plate source. The power supplied is equal to the product of 1,000 d-c v and 100 d-c ma, or 100 w. The peak value of *a-f* voltage applied by T_M is 1,000 v, which produces 100 per cent modulation.

Recall that the *a-f* voltage in Fig. 17–7(a) is sinusoidal and has a peak current value of 100 ma. We know that the current value is the same anywhere in a series circuit; the output winding of the modulation transformer T_M is connected in series with the plate tank and E_{BB}. Of course, the *a-f* power supplied to this circuit is equal to one-half the product of the maximum voltage and the maximum current values. Therefore, the power supplied by the *a-f* modulator is formulated

$$P_m = \frac{V_o I_o}{2} \quad \textbf{(17–4)}$$

where V_o = output voltage, and I_o = output current.

Example 2. What is the modulated power in the circuit in Fig. 17–7 when $e_c = 1{,}000$ v and $i_b = 100$ ma.

$$P_m = \frac{1000 \times 0.1}{2}$$

$$= 50 \text{ w} \qquad \text{(answer)}$$

Observe that the transmitter output power (neglecting losses) before modulation is 100 w, and after 100 per cent modulation is equal to 150 w. The antenna current before modulation is formulated

$$I_a = \sqrt{\frac{P}{R}} \quad \textbf{(17–5)}$$

Example 3. What is the antenna current before modulation in Example 2 if the antenna-load resistance is 100 Ω?

$$I_a = \sqrt{\frac{100}{100}}$$

$$= 1 \text{ amp} \qquad \text{(answer)}$$

With 100 per cent modulation, the power increases to 150 w. The antenna current in Example 2 is then increased.

Example 4. What is the internal current in the problem given in Example 3 when the modulation is increased to 100 per cent?

$$I_a = \sqrt{\frac{P}{R}} = \sqrt{\frac{150}{100}} = 1.224 \text{ amp}$$

From Example 3 we see that an increase of modulation from 50 per cent to 100 per cent gives an increase of 22.4 per cent in antenna current. Next, suppose that modulation is less than 100 per cent. Then, Formula 17–1 is evidently written

$$e = E_c(1 + m \sin 2\pi f_m t) \sin 2\pi f_c t \tag{17–6}$$

where m corresponds to the percentage of modulation, such as 0.50, 0.25, 0 or any value determined by the audio level. Formula 17–2 is written in the more general form

$$e = E_c \sin 2\pi f_c t + 0.5mE_c \cos 2\pi(f_c - f_m)t - 0.5mE_c \cos 2\pi(f_c + f_m)t \tag{17–7}$$

As previously noted, the carrier amplitude remains the same when the percentage of modulation has any value from 0 to 100 per cent; only the amplitude of the sideband frequencies is affected. When the carrier is 100 per cent modulated, one-sixth of the total power is contained in each sideband, or one-third of the power is contained in the pair of sidebands. During 100 per cent modulation, the *r-f* amplifier must handle peak currents that are twice the unmodulated value. In other words, the amplifier handles four times as much power from the unmodulated condition to the 100 per cent modulated condition.

The transistor modulator depicted in Fig. 17–7(b) operates in the common-emitter configuration in either Class-*B* or Class-*C* operation. In Class-*B* operation, the transistor conducts in the positive half cycle of the carrier signal. The modulation signal is applied in series with the collector circuit and the battery voltage. The relation of the carrier and the modulation signals is illustrated in Fig. 17–8. The output waveform is rich in harmonics from which we obtain the modulation index in Equation 17–2 resulting in an AM signal.

A small-signal or square-law modulator is illustrated in Fig. 17–9. Both signals can be connected in series to the base; however, the same result may be obtained by another method in the circuit by connecting one signal in series with the base and the other signal in series with either the emitter or the collector. In Fig. 17–9, the transistor is biased in the conduction region. The conduction is varied by the algebraic sum of the carrier signal at the base and the modulation signal at the emitter. Capacitor C_E bypasses the *r-f* carrier signal, but it does not appreciably reduce the *a-f* modulating signals.

With 100 per cent modulation, distortion is very high. For this reason, 60 or 70 per cent modulation, which requires about 0.1 v for each signal, is considered optimum.

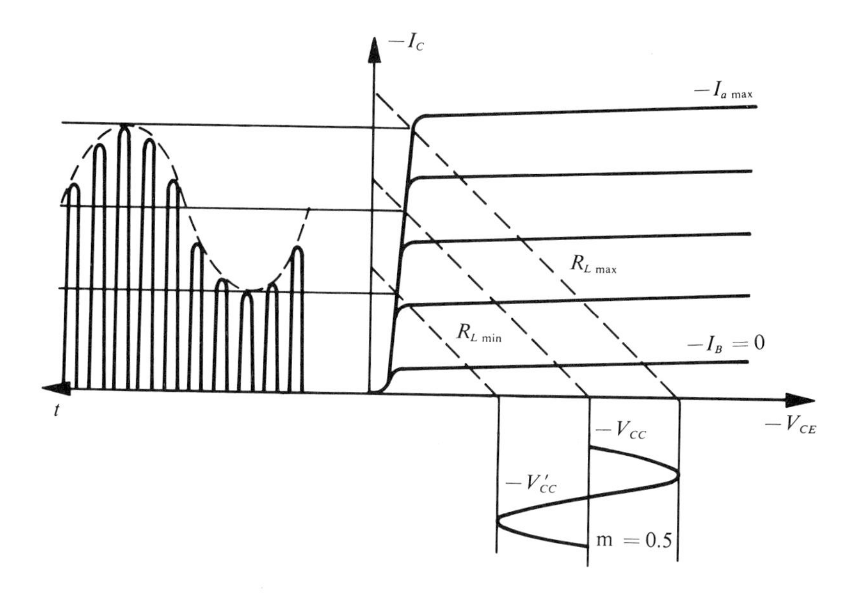

Figure 17-8. Output characteristics for high level linear modulators depicted in Figures 17-7 (a) and (b).

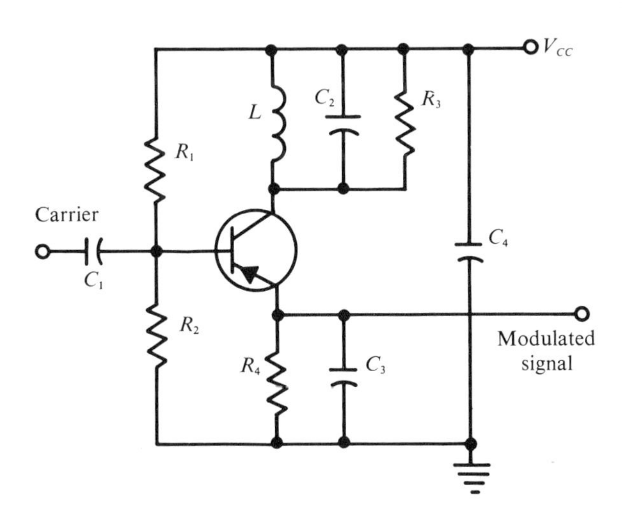

Figure 17-9. Small-signal or square-law modulator.

17-4. Amplitude Demodulation

Demodulation, or *detection*, is the process of recovering the original information from a modulated waveform. Demodulation of an AM wave produces currents or voltages that vary with the envelope of the modulated wave. It follows from previous discussion that a modulator is basically a nonlinear circuit system. Similarly, a demodulator is a nonlinear device. In other words, the plate resistance of *V*1 in Fig. 17-7(a) does not remain constant during modulation; the plate resistance varies from a low value to infinity over the modulation cycle. This is just another way of saying that the voltage-current characteristic of the system is not a straight line. At the positive peak of modulation, the voltage-current ratio is small, but at the negative peak of modulation, the voltage-current ratio is infinite for 100 per cent modulation.

This fact will become clearer to the student as we proceed with our analysis. At the receiver, let us first assume that the modulated wave is passed through a linear component such as an ordinary fixed resistor. Obviously, the current flow will contain the same frequencies as the voltage waveform; the modulated waveform is not changed by passage through a linear resistance. By the same token, let us consider the effect of passing an *r-f* current and an *a-f* current through a fixed resistor, instead of a vacuum tube at the transmitter. Clearly, the current flow through the resistor will not be a modulated waveform, but simply a mixture of the *r-f* and *a-f* sine waves, as depicted in Fig. 17-10(a). To make certain that we fully understand the nature of nonlinear mixing, observe Fig. 17-11. At the input, the *r-f* wave "rides on" the *a-f* wave to determine the peak excursions. At the positive peak, the nonlinear device has low resistance and passes a maximum amount of signal. On the other hand, at the negative peak, the nonlinear device has infinite resistance and passes no signal. At intermediate values, a part of the signal is passed.

At the receiver, the modulated waveform is passed through a nonlinear device called a demodulator. Demodulator action is depicted in Fig. 17-12. We observe that the average output (signal component) follows the envelope of the applied modulated wave. This average output current is pulsating d-c. The a-c component of the pulsating d-c waveform is the audio frequency that is recovered by the demodulator. If this average output current is passed through an earphone, the original audio information is reproduced as sound waves from the earphone. Note that if the modulated waveform were not demodulated, and if the modulated waveform were passed through an earphone, we

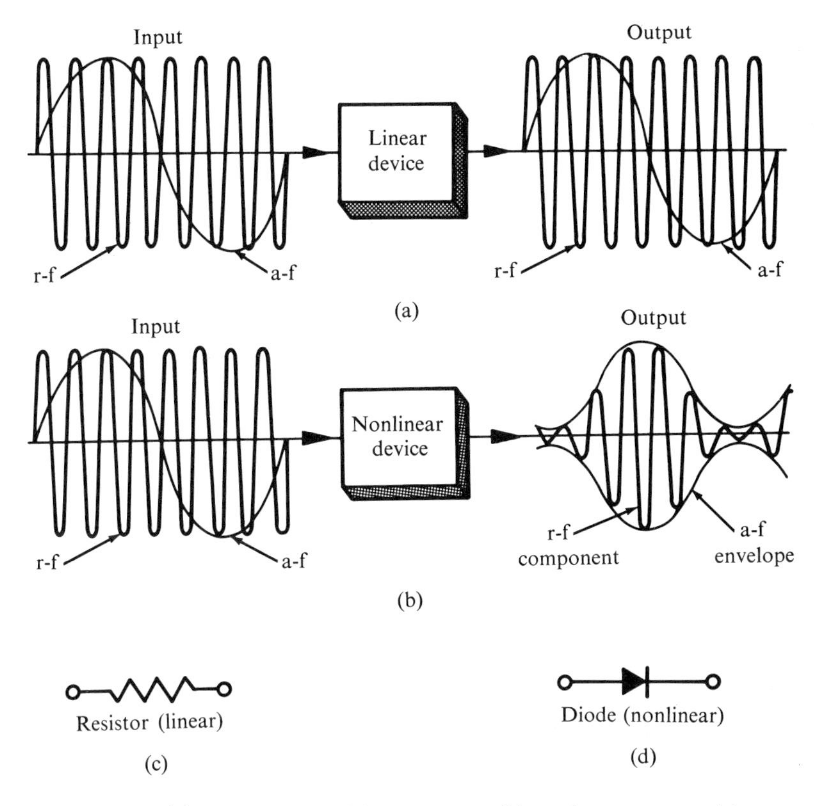

Figure 17–10. (a) Visualization of linear mixing; (b) Nonlinear mixing; (c) A conventional resistor is a linear device; (d) A diode is a nonlinear device.

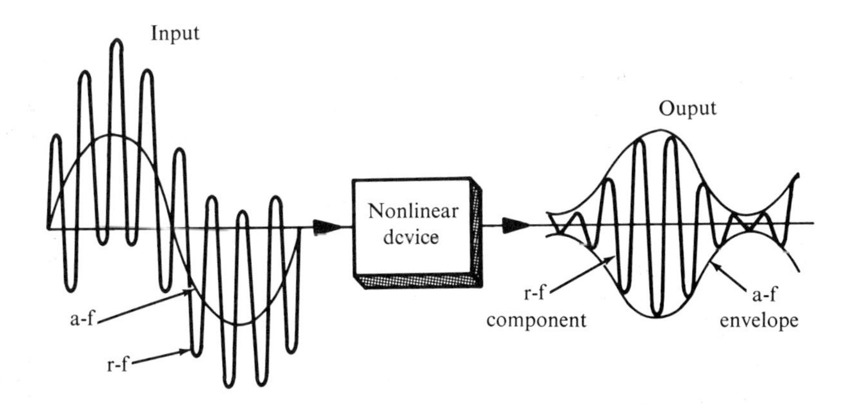

Figure 17–11. Another visualization of nonlinear device operation.

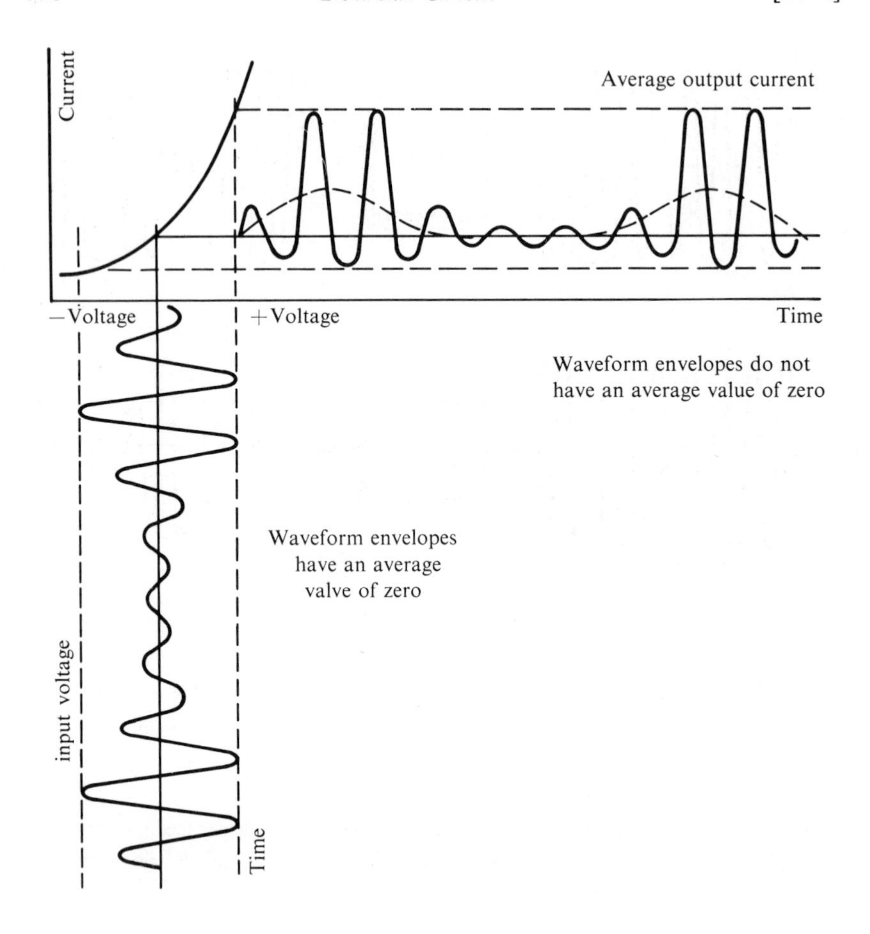

Figure 17–12. **Demodulating action of a nonlinear device.**

would hear no sound, simply because the average value of the modulated waveform is zero. On the other hand, the average output current in Fig. 17–12 has an average value that is *not* zero. Its average value varies in accordance with the envelope of the modulated waveform.

This is such an important circuit action that we will examine some of its details. With reference to the modulated waveforms depicted in Fig. 17–4, we observe that the upper and lower modulation envelopes have equal instantaneous values. Therefore, if *both* envelopes are permitted to pass through an earphone, the envelope values cancel out, and no sound is produced. However, in Fig. 17–12, *both* envelopes are *not* permitted to pass; instead, the lower envelope is greatly weakened by the nonlinear detector. The resultant of this circuit action is dominated by the upper envelope, and the upper envelope is thereby *recovered.* This is just another way of saying that the envelopes do not cancel each other when passed through an earphone.

Now, let us consider a diode detector circuit, as shown in Fig. 17–13. This is one of the simplest and most widely used detector circuits. The modulated signal voltage [Fig. 17–13(c)] is developed across the tuned circuit L_2C_1 of the detector configuration. Signal current flows through the diode only when its anode is positive with respect to the cathode—that is, only on the positive half cycles of the *r-f* voltage wave. The rectified signal flowing through the diode [Fig. 17–13(d)] actually consists of a series of *r-f* pulses, and is not a smooth outline or envelope. Note that the average value of these pulses, with little or no filtering, does increase and decrease at the *a-f* rate, as shown by the dotted line. There is an audio voltage output even if no filtering is employed. However, some stray capacitance exists, and consequently some filtering action takes place.

If a capacitor [C_2 in Fig. 17–13(a)] of suitable value is used as a filter, the output voltage from the detector is increased and its outline more nearly follows the envelope. On the first quarter cycle of applied *r-f* voltage, C_2 charges up to nearly the peak value of the *r-f* voltage [point A in Fig. 17–13(e)]. The small voltage drop in the tube prevents C_2 from charging up completely. Then, as the applied *r-f* voltage falls below its peak value, some of the charge on C_2 leaks through R, and the voltage across R drops only a slight amount to point B. When the *r-f* voltage applied to the plate on the next half cycle exceeds the potential at which the capacitor holds the cathode (point B), diode current again flows and the capacitor charges up to almost the peak value of the second positive half cycle at point C.

Thus, the voltage drop across the capacitor follows very nearly the peak value of the applied *r-f* voltage and reproduces the *a-f* modulation. This detector output, after rectification and filtering, is a pulsating d-c voltage that varies at an audio rate, as shown by the solid line in Fig. 17–13(e). The curve of the output voltage across the capacitor is shown somewhat jagged. Actually, the *r-f* component of this voltage is negligible and, after amplification, the speech or music originating at the transmitter is faithfully reproduced. Correct choice of R and C_2 values [Fig. 17–13(a)] in the diode-detector circuit is very important if maximum sensitivity and fidelity are to be obtained. The load resistor, R, and the plate resistance of the diode act as a voltage divider for the received signal. Therefore, the load resistance should be high

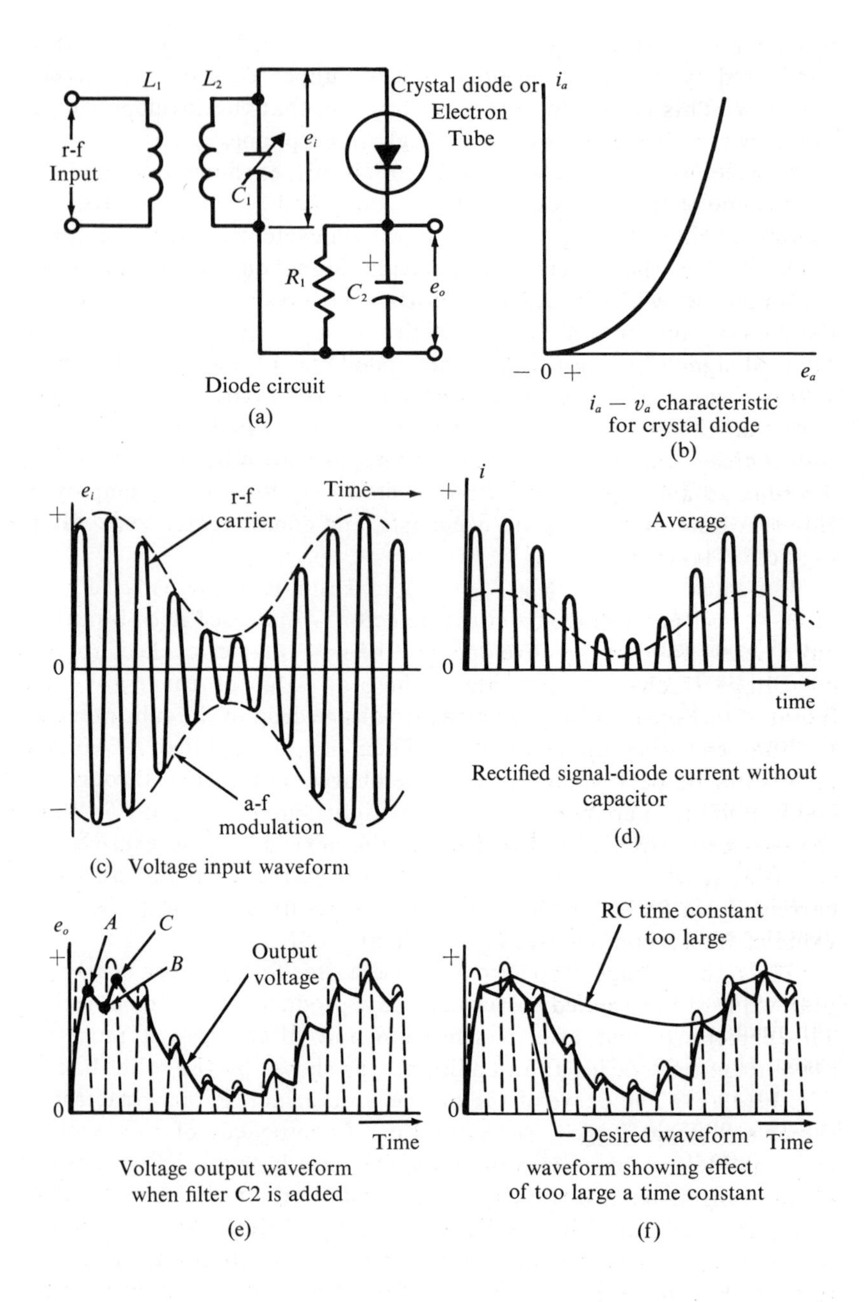

Figure 17–13. Diode detector circuit and operating waveforms.

compared with the plate resistance of the diode, so that maximum output voltage will be obtained.

The value of C_2 should be such that the RC time constant is long compared with the period of one *r-f* cycle. This is necessary because the capacitor must maintain current flow. Also, the time constant must be short compared with the time of one *a-f* cycle, in order that the capacitor voltage can follow the modulation envelope. Values of R_1 and C_2 thus place a limit on the highest modulation (audio) frequency that can be detected. Figure 17–13(f) shows the type of distortion that occurs when the RC time constant is too large. At the higher modulation frequencies, the capacitor does not discharge as rapidly as required, and negative peak clipping of the audio signal results.

Efficiency of rectification in a diode circuit is the ratio of peak voltage appearing across the load to the peak input signal voltage. Efficiency increases with the value of R, compared with the diode plate resistance, because R and the diode are connected in series across the input circuit, and their voltages divide in proportion to their resistances. At audio frequencies, a large value of R may be used (in the order of 100,000 Ω), and consequently the efficiency is comparatively high (95 per cent). When high modulation frequencies such as those used in television are to be detected, the value of R must be reduced to keep the RC time constant low enough to follow the modulation envelope. In turn, efficiency is reduced.

A diode detector can handle large signals without overloading, and it can provide high fidelity. However, it has the disadvantage of drawing power from the input tuned circuit, because the diode and its load form a low-impedance shunt around the circuit. Consequently, the circuit Q, sensitivity, and selectivity are reduced. Interelectrode capacitance of the diode limits its usefulness at very high carrier frequencies, and the lower portion of the current-voltage characteristic entails distortion of weak signals. Therefore, a diode detector provides high fidelity only if the input signal has a large amplitude; this is the reason that considerable amplification of the modulated signal is provided in most receivers before the signal is demodulated by a diode detector. Thus, linear-rectifier demodulation occurs.

A transistor can be used as a detector of very small signals. The detection occurs in the nonlinear portion of the input emitter base diode characteristics. The transistor, which operates as both a detector and amplifier, is biased at the most nonlinear portion of the emitter-base diode. The current gain is rather low due to the low bias point on the collector characteristic curves.

SUMMARY

1. Modulation is the process of inserting information into a sine waveform called a *carrier*.

2. Demodulation or detection is the process by which the modulating signal (intelligence) is recovered from the modulated carrier at the receiver.

3. Amplitude modulation is defined as variation in amplitude of an *r-f* carrier waveform at an audio-frequency rate.

4. The bandwidth of a modulated wave in the *r-f* spectrum is double the frequency of the modulating signal.

5. In the case of 100 per cent modulation, the carrier contains two-thirds of the transmitted power.

6. When a carrier is overmodulated, there are intervals called sideband splatter during which gaps occur in transmission. The sideband splatter can result in an increased bandwidth and possible interference with other transmissions operating on nearby carrier frequencies.

7. A buffer amplifier is used to isolate the oscillator from the power stage to prevent frequency shifting of the oscillator.

8. Demodulation or detection is a process of recovering the original intelligence from a modulated waveform.

9. Demodulation of an AM wave produces a voltage or current waveform that varies with the envelope of the modulated waveform.

10. The upper and lower envelopes of an amplitude-modulated waveform have the same instantaneous value, and if both are passed through an earphone or speaker, the waveforms cancel.

11. In diode detection, the diode rectifies the modulated waveform, and a capacitor across the output filters the *r-f* frequency from the audio frequency.

12. Efficiency of rectification in a diode circuit is the ratio of the peak voltage appearing across the load to the peak-input signal voltage.

13. Square-law or small-signal modulation is useful for small output power levels.

14. Linear or high-level modulation is useful for large output power levels.

Questions

1. State the basic definition of modulation.
2. State the basic definition of demodulation.
3. Give four reasons why it is impractical to transmit electromagnetic waves at audio frequencies.
4. Suppose a 1-MHz sine wave is modulated by a 1-kHz sine wave. What frequencies are present in the modulation envelope?
5. Why is the carrier necessary for transmission of information even though the carrier contains no information?
6. What is the bandwidth of a 1-MHz *r-f* wave that is amplitude modulated by a 4-kHz wave?
7. How can you prove in the laboratory that sideband frequencies are actually generated when an *r-f* carrier is amplitude-modulated?
8. Can amplitude modulation be improved by more than 100 per cent modulation? Explain.
9. Explain the term "sideband splatter" and its undesirable effects.
10. With reference to Fig. 17-6, explain the function of each block in the diagram.
11. Explain the term "high-level" modulation.
12. Describe the functions of a buffer amplifier.
13. With reference to Fig. 17-7, explain the operation of the transmitting circuit.
14. With reference to Fig. 17-7, why is the antenna current proportional to the percentage of modulation?
15. What is the relationship between carrier amplitude and percentage of modulation?
16. Define the term "detection."
17. Why must a modulator be a nonlinear device?
18. Why must a demodulator be a nonlinear device?
19. With respect to Fig. 17-12(a), what is the function of the diode; what is the function of the capacitor C_2?
20. What is the relationship of the *RC* time of R_1C_2 in the circuit in Fig. 17-11 and the time of one cycle of the *r-f* waveform?
21. What is the relationship of the *RC* time of R_1C_2 in Fig. 17-11 and the time of one cycle of the *a-f* waveform?
22. Define the efficiency of a diode in a rectifier circuit and relate diode efficiency to the value of the load resistance.

23. What are the advantages of a diode detector?

24. Which type of modulation, square-law or linear-rectifier, is useful at high output power levels?

25. Compare square-law and linear-rectifier modulation.

26. Explain how a transistor may be employed as both a detector and an amplifier.

PROBLEMS

1. Suppose the waveform depicted in Fig. 17-4(d) is observed on an oscilloscope to have a crest voltage of 110 v and a trough voltage of 68 v. What is the percentage of modulation?

2. What value of trough voltage is necessary to establish 42 per cent of modulation of the waveform in Problem 1 if the crest voltage remains at 110 v?

3. With reference to the circuit in Fig. 17-7, if the supply voltage is reduced to 600 v, what value of plate current will be necessary to develop the same output power?

4. Suppose the peak-input signal to a diode detector is 15 v and the peak-output signal is 12 v. What is the efficiency of the diode?

CHAPTER 18

Frequency Modulation and Demodulation

18–1. Introduction

Information may be transmitted in the form of electromagnetic waves by varying the *frequency* of a carrier. This technique is called *frequency modulation* and is abbreviated FM. Frequency modulation of a signal can provide a better signal-to-noise ratio than amplitude modulation. The chief disadvantage of FM in many practical applications is the necessity for greater bandwidth than for AM. We know that an AM waveform comprises one upper sideband frepuency and one lower sideband frequency for each modulating frequency. An AM waveform occupies a bandwidth that is twice the value of the modulating frequency. However, we will find that an FM waveform may contain more than one pair of sideband frequencies for each modulating frequency.

As depicted in Fig. 18–1, an FM wave comprises a carrier wave with a frequency f_o and associated sideband frequencies of $f_o \pm f_m$,

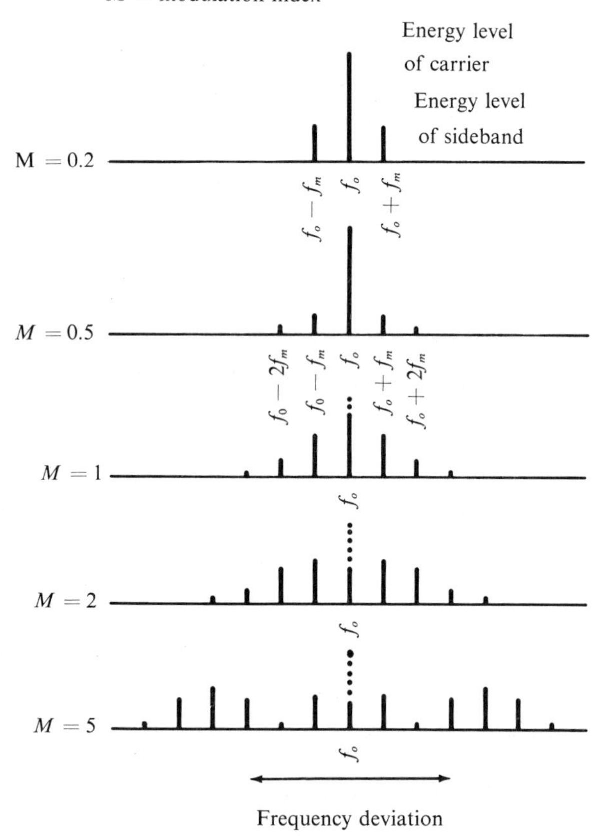

Figure 18-1. Sideband frequencies in a frequency-modulated wave.

$f_o \pm 2f_m$, $f_o \pm 3f_m$, and so on, where f_m is the modulating frequency and f_o is the carrier frequency. In Fig. 18-1, each line beside the center line represents a particular component of the FM wave. The center line represents the carrier. Lines to the right of center represent the upper sideband frequency components, and lines to the left of center represent the lower sideband frequency components. The length of each line denotes its energy level. Note that the horizontal distance between the center line and the last significant sideband (farthest removed from the center line) is proportional to the *deviation frequency* of the carrier, which depends in turn on the *amplitude* of the modulating frequency f_m. Sideband frequency components are spaced apart by a number of cycles equal to the modulating frequency.

Figure 18–2 shows the formation of an FM wave. We observe that

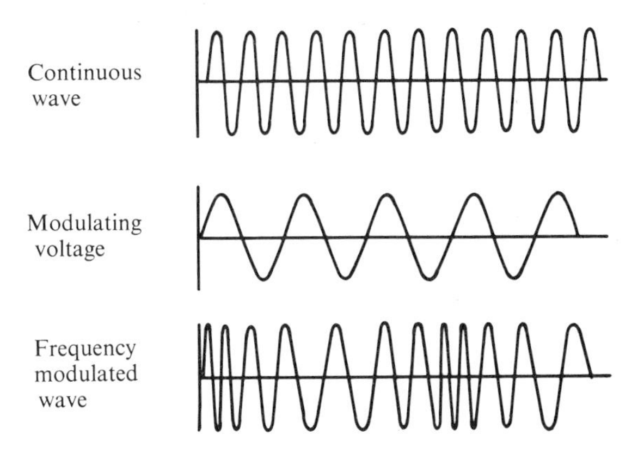

Figure 18–2. Formation of an FM wave.

the instantaneous frequency of the carrier becomes higher as the modulating voltage increases in a positive direction. On the other hand, the instantaneous frequency of the carrier becomes lower as the modulating voltage increases in a negative direction. In frequency modulation the carrier frequency is varied by an amount depending on the signal amplitude, and the rate of the carrier variation depends on the frequency of modulation. With reference to Fig. 18–1, the number of sideband frequencies that contain sufficient energy to be significant in information transfer depends on the frequency deviation imposed on the carrier by the modulating signal. For example, if the modulating frequency f_m causes the carrier to deviate by an amount equal to f_m and no more, the first pair of sidebands, $f_o \pm f_m$, is the only pair of importance. Note that no additional energy is supplied to an FM wave during modulation. By way of comparison, we recall that additional energy is applied to an AM wave during modulation. Observe that the energy of the carrier is merely *redistributed* during FM modulation, as depicted in Fig. 18–1.

In other words, the carrier energy is significantly reduced when the *modulation index* (M) exceeds 0.5. The modulation index is equal to f_d/f_m, where f_d is the frequency deviation of the carrier, and f_m is the modulating frequency. During frequency modulation, energy is taken from the carrier and is redistributed in the sideband components. For the cases $M = 2$ and $M = 5$, some of the sideband frequency components contain more energy than the carrier. During strong modulating signals, the energy level of the carrier approaches zero. Starting at the carrier and counting the sideband component consecutively in each direction, each upper odd-numbered sideband frequency component is 180° out of phase (these phase relations are not shown in Fig. 1-1) with its associated lower sideband component. Thus, the upper odd-numbered sideband component $f_o + f_m$ is 180° out of phase with the odd-numbered lower sideband component $f_o - f_m$. All even-numbered sideband frequency components are in phase with each other; $f_o + 2f_m$ is in phase with $f_o - 2f_m$. Furthermore, energy levels of each pair equally spaced from the carrier are equal.

As in the case of amplitude modulation, the bandwidth required for frequency modulation is determined by the sidebands associated with the carrier. An FM wave single-tone modulation theoretically has an infinite number of sideband pairs, instead of just one pair, as in amplitude modulation. Fortunately, only a limited number of sidebands contain sufficient energy to be significant. An approximation of the number of significant sideband frequencies may be made by assuming that the important sideband components extend over a frequency range on each side of the carrier by an amount equal to the sum of the modulation frequency and the carrier frequency deviation.

For example, suppose the frequency deviation of the carrier f_o in Fig. 18-1 is assumed to be 50 kHz when the modulating frequency f_m is equal to 10 kHz ($M = 5$) and the bandwidth on each side of the carrier is approximately 50 + 10, or 60 kHz, making a total bandwidth of 60 × 2, or 120 kHz. This bandwidth is an approximation derived from the "rule-of-thumb" noted previously. Because the sideband components are spaced by an amount equal to the modulation frequency, the product of the number of significant sideband components and the modulation frequency is equal to the bandwidth. In this example there are eight significant sideband frequencies on each side of the carrier, as shown at the bottom of Fig. 18-1. Thus, the bandwidth is 8 × 10, or 80 kHz, on each side of the carrier, or a total of 2 × 80, or 160 kHz. The bandwidth thus depends both upon the modulation frequency and the total frequency deviation of the carrier.

We perceive that the bandwidth requirement for an FM system is greater than twice the frequency deviation of the carrier by an amount equal to at least twice the modulating frequency. For most FM systems, the bandwidth is greater than that required of AM systems. FM transmission is made on higher carrier frequencies (88 to 108 MHz for commercial channels) to obtain spectrum space necessary for a number of wideband channels. In commercial high-fidelity broadcast transmission, 15 kHz is the highest modulating frequency. Maximum frequency deviation of the carrier is limited by the Federal Communications Commission (FCC) to 75 kHz on each side of the carrier frequency. The ratio of frequency deviation to modulation frequency, or modulation index M, is accordingly $75/15 = 5$. As previously explained for a modulation index of 5, there are 8 important sidebands. Because the sidebands are spaced 15 kHz apart, the bandwidth requirement is 8×15, or 120 kHz, on each side of the carrier.

Although the FCC regulation limits the carrier shift to ± 75 kHz, some significant sidebands may extend beyond this frequency. A guard band (Fig. 18–3) of 25 kHz on each side of the permissible frequency

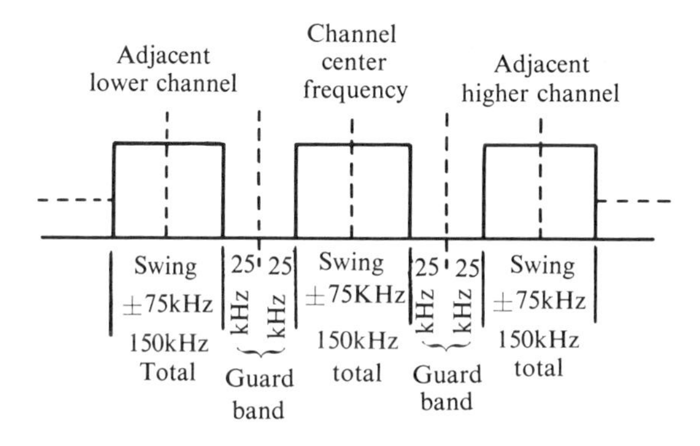

Figure 18–3. **FM broadcast channel allocations.**

swing of ± 75 kHz is assigned to accommodate most of the significant sidebands beyond the established limits. In the armed services, communication systems do not need high-fidelity response and accordingly use modulating frequencies up to only a few thousand Hertz. With a carrier frequency deviation of 12 kHz and a modulating frequency of 3 kHz, the approximate bandwidth is $2(3 + 12) = 30$ kHz for narrow-band FM, abbreviated NFM.

18–2. Degree of Frequency Modulation

To explain the meaning of 100 per cent modulation in an FM system, it is helpful to review the same condition for an AM wave. We know that 100 per cent modulation exists when the amplitude of the carrier varies between zero and twice its normal unmodulated value. There is a corresponding increase in power of 50 per cent. The amount of power increase depends upon the degree of modulation; because the degree of modulation varies, the tubes cannot be operated at maximum efficiency continuously.

In frequency modulation, 100 per cent modulation has a different meaning. The *a-f* signal varies the frepuency of the oscillator, but not its amplitude. Therefore, tubes operate at maximum efficiency continuously, and the FM signal has a constant power input at the transmitting antenna regardless of the degree of modulation. A modulation of 100 per cent simply means that the carrier is deviated in frequency by the full permissible amount. For example, an 88-MHz FM station has 100 per cent modulation when its audio signal deviates the carrier 75 kHz above and 75 kHz below the 88-MHz value, when this value is assumed to be the maximum permissible frequency swing. For 50 per cent modulation, the frequency would be deviated 37.5 kHz above and below the resting frequency.

18–3. FM Systems

A practical FM transmitter must fulfill two basic requirements: (1) the frequency deviation must be symmetrical about a fixed frequency, and (2) the deviation must be directly proportional to the amplitude of the modulation and independent of the modulation frequency. There are several systems of frequency modulation that meet these requirements. A *mechanical modulator* employing a capacitor microphone is the simplest means of obtaining frequency modulation, but it is seldom used. Two of the most important systems are called *reactance-tube* modulation, and *angle*, or *phase-angle* modulation. The chief distinction between these two systems is that in reactance-tube modulation, the *r-f* wave is modulated at its source (the oscillator), while in phase modulation, the *r-f* wave is modulated in some stage following the oscillator. In each of these systems, the end result is the same—the FM wave generated by either system can be reproduced by the same type of FM receiver.

One of the simplest forms of frequency modulation employs a capacitor microphone, which shunts the oscillator tank circuit *LC*, as shown in Fig. 18–4. A capacitor microphone is equivalent to an air-dielectric capacitor, one plate of which forms the diaphragm of the microphone. Sound waves striking the diaphragm compress and release it, thus causing the capacitance to vary in accordance with the instantaneous spacing between the plates. This type of FM transmitter is not very practical (the frequency deviation is limited at conventional carrier frequencies), but it is an instructive basic system. The

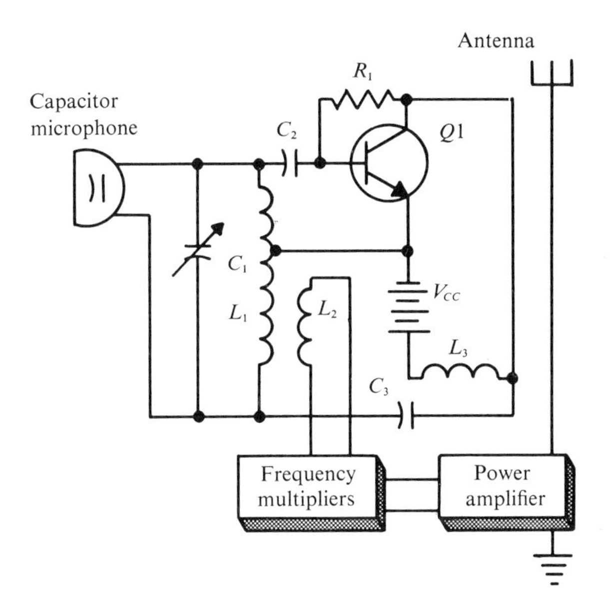

Figure 18-4. An FM transmitter arrangement modulated by a capacitor microphone.

oscillator frequency depends on the inductance and capacitance of the tank circuit *LC*, and accordingly varies in response to the changing capacitance of the capacitor microphone.

If the capacitor microphone is replaced by a varactor (varicap) diode, as depicted in Fig. 18-5, full frequency deviation is easily obtained at conventional carrier frequencies. Recall that a reverse-biased diode has a depletion layer that operates as a capacitor. The capacitance across this depletion layer varies with the value of applied voltage. Junctions can be designed to have a large value of capacitance that can be varied over a wide range by comparatively small voltage changes. This type of FM modulator is widely used in FM generators of various classes.

When sound waves vibrate a capacitor microphone diaphragm (or a conventional microphone output varies the reverse bias voltage of a varicap diode), and the vibrations have a low frequency, the oscillator frequency is changed only a few times per second. Again, if the sound frequency is higher, the oscillator frequency is changed more times per second. When the sound waves have low amplitude, the extent of the oscillator frequency change from the no-signal, or resting frequency, is small. On the other hand, a loud *a-f* signal changes the capaci-

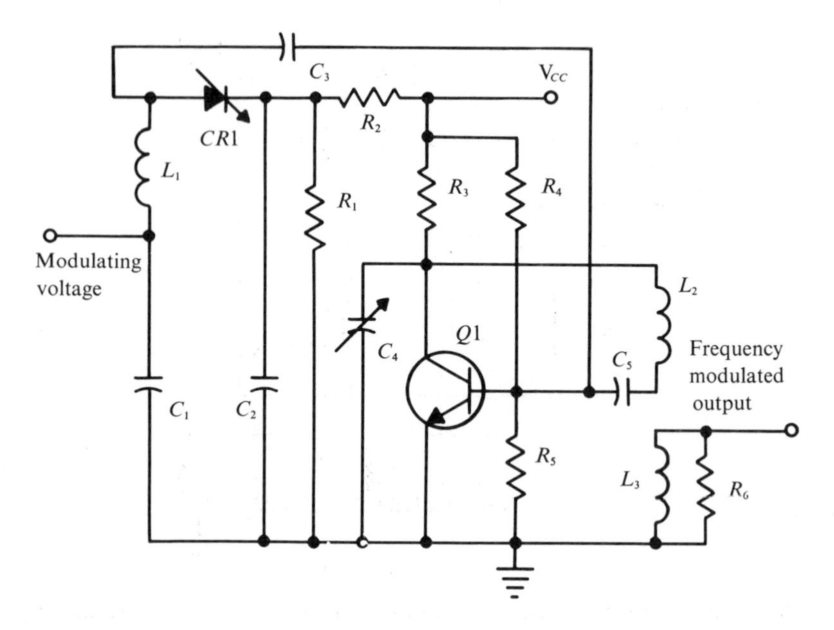

Figure 18–5. Frequency-modulated 100-MHz oscillator in an FM generator; the capacitance of the voltage variable capacitor *CR*1 varies in step with the modulating voltage.

tance a greater amount, and this deviates the oscillator frequency to a greater degree. Hence, the deviation frequency of the oscillator tank circuit depends upon the amplitude of the modulating signal.

To increase the initial deviation frequency of the oscillator (which is greatly restricted in the case of the capacitor-microphone modulator at conventional frequencies) to a suitable value, a system of frequency multiplication may be employed. The circuits used to accomplish this frequency multiplication are contained in the block diagram marked "frequency multipliers" in Fig. 18–4. One method utilizes a broadly tuned plate-tank circuit in a Class-*C* amplifier. The tank is tuned to the second harmonic of the grid-input signal and thus develops a tank current and output signal at double frequency. Output from the first doubler is fed into another similar doubler. Several doubler stages may be utilized. For example, a 5-MHz signal fed into a 3-stage doubler becomes a 40-MHz signal at the output. An initial deviation of 1 kHz produced by an audio-modulation signal becomes a frequency deviation of 8 kHz at the output of the third doubler stage.

18-4. Reactance-Tube System

The reactance-tube system of frequency modulation is depicted in Fig. 18-6. A reactance tube is a vacuum tube operated in a circuit in

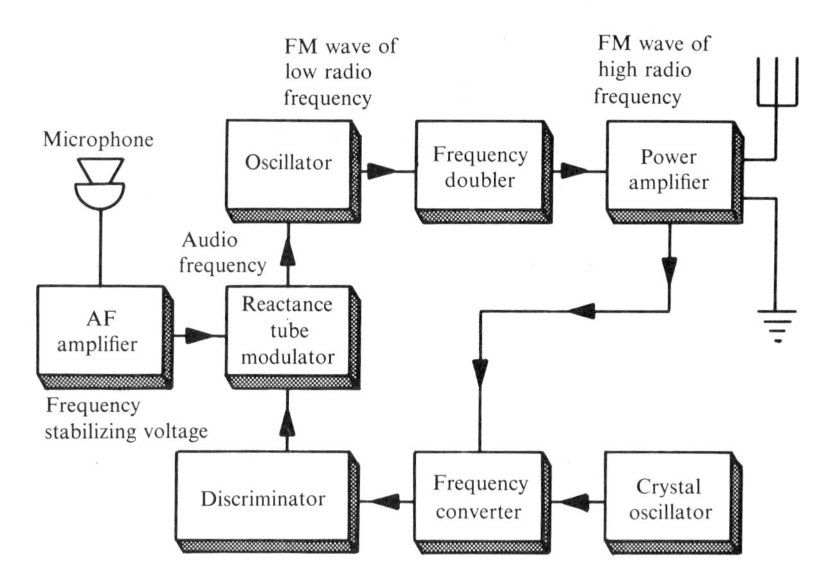

Figure 18-6. **Block diagram of a reactance tube FM transmitter.**

such a manner that the circuit's terminal reactance varies with the modulation signal, and thereby varies the frequency of the oscillator stage. The reactance tube is connected in parallel with the oscillator tank and functions like a capacitor; its capacitance value is varied in accordance with the audio signal, as in the capacitor-microphone system of FM. In turn, the oscillator frequency is changed, and the resulting FM signal is passed through a frequency doubler to increase the carrier frequency and the deviation frequency. A power amplifier feeds the final signal to the antenna.

Figure 18-7 depicts the operation of a reactance-tube circuit. Reactance tube $V1$ is effectively in shunt with the oscillator tank LC, and with the phase-shift circuit R_gC_1. At the operating frequency, the capacitive reactance of the capacitor is large, compared with the resistance of R_g. Current i leads voltage e_b by approximately 90°. Voltage e_b is the a-c component of the plate-to-ground voltage appearing simultaneously across the reactance tube, the phase-shift circuit, and the oscillator tank.

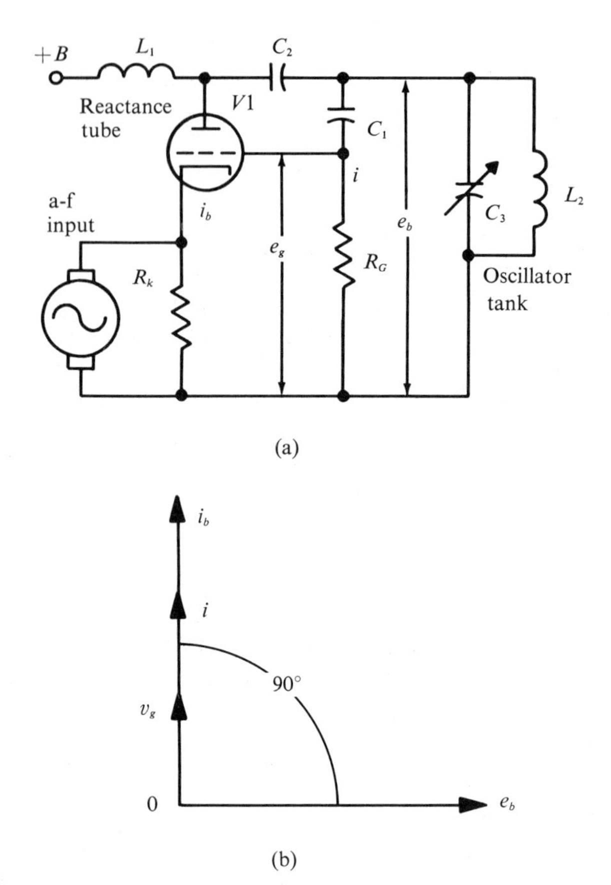

Figure 18-7. Frequency modulation with a reactance-tube modulator. (a) Circuit diagram; (b) Vector diagram.

Coupling capacitor C_2 has relatively low capacitive reactance to the a-c component of current that flows through it; at the same time, it blocks d-c voltage from the phase-shift circuit and the tank. Reactance tube V_1 receives its a-c grid-input voltage e_g from the voltage drop across R_g. Grid voltage e_g is, of course, in phase with plate current i_b inasmuch as i_b is "valved" by e_g. Because e_g is in phase with both i and i_b, and e_g leads e_b by approximately 90°, both i and i_b lead e_b by approximately 90°. These relations are seen in the vector diagram in Fig. 18-7(b). Both i and i_b are supplied by the oscillating tank circuit, and because both are leading currents with respect to the tank voltage e_b, their circuit significance is the same as the current in tank capacitor C. The effect of these currents on the tank frequency is the same as though additional capacitance were connected across the tank.

Now let us consider the effect of applying an audio signal across R_k. With zero audio voltage, the *r-f* plate current i_b comprises a succession of rapid pulses at constant amplitude, and the oscillator tank operates at a constant frequency called the no-signal, or resting frequency. When the audio voltage rises in a polarity that swings the cathode negative with respect to the grid, the pulses of plate current increase proportionally in amplitude. This leading *r-f* plate current is drawn through the oscillator tank and is equivalent to an increasing value of tank capacitance. As a result, the oscillating frequency is lowered. Conversely, when the audio signal swings the grid of the reactance tube negative with respect to the cathode, the *r-f* plate-current pulses decrease proportionally in amplitude and the oscillating frequency increases.

We perceive that the frequency of the *a-f* signal determines the number of times per second that the oscillator-tank frequency changes. On the other hand, the amplitude of the *a-f* signal determines the extent of the oscillatory frequency change or determines the deviation. In summary, a reactance-tube modulator with a reference audio-signal input generates an FM output signal that has the same characteristics as that of a capacitor-microphone modulator, or the same as that of a semiconductor varicap modulator.

18-5. Equivalent FM Generation by Phase Modulation

Equivalent FM generation denotes an arrangement that basically produces phase modulation. The audio-modulating signal is modified so that an FM output is obtained instead of a PM output. One arrangement is shown in Fig. 18-8(a). Phase modulation is produced by amplitude modulating a carrier wave and also by combining this amplitude-modulating wave with a large carrier voltage that has been shifted 90° in phase. Figure 18-8(b) shows the upper and lower sideband vectors in the AM wave; these vectors rotate in opposite directions with respect to the carrier and produce a resultant that varies in amplitude. When the AM wave is added to a large carrier voltage that has been shifted 90° in phase, the resultant is essentially a phase-modulated wave, as shown in Fig. 18-8(c). If the shifted-carrier vector is made comparatively large, the varying phase angle θ will be practically proportional to the amplitude of the modulating voltage.

Since there is some residual envelope variation in this PM wave, it is passed through a limiter, as shown in Fig. 18-8(a). The limiter

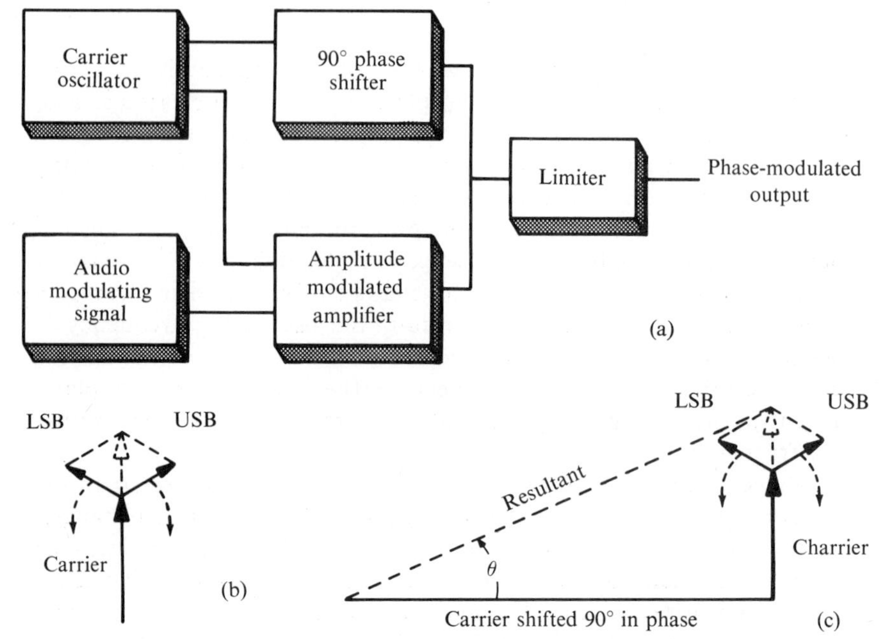

Figure 18–8. (a) Arrangement for generating a phase-modulated wave; (b) Conventional AM waveform vectors; (c) Vector relations with carrier shifted 90° in phase.

operates as a clipper and removes residual amplitude modulation; the output from the limiter is then a phase-modulated wave. Since equivalent FM output is desired, instead of PM output, the audio-modulated signal in Fig. 18–8(a) is first passed through an *RC* network that provides an output voltage which is inversely proportional to frequency. This makes the modulation index inversely proportional to the modulating frequency and provides equivalent FM output from the limiter. This arrangement has an advantage in that the oscillator can be crystal controlled, so that the center frequency is highly stable. On the other hand, it has the disadvantage of a small modulation index and thus requires the use of several frequency-multiplying stages to obtain a large modulation index.

18–6. Demodulation of the FM Signal

We know that in FM transmission, the information is contained as a variation in the instantaneous frequency of the carrier either above or below the center (resting) frequency. Accordingly, an FM detector must be so constructed that its output will vary linearly with the instantaneous frequency of the incoming signal. Also, the FM detector must be insensitive to amplitude variation produced by interference or by receiver nonlinearities. To fulfill this requirement, a limiting (clipper) device called the *limiter* precedes the FM detector. There are various types of FM detectors, and each type has certain advantages and disadvantages. Two of the most widely used types are the *discriminator* and the *ratio detector*. Another type of FM detector, called the *locked* oscillator, is sometimes utilized. Economy-type receivers may utilize a *slope* detector. Let us consider the operation of a slope detector, inasmuch as this is the simplest form of FM detector.

We will find that even an AM receiver may develop a distorted reproduction of an FM signal under certain operating conditions. When the carrier frequency of the FM signal falls on the sloping side of the *r-f* response curve in an AM receiver, the frequency variations of the carrier signal are converted into equivalent amplitude variations. This conversion results from the unequal response above and below the carrier center frequency (point *B*) shown in Fig. 18–9. We perceive that when the incoming FM signal is less than the carrier frequency—for example, at point *A*, which is the minimum value—the output voltage is at a minimum in the negative direction. When the incoming signal swings to point *C* (the maximum value), the output voltage is maximum in the positive direction. The resultant AM signal may be coupled to a conventional AM detector, in which the original audio voltage is recovered.

It is evident that one disadvantage of slope detection is the nonlinearity of the response curve. If we should operate a slope detector over the most linear portion of the response curve depicted in Fig. 18–9(a), only a narrow frequency range is available, and substantial deviation of an FM signal cannot be accommodated. Another disadvantage of this simple arrangement is that AM interference is not rejected. Note that if the slope-detector circuit were tuned to place the center frequency at point *D*, and the maximum frequency swing were from point *C* to point *E*, the output signal would be highly distorted. Accordingly, point *B* is the optimum operating point. If the slope detector is preceded by a clipper circuit that slices off the positive and negative peaks from the FM signal, AM interference will be rejected as visualized in Fig. 18–10. Of course, a limiter does not improve the linearity of demodulation.

A discriminator provides good demodulation linearity. The simplest form of discriminator, the Round-Travis discriminator, is shown in Fig. 18–11(a). The resonant circuit *A* tuned to the carrier frequency

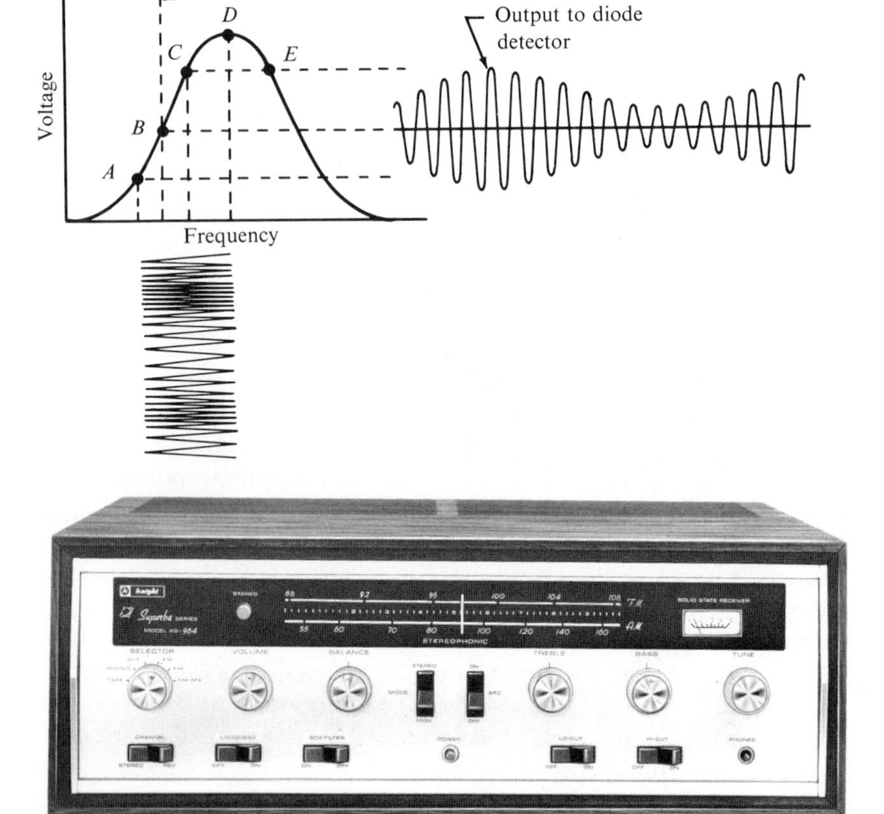

Courtesy of Knight Electronics

Figure 18–9. (a) The process of slope detection; (b) View of an FM and AM transistor stereo tuner.

(f_m) forms the primary of the transformer. The tank circuit B is tuned at f_2 above f_m, and the tank circuit C is tuned at f_1 an equal amount below f_m. As the carrier current in the primary fluctuates, the currents in the detuned circuits vary. When there is no deviation of the carrier (carrier at f_m), the voltage developed across tank B is equal to the voltage developed across tank C. Conversely, the voltages across capacitors C_1 and C_2 are equal and opposite and develop zero volts at the output. When the FM frequency is above the carrier, at f_2, the voltage across tank B is much larger than the voltage across tank C. Conversely, the voltage developed across C_1 is much greater than that developed across C_2 and develops a positive output voltage. When the

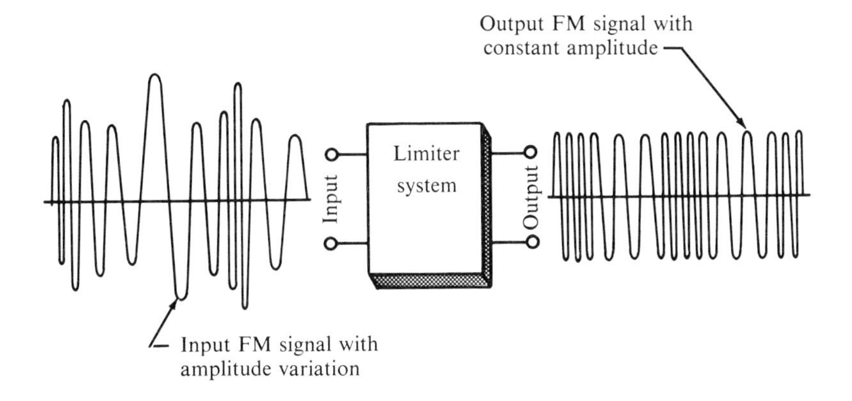

Figure 18–10. Visualization of limiter action.

FM frequency is below the carrier, at f_1, the voltage across tank C is much larger than the voltage developed across tank B. Likewise, the voltage developed across C_2 is larger than that across C_1 and produces a negative output voltage.

By careful adjustment of the circuit components and limitation of the frequency deviation between f_1 and f_2, we obtain an output voltage that is an approximate linear function of the input FM signal. We should note here that since the output is a rectified voltage, adjustment of the tuned circuits to f_2 and f_1 is accomplished with an input *r-f* generator and a *d-c* voltmeter.

Another commonly used discriminator type is the Foster-Seely discriminator shown in Fig. 18–12(a). In practice, this configuration is always preceded by one or more limiter stages. Basically, an output voltage e_{10} is developed by the discriminator circuit that varies in amplitude according to the instantaneous frequency of the FM signal. Input voltage e_1 is applied across the tuned input circuit. Current i_1 lags e_1 by 90°. The mutually induced voltage e_2 lags i_1 by 90°. Thus, e_2 is 180° out of phase with e_1, as shown in (1) of Fig. 18–10(b). Inductor L_4 is shunted across the input tuned circuit via C_2 and C_5, which have negligible reactance at the resonant frequency. Thus, e_1 is also applied across L_4.

Let us assume first that the incoming signal is at its resting frequency. The induced current i_2 is in phase with e_2, as shown in (2) and (3) of Fig. 18–12(b). Voltages e_3 and e_4 are the iX_L drops across L_2 and L_3, respectively. From Fig. 18–12(a) and Section (3) of Fig. 18–12(b), we observe that e_6, the voltage applied to $CR1$, is the vectorial sum

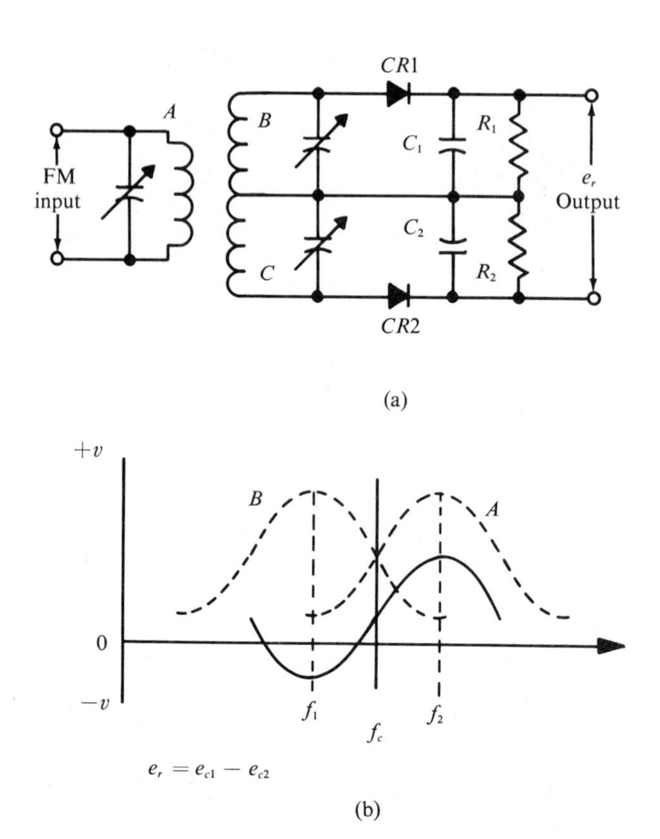

Figure 18-11. **A discriminator employing two stagger-tuned circuits. (a) Round-Travis configuration; (b) Voltage waveforms.**

of e_1 and e_3; e_7, the voltage applied to $CR2$, is the vectorial sum of e_1 and e_4. The rectified output from $CR1$ is e_8, and the rectified output from D_2 is e_9. Output voltage e_{10} is the algebraic sum of e_8 and e_9. In Section (3) of Fig. 18–12(b), e_6 and e_7 are equal, because the incoming signal is at its resting frequency. Therefore, e_8 is equal to e_9, and since they are in opposite directions, the output voltage is zero.

Below the resting frequency, i_2 leads e_2 because X_C is greater than X_L. Voltages e_3 and e_4 are still in phase opposition, but each is 90° out of phase with i_2, as shown in Fig. 18–11(c). Consequently, e_7 is greater than e_6, and e_9 is greater than e_8. Point A becomes negative with respect to ground, and an output signal is developed. Above the resting frequency, i_2 lags e_2 because X_L is greater than X_C. Voltages e_3 and e_4 bear the same phase relation with each other and with i_2 as they did

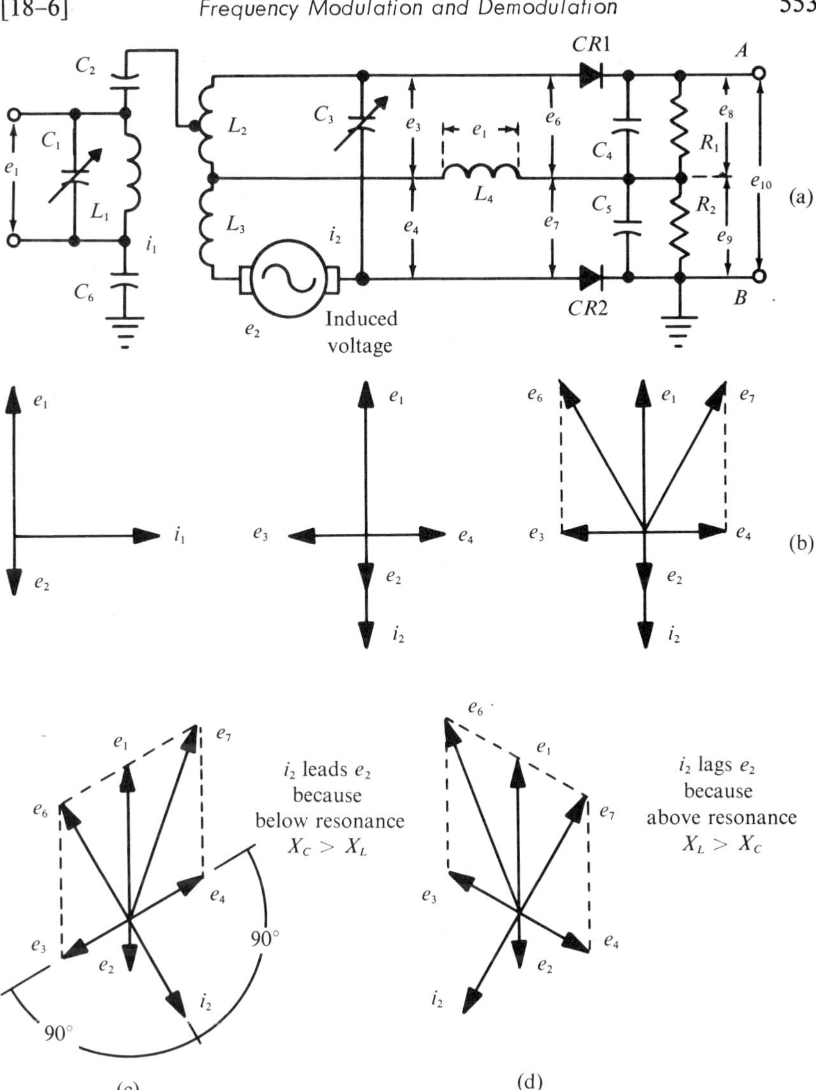

Figure 18–12. Discriminator configuration and vector diagrams.

for each of the foregoing conditions. However, from Fig. 18–11(d), e_6 is now greater than e_7. Therefore, e_8 is greater than e_9, point *A* becomes positive with respect to ground, and an output voltage is developed that is the other half cycle of the *a-f* signal.

A ratio detector is another configuration that provides good demodulation linearity. One form of ratio detector is depicted in Fig.

18-13(a). It differs from the discriminator configuration that has been described in that the diodes are connected in series across the transformer secondary. There is a difference in the method of obtaining output voltage and in the amount of limiting that precedes the detector. However, the vector analysis is essentially the same for a ratio detector as for a discriminator.

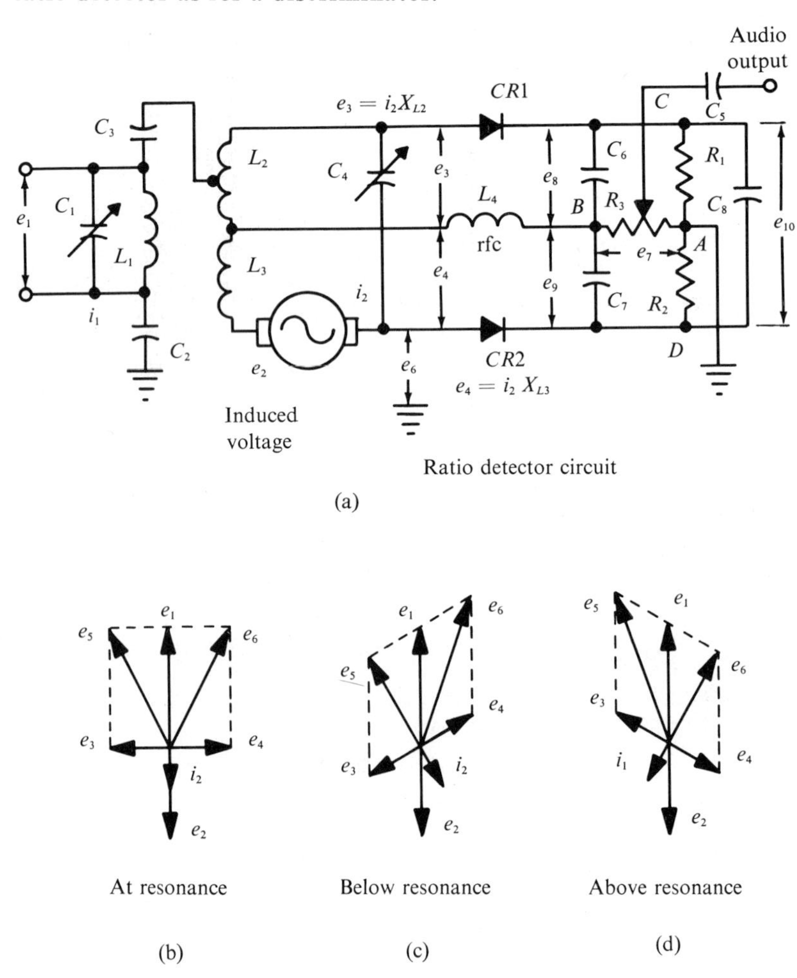

Figure 18-13. Ratio-detector configuration and vector diagrams.

With reference to Fig. 18-13, the induced voltage e_2 is 180° out of phase with e_1, as indicated by the vector diagrams. At resonance (Fig. 18-13), i_2 is in phase with e_2, as in all tuned circuits. Voltages e_3 and e_4, developed by the i_2X_{L2} and i_2X_{L3} drops, respectively, are 180° out of phase with each other and 90° out of phase with i_2. This out-of-phase relation holds true at resonance, as well as below or above resonance. From Fig. 18-13(b), e_5, the a-c voltage applied to $CR1$, is the vector sum of e_1 (coupled through C_3 between the center tap and ground) and e_3. Also, e_6, the voltage applied to $CR2$, is the vector sum of e_1 and e_4. At resonance, e_5 and e_6 are equal.

The angle of current flow through $CR1$ and $CR2$ may be in the order of 40° or less. In Fig. 18-13(b), vector e_5 leads e_6 by approximately 50°. Thus, $CR1$ will conduct for 40° and be cut off for 10° of the input cycle before $CR2$ begins conduction. The current path of $CR1$ is from C to A, B, L_4, L_2, $D1$, and back to C. Again, the current path of $CR2$ is from D to $D2$, L_3, L_4, B, A, and back to D. Hence, *r-f* rectified current flows via $CR1$ from A to B, and via $CR2$ from B to A. Without capacitors C_6 and C_7, the *r-f* voltage developed across R_2 as a result of these current flows is small, because of *r-f* choke L_4 in series with R_3.

C_6 and C_7 will charge up to *d-c* voltage values e_8 and e_9, respectively, and will keep the voltage across R_3 at the average value of these two currents which is evidently equal to zero at the resting (resonant) frequency. In other words, when $CR1$ conducts, the path for current flow is via C_6 instead of R_1 and R_3. Again, when $CR1$ conducts, the path for current flow is via C_7 instead of R_2 and R_3. During the interval when neither diode is conducting, C_6 and C_7 will discharge only slightly. Thus, the output voltage e_7 across R_3 is zero at the resting (resonant) frequency, because $CR1$ and $CR2$ conduct equal current values.

Below resonance in Fig. 18-12(c), i_2 leads e_2 because X_C is greater than X_L. Therefore, e_6, the vector sum of e_1 and e_4, is greater than e_5, the vector sum of e_1 and e_3. The current paths are the same as explained previously, but because e_6 is greater than e_5, the current flow through $CR2$ is greater than that through $CR1$. As a result, the current through R_3 from B to A ($D2$ conducting) will be greater than from A to B ($D1$ conducting). Point B accordingly swings negative with respect to point A to develop the audio output voltage e_7.

Above resonance in Fig. 18-13(d), i_2 lags e_2 because X_L is greater than X_C. In turn, e_5 is greater than e_6. Current flow through $CR1$ exceeds that through $CR2$. Current flow through R_3 from A to B is greater than from B to A, and the potential at point B swings positive with respect to point A. The other half of the audio output cycle appears across R_3. If there is a sudden undesired increase in amplitude of the input signal to the ratio detector due to noise or static, capacitor C_8 resists any change in voltage amplitude across terminals CD because the capacitor has a large capacitance value. This capacitor effectively "absorbs" all but very low frequency amplitude variations.

Ratio detectors are widely used for reasons of economy; fewer amplifier stages are required and less limiting action is required in the driver stage. It is, however, more difficult to align (tune the resonant circuits) for optimum demodulator action in a ratio-detector circuit than in a discriminator circuit. Even with optimum alignment, the audio output signal may be distorted somewhat at high

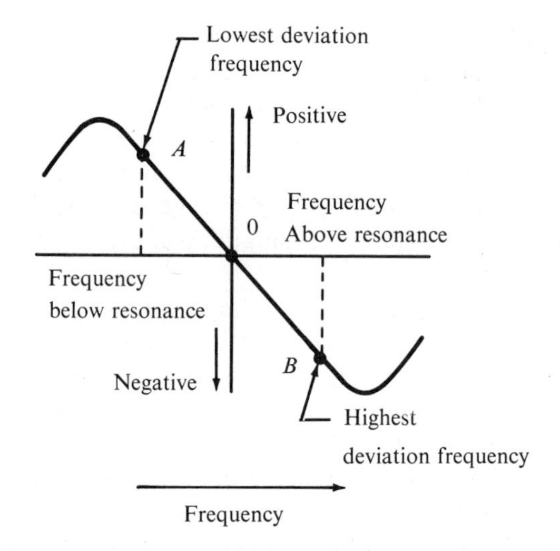

Figure 18-14. **Frequency-response curve of an FM detector aligned for linear demodulation.**

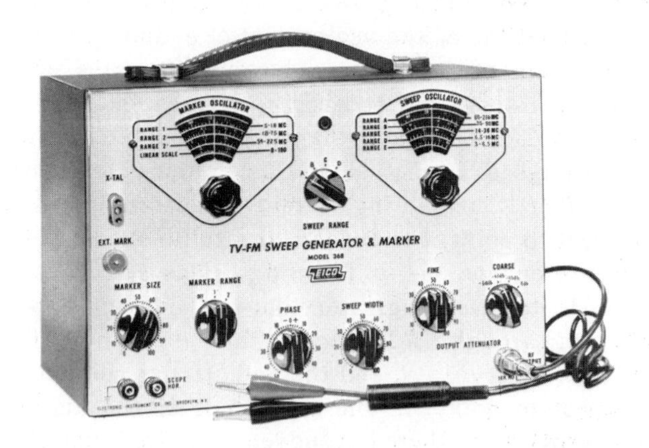

Courtesy of EICO, Electronic Instrument Co., Inc.

Figure 18-15. **A typical FM generator.**

input signal levels, unless appreciable limiting is employed in the driver stage. Figure 18–14 shows the frequency response curve for a properly aligned FM detector. FM systems are commonly aligned with the aid of an oscilloscope and an FM (sweep) generator, such as illustrated in Fig. 18–15. These instruments provide a display of the frequency-response curve on the oscilloscope screen.

SUMMARY

1. Frequency modulation is the technique of transmitting information in the form of electromagnetic waves by varying the frequency distribution of the wave.
2. The chief disadvantage of FM transmission in many practical applications is the necessity of a greater bandwidth than for AM.
3. Frequency modulation may have many sideband frequencies for each modulating frequency frequency, whereas amplitude modulation has only two sideband frequencies for each modulating frequency.
4. The deviation frequency of the carrier depends upon the amplitude of the modulating frequency f_m, and sidebands are spaced apart by a number of cycles equal to the modulating frequency. The rate of change of the carrier frequency is determined by the frequency of the modulation.
5. During modulation, no additional energy is applied to an FM waveform.
6. The maximum shift of an FM carrier as regulated by the FCC is ± 75 kHz; however, a guard band of 25 kHz separates each channel to prevent interference.
7. In frequency modulation, 100 per cent modulation means that the carrier is deviated in frequency by the full permissible amount.
8. In a practical FM transmitter, the frequency deviation must be symmetrical about a fixed frequency, and the deviation must be directly proportional to the amplitude of the modulation and independent of the modulation frequency.
9. A reactance tube is a vacuum tube operated in such manner that its reactance varies with the modulating frequency, thereby varying the frequency of the parallel oscillator tank circuit.
10. The output of an FM detector must vary linearly with the instantaneous frequency of the incoming signal.
11. A limiter is used to clip amplitude variations from the incoming FM signal.
12. The disadvantage of a slope detector is in the nonlinearity of the response curve.

Questions

1. Define frequency modulation as referred to information transmission.
2. What are the advantages and disadvantages of FM transmission in relation to AM transmission?
3. Define the deviation frequency of an FM carrier.
4. What determines the spacing between the sidebands of an FM waveform?
5. What effect does the frequency of modulation have on the carrier frequency?
6. Does a modulating frequency add energy to an FM carrier? Explain your answer.
7. Explain the phase relationship between the odd and even sidebands of an FM waveform above and below f_o.
8. Compare reactance-tube modulation and phase-angle modulation.
9. How many sidebands will a single-tone modulation produce in an FM waveform?
10. What is the maximum allowable frequency deviation of the carrier in commercial FM broadcasting?
11. What is a "guard band"?
12. Why does military FM transmission require less bandwidth than commercial transmission?
13. Compare the meaning of 100 per cent modulation for an AM carrier with the meaning of 100 per cent modulation for a practical FM transmitter?
14. Explain the purpose of the voltage variable capacitor in Fig. 18-5.
15. What is the relationship of total capacitance change in the tank circuit of Fig. 18-5 to signal frequency and signal amplitude?
16. Suppose a modulation waveform varies an oscillator frequency by ± 5 kHz. What is the deviation of the output if it is amplified by a three-stage doubler?
17. Discuss the operation of the FM oscillator in Fig. 18-7.
18. What is the function of a limiter in an FM receiver?
19. What are the disadvantages of a slope detector?
20. With reference to Fig. 18-11, what are the output symptoms if the following troubles occur: (a) tank *B* detuned, (b) diode *CR*1 opens, (c) tank *A* detuned, and (d) capacitor C_2 shorts?

PROBLEMS

1. With reference to Fig. 18-1, suppose the carrier is 88 MHz and the modulation index is 0.5. What is the approximate bandwidth when f_o is 10 kHz?

2. An FM carrier of a commercial station is being modulated 30 per cent. What is the frequency deviation?

3. With reference to Fig. 18-11, the carrier frequency is 88 MHz and the deviation frequency is to be ± 15 kHz. What is the maximum value of f_1 and the minimum value of f_2?

4. Suppose that the carrier frequency in Problem 1 is 108 MHz. What is the approximate bandwidth when (a) $M = 0.1$, (b) $M = 0.2$, (c) $M = 8$, and (d) $M = 10$?

5. What is the maximum allowable deviation (by FCC Regulations) of the carrier in Problem 2?

CHAPTER 19

RC Nonsinusoidal Oscillators

19–1. Introduction

RC oscillators that provide nonsinusoidal output waveforms are utilized in a great number of electronics technology areas. For example, *RC* nonsinusoidal oscillators are found in television receivers, radar systems, industrial-electronics devices, electronic instruments, navigation systems, and so on. Such oscillators may generate square waves, pulses, sawtooth waveforms, triangular waves, ramps, staircase waveforms, trapezoidal waveforms, exponential waves, or a wide variety of shaped complex waveforms. Nonsinusoidal *RC* oscillators operate on the basis of charge and discharge of a capacitor or capacitors in series and/or shunt with a resistor or resistors. To obtain specialized operating characteristics, one or more of the resistors may have a nonlinear resistance characteristic; similarly, a capacitor might have a nonlinear capacitance characteristic. The active component is commonly a vacuum tube, gas tube, or transistor;

however, negative-resistance devices such as tunnel diodes are also utilized. This chapter discusses the more basic types of *RC* nonsinusoidal oscillators.

One of the simplest types of *RC* nonsinusoidal oscillators is the neon-tube sawtooth generator. Hence, we will begin our analysis with this device. Small neon bulbs are used in sawtooth-generator circuits for toys or decorative devices that are commonly called "electronic fireflies." Large neon bulbs are used in the simpler types of stroboscopes to measure speeds of rotation or frequencies of vibration. With reference to Fig. 19–1, capacitor *C* is charged through resistor *R* until

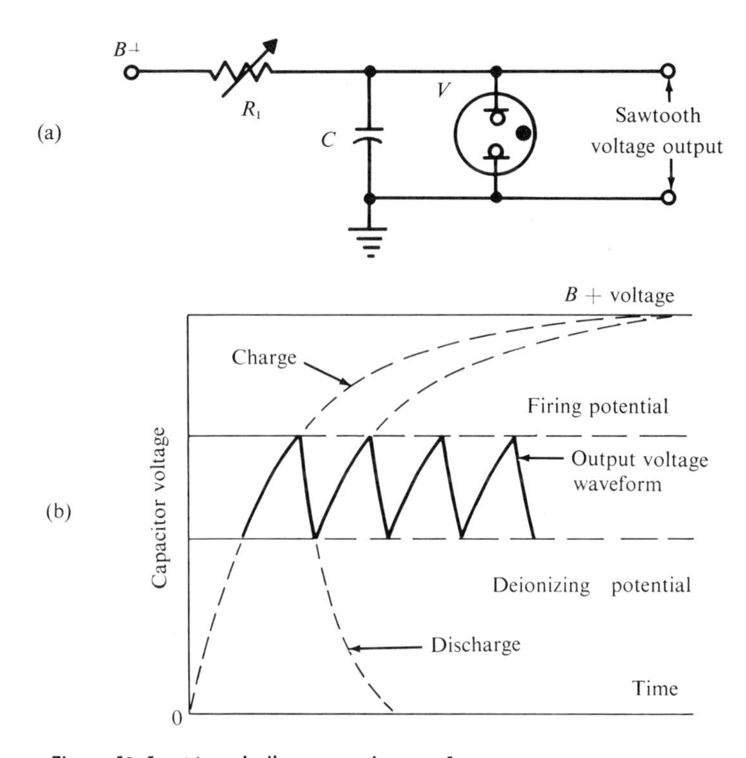

Figure 19–1. Neon-bulb sawtooth waveform generator.

the potential across *C* is sufficiently high to ionize the gas in the neon bulb. The bulb has a very high resistance until the ionization potential is reached; at the ionization point, however, its internal resistance drops suddenly to a low value, and *C* discharges rapidly through the bulb. When the voltage across *C* falls below the deionizing potential, the initial high resistance is re-established in the bulb. The capacitor then stops discharging because the voltage across C is less than the value required to ionize the neon gas. Thereupon, the capacitor again starts to charge. A neon bulb typically ionizes at 64 v and deionizes at 58 v.

Example 1. The sawtooth generator in Fig. 19–1 has a firing potential of 72 v and a deionizing potential of 12 v. What is the oscillator frequency when $R = 68\ \text{k}\Omega$, $C = 1.0\ \mu\text{f}$, and $B+ = 152$ v?

The sawtooth waveform varies from 12 v to 72 v across the capacitor. However, the capacitor is allowed to change only 60 v before the neon bulb fires, discharging the capacitor to 12 v.

$$\therefore \quad \Delta E_C = 72 - 12\ \text{v} = 60\ \text{v}$$

From the universal time constant chart (Fig. 19–2),

$$T = 0.7\ RC$$

$$\begin{aligned} T &= 0.7\ RC \\ &= 0.7 \times 68 \times 10^3 \times 1 \times 10^{-6} \\ &= 47.6\ \text{m sec} \\ f &= \frac{1}{T} = \frac{1}{47.6} \\ &= 20.2\ \text{Hz} \qquad \text{(answer)} \end{aligned}$$

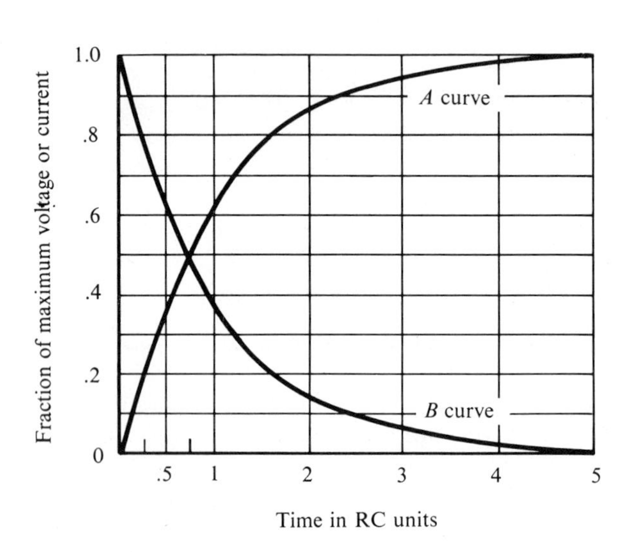

Figure 19–2. Universal *RC* time-constant chart.

For a given d-c supply voltage, the frequency of the generated sawtooth waveform depends upon the *RC* time constant, and the frequency is varied by adjusting the value of *R*. It is also possible to vary the frequency of the sawtooth waveform by adjusting the value of the d-c supply voltage. Consideration of Fig. 19–1 reveals that the sawtooth voltage varies between the deionizing potential and the firing potential of the neon bulb. The full d-c supply voltage is not applied across *C* because the firing potential has a lower value then the d-c supply, and the difference appears across *R*. Similarly, *C* does not completely discharge because it must stop discharging when the deionizing potential is reached. The capacitor voltage follows a normal *RC* charging curve between these two limits (curve *A* in Fig. 19–2). The discharge interval follows a similar curve (curve *B* in Fig. 19–2), except that the discharge time is only a small fraction of the charge time. The resistance in the discharge path is only a small fraction of the resistance in the charge path.

Observe in Fig. 19–3 the general results of increasing the *RC* time constant and of increasing the d-c supply voltage. Note the following points.

1. The amplitude of the generated sawtooth waveform is independent of the *RC* time constant and of the d-c supply voltage value.
2. Reduction of the d-c supply voltage or increase of the *RC* time constant reduces the frequency of the generated sawtooth waveform.
3. Linearity of the generated waveform is improved by an increase in the d-c supply voltage.

The third point is of considerable importance in applications that require a generated waveform that approaches a true sawtooth shape. Although the waveform generated by the configuration in Fig. 19–1 always has an exponential rise and fall, the initial portion of an exponential waveform is more nearly linear than following portions. Therefore, when we wish the generated waveform to rise as linearly as possible, we employ a d-c supply voltage that is as high as practical. The generated frequency will be the same if *R* is increased suitably when a higher value of d-c supply voltage is used. This is just another way of saying that the generated waveform would have a perfectly linear rise if the capacitor were charged from a constant-current source. Observe the following formulas.

$$Q = It \qquad \textbf{(19–1)}$$

$$Q = Ce \qquad \textbf{(19–2)}$$

or,

$$e = \frac{It}{C} \qquad \textbf{(19–3)}$$

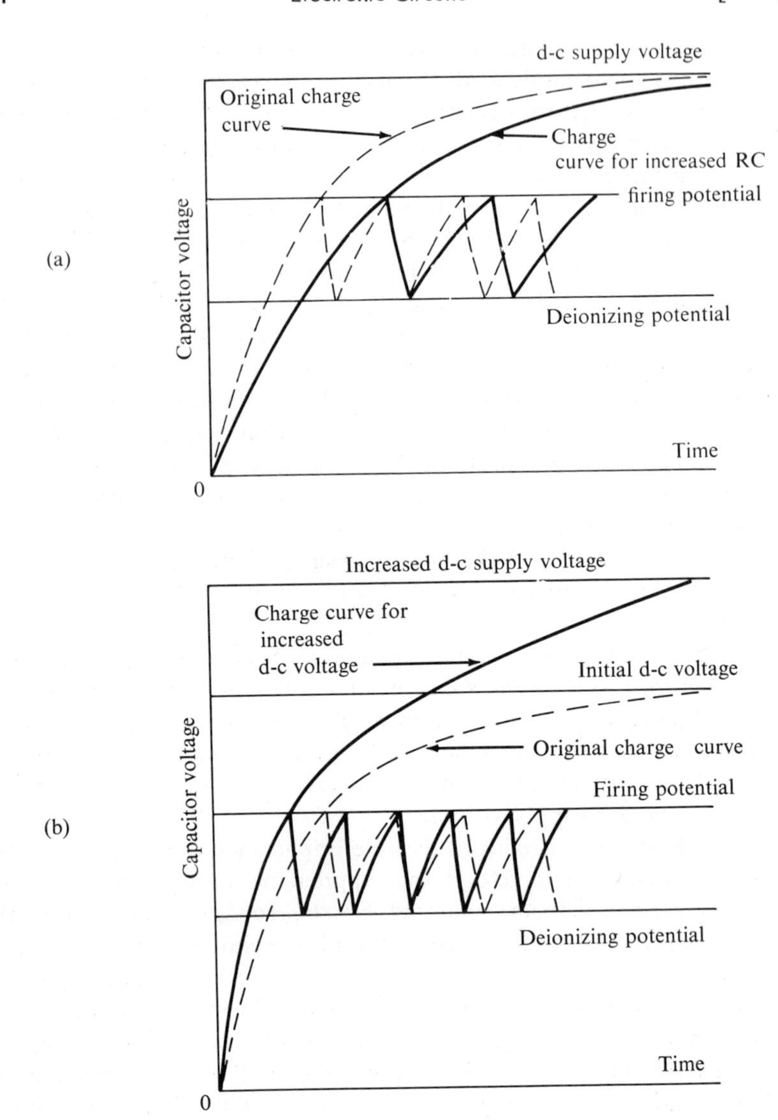

Figure 19–3. **Results of increasing the *RC* time constant and increasing the supply voltage.**

Formula 19–3 states that e will be directly proportional to t, if I is maintained constant. We recall that a source with high internal resistance approximates a constant-current source. The linearizing effect of a high d-c supply voltage is evident from Coulomb's law for capacitors, as well as from the universal *RC* time-constant chart.

Next, observe the configuration depicted in Fig. 19–4. If capacitor

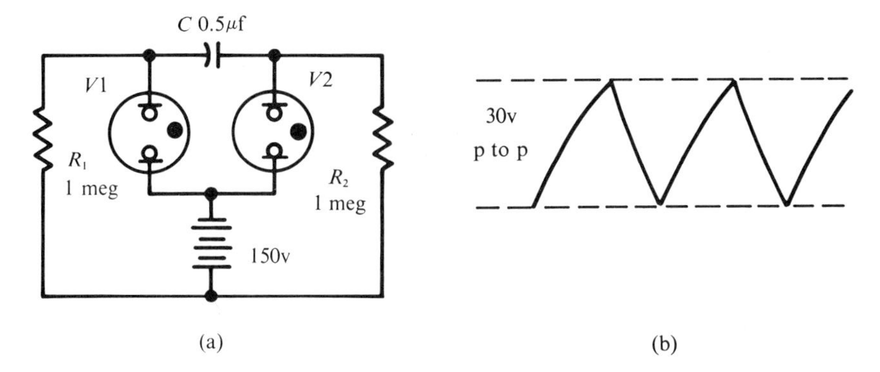

Figure 19–4. (a) The neon bulbs conduct alternately; (b) Waveform across C.

C is omitted, it is obvious that *V*1 and *V*2 will conduct equally and continuously. On the other hand, when *C* is connected between the neon bulbs, the bulbs conduct alternately in see-saw fashion. Let us see why. No two neon bulbs have *exactly* the same ionization potential. Accordingly, at the instant that the voltage is applied in the circuit, one of the bulbs will conduct first; let us suppose that *V*1 conducts first. We will find that conduction in *V*1 suppresses conduction in *V*2 for a period of time during which *C* charges. That is, we may consider that the tube *V* represents a very low resistance, or practically a short circuit, while it conducts. Evidently, *C* must immediately start charging through R_2 and the *V*1 short circuit. The left-hand plate of *C* charges negative, and the right-hand plate charges positive.

This initial sequence is diagrammed in Fig. 19–5. Observe that at the instant that *V*1 is replaced by a short circuit (conduction), the top terminal of R_1 becomes 150 v negative. Charging current I_C flows into *C* and back through R_2 to the 150-v source. Hence, at the first instant, the top terminal of R_2 is also 150 v negative. *V*2 cannot conduct because there is zero volts drop across *V*2 in accordance with Kirchhoff's voltage law. Thus, *V*1 conducts and *V*2 cannot conduct as *C* charges. This state of affairs cannot persist indefinitely, because I_C is decreasing in accordance with Fig. 19–2, and the voltage drop across R_2 is also decreasing. When *C* becomes charged sufficiently, the drop across *V*2 ($150—I_C R_2$) rises to the ionization potential of *V*2;

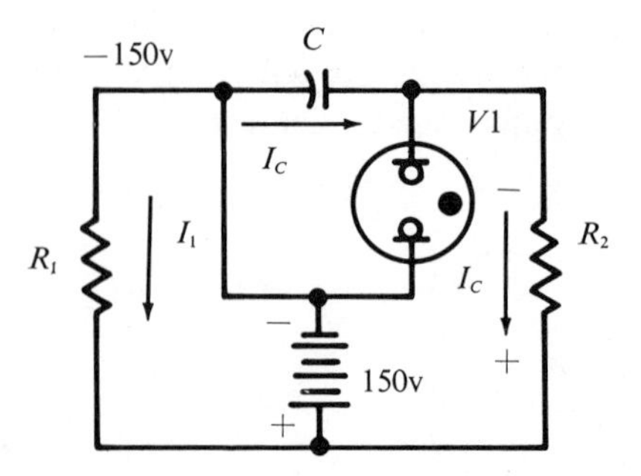

Figure 19-5. Charging current I_C suppresses conduction by V1.

thereupon $V2$ suddenly starts to conduct. We will perceive that the onset of conduction by $V2$ is accompanied by extinguishing of $V1$.

What happens at this instant is seen in Fig. 19-6; although both $V1$ and $V2$ are conducting at the first instant, C has been charged to a

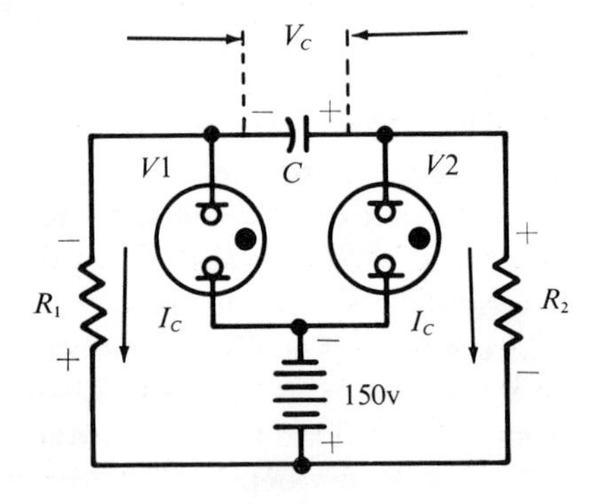

Figure 19-6. I_C increases the drop across V2 and decreases the drop across V1.

voltage E_C, and the conduction of $V2$ prevents further charge. In turn, E_C causes a discharge current flow I_C through R_1 and R_2. Note that $I_C R_1$ reduces the voltage drop across $V1$, and that $I_C R_2$ increases the voltage drop across $V2$. Therefore, conduction in $V2$ is immediately accompanied by extinguishing of $V1$. Now, $V1$ is an open circuit, and $V2$ represents a very low resistance, or practically a short circuit. Hence, C proceeds to charge in opposite polarity via $V2$ and R_1; the right-hand plate of C rises in negative polarity and the left-hand plate rises in positive polarity. Eventually, the drop across $V1$ increases to the ionization potential, and the cycle of operation repeats. We call this configuration a two-tube neon relaxation oscillator.

The same basic principle applies to the three-tube neon ring oscillator (relaxation oscillator) depicted in Fig. 19-7. Only one neon

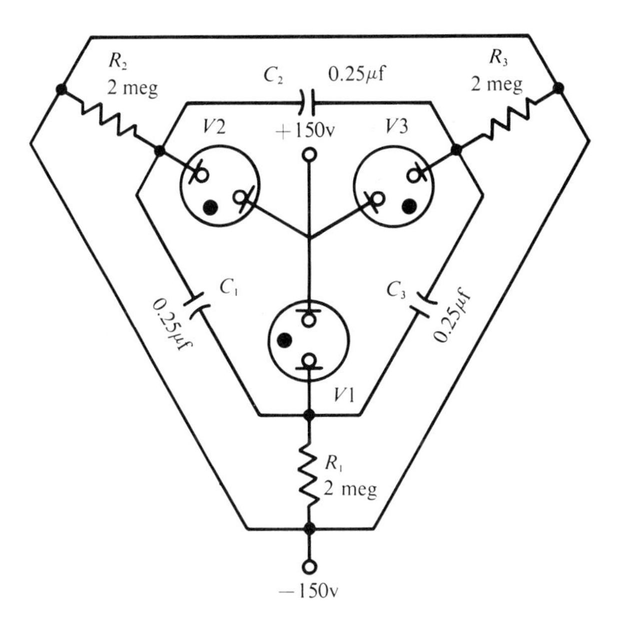

Figure 19-7. A three-tube neon ring oscillator.

bulb can conduct at a time. Conduction is in rotation around the ring, either clockwise or counterclockwise, depending upon initial conditions. Conduction in any one bulb suppresses conduction in both of the other bulbs until the capacitor charges change sufficiently to cause switching. When voltage is first applied in the circuit, the firing sequence is determined by component tolerances comprising ionization potentials of the individual bulbs, resistive values, and capacitive values. Unless bulbs, resistors, and capacitors are reasonably well matched, only two of the bulbs will conduct alternately, and the third bulb will not conduct at any time. On the other hand, with well-matched components, conduction occurs clockwise or counterclockwise around the ring, as the case may be.

You can reverse the direction of conduction around the ring in Fig. 19-7 by touching any one of the bulbs at a suitable instant in the operating sequence. This introduces body capacitance into the circuit, and also couples stray-field voltage into the bulb that you touch. In other words, your body is normally above ground potential, as you can demonstrate by touching the vertical-input terminal of an oscilloscope. When you couple stray-field voltage into one of the bulbs at a suitable instant, the voltage distribution in the circuit is changed; this change can cause a bulb to fire out of the established order, and reverse the sequence of operation. A three-tube neon ring oscillator is comparatively easy to design, but a four-tube configuration becomes more difficult because of the tight component tolerances that are imposed. Nevertheless, this makes an interesting experiment.

Note that the capacitors in Fig. 19–7 operate in a very high impedance circuit when no bulb is conducting. Each capacitor is connected to the remainder of the circuit through a 2-MΩ resistor. Suppose that the d-c voltage is reduced to the point that no bulb is quite at the ionization potential. You can then make one of the bulbs fire by touching it with your finger; the stray-field voltage that you couple into the bulb brings it up to ionization potential. The bulb will continue to flash on and off as long as you hold your finger against it. When you remove your finger, the bulb no longer fires.

19–2. Four-Layer Diode Oscillators

The four-layer diode is a semiconductor device which can replace the neon bulb in a relaxation oscillator. It is a solid-state silicon device which has a property corresponding to a gas diode. Like neon bulbs, four-layer diodes have two stable states, conducting and non-conducting, and they switch rapidly from one state to the other. The schematic of the construction of the four-layer diode in Fig. 19–8 shows that the diode has four alternate PN layers. Because the characteristics of these layers are closely controlled by the diffusion process, the characteristics of these diodes have a close tolerance. This close tolerance factor, along with the fact that the diode is a solid state device and may be obtained in a wide range of switching voltages, makes the four-layer diode a natural choice as a switching device.

The voltage-current characteristic curve in Fig. 19–9 gives a quantitative description of the electrical performance of the four-layer diode at low frequency operation. The following definitions specify the electrical current-voltage characteristics of the four-layer diode.

I_s = switching current
V_s = switching voltage
V_h = holding voltage
I_h = holding current

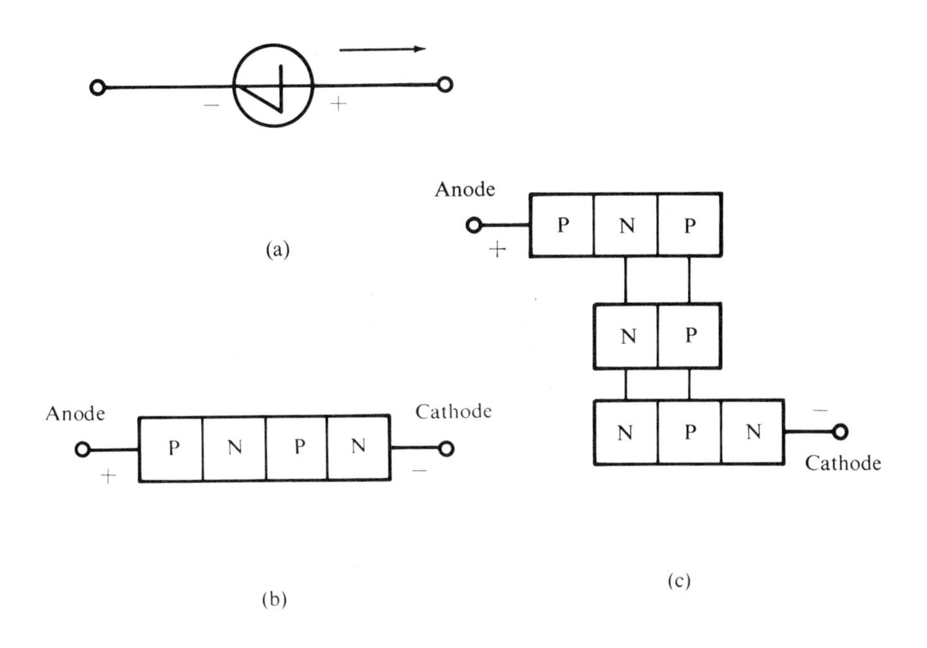

Figure 19–8. Four-layer diode. (a) Circuit symbol showing current and voltage relations; (b) Schematic representative; (c) Equivalent circuit representation with a PNP transistor, an NPN transistor, and a back-biased diode.

R_{off} = dynamic resistance when off
R_{on} = dynamic resistance when on

Three regions of operation are noted in the four-layer voltage-current characteristic in Fig. 19–9. In Region I, the resistance of the diode is extremely high up to the point where an increasing voltage across the diode reaches the switching voltage V_s. When the voltage across the diode reaches or exceeds V_s, transistor action occurs in the diode in the form of an internal feedback, producing the negative resistance effect in Region II. As the current through the diode exceeds the holding current I_h, the diode moves into saturation, Region III. The device will remain in Region III as long as an external source, such as a capacitor, can maintain a current flow greater than I_h. The moment the current falls below I_h, the diode switches back into Region I.

Let us compare the four-layer diode sawtooth oscillator in Fig. 19–10 with the neon-bulb sawtooth oscillator in Fig. 19–1. Only two

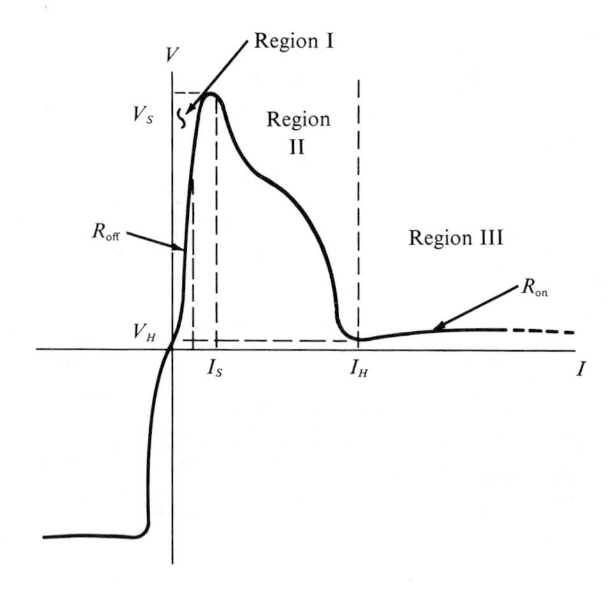

Figure 19–9. Characteristics of a 4-layer diode showing the current-voltage characteristic curve.

differences are apparent: first, the four-layer diode voltage returns essentially to zero; second, a resistor R_2 must be added to the four-layer diode circuit as a current limiter to prevent destruction of the diode. The value of R_2 is less than 50 Ω.

Example 2. Determine the values of the components in the circuit Fig. 19–10: R_1 is 22 kΩ, R_2 is 50 Ω, and C is 0.01 μf. What is the frequency of oscillation when the supply voltage is 50 v using a diode with a switching voltage of 12 v and a holding current of 50 μa?

The output waveform varies from approximately 0 to 12 v, since the product of R_2 times the holding current is small.

$$V = 12 \text{ v}$$

$$\% \text{ of } V_o = \frac{12}{50} = 24$$

From the universal time-constant chart,

$$24\,\% = T = 0.28\ TC = 0.28\ RC$$

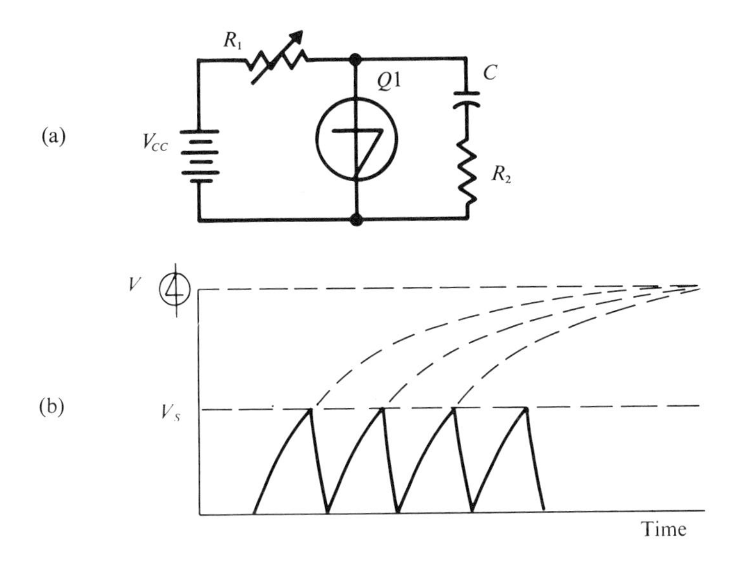

Figure 19–10. Four-layer diode sawtooth generator. (a) Circuit; (b) Output sawtooth waveforms.

$$f = \frac{1}{0.28 \times 2.2 \times 10^{4} \times 1 \times 10^{-8}}$$

$$f = 16.2 \text{ kHz} \qquad \text{(answer)}$$

Four-layer diodes are manufactured with switching voltages of about 2 to 1,000 v and switching currents up to several hundred amperes. Higher switching voltages may be obtained by placing several diodes in series, and higher switching currents may be obtained by paralleling similar diodes.

19–3. Thyratron Sawtooth Oscillators

The thyratron, or gas-filled triode, is widely used to generate sawtooth waveforms, and has certain advantages over the neon-tube sawtooth arrangement. A thyratron is more stable, has lower internal resistance and faster deionization time, can conduct comparatively large current, and can provide partial grid control of conduction. In other words, conduction can be initiated by decreasing the grid bias, but conduction can be stopped only by reducing the anode potential almost to zero. This is just another way of saying that the grid loses control after conduction is initiated. Thyratron *RC* oscillators are used in various instruments, such as the capacitor tester illustrated in Fig. 19–11. Thyratron *RC* oscillators are also used to some extent

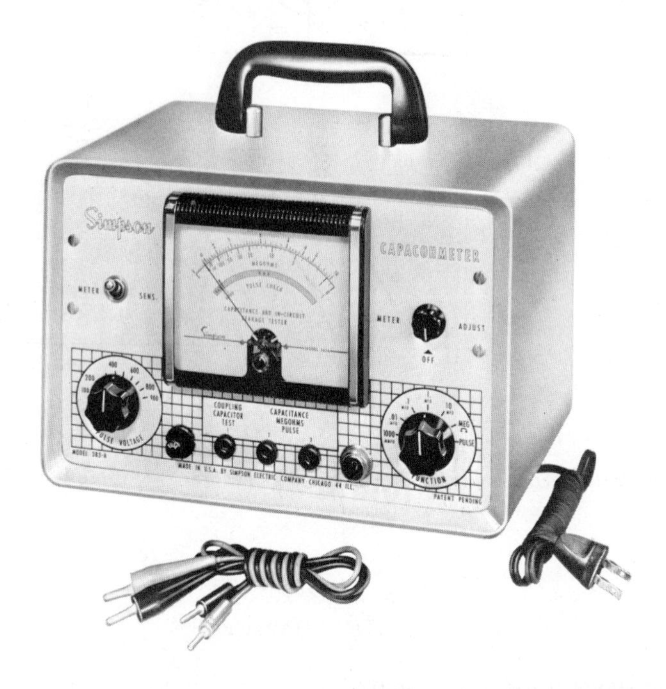

Courtesy of Simpson Electric Co.

Figure 19–11. A capacitor tester that utilizes a thyratron *RC* oscillator.

in oscilloscopes, although they have been largely supplanted by vacuum-tube relaxation oscillators. Thyratrons are in extremely wide use in industrial-electronic equipment. Due to their low anode resistance, thyratrons are more efficient than high-vacuum tubes, and efficiency becomes an important consideration when large current values must be switched.

We will find that a thyratron tube operates in the same basic manner as a neon bulb, except for grid control of ionization potential. As previously noted, the grid has practically no control over the deionization potential. If the grid of a thyratron is highly negative, the tube cannot fire, regardless of the anode potential. That is, the more negative that we bias the grid, the higher the ionization potential becomes. This relation for a typical thyratron is depicted in Fig. 19–12. For example, if the anode is operated at a potential of 350 v, the thyratron will fire when the grid-bias voltage is reduced to 4 v.

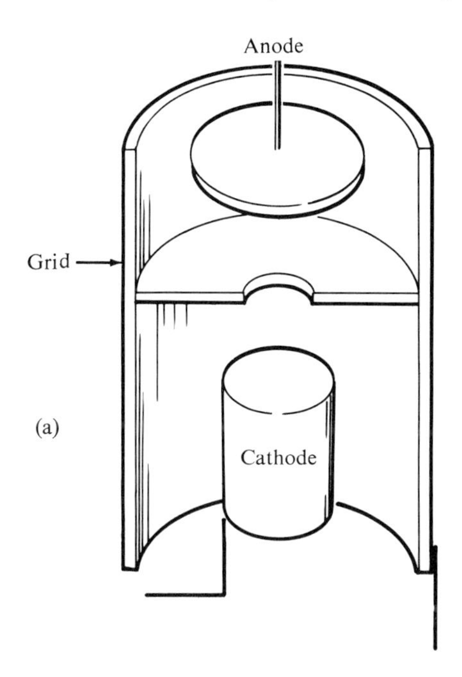

Figure 19–12. (a) Construction of a thyratron triode.

Thereupon, the grid loses control, and the thyratron will not deionize until the anode voltage is reduced to practically zero. This removal of plate voltage is accomplished automatically in a thyratron sawtooth oscillator circuit.

A simple thyratron sawtooth generator circuit and the output waveforms with high and low values of grid potential are shown in Fig. 19–13. The approximate frequency of the sawtooth waveform is formulated

$$f = \frac{1}{2.3\, RC \log_{10}(E_b - E_2/E_b - E_1)} \text{ Hz, approximately} \qquad \textbf{(19–4)}$$

where R denotes the total charging resistance in megohms, C denotes the total capacitance in microfarads, E_b denotes the anode supply voltage, E_2 denotes the deionization potential, and E_1 denotes the ionization potential in volts.

(b)

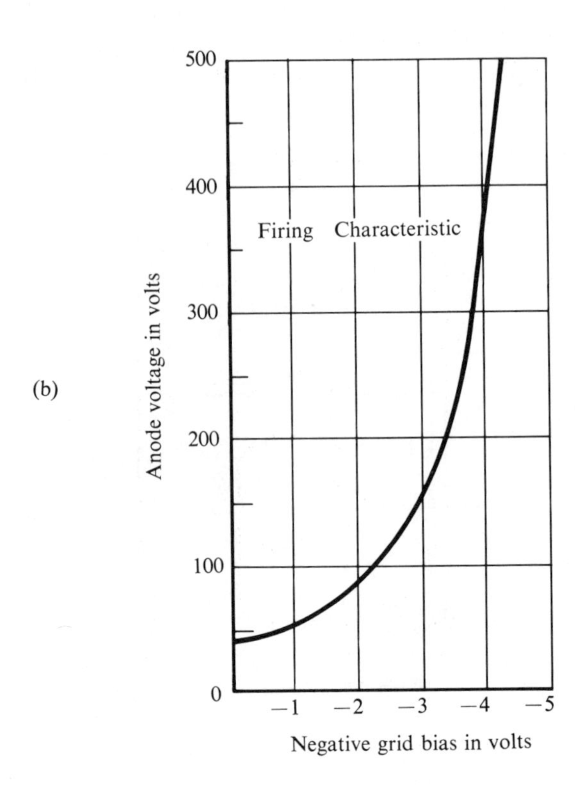

Figure 19–12. (b) Grid voltage versus anode voltage for ignition or conduction.

Example 3. The circuit in Fig. 19–13 uses a tube with a firing potential of 115 v and a deionizing voltage of 56 v. What is the oscillating frequency when $R = 100\ \mathrm{k\Omega}$, $C = 0.1\ \mu\mathrm{f}$, and the supply voltage is 300 v?

$$f = \frac{0.435}{1 \times 10^{3} \times 1 \times 10^{-8} \log (300 - 115/300 - 56)}$$

$$f \simeq 32.2\ \mathrm{kHz} \qquad \text{(answer)}$$

In the circuit of Fig. 19–13, the supply voltage must of course be much larger than the ionization potential of the tube. When a comparatively low supply voltage is used, the output waveform has greater nonlinearity, as we have seen in the neon-bulb sawtooth oscillator circuit. No gas tube relaxation oscillator generates a highly stable

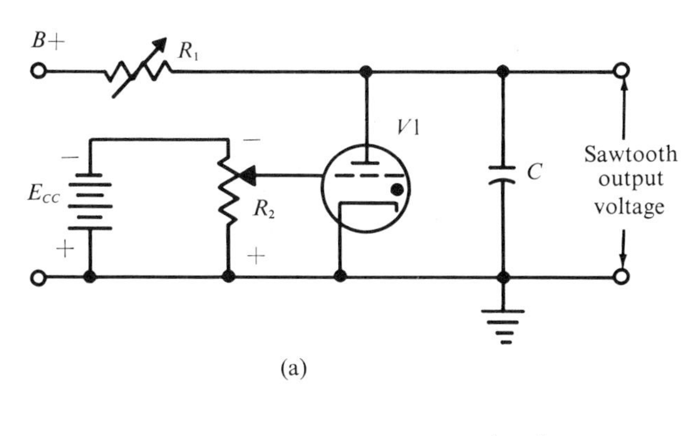

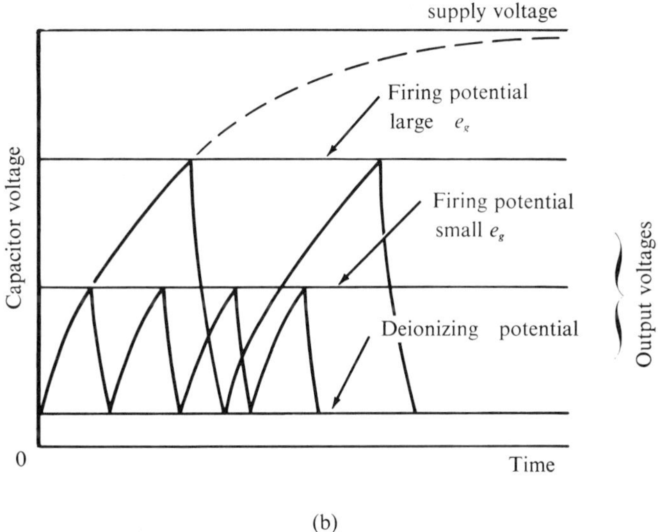

Figure 19-13. Thyratron sawtooth generator circuit and output waveforms.

output frequency; however, the output frequency can be synchronized by a stable reference frequency. This is done by injecting a small amplitude of the reference or synchronizing frequency into the grid circuit to accurately fix the instant of ionization. A thyratron sawtooth oscillator stabilized by a sync voltage is depicted in Fig. 19-14.

Circuit operation in Fig. 19-14 is analyzed as follows. The grid bias is adjusted to make the free-running frequency somewhat lower than the frequency of the sync signal. This free-running condition is indicated by the dotted sawtooth curve (without sync signal) in Fig. 19-14(b). Without the sync signal, the thyratron fires at points *A*, *C*, and so on; but with the sync signal applied, the firing potential varies

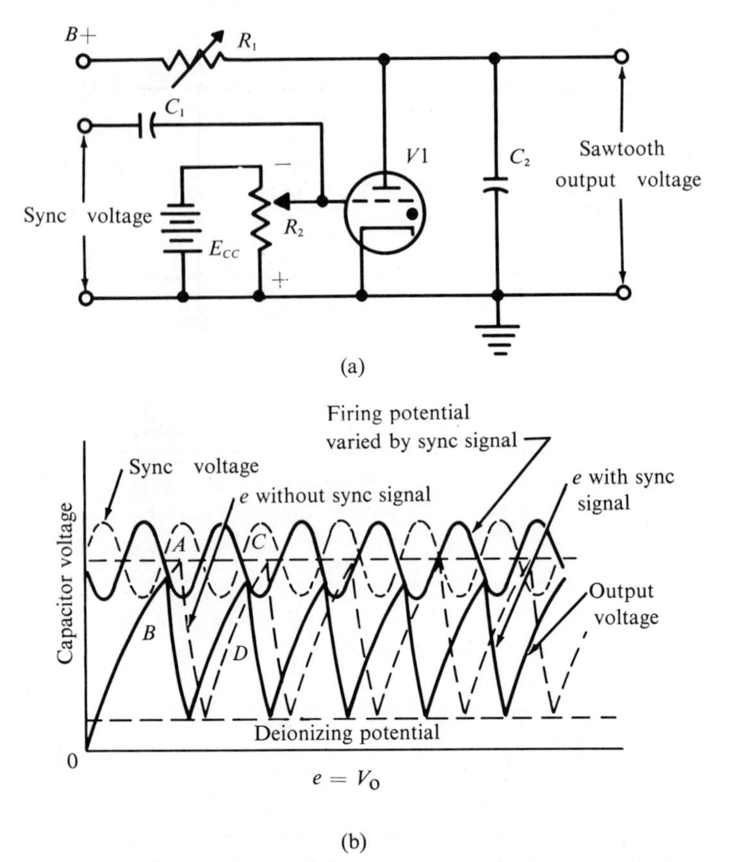

Figure 19–14. Synchronized thyratron sawtooth generator circuit.

in accordance with the instantaneous value of the grid potential. In other words, when the positive half cycle of the sync signal is applied to the grid, the firing potential is reduced, and the thyratron fires at points B, D, and so on. When the negative half cycle of the sync signal is applied, the firing potential is increased. Thus, if the sync voltage is applied, the time for each oscillation is reduced from AC to BD, and the oscillator is *locked* to the frequency of the sync voltage. We will perceive that the oscillator may also be locked to a multiple or submultiple of the sync-voltage frequency.

19–4. Silicon Controlled Rectifiers

The silicon controlled rectifier (SCR) is a four-layer semiconductor device with a control gate (G). A change of voltage on the control gate shifts the bias on the internal PN junction, as illustrated in Fig. 19–15. Increasing the gate bias decreases the switching voltage (V_s)

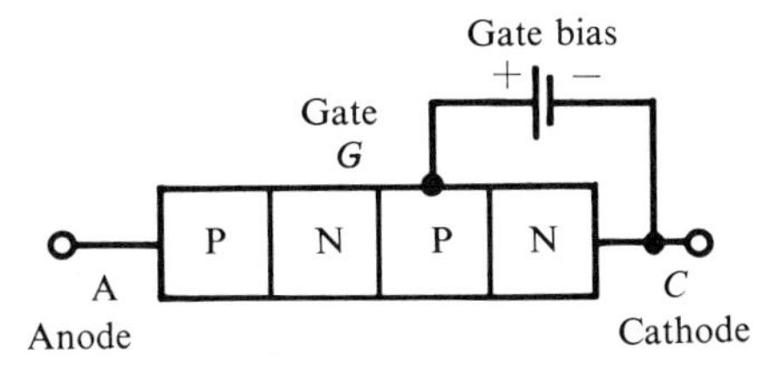

Figure 19–15. Representation of a controlled rectifier showing the electrodes: gate (G), anode (A), and cathode (C).

level. Once the switching voltage is exceeded, the gate loses control and the device can only be returned to the nonconducting state by a decrease of the external voltage and a corresponding decrease of the diode current. We will note here that these are also typical characteristics of a thyratron vacuum tube.

Generally speaking, the SCR can be utilized in any operation that requires the application of a high-speed, high-current switch. A useful application of the SCR is as a rectifier with a variable output level

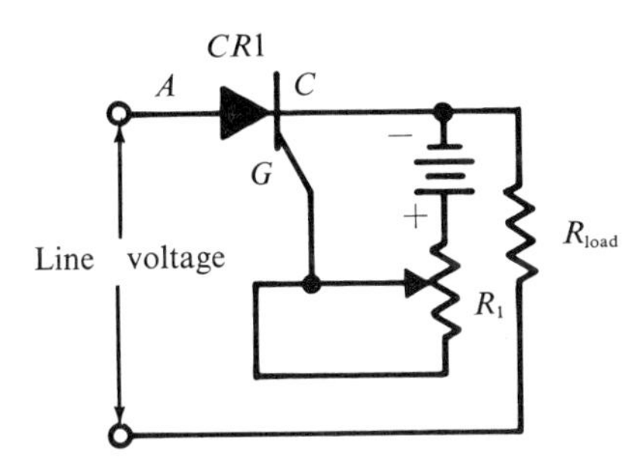

Figure 19–16. Half-wave SCR rectifier—the output voltage pulses are controlled by the gate bias.

(see Fig. 19–16). The conduction angle is controlled by the gate bias which controls the average load current. Figure 19–17 depicts the current-voltage characteristics of the SCR for three values of gate bias. In most applications, the SCR is triggered ON by forward bias pulses at the gate. Once the rectifier is conducting, it can be turned OFF only by the removal or reversal of the anode voltage.

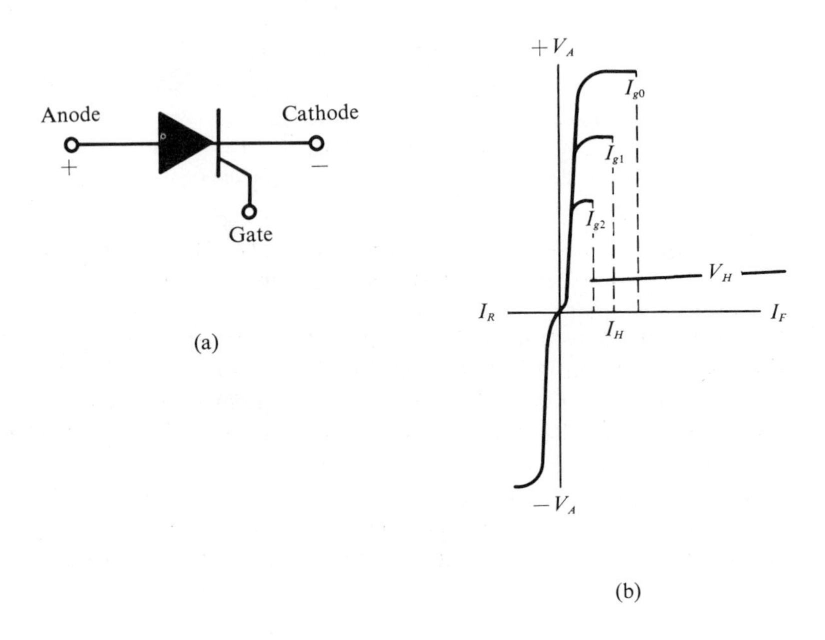

Figure 19–17. Silicon controlled rectifier. (a) Circuit symbol; (b) Characteristics in relation to gate bias current.

19–5. Multivibrator Waveform Generators

A multivibrator is a vacuum-tube or transistor oscillator that utilizes two tubes or two transistors to feed the output from one tube to the input of the other tube (and vice versa) by means of *RC* coupling networks. Output waveforms from multivibrators are basically square waves or pulses; however, a multivibrator may be elaborated to provide a sawtooth waveform, as in many television receivers. Free-running multivibrators generate a frequency (operate at a repetition rate) that is determined by the *R* and *C* values of the coupling networks. The repetition rate can be synchronized by a reference signal frequency, and the configuration is then called a *driven multivibrator*. It will be helpful at this point to briefly review the following principles of electron-tube circuit operation.

1. When the grid of a vacuum tube is made less negative (more positive) with respect to the cathode, the plate current increases, and vice versa.

2. An increase in current flow through the plate-load resistor results in a greater *IR* drop across the resistor and a reduction in the plate voltage. Conversely, lower current results in higher plate voltage.

3. There is a 180° phase shift between the grid signal voltage and the a-c component of plate voltage.

4. The voltage across a capacitor cannot build up or decay instantaneously in an *RC* circuit. (see Fig. 19–2.)

5. Current flow through a resistor is from the negative terminal to the positive terminal.

Basic multivibrator operation will be more easily understood if the action of the Eccles-Jordan *trigger circuit* is considered first. This type of circuit is used when electronic-switching action is desired in response to reference signals or pulses. This circuit is also widely used in electronic counters or scaling devices. In the strict sense of the term, the Eccles-Jordan trigger circuit depicted in Fig. 19–18 is not an oscillator. Instead, it is a circuit that has two conditions of equilibrium. One condition exists when $V1$ is conducting and $V2$ is cut off; the other condition of equilibrium exists when $V2$ is conducting and $V1$ is cut off. Both conditions are quiescent, in that there is no change in current or in any of the potentials until the circuit is triggered by an external signal. When a trigger pulse is applied, the nonconducting tube is suddenly driven into conduction, and the formerly conducting tube ceases to conduct. On the next trigger pulse, the reverse operation occurs. This circuit is often called a *flip-flop* circuit, or a bistable multivibrator.

The grids of $V1$ and $V2$ in Fig. 19–18 are connected to voltage-divider networks. Note that the voltage-divider network for $V1$ includes R_2, R_4, R_5, and E_{CC}. The voltage-divider network for $V2$ includes R_1, R_3, R_6, and E_{CC}. We observe that the voltage drops across R_5 and R_6 are individually less than E_{CC}, and because these voltages subtract from E_{CC}, the grids are always negative with respect to the cathodes. The action of the Eccles-Jordan circuit is analyzed as follows.

1. Assume that the cathodes are heated and that d-c supply voltage is applied to both tubes. If both tubes and their corresponding elements were exactly alike, equal currents would flow through the plate circuits. It is not likely, however, that the two tubes and their circuit elements will be balanced so exactly that this critical state would occur. Actually, one tube starts to conduct an instant before the other, or conducts more heavily than the other. We will assume that $V1$ in Fig. 19–10 conducts more current than $V2$.

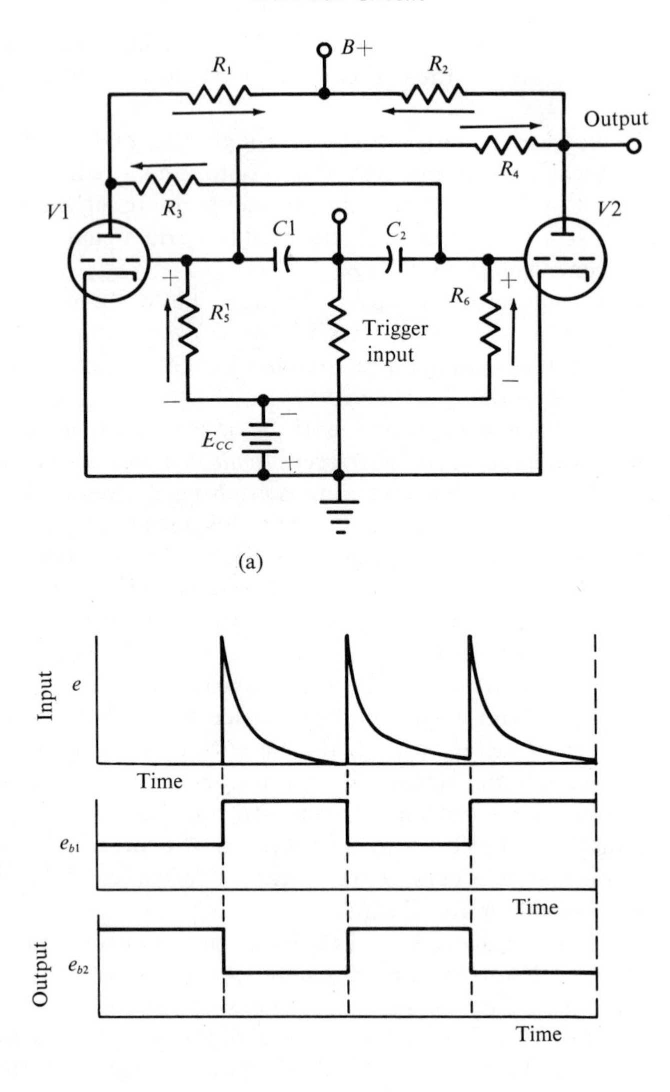

Figure 19–18. Basic Eccles-Jordan trigger (bistable or flip-flop) circuit and waveforms.

2. The voltage drop across R_1 is then greater than the drop across R_2, and the voltage at the plate of $V1$ is lower than the voltage at the plate of $V2$.

3. Note that lower voltage on the plate of $V1$ reduces the voltage across R_6 and increases the negative bias on $V2$. In turn, the current through $V2$ is further reduced.

4. Accordingly, the voltage at the plate of *V*2 is increased, and the drop across R_5 is increased, reducing the negative bias on the grid of *V*1. Current flow in the plate circuit of *V*1 is further increased and the plate voltage is further decreased.

5. This action is cumulative and very quickly reaches a condition where the plate current of *V*1 reaches a maximum and the plate current of *V*2 is cut off. This is one condition of stable equilibrium. During this quiescent period, the voltage drop across R_5 is larger than the drop across R_6.

6. Assume that a positive-going voltage is applied simultaneously to the grids of both tubes at the trigger input. Since *V*1 is already passing a heavy current, the positive pulse on its grid has little effect on the flow of current through the tube. Tube *V*2, however, is cut off; the positive pulse on its base, if of sufficient amplitude, removes the negative bias momentarily. Current then flows in the plate circuit of *V*2.

7. The plate voltage of *V*2 is reduced, and the reduced voltage across R_5 makes the grid of *V*1 more positive.

8. The plate current in *V*1 falls, and the plate voltage of *V*1 increases.

9. Increased plate voltage at *V*1 increases the drop across R_6, further reducing the bias of *V*2, and its plate current continues to increase, thus applying more negative bias to *V*1.

10. Plate current through *V*1 thus quickly ceases, and at the same time, plate current through *V*2 reaches saturation.

If negative-going pulses had been employed, the conducting tube would have been affected first. Its plate current would have been decreased with the same end result (*V*1 cut off and *V*2 conducting). One alternation is thus completed for each pulse, and two pulses are necessary to complete a full cycle. It is possible, however, to bring about a complete cycle of operation with a single trigger pulse by making suitable circuit changes. This will be discussed next.

19-6. One-Shot Multivibrator

A one-shot, or monostable, multivibrator circuit is a modified form of the bistable trigger configuration that has been described. As seen in Fig. 19-19, the one-shot multivibrator is essentially a two-stage *RC*-coupled amplifier. In its normal or balanced state, one tube is cut off and the other tube is in conduction. This state is established by the biasing arrangement for the tubes. When a trigger pulse is applied, the conducting tube is suddenly cut off, and the other tube

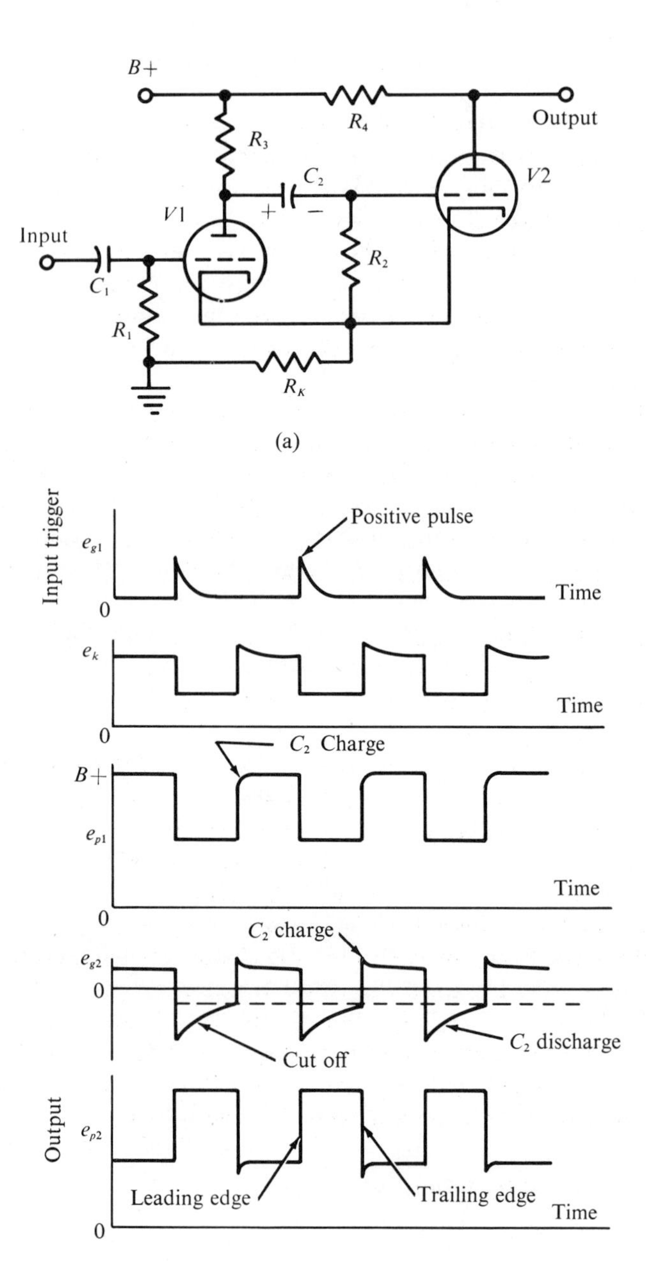

Figure 19-19. One-shot (monostable) multivibrator configuration and operating waveforms.

immediately conducts. After a definite period, tubes automatically revert to their original balanced state, and the circuit is ready to respond to another trigger pulse. We observe in Fig. 19–19 that the grid of *V*2 is returned to its cathode and that no current normally flows through R_2; accordingly *V*2 normally has zero grid bias. With this zero bias, *V*2 conducts and plate current flows through cathode resistor R_k. The resulting voltage drop across R_k biases *V*1 to cutoff. This is the normal condition of the circuit when no trigger pulse is applied. Note that when *V*2 is *not* conducting, *V*1 cannot be cut off by the self bias developed across R_k.

Next, a positive pulse, e_{g1} in Fig. 19–19(b), applied to the *V*1 grid via C_1 will raise the *V*1 grid above cutoff and cause *V*1 to conduct. The resulting drop in *V*1 plate voltage is applied through C_2 as a negative-going signal that quickly cuts off *V*2. The accompanying rise in plate voltage generates the leading edge of the output waveform. Capacitor C_2 discharges through R_2 toward the lowered value of *V*1 plate voltage. The drop across R_2 holds *V*2 cut off for a certain length of time which depends on the time constant R_2C_2 and the cutoff bias of *V*2. *V*1 continues to conduct after the input trigger has passed, because as previously noted, the drop across R_k is insufficient to cut off *V*1. The circuit remains with *V*1 conducting and *V*2 cut off, while C_2 discharges toward the lowered *V*1 plate voltage until the *V*2 grid voltage rises above cutoff.

Thereupon, *V*2 conducts and the drop in *V*2 plate voltage generates the trailing edge of the output waveform. The accompanying increase in voltage across R_k cuts off *V*1. Consequently, the *V*1 plate voltage immediately rises and the positive-going signal applied to the *V*2 grid via C_2 causes a further rise in *V*2 plate current as *V*2 returns to its original conducting state. This action is very rapid, so that *V*2 conducts at practically the same time that *V*1 is cut off. The circuit is now in its original quiescent state. It is important to recognize the following operating characteristics of a monostable multivibrator.

1. Although the output waveform depicted in Fig. 19–19 is essentially a square wave, it is evident that if the trigger pulses occur at longer intervals of time, the ouput will be a rectangular wave, as depicted in Fig. 19–20(a).

2. If the trigger pulses occur at the same rate as before, but the *RC* time constant for *V*2 is reduced, the output will be termed a pulse waveform, as shown in Fig. 19–20(b).

3. In case trigger pulses are applied at nonuniform time invervals, the output will comprise pulses of uniform width that occur at nonuniform intervals, as depicted in Fig. 19–20(c).

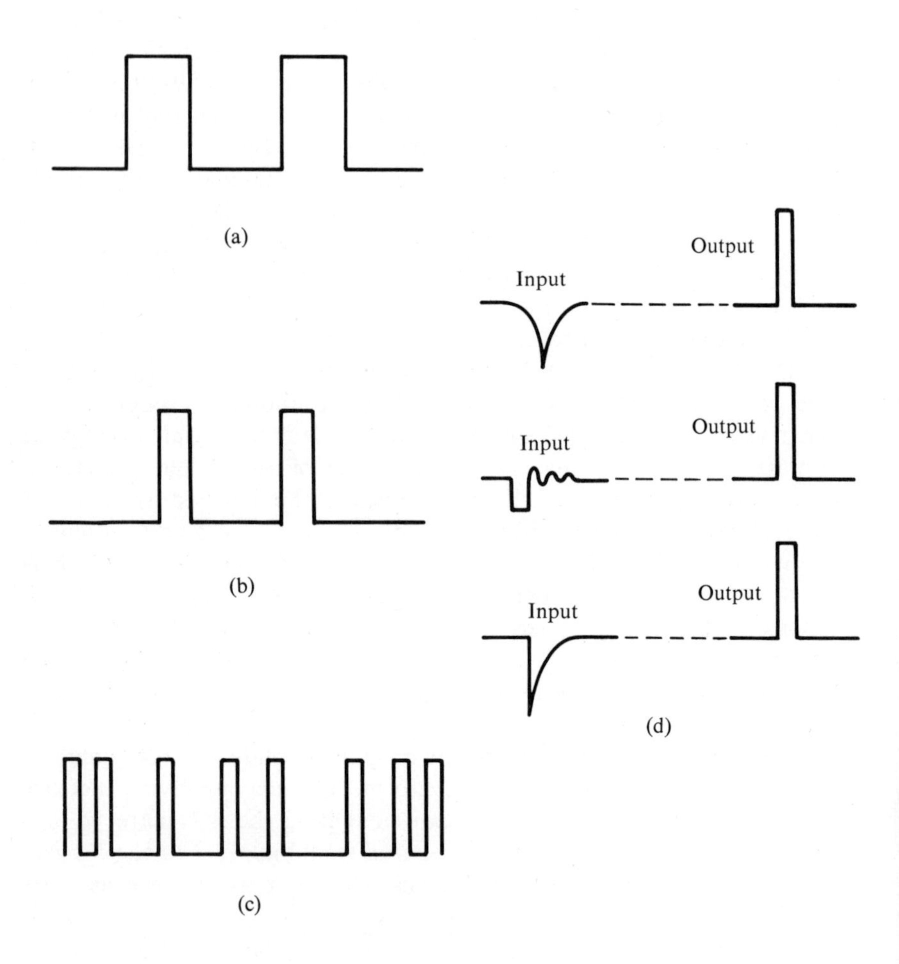

Figure 19–20. (a) Rectangular waveform; (b) Pulse waveform; (c) Uniform pulses occuring at nonuniform intervals; (d) Output pulse shape is independent of the input pulse shape.

4. Regardless of the shape of the trigger pulse, the generated output waveform always has the same shape and amplitude, as shown in Fig. 19–20(d).

These operating characteristics all have useful applications. For example, consider the fact that the generated waveform always has the same shape and amplitude, regardless of the shape of the trigger pulse. In computer circuitry, pulses must pass through many components which progressively distort and attenuate the wave-shapes.

Therefore, it often becomes desirable or necessary to convert the pulse back into standard shape and amplitude. For this purpose, the monostable multivibrator is well adapted; Fig. 19–20(d) depicts this capability as a *pulse regenerator.*

19–7. Free-Running Multivibrator

A basic form of the free-running or astable multivibrator circuit is shown using vacuum tubes in Fig. 19–21. It is simply a two-stage *RC*-coupled amplifier, with the output from the second stage coupled

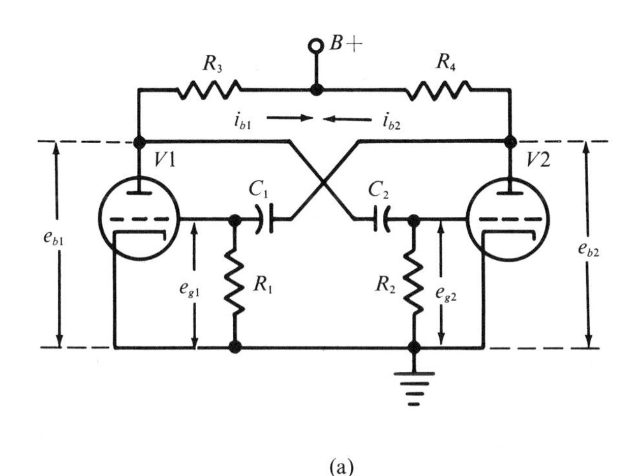

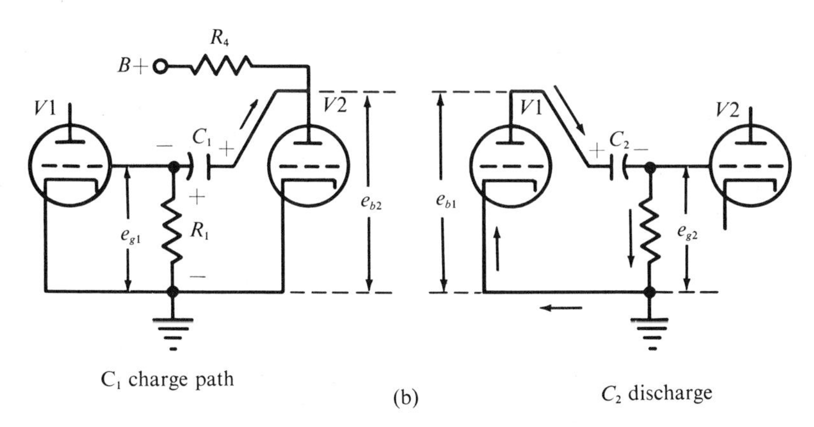

Figure 19–21. The basic free-running or astable multivibrator. (a) Circuit diagram; (b) Charge circuit for capacitor C_1 and discharge circuit for capacitor C_2.

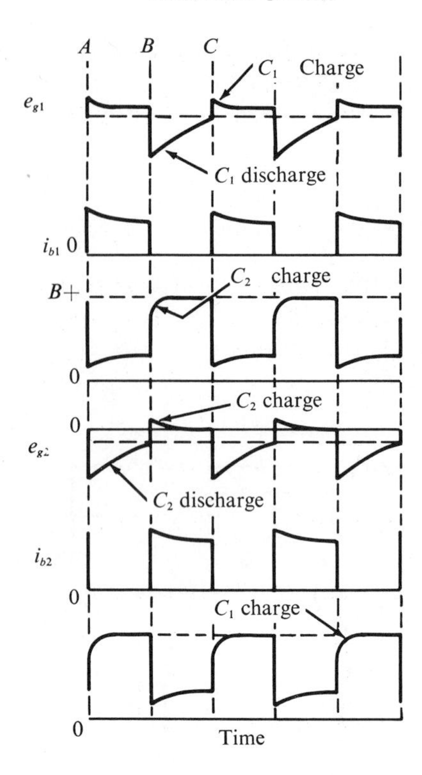

Figure 19–21. The basic free-running or astable multivibrator. (c) Circuit waveforms.

through C_1 to the input of the first stage, and the output from the first stage coupled through C_2 to the input of the second stage. Free-running multivibrators are used as square-wave generators and electronic switches, and have many other applications. A typical electronic switch is illustrated in Fig. 19–22. Because the voltage that is fed back in each case is of proper polarity to reinforce the voltage at the grid of the tube receiving the feedback voltage, signals are reinforced and oscillation takes place. The operation of the basic free-running multivibrator depicted in Fig. 19–21 is analyzed as follows.

When the cathodes are heated and plate potential is applied, both tubes begin to conduct. Initially, the plate currents are nearly equal, but there is always a difference between them. This slight initial unbalance brings about a cumulative or regenerative switching action, which in this example is assumed to end with i_{b1} increased to a maximum value and i_{b1} reduced to zero. Although we have described this switching action as through it occurred slowly, it actually occurs

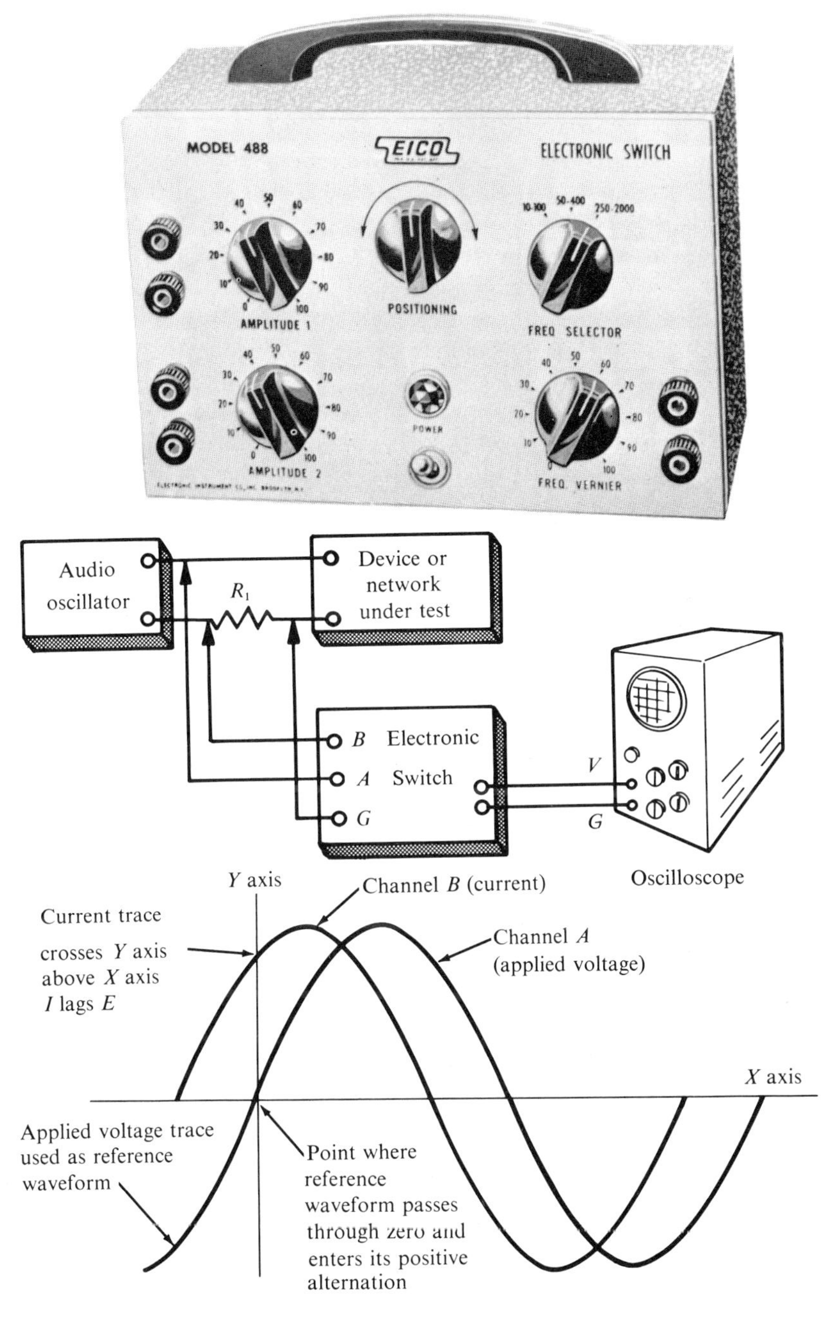

Courtesy of EICO, Electronic Instrument Co., Inc.

Figure 19-22. (a) An electronic switch; (b) Typical application of an electronic switch to display two waveforms simultaneously on an oscilloscope screen.

with extreme rapidity—in a very small fraction of a second in a well-designed multivibrator. This action is followed by a relatively long period during which the tubes are quiescent. Note that during this interval, one capacitor charges and the other capacitor discharges.

We may assume that initially i_{b1} rises more rapidly than i_{b2}. Plate voltage e_{b1} falls due to the increased drop across R_3; C_2 discharges through R_2, making the grid of $V2$ negative, thus reducing i_{b2}. Plate voltage e_{b2} rises because of the decreased voltage drop across R_4; C_1 charges through R_1, thus applying a positive bias to the grid of $V1$. Hence, the plate current of $V1$ rises to a maximum value, and $V2$ is cut off. The charge path for C_1 and the discharge path for C_2 are shown in Fig. 19–21(b). Waveforms of plate current and plate and grid voltages are shown in Fig. 19–21(c).

The negative grid voltage applied to $V2$ results from the discharge of C_2 through R_2, and the grid potential returns to zero as the capacitor discharge is completed. When the bias is reduced to the cutoff point, plate current i_{b2} begins to flow, and a second switching action takes place. This switching action is like the first, except that i_{b2} is increasing and i_{b1} is decreasing. Plate voltage e_{b2} decreases because of the increased drop across R_4; C_1 discharges through R_1, making the grid of $V1$ negative and thus reducing plate current i_{b1}. Plate voltage e_{b1} rises because of the decreased drop across R_3; C_2 charges through R_2, making the grid of $V2$ positive. Thus, the second switching action ends with $V2$ conducting maximum current and with $V1$ cut off. During the cycle of operation, current is maintained at a relatively steady value in one tube during the interval that the other tube is cut off. This action repeats indefinitely, with first one tube and then the other conducting.

The frequency of oscillation depends on the time constants of the coupling networks, R_1C_1 and R_2C_2. For a square-wave output, $R_1 = R_2$ and $C_1 = C_2$, the approximate frequency of oscillation is formulated

$$f = \frac{1{,}000}{R_1C_1 + R_2C_2} = \frac{1{,}000}{2R_1C_1} \qquad \textbf{(19–5)}$$

where f denotes kHz, R_1 is in ohms, and C_1 is in microfarads.

Example 4. Determine the free-running frequency of the circuit in Fig. 19–21 when $R_1 = R_2 = 220$ kΩ and $C_1 = C_2 = 0.005$ μf.

$$f = \frac{1}{2 \times 2.2 \times 10^5 \times 5 \times 10^{-9}}$$

$$f \simeq 450 \text{ Hz} \qquad \text{(answer)}$$

We will recognize that if both grid-circuit time constants are the same in a multivibrator, the output is a square wave and is called a *symmetrical multivibrator*. On the other hand, if the grid-circuit time constants are unequal, the multivibrator becomes a pulse generator and is then called an *asymmetrical multivibrator*.

19-8. The Transistor as a Switch

Any device which has two distinct stable states may be used as a switch. Among the numerous devices which fit into this category, only semiconductors have so many desirable features. Among these features are small size, low power consumption, low cost, fast speed, rugged construction, no filament power, and long life. Even with all these excellent features, the transistor is not an ideal switch; in fact, transistors have characteristics such as temperature instability and carrier storage time that are not encountered in vacuum-tube switching circuits.

The most common application of the transistor as a switch is in analogy to the mechanical switch. That is, the transistor has a low resistance in the closed state (ON) and a high resistance in the open state (OFF). These two points are illustrated on the characteristic curves in Fig. 19-23. The characteristic curves may be divided into three regions: the saturation region (*A*), the cutoff region (*B*), and the active region (*C*). Region *A* represents the conditions of a closed switch when the transistor is on or saturated. For the transistor represented in Fig. 19-18, the saturation voltage (V_{sat}) at point *A* is approximately 0.25 v with a saturation current (sat) of 100 ma giving a saturation resistance of 2.5 Ω. The open-circuit current represented at point *B* is very small, possibly 2 namp at 20 v for an open-circuit resistance (r_{off}) of several thousand megohms in a silicon transistor.

The transient response or, in other words, the speed with which the transistor switches from one condition to another is of great importance. The transition time should be as short as possible to minimize pulse distortion and pulse delay.

Pulse distortion is a result of both transition time and pulse delay, as we will observe by an analysis of the waveforms in Fig. 19-24. The *pulse-delay time* (t_d) is the time delay between the application of an input pulse and a change of 10 per cent of the maximum change. This is the result of the junction capacitance between the emitter and the base, which must be charged, and the transit time of the initial emitter current to diffuse through the base to the collector.

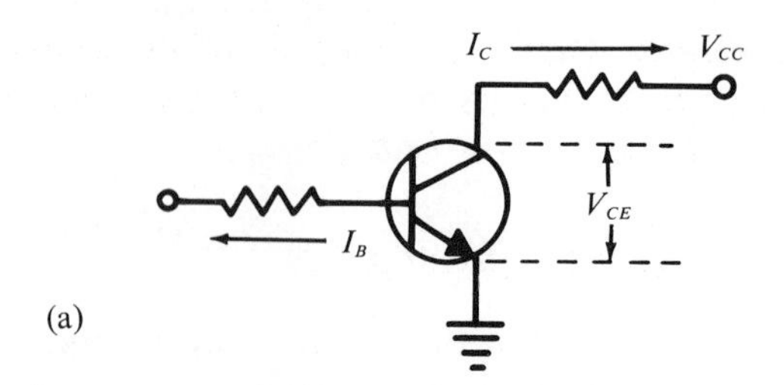

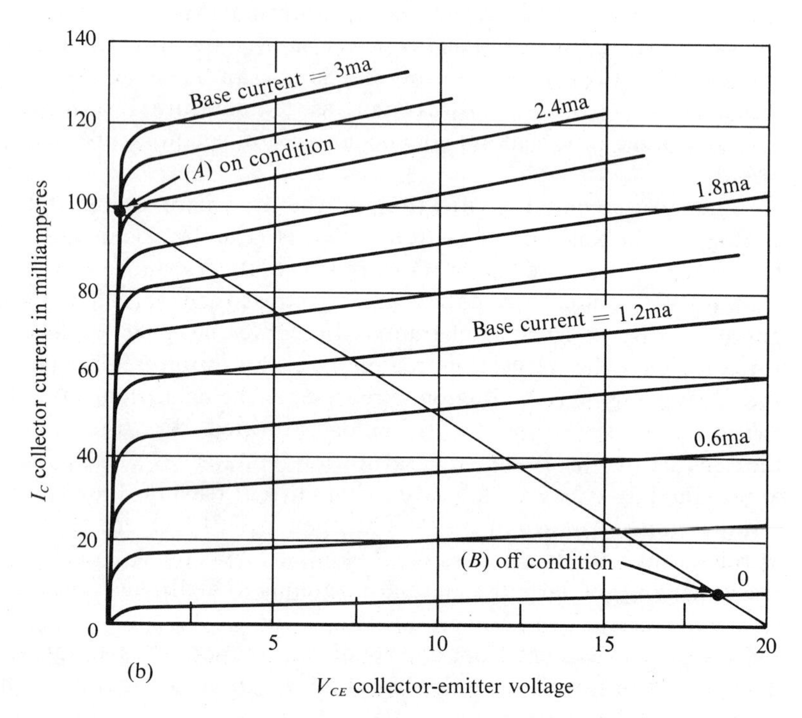

Figure 19–23. Transistor as a switch. (a) Transistor and current indication; (b) Output characteristics curves: Point *A* is the ON condition of the transistor (saturated); Point *B* is the OFF condition (base current 0).

The *pulse-rise time* (t_p) is the time required for the pulse to rise from 10 per cent to 90 per cent of its maximum value. The pulse-rise time is dependent upon the frequency response of the transistor, the magnitude of the applied base pulse, and the circuit capacitance.

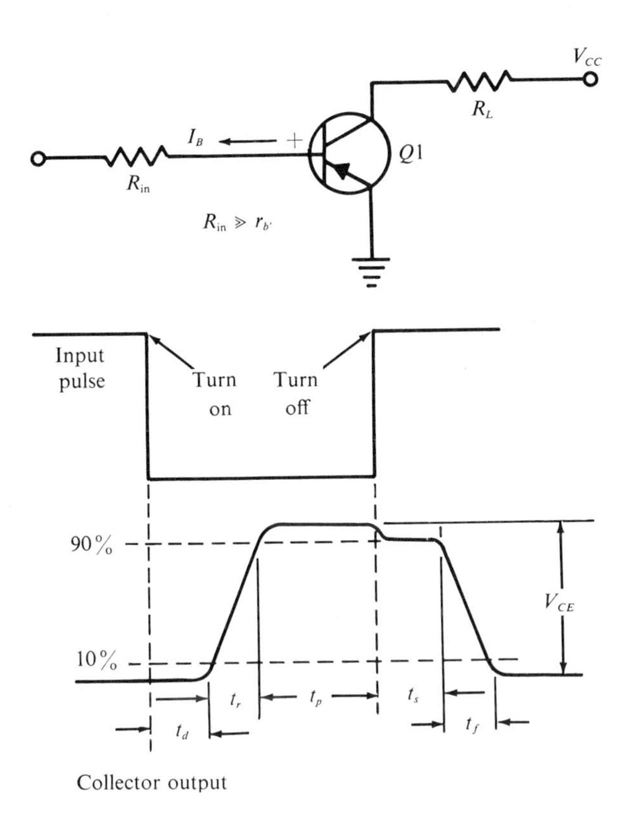

Figure 19-24. Transistor switching characteristics.

The *storage time* (t_s) is the result of diffusion of large numbers of minority carriers into the base region. The storage time is the time in which the output pulse remains near its maximum value after the input pulse has been removed. Accumulation of minority carriers into the base region occurs when the transistor is in the saturation region (ON), and the base is forward biased in respect to *both* the emitter and collector. Storage time or storage delay occurs when the base signal tries to turn the transistor off, and these minority carriers must be pulled from the base region into the collector circuit before the pulse can begin to change states.

To achieve faster switching rates, it is desirable to make the turn-on and turn-off rates as short as possible. The turn-on time is the sum of the delay time and rise time. The turn-off time is the sum of the storage time and the fall time. Obtaining a fast switching speed begins by the selection of a high speed switching transistor. Then the rise and fall time of the transistor may be improved by adding a *speed-up* capacitor (C_B) as shown in Fig. 19–25. This capacitor draws

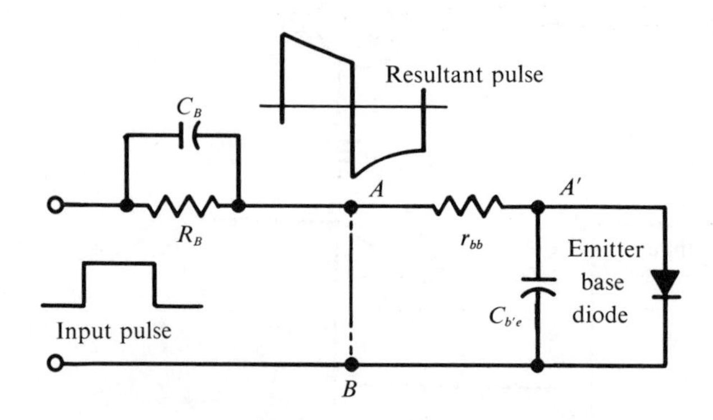

Figure 19–25. **Speed-up network C_BR_B shown with the equivalent input circuit of a transistor.**

a charging current during the switch-on time and returns the stored energy to the base in the turn-off time. The charging and discharging of C_B produces a larger base drive and therefore a faster collector rise and fall current. The value of C_B is chosen to form a voltage divider between C_BR_B and $C_{be}r_{bb}$ which effectively reduces the reactive current into the base circuit when the product of C_BR_B is equal to the product of $r_{bb}C_{be}$.

$$C_BR_B = r_{bb}C_{be} \tag{19–6}$$

To calculate the value of C_B, we must first determine the largest value of R_B which will produce saturation in the collector circuit. We will assume that the value of the collector voltage across the switching transistor in Fig. 19–26 is V_{sat} (0 v) and that the base voltage at the time-base current is maximum is V_{BE} is 0v. With these approximations and the transistor's current gain beta (β), we can develop the formulas for I_C, I_B, and R_B.

$$I_C \simeq \frac{V_{CC}}{R_C} \tag{19–7}$$

Since $I_C = \beta I_B$, we may state

$$I_B = \frac{V_{CC}}{\beta R_C} \tag{19–8}$$

Noting that

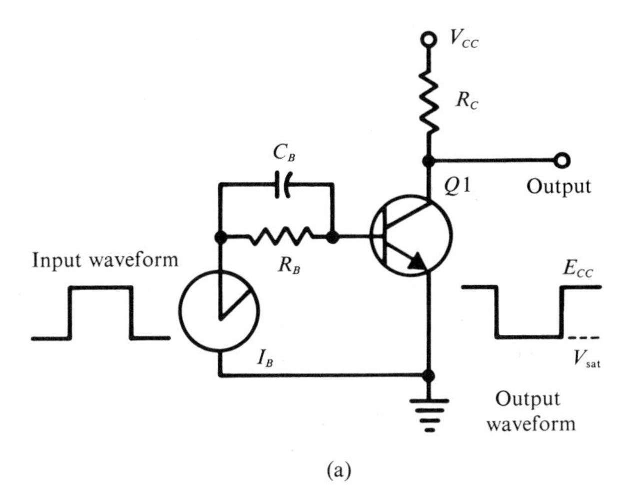

(a)

Figure 19–26. **Example of a switching circuit using a speed-up capacitor.**

$$I_B \simeq \frac{V_{BB}}{R_B} \tag{19–9}$$

and in most practical switching circuits $V_{BB} = V_{CC}$, then

$$I_B = \frac{V_{CC}}{R_B} \tag{19–10}$$

Substituting the value of I_B in Formula (19–8) into Formula (19–8),

$$\frac{V_{CC}}{R_B} \simeq \frac{V_{CC}}{\beta R_C}$$

and finally

$$R_B \simeq \beta R_C \tag{19–11}$$

To produce saturation in a transistor switch when $V_{CC} = V_{BB}$, the value of R_B must be *less* than βR_C.

Example 5. Determine the values of R_B and C_B for the circuit in Fig. 19–26 when $V_{CC} = V_{BB} = 20$ v, $r'_{bb} = 300\ \Omega$, $C'_{be} = 1{,}000$ pf, $\beta = 100$, and $R_C = 100\ \Omega$.

$$\begin{aligned} R_B &\leqq \beta R_C \\ &\leqq 100 \times 100 \\ &\leqq 10\ \text{k}\Omega \qquad \text{(answer)} \\ C_B R_B &= C'_{be} r'_{bb} \\ C_B &= \frac{1 \times 10^{-9} \times 3 \times 10^2}{1 \times 10^4} \\ &= 30\ \text{pf} \qquad \text{(answer)} \end{aligned}$$

19–9. Transistor Multivibrators

Transistor flip-flops (bistable multivibrators) are basic building blocks in many computer and switching-circuit applications. A basic configuration is shown in Fig. 19–27(a). The input trigger pulses make the two sides of the flip-flop conduct alternately (compare the transistor configuration with the Eccles-Jordan vacuum tube flip-flop depicted in Fig. 19–18). When $Q1$ conducts, $Q2$ is off, or when $Q2$ conducts, $Q1$ is off. In this example, the circuit is assumed to be symmetrical; that is, $R_{C1} = R_{C2}$, $R_1 = R_2$, $R_3 = R_4$, $C_1 = C_2$, $C_3 = C_4$, and transistor $Q1$ is similar to transistor $Q2$. Operating waveforms are depicted in Fig. 19–27(b).

We will assume that when battery power is applied, $Q1$ conducts more rapidly than $Q2$. Collector voltage v_{C1} falls because of the increased drop across R_{C1}. This action reduces the voltage applied to the divider R_1, R_4, and V_{BB}, with the result that V_{BB} cuts off $Q2$ [V_{BB} is greater than the drop across R_4, as shown in Fig. 19–27(c).] Collector voltage v_{C2} rises almost to the collector supply voltage V_{CC} because there is only a small drop across R_{C2}. This action supplies increased voltage across divider $R_2 - R_3$. The drop across R_3 exceeds V_{BB} to place a forward bias on the base of $Q1$. See Fig. 19–27(d). In turn, transistor $Q1$ conducts heavily. Thus, prior to time t_0 [Figure 19–27(b)], $Q1$ conducts and $Q2$ is cut off.

At time t_0, a positive pulse is applied between the base and ground of $Q2$ which increases the current through R_2. This produces a large base current which causes an even larger collector current through $Q2$. Then v_{C2} drops (there is more voltage drop across R_{C2}) and the voltage supplied to divider $R_2 - R_1$ decreases. The drop across R_3 decreases to less than V_{BB} to produce a reverse bias on the base of $Q1$. This reverse base bias decreases the conduction of $Q1$, resulting in a collector voltage rise of $Q1$ (due to a decrease in voltage drop across R_{C1}). As a result more voltage is supplied to the divider $R_1 - R_4$. The drop across R_4 becomes greater than V_{BB} to hold the base of $Q2$ positive and thus to hold $Q2$ in the conducting state. Thus, a positive pulse to the base of $Q2$, the OFF transistor, turns $Q2$ on and $Q1$ off. This condition is stable. Note that if the same positive pulse were applied to the ON transistor, $Q1$ at time t_0 would have no effect on $Q1$ because $Q1$ is already in the conducting (saturated) state.

At time t_1, a positive pulse applied to the base of $Q1$ ($Q1$ OFF) turns $Q1$ on. The drop in collector voltage v_{C1} will apply a back bias to the base of $Q2$ (V_{BB} will exceed R_4), and $Q2$ will decrease conduction with a resulting increase in v_{C2}, caused by the decrease across R_{C2}.

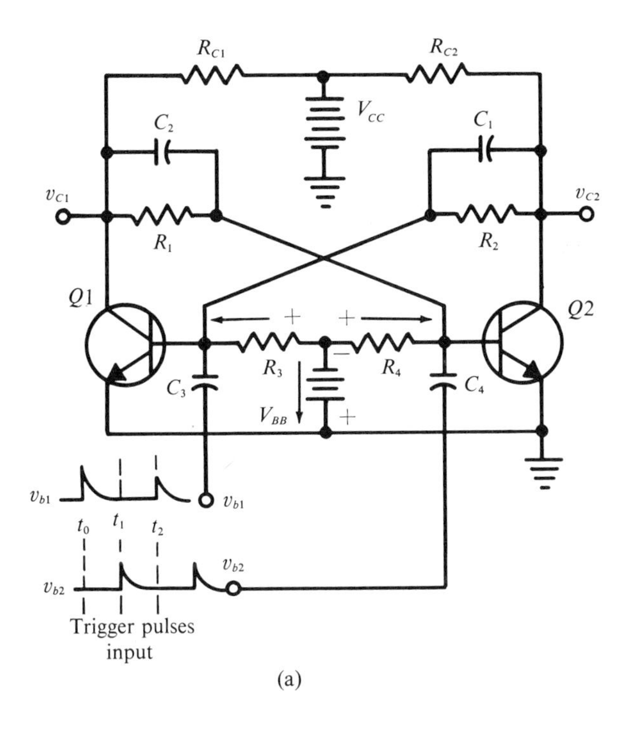

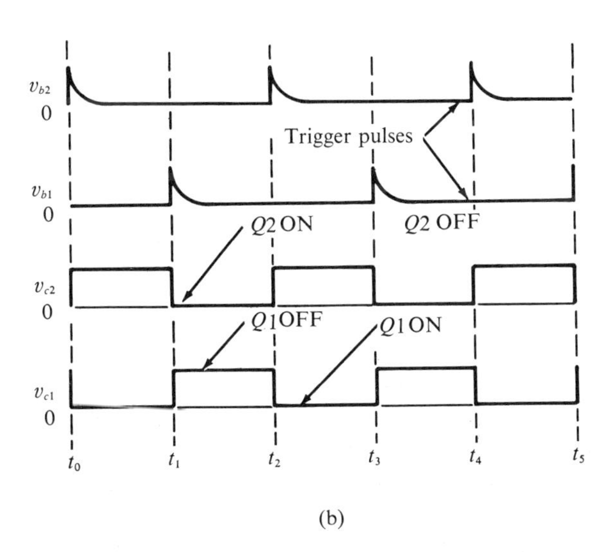

Figure 19-27. (a) Basic transistor bistable multivibrator circuit; (b) Operating waveforms.

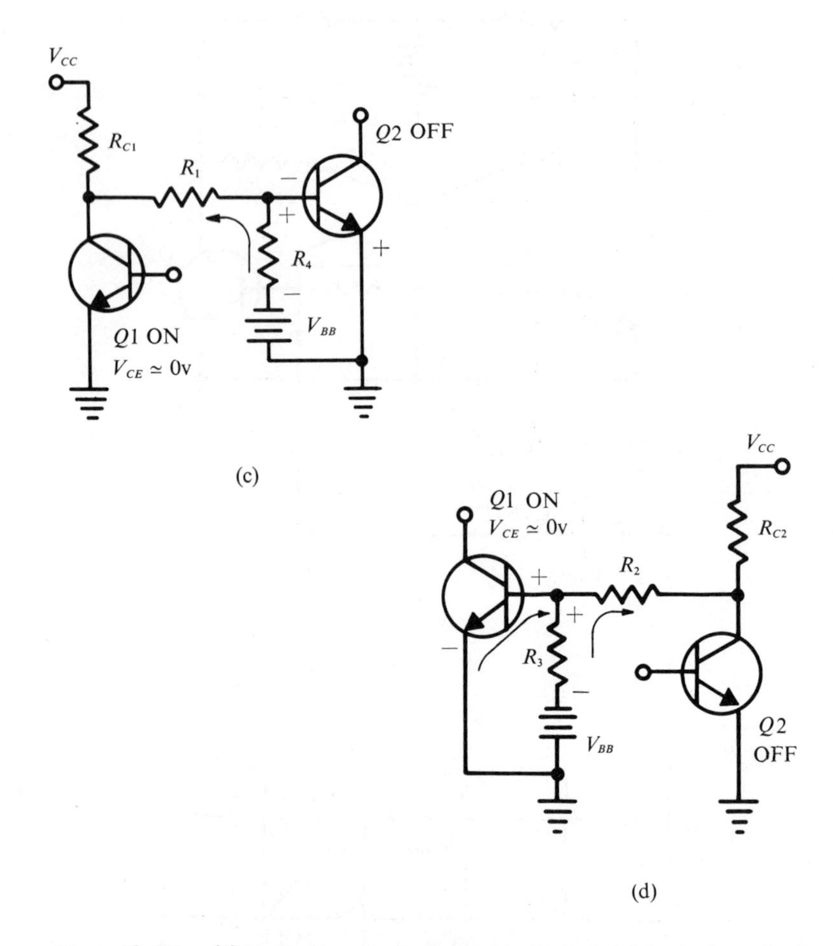

Figure 19–27. (c) Equivalent circuit that holds Q1 ON; (d) Equivalent circuit that holds Q2 OFF.

Thus, a second stable state exists at time t_1 when $Q2$ conducts and $Q1$ is cut off. Note that the repetition rate of the output waveform between collector and ground is one-half that of the input pulses. In other words, there are two input pulses for each output pulse.

A change of the input trigger coupling circuit, as shown in Fig. 19–28, will allow the flip-flop circuit to be triggered from one source. Diodes $CR1$ and $CR2$ (isolation diodes) are used to couple the input trigger to the base of the OFF transistor. This type of connection is called a symmetrical base-triggering circuit. In the event collector triggering is to be used, as indicated by the dashed lines of diodes $CR3$

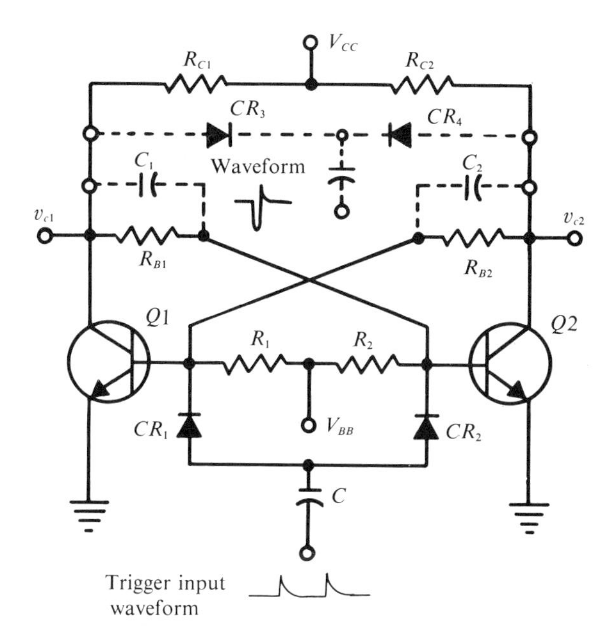

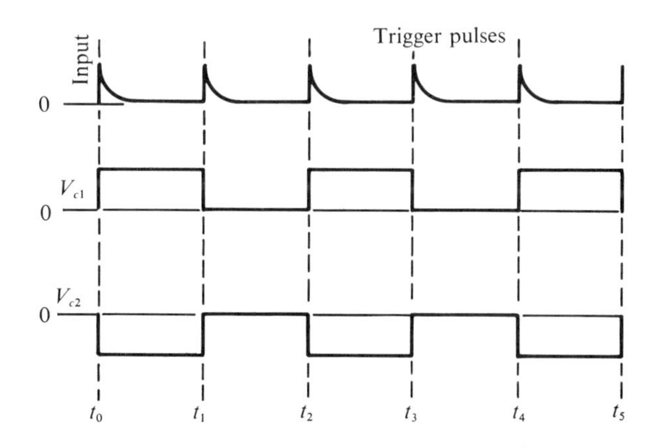

Figure 19-28. **Basic Eccles-Jordan multivibrator. (a) Circuit diagram; (b) Circuit waveforms.**

and *CR*4, negative pulses should be used with NPN transistors to prevent overloading of the ON transistor. The negative pulse is transferred from the collector of the OFF transistor to the base of the ON transistor, decreasing the base current and thereby switching the circuit into another stable state.

Base triggering requires less trigger amplitude than collector triggering; however, there is the undesirable possibility of a base triggered by spurious circuit noise. Collector triggering, on the other hand, has the desirable chracteristic of a slightly faster switching time.

We calculate the values of the collector resistors, base resistors, and speed-up capacitors for the Eccles-Jordan multivibrator as in Section 19–6 (The Transistor as a Switch). However, the value of R_B in Formula 19–10 is now the sum of R_B and R_C.

Example 6. The flip-flop circuit in Fig. 19–27 uses transistors with a beta of 50 and collector resistors with values of 500 Ω. Determine the maximum value of R_B to produce saturation.

$$R_{B1} \leq \beta R_{C2} - R_{C1}$$
$$R_{B1} \leq 50 \times 500 - 500$$
$$R_{B1} = R_{B2} \leq 24.5\ \text{k}\Omega \qquad \text{(Answer)}$$

Next, let us observe the transistor monostable (one-shot) multivibrator configuration depicted in Fig. 19–29. This circuit may be compared with its vacuum-tube counterpart in Fig. 19–19. The transistor one-shot multivibrator shown in Fig. 19–29 has a single stable condition of equilibrium. Like its tube counterpart, it requires a trigger pulse to cycle; it then returns to its initial state. In the absence of a trigger, the base-emitter circuit of $Q2$ is forward-biased and $Q2$ conducts. The base-emitter circuit of $Q1$ is reverse-biased (the drop across R_E exceeds that across R_2) and $Q1$ is cut off. A negative trigger [Fig. 19–29(b)] is applied to the base circuit of $Q1$ to start the cycle. The voltage drop across R_2 exceeds that across R_E to momentarily forward-bias the input of $Q1$; the collector current through $Q1$ causes the collector voltage of $Q1$ to drop; v_{C1} falls due to the drop across R_{C1}. C_2 discharges via R_B, the collector-emitter circuit of $Q1$ and voltage V_{CC}.

This action biases the base-emitter circuit of $Q2$ in the reverse direction to cut off $Q2$; $Q2$ is held cut off until C_2 discharges to zero. Now, $Q2$ is no longer reverse-biased and $Q2$ collector current starts to flow. The drop across R_E exceeds that across R_2 to place reverse bias on the input of $Q1$. Transistor $Q1$ is cut off. As a result v_{C1} rises, and C_2 charges via R_{C1} and the $Q2$ input to place a forward-bias current through the $Q2$ input. The collector current of $Q2$ goes to saturation, and the voltage drop across R_E holds $Q1$ cut off until the next input trigger via C_1. The input trigger is of short duration and lasts only

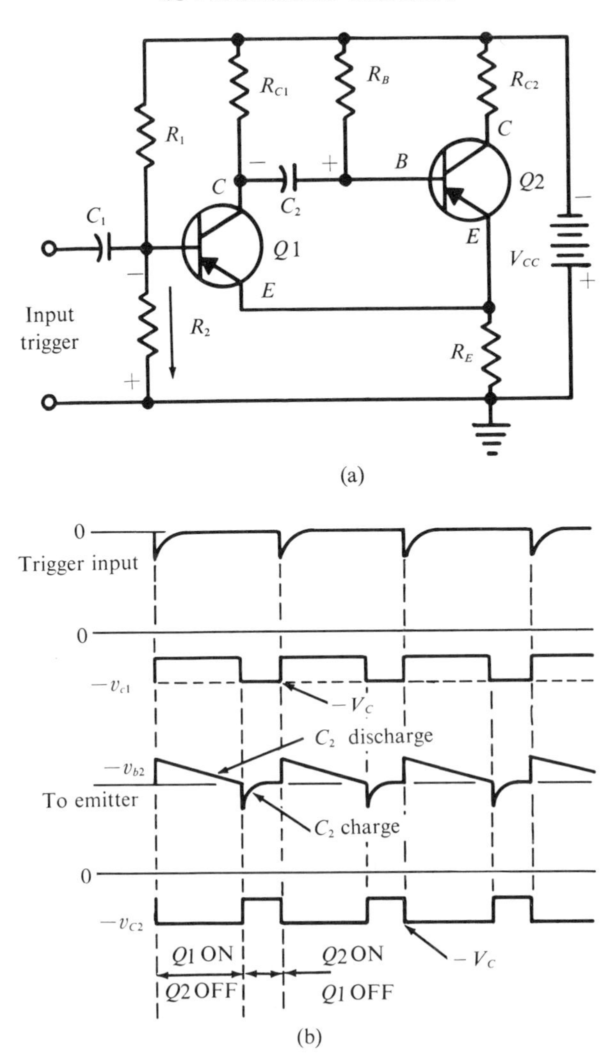

Figure 19-29. **A simple monostable multivibrator configuration and operating waveforms.**

long enough to initiate the cycle. Note that the relatively long interval when $Q2$ is off and $Q1$ is conducting is caused by the relatively large value of R_B (in the C_2 discharge path) compared with that of R_{C1} (in the C_2 charge path). The repetition rate of the pulse circuit is that of the external trigger. In summary, the one-shot action is seen to be that of temporarily turning $Q1$ ON and turning $Q2$ OFF; after C_2 discharges, $Q2$ reverts to its initial conducting state and $Q1$ returns to cutoff.

The output pulse width of the transistor one-shot multivibrator when $V_{CC} \gg V_E$, as shown in Fig. 19–29, is approximately formulated

$$T_p = 0.69\, R_B C \qquad \textbf{(19–12)}$$

Example 7. What is the output pulse width of the circuit in Fig. 19–29 when $R_B = 47\text{ k}\Omega$ and $C_2 = 1\mu$?

$$T_p = 0.69 \times 4.7 \times 10^3 \times 1 \times 10^{-6}$$
$$T_p \simeq 3.25\ \mu\text{sec} \qquad \text{(answer)}$$

Finally, let us consider the free-running (astable) transistor multivibrator shown in Fig. 19–30(a). This arrangement, which uses two NPN transistors, may be compared with its vacuum tube counterpart shown in Fig. 19–21.

To begin the circuit analysis, assume that when the supply voltage is applied, $Q2$ conducts more heavily than $Q1$. The collector voltage v_{C2} falls and approaches zero faster than the collector voltage v_{C1}, due to the large drop across R_{C2}. Capacitor C_1 discharges through the collector-emitter circuit of $Q2$ through R_{B1} and the power supply. This action places reverse bias on the base-emitter circuit of $Q1$, holding $Q1$ cut off. From these conditions, $Q2$ ON and $Q1$ OFF, we will make an analysis of the free-running multivibrator.

Capacitor C_2 in Fig. 19–30(a) has charged to a value of approximately E_{CC} through the emitter-base resistance (r_{be2}) and the collector resistor (R_{C1}). Capacitor C_1 has discharged through R_{B1} until the base circuit of $Q1$ is forward-biased. At this point, the waveforms are as shown in Fig. 19–30(d) and (e) at time t_0. The switching action starts by transistor $Q1$ going into conduction. Collector current i_{c1} through R_{C1} results in a drop of voltage v_{ce1} which is coupled through capacitor C_2 to develop a negative voltage at the base of transistor $Q2$. This results in a decreased collector current and an increased collector voltage for $Q2$, as shown in Fig. 19–30(e). As voltage v_{c2} increases, capacitor C_2 [Fig. 19–30(b)] begins to charge through the emitter-base resistance (r_{eb}) of $Q1$, and the collector resistor R_{C2} to the value of E_{BB}, as shown in Fig. 19–30(e) as time (T_2').

$$T_2' = 5C\,(r_{ie1} + R_{C2}) \qquad \textbf{(19–13)}$$

Example 8. The rise time of the waveform in Fig. 19–30(d) when $r_{eb} = 100\ \Omega$, $R_C = 1\text{ k}\Omega$, and $C = 0.001\ \mu\text{f}$ is figured on page 602.

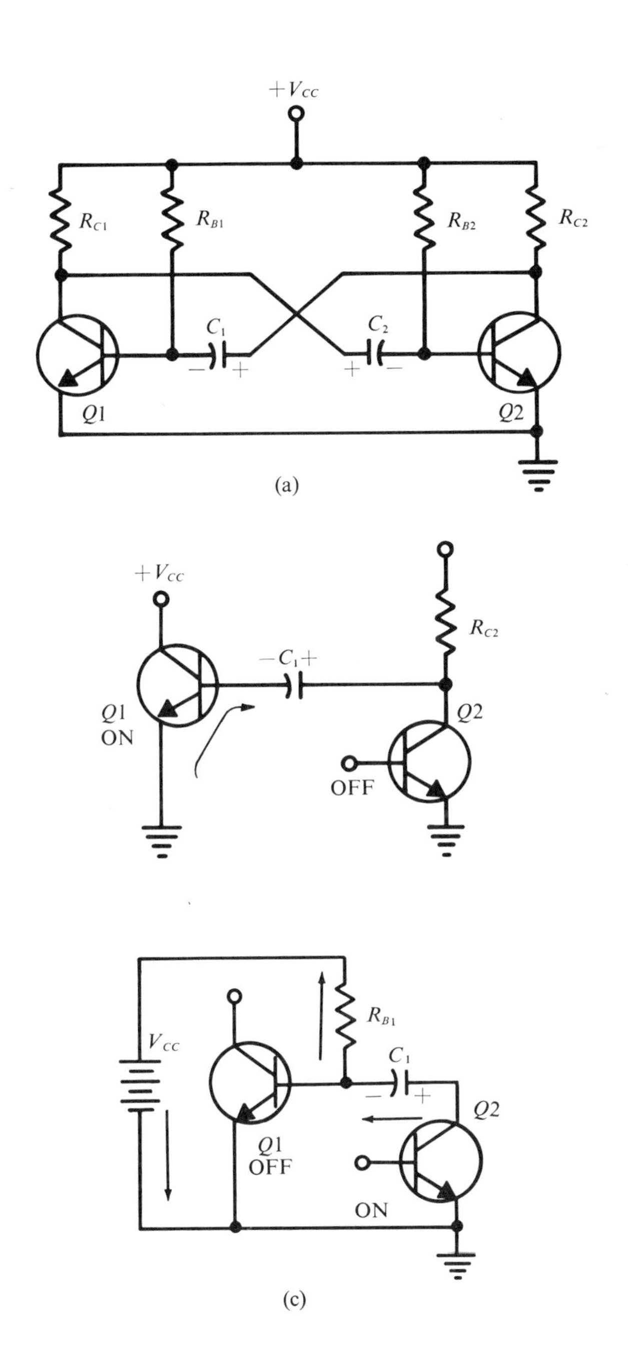

Figure 19-30. (a) Transistor astable multivibrator circuit diagram; (b) Charge path of C_1; (c) Discharge path of C_2.

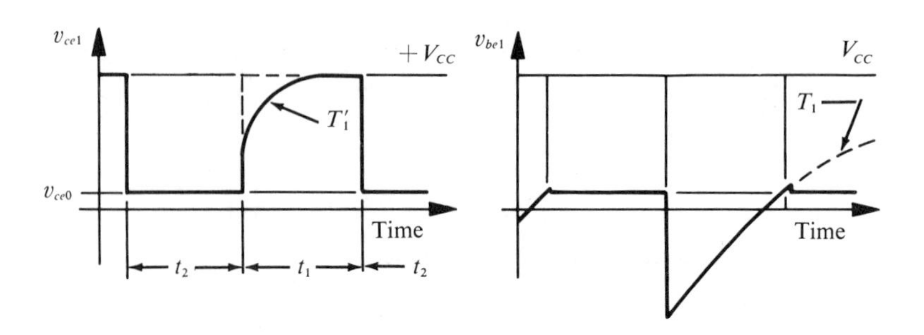

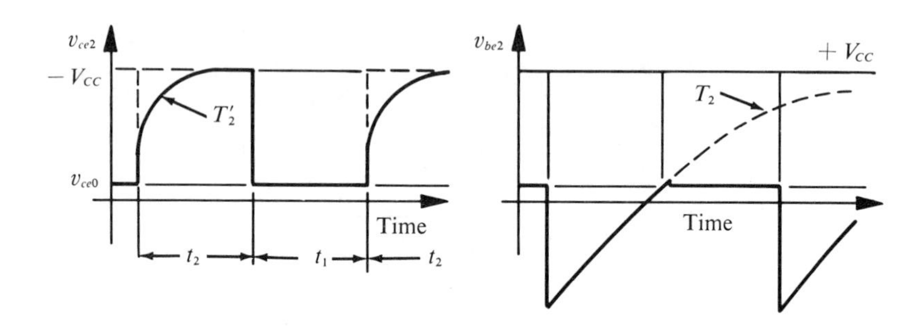

Figure 19-30. (d) Voltage waveform for Q_1; (e) Voltage waveforms for Q_2.

$$T_2' = 5 \times 1 \times 10^{-9} \times 1 \times 10^2 \times 1 \times 10^3$$
$$\simeq 500\ \mu\text{sec} \qquad \text{(answer)}$$

We should note that the switching time of the transistor is much less than the rise time of the circuit, and that the rise time measured on an oscilloscope will be only 2.2 time constants (10 per cent to 90 per cent).

At time t_0 capacitor C_1, which is charged to E_{CC}, begins to discharge through transistor $Q1$ (r_{ec}), resistor R_{B2}, and the supply (V_{CC}), as shown in Fig. 19–30(c). The discharge of the voltage across capacitor C_1 results in a negative base voltage (v_{be2}), shown in Fig. 19–30(e), which holds transistor $Q2$ in the OFF condition until its base voltage exceeds the potential barrier and becomes forward-biased. Careful inspection of the discharge circuit in Fig. 19–30(c) shows that capacitor C_2 appears to be discharging from $+V_{CC}$ and charging to a value of $-V_{CC}$. Capacitor C_2 appears to "see" a voltage change of $2\,E_{CC}$ as indicated by the dashed lines in Fig. 19–30(e). However, at approximately one-half the expected voltage change, transistor $Q2$ conducts, ending the period t_2, and capacitor C_2 quickly charges through the low resistance of the emitter-base diode.

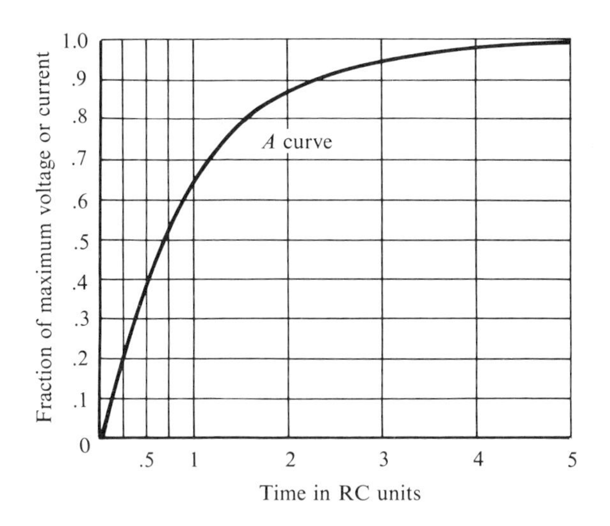

Figure 19–31. Fifty per cent of the total change shows 0.69 time constant.

Pulse time, t_2, may be determined by examination of the *RC* time-constant curve (Fig. 19–31) and the discharge circuit of C_2 [Fig. 19–30(c)]. We observe from Fig. 19–31 that a voltage of 50 per cent results in a time duration of approximately 0.69 time constants. The pulse time for t_2 is formulated

$$t_2 = T_{OFF} \simeq 0.69\,(r_{eb} + R_{B2})\,C_2 \tag{19–14}$$

When resistor $R_B \gg r_{ce}$, the formula 19–14 may be simplified.

$$t_2 \simeq 0.69\,R_B C \tag{19–15}$$

The analysis of the output voltage waveform t_1 and the rise time T'_1 follows a similar logic, and t_1 is formulated

$$t_1 = T_{OFF} = 0.69\,(r_{eb} + R_{B1})\,C_1 \tag{19–16}$$

and the rise time T'_1 is formulated

$$T'_1 \simeq 5\,(r_{eb} + R_{C1}) \tag{19–17}$$

The frequency of oscillation depends upon the OFF time and is equal to the reciprocal of the total time.

$$f = \frac{1}{t_1 + t_2} \tag{19-18}$$

When output from the multivibrator is symmetrical and the resistance $R_{BB} \ll r_{b2}$, the frequency may be calculated by the formula

$$f \simeq \frac{2.1}{R_B C} \tag{19-19}$$

Example 9. What is the rise time of the output pulse in Example 7, assuming that the switching time of the transistor is much faster than the rise time of the circuit?

$$\begin{aligned} T_{\text{rise}} &= 5TC = 5R_C C \\ &= 5 \times 1 \times 10^3 \times 10.6 \times 10^{-9} \\ &= 53\ \mu\text{sec} \qquad \text{(answer)} \end{aligned}$$

19-10. Bandwidth Considerations

At this point, we should clearly recognize that a multivibrator has a certain bandwidth or frequency response. In other words, the bandwidth of the circuitry comprising a square-wave or pulse generator is a limiting factor in rise time of the output waveform. We recall that a multivibrator is basically an *RC*-coupled amplifier with positive feedback from output to input. To calculate its frequency response (rise time), we must regard the multivibrator from the viewpoint of amplifier theory. Let us consider an *RC*-coupled vacuum-tube audio amplifier with a cutoff frequency of 20 kHz. The plate-load resistors in this amplifier will have values of about 0.1 MΩ. The frequency response is limited to 20 kHz, as we have learned, because there are stray capacitances in the amplifier circuit. The rise time of this amplifier will be approximately 17 μsec.

Next, if we reduce the plate-load resistance value to 10,000 Ω, the rise time of the amplifier will be much faster. Of course, the amplitude of the output waveform is considerably reduced. Recall that bandwidth must be paid for in terms of gain. These considerations apply in the same manner to multivibrators as to *RC*-coupled amplifiers. Hence, if we wish to obtain a fast-rise output waveform from a multivibrator, we must utilize low-value plate-load resistors. The alert student will perceive that high-frequency compensation may be applied to multivibrator circuits, just as to *RC*-coupled amplifier circuits, to obtain maximum bandwidth for a given value of plate-load resistor. Thus, peaking coils may be inserted into the plate-load circuits of a multivibrator as shown in Fig. 19-32. In turn, the rise time of the output waveform is reduced considerably.

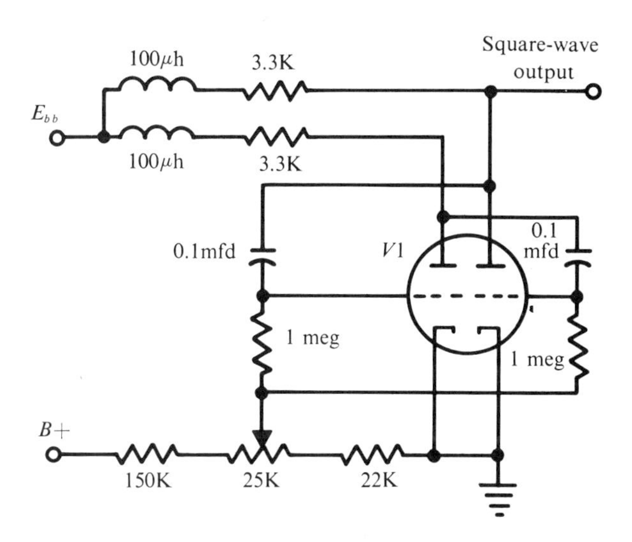

Figure 19-32. Series peaking coils provide faster rise of the square-wave output.

It is customary practice to use series peaking coils only to minimize ringing. Series peaking coils cause overshoot of the generated waveform, but ringing is negligible. If the output waveform were used

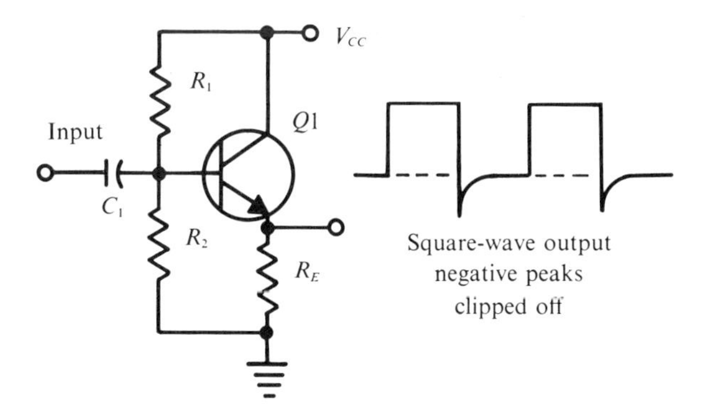

Figure 19-33. Bottom clipping is provided by an overdriven cathode follower.

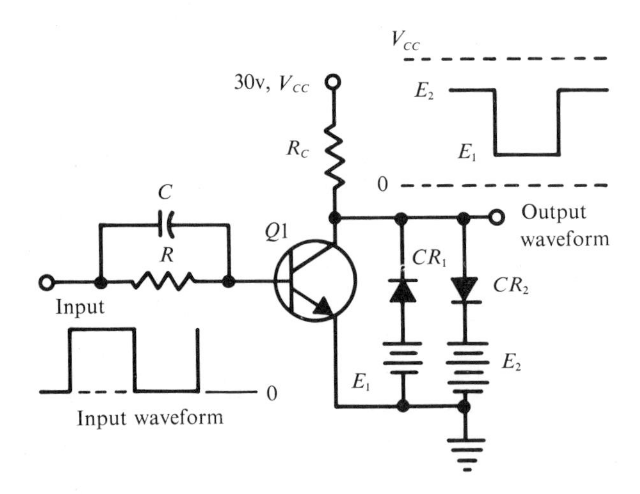

Figure 19–34. **A transistor switching circuit with limiter diodes CR1 and CR2 that improve the switching time.**

directly, this overshoot would be objectionable, because most test procedures with square waves are referenced to a source waveform that is as nearly ideal as possible. However, the square-wave output from a multivibrator is seldom used directly. Instead, a cathode follower is customarily placed between the multivibrator and the output terminals of the generator. A cathode follower provides isolation so that multivibrator operation is not disturbed by loads imposed at the generator output terminals. Note that a cathode follower, if overdriven, clips the waveform. In this way, both isolation and elimination of overshoot are provided. Figure 19–33 depicts bottom clipping of a waveform by an emitter follower. A method to improve the rise time of a transistor switch is shown in Fig. 19–34. Diodes *CR*1 and *CR*2 operate as conventional diodes and prevent the transistor from going into saturation or cutoff; the clamping levels are determined by the values of E_1 and E_2.

SUMMARY

1. A neon bulb has the advantage of visual indication over the four-layer diode as a sawtooth generator.
2. The oscillating frequency of a neon bulb sawtooth generator is determined by the values of the series resistor *R*, the shunt capacitance *C*, the ionizing potential of the bulb, and the value of *B*+ voltage.
3. A linear voltage can be produced across a capacitor by an energy source which furnishes a constant current.

4. The four-layer diode is a solid state device which conducts heavily when the switching voltage V_S is exceeded and which becomes nonconducting when the current value drops below the holding current I_H.

5. The switching current of a four-layer diode must be limited by a series resistance to prevent overheating of the diode.

6. The control grid in a thyratron tube may be used to vary the tube's firing potential. Once the tube has ionized, however, the grid loses control, and the tube may be deionized only by reduction of the plate-to-cathode voltage below a predetermined level.

7. A thyratron tube may be synchronized by an external trigger voltage which has a repetition rate greater than that of the free-running frequency of a thyratron oscillator.

8. An SCR is a solid-state device with characteristics similar to those of a thyratron.

9. Multivibrator circuits are commonly used to generate square-wave signals.

10. An Eccles-Jordan or flip-flop multivibrator is a trigger switching circuit. Its output frequency is one-half the input frequency.

11. The rise time of the output pulse of an Eccles-Jordan multivibrator is determined by the stray capacitance and the switching time of the control device.

12. When an input trigger is applied, the one-shot multivibrator gives an output pulse that may be used to generate a delay.

13. The free-running frequency of a multivibrator is determined primarily by coupling capacitors and bias resistors.

14. The switching speed of a transistor is limited by the input resistance and capacitance, the storage time, the output capacitance, and the cutoff frequency of the transistor.

15. Vacuum tubes, junction transistors, or field-effect transistors may be utilized as control devices in multivibrator circuits.

16. The switching times of the output pulse of a multivibrator may be improved by various devices such as a peaking coil, a limiter diode, or an overdriven amplifier.

Questions

1. Explain the operation of a neon bulb in a sawtooth generator.
2. What three factors may be used to increase the frequency of a sawtooth generator which uses a neon bulb or a four-layer diode as a switch?
3. What factor may be used to increase the linearity of a neon bulb or a four-layer diode sawtooth generator?
4. What is the necessary requirement for an energy source to produce a linear voltage rise across a capacitor?
5. What are the physical differences between a neon bulb and a four-layer diode?
6. Compare the operation of a neon bulb to that of a four-layer diode.
7. What are the functions of R_1 and R_2 in the four-layer diode circuit in Fig. 19-10?
8. What function does the control grid serve in a thyratron tube?
9. What are the requirements to synchronize a thyratron oscillator to an external signal?
10. Compare the operation of an SCR to that of a thyratron tube.
11. The values of R and C of a thyratron oscillator determine the free-running frequency. What are the limits on the value of R?
12. Define the terms monostable, bistable, and astable as they are used in connection with multivibrators.
13. Define the terms symmetrical and asymmetrical as they are used with multivibrators.
14. What are the characteristics of an Eccles-Jordan multivibrator?
15. Explain the switching action of the Eccles-Jordan multivibrator when an input positive-trigger pulse is applied to the input of the OFF control device.
16. The input trigger frequency to an Eccles-Jordan multivibrator is 10 kHz. What is the frequency of the output signal?
17. Specify three possible uses of the one-shot multivibrator.
18. What is the basic difference between the operation of a vacuum tube multivibrator and a transistor multivibrator?
19. When the transistor is used as a switch, what factors limit its on and off switching times?
20. What is the purpose of a speed-up capacitor in an Eccles-Jordan multivibrator?
21. What causes the storage time delay in a transistor switch, and how can this time be decreased?

22. List three ways in which the output pulse rise time of a multivibrator can be improved.

PROBLEMS

1. A neon bulb used in a sawtooth oscillator has a firing potential of 78 v and a deionizing potential of 52 v. What is the oscillator frequency when $R = 100\ \text{k}\Omega$, $C = 0.01\ \mu\text{f}$, and $B+ = 200$ v?
2. The sawtooth generator in Fig. 19-10 has the following values of circuit components: $R_1 = 10\ \text{k}\Omega$, $R_2 = 50\ \Omega$, $C = 1\ \mu\text{f}$. A type 1N24 four-layer diode has a switching voltage (V_S) of 22 v and a holding current of 20 μa. What is the frequency of output waveform where a supply voltage of 60 v is applied to the circuit?
3. The values for the circuit shown in Fig. 19-13 are $B+ = 300$ v, $R = 22\ \text{k}\Omega$, $C = 1\ \mu\text{f}$, firing potential $= 150$ v, and the deionizing potential $= 25$ v. What is the oscillating frequency?
4. What value of capacitance would be necessary to establish the oscillating frequency of the circuit in Problem 3 at 10 kHz?
5. Determine the output pulse width of $V2$ in the circuit in Fig. 19-19 when $R_k = 100\Omega$, $R_2 = 220\ \text{k}\Omega$, $E_k = 15$ v, and $C_2 = 0.01\ \mu\text{f}$. (Note that R_2 returns to the cathode of $V1$.)
6. Determine the free-running frequency of the multivibrator in Fig. 19-21 when the circuit values are $R_1 = R_2 = 100\ \text{k}\Omega$ and $C_1 = C_2 = 0.1\ \mu\text{f}$.
7. Determine the values of C_1 and C_2 in Problem 6 which would be needed to develop a frequency of 20 kHz.
8. The flip-flop circuit in Fig. 19-27 uses transistors with a beta of 100 and 220-Ω collector resistors. What are the maximum values of R_1 and R_2 that will produce saturation and what are the correct values of C_1 and C_2 when $r_i = 100\ \Omega$?
9. Determine the pulse width of the circuit in Fig. 19-29 when $R_B = 33\ \text{k}\Omega$, $V_{CC} = 10$ v, $V_E = 1$ v, and $C = 0.1\ \mu\text{f}$.
10. The circuit in Fig. 19-29 has the following circuit values: $R_{C1} = 2\ \text{k}\Omega$, $R_{C2} = 1\ \text{k}\Omega$, $R_E = 100\ \Omega$, $V_{CC} = 20$ v, and $\beta = 100$. What are the values of R_{B1}, and C_2, and R_2 to develop an output of 30 μsec duration when $V_E = 1$ v?
11. Determine the free-running frequency of the multivibrator depicted in Fig. 19-30 when $R_{B1} = R_{B2} = 47\ \text{k}\Omega$, and $C_1 = C_2 = 2.2\ \mu\text{f}$.
12. Determine the values of C_1 and R_{B1} for a symmetrical 1 kHz output of the circuit in Fig. 19-30 when $R_{C1} = R_{C2} = 470\ \Omega$, $V_{CC} = 25$ v, and the beta of each transistor is 70.
13. What is the output pulse rise time of the circuit in Problem 12 without the effect of the transistor characteristics?

CHAPTER 20

Synchronization of Complex Waveform Oscillators

20–1. Introduction

Practical applications often require that a complex waveform oscillator be free-running in the absence of a synchronizing signal and that locked operation ensue in the presence of a synchronizing signal. A familiar example is the vertical oscillator in a television receiver. Unless the vertical oscillator were free-running in the absence of a station signal, the scanning lines would collapse to a single horizontal line, and the picture-tube screen would be burned. On the other hand, when a station signal is tuned in, the picture would roll up or down on the screen unless the vertical oscillator were locked. Similarly, complex waveform oscillators in electronic test instruments, such as the square-wave generator illustrated in Fig. 20–1, may operate either as free-running oscillators or as synchronized oscillators, depending upon the particular instrument application. Accordingly, we

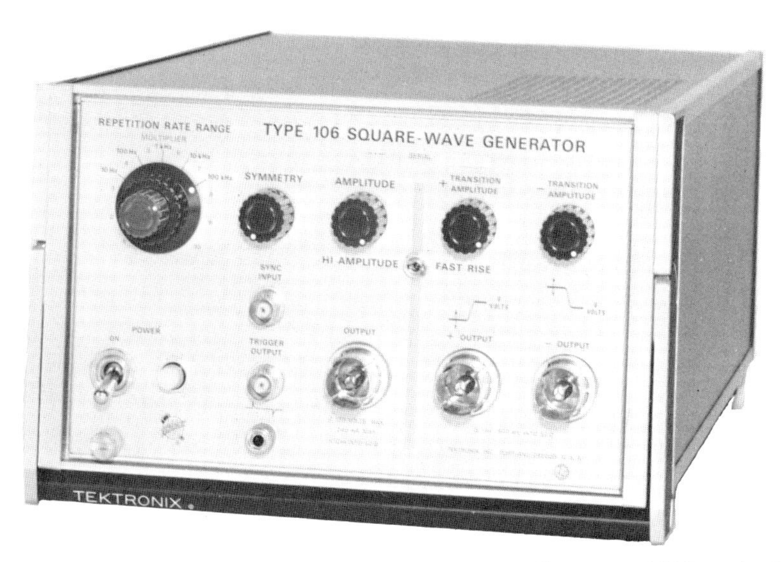

Courtesy of Tektronix, Inc.

Figure 20–1. A multivibrator-type square-wave generator.

will now analyze the circuit action of nonsinusoidal oscillators that operate in either mode.

When a free-running multivibrator generates an output frequency that is locked to (exactly the same as) a synchronizing frequency, the multivibrator is said to be driven by the synchronizing signal. Sine waves or pulses are commonly used to synchronize an otherwise free-running multivibrator, although almost any sync waveshape could be utilized. We will consider pulse-trigger synchronization as depicted in Fig. 20–2(a). Positive pulses are applied to the base circuit of one transistor through resistor R_S. The value of R_S should be large so as not to change the RC time constant of the base circuit.

Figure 20–2 depicts the effect of positive pulses on the multivibrator base-voltage waveform. A positive pulse of insufficient amplitude to drive the base above cutoff (zero volts) when applied to a nonconducting transistor at instant A does not cause switching action. Its only effect is to reduce the negative bias slightly, as shown at A'. A positive pulse applied to a transistor that is already conducting (points B or C) serves only to increase the base voltage momentarily; it has no effect on multivibrator action. With the exceptions of the variations in the v_b waveform at times A', B', and C', the multivibrator is essentially free-running. If applied at instant D, however, the positive

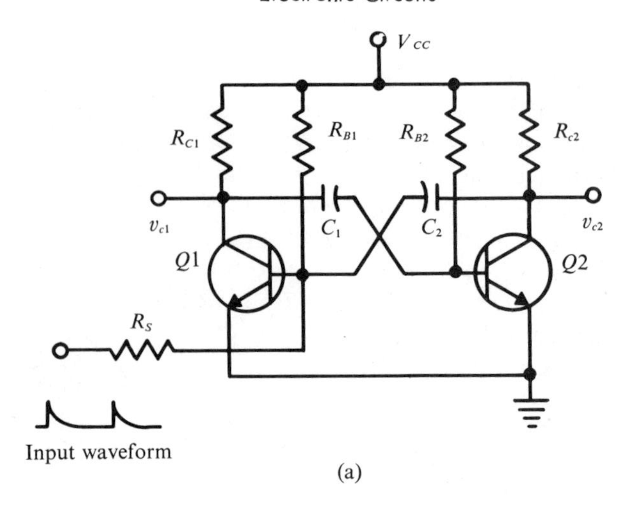

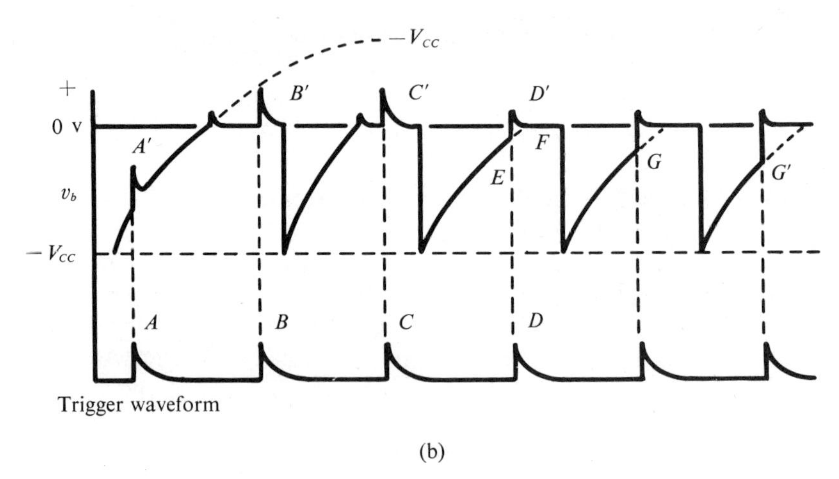

Figure 20–2. **(a) Transistor synchronized multivibrator; (b) Waveforms on the base of a synchronized free-running multivibrator driven by positive pulses.**

trigger pulse has sufficient amplitude to overcome the negative voltage on the base, and it drives the base above zero volts. The period of the multivibrator is thereby shortened by an amount *EF*.

For proper synchronization, the natural period of the free-running multivibrator must be greater than the time interval between pulses. In these circumstances, the positive trigger pulses cause the switching action to occur earlier in the cycle than it would in the free-running state. Thus, the transistor conducts at *E*, *G*, and *G′* in Fig. 20–2, whereas it would have conducted later in each instance had the pulses not been applied. In this way, the frequency of the multivibrator is forced to become the same as the repetition rate of the trigger pulses. The multivibrator may be synchronized to a *submultiple* of the trigger frequency if both frequencies are such that every second, third, fourth, etc. sync pulse occurs at the right time to drive the base voltage of the nonconducting transistor above cutoff. Furthermore, a free-running multivibrator may be synchronized by positive or negative pulses applied to either the collector or base of a transistor circuit or the grid or plate of a vacuum-tube circuit.

The sine-wave sync signal for the vacuum tube in Fig. 20-3 may

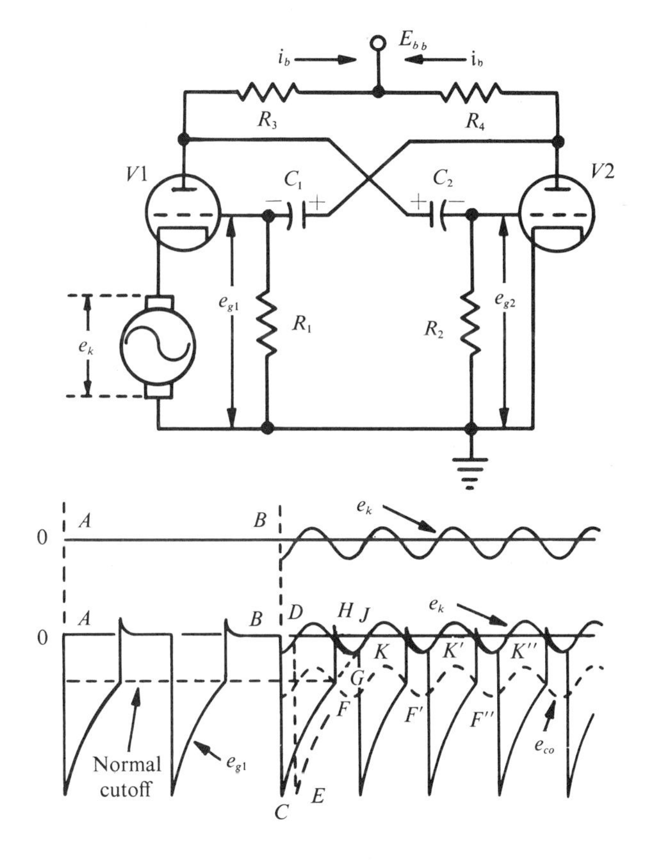

Figure 20-3. Multivibrator circuit action with a sync signal applied to the cathode of a tube.

be injected at the cathode or between the grid and cathode; it is instructive to analyze cathode injection at this time. Figure 20–3 depicts a multivibrator configuration with the synchronizing sine-wave voltage e_k applied to the cathode. A set of voltage curves is also shown to clarify the analysis. The actual grid-to-cathode voltage of *V*1, which is the voltage controlling the flow of plate current, is the algebraic sum of e_{g1} and e_k. Note that the source of synchronizing voltage should have a low internal impedance, so that the flow of i_{b1} through this source will not cause an alteration in waveshape.

If the multivibrator is properly balanced and running freely, the voltage curve of e_{g1} is that shown between time *A* and time *B* in Fig. 20–3. Because there is no synchronizing voltage on the cathode, its voltage remains constant at ground potential. At time *B*, the synchronizing voltage is applied, and e_k begins to vary sinusoidally. Although e_{g1} is not affected by this variation, the grid-to-cathode potential now contains this sinusoidal voltage component, so that the effective cutoff voltage of the tube varies sinusoidally about its normal cutoff value and in phase with the synchronizing voltage on the cathode. The cathode voltage curve, e_k, and the effective cutoff voltage curve, e_{co}, are shown with the e_{g1} curve in order to explain the synchronizing action.

At the instant when *V*1 conducts, the e_{g1} curve crosses the e_{co} curve. At instant *B*, e_k starts to rise in a positive direction, and i_{p1} is decreased. The positive-going voltage resulting at the plate of *V*1 initiates the switching action via C_2 to the grid of *V*2, and from the plate of *V*2 via C_1 back to the grid of *V*1. Then, e_{g1} drops along line *BC* instead of along *DE*, as it would in the free-running state, and *V*1 is quickly cut off. C_1 discharges along curve *CFG*; but since the e_{g1} curve intersects the e_{co} curve at *F*, the switching action by which *V*1 is made conducting and *V*2 is cut off takes place at *F*, instead of *G* as it would in the free-running state.

This switching action drives the grid of *V*1 positive; but the resulting grid current quickly charges C_1, and the grid returns to cathode potential. Grid voltage e_{g1} does not follow curve *HJ*, as it would in the free-running state. Instead, the grid draws current because the synchronizing voltage causes the cathode to be negative with respect to ground at this time, and e_{g1} follows the cathode voltage along curve *HK*. When the cathode voltage begins to rise in a positive direction, the plate current of *V*1 starts to decrease. At instant *K*, the rise in voltage at the plate of *V*1, resulting from the decrease in i_{b1}, is large enough to drive *V*2 into conduction, and the tubes are rapidly switched.

Note that this action of the sine-wave sync voltage forces the period of one cycle to be shorter (frequency higher) than it would be in absence of the sync signal. Switching in one direction occurs at instants F, F', F'', etc.; switching in the other direction occurs at instants K, K', K'', etc. With the exception of the short transition time, the period of the multivibrator is equal to the period of the sync voltage. Thus, the multivibrator circuit action is locked or synchronized with the repetition rate of the sync signal. The sync voltage can make the multivibrator operate above or below its free-running frequency. If, however, an attempt is made to pull the multivibrator frequency to a value that is too high, it will synchronize at a frequency that is one-half (or some other division) of the sync frequency, and *frequency division* is obtained. Multivibrators are utilized as frequency dividers in numerous practical applications.

If the square-wave output from a multivibrator is passed through an integrating circuit that has suitable time constants, the integrator provides an essentially sawtooth-shape waveform. This arrangement is used as a sweep-frequency (horizontal-deflection) generator for cathode-ray tube applications, as in oscilloscopes. Figure 20–4 depicts the plan

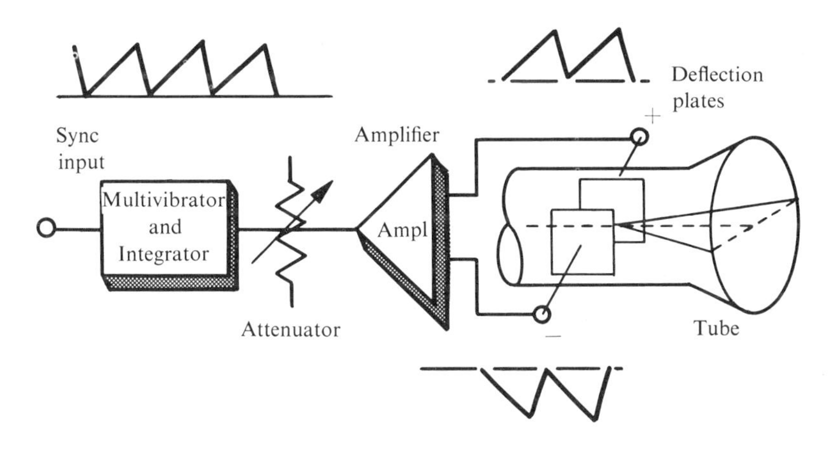

Figure 20–4. Plan of a horizontal-deflection system for an oscilloscope.

of a horizontal-deflection system for an oscilloscope. It is evident that a multivibrator can also be used to introduce a time delay between the operation of two circuits by using the leading edge of the square wave to trigger one circuit, and the trailing edge to trigger another circuit. In turn, the time delay can be controlled by varying the RC time constant of the multivibrator circuits which control the pulse width. Trigger pulses are obtained from a square wave by a differentiating circuit, as depicted in Fig. 20–5.

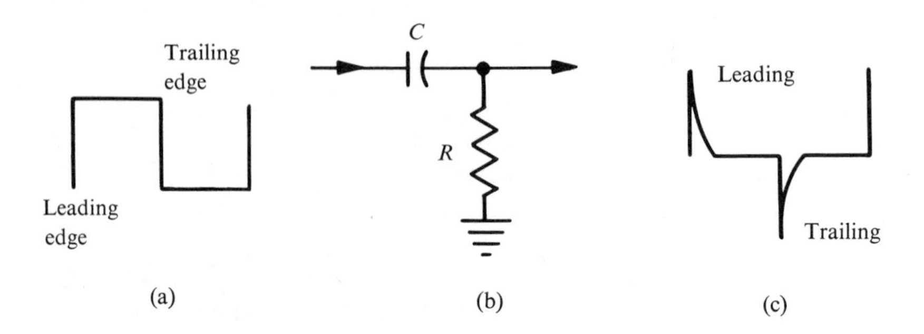

Figure 20–5. **Trigger pulses are formed from a square wave by a differentiating circuit.**

20–2. Blocking Oscillators

Blocking oscillators that are either free-running or locked if a sync signal is applied are widely utilized in electronics technology. For example, the vertical deflection oscillator in a television receiver is often of this type. A blocking oscillator periodically cuts itself off after one or more cycles of operation as a result of a gradual charge accumulation on the input capacitor which makes the input terminal reverse-biased into cutoff.

Figure 20–6 shows a simple transistor-blocking oscillator. This circuit is caused to block or be cut off periodically by means of capacitor C_1 and resistor R_1 connected across the base-emitter circuit. The operation of the transistor-blocking oscillator and the vacuum-tube oscillator is quite similar. We shall follow the operation of the transistor from the moment the supply voltage E_{CC} is switched on.

At the first instant, the transistor is cut off because the base is at ground potential and only a small collector-leakage current flows. This small collector current induces a voltage into the secondary winding, forward-biasing the base-to-emitter junction, and starts base current flow. (Resistor R_B may be added if the collector-leakage current is too small to start this action.) This small base current flow causes an increase in collector current which, in turn, induces an increasing voltage to the base circuit to further increase base current. This accumulative action (positive feedback) continues until the transistor reaches the saturation region. During this time, capacitor C_1 has charged

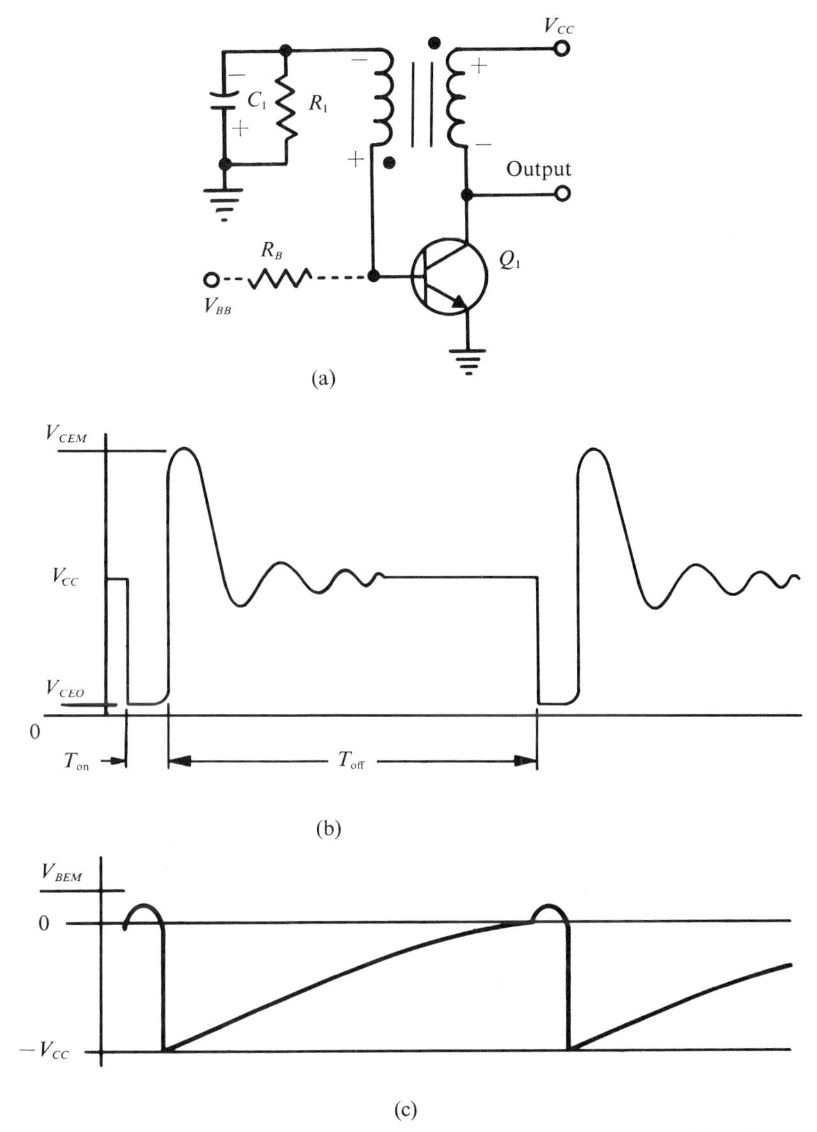

Figure 20–6. **Transistor-blocking oscillator. (a) Circuit diagram; (b) Collector voltage waveform; (c) Base voltage waveform.**

through the low base-to-emitter impedance to the voltage induced across the secondary of the transformer.

As the collector current stabilizes at a value I_{sat}, the voltage induced in the secondary winding drops to zero. (Only a changing current, and thereby changing magnetic flux lines, induce a voltage.) Capacitor C_1 starts to discharge through R_1 biasing the base negative, resulting in a decrease in collector current and an increase in collector voltage. Almost instantaneously this decreasing collector current in the secondary winding drives the base further into cutoff. The positive swing of the collector voltage in Fig. 20–6 is produced by the collapsing magnetic field in the primary winding of the transformer.

One cycle of operation of the blocking oscillator consists of two periods. The first period, (T_{on}), is determined by the pulse transformer, the transistor switching time, and the circuit capacitance. This period is rather difficult to calculate and is usually set by the pulse transformer, but it can be varied slightly by a careful analysis of the formula

$$\text{Pulse width} \propto L/R \qquad \textbf{(20–1)}$$

$$PW = KL/R \qquad \textbf{(20–2)}$$

where L is transformer primary inductance, R is effective series resistance, and K is a constant for the circuit.

In other words, any action which will change the time for the transistor to reach saturation will have a slight effect on pulse width. The duration of the second period, (T_{off}), is determined by the RC time constant of the base circuit. The repetition rate of the oscillator depends chiefly upon the time constant of R_1C_1.

The blocking oscillator in Fig. 20–6 has the transistor connected in the common-emitter configuration. A common-base configuration, such as shown in Fig. 20–7, can be used, but in this case the polarity of the feedback voltage is not reversed through the transformer. Bias battery E_{EE} has been added to provide base bias and to assure enough base current to start operation. The resistor R_S has been added in the base to provide for synchronized operation. The requirements for synchronization of the blocking oscillator are the same as for other pulse circuits.

The reverse voltage at the collector (V_{CEM}) in Fig. 20–6(b) approaches 2 E_{CC} and can cause breakdown of the transistor. This peak pulse can be limited by the action of diode $CR1$ in Fig. 20–7(b). When the transistor is conducting (T_{on}), diode $CR1$ is reverse-biased to offer a very high impedance across the primary windings. When a high positive voltage is induced by the energy stored in the magnetic field of the transformer, $CR1$ conducts and offers a low impedance to this induced voltage. This "damping" action eliminates the overshoot and ringing from the output pulse waveform.

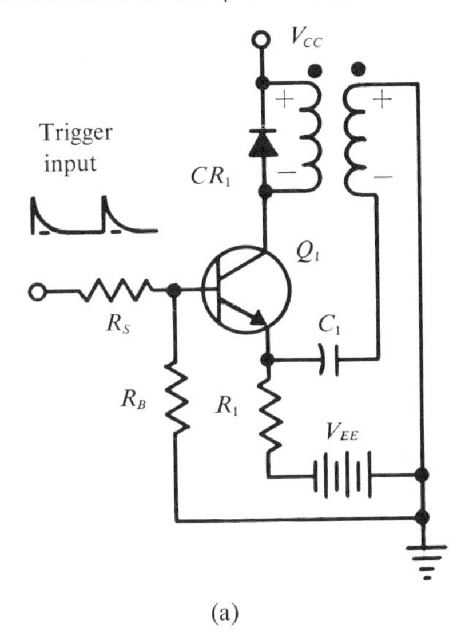

(a)

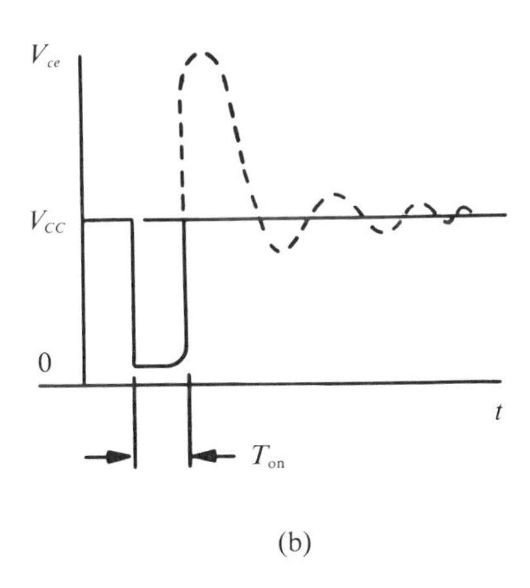

(b)

Figure 20-7. (a) Blocking oscillator with the transistor in the CB configuration; reversing the emitter bias results in a monostable circuit; (b) Output waveform as a result of diode CR1.

The circuits in Fig. 20–8 include a tuned-grid branch that has a high *LC* ratio. Typically, the capacitance is solely the distributed capacitance of the winding and the interelectrode capacitance of the triode. Coupling is tight, and there are many more turns on the secondary than on the primary, so that the grid drive is relatively large compared with that for conventional oscillator operation. Grid-

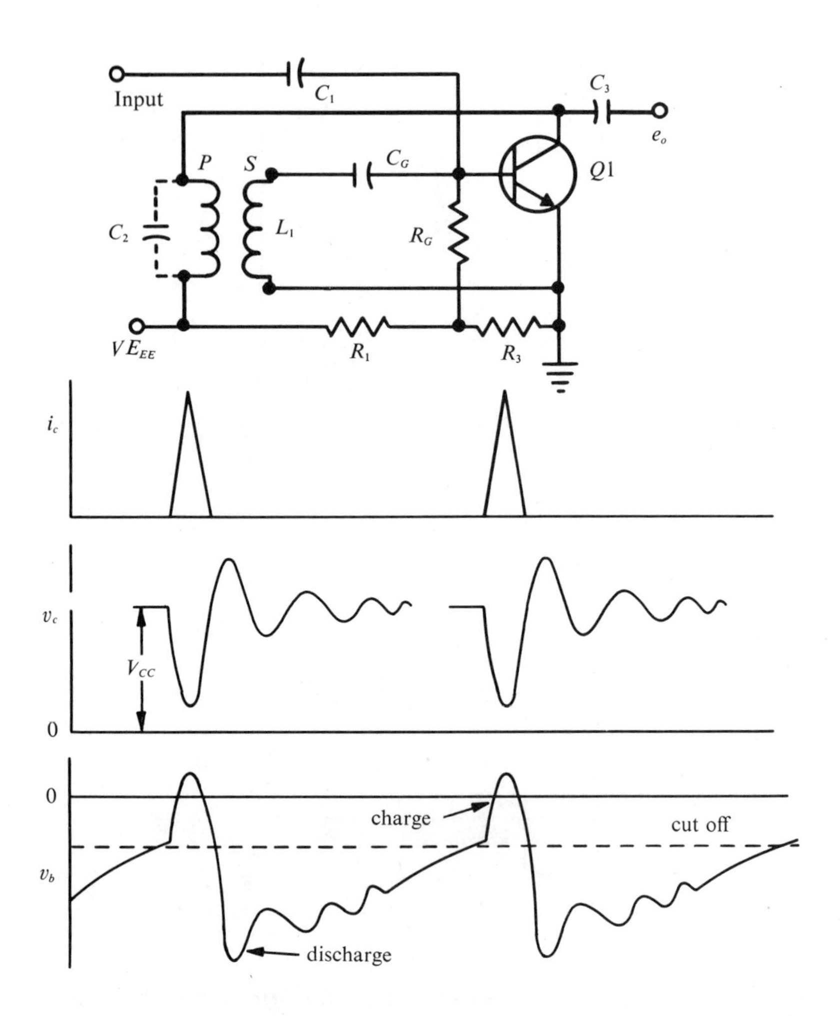

Figure 20–8. A single-swing blocking oscillator and operating waveforms.

leak resistor R_g has a large value, and the R_gC_g time constant is long compared with the period of the resonant tank circuit.

In the absence of bias, oscillation starts and quickly builds up in the circuit in Fig. 20–8. The circuit action is similar to that of a Class-*C* amplifier, in which plate current flows during the time that the plate voltage is decreasing and the grid voltage is swinging positive. Oscillations start to build up rapidly until the voltage drop across the plate coil is almost equal to the d-c supply voltage. The large grid-driving voltage causes C_G to quickly charge up to this peak voltage; during the portion when the a-c voltage across the tank coils is zero, the voltage across C_G biases the triode well beyond cutoff. Because of energy losses in the coils, the positive peak in the next cycle of operation will not be sufficient to raise the grid above cutoff, and succeeding oscillations will quickly die out as indicated by the damped oscillations in the diagram.

When the grid swings positive with respect to the cathode (positive feedback), grid current flows and charges the grid capacitor with the grid terminal negative. On the positive portion of the cycle, the grid capacitor discharges through the grid-leak resistor; if the *RC* time constant is relatively long in relation to the period of the cycle, this action will bias the tube to cutoff at or before completion of one cycle (single-swing type), or after several cycles (self-pulsing type). When the grid-capacitor discharge has progressed sufficiently to raise the bias above cutoff, the tube again conducts and the action repeats itself periodically. Thus, the circuit becomes an intermittent oscillator.

The period of the negative-going output pulse is approximately equal to one-half of the period of the resonant tank coil. We recognize that the cutoff period varies with the $R_G C_G$ time constant. As C_G gradually discharges through R_G, the grid voltage is positive-going, and at cutoff the plate current starts to flow again; the cycle of operation then repeats. In practical applications, a blocking oscillator is usually synchronized. The synchronized vacuum-tube blocking-oscillator configuration shown in Fig. 20–9 is typical. This arrangement can be used as a timing oscillator to drive auxiliary equipment, such as a sawtooth-shaper and output amplifier. Normally, in the absence of an input sync signal, the circuit generates its own repetition rate, or is free-running as indicated by the waveshapes marked "e_g, no sync." The natural frequency of the circuit is 300 pulses per second (pps) in this example.

A sync pulse can be introduced at the plate, without any circuit alterations that will cause the pulse-repetition frequency (prf) to shift as high as 1,000 pps. In this example, the sync pulses occur at 700 pps. Their amplitude is such that although one pulse makes the circuit operate, the next pulse does not bring the grid above cutoff; however, the third pulse again triggers the tube into conduction. Since the oscillator is synchronized on alternate pulses, its prf becomes equal to 700/2, or 350 pps. We will perceive that the amplitude of the synchronizing pulse has great significance on the operation of this circuit. If the amplitude of the sync pulse is increased 50 per cent, the first

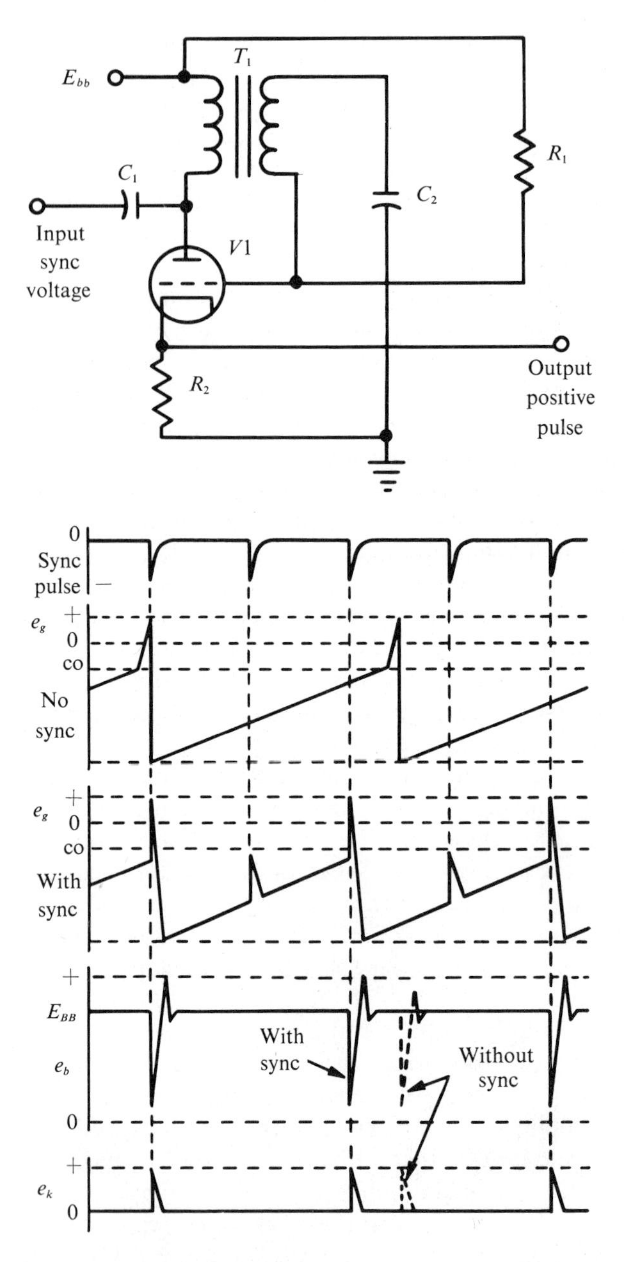

Figure 20-9. A synchronized vacuum-tube blocking-oscillator circuit and operating waveforms.

pulse that occurs after the circuit begins operation will bring the grid above cutoff, and the circuit will be synchronized on every pulse.

In each of the foregoing analyses, it was assumed that the natural period of the circuit was longer than the period of the trigger pulses. If the period of the blocking oscillator is shorter than that of the sync pulse, long and short output cycles will be generated because the circuit will be triggered by some pulses but will conduct at its natural period in between. Maximum frequency stability is achieved by connecting the grid-leak resistor to the d-c supply voltage. This stabilization method is also used in *RC* multivibrator circuits. Injection of a substantial positive d-c voltage causes the grid-decay waveform to be more linear and to make a greater angle of intersection with the cutoff axis. In turn, random noise voltages in the grid circuit have less tendency to produce erratic triggering. The variation in output frequency due to noise voltages is called *jitter*.

Note that the output from the configuration in Fig. 20-9 is taken from a small (680-Ω) resistor in the cathode circuit. This resistor has a negligible effect on circuit operation but develops a useful output waveform from the plate current flow. In turn, the output impedance is low, and the output waveform has a comparatively small amplitude. Operation of the blocking oscillator is thereby made practically independent of output current demand. A blocking oscillator is rich in harmonic content. Hence, the output from a blocking oscillator may be used as a generalized type of test signal for preliminary troubleshooting. In this application, the blocking oscillator serves as a primitive form of signal generator. A typical harmonic generator is illustrated in Fig. 20-10.

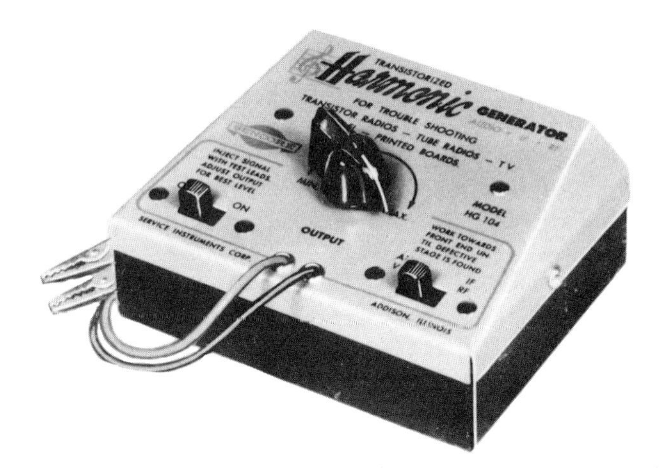

Courtesy of Sencore

Figure 20-10. A harmonic generator that consists of a free-running blocking oscillator.

20–3. Synchronized Blocking-Oscillator Sawtooth Generator

In various practical applications, such as television-receiver vertical-deflection oscillators, generation of sawtooth waveforms is required. It was previously noted that an integrating circuit with suitable time constants can be used to generate a sawtooth waveform from the output of a nonsinusoidal oscillator. Let us consider the modification of a synchronized blocking oscillator for this purpose. The essential circuit elaboration consists of a capacitor C [Fig. 20–11 (a)] connected from plate to ground. Observe that in the absence of shunt capacitor C, the plate waveform e_b would appear as depicted in Fig. 20–11(b). This is basically a pulse waveform; the plate voltage e_b suddenly falls almost to zero each time that the grid is driven positive. Then, e_b suddenly rises to the $B+$ value because of the cutoff self bias that develops from grid-current flow.

Next, let us consider the plate-voltage waveform when shunt capacitor C is connected into the circuit [Fig. 20–11(c)]. After the grid is driven positive, the plate voltage cannot suddenly rise to the $B+$ value, because C must charge gradually through the series resistors in the plate circuit. Hence, the plate voltage E_b rises slowly, as seen in Fig. 20–11(c). When a trigger pulse arrives, the blocking oscillator suddenly conducts, and C discharges rapidly through the low plate resistance of the triode. Again, C must charge slowly through the series resistors in the plate circuit. We observe that a sawtooth waveform is formed because C is associated with two different time constants. The charging time constant is determined by the two resistors in the plate circuit, but the discharge time constant is determined by the plate resistance of the triode during the time that the grid is driven positive.

Note that the 1-MΩ potentiometer in the grid circuit in Fig. 20–11 (a) determines the free-running frequency of the sawtooth generator. The potentiometer is adjusted to make the free-running frequency slightly lower than the repetition rate of the sync pulses. Thereby, the oscillator is triggered by each sync pulse. We perceive that the amplitude of e_b is determined by the setting of the 2.5-MΩ potentiometer in the plate circuit. If the total plate-load resistance has a high value, C can charge up to only a small voltage before it is discharged. On the other hand, if the total plate-load resistance has a low value, C will charge almost to the $B+$ value before it is discharged. The amplitude of the output waveform can be adjusted as required.

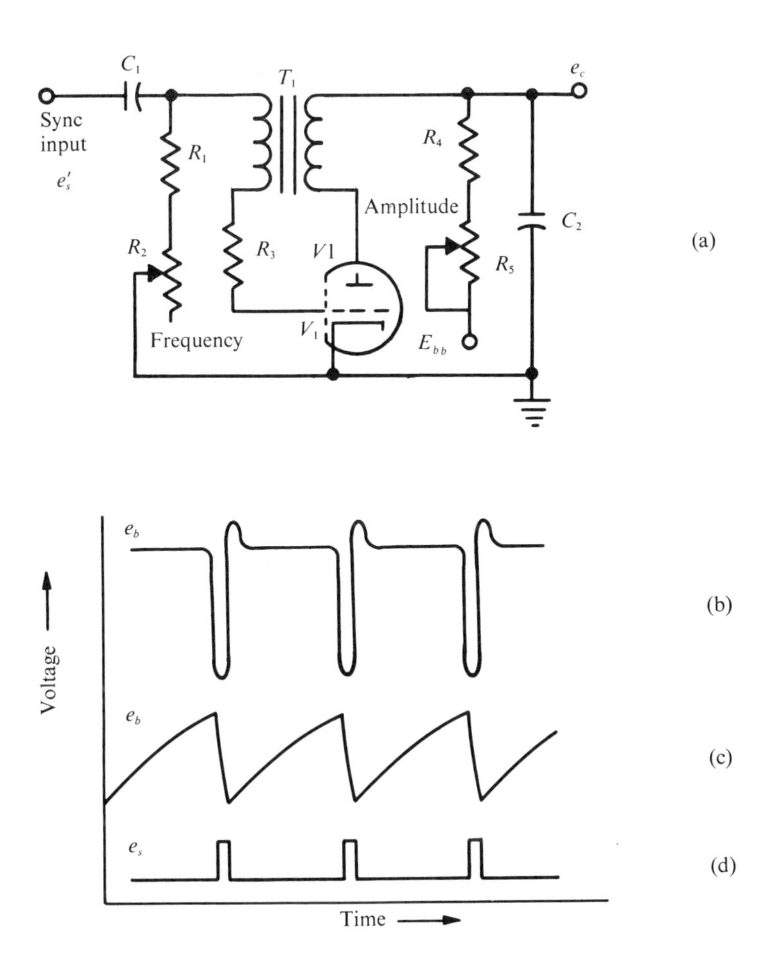

Figure 20–11. **(a) Blocking oscillator modified for sawtooth generation; (b) Output waveforms without capacitor C_2 connected; (c) Output waveforms with capacitor C_2 connected; (d) Input trigger.**

Of course, the charge and discharge intervals are not linear but are more or less exponential. To linearize the sawtooth waveform within practical limits, it is usual practice to use a high value of $B+$ voltage. In turn, the output waveform consists of only a small portion of the total available exponential curve, as previously explained.

Although a good approximation to a sawtooth waveform can be obtained from the configuration depicted in Fig. 20–11, practical applications often require greater linearity than can be obtained simply by using a high value of $B+$ voltage. Hence, a nonlinear resistance is employed to obtain the required linearity. Recall that a triode has a nonlinear transfer characteristic. The nonlinear plate resistance of a triode amplifier stage is often used to linearize a sawtooth waveform, as shown in Fig. 20-12. Observe that the input waveform has a cur-

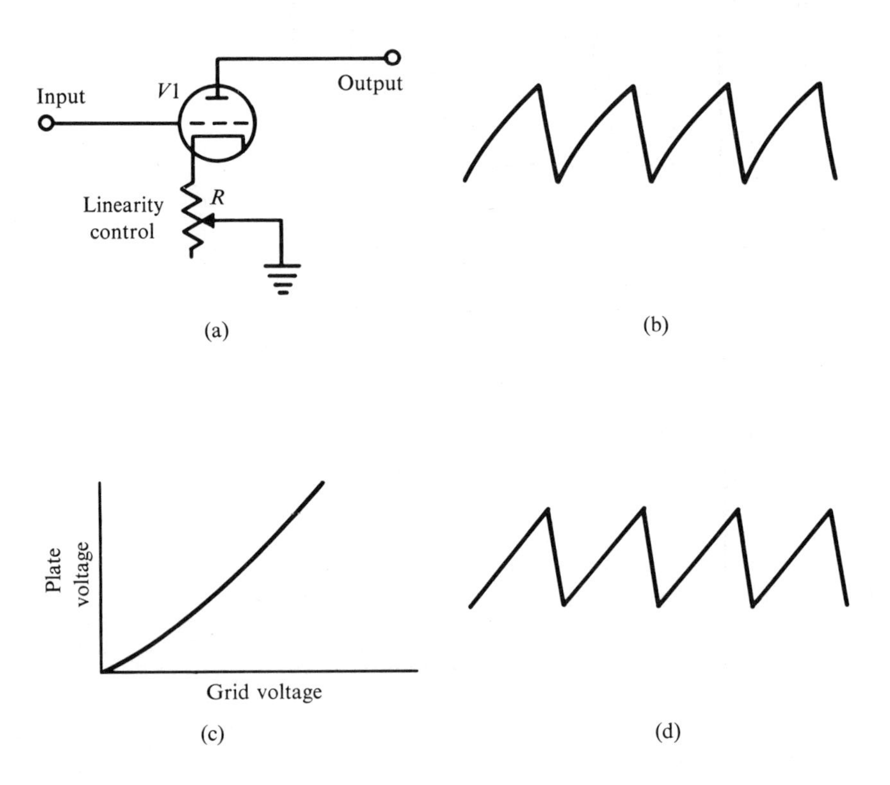

Figure 20-12. (a) Amplifier stage with adjustable cathode bias; (b) Input waveform; (c) Tube transfer characteristic; (d) Linearized output waveform.

vature that is opposite to that of the triode transfer characteristic. We can adjust the amount of curvature in the tube characteristic by varying the cathode bias of the tube. Thus, when the linearity control in Fig. 20-12 is suitably adjusted, the output sawtooth waveform is effectively linearized.

We will now observe operation of pulse circuits in a system by the analysis of the vertical-deflection unit of a television receiver, depicted in Fig. 20–13. The first transistor operates as a free-running syn-

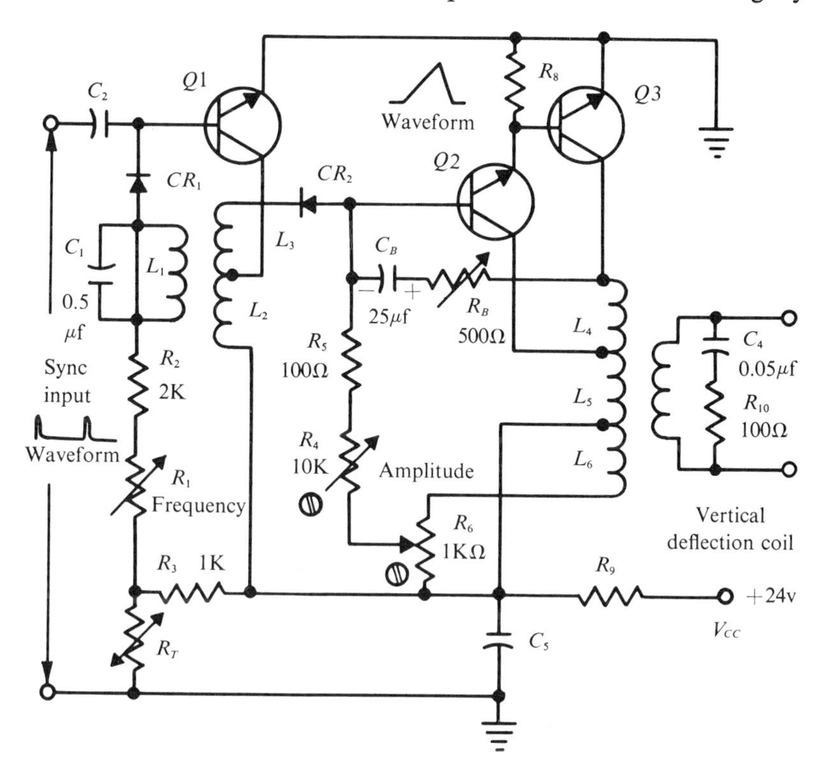

Figure 20–13. Vertical deflection circuit of a television receiver.

chronized blocking oscillator. The value of R and C in the base circuit are selected for a free-running frequency of approximately 50 Hz, and the pulse transformer is selected for a pulse width of about 1.2 msec.

The oscillator is synchronized by positive pulses from the differentiated video signal coupled through capacitor C_2. The free-running frequency of the oscillator is adjusted by R_1 in the base circuit. Resistor R_T is a thermistor to prevent oscillator drift with a temperature change. Capacitor C_1 is a filter to keep the horizontal frequency (15.75 kHz) out of the vertical oscillator. Capacitor C_5 and R_9 form a decoupling network to keep the vertical signal out of the power supply. Finally, diode $CR1$ eliminates ringing and overshoot of the output waveform of $Q1$.

The driver and output stage ($Q2$ and $Q3$) generates a sawtooth of current to drive the vertical deflection coil of the picture tube. In the absence of a pulse from $Q1$, capacitor C_3 charges through the base of $Q2$, R_7, and the primary to E_{CC}. The exponential rise of current in C_3 produces an almost linear change of current across R_8, causing a linear current in the collector of $Q3$ and finally a linear output to the vertical deflection coils. This rise of current is suddenly stopped, and a pulse at the collector of $Q1$ which clamps the left end of $CR2$ to almost zero volts to discharge C_B.

Capacitor C_B, the 25 μf capacitor, charges through R_7. The value of R_B is limited by the maximum collector current, as we have previously seen.

$$R_{B\,\max} \simeq \frac{\beta\, E_{CC}}{I_{C\,\max}} \tag{20–3}$$

The smaller the value of R_B the better the linearity; however, the size of C_B also depends upon the value of R_B, as C_B must appear very large to maintain a constant current during the charge cycle. A large current gain beta is required if we wish R_B and C_B to be normal size components.

Example 1. Suppose a maximum collector current of 3 amp is required with $\beta = 30$ and $E_{CC} = 24$ v. The maximum value of R_B would be

$$R_{B\,\max} \simeq \frac{30 \times 24}{3} \simeq 240\ \Omega \qquad \text{(answer)}$$

Two advantages are gained by direct connection of the transistor and connection of the capacitor to the collector of $Q3$ instead of E_{CC}. First, the capacitor may be smaller by a factor of $(1 + \beta)$ where β is $\beta_1 + \beta_2$. Second, the changing voltage at the collector of $Q3$ increases the linearity of the current charging capacitor C_B.

20–4. Overshoot and Corner Rounding

We have seen that although a multivibrator is a square-wave (or pulse) generator, it does not generate ideal waveforms. One departure from the ideal has been discussed—the generated waveform does not rise in zero time. Thus, the idealized square waveforms depicted in Fig. 20–14(b) have zero rise time, but the actual waveforms shown at (c) have finite rise time. Furthermore, the waveforms shown at (c) have rounded corners and overshoots. We will now analyze the circuit

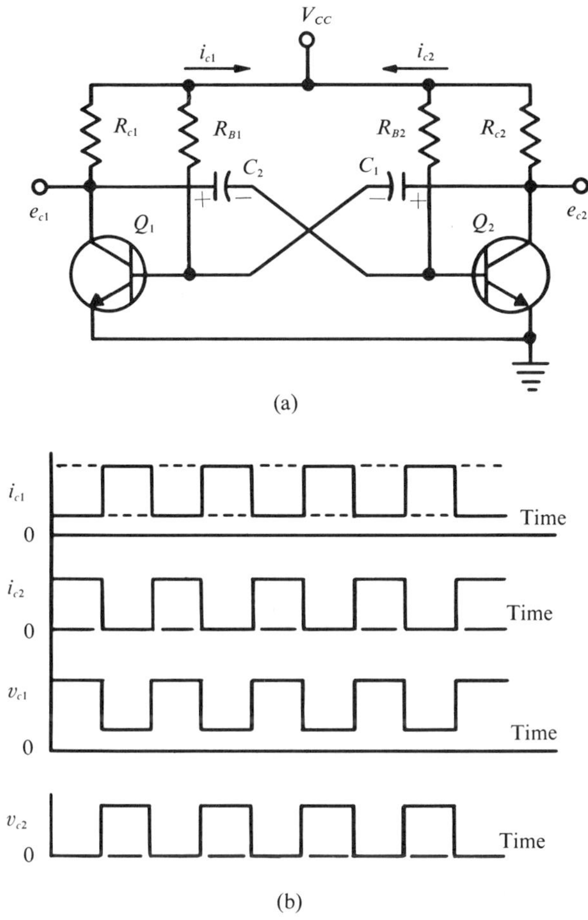

Figure 20–14. (a) Basic multivibrator circuit; (b) Idealized operating waveforms.

action to determine the source of these distortions. With reference to Fig. 20–14(a), a tabulation is given of sequential circuit action in the first switching mode. Observe the following relation.

First, current flow increases through $Q1$ in Fig. 20–14(a). This current flows through R_{C1}, and the resulting voltage drop across R_{C1} lowers the collector voltage suddenly. In other words, $Q1$ has been cut off, and $Q2$ has been conducting. At the instant that the first switching action occurs, $Q1$ is driven into conduction and $Q2$ is driven to cutoff. Accordingly, as the current flow through $Q1$ suddenly increases, the current flow through $Q2$ decreases simultaneously. This is just another way of saying that the base of $Q1$ is being driven positive, and the

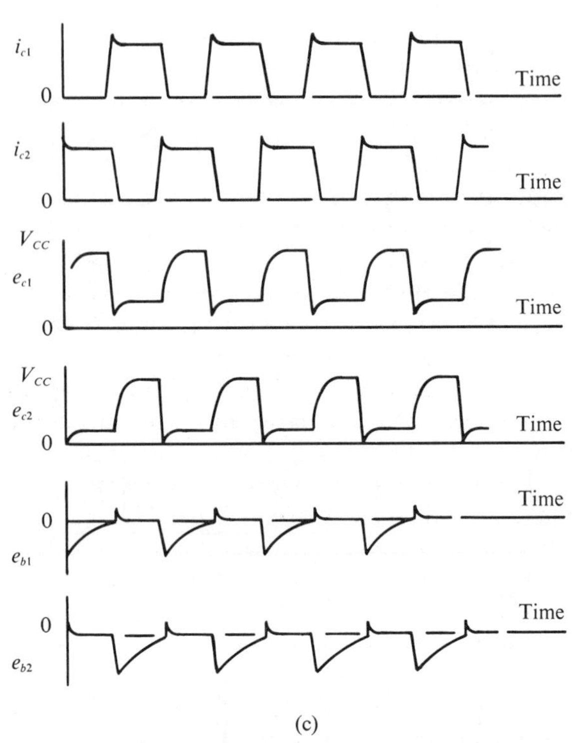

Figure 20–14. (c) Actual waveforms.

base of $Q2$ is being driven negative. Since the base of $Q1$ is driven positive, the base draws current; this current charges C_1 through the base-emitter resistance of $Q1$. This time constant is quite short, and C_1 becomes charged in a small fraction of the operating period. But while C_1 is being charged, a pip, or overshoot, is generated on the base voltage waveform e_{b1} [Fig. 20–14(c)].

During the brief charging interval of C_1, the pip on the e_{b1} waveform corresponds to *two* paths of current flow for $Q1$. In other words, $Q1$ supplies current during this brief interval to both R_{C1} and to C_1. Therefore, the total current flow through $Q1$ during this brief interval is greater than normal; a pip is generated on the i_{c1} waveform [Fig. 20–14(c)]. After C_1 is charged, $Q1$ then supplies current to R_{C1} only. We observe in Fig. 20–14(c) that the e_{c1} voltage waveform also develops a pip. This pip develops because the collector resistance of $Q1$ becomes abnormally low during the brief interval that excess base current flows. Stated otherwise, the collector resistance of $Q1$ and R_{C1} comprises a voltage divider for E_{CC}. Thus, when r_{ce} is abnormally low during C_1 charge-current flow, e_c undershoots its resting level, thereby developing a pip. We perceive that pips are developed in the i_{c2}, e_{c2}, and e_{b2} waveforms in the same manner.

Let us investigate the alternate corner rounding seen in the waveforms e_{c1} and e_{c2} in Fig. 20–14(c). The term alternate corner rounding refers to the fact that the upper left-hand corner of e_{c1} is rounded, but the upper right-hand corner of e_{c1} is not rounded. Corner rounding occurs because $Q1$ drives the base of $Q2$ very positive during the brief interval that $Q1$ is switching from conduction to nonconduction. During the extra base-current interval, the total collector load is abnormally low, and e_{c1} cannot rise to its final value. The base-current flow soon decreases, and the total collector load increases; e_c then rises to its resting value. This delayed rise to the final e_{c1} value is seen as a rounding of the upper left-hand corner in the e_{c1} waveform.

Now let us consider the development of the upper right-hand corner in the e_{c1} waveform [Fig. 20–14(c)]. At this time, $Q1$ is switching from nonconduction into conduction. Waveform e_{c1} suddenly falls, driving $Q2$ to cutoff; of course, no base current flows in $Q2$ at this time, because the base of $Q2$ is being driven negative. Therefore, the total collector load for $Q1$ is simply R_{C1}. Also, no corner rounding occurs. In summary, a fixed collector-load resistance is present as $Q1$ switches from nonconduction into conduction. On the other hand, a rapidly changing total collector-load resistance is present as $Q1$ switches from conduction into nonconduction.

The leading edge of e_{c1} has a comparatively slow rise time, and the trailing edge of e_{c1} has a comparatively fast fall time. However, the fall time of the trailing edge is not zero, because of the stray capacitances in the circuit. As previously noted, the stray capacitances have the basic effect of integration, which limits the speed of the switching action. We recall that the rise and fall times can be improved by reducing the values of the plate-load or collector-load resistors, and that this improvement in switching time is obtained at the cost of reduced output amplitude. We will recall also that the rise and fall times can be improved by insertion of series peaking coils; this improvement in switching time is obtained at the cost of increased overshoot (higher amplitude pips). However, an overshoot excursion can be eliminated by passing the multivibrator output waveform through an overdriven cathode follower.

20–5. Waveform Clipping

At this point, it is instructive to consider the action of an overdriven emitter or cathode follower in detail. With reference to Fig. 20–15

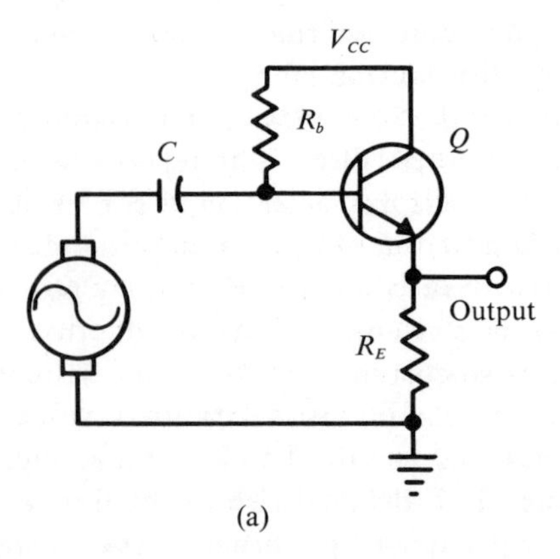

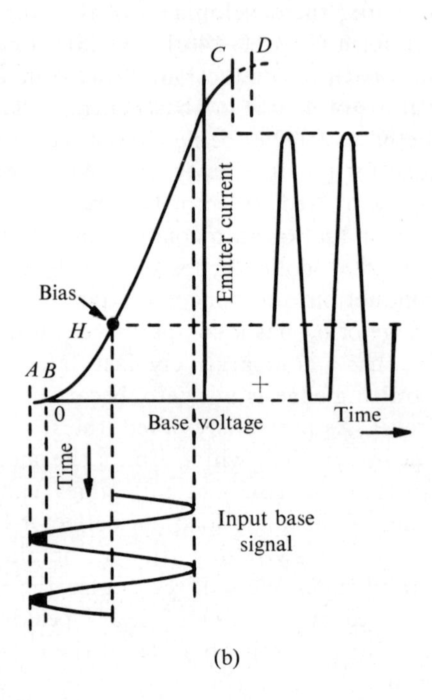

Figure 20-15. **(a) Emitter follower circuit; (b) Bottom of waveform is clipped.**

(a), a simple emitter-follower configuration is driven from a sine-wave source. We will assume that R_E and R_B provide a small base bias current, and that a comparatively large sine-wave signal is applied. The associated circuit relations are depicted in Fig. 20–15(b). With the quiescent point located at point H on the transfer characteristic, the negative peaks of the sine-wave signal drive the transistor beyond collector-current cutoff. No emitter current flows during the shaded negative peaks of the input base signal; hence, the output emitter-current waveform has clipped negative peaks, as shown in the diagram. Cutoff is complete over the clipped interval, and the bottom excursion of the current waveform is quite flat. Of course, the output voltage waveform has the same shape as the current waveform.

Let us next investigate the possibility of clipping the top of the sine-wave signal. In an attempt to produce top-clipping, we may shift the quiescent point in Fig. 20–15(b) up the transfer characteristic near the saturation region of the transistor, where the curve begins to droop. In this case, the current waveform will become *compressed* at the top, but the waveform will not be sharply clipped. The reason for compression is that the transistor characteristic continues to rise (if at a slower rate) into the saturation region. This is just another way of saying that the cutoff interval from A to B is completely flat, but that the saturation interval from C to D is not flat. Therefore, saturation clipping is not practical in an overdriven cathode follower. Cutoff clipping, though, is very effective and is employed extensively. Comprehensive analysis of clipper circuits and other waveshaping circuits is reserved for subsequent discussion.

20–6. Basic Oscilloscopes

Many of the basic principles that we have learned are exemplified to good advantage in oscilloscope circuitry. The cathode-ray tube (Fig. 20–16) is a special type of vacuum tube in which electrons emitted from the cathode are shaped into a narrow beam and accelerated to a high velocity before striking a phosphor-coated viewing screen. A cathode-ray oscilloscope is a test instrument which uses the cathode-ray tube as an indicator. Waveforms can be positioned on the screen and appear stationary because the electron beam repeatedly reproduces the pattern in the same location; this requires that the horizontal motion of the beam be synchronized with the frequency of the signal under test.

The electrostatic type of cathode-ray tube used in oscilloscopes has two pairs of deflection plates mounted at right angles to each other as seen in Fig. 20–16(b). Vertical-deflection plates YY' deflect the electron beam vertically, and horizontal-deflection plates XX' deflect the beam horizontally. Both pairs usually operate simultaneously in attracting and repelling the electron beam by electrostatic fields. If a sine-wave voltage is applied across the vertical-deflection plates, and a

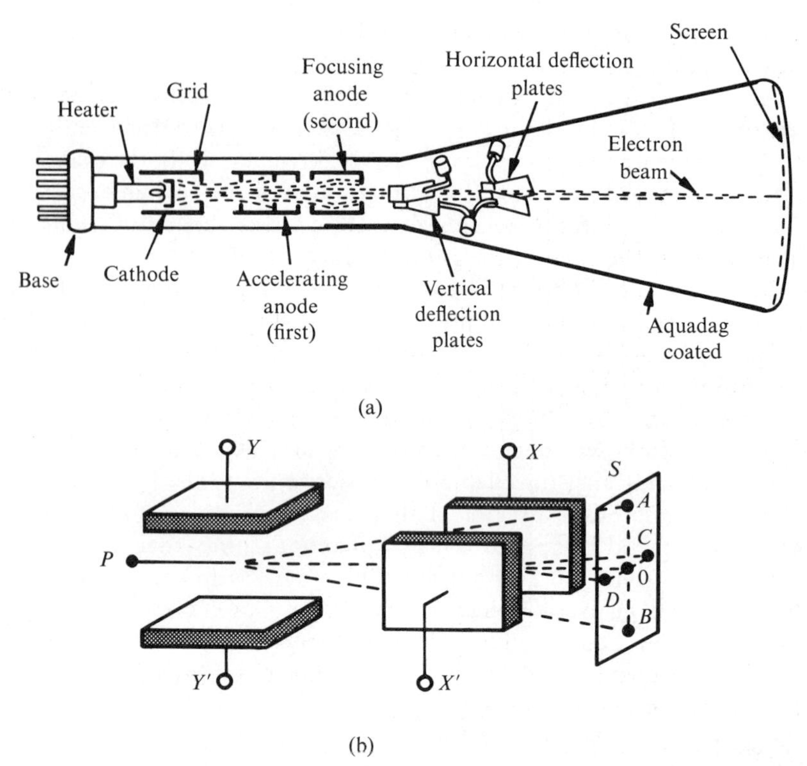

Figure 20-16. (a) Plan of a cathode-ray tube; (b) Action of deflecting plates in a CRT.

sawtooth voltage of suitable frequency is applied to the horizontal deflection plates, a sine waveform is displayed on the CRT screen, as depicted in Fig. 20–17. Note that a sawtooth voltage increases in amplitude linearly with time; the circuitry that generates the sawtooth voltage is called a *time base*.

A block diagram for an oscilloscope is shown in Fig. 20–17(b). The horizontal-deflection amplifier is a high-gain *RC*-coupled Class-*A* wideband voltage amplifier that increases the amplitude of the horizontal-input voltage and applies it to the horizontal-deflection plates. The sweep generator, or time base, supplies a sawtooth voltage to the input of the horizontal amplifier through a switch that provides an optional external connection. Similarly, the vertical-deflection amplifier increases the amplitude of the arbitrary vertical-input voltage waveform before applying it to the vertical-deflection plates. In usual application, the horizontal-deflection plates are always energized by a sawtooth voltage waveform, the frequency of which may be adjusted by the

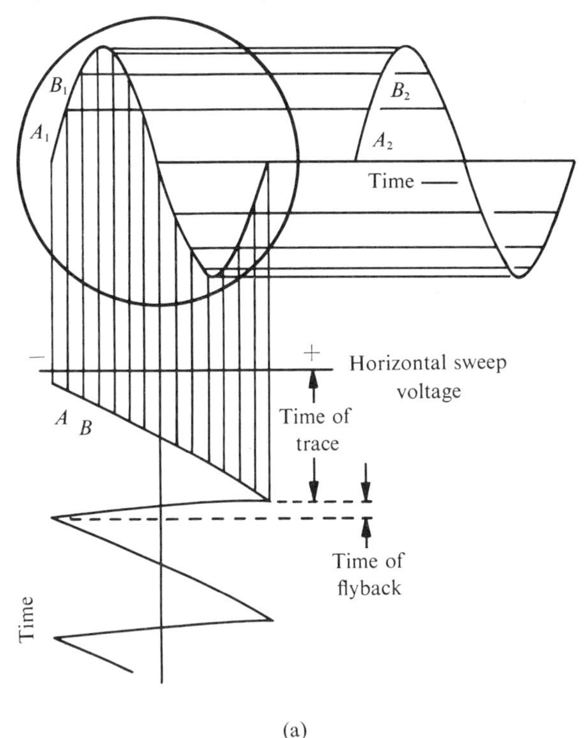

Figure 20-17. **(a) Development of a sine-wave display on a CRT screen.**

operator. On the other hand, the vertical-deflection plates may be energized by a sine wave, square wave, pulse, or other simple or complex waveform.

It is evident that an oscilloscope can be used as a voltmeter; the amplitude of the pattern displayed on the screen is directly proportional to the signal amplitude applied at the vertical-input terminals. Thus, after an oscilloscope has been calibrated, the peak-to-peak voltage of a displayed waveform may be measured on the screen. A schematic diagram for an elementary oscilloscope is shown in Fig. 20-18. *V*1 is the vertical-amplifier tube, *V*2 is the horizontal-amplifier tube, and *V*2 is a thyratron sawtooth-generator tube. Potentiometer R_1 is a manual vertical-gain control, R_2 is a manual horizontal-gain control, S_3 is the coarse-frequency adjustment for *V*3, and R_{10} is thc fine-frequency adjustment. An external synchronizing signal may be applied from an external source to the grid of *V*3 when S_2 is in the external sync position. Alternatively, the synchronizing signal is obtained from the plate of *V*1, and S_2 is in the internal sync position. R_3 provides manual control of the sync-signal amplitude.

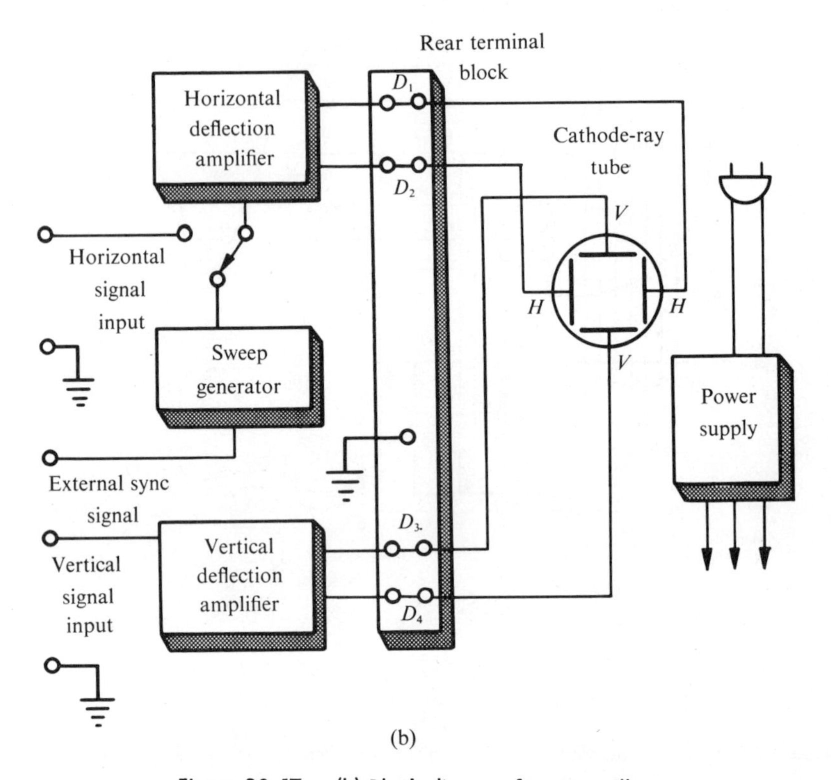

Figure 20–17. **(b) Block diagram for an oscilloscope.**

Note that the low-voltage power supply provides plate-supply potential (400 v) for *V*1, *V*2, and *V*3. The high-voltage power supply provides accelerating potential (−1,100 v) for the cathode-ray tube. In the interest of operating safety, the CRT screen is maintained at approximately ground potential, and −1,100 v are applied to the cathode. R_{17} and R_{18} are positioning (centering) controls that provide manual adjustment of the low d-c voltage applied across the pairs of deflection plates. The visual effect of varying the settings of R_{17} and R_{18} is illustrated in Fig. 20–19. Next, note that C_9 couples a blanking pulse to the grid of the CRT which blanks out the return trace (retrace, or flyback interval) as the sawtooth voltage completes one cycle and starts another. Unless a blanking circuit is used, a visible retrace line appears in waveform displays, as illustrated in Fig. 20–19.

Let us analyze the generation of the blanking pulse. Figure 20–20 (a) shows a familiar *RC* differentiating circuit. The output voltage is basically proportional to the *rate of change* of the input voltage. A sawtooth wave [Fig. 20–20(b)] changes with comparative slowness

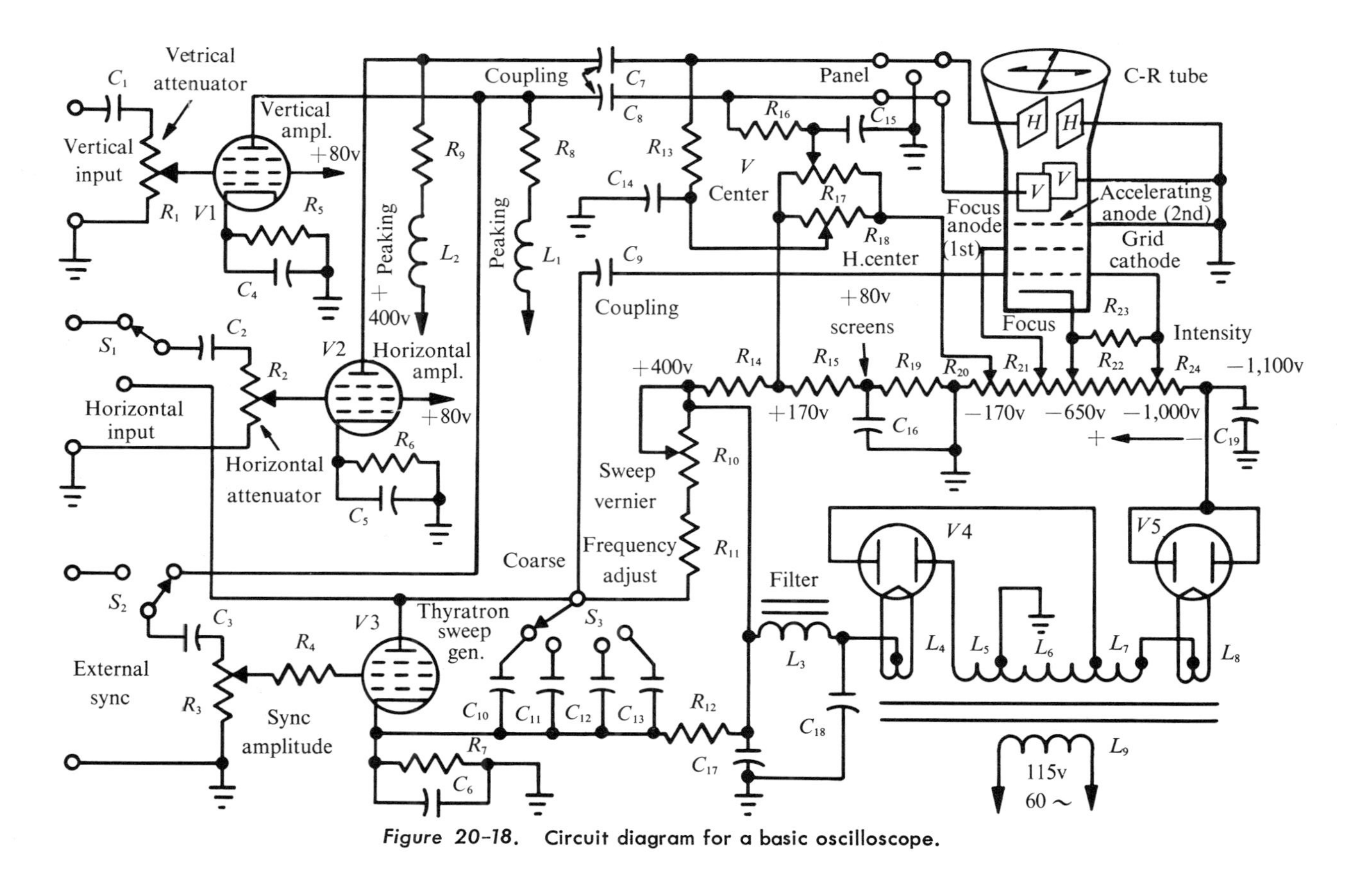

Figure 20-18. **Circuit diagram for a basic oscilloscope.**

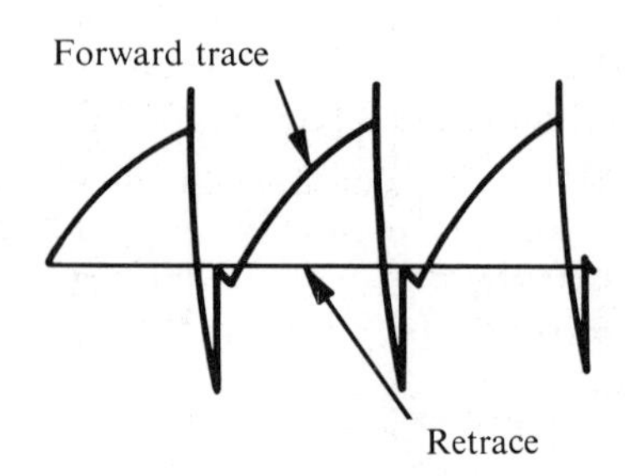

Figure 20–19. **Visible retrace line in a waveform display.**

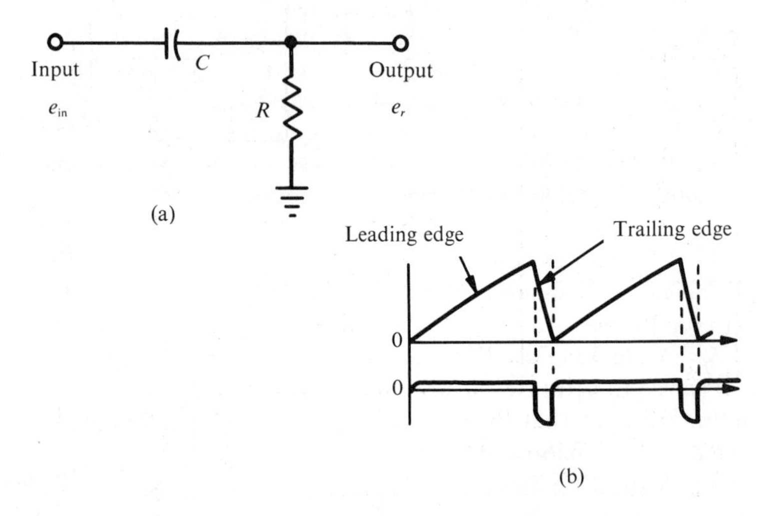

Figure 20–20. **(a) Differentiating circuit; (b) Sawtooth wave shaped into a pulse.**

along its leading edge, but falls rapidly in voltage along its trailing edge. Therefore, e_R has only a small amplitude above the zero-volt axis, but has a large amplitude below the zero-volt axis. Both the leading edge and the trailing edge of the sawtooth wave supply a steadily changing voltage to produce steady current flow through R; however, the trailing edge applies a steadily changing voltage at a greatly increased rate. Therefore, the output voltage has essentially a pulse waveshape.

In normal operation, the free-running frequency of the thyratron sawtooth generator is adjusted to a slightly lower value than the vertical-input signal frequency. The synchronizing action speeds up this free-running frequency slightly, and thereby keeps the sawtooth wave in step with the displayed signal waveform. If the thyratron is synchronized at a submultiple of the signal frequency, two or more cycles of the signal will be displayed on the screen. For example, let us suppose that the signal frequency is 1,000 Hz. If the sawtooth repetition rate is 1,000 Hz, one cycle of the signal is displayed; again, if the sawtooth repetition rate is 500 Hz, two cycles of the signal are displayed, and so on.

SUMMARY

1. Practical applications often require that a complex waveform oscillator be free-running in the absence of a synchronizing signal.
2. When a free-running oscillator produces an output frequency that is the same as a synchronizing frequency, the multivibrator is said to be "locked" to the synchronizing frequency.
3. By proper selection of the trigger amplitude and frequency, a free-running multivibrator may be developed into a frequency divider.
4. When a free-running multivibrator is synchronized by an external waveform, one-half of the multivibrator's output waveform decreases in length.
5. A blocking oscillator is an oscillator in which the feedback is so great as to saturate the amplifying device almost instantaneously, and so arranged that the circuit cuts off shortly after saturation to periodically develop a very narrow output pulse.
6. A limiting diode may be placed across the primary (Fig. 20–7) or secondary (Fig. 20–14) of a blocking-oscillator transformer to eliminate overshoot and ringing of the output waveform.
7. The primary differences between a vacuum tube and a transistor free-running blocking oscillator are the values of the supply voltages and the fact that the transistor input must be forward-biased to start the regenerative oscillator action.
8. A blocking oscillator may be utilized as a switch to discharge the capacitor of an *RC* circuit which develops a sawtooth output waveform (Fig. 20–12).
9. The linearity of a sawtooth waveform may be improved by proper employment of the nonlinear characteristics of a transistor or vacuum tube to counteract the exponential rise of waveform.
10. All pulse waveforms have a rise time and a decay time determined by the capacitance, inductance, and resistance in the circuit.
11. The output waveform of an overdriven cathode or emitter

follower is spiked on the one extreme due to the change of the amplifier-device characteristics in the saturation region.

12. An oscilloscope uses an electrostatic-type cathode-ray tube.

13. An oscilloscope has five basic circuits: the power supply, the CRT, the sweep generator, the vertical amplifier, and the horizontal amplifier.

14. The oscilloscope can be used to measure a-c and d-c voltage, time, frequency, and phase, to compare waveforms, and to evaluate many other electrical quantities.

Questions

1. A free-running multivibrator is to be synchronized by an external signal. What is the necessary relationship between the external signal and the free-running frequency?
2. Give two applications in which it is desirable to have a complex waveform oscillator free running in the absence of an external synchronizing signal.
3. Explain the term "locked" as applied to a synchronized oscillator.
4. In reference to the circuit in Fig. 20-2, state three different points at which the circuit may be synchronized and the trigger requirements for each point.
5. In reference to the waveforms in Fig. 20-2, explain how a multivibrator may be made to divide the input trigger frequency by three.
6. Draw the output of a multivibrator which free-runs at 1-kHz rate, and compare this with the resultant output when the multivibrator is triggered by a 1.2 kHz pulse.
7. In respect to Fig. 20-6, draw the base and collector waveforms for the blocking oscillator circuit employing a PNP transistor.
8. What causes overshoot and ringing in the output waveform of a blocking oscillator?
9. Explain a method that may be employed to eliminate overshoot and ringing in the output waveform of a blocking oscillator.
10. In Fig. 20-7, what is the purpose of the voltage supply E_{EE}? What is the purpose of the resistor R_B? Which component determines the free-running frequency? What is the function of diode *CR*1? What is the phase relation between a voltage change at the collector and one induced at the base?
11. If the circuit in Fig. 20-7 is connected but does not oscillate, what are the two most probable problems?
12. Compare a vacuum-tube and a transistor-blocking oscillator.
13. What determines the lowest frequency to which a given blocking oscillator may be synchronized?
14. What determines the maximum frequency to which a blocking oscillator may oscillate?
15. With reference to the blocking oscillator circuit in Fig. 20-9, what is the function of the 680-Ω resistor in the cathode? What components determine the free-running frequency of the blocking oscillator?
16. With reference to Fig. 20-12, what is the function of the blocking oscillator? What determines the rise time of the output sawtooth waveform? What determines the discharge time of the output pulse?

17. What is the purpose of each of the following components in the circuit of Fig. 20-14: resistor R_T; diode $CR1$; diode $CR2$; capacitor C_5 and resistor R_9; resistor R_3, and capacitor C_1?

18. In the circuit in Fig. 20-14(a), what determines the rise time of output waveform e_{c1}? What causes the small pip at the leading edge of the saturation current i_{c2}? What components cause the pulse width of the OFF time of $Q1$?

19. Explain why an overdriven emitter or cathode follower produces an output waveform clipped on one peak and compressed on the other peak?

20. Referring to Fig. 20-16, explain how an electrical signal is developed on the CRT screen.

21. What are the five basic blocks of an oscilloscope, and what is the function of each block?

22. State five quantitative measurments that may be made with an oscilloscope, and explain the operation of the oscilloscope in each measurement.

CHAPTER 21

Wave-Shaping Circuits

21–1. Introduction

An extremely large number of wave-shaping circuits are utilized in electronics technology. This chapter is concerned with some of the more basic types of wave-shaping arrangements. We have seen that an overdriven cathode follower operates as a squaring circuit and clips the negative peak from a waveform. We have also seen that the curvature in a tube characteristic may be employed to linearize an opposite curvature in a waveform. We recall that a monostable multivibrator is often used to regenerate pulse waveforms. These techniques serve as an instructive introduction to the general topic of wave-shaping circuits.

Several methods of square-wave generation are in use. We know that a symmetrical multivibrator serves this purpose. Another way to generate a square waveform is to shape a sine wave that has a frequency equal to the repetition rate of the desired square wave. Clipper and amplifier stages, or clipper-amplifier stages, are commonly employed for this purpose; diode circuits are also utilized. To anticipate subsequent discussion, a clipper will slice off the top and bottom from a sine wave, but the rise and fall times of the output waveform will evidently be comparatively slow. That is, the output from the clipper is only a rough approximation to an ideal square waveform. We will find, however, that if this semi-square wave is amplified and again clipped, the output waveform then has a faster rise and fall. This shaping technique can be repeated as many times as may be required to obtain a square-wave output with specified rise and fall time.

Let us analyze the operation of a peak diode clipper. Figure 21–1(a) depicts the configuration for a negative-peak clipper; the voltage-current characteristic for this circuit is shown in Fig. 21–1(e). The diode operates as an electronic switch. The "switch" is closed during the positive half cycle of the input signal and is open during the negative half cycle. This is called a series diode-clipper circuit. A clipper circuit may also employ a shunt-connected diode, as shown in Fig. 21–1(b). The value of resistor R is large, compared with the a-c plate resistance of the diode. Note that the diode is polarized to clip the positive half cycle of the input waveform. Over the positive half cycle, there is only a slight output voltage, which stems from the plate resistance of the diode; the plate resistance operates as a voltage divider in combination with R. Over the negative half cycle, however, the diode is effectively out of the circuit, and the negative half cycle is passed to the output terminals of the circuit.

Observe that a low-resistance load may be connected across the output terminals of the circuit in Fig. 21–1(a), and clipping action occurs much as with an open-circuited output. The only difference is that the diode voltage drop e_b is increased to some extent when a low-resistance load is connected. On the other hand, if we connect a low-resistance load across the output terminals of the circuit in Fig. 21–1(b), most of the positive half cycle will be passed as well as the negative half cycle. The shunt diode-clipper circuit is not well adapted for low-resistance loads; this configuration is used only when the load impedance is very high, such as the grid of a following amplifier tube. The shunt diode-clipper circuit depicted in Fig. 21–1(c) is the same as shown in (b), except that the diode polarity has been reversed. Note that we may reverse the diode polarity in (a) to obtain negative peak clipping.

Of course, the vacuum diodes in Fig. 21–1 may be replaced by semiconductor diodes, and circuit operation will remain the same. Let

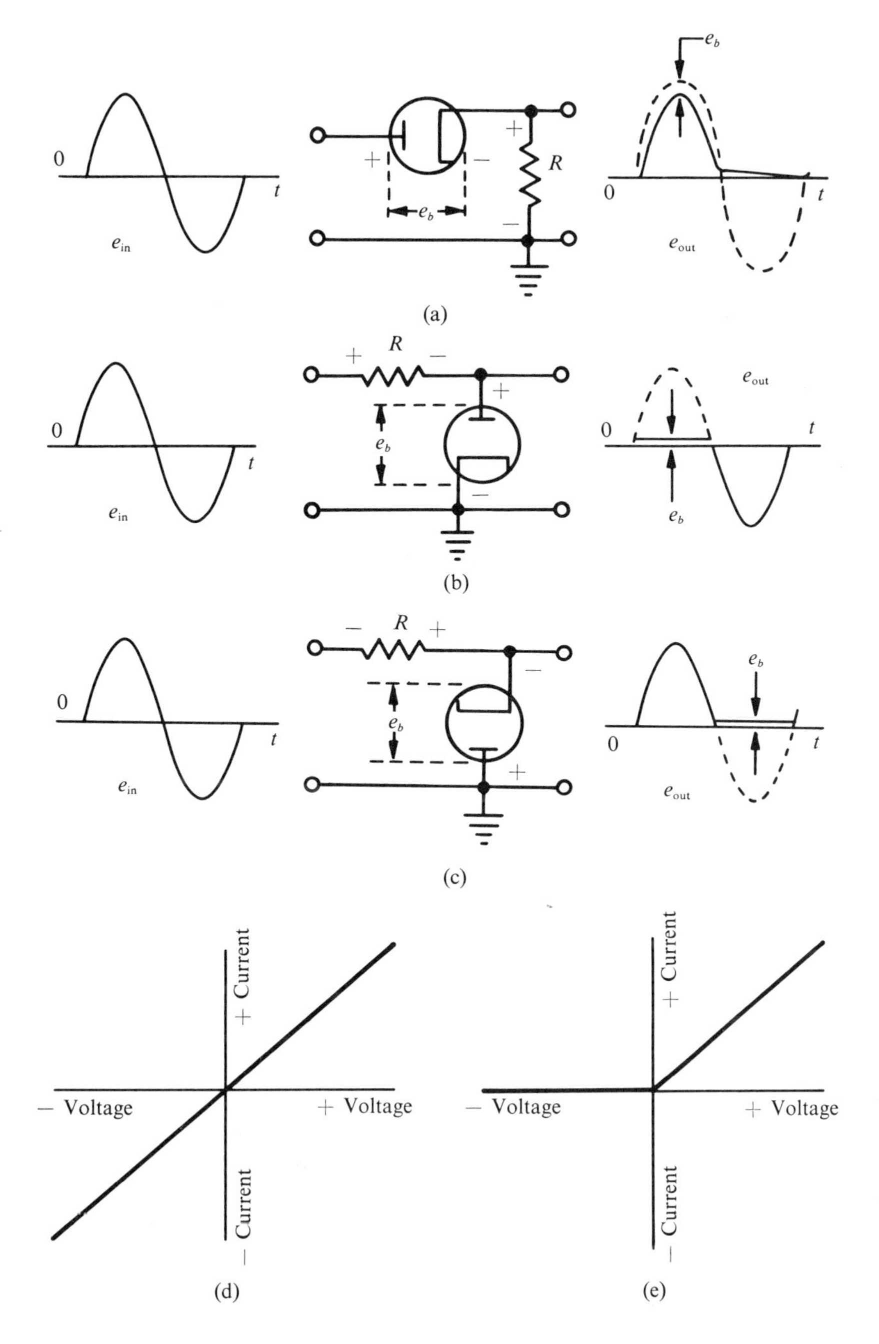

Figure 21-1. (a) Series diode clipper; (b) Shunt diode clipper; (c) Shunt diode clipper (opposite polarity); (d) Linear resistance characteristic; (e) Resistance characteristic of circuit (a).

us see how we can elaborate a shunt diode-clipper circuit to obtain a semi-square waveform from a sine wave. Observe the configuration depicted in Fig. 21–2(a). Bias batteries are connected in series with a

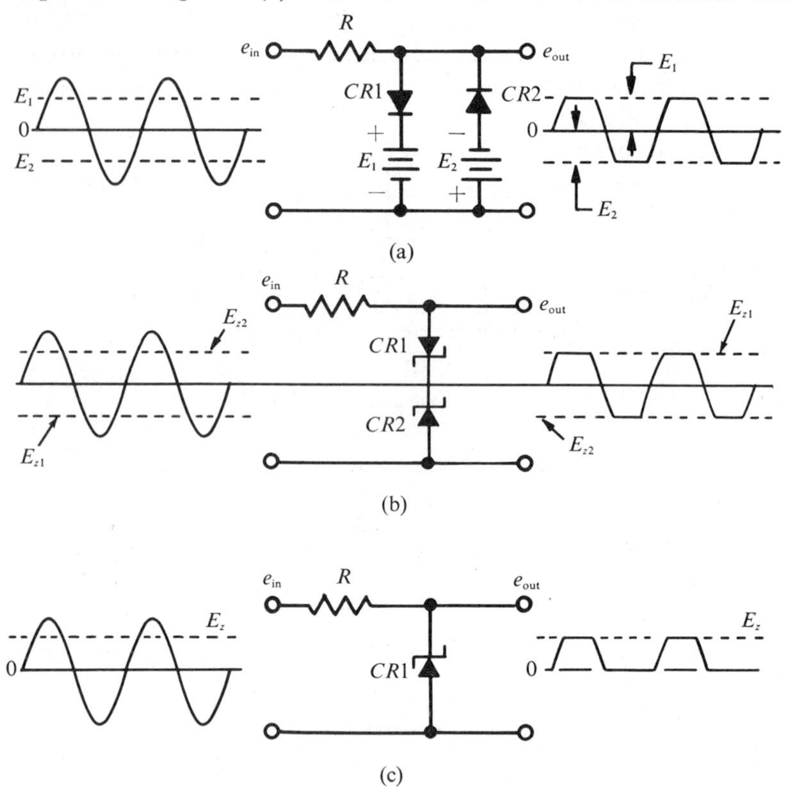

Figure 21–2. **A diode clipper circuit. (a) Double-shunt diode clipper; (b) Zener diode clipper; (c) Double clipping with a Zener diode.**

pair of oppositely polarized shunt diodes. Each diode is reverse-biased and cannot conduct until the sine-wave input voltage exceeds the bias voltages E_1 and E_2. The resultant output waveform is a semi-square wave. The rise and fall times are slow in this example; faster rise and fall can be obtained by applying a larger amplitude of input signal. Alternatively, the output from the circuit in Fig. 21–2(a) can be amplified and then passed through another clipper circuit; this approach will also provide faster rise and fall.

The back-to-back Zener diode circuit in Fig. 21–2(b) may be used to develop a semi-square waveform. Diode *CR*1 cannot conduct until the positive input signal exceeds the Zener breakdown voltage of $CR2$. Diode $CR2$ then limits the output voltage to a maximum value of E_{z2}. In the negative half cycle, diode $CR2$ cannot conduct until the input exceeds the Zener breakdown voltage of $CR1$. Diode $CR1$ then limits the output to a minimum value of E_{z1}. Again, we may limit the output to zero and some positive or negative value by use of a Zener diode, as illustrated in Fig. 21-2(c).

Next, we will perceive that a triode grid-clipper circuit such as shown in Fig. 21-3(a) provides both clipping and amplifying action.

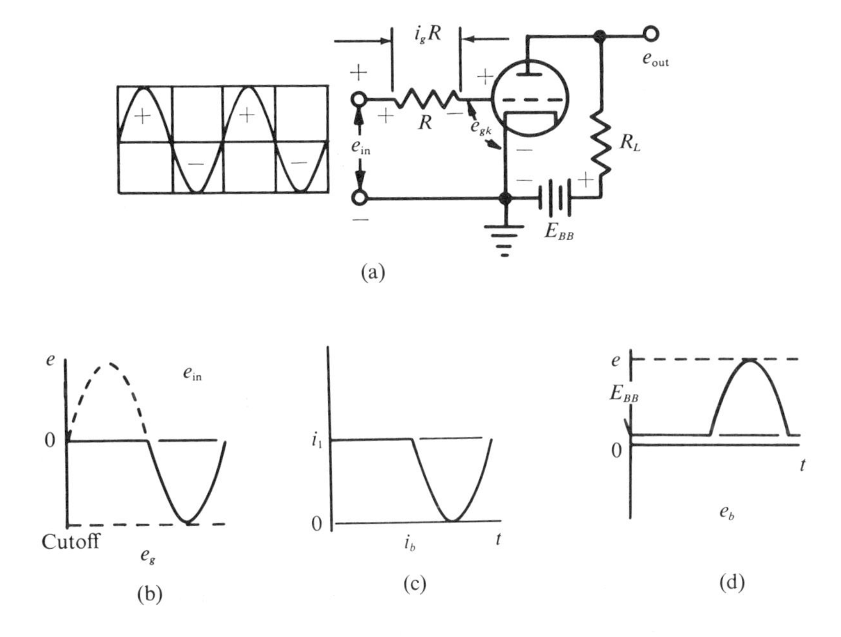

Figure 21-3. (a) Basic triode grid-clipper circuit; (b) Grid-voltage waveform; (c) Plate-current waveform; (d) Plate-voltage waveform.

The grid-cathode circuit clips the input waveform in the same manner as a diode; the grid-voltage waveform is depicted in Fig. 21-3(b). On the positive half cycle of the input signal, the signal voltage is dropped across R because grid-current flow is associated with a very low grid-cathode resistance. On the negative half cycle, the triode operates as a Class-B amplifier. Plate current flows as shown in Fig. 21-3(c), and the amplified plate-output voltage has inverted phase, as seen in (d). Note that the grid-cathode resistance of a triode may change from a practically infinite value (when the grid is driven negative) to a resistance of 1,000 Ω or less (when the grid is driven positive). If R has a value of 1 MΩ, it is evident that e_{gk} is practically zero over the positive half cycle of the input signal.

We can also obtain clipping action by the use of an unbiased transistor as illustrated in Fig. 21-4. The action is similar to that of

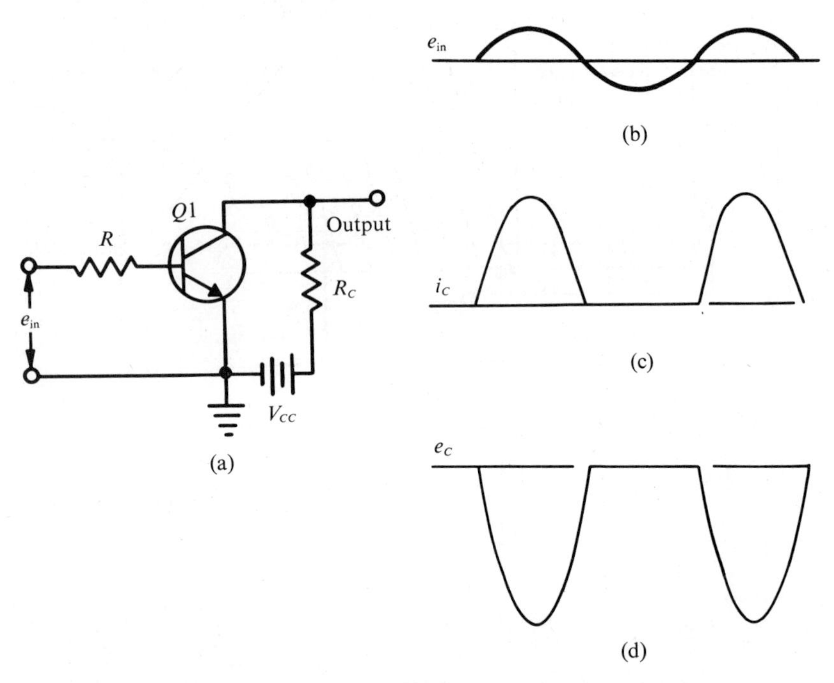

Figure 21-4. (a) Basic transistor clipper circuit; (b) Input signal; (c) Collector-current waveform; (d) Collector-voltage waveform.

the vacuum-tube limiter in Fig. 21-3, except that the series resistor in a transistor limiter is very small (only large enough to limit the base current to a safe value). Limiting action in a transistor is the result of the transistor's being cut off during the negative half cycle of the input waveform. During the positive half cycle of the input signal, base current flows, and the amplified output is developed across the collector resistor (R_C).

We can also obtain clipping action by utilizing plate-current cutoff in a triode, as seen in Fig. 21-5. In this circuit, cathode bias is used, and the quiescent point occurs at −5 v grid potential. For the plate-load resistance value and the e_b value depicted, plate-current cutoff occurs at a grid potential of −7 v. The sine-wave input voltage (3) has

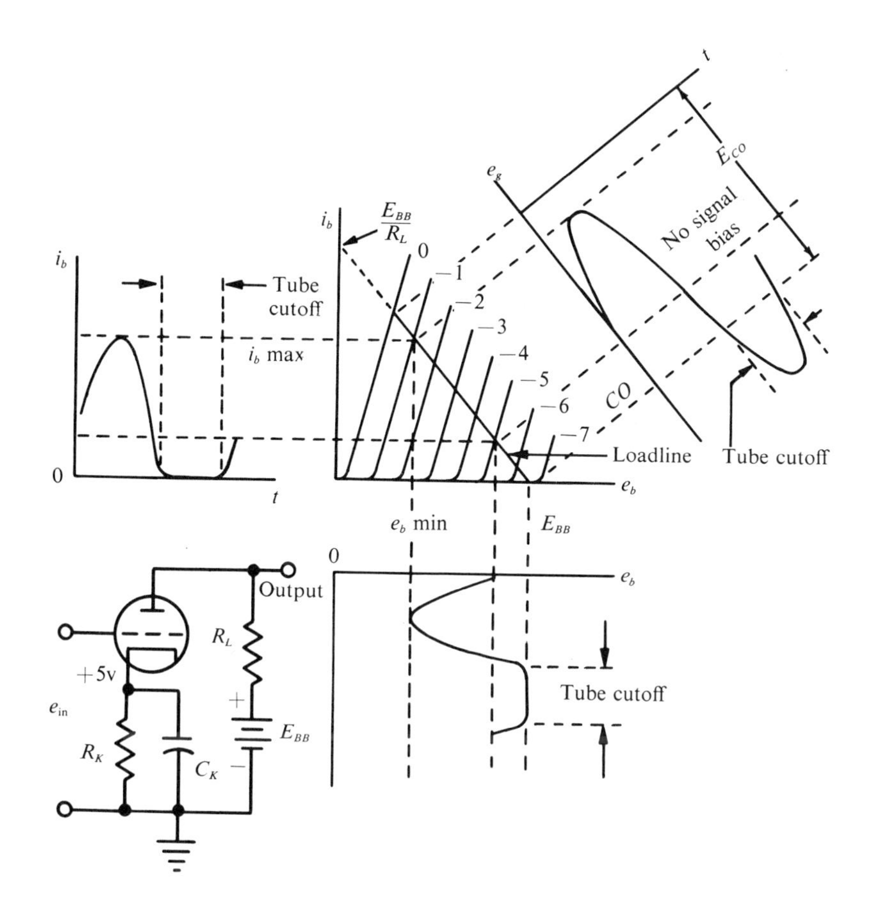

Figure 21–5. Plate-current cutoff action clips the peak from the output waveform.

an amplitude such that the grid swings from —1 to —9 v. The stage operates as an overdriven Class-*AB* amplifier. A full positive half cycle (4) of plate current flows, but the negative half cycle of plate current flow is clipped because the triode cuts off while the grid swings from —7 to —9 v. In turn, the plate-output voltage waveform (5) has its positive half cycle clipped.

As would be anticipated, a transistor or a vacuum tube can be used to provide both input clipping and cutoff clipping in order to produce a semi-square output waveform from an input sine wave. A typical configuration is shown in Fig. 21–6(a). Grid resistor R has a value of 1 MΩ. The positive half cycle of the input sine-wave voltage, e_g, is

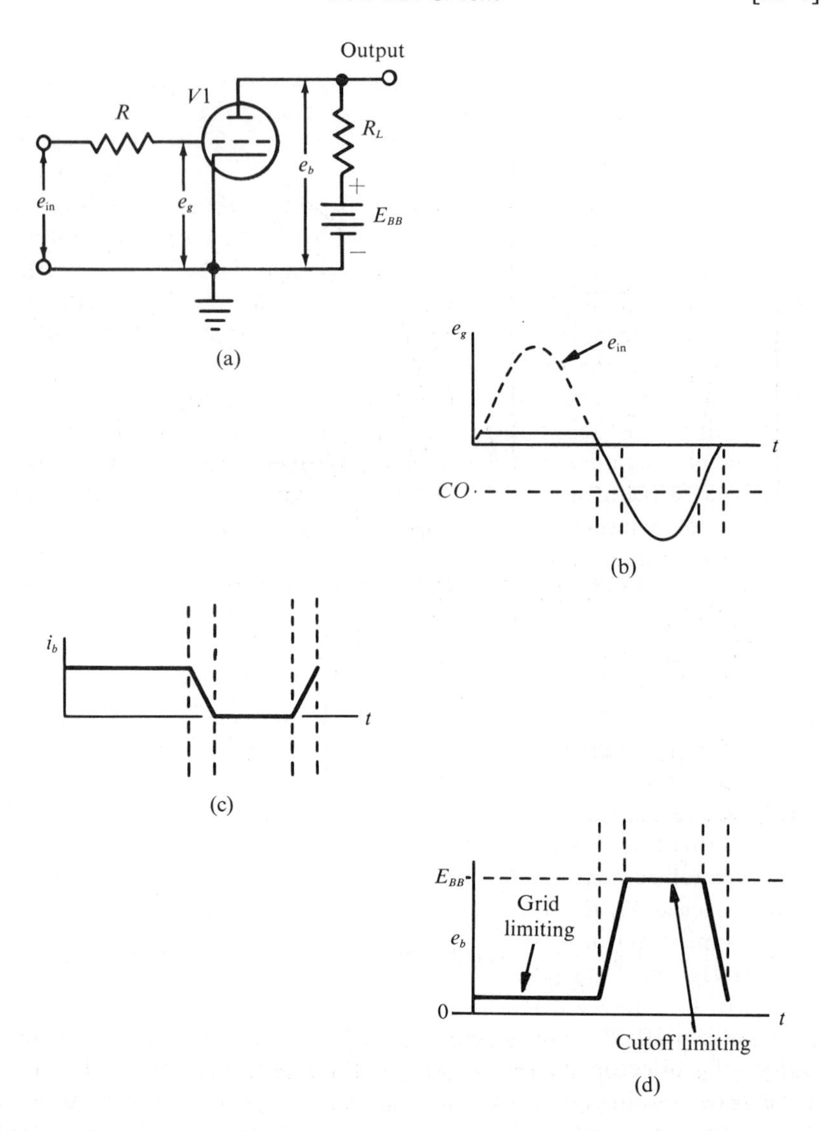

Figure 21-6. (a) Combined grid and cutoff clipper circuit; (b) Grid-voltage waveform; (c) Plate-current waveform; (d) Plate-voltage waveform.

clipped due to grid-current flow. However, the stage operates as a Class-*B* amplifier over the negative half cycle of the input waveform. Note that if the tube is overdriven, plate-current cutoff ensues, and the negative peak of plate-current flow is clipped as depicted in Fig.

21-6(c). The plate-voltage waveform appears as seen in (d). The end result of combined grid and cutoff clipping is a semi-rectangular wave; it is not a semi-square wave because the grid-clipping interval is longer than the cutoff interval. Rise and fall times can be improved by driving the grid harder.

It is evident that two stages similar to the configuration shown in Fig. 21-6 can be connected in cascade and that the resulting amplifying and clipping action will improve the rise and fall times. In other words, let us suppose that a semi-rectangular wave has a rise time of 10 μsec. If we amplify this waveform 20 times and then clip off 0.9 of its amplitude, the resulting clipped waveform will have a rise time of 1 μsec and double the amplitude of the original waveform. Assuming that *RC*-coupling is used between the first and second stages, the student may analyze the second clipping actions to show whether the output waveform from the second stage is a semi-rectangular wave as before, or whether the waveform becomes an approximation of a semi-square wave.

Another type of clipper employs *saturation limiting* or collector-current saturation limiting. If we employ a high-resistance collector load and a comparatively low d-c supply voltage as depicted in Fig. 21-7(a), saturation limiting or clipping occurs over the positive half cycle of the base-driving current. Observe in Fig. 21-7(b) that the base-driving current is not of sufficient amplitude to drive the transistor to cutoff on its negative swing. However, on its positive swing, the positive half cycle of collector-current flow is clipped. Let us see how this clipping action occurs. Note that the load line intersects the collector-current axis at point $V_{CC}R_C$. Obviously, the plate current cannot exceed $V_{CC}R_C$ no matter how positive the base is driven. Therefore, clipping action ensues.

Strictly speaking, the maximum collector-current value in Fig. 21-7 will be somewhat less than $V_{CC}R_C$ because the collector resistance is in series with R_C, and their sum determines the maximum possible value of collector current. This maximum collector-current value determines the lowest value to which the collector voltage can fall. We perceive that saturation clipping develops approximately the same output waveshape as grid or base clipping. However, saturation clipping provides a greater output amplitude. This increased output amplitude is obtained at the cost of a higher-amplitude input-drive waveform. Engineers often refer to the collector-saturation action as "bottoming."

We have seen that saturation clipping can be combined with cutoff clipping as depicted in Fig. 21-8; however, this action does not

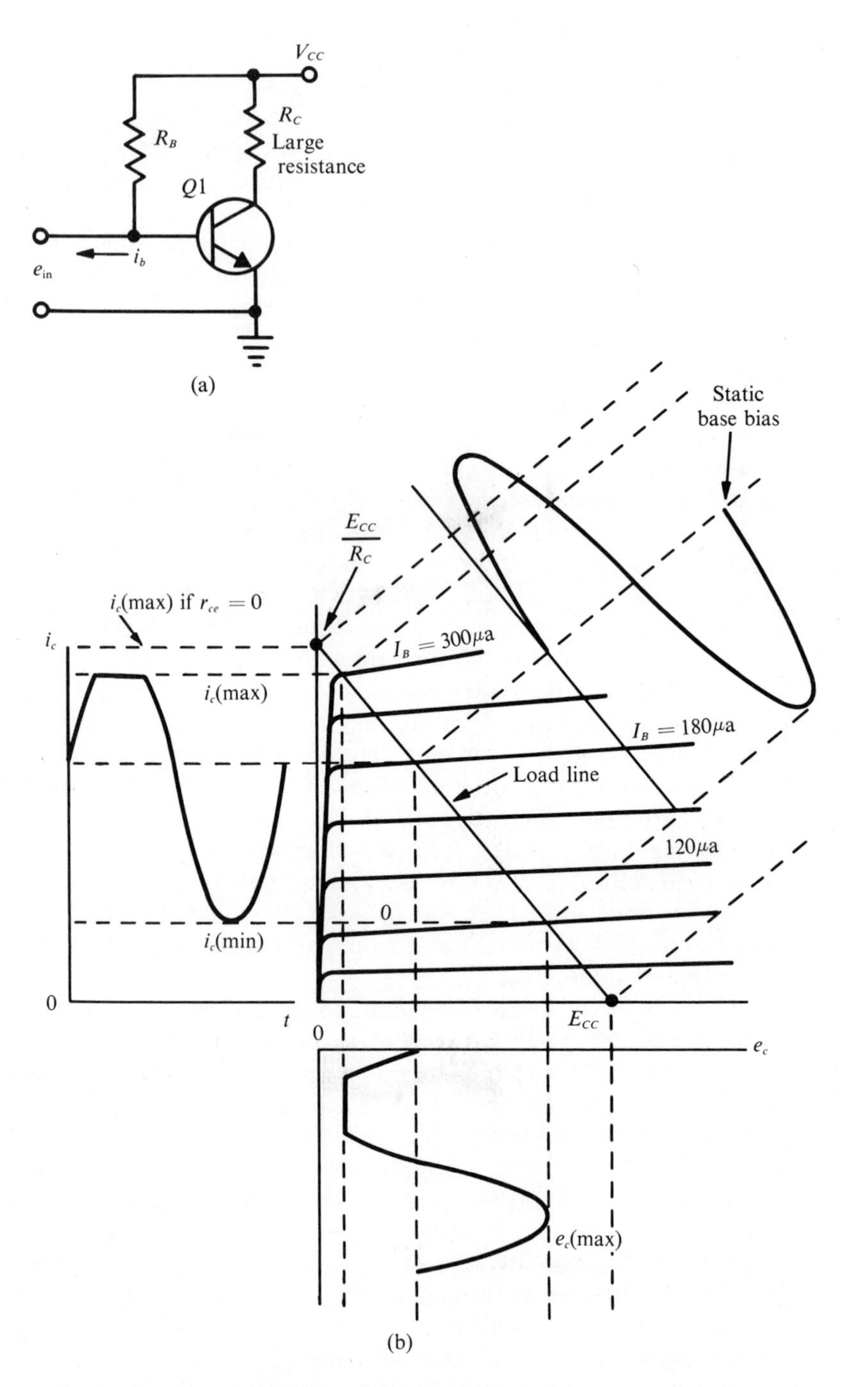

Figure 21-7. **(a) Saturation clipper circuit; (b) Operating waveforms.**

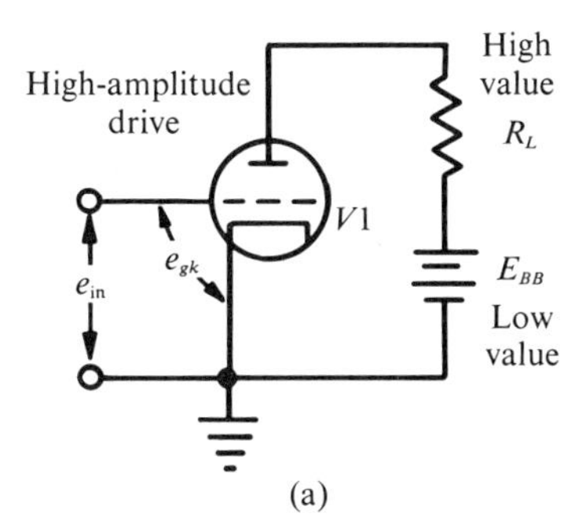

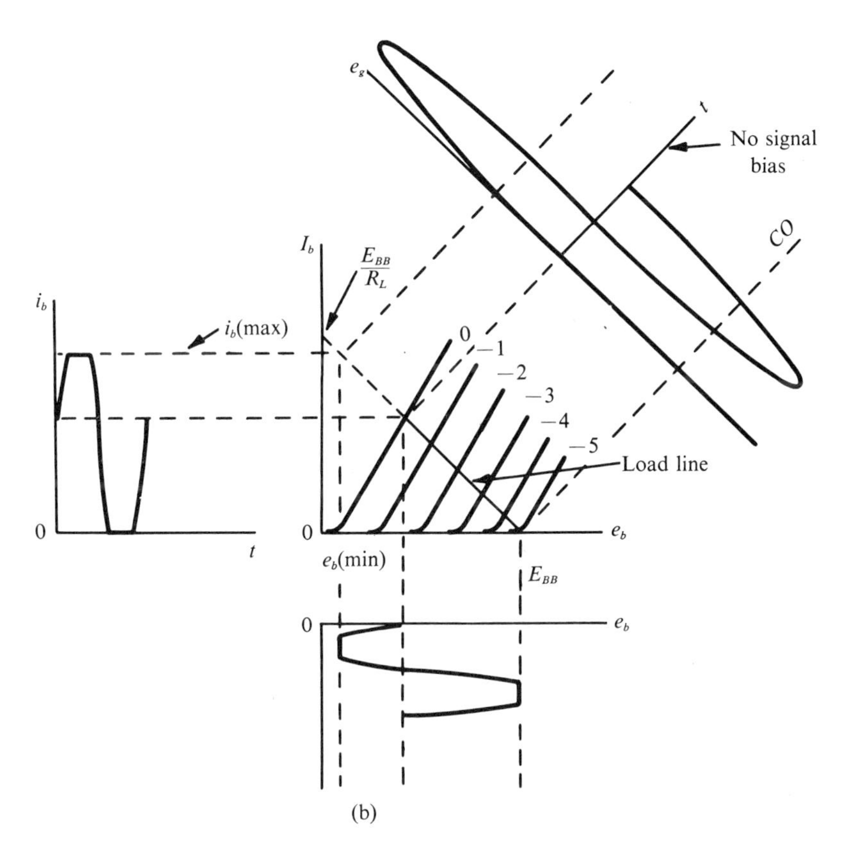

Figure 21–8. (a) Saturation and cutoff clipper circuit; (b) Operating waveforms.

develop a rectangular wave. The output waveform is a semi-rectangular wave, inasmuch as the clipping action is unsymmetrical. To improve the rise time and fall time of the output waveform, two or more such stages can be cascaded. This arrangement is called an *overdriven amplifier*. Assuming *RC* coupling between a pair of cascaded stages, the student may analyze the clipping actions in the second stage to determine whether the output waveform from the second stage is a semi-rectangular wave as before, or whether the output waveform more nearly approximates a semi-square wave.

21–2. Oscilloscope Calibrator

Let us observe how a duo-diode clipper circuit operates in an oscilloscope calibrator illustrated in Fig. 21–9. This instrument pro-

Courtesy of EICO, Electronic Instrument Co., Inc.

Figure 21–9. An oscilloscope calibrator.

vides a semi-square wave of predetermined amplitude for use in measuring the vertical-amplifier sensitivity of an oscilloscope. A type 6AL5 tube is used as a clipper. The heaters are not shown in the diagram of Fig. 21–10; note that the yellow lead from the power transformer connects to pin 3 of the tube socket to provide heater voltage. Since accurately known calibrating voltages are required and line voltage is subject to fluctuation, an 0C3 voltage-regulator tube is utilized for voltage reference. This regulator tube stabilizes the positive d-c voltage that is applied to the cathode (pin 5) of the 6AL5 tube. In the figure, a-c voltage is taken from the red lead of the power transformer, and applied through R_2 and C_2 to the waveshaping circuitry.

Let us consider the circuit action in Fig. 21–10 when the positive half cycle of a-c voltage is applied to the diode. The pin-1 cathode is driven positive, and this diode is effectively an open circuit since the

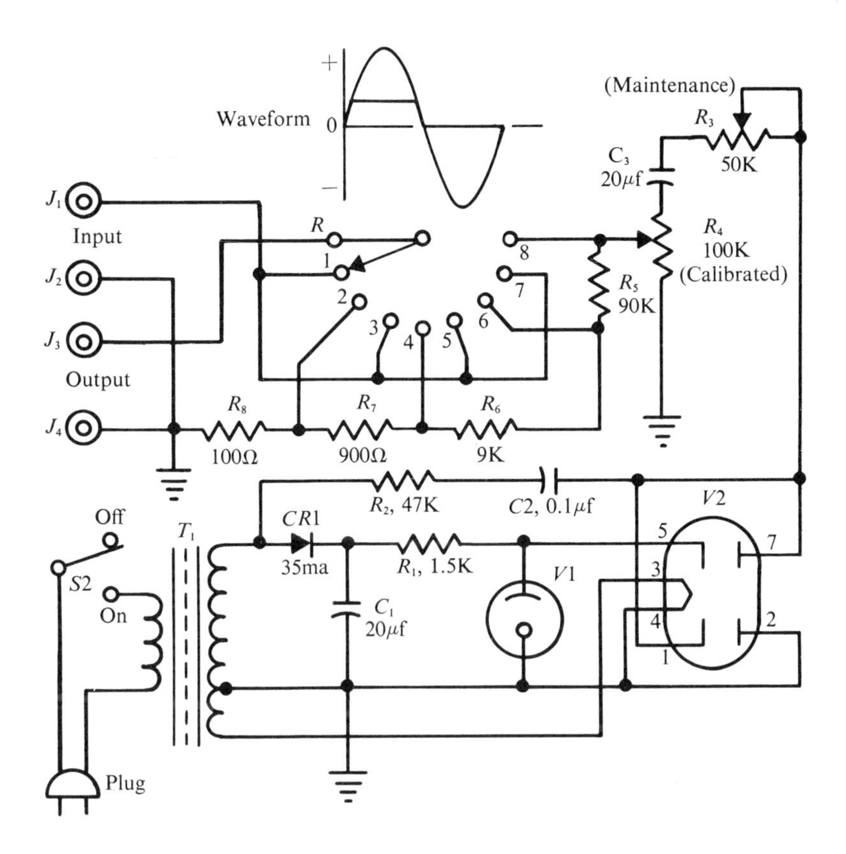

Courtesy of EICO, Electronic Instrument Co., Inc.

Figure 21–10. **Circuit diagram for oscilloscope calibrator.**

diode plate is at ground potential. At the same time, the pin-2 plate is driven positive, and the diode conducts; the plate voltage can rise up to but only slightly above the conduction level, because of R_2, C_2, and the low resistance to ground via the diode and $V1$. During the conduction interval, the voltage across $V1$ is effectively applied to R_3, C_3 and C_4. Next, when the negative half cycle of a-c voltage is applied, the pin-7 diode conducts, and the pin-2 diode is cut off. Thus, the negative half cycle of a-c voltage flows to ground via the pin-7 diode.

We perceive that the output waveform from R_4 is a semi-square wave and has a peak-to-peak amplitude that is determined by the d-c voltage drop across $V1$. An 0C3 voltage-regulator tube maintains a constant drop of 105 v. Accordingly, the 60-Hz semi-square voltage waveform applied to R_3 has a peak-to-peak amplitude of 105 v. R_3 is a maintenance adjustment that is set to provide correct indication by the calibration control R_4. As seen in Fig. 21–9, R_4 is calibrated for voltage values from 0 to 100 peak-to-peak v. R_3 and R_4 form a voltage divider; thus, R_3 may be set to provide a 100 p-p v waveform when R_4 is at the top limit of its travel. Once R_3 is correctly adjusted, it requires no further attention.

Because oscilloscope calibration requires a very wide range of reference voltage values, it is not practical to provide attention by means of R_4 only. For example, if a 50-mv reference should be required, R_4 would have to be set so near to its lower limit that the fractional scale divisions could not be read with any reasonable accuracy. Therefore, a step attenuator comprising R_5 through R_8 is also provided. Contact 8 provides a maximum reference voltage of 100 p-p v; contact

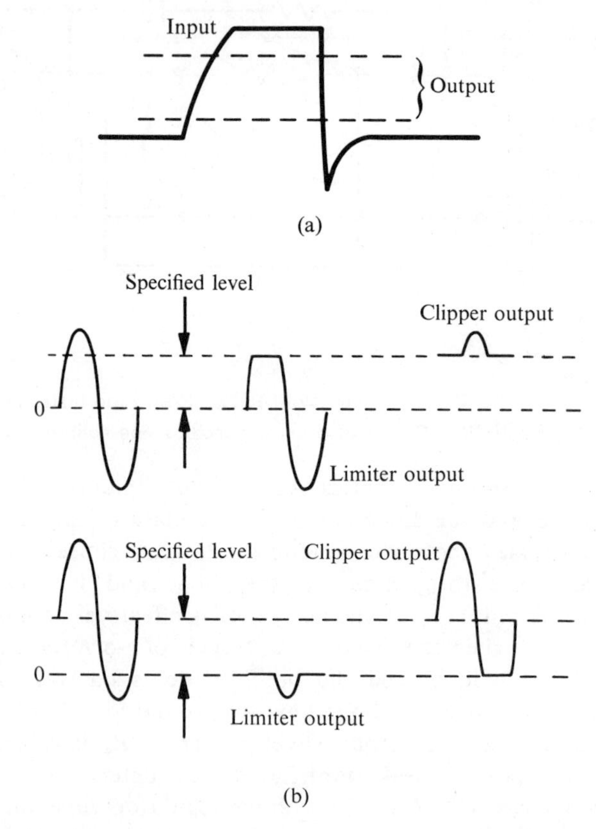

Figure 21–11. (a) A clipper may slice a voltage at any desired level; (b) A limiter provides an output up to a specified value.

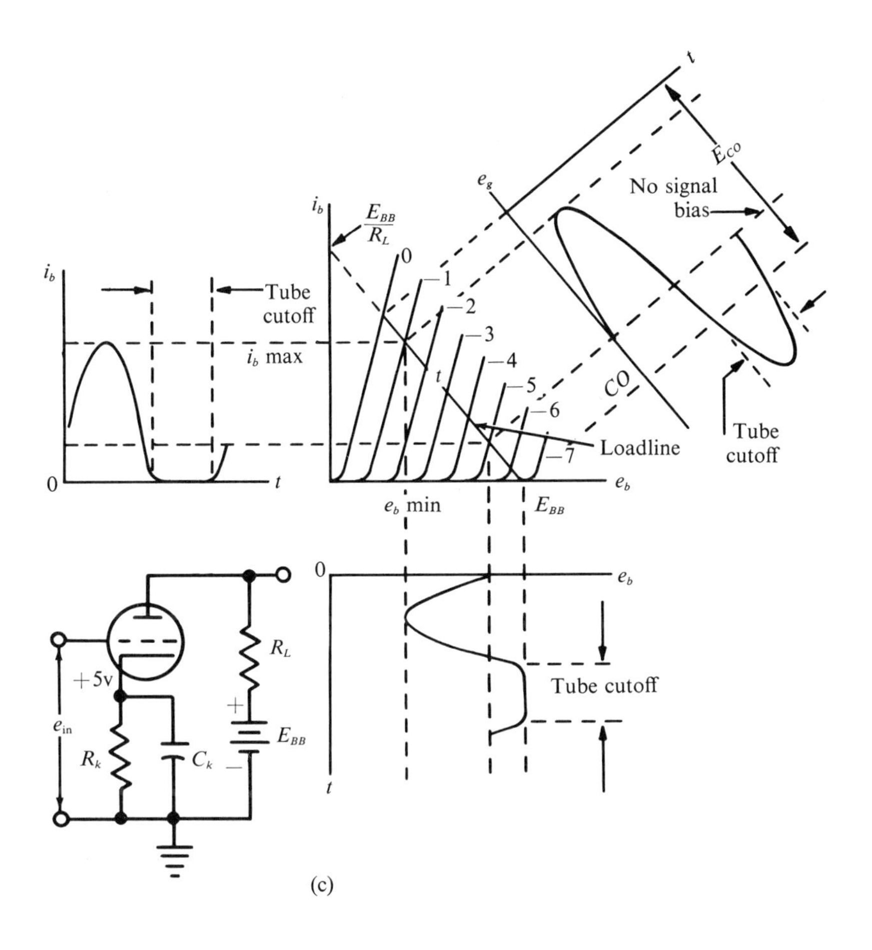

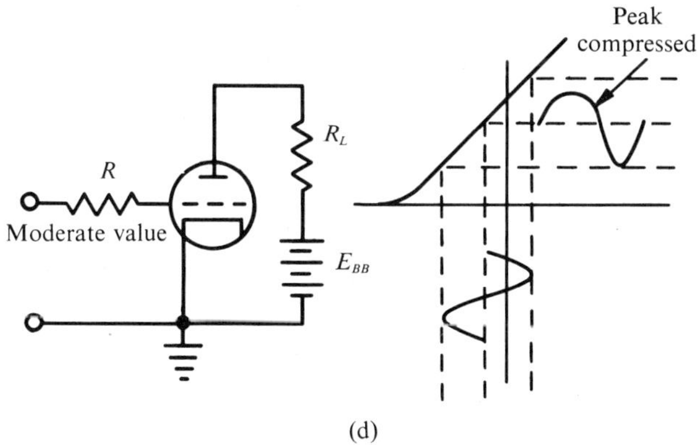

Figure 21-11. (c) Cutoff clipping action; (d) Incomplete clipping is called peak compression.

6 provides 10 p-p v; contact 4 provides 1 p-p v; contact 2 provides 0.1 p-p v. Suppose that S_1 is set to contact 1, and R_4 is set to "10" on its 0–100 scale; then, the reference voltage is evidently equal to 10 mv p-p.

Switch contacts 1, 3, 5 and 7 permit convenient switching of the test signal to the scope for comparison with the reference semi-square wave amplitude. The test signal is applied to J_1 and J_2; the vertical-input terminals of the oscilloscope are connected to J_3 and J_4. Therefore, if S_1 is set to contact 1, the test signal is displayed on the oscilloscope screen, but 'f S_1 is set to contact 2, the reference semi-square wave is displayed on the screen.

Semiconductor diodes may be used instead of high-vacuum diodes in circuits such as depicted in Fig. 21–10; the essential requirement is that the diode have an ample peak-inverse voltage rating. A clipper arrangement of this type that clips both the top and the bottom of a waveform is termed a *slicer*, *amplitude gate*, or *clipper-limiter*. Bias voltages may be chosen to slice a waveform at any desired levels [see Fig. 21–11(a)]. Strictly speaking, a limiter provides a waveform output up to a specified value, as seen in Fig. 21–11(b). Similarly, a *clipper* is defined as a circuit that provides a waveform output above a specified value, as depicted in Fig. 21–11(b). However, engineers tend to use the word "clipper" to denote either clipping action or limiting action. An a-c waveform, of course, has both a positive peak and a negative peak; it is customary (although not universal) to graph the positive excursion upward and the negative excursion downward.

Cutoff clipping action [Fig. 21–11(c)] entails a clipped plate-current waveform and a limited plate-voltage waveform. When clipping or limiting action is incomplete, as seen in Fig. 21–11(d), the circuit is called a *peak compressor*. Of course, circuits may be devised for compression of both positive and negative peaks. A *speech compressor* is an amplifier that reduces the amplitudes of both positive and negative peak regions. Although the audio waveform is distorted, it retains acceptable intelligibility, and the average power of the waveform is increased. Hence, judicious compression can increase the effectiveness of a speech waveform under conditions of interference; greater average power enables the speech waveform to compete more effectively against the interference level.

21–3. *RC* Peaking Circuits

Another widely used waveshaping circuit is called the *RC* peaking configuration and is commonly utilized to add a *peaking pulse* to a sawtooth waveform in television deflection networks. With reference to Fig. 21–12, the triode is called a *discharge tube*; the triode is biased to cutoff and is driven into conduction for brief intervals by positive pulses applied to the grid. During the comparatively long period that the triode is cut off, capacitor C charges through R and generates an approximate sawtooth waveform (exponentially curved sawtooth).

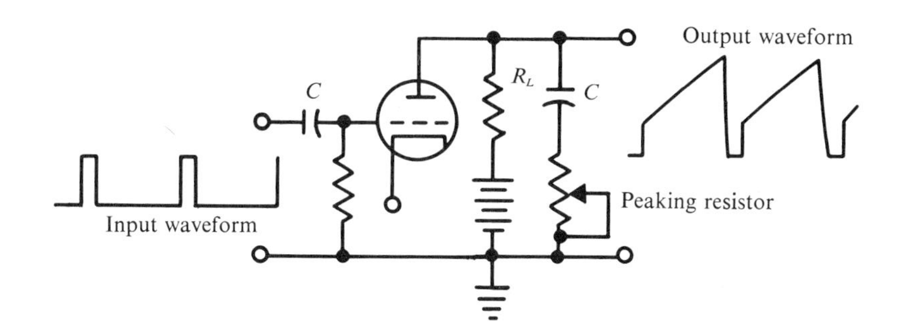

Figure 21–12. Peaked-sawtooth waveshapes.

When a positive pulse arrives at the grid, the triode suddenly presents a very low value of plate resistance to the charged capacitor. Hence, C quickly discharges through the low plate resistance of the triode. Note carefully that a *peaking resistor* is connected in series with C. Its circuit action may be analyzed in the following way.

We will analyze the plate-load circuit in Fig. 21–12 by components. If the peaking resistor were short-circuited, the voltage developed across C would be an approximate sawtooth waveform. If the peaking resistor were operating with C short-circuited, an amplified pulse waveform would be developed. Finally, with both C and the peaking resistor operating in the plate circuit, it is evident that a combination pulse and sawtooth waveform will be developed. This combination waveform is called a peaked sawtooth. The amplitude of the peaking pulse is determined by the value of the peaking resistor.

Let us observe the utility of a peaked-sawtooth waveform in producing a sawtooth current flow through an impedance. As shown in Fig. 21–13, voltage and current waveform are the same for a resistive load; on the other hand, voltage and current waveforms are different for reactive loads and for impedance loads. Thus, a sawtooth-voltage waveform drives sawtooth-current waveforms through a resistor. However, a rectangular waveform drives a sawtooth-current waveform through a pure inductor. It follows that if a load consists of inductance and resistance connected in series, a peaked-sawtooth voltage will drive

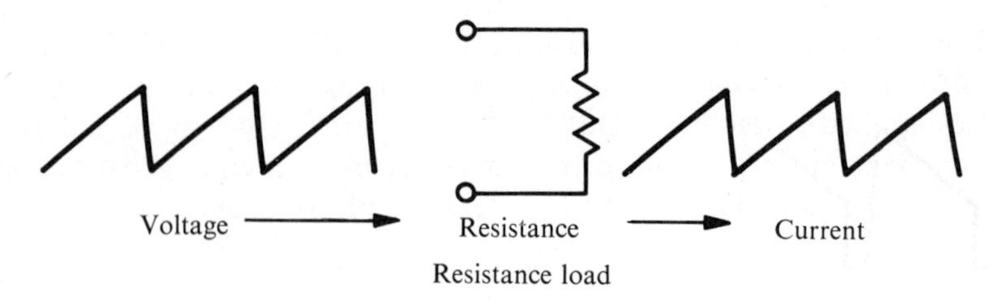

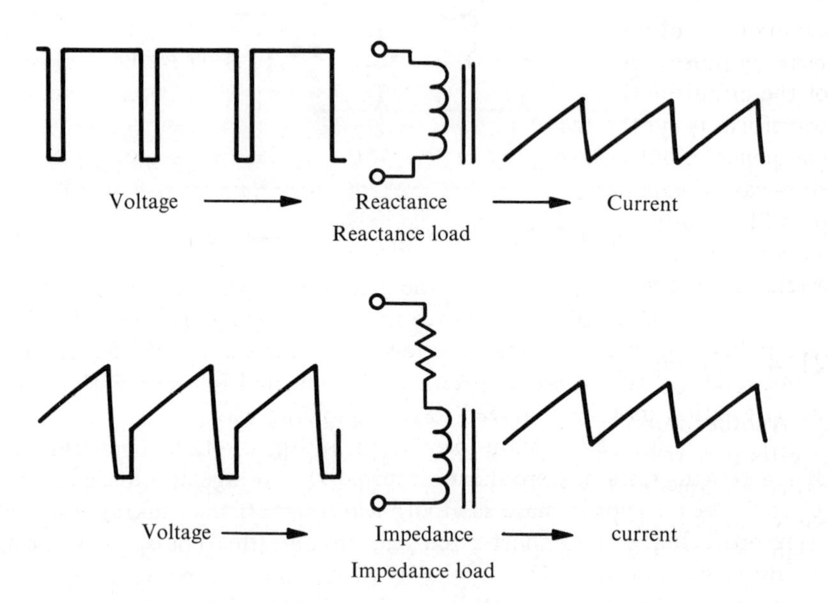

Figure 21–13. **Voltage and current waveforms for resistance, reactance, and impedance.**

a sawtooth-current waveform through impedance. The deflection coils used with a television picture tube have winding resistance and present an impedance load. It is necessary to pass a sawtooth-current waveform through the deflection coils; hence, a peaked-sawtooth voltage waveform is applied across the deflection coils.

We have learned that a differentiating circuit changes a square wave into spikes or modified pulses. Now, we will observe how a differentiating circuit changes a peaked-sawtooth waveform into a pulse waveform. This circuit action is shown in Fig. 21–14. When the time constant of the differentiating circuit is suitably chosen, the sawtooth

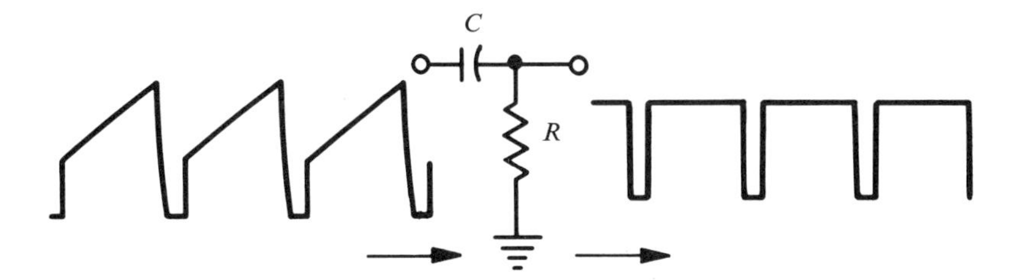

Figure 21–14. A differentiating circuit changes a peaked-sawtooth wave into a pulse waveform.

excursion is eliminated from the peaked-sawtooth waveform, and the peaking pulses only appear at the output terminals. An exact analysis of the circuit action can be made by Fourier analysis. A sawtooth waveform is synthesized from comparatively low-frequency sine-wave components, but a narrow pulse is synthesized from high-frequency sine-wave components. Since a differentiating circuit is a form of high-pass filter, only the high-frequency sine-wave components are passed. This waveshaping circuit is commonly used in the blanking section of a television receiver.

21–4. Clamper Waveshaping

Another important type of waveshaper employs a clamper circuit. A circuit that holds either peak of a waveform to a given d-c level is called a clamping configuration, or a clamper. This circuit action is shown in Fig. 21–15. We should note that the reference voltage level might be positive, or negative, or zero in various applications. The

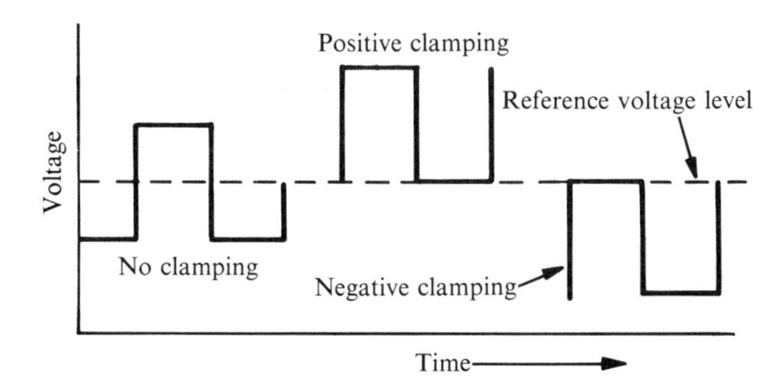

Figure 21–15. Visualization of clamping action.

simplest type of clamping circuit utilizes a diode in combination with an *RC* coupling circuit, as depicted in Fig. 21–16; this is a positive clamping circuit. In this example, the capacitor voltage is maintained at approximately the minimum applied voltage. Note the following points.

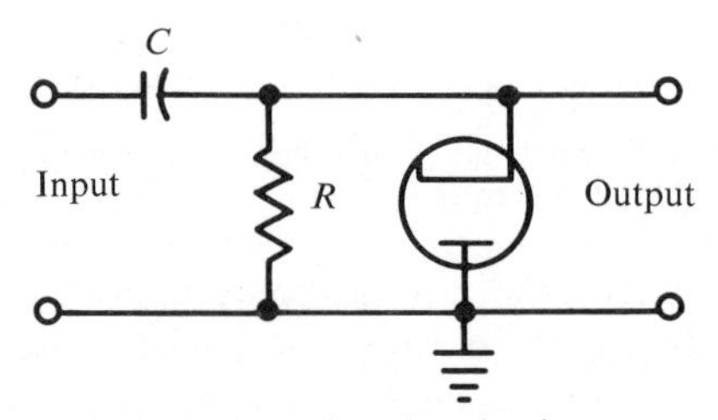

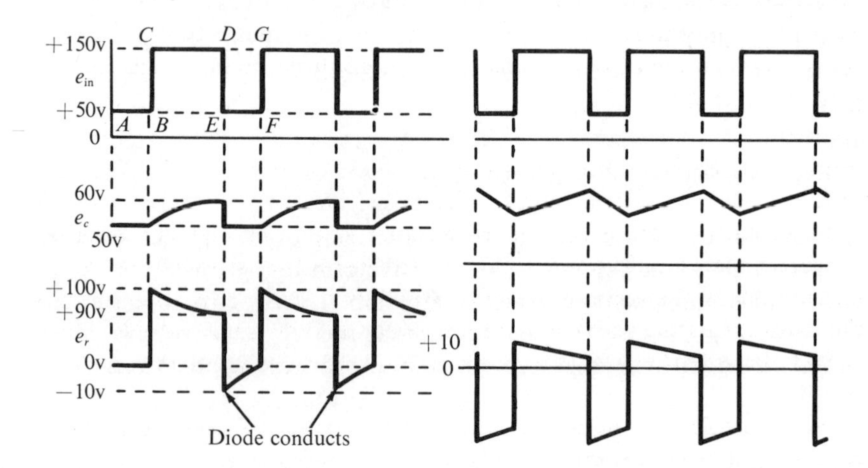

Figure 21–16. **Positive clamping circuit and voltage waveforms.**

1. If the cathode of the diode is driven negative with respect to the plate, electrons flow from cathode to plate and the tube can be regarded as practically a short circuit.

2. If the cathode is driven positive with respect to the plate, no current flows and the tube can be regarded as practically an open circuit.

The plate-voltage variation in a circuit generating a square-wave voltage is typical of the input (e_{in}) applied to the clamping circuit in Fig. 21–16. In this clamping circuit, capacitor *C* charges gradually through the high resistance R. After a period of time, depending on the RC time constant, the charge on the capacitor reaches 50 v, which corresponds to the base of the input waveform. Our problem is to maintain the charge at this value in spite of the tendency of the capacitor to charge to a higher level when the applied voltage rises to 150 v. The waveforms at the right in Fig. 21-16 will be obtained if no diode clamper is used.

Assuming that a steady voltage equal in magnitude to that at time A has been applied for some time, the capacitor may then be considered to be charged to 50 v. During the time interval between A and B, the charge on the capacitor is equal to the applied voltage, and no current flows through R. Then, at point B, the applied potential suddenly increases to 150 v. Because it is impossible for the charge on the capacitor to change instantaneously, the difference between the 150 v applied and the 50 v across the capacitor must appear across R. This difference of 100 v is the output voltage e_r.

It follows that when a voltage exists across R, that current must flow through the resistor. This current adds to the charge on C. Generally, the RC time constant is very long, and the charge added to C is small. For simplicity, we will assume that the 150 v potential is applied for a time equal to 0.1 RC—that is, the interval from point C to point D. Because the cathode of the diode is positive with respect to the anode (which is at ground potential), the tube is in effect an open circuit. During a time equal to 0.1 RC, the charge on the capacitor increases exponentially by 10 per cent of 100 v, or 10 v, making the total charge on the capacitor 60 v. During this same period, the drop across the resistor decreases exponentially by 10 v to a value of 90 v, leaving the sum of e_r and e_c still equal to the applied potential of 150 v.

At point D, the applied voltage suddenly drops back to 50 v. The capacitor, however, is charged to 60 v. This would leave an output voltage across R of 10 v negative with respect to ground—a condition that must be avoided. In order for the output to return to zero very quickly, the capacitor must discharge the extra 10 v through a path having a very short time constant. In Fig. 21-16, the cathode of the diode is connected to the high side of R, and the plate is grounded. Any output voltage that is negative with respect to ground makes the cathode negative with respect to the plate. Under this condition, the diode conducts. It becomes, in effect, a very low resistance discharge path for the capacitor until the charge is again equal to the applied voltage, and the output voltage returns to zero. At this time, the diode becomes nonconducting.

To illustrate the operation of the positive clamping circuit further, assume that the negative-going waveform shown in Fig. 21–17 is applied to the input of the clamping circuit. Because at point *A* the input voltage is zero, the output voltage is zero and remains so until

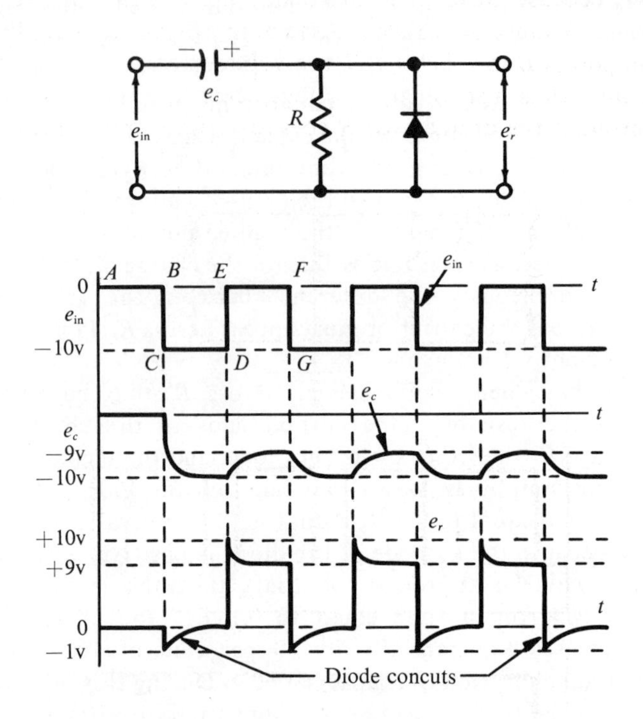

Figure 21–17. **Negative-going input signal applied to a positive clamping circuit.**

point *B* is reached. At this time, the input voltage drops suddenly toward −10 v at point *C*. Because the capacitor cannot change its charge instantaneously, the output voltage across *R* also drops suddenly toward −10 v. When the cathode of the diode is sufficiently negative with respect to the anode, the diode conducts, charging the capacitor very rapidly through the short *RC* time constant of the conducting diode and capacitor, and reducing the voltage across *R* to that across the conducting diode (almost zero). When the capacitor voltage becomes equal to the applied voltage, the diode becomes nonconducting. As long as the input remains −10 v, from points *C* to *D*, the output voltage remains at zero potential.

At point D in Fig. 21–17, the input voltage changes back to zero, a rise of 10 v in the positive direction (−10 to 0). This rise produces a drop of 10 v (0 to +10) across R, because again the capacitor cannot change its charge instantaneously. The capacitor must now discharge very slowly because the diode is nonconducting, and the high-resistance path through R must be utilized. Assuming again that the discharge time from points E to F is 0.1 RC, the voltage across the capacitor at point F, and thus the output voltage, decreases to 9 v (point G). Instantaneously, the drop across R goes to −10 v (input minus e_c). The

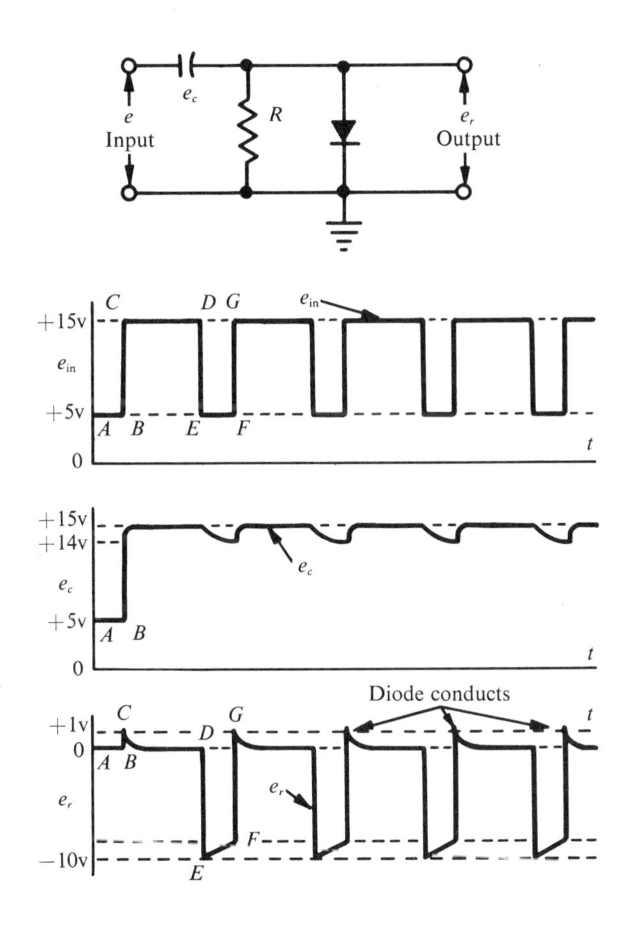

Figure 21–18. Negative clamping circuit and voltage waveforms.

diode conducts quickly, returning the charge on the capacitor to 10 v and the output to zero. The output voltage waveform is shown in the lower part of the diagram. Note that no portion of the waveform is lost after the first cycle. The function of the clamping circuit is merely to shift the waveform from above to below the zero-voltage reference level (ground).

A negative clamping circuit and its associated waveforms are illustrated in Fig. 21–18. This diode clamping circuit is capable of causing the output voltage to vary between some negative value (−10 v in this example) and the zero reference voltage. The only difference between this circuit and the one illustrated in Fig. 21–16 is in the manner in which the diode is connected. In Fig. 21–16, the anode is grounded, and the diode conducts if the cathode is made negative with respect to the anode. In Fig. 21–18, the cathode is grounded, and the diode conducts whenever the anode voltage rises above ground.

Next, let us consider the baseline leveling circuit depicted in Fig. 21–19. This is basically a clamper circuit that is used to shape the input waveform into a series of pulses positioned on a straight baseline.

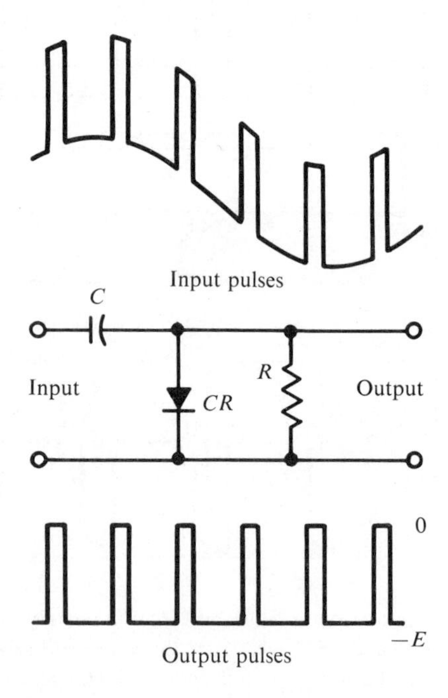

Figure 21–19. A baseline leveling circuit.

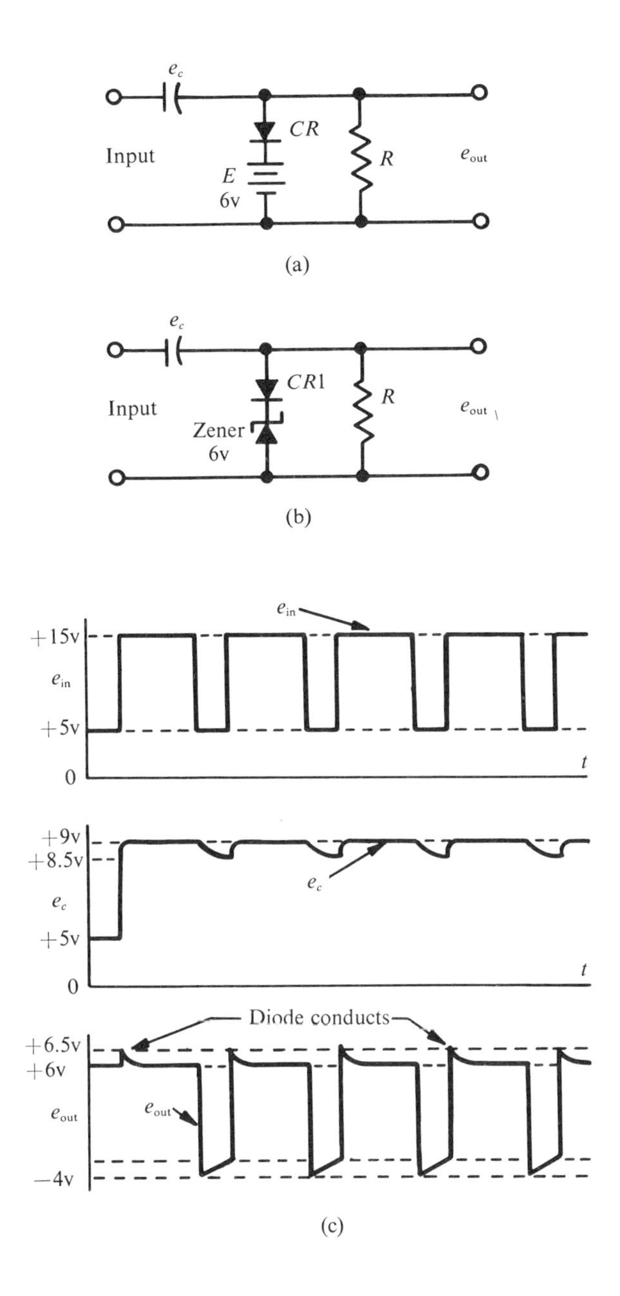

Figure 21-20. Biased negative clampers. (a) The battery voltage determines the peak positive output; (b) A Zener diode is employed to limit the output signal to +6 v; (c) Circuit voltage waveforms.

Again a semiconductor diode is used in this example instead of a vacuum diode. Since the input waveform is positive-going, the diode conducts on the positive excursion, and the capacitor C charges with its right-hand plate negative. Since the RC time constant is comparatively long, the d-c voltage across R is maintained at virtually a fixed value, and the pulse component of the input waveform appears across R. Next, consider the processing of the low-frequency baseline waveform; since R and C form a low-pass filter, the low-frequency component is greatly attenuated. The end result of the combined clamper and filter action is an output pulse waveform that is clamped to a practically straight baseline.

The clamper circuits depicted in Fig. 21–20 develop an output at some reference level other than zero volts. In the circuit in Fig. 21–20 (a), the battery voltage determines the peak positive output. The positive input signal charges capacitor C to the average value of peak input minus the battery voltage. On the negative-going cycle, the input opposes the capacitor voltage developing -4 v across R. Battery E may be reversed to clamp the output to a maximum value of -6 v.

The clamper circuit in Fig. 21–20(b) employs a 6 v Zener diode as a battery to limit the output signal to a maximum of $+6$ v. The Zener diode prevents diode $CR1$ from conducting until the input level exceeds V_z (6 v). As the input level exceeds V_z, capacitor charges to 9 v. The remainder or the circuit operation is identical to that of the clamper in Fig. 21–2(a). However, the Zener diode cannot simply be turned over to develop a clamper with a maximum of -6 v.

SUMMARY

1. A square wave may be developed by shaping a sine wave with clipper and amplifier stages.
2. In a series clipper the ratio of the series resistor to the diode ON resistance must be high and the ratio of the series resistor to the diode OFF resistance must be low.
3. A back-to-back Zener diode may be utilized to develop a square wave. Zener diodes in this configuration are manufactured in one case with two leads.
4. Grid or base clipping has the advantage of amplifying one-half of the input signal.
5. The output signal of an amplifier can be limited to zero volts by producing saturation. Saturation is accomplished by using a very high plate or collector-load resistor and a low value of supply voltage.

6. Clipping of the top or bottom of a waveform may be accomplished by locating the operating point of the control device near cutoff or saturation.

7. Both the top and bottom of a waveform may be clipped by overdriving an amplifier with a large input signal.

8. An oscilloscope calibrator develops a square-wave output at a particular amplitude to be used for calibration of the gain of the vertical amplifier so that the oscilloscope can be used as a voltmeter. Better quality laboratory oscilloscopes often have an internal calibrator.

9. A limiter provides a waveform output up to a specific value.

10. A clipper provides a waveform output above a specific value.

11. A compressor circuit reduces the peak amplitude of a waveform without clipping or limiting the waveform.

12. An *RC*-peaking circuit may be used in a waveshaper circuit to develop a linear current flow through an inductor.

13. A clamper or d-c restorer is used to hold either peak of a waveform to a given d-c level.

Questions

1. Explain the process of developing a square wave from a sine wave with clipper and amplifier stages.
2. Construct a schematic of two positive clipper circuits using semiconductor diodes, and explain the operation of each circuit.
3. Draw a schematic of a back-to-back Zener diode clipper which will clip a 100 v peak-to-peak sine wave (± 50 v) to an output of $+5$ v and -8 v. Discuss the operation of the clipper.
4. Show two methods to limit a sine wave which varies about zero to a value of 0 and $+10$ v.
5. Show a vacuum tube or transistor in an input clipper circuit which will produce a negative-going output signal limited at 0 v.
6. What is the approximate value of grid-to-cathode resistance of a triode when the grid draws current?
7. Explain the operation of a grid clipper.
8. In reference to the circuit in Fig. 21-5, explain how the amplifier clips the peak from the output waveform.
9. With reference to the circuit in Fig. 21-5, discuss the component and input signal values to produce square-wave output.
10. With reference to the circuit in Fig. 21-7, what are the necessary relative values of R_C and E_{CC} to produce collector-current saturation limiting?
11. With reference to the circuit in Fig. 21-8, what are the relative values of R_L, E_{BB}, and e_{in} to produce saturation and cutoff limiting?
12. Refer to the circuit diagram in Fig. 21-10(a). What type of limiter is used to develop a square wave? What is the function of S_1? What is the function of R_3? What is most probably wrong in the circuit if the output signal is limited only in negative half cycles?
13. What are three names for a circuit which clips some portion of both the positive and negative half cycles of a waveform?
14. Compare the output of a clipper to the output of a limiter.
15. Disuss the operation of a compressor circuit, and give an example of one use for the circuit.
16. With reference to the circuit in Fig. 21-18, what is the triode tube called, and how does the circuit develop the trapezoid waveform?
17. Refer to the waveforms in Fig. 21-13, and explain why a trapezoid waveform is required to develop a linear current flow through a practical inductor?

18. Referring to Fig. 21-16, give the average value of the input signal, the average charge on the capacitor, and the average value of the output waveform.

19. Design a Zener diode clamper to limit the positive-peak output to +8 when the input is a 20 v square wave varying about 0 v.

20. What is the function of the base-line clamper?

21. Compare the function of a clamper circuit to that of a limiter circuit.

CHAPTER 22

Fundamentals of Electronic Computers

22–1. Introduction

An electronic computer is often described as "a machine for doing mathematics," in the same manner as an adding machine (desk calculator). However, an electronic computer is much more than a "mathematics machine" in the traditional sense of the term. Not only does an electronic computer operate at extremely high speed—it is also a data processor. For example, it can operate as a sorting device and can be designed to perform rote translation of languages. A computer can perform similar mathematical operations hundreds, thousands, or millions of times without becoming tired. If the computer is well designed, the probability of its making an error in calculation is very small, even after a trillion operations. Modern computers can solve a problem millions of times faster than a skilled mathematician.

Special-purpose computers are designed to solve only one type of problem; on the other hand, general-purpose computers are designed to solve a range of mathematical operations. Although a computer might occasionally give an inaccurate answer, due to a component defect for example, many computers are designed to check their own answers; they thereby provide an extremely high probability of accuracy. Popular articles and books sometimes convey the impression that a computer is a "thinking machine". However, it has not been found possible to design a computer that thinks. In other words, a computer can only follow mechanical orders (its *program*); it cannot consider any new information (outside the program that is fed into it) no matter how important the new information might be. For example, a navigational computer will fly its damaged plane and wounded crew directly into enemy territory if it was previously programmed for this course of flight. A human pilot in this situation forms an opinion or judgment on the basis of the new information and attempts to bring the plane to a safe landing in friendly territory.

Similarly, an electronic computer has no intuition and operates in rote manner exactly as directed. On the basis of intuition, a student occasionally experiences a sudden insight into a problem and can state the correct answer without working out the details in traditional rote fashion. A computer, however, cannot take advantage of intuition and new insights into modes of problem attack; it can only proceed step by step in rote fashion according to the program that has been prepared for its operation. In a broad context, an electronic computer is a physical model of a mathematical operation or of a group of mathematical operations. For a simple example, a child's set of building blocks is a physical model of a set of counting numbers, although we recognize that building blocks are not numbers but only physical models of numbers. Analogously, electronic circuits are physical arrangements, and circuit actions are physical actions which engineers can arrange to model operations on mathematical ideas.

Recall that a formula, such as Ohm's law, is a mathematical model of a physical situation. An equation, such as a quadratic equation, expresses a mathematical operation on pure numbers. Conversely, if a computer is programmed to solve a quadratic equation, the computer operates as a physical model of a quadratic equation, and the computer circuit action serves as a physical model of the human operator upon numbers. Hence, we perceive that electronics technology entails both mathematical models of physical situations, and physical models of mathematical situations.

22–2. Basic Types of Computers

Computers are classified into the two basic types: analog and digital. Hybrid computers have both analog and digital characteristics.

Let us consider some fundamentals of analog computation. A comparison of Newton's law and Ohm's law shows that these formulas have a similar form:

$$F = MA \tag{22-1}$$

$$E = IR \tag{22-2}$$

In turn, we state that an electric circuit described by Ohm's law is analogous to a moving object described by Newton's law. This analogy permits us to solve Formula (22-1) by comparison with answers obtained in solution of Formula 22-2. For example, with reference to Fig. 22-1, if we set the circuit values as indicated, we can read the

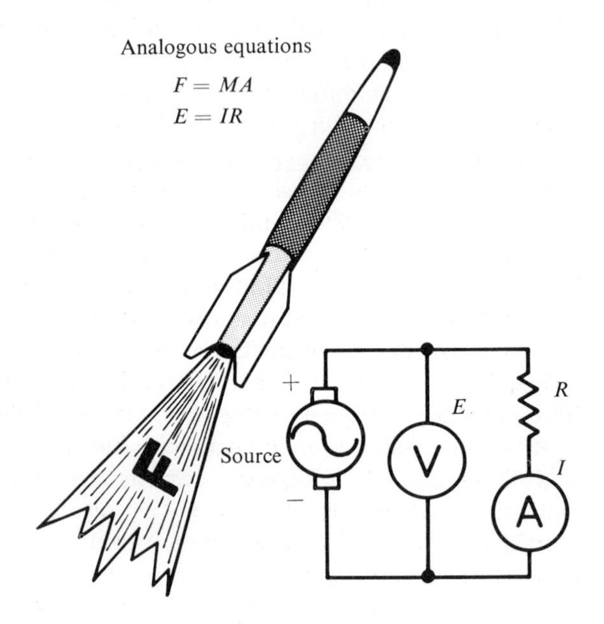

Figure 22-1. An example of analog computation.

voltmeter and state by comparison the value of force exerted by the rocket. In this analogous electric circuit, voltage E is proportional to force F; resistance R is proportional to mass M; current I is proportional to acceleration A. Finally, the product IR is proportional to the product MA.

We can call the electric circuit an analog computer. In engineering language, an analog computer is actually two devices that operate as a system. It is a *simulating* machine, because in a mathematical sense it is a model, or *analog*, of another system. It is also a mathematical machine in that it can be programmed to yield an answer to a mathematical problem. It is clear that the analog circuit depicted in Fig. 22–1 cannot think in the way that thought processes take place in a human mind. For example, human beings have a wide range of beliefs concerning truth and falsehood; we believe that the set of counting numbers can be extended indefinitely and that there is no "largest number." On the other hand, the circuits in a computer cannot state any belief in this matter; the circuits can only respond rotely according to their program.

We will find that a *digital computer* can also process Formulas 22–1 and 22–2, but in a different way. Instead of looking for some analogous physical system, the programmer regards $F = MA$ as a multiplication problem, and the particular numerical values are utilized immediately. In digital form, the problem is written

$$75 \times 2 = ? \qquad \textbf{(22–3)}$$

An office adding machine can be used as a digital computer to solve this problem; the digits 75 are punched twice, and the answer is read as the three digits 150. We perceive that the *input* to an analog computer is never an absolutely exact number; it is an approximate position on a continuous scale (for example, the position of a potentiometer arm) as shown in Fig. 22–2(a). Hence, there will always be a residual error, no matter how small it might be. On the other hand, the number fed into a digital computer that is operating normally is precisely the number desired. Thus, in its *processing*, an analog computer works with continuous values, while a digital computer works with discrete individual digits [Fig. 22–2(b)]. The *output* from an analog computer, accordingly, is never an absolutely exact number; it is merely an approximate position on a continuous scale. However, the output from a digital computer is a specific series of digits [Fig. 22–2(c)].

The *speed* of an analog computer is extremely great; most analog computers provide an answer almost instantaneously. A digital computer does not work quite as fast, although its speed is also very great; for example, a digital computer can solve a fairly simple problem in a few millionths of a second. *Presentation* differs for the two basic types of computers. Once an analog computer is set up to solve the previous problem, it is possible to do much more than multiply 75 × 2. By varying the voltage, the answers to a number of problems can be rapidly obtained, such as 75 × 2, 75 × 2.1, 75 × 2.2, and so on. However, it is much more difficult to obtain a spectrum of answers with a digital computer. It can be done, but it would require a separate program for each answer. Automatic digital computers multiply one

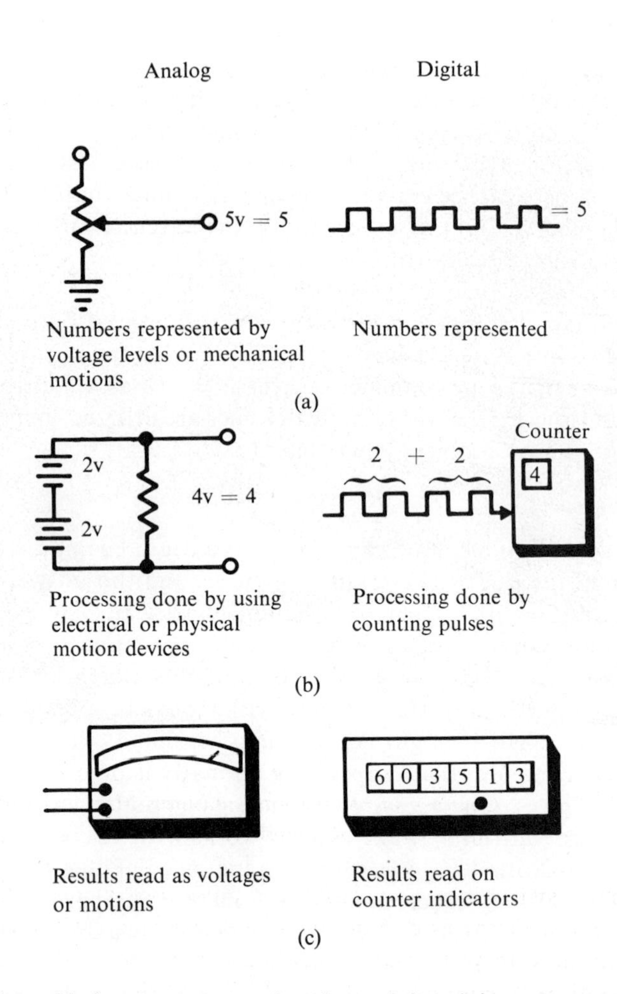

Figure 22-2. **Comparisons of analog and digital data processing.**

pair of numbers, record the answer, multiply another pair of numbers, record the answer, and so on. Obviously, somewhat more time is required than if an analog computer were connected to a recording voltmeter. The digital computer does have a great advantage in that it can store information for immediate or future reference.

22-3. Digital Computers

Electronic circuits used in digital computers (Fig. 22-3) operate on the basis of "off" or "on." In other words, computation is accomplished

Courtesy of International Business Machines Corporation
Figure 22–3. A digital computer.

by electronic switching, and basic computer operation entails the binary system of arithmetic. You have learned the fundamentals of binary arithmetic in your mathematics courses and will recall that two digits only are employed: 0 and 1. These digits correspond to "on" and "off" states of circuits in a digital computer. Binary digital computers can add, subtract, divide, and multiply. For simplicity, let us briefly review binary addition only. We will first consider the following pencil-and-paper addition of two binary numbers.

$$\begin{array}{r l} \underline{1} & \text{carried} \\ 10 & = 2 \\ \underline{11} & = \underline{3} \\ 101 & = 5 \end{array} \qquad \textbf{(22–4)}$$

We recall that this addition is performed by observing the following rules at the top of page 678.

$$
\begin{aligned}
&0 \text{ and } 0 = 0\\
&0 \text{ and } 1 = 1\\
&1 \text{ and } 0 = 1\\
&1 \text{ and } 1 = 0, \text{ and carry } 1
\end{aligned}
\qquad \textbf{(22–5)}
$$

A digital computer is an electronically operated machine or system that receives data and processing instructions and then responds by circuit actions to produce the result of the instructions. The computer rapidly counts electrical pulses that represent numbers (binary digits). These pulses are d-c voltages that rise quickly from zero to a maximum value, and then quickly fall to zero; the pulse width is commonly a few microseconds or less, and the pulse repetition rate may be 100 kHz or more pps (pulses per second).

One of the basic circuits employed as a building block in a digital computer is the flip-flop or bistable multivibrator that we studied in Chapter 16. A typical configuration is shown in Fig. 22–4.

To review briefly, when the right hand transistor $Q3$ is off, its collector voltage is almost equal to the supply voltage (a small current flows through R_1, R_{B2}, and R_{C2}). The voltage division by R_1, R_{B2} and R_{C2}, at the base of $Q2$, is slightly more than the voltage developed across R_E, and $Q2$ is biased ON. In turn, the collector voltage of $Q2$ is low due to the relative drop across R_{C1}, resulting in approximately zero volts at the base of $Q2$. The voltage across R_E results in $Q3$ being biased OFF.

We find that the driver transistor $Q1$ depends upon the collector voltage of $Q2$ for bias. As a result, the lamp is OFF indicating a "0" (zero) condition. Application of a negative input pulse to the junctions of diodes $CR1$ and $CR2$ produces conduction through diode $CR1$ which results in a negative-going pulse at the collector of $Q2$. This negative-going pulse is coupled through R_{B2} to the base of $Q2$ decreasing the conduction of $Q2$. The resulting increase in collector voltage is coupled back to the base of $Q3$ to drive $Q3$ into saturation.

The very low voltage at the collector is coupled through the voltage divider R_{B2} and R_1 to bias $Q2$ in the off condition. In this stable state, transistor $Q1$ is biased on and the lamp indicates a "1" condition of the flip-flop.

A second trigger applied to the input of the flip-flop circuit will cause it to go from the "set" (or "1") condition ($Q2$ OFF and $Q3$ ON) to the "reset" or "0" state ($Q3$ OFF and $Q2$ ON). This flip-flop has useful positive output from either A ("1") or $\bar{A}$ ("0"). The $\bar{A}$ is called "not A." Either of these outputs may be utilized to control other gates or flip-flops.

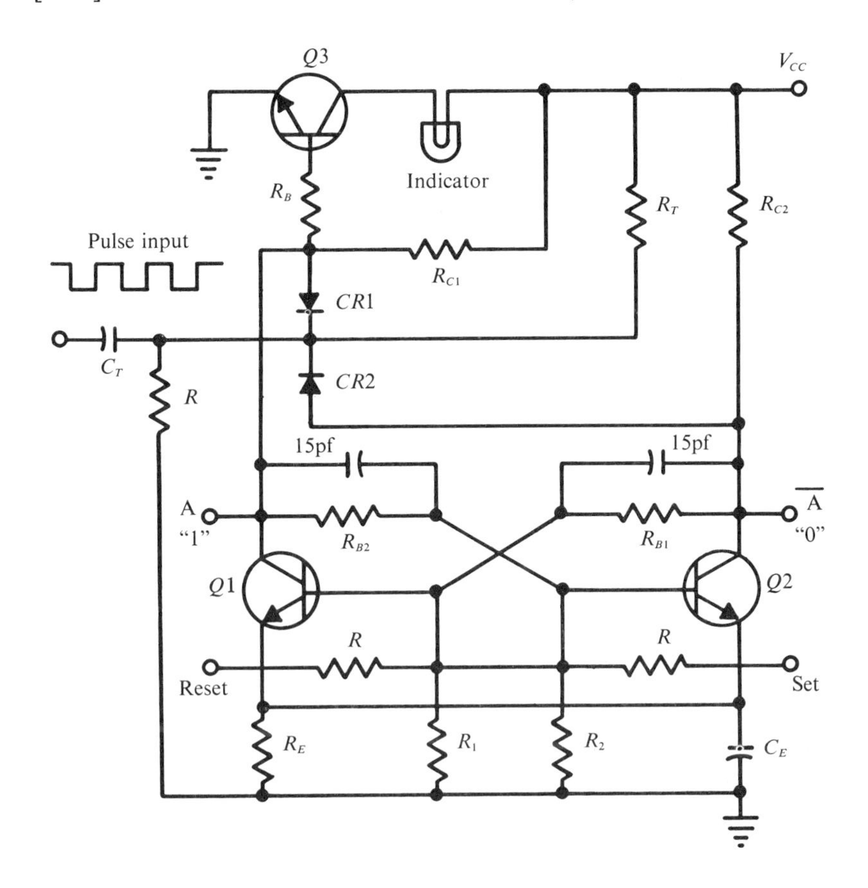

Figure 22-4. A flip-flop configuration for a circuit section in a digital computer.

In brief, the flip-flop (FF) has two stable states, and changes from one state to the other when an input is applied. For each two input triggers, either side of the FF produces an output pulse. In other words, the flip-flop divides the input by "2" and, therefore, a series of flip-flops will add pulses and show their sum in binary form.

Assume that four flip-flop circuits as depicted in Fig. 22-4 are connected in cascade to form a four-digit binary counter, as shown in Fig. 22-5. Neon indicators may be used to indicate a "1" or "set" condition when they are glowing, or a "0" or "reset" condition when they are off. The four indicators may be read from left to right as a four-digit binary number. Note that input pulses to be counted are applied to the right-hand flip-flop, which represents the least signifi-

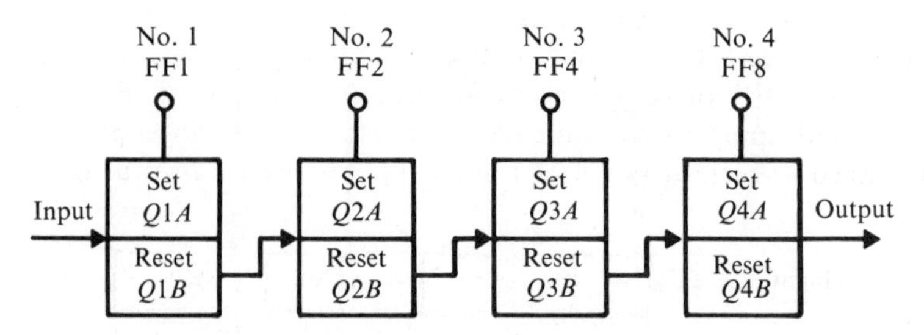

Figure 22-5. **Block diagram for a binary counter comprising cascaded flip-flops.**

cant digit. We will assume that all four flip-flops are in the "0" (reset) condition (upper triodes OFF), and that a negative pulse is applied at the input. FF_1 changes state from reset to set, and its indicator is energized; however, FF_1 produces no output, and the other three flip-flops continue to indicate zero. The four indicators may be read as the binary number: 0001. Upon application of a second negative pulse at the input, FF_1 returns to zero, and provides a negative-going input to FF_2. Thereupon, FF_2 changes to the "1" state, but produces no output. The binary indication is now 0010 (the binary equivalent of 2 in decimal notation), indicating that two input triggers have been counted.

With the third input trigger applied to the counter in Fig. 22-5, FF_1 changes to the "1" (set) state; however, it produces no output. FF_2 therefore remains in the "1" state, and the binary indication is now 0011 (the binary equivalent of 3 in decimal notation). On the fourth trigger input, FF_1 returns to zero (set to reset) and produces an output. FF_2 returns to zero (set to reset) and also produces an output. FF_3 changes to the "1" state (reset to set), but produces no output. The binary indication is now 0100 (the binary equivalent of 4 in decimal notation).

After fifteen input triggers, the neon tubes will indicate the binary number 1111 (decimal 15). On the sixteenth input trigger, FF_1 changes to zero and produces an output: FF_4 therefore changes to zero and produces an output, which changes FF_8 to zero. The counter accordingly returns to its original state (0000) after sixteen input triggers. If a fifth flip-fiop were added, receiving its input from FF_8, the counter would count up to 11111 (decimal 31), and return to the original state on the 32nd input. To summarize briefly: in the "0" condition, all the flip-flops are OFF, or in the reset condition (see part A of Fig. 22–6). Next, we will apply three input trigger pulses to the flip-flops, which are arranged in series (part B of Fig. 22–6). The first pulse turns No. 1

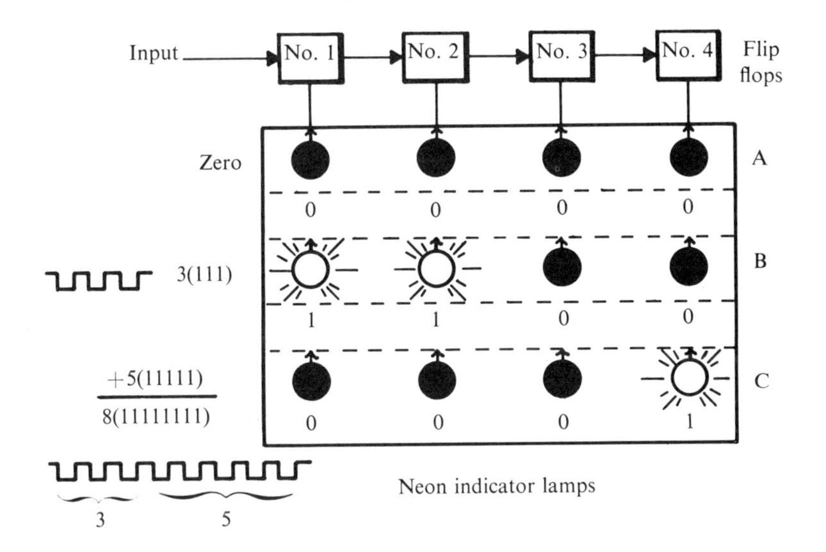

Figure 22–6. A chain of flip-flops can perform the operation of addition.

from nonconducting, OFF, to conducting, ON. This action energizes the neon lamp associated with the first flip-flop. The second pulse to No. 1 causes it to go from "1" to "0" and to produce an output to No. 2, switching it ON. The third pulse switches No. 1 ON again.

After three trigger pulses, the indicator lamps for flip-flops Nos. 1 and 2 are "on", while Nos. 3 and 4 are dark. If a glowing lamp equals 1 and a dark lamp equals 0, the binary indication is 0011. This indication corresponds to 3 in decimal notation. Now, we will apply 5 more pulses to the system. The first pulse of this group produces outputs from Nos. 1 and 2 since they are ON, thereby switching No. 3 to ON, and in passing turning Nos. 1 and 2 to OFF. The second pulse in the series of 5 merely turns No. 1 ON (no output), so that the indication is 0011, which denotes the number 5 in decimal notation. If we proceed in this manner to 8 triggers, the flip-flop will indicate 1000, as shown in part C of Fig. 22–6.

The foregoing example illustrates the counting function possible with a series of flip-flops. With modifications and elaborations of this and the addition of other circuits, it is possible to perform all the fundamental operations in arithmetic. Note too that the flip-flops are capable of storing information. In other words, a flip-flop chain is a memory. In this case, they remembered the binary number 3 until the pulse train representing 5 was applied. These two numbers were added and the binary sum was stored. In a digital computer, the unit that stores a number, then adds another number to the stored number, and then stores the sum of the two numbers is called an *accumulator*.

22–4. Basic Logic Circuits

To perform arithmetical operations, digital computers employ circuitry to make logical decisions (but not to reason) about physical situations which can be logically decided upon. In other words, performance of arithmetical operations entails more than flip-flop counter circuits alone. Some logical decisions that a computer may be required to make are (1) when to perform an operation; (2) what operation to perform; (3) which of several ways to perform the operation.

These circuit actions are called *machine logic*. We recall that circuits never reason why; they merely operate from instructions prepared by a programmer who orders the machine to make certain decisions. The computer is incapable of making any decision until its program is fed into it. The program is a series of coded instructions which are stored in the computer's memory; these instructions tell the computer how to store, locate, process, and present the data it receives. In brief, solutions to various problems are solved on the basic of *logical truths* that are modeled by circuit actions. We employ logical truths in everyday life without realizing it. Our simpler logical patterns are distinguished by words such as AND, OR, NOT, IF, ELSE, THEN. In mathematical terms, these logical patterns may be indicated by plus, minus, times, divided by, or combinations of these terms.

We will find that our physical model of binary mathematical operations—a digital computer—can be constructed more economically if designed with switching circuits that are physical models of the words AND, OR, and so on. If a set of statements is made, it is then possible to establish a table that will determine what is called the truth values of these statements. In compiling this table, it is necessary to use some of the logical connectives, such as AND, OR, NOT, and so on. We can then tabulate the truth or falsity of each of the conditions for the statements under consideration. Furthermore, it is possible to set up mathematical relations of the truth values of statements and to perform mathematical operations on these statements. This study of mathematical logic is called the *algebra of logic*. It is of great importance for the student to recall that *machine logic* is devoid of reason and that only the programmer can reason.

A mathematician named George Boole demonstrated in 1847 that when logical statements are written in certain symbolic forms, these symbols can be manipulated very much like algebraic symbols. There are many types of symbolic logic, but we will need to consider only one type. This is commonly called *symbolic logic*, or *Boolean algebra*. Machine logic entails a physical modeling of symbolic logic. In the arithmetical operations that we learned in primary school, there are four basic operations: addition, subtraction, multiplication, and division. However, in Boolean algebra there are only three basic operations: AND, OR, NOT. If these words do not sound mathematical, it is because we have become sophisticated in our thought processes. For example, a small child understands that "two and two are four" before he makes the more sophisticated assertion that $2 + 2 = 4$. Just as "and" means "plus", so do "or" and "not" mean various steps in mathematical operations.

For convenience, the familiar arithmetic symbols are used for two of the Boolean algebra operations. Thus,

$$\text{AB means A and B} \tag{22-6}$$

$$\text{A + B means A OR B} \tag{22-7}$$

Note that the OR operation is indicated by the addition symbol, while the AND operation is indicated by the multiplication symbol. If desired, we can use any of the multiplication symbols, such as the parentheses, or the dot, to indicate the AND operation. Thus,

$$\text{(A + B)(C) means A OR B, AND C} \tag{22-8}$$

$$\text{A} \cdot \text{B means A AND B} \tag{22-9}$$

The NOT operation is indicated by a bar over the letter. Thus,

$$\bar{A} \text{ means A NOT} \tag{22-10}$$

$$\text{A } \bar{B} \text{ means A AND B NOT} \tag{22-11}$$

Let us consider the physical model for the AND operation. It is a simple electric circuit, as depicted in Fig. 22-7. Here, we see a circuit that lights a lamp. In this example, series switching is used to represent the AND operation. The condition of each switch may be denoted by 0. Again, the closed condition of each switch may be denoted by 1. Absence of current flow through the lamp is denoted by 0, and current flow is denoted by 1. A *truth table*, which is a shorthand

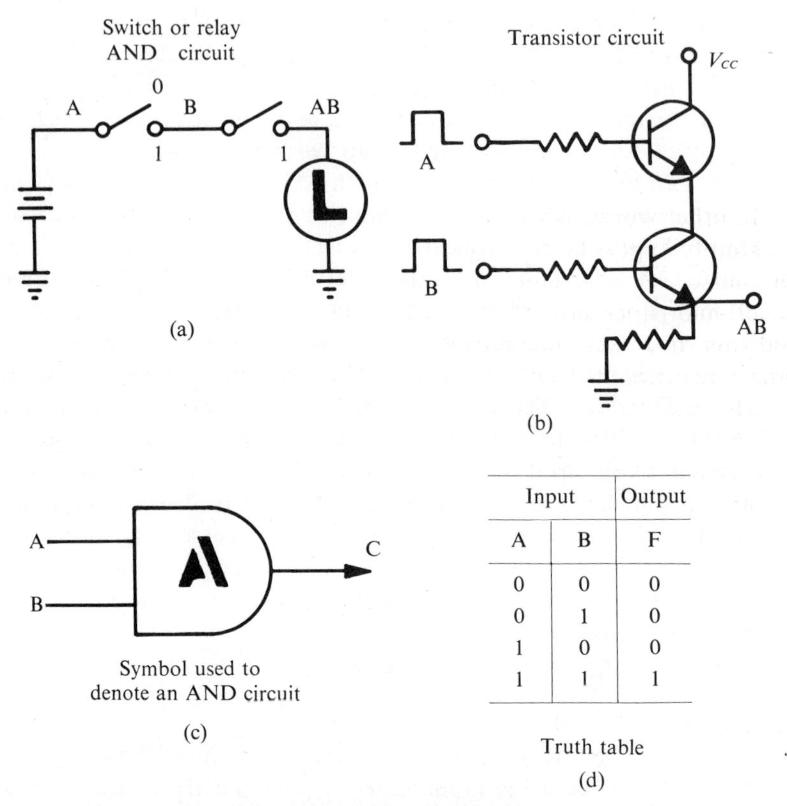

Input		Output
A	B	F
0	0	0
0	1	0
1	0	0
1	1	1

Figure 22–7. **Basic AND circuit. (a) Switch symbolism; (b) Transistor circuit; (c) Circuit diagram; (d) Truth table.**

method of describing logic-circuit action, may be employed to list the combinations of conditions and results. If you observe the truth table in Fig. 22–7, you will see that the AND circuit represents *multiplication.* In this circuit, if a switch is open, it is tabulated in the truth table as 0; if a switch is closed, it is tabulated in the truth table as 1. A 0 in the output column indicates no-flow of current, and a 1 in the output column indicates current flow. The truth table shows that there is just one condition of the switches that permits current flow through the lamp (output). This condition exists when switches "A" and "B" are both closed.

Example 1. To make certain that we clearly understand this physical model of multiplication, let us briefly review a simple problem. We will multiply 10011 by 1011.

$$\begin{array}{r}
10011 \\
1011 \\
\hline
10011 \\
10011 \\
00000 \\
10011 \\
\hline
11010001
\end{array}$$

In other words, when we calculate the product of 10011 and 1011, we simply apply the rules tabulated in the truth table of Fig. 22–7. Of course, we also employed the rules for binary addition in our pencil-and-paper calculation. Recall that an OR circuit represents addition. If it were not necessary to perform the carry operation in binary addition, the OR circuit could be used alone. However, this is not the case, so we must use OR circuits in combination with AND and NOT circuits to perform addition of numbers that require a carry in an adder circuit.

We can represent an OR circuit electrically by using two single-pole knife switches connected in parallel as shown in Fig. 22–8. There

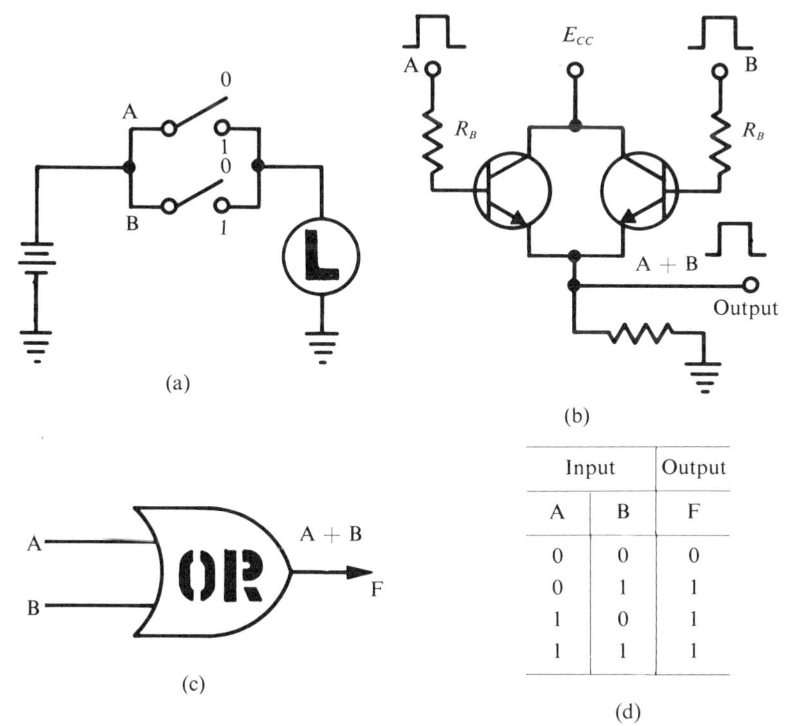

Input		Output
A	B	F
0	0	0
0	1	1
1	0	1
1	1	1

(d)

Figure 22–8. Basic OR circuit. (a) Simplified as switches; (b) Transistor configuration; (c) Circuit symbol; (d) Truth table.

is an output signal at F if an input signal is placed at A or B. In this case, placing a signal at a point in the circuit means to close a switch. We have used only two inputs to keep the illustration simple; however, there can be any number of input signals to an OR circuit. Any one or more of these input signals can produce an output signal.

The last of the basic logic circuits that we will describe is the NOT circuit. The requirements for a NOT circuit are that a signal injected at the input shall produce no signal at the output, and that no signal at the input shall produce a signal at the output. A switch and a relay, for example, can be connected as shown in Fig. 22–9(a) to form a simple NOT circuit. If we identify a closed contact as a signal, and an open contact as NO signal, we see that this connection fulfills the binary requirements for a NOT circuit. A signal (closed contact)

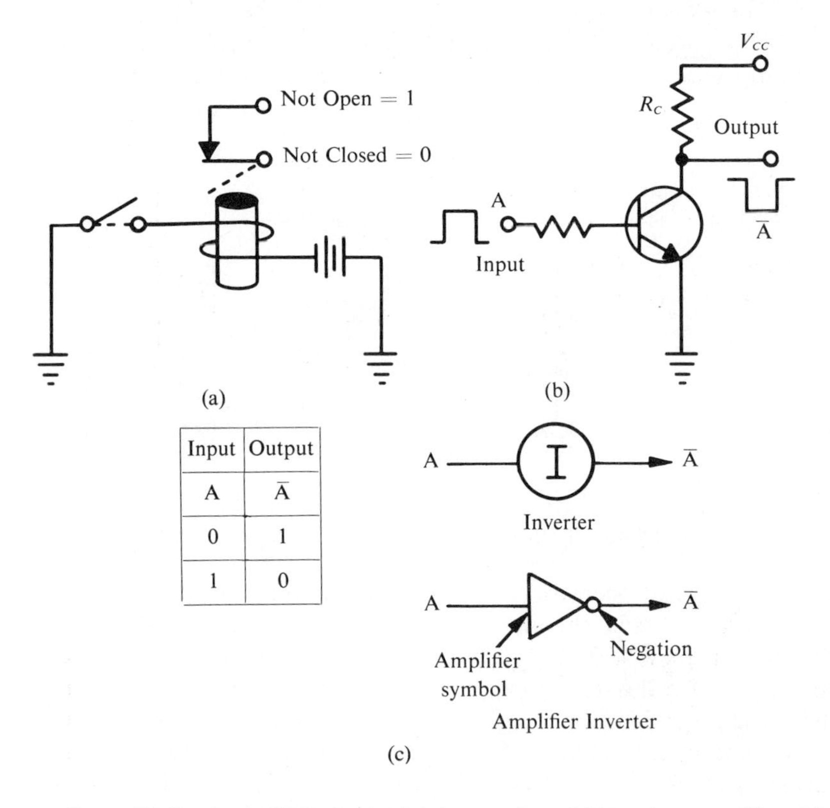

Input	Output
A	$\bar{A}$
0	1
1	0

Figure 22–9. Basic NOT circuit. (a) As a relay; (b) Transistor amplifier; (c) Truth table.

at its input produces NO signal (open contact) at its output, and vice versa. The truth table shown in Fig. 22-9 illustrates the shorthand method of describing the NOT circuit's function. The symbol "$\bar{A}$" [which we read "NOT A" (or "bar A")]means the opposite of A. This opposite or inverted condition of A points out the reason why the NOT circuit is sometimes called an INVERTER. The transistor inverter in Fig. 22-9(b) represents another NOT circuit; when a positive input ("1") is applied to the base zero ("0") volts, output is developed at the collector.

The half adder circuit shown in Fig. 22-10 is an example of the

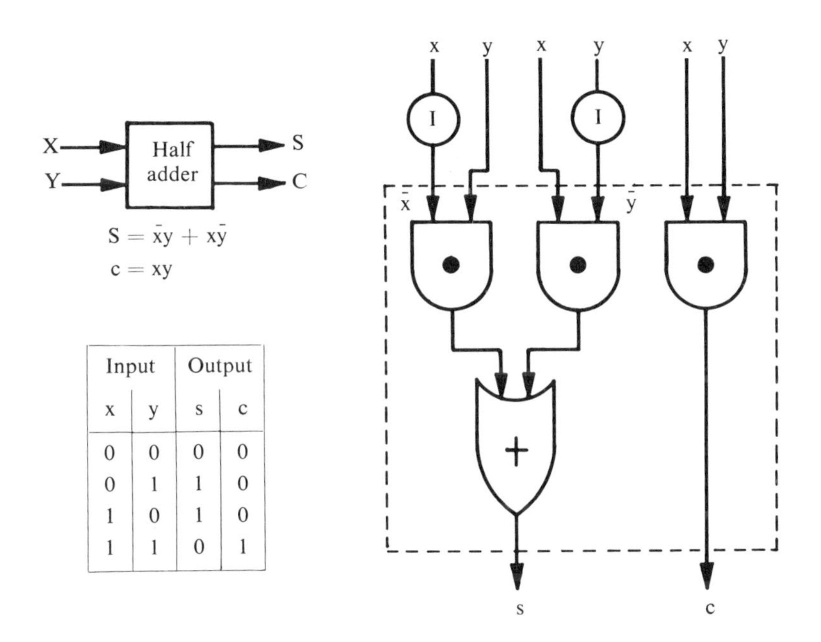

Input		Output	
x	y	s	c
0	0	0	0
0	1	1	0
1	0	1	0
1	1	0	1

Figure 22-10. Half-adder block diagram and truth table.

basic logic circuit operating to perform an arithmetic operation.

In the last decade, the physical size of computers has decreased by a factor of at least 100 : 1. The first step in this evolution was the application of transistors to computer circuits in place of vacuum tubes. The second step, which is still in process, is the development of micro-logic circuits in which each small module will replace many transistors, resistors, and diodes. Figure 22-11 illustrates a transistor 0-10-0 counter which is mounted in a standard TO-5 transistor case with 8 leads. The photograph of the micro-logic circuit layout in

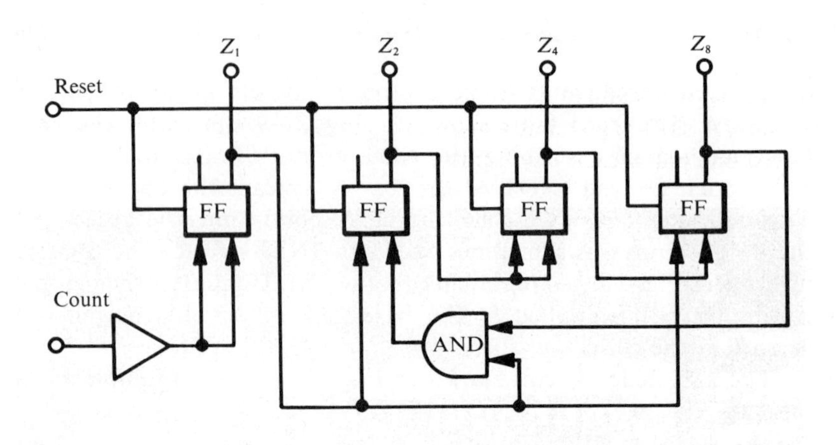

(b)

Count H = High, L = Low

	0	1	2	3	4	5	6	7	8	9
Z_1	H	O	H	O	H	O	H	O	H	O
Z_2	H	H	L	O	H	H	L	L	H	H
Z_4	H	H	H	H	O	O	O	O	H	H
Z_8	H	H	H	H	H	H	H	H	O	O

(c)

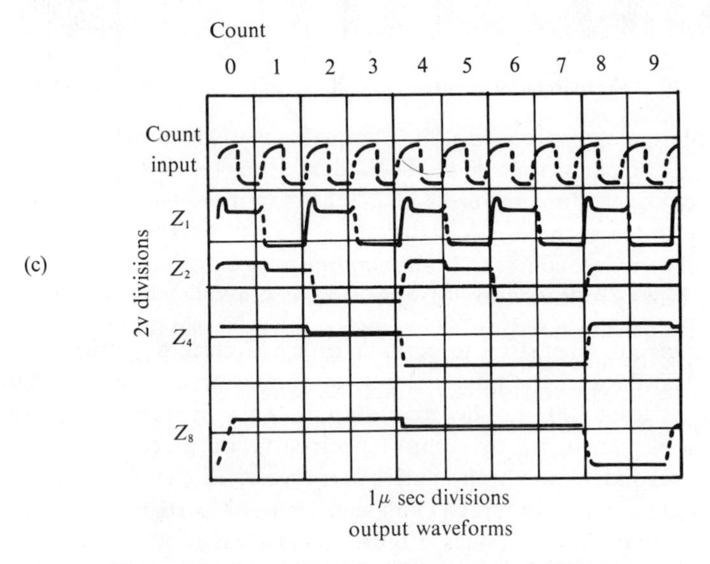

Figure 22-11. **Micro-logic 0-10-0 counter circuit. (a) Circuit diagram; (b) Truth table; (c) Output waveforms.**

Fig. 22-12 is expanded 26 times to show the detail of the 48-bit shift register. A shift register is a storage register in which a binary bit is shifted one place each clock pulse.

Figure 22-12. Photograph of a micro-logic 48-bit shift expanded 26 times.

A complete electronic computer comprises many sections comprising the basic circuits that have been described to perform binary mathematical operations in terms of machine logic. You will have an opportunity to learn the organization of a digital computer in your advanced electronics courses. Many computers contain circuits for converting binary numbers to decimal numbers, so that answers are displayed in decimal notation. The programmer must have a clear understanding of machine logic, so that he can convert a problem that is stated in decimal notation into machine language. Some computers contain circuits for automatic programming; these circuits minimize the labor of program preparation.

SUMMARY

1. A digital computer is a member of a family of machines known as data processors. There are three characteristics of a modern

digital computer: extremely high speed, possibly over a million operations per second; an addressable storage unit to hold the sequences of operations to be carried out automatically; and the ability to choose one of several paths in a sequence by the information present.

2. An electronic computer is not a brain in that it cannot think, it cannot feel, and it has no intuition.

3. An analog computer works with continuous values, while a digital computer works with discrete individual digits.

4. A series of flip-flop circuits may be used as a frequency divider or binary counter.

5. The action of circuits which are required to make decisions in a computer is called machine logic.

6. A truth table is a tabulation of a condition of a circuit.

7. The three basic operations in Boolean algebra are AND, OR, NOT.

Questions

1. What is the general function of a digital computer?
2. What are the limitations of a computer?
3. Compare the operating principles of an analog computer to those of a digital computer.
4. Compare the speed of an analog computer to that of a digital computer.
5. Show by a block diagram how bistable multivibrators may be connected to count in the binary system.
6. What are three logical decisions that a computer may be required to make?
7. Explain the terms AND, OR, and NOT as used in Boolean algebra.
8. Add the binary numbers 10111 11001.
9. How is a NOT function indicated in a Boolean equation?
10. Why is the NOT circuit sometimes called an inverter?

Appendix

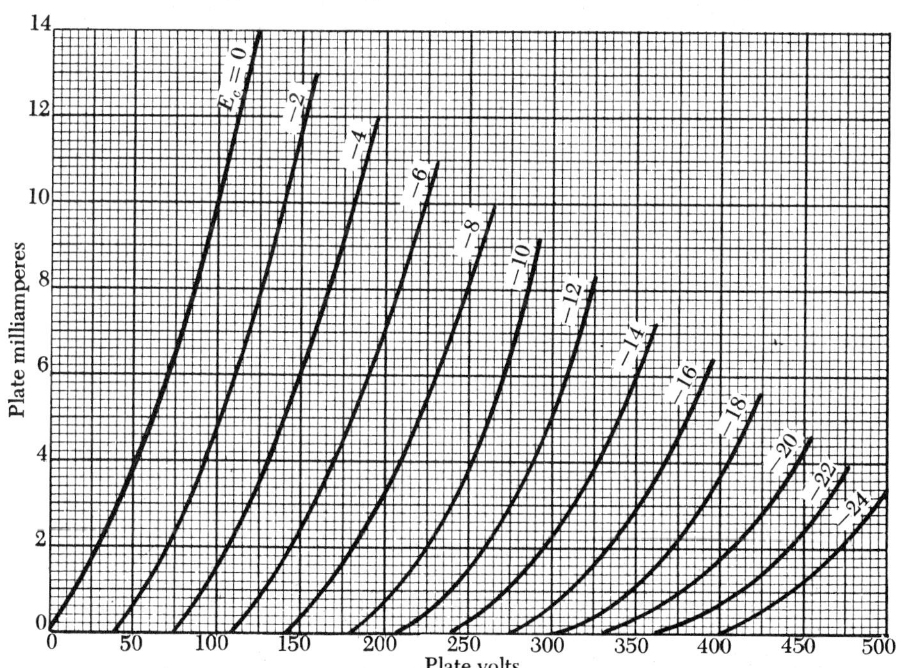

Plate characteristics, 6J5 triode, with the operating conditions $E_{bo} = 250$ v, $I_{bo} = 9$ ma, $E_c = -8$ v, $r_p = 7700\ \Omega$, $\mu = 20$, $g_m = 2600\ \mu\mho$.

Figure A-1

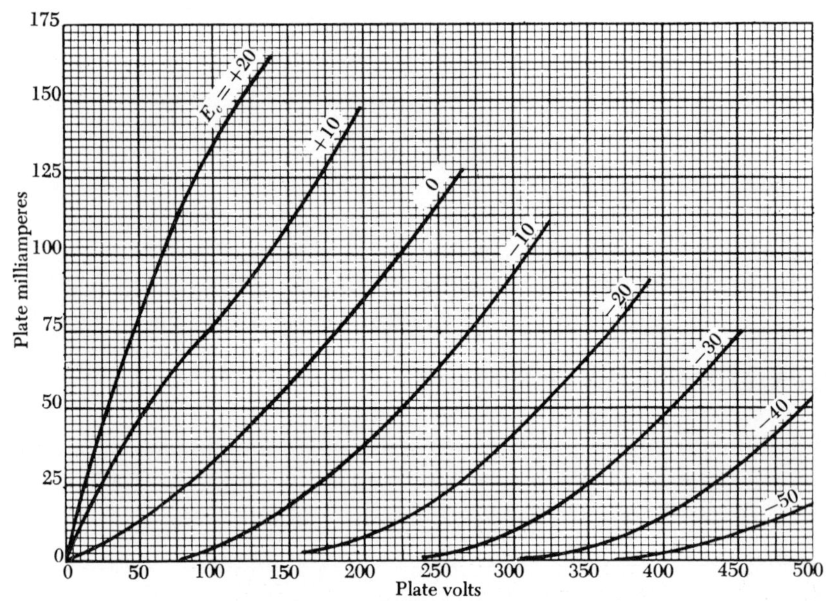

Plate characteristics, 6V6 triode connection.

Figure A-2

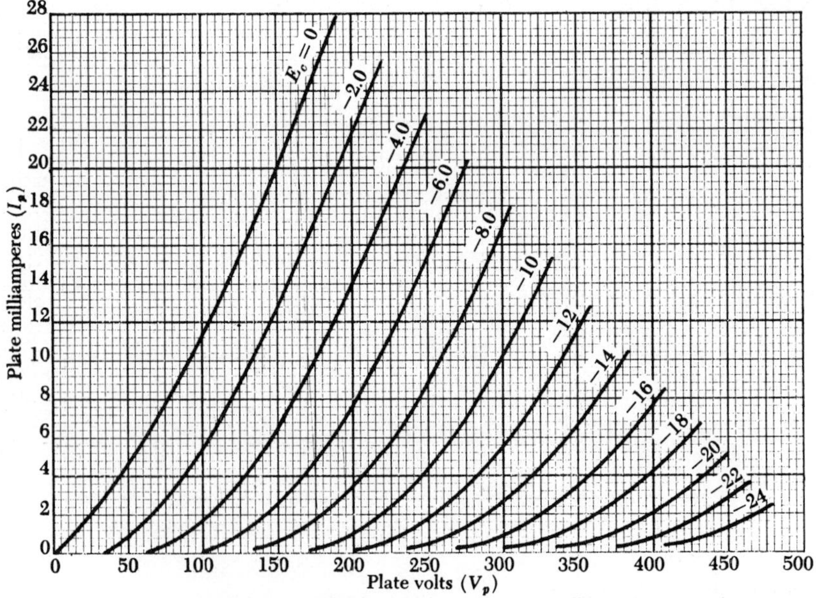

Plate characteristics, 6SN7GT, with the operating conditions E_f = rated value, $E_{bo} = 250$ v, $E_c = -8$ v, $I_{bo} = 9$ ma, $r_p = 7700\ \Omega$, $g_m = 2600$ μ℧, $\mu = 20$.

Figure A-3

Symbols and Formulas

ELECTROSTATICS

$F = \dfrac{Q_1 Q_2}{Kd^2}$	Force between charges
$E = \dfrac{Q}{Kd^2}$	Electric field intensity
$E = \dfrac{V}{d}$	Field between plates
$F = QE$	Force on Q in field E
$W = QV = eV$	Work
$C = \dfrac{Q}{V}$	Capacitance
$V = \dfrac{Q}{d}$	Potential

DIRECT CURRENT

$E = IR$	Ohm's Law
$P = EI = I^2R = \dfrac{E^2}{R}$	Power

Series Circuits

$R_t = R_1 + R_2 + R_3 + \ldots$ Total resistance in a series circuit

$E_t = E_1 + E_2 + E_3 + \ldots$ Total voltage in a series circuit

$I_t = I_1 = I_2 = I_3 = \ldots$ Total current in a series circuit

$P_t = P_1 + P_2 + P_3 + \ldots$ Total power in a series circuit

Parallel Circuits

$R_t = \frac{R_1 R_2}{R_1 + R_2}$ Total resistance — Two resistors in parallel

$R_2 = \frac{R_t R_1}{R_1 - R_t}$ Unknown resistance — Two resistors in parallel with total resistance known

$G_t = \frac{1}{R_t}$ Total conductance

$G_t = G_1 + G_2$ Total conductance

$R_t = \frac{1}{\frac{1}{R_1} + \frac{1}{R_2} + \frac{1}{R_3} + \ldots}$ Total resistance with any number of resistors in parallel

$E = E_1 = E_2 = E_3 = \ldots$ Total voltage in a parallel circuit

$I_t = I_1 + I_2 + I_3 + \ldots$ Total current in a parallel circuit

$P_t = P_1 + P_2 + P_3 + \ldots$ Total power in a parallel circuit

Resistance

$R = P \frac{l}{d^2}$ Value of resistance as a function of the material and its physical dimensions

where R = resistance in ohms
P = resistance in ohms per *circular-mil-foot* of the material (specific resistance)
l = length of conductor *in feet*
d = diameter of wire *in mils*

Meters

$$R_{\text{shunt}} = \frac{I_m R_m}{I_t - I_m}$$ Shunt required to extend range of ammeter

where I_m = current for full scale deflection
R_m = meter resistance
I_t = extended range of cuttent

$$\text{Ohms per volt} = \frac{\text{Resistance of meter}}{\text{Full scale reading in volts}}$$ Sensitivity of a meter

$$\text{or} = \frac{1}{\text{full scale current}}$$

$$R_m = (E \times (E \text{x} \Omega/V) - R_v$$ Multiplier required to extend range of voltmeter

where E = extended range of voltage
Ω/V = sensitivity of meter
R_v = meter resistance

Networks

$$R_1 = \frac{R_a R_b}{R_a + R_b + R_c}$$

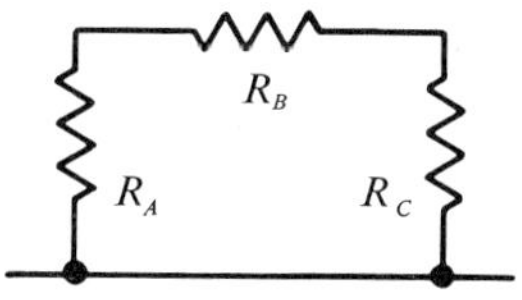

R_B R_A R_C

π to T transformation

$$R_a = \frac{R_1 R_2 + R_2 R_3 + R_3 R_1}{R_2}$$

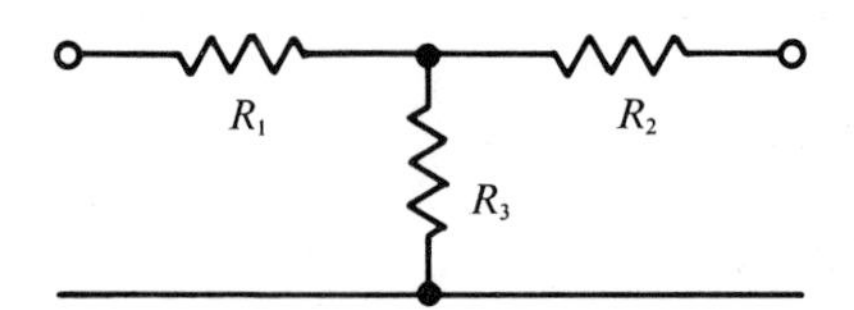

T to π transformation

MAGNETISM AND ELECTROMAGNETS

$F = \dfrac{m_1 m_2}{\mu d^2}$ Force between two magnetic poles

where μ = permeability of the medium

$H = \dfrac{F}{m}$ oersteds Magnetic field intensity

$H = \dfrac{2I}{10r}$ oersteds Magnetic intensity about a conductor

where I = current in amperes
r = radius in centimeters

$H = \dfrac{4\pi NI}{10l}$ oersteds Magnetic intensity for long coils

$H = \dfrac{2\pi NI}{10r}$ oersteds Magnetic intensity for short coils

$\text{mmf} = \dfrac{4\pi NI}{10}$ gilberts Magnetomotive force

$\phi = \dfrac{\text{mmf}}{R}$ maxwells Flux

$R = \dfrac{L}{\mu A}$ rels Reluctance

$B = \dfrac{\phi}{A}$ gauss Flux Density

$F = \dfrac{BIl}{10} \sin\theta$ dynes Force exerted on conductor

$\mu = \dfrac{\beta}{H}$ Permeability

Induction

$e = -N\dfrac{\Delta\phi}{\Delta t} \times 10^{-8}$ Average Induced voltage

$e = -M\dfrac{\Delta i}{\Delta t}$ Induced voltage

where M is mutual inductance

$M = K\sqrt{L_1 L_2}$ Mutual Inductance

where K is the coefficient of coupling

$L = \dfrac{N\phi}{I} \times 10^{-8}$ Inductance

where N is number of turns
I is current in amperes

$e_L = -L\dfrac{\Delta i}{\Delta t}$ Voltage across inductance

$e_L = -N\dfrac{\Delta\phi}{\Delta t} \times 10^{-8}$ Voltage across inductance

ALTERNATING CURRENT

General

$e = E_{max} \sin \omega t$ Instantaneous voltage

where $\omega = 2\pi f$

$i = I_{max} \sin \omega t$ Instantaneous current

where $\omega = 2\pi f$

$P = \dfrac{E^2}{R} = I^2 R = EI \cos \theta$ Power in AC circuits

where θ is the phase angle

$PF = \frac{R}{Z} = \cos \theta$ Power factor

$\theta = \text{arc tan } \frac{X}{R} =$ angle of lead or lag (degrees) Phase angle

$C = K \frac{A}{d}$ Capacitance in micromicrofarads

where A is area of plates
d is distance between plates
K is dielectric constant

$X_L = 2\pi f L$ Inductive reactance

$X_C = \frac{1}{2\pi f C}$ Capacitive reactance

$Z = \sqrt{R^2 + (X_L - X_C)^2}$ Impedance — triangular form

$Z = R \pm jX$ Impedance — rectangular form

$f_r = \frac{1}{2\pi\sqrt{LC}}$ Frequency at resonance

$Q = \frac{X_L}{R}$ Merit of a coil or tank

where R is total resistance in series with coil

$Q = \frac{R}{X_L}$ Merit of a coil or tank

where R is total resistance in parallel with coil

$L_T = L_1 + L_2 \pm 2M$ Total inductance of two coils

where M is the mutual inductance

$L_t = L_1 + L_2 + L_3 \ldots$ Inductance in series

$1/L_t = 1/L_1 + 1/L_2 + 1/L_3 \ldots$
Inductance in parallel

$C_t = C_1 + C_2 + C_3 \ldots$
Capacitances in parallel

$1/C_t = 1/C_1 + 1/C_2 + 1/C_3 \ldots$
Capacitances in series

$Q_t = Q_1 = Q_2 = Q_3 \ldots$
Total charge on condensers in series

$Q_t = Q_1 + Q_2 + Q_3 \ldots$
Total charge on condensers in parallel

+ is used when the fields add
— is used when the fields oppose

Transformers

$$\frac{E_p}{E_s} = \frac{N_p}{N_s} = \frac{I_s}{I_p}$$ Relationships among voltages, turns, and currents in transformers

$$\frac{Z_p}{Z_s} = \left(\frac{N_p}{N_s}\right)^2$$ Impedances and turns

$$Z = Z_1 + Z_p - \frac{Z_{m^2}}{Z_2 + Z_s}$$ Impedance when looking into primary of a transformer

where Z_1 is in series with primary
Z_p is impedance of primary windings
Z_m is mutual impedance
Z_s is impedance of secondary windings
Z_2 is impedance in series with secondary

$$Z_r = \frac{-Z_{m^2}}{Z_2 + Z_s}$$ Reflected impedance

Transients

$t = R \times C$ Time constant for *RC* circuits

where t is in seconds
R is in ohms
C is in farads

$t = \frac{L}{R}$ Time constant for L/R circuits

where t is in seconds
R is in ohms
L is in henries

$e_c = E(1 - \epsilon^{-t/RC})$ Voltage change across a capacitor in an RC circuit

where E is the applied voltage

$e_R = E\epsilon^{-t/RC}$ Voltage across a resistor in an RC circuit

$i = \frac{E}{R}\epsilon^{-t/RC}$ Current in an RC circuit

$e_L - E\epsilon^{-Rt/L}$ Voltage across coil in an L/R circuit

$e_R = E(1 - \epsilon^{-Rt/L})$ Voltage change across a resistor in an L/R circuit

$i = \frac{E_a}{R}(1 - \epsilon^{-Rt/L})$ Current in an L/R circuit

VACUUM TUBES AND AMPLIFIERS

$\mu = \frac{\Delta E_p}{\Delta E_g}(I_b \text{ constant})$ Amplification factor

$r_p = \frac{\Delta E_b}{\Delta I_b}(E_g \text{ constant})$ Plate resistance

where r_p is in ohms
E_b is in volts
I_b is in amperes

$g_m = \frac{\Delta I_p}{\Delta E_g}(E_b \text{ constant})$ Transconductance

where g_m is in mhos
I_b is in amperes
E_g is in volts

$\mu = g_m r_p$ — Relationship among g_m, r_p and μ

where g_m is in mhos
r_p is in ohms

$$A = \frac{E_o}{E_s} = \frac{E_o}{E_g} = \frac{\mu Z_L}{r_p + Z_L}$$ Voltage gain of triode operated Class A

where E_o is the output voltage
E_s is the input voltage
E_g is the grid-to-cathode voltage
Z_L is the equivalent load impedance
r_p is the plate resistance

$$\text{db} = 20 \log \frac{E_o}{E_{\text{in}}} = 20 \log \frac{I_o}{I_{\text{in}}}$$ Voltage or current gain in decibels, if R_{in} equals R_{out}

$$\text{db} = 10 \log \frac{P_0}{P_{\text{in}}}$$ Power gain of stage operated Class A

where P_o = output power
P_{in} = input power

$$A_t = A_1 \times A_2 \times A_3 \ldots\ldots A_n$$ Total voltage gain of stages in cascade

$$\text{db}_t = \text{db}_1 + \text{db}_2 + \text{db}_3 \ldots \text{db}_n$$ Total power gain of stages in cascade

$$f_h = \frac{1}{2\pi R_L C_s}$$ Highest frequency in band passed by a video amplifier

where R_L is the total load resistance
C_s is total shunt capacitance

$$f_l = \frac{1}{2\pi R_G C_C}$$ Lowest frequency in band passed by a video amplifier

where R_G is the grid resistor
C_C is the coupling capacitor

$$A' = \frac{E_o}{E_g} = \frac{A}{1 - BA}$$ Gain in a feedback amplifier

where A is the gain without feedback
B is the fraction of the output voltage that is fed back

Cathode Followers

$$Z_o = \frac{R_k r_p}{r_p + R_k(\mu + 1)}$$ Output impedance of a cathode follower

$$A = \frac{E_o}{E_s} = \frac{\mu R_k}{r_p + R_k(\mu + 1)}$$ Gain of a cathode follower

$$A_p = A_v A_i$$ Power gain

$$Z_o = \frac{h_{ie} + Z_{\text{gen}}}{(h_{ie}h_{oe} - h_{re}h_{fe}) + Z_{\text{gen}} h_{oe}}$$ Transistor output impedance

FIELD-EFFECT TRANSISTORS AND CIRCUITS

$$g_m = \frac{\Delta I_D}{\Delta V_{GS}} (E_{DS} \text{ constant})$$ Transconductance

$$\mu = \frac{\Delta E_D}{\Delta V_{GS}} (I_D \text{ constant})$$ Amplification factor

$$r_{ds} = \frac{\Delta V_{DS}}{\Delta I_D} (V_{GS} \text{ constant})$$ Drain resistance

$$I_{DSS} = \frac{1}{I_D}\left(1 - \frac{V_{GS}}{V_p}\right)^2$$ Drain current at $V_{GS} = 0$

$$A_v = g_m R_L$$ Voltage gain

$$A_v = \frac{\mu R_L}{r_{ds} + R_L}$$ Voltage gain

JUNCTION TRANSISTORS AND AMPLIFIERS

$$\beta = \frac{\Delta I_C}{\Delta I_B}$$ Amplification factor, Common Emitter

$\alpha = \dfrac{\Delta I_C}{\Delta I_E}$	Amplification factor, Common Base
z_i	Transistor input impedance
Z_i	Circuit input impedance
$\beta = \dfrac{\alpha}{1 - \alpha}$	Conversion
$z_i = h_{ie} - \dfrac{h_{re}h_{fe}}{h_{oe} + Y_L} \simeq h_{ie}$	Transistor input impedance
$Y_o = h_{oe} - \dfrac{h_{re}h_{fe}}{h_{ie} + Z_i}$	Output admittance
$A_i = \dfrac{h_{fe}}{1 + h_{oe}R_L} \simeq h_{fe}$	Current gain
$A_v = A_i \dfrac{R_L}{Z_i} \simeq A_i \dfrac{R_L}{h_{ie}}$	Voltage gain

Answers to Odd-numbered Problems

Chapter 1

1. $r_{int} = 0.0967\ \Omega$
3. $E_{avg} = 49.45$ v
5. $V_{pp} = 1{,}190$ v
7. $R_{int} = 29.7\ k\Omega$

Chapter 2

1. V_{out} of supply A is approximately 64 per cent of maximum; V_{out} of supply B is approximately 58 per cent of maximum
3. R_{int} of gas diode $\simeq 25\ \Omega$; R_{int} of vacuum diode $\simeq 163\ \Omega$
5. Ripple voltage $= 33.25$ v
7. $E_r = 0.97$ v
9. $L_{cr} = 0.75$ h
11. $\theta = 216°$
13. Loss of V_1 or V_2 which charges operation from full-wave to half-wave operation
15. L_1 shorted

Chapter 3

1. $V_{R3} = 105$ v; $V_{R2} = 210$ v; $V_{R1} = 315$ v
3. $R = 7.28\ k\Omega$
5. $V_{applied} = 311.5$ v; $I_3 = 4.51$ ma; $I_2 = 5.98$ ma; $I_1 = 3.96$ ma
7. $r_{int} = 4\ \Omega$

Chapter 4

1. $I_b \simeq 8$ ma; $E_b \simeq 180$ v
3. $E_c = -6$ v
5. $R_k = 220\ \Omega$

Chapter 5

1. (a) 3 db
   (b) 2.8 watts increase
3. db = 80
5. $E = 675$ mv
7. From chart, Fig. 5-2(b), -3 db
9. (a) db = 36.99
   (b) db = 36.99
11. $P_{in} = 1.59\ \mu$w
13. $P_{in} = 2.75\ \mu$w, db = 63.38
15. For 68 db, $P_0 = 6.31$ kw; for 32 db, $P_0 = 1.59$ w; for 22 db, $P_0 = 0.159$ w; for -15 db, $P_0 = 31.6\ \mu$w
17. db = 61.48
19. 20 Hz, -2.3 db; 100 Hz, -1.5 db; 400 Hz, -0.52 db; 2 kHz, -0.4 db; 30 kHz, -1.16 db
21. $R_{int} = 25$ kΩ
23. 1 v = 9 kΩ; 3 v = 29 kΩ; 10 v = 99 kΩ; 15 v = 149 kΩ; 50 v = 499 kΩ; 300 v = 2.999 MΩ; 1,000 v = 9.999 MΩ
25. For 1 ma, $R_s = 83.4\ \Omega$; for 5 ma, $R_s = 16.1\ \Omega$; for 10 ma, $R_s = 8.04\ \Omega$; for 25 ma, $R_s = 3.21\ \Omega$; for 100 ma, $R_s = 0.8\ \Omega$; for 250 ma, $R_s = 0.32\ \Omega$; for 500 ma, $R_s = 0.16\ \Omega$; for 1 a, $R_s = 0.08\ \Omega$

Chapter 6

1. 17.6 times
3. Reduction of high-frequency response; requirement of excessively high supply voltage
5. $I_b = 2$ ma, $A_v = 5.15$; $I_b = 4$ ma, $A_v = 5.15$; $I_b = 6$ ma, $A_v = 4.67$; $I_b = 8$ ma, $A_v = 5.68$
7. 21 watts, approximately
9. 2.5 ma, approximately
11. 1.6 mw, approximately
13. 94 times, approximately

Chapter 7

1. 70.7 per cent; 89.4 per cent; 44.7 per cent
3. 61 MHz, approximately
5. 4,560 μv, approximately
7. $R \simeq 750\ \Omega$

Chapter 8

1. 3,400 μf
3. $R_1 \simeq 27.7$ kΩ
5. $I_B \simeq 25$ μa
7. $I_B \simeq 22.3$ μa
9. 9.6 μMhos
11. 1.4 MHz
13. 20 MHz

Chapter 9

1. 99, approximately
3. 45, approximately
5. 825 Ω, approximately
7. 15 μv, approximately
9. $I_B \simeq 108$ μa

Chapter 10

1. $P_o = 40$ w
3. $R_c = 5\ \Omega$
5. $A_p = 41.6$
7. Yes, allowable dissipation = 60 w
9. $A_p \simeq 160$, $P_o = 50$ w
11. $P_{\text{diss}} = 11.42$ w, $V_{CE\ \max} = 91.2$ v

Chapter 12

1. $H_2 \simeq 4.5\ \%$
2. $I_B = 175$ μa, $H_2 \simeq 0\ \%$; $I_B = 150$ μa, $H_2 \simeq 7.1\ \%$; $I_B = 125$ μa, $H_2 \simeq 4.55\ \%$; $I_B = 100$ μa, $H_2 \simeq 3.84\ \%$; $I_B = 75$ μa, $H_2 \simeq 0\ \%$

Chapter 13

1. $A_v' = 5$
3. $r_o = 127\ \Omega$
5. Because of the value of g_m, no value of R_s will allow the output resistance to equal 100 Ω. In other words, the value of g_m limits the output resistance to less than 100 Ω.
7. $C_{\text{in}} = 139.8$ pf
9. $C_{\text{in}} = 218$ pf
11. $C_{\text{in}} = 84$ pf

Chapter 14

1. $f_o = 890$ kHz
3. $f_L = 1.13$ MHz, $f_h = 3.38$ MHz
5. $\dfrac{L_2}{L_1} = \dfrac{2{,}200}{1}$

7. $f = 9.65 \times 10^6$ Hz
9. $f_o \simeq 42.2$ MHz

Chapter 15

1. $R \simeq 10$ kΩ
3. $e_g \simeq 3.2$ v
5. $C \simeq 800$ pf
7. $h_{fe} \geq 69.3$
9. $f = \dfrac{1}{2\sqrt{3}\,\pi RC}$
11. $f \simeq 4$ kHz
13. $C \simeq 2.25$ nf
15. $C_1 = 1.39$ nf

Chapter 16

1. $f_o \simeq 875$ kHz, $Q = 108$, $BW = 8.1$ kHz
3. $Z \simeq 600$ kHz, $Q = 280$, $BW = 8.05$ kHz, $Z = 988$ kΩ, $f_o \simeq 8.05$ kHz; Z is the ratio of L/RC; as C decreases, Z increases, all other factors equal.
5. $C_N = 0.83$ pf, $R'_N = 540\ \Omega$, $R_N = 360\ \Omega$
7. $C_N = 1.5$ pf, $R_N = \infty$
9. $K = 0.2$, $L_M = 20\ \mu$h; $K = 0.05$, $L_M = 5\ \mu$h
11. $BW = 37$ kHz
13. $BW = 617$ kHz

Chapter 17

1. $M = 23.6$ per cent
3. $I_o = 333$ ma

Chapter 18

1. $BW = 30$ kHz
3. $f_1 = 87{,}995$ kHz, $f_2 = 88{,}015$ kHz
5. ± 75 kHz

Chapter 19

1. $T = 0.2 \times 10^{-3}$ sec, $f = 5$ kHz
3. $T = 13.2 \times 10^{-3}$, $f = 75.8$ Hz
5. $T = 6.82$ msec
7. Time each pulse $= 2.5 \times 10^{-5}$ sec, $C = 250$ pf
9. $T \simeq 2.27$ msec
11. $f \simeq 7$ Hz
13. $T_{rise} \simeq 51.7\ \mu$s, measured $23.7\ \mu$s

Index